GLENCOE

Pre-Algebra

An Integrated Transition to Algebra & Geometry

GLENCOE
McGraw-Hill

New York, New York   Columbus, Ohio   Woodland Hills, California   Peoria, Illnois

Visit the Glencoe Mathematics Internet Site for
Pre-Algebra at

www.glencoe.com/sec/math/prealg/mathnet

You'll find:

Chapter Review

Test Practice

Data Collection

Games

*inter*NET CONNECTION links to websites relevant to

Chapter Projects, Investigations, exercises

and much more!

Glencoe/McGraw-Hill

A Division of The McGraw·Hill Companies

Send all inquiries to:
Glencoe/McGraw Hill
8787 Orion Place
Columbus, OH 43240-4027

ISBN: 0-07-822873-5

8 9 10 027/043 08 07 06 05

WHY IS MATHEMATICS IMPORTANT?

Why do I need to study mathematics? When am I ever going to have to use math in the real world?

Many people, not just mathematics students, wonder why mathematics is important. **Pre-Algebra** is designed to answer those questions through **integration**, **applications**, and **connections**.

Did you know that algebra and geometry are closely related? Topics from all branches of mathematics, like geometry, probability, and statistics, are integrated throughout the text.

You'll learn how to use proportions to relate the measures of similar triangles. (Lesson 11-6, pages 578–583)

How do they make those dinosaurs in movies? They used rational numbers! In **Pre-Algebra**, you'll study many ways that math is used in the real world.

The size of a model is related to the size of a real object by using a rational number. (Lesson 5-1, page 224)

What does mathematics have to do with biology? Mathematical topics are connected to the other subjects that you study.

Scientists use scientific notation to describe the size of very large distances, like the distance from Earth to the Moon, and very small distances, like the diameter of a red blood cell. (Lesson 4-9, page 211)

Authors

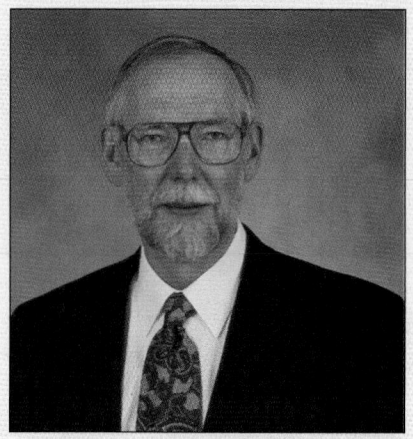

William Leschensky is a former teacher of mathematics at Glenbard South High School and the College of DuPage in Glen Ellyn, Illinois. He has taught mathematics at every level of the high school curriculum, served as mathematics department chairperson, and is a former textbook editor and editorial director. Mr. Leschensky received his B.A. in mathematics from Cornell College and his M.A. in mathematics from the University of Northern Iowa. He has participated in special institutes in mathematics, physics, and computer science and is a frequent speaker at local, state, and regional conferences. Mr. Leschensky is a question writer for MATHCOUNTS, a nationwide series of competitions designed to stimulate student interest and achievement in mathematics.

This program builds on the mathematical background students obtain in earlier grades and prepares them to succeed in algebra. Technology is included wherever appropriate, and links between mathematics and life situations abound. Math is fun!

Wm D Leschensky

Carol Malloy is an assistant professor of mathematics education at the University of North Carolina at Chapel Hill. Dr. Malloy previously taught middle and high school mathematics. She received her B.S. in mathematics and education from West Chester State College, her M.S.T. in mathematics from Illinois Institute of Technology, and her Ph.D. in curriculum and instruction from the University of North Carolina at Chapel Hill. Dr. Malloy is an active member of numerous local, state, and national professional organizations and boards, including National Council of Teachers of Mathematics (NCTM) and Association for Supervision and Curriculum Development (ASCD). Dr. Malloy also serves as an editor of the Benjamin Banneker Association newsletter. She frequently speaks at conferences and schools on the topics of equity and mathematics for all students.

Carol E Malloy

Mathematics teachers are excited about implementing the NCTM Standards in their mathematics classrooms. Glencoe's Pre-Algebra affords students with a strong foundation to high school mathematics through comprehensive real-world explorations, technology, and hands-on activities.

Jack Price received his B.A. from Eastern Michigan University and his Doctorate in Mathematics Education from Wayne State University. Dr. Price has been active in mathematics education for over 40 years, 38 of those years at grades K through 12. In his current position as co-director of the Center for Education and Equity in Mathematics, Science, and Technology at California State Polytechnic University, he teaches mathematics and methods courses for preservice teachers and consults with school districts on curriculum change. Dr. Price is a past president of the National Council of Teachers of Mathematics, is a frequent speaker at professional conferences, conducts many in-service workshops, and is author of numerous mathematics and science instructional materials. Dr. Price is co-author on Glencoe's *Mathematics: Applications and Connections* and *Interactive Mathematics: Activities and Investigations*.

With the study of pre-algebra, you are taking your first steps toward a mathematics education that will be valuable to you throughout your lifetime. I know you will find it exciting and challenging.

Jim Rath has 30 years of classroom experience in teaching mathematics at every level of the high school curriculum. He is a former mathematics teacher and department chairperson at Darien High School in Darien, Connecticut. Mr. Rath earned his B.A. in philosophy from the Catholic University of America and his M.Ed. and M.A. in mathematics from Boston College. He held an appointment as a Visiting Fellow in the Mathematics Department at Yale University in New Haven, Connecticut for the academic year 1994-1995. Mr. Rath is a co-author on Glencoe's *Algebra 1* and *Algebra 2*.

What you learn today is a building block for what comes tomorrow. That is the reason why a student needs to do his or her best each day . . . what you learn in pre-algebra is building a strong foundation for you.

Yuria Alban received her B.S. in Mathematics from Armstrong State College, her M.A. in Mathematics from the University of Georgia, and her Educational Specialist degree in Computer Education from Nova University. Ms. Alban has more than twenty years of teaching experience in middle school, high school, and college mathematics. She is currently a District Mathematics Supervisor for the Dade County Public Schools in Florida. Ms. Alban is a past President of the Dade County Council of Teachers of Mathematics and has received numerous teaching awards at the local and state levels. Recently, Ms. Alban participated in U.S. Justice Department programs to assess students' needs and advise on mathematics curriculum in Guantanamo Bay, Cuba and Venezuela.

Mathematics today is a thrilling mix of present reality and investigation. Students will travel on paths of exploration to reach into the future. This textbook will make the journey a unique and exciting experience.

Consultants and Reviewers

Consultants

Gilbert J. Cuevas
Professor of Mathematics
  Education
University of Miami
Coral Gables, Florida

Dr. Luis Ortiz-Franco
Consultant, Diversity
Associate Professor of
  Mathematics
Chapman University
Orange, California

David Foster
Glencoe Author and
  Mathematics Consultant
Morgan Hill, California

Daniel Marks
Consultant, Real-World
  Applications
Associate Professor of
  Mathematics
Auburn University at
  Montgomery
Montgomery, Alabama

Melissa McClure
Consultant, Tech Prep
Mathematics Consultant
Teaching for Tomorrow
Fort Worth, Texas

Cleo M. Meek, Ed. D.
Mathematics Consultant
NC Dept. of Public Instruction,
  Retired
Raleigh, North Carolina

Jim Zwick
Mathematics Department
  Chairman
Guion Creek Middle School—
  MSD of Pike Township
Indianapolis, Indiana

Reviewers

Robert M. Allen
Mathematics Teacher
Vista Verde Middle School
Phoenix, Arizona

Bruce I. Althouse, Sr.
Mathematics Department
  Chairperson
Northern Lebanon
  Junior/Senior High School
Fredericksburg, Pennsylvania

Glenn Aston-Reese
Mathematics Department
  Chairperson
Trinity Middle School
Washington, Pennsylvania

Mary R. Beard
Mathematics Teacher
Southeast Guilford Middle
  School
Greensboro, North Carolina

Ronald L. Bow
Mathematics Teacher
Henry Clay High School
Lexington, Kentucky

Alan Brown
Registrar, Adult Education
Springfield Adult Learning
  Center
Springfield, Massachusetts

Sharon A. Cichocki
Secondary Math Coordinator
Hamburg High School
Hamburg, New York

Debra M. Cline
Mathematics Teacher
Atkins Middle School
Winston-Salem, North Carolina

Linda Coutts
Mathematics Coordinator
Columbia Public Schools
Columbia, Missouri

Jean E. Ellis
Mathematics Teacher
Ruddiman Middle School
Detroit, Michigan

Gary A. Graves
Coordinator of Mathematics
  K-12
St. Joseph Public Schools
St. Joseph, Missouri

Earlene J. Hall
Administrator
Detroit Public Schools
Detroit, Michigan

Eileen D. Harris, Ph.D.
Mathematics/Science
  K-12 Specialist
Charlotte County School
  District
Port Charlotte, Florida

Richard Hedlund
Mathematics Teacher
South Broward High School
Hollywood, Florida

James P. Herrington
Mathematics/Science
  Department Chairperson
O'Fallon Township High School
O'Fallon, Illinois

Deborah D. Johnson
Supervisor of Mathematics
Camden City Public Schools
Camden, New Jersey

Nancy L. Jones
Mathematics Teacher
Aycock Middle School
Greensboro, North Carolina

Stephen R. Lightner
Mathematics Department
  Chairperson
D.S. Keith Junior High School
Altoona, Pennsylvania

Gaye S. McKinnon
Mathematics Teacher
Columbiana Middle School
Columbiana, Alabama

Bernadette M. Meglino
Mathematics Teacher
Lake Weir High School
Ocala, Florida

Nance Marie Minnick
Mathematics Teacher
Denver Public Schools
Denver, Colorado

Rhonda C. Niemi
Mathematics Teacher
Meyzeek Middle School
Louisville, Kentucky

Georgia Thielen O'Connor
Mathematics Instructor
Stevens High School
Rapid City, South Dakota

Pamela H. Ostrosky
Mathematics Teacher
Dake Junior High School
Rochester, New York

Marcia Perry
Mathematics Teacher
Palm Beach Gardens High
  School
Palm Beach Gardens, Florida

Odelia Schrunk
Mathematics Department
  Chairperson
Clinton High School
Clinton, Iowa

Sharon Ann Schueler
Mathematics Department
  Chairperson
Roosevelt High School
Sioux Falls, South Dakota

Sara P. Shaw
Mathematics Teacher
Adams Middle School
Saraland, Alabama

Mary Eleanor Wood Smith
Mathematics Teacher
North Augusta High School
North Augusta, South Carolina

Diane Stilwell
Mathematics Department
  Chairperson
South Junior High School
Morgantown, West Virginia

Janice M. Stricko
Mathematics Teacher
Gateway Middle School
Monroeville, Pennsylvania

Jeanette Tomasullo
Assistant Principal Supervision
  —Mathematics
Eastern District High School
Brooklyn, New York

Ivan T. Van Dyke
Mathematics Teacher
Norfolk Junior High School
Norfolk, Nebraska

Kathy Vielhaber
Mathematics Teacher
Parkway East Middle School
Creve Coeur, Missouri

Marlene T. Vitko
Mathematics Department
  Chairperson
Godwin Middle School
Dale City, Virginia

Carol Welsch
Learning Coordinator
Toki Middle School
Madison, Wisconsin

Table of Contents

Introduction to the Graphing Calculator xx

Chapter 1 Tools for Algebra and Geometry 4

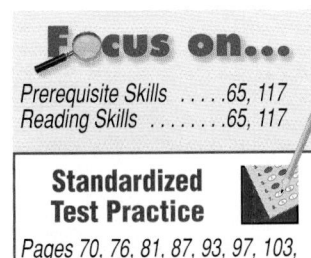

F☉cus on...

Prerequisite Skills 65, 117
Reading Skills 65, 117

Standardized Test Practice

Pages 70, 76, 81, 87, 93, 97, 103, 108, 114–115, 122, 128, 133, 137, 144, 150, 155, 159

*inter*NET
CONNECTION

Chapter Review 109, 161
Test Practice 115

In addition to the Graphing Calculator Activities, Modeling with Technology is included on pages 83, 94, and 124.

In addition to the Hands-On Activities, Modeling Math exercises are included on pages 68, 91, 102, 120, and 135.

LONG-TERM PROJECTS
Investigation

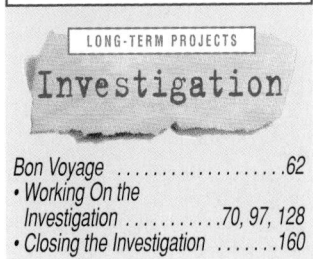

Bon Voyage 62
• Working On the
   Investigation 70, 97, 128
• Closing the Investigation 160

Content Integration

Why are topics like geometry and statistics in a pre-algebra book? Believe it or not, you can usually study an algebra topic from a geometric point of view and vice versa. Some examples are described below.

INTEGRATION

Probability You'll learn more about your chances of making that winning move in backgammon. (Lesson 9-3, pages 440-443)

Geometry You'll use ordered pairs to develop algebraic expressions. (Lesson 1-7, pages 36-40)

Statistics You'll learn how to use mean, median, and mode. (Lesson 6-6, pages 301-306)

Discrete Mathematics You'll learn to represent an arithmetic sequence algebraically. (Lesson 5-9, pages 258-262)

Problem Solving You'll learn how to solve problems by looking for a pattern. (Lesson 2-6, pages 94-97)

You can review dividing fractions in Lesson 6-4.

Look Back features refer you to skills and concepts that have been taught earlier in the book. Lesson 9-7, page 459

Chapter 4 Exploring Factors and Fractions 168

Standardized Test Practice

📖interNET
CONNECTION

In addition to the Graphing Calculator Activities, Modeling with Technology is included on pages 205, 210, 234, and 258.

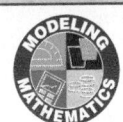

In addition to the Hands-On Activities, Modeling Math exercises are included on pages 173, 182, 192, 198, 202, 227, 246.

LONG-TERM PROJECTS

Investigation

Chapter 5 Rationals: Patterns in Addition and Subtraction 222

Real-Life Applications

Every lesson in this book is designed to help show you where and when math is used in the real world. Since you'll explore many interesting topics, we believe you'll discover that math is relevant and exciting. Here are some examples.

Real World APPLICATION

Recycling You'll discuss recycling as you learn about solving inequalities. (Lesson 3-7, page 151)

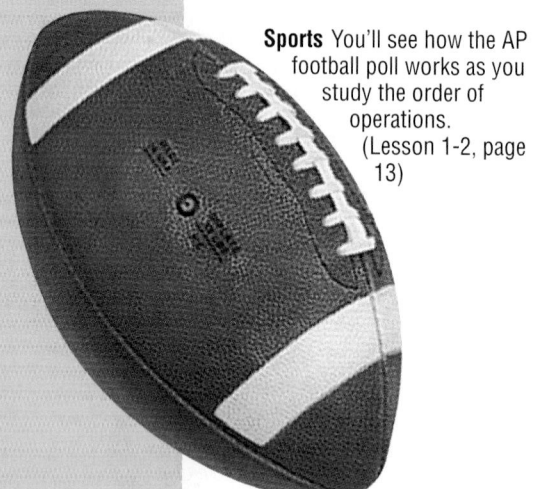

Sports You'll see how the AP football poll works as you study the order of operations. (Lesson 1-2, page 13)

FYI

The earliest date for a hurricane to strike the U.S. was Hurricane Allison on June 5, 1995. The previous early hurricane record holder was Hurricane Alma on June 9, 1966, which like Allison, struck the Florida panhandle.

FYI features contain interesting facts that enhance the applications. (Lesson 5-6, page 248)

Games
Rolling dice in the game of parcheesi, which originated in India, is related to exponents. (Lesson 4-2, page 175)

From the FUNNY PAPERS
When is mathematics funny?
An actual comic illustrates how mathematics is a part of our society. (Lesson 3-5, page 144)

Health and Medicine You'll study a health-related application that involves heart rates and algebraic expressions. (Lesson 1-3, pages 16-17)

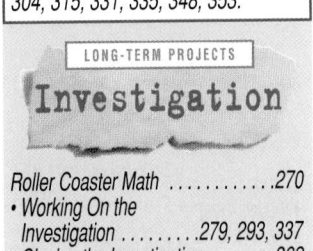

Chapter 8 Functions and Graphing 370

Chapter 9 Ratio, Proportion, and Percent 430

Sidebar

F⊙cus on...

Standardized Test Practice

interNET CONNECTION

In addition to the Graphing Calculator Activities, Modeling with Technology is included on pages 379, 412, 440, and 467.

MODELING MATHEMATICS

In addition to the Hands-On Activities, Modeling Math exercises are included on pages 387, 404, 414, 434, 442, 451, 460.

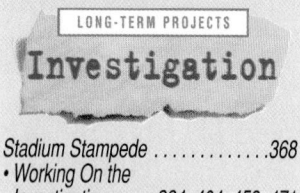

LONG-TERM PROJECTS

Investigation

Connections

Did you realize that math can be used to model many real-world applications? Do you know that mathematics is used in biology? in history? in geography? Yes, it may be hard to believe, but mathematics is frequently used in subjects besides math.

Connection to > Algebra

You'll use algebra as a tool to solve problems from math topics such as geometry and arithmetic. (Lesson 7-4, page 342)

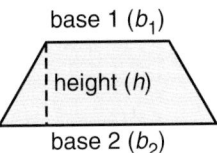

base 1 (b_1)

height (h)

base 2 (b_2)

CULTURAL CONNECTIONS

Pythagoras is credited with the proof of the theorem that bears his name. However, the theorem may have been known to Egyptian, Babylonian, and Chinese scholars as much as one thousand years earlier.

Connection to > Geometry

You'll learn how ratios can be used to describe the dimensions of the pyramids of Egypt. (Lesson 9-1, page 434)

Cultural Connections introduce you to a variety of world cultures. (Lesson 13-4, page 678)

CONNECTIONS   Interdisciplinary Connections

Geography You'll study how to write and solve an equation involving the highest and lowest points in Africa. (Lesson 2-5, page 90)

Music You'll solve equations that involve the number of vibrations on a guitar for a particular musical note. (Lesson 6-7, page 308)

Ecology You'll learn to model data about household waste as ordered pairs. (Lesson 8-1, page 372)

CAREER CHOICES

Oceanographers are needed in all parts of the world. For more information, contact: National Ocean Industries Assn. 1050 17th Street NW, Suite 700 Washington, D.C. 20036

Health You'll relate pollen count to comparing integers on a number line. (Lesson 2-3, page 78)

Math Journal

Math Journal exercises give you the opportunity to assess yourself and write about your understanding of key math concepts. (Lesson 5-2, page 231)

Career Choices features include information on interesting careers. (Lesson 8-4, page 393)

Chapter 10 More Statistics and Probability   484

Chapter 11 Applying Algebra to Geometry   546

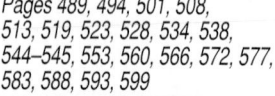

F⚲cus on...

Prerequisite Skills 485, 547
Reading Skills 485, 547

Standardized Test Practice
Pages 489, 494, 501, 508, 513, 519, 523, 528, 534, 538, 544–545, 553, 560, 566, 572, 577, 583, 588, 593, 599

interNET CONNECTION

Chapter Review 539, 601
Test Practice 545
Data Collection 483

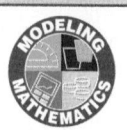

In addition to the Graphing Calculator Activities, Modeling with Technology is included on pages 490, 530, 573, and 589.

MODELING MATHEMATICS

In addition to the Hands-On Activities, Modeling Math exercises are included on pages 488, 498, 511, 552, 575, 581, 586, and 597.

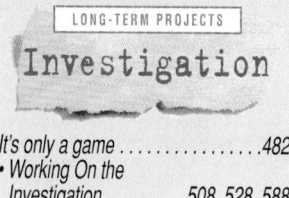
LONG-TERM PROJECTS
Investigation

It's only a game 482
• *Working On the Investigation* 508, 528, 588
• *Closing the Investigation* 600*

Modeling Mathematics

Using Manipulatives and Technology

One of the best ways to learn is to learn by doing. Here are two examples.

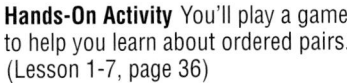

Hands-On Activities

In Lesson 9-4B, you'll learn how to use the capture-recapture method of counting deer.

Hands-On Activity You'll play a game to help you learn about ordered pairs. (Lesson 1-7, page 36)

Do you know how to use high-tech machines like graphing calculators? We know that such knowledge is vital to your future success. That is why graphing calculator technology is integrated throughout this book.

There are several ways in which graphing calculators are integrated.

▶ **Introduction to Graphing Calculators** On page xx and page 1, you'll get acquainted with the basic features and functions of a graphing calculator.

▶ **Graphing Calculator Activities** In Lesson 3-5B, you'll use a program to find the area of quadrilaterals given the coordinates of the vertices.

▶ **Graphing Calculator Exercises** Many exercises are designed to be solved using a graphing calculator. For example, Exercise 23 in Lesson 8-2 involves making a scatter plot of keyboarding speeds.

▶ **Technology Activity** On page 467, Lesson 9-9 begins with an application involving a spreadsheet.

Option	Minimum Investment	Monthly Interest Rate (Percent)	Term (Months)	Interest for Term
1	$2500	0.375	6	$56.25
2	$2500	0.45	12	$135.00
3	$5000	0.54	36	$972.00
4	$7500	0.625	60	$2812.50

Techno Tips are designed to help you make more efficient use of technology through practical hints and suggestions.

TECHNO TIP

Some calculators use a
$+/-$ key, called the
"change sign" key instead
of the $(-)$ key.

(Lesson 2-4, page 83)

Chapter 12 Measuring Area and Volume 608

Chapter 13 Applying Algebra to Right Triangles 662

Chapter 14 Polynomials   704

Student Handbook

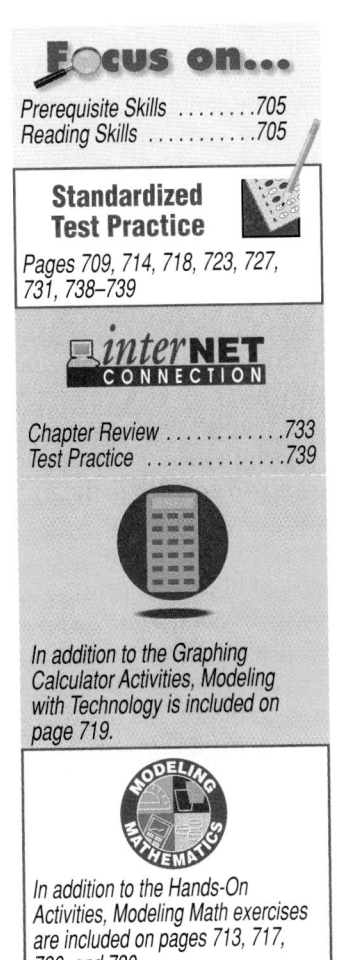

F⚲cus on...

Prerequisite Skills 705
Reading Skills 705

Standardized Test Practice

Pages 709, 714, 718, 723, 727, 731, 738–739

inter NET CONNECTION

Chapter Review 733
Test Practice 739

In addition to the Graphing Calculator Activities, Modeling with Technology is included on page 719.

MODELING MATHEMATICS

In addition to the Hands-On Activities, Modeling Math exercises are included on pages 713, 717, 726, and 730.

Introduction to the Graphing Calculator

What is it?
What does it do?
How is it going to help me learn math?

These are just a few of the questions many students ask themselves when they first see a graphing calculator. Some students may think, "Oh, no! Do we *have* to use one?", while others may think, "All right! We get to use these neat calculators!" There are as many thoughts and feelings about graphing calculators as there are students, but one thing is for sure: a graphing calculator *can* help you learn mathematics.

So what is a graphing calculator? Very simply, it is a calculator that draws graphs. This means that it will do all of the things that a "regular" calculator will do, *plus* it will draw graphs of simple or very complex equations. In pre-algebra, this capability is nice to have as you learn to graph equations and solve equations by graphing.

But a graphing calculator can do more than just calculate and draw graphs. For example, you can program it and work with data to make statistical graphs and computations. If you need to generate random numbers, you can do that on the graphing calculator. It's really a very powerful tool—so powerful that it is often called a pocket computer. A graphing calculator can save you time and make doing mathematics easier.

As you may have noticed, graphing calculators have some keys that other calculators do not. The keys located on the bottom half of the calculator are probably familiar to you as they are the keys found on basic scientific calculators. The keys located just below the screen are the graphing keys. You will also notice the up, down, left, and right arrow keys. These allow you to move the cursor around on the screen, to "trace" graphs that have been plotted, and to choose items from the menus. The other keys located on the top half of the calculator access the special features such as statistical computations and programming features.

There are some keystrokes that can save you time when using the graphing calculator. A few of them are listed below.

- Any yellow commands written above the calculator keys are accessed with the [2nd] key, which is also yellow. Similarly, any green characters above the keys are accessed with the [ALPHA] key, which is also green.
- [2nd] [ENTRY] copies the previous calculation so you can edit and use it again.
- Pressing [ON] while the calculator is graphing stops the calculator from completing the graph.
- [2nd] [QUIT] will return you to the home (or text) screen.
- [2nd] [A-LOCK] locks the [ALPHA] key, which is like pressing "shift lock" or "caps locks" on a typewriter or computer. The result is that all letters will be typed and you do not have to repeatedly press the [ALPHA] key. (This is handy for programming.) Stop typing letters by pressing [ALPHA] again.
- [2nd] [OFF] turns the calculator off.

Some commonly-used mathematical functions are shown in the table below. As with any scientific calculator, the graphing calculator observes the order of operations.

Mathematical Operation	Example	Keys	Display
evaluate expressions	Find 2 + 5.	2 [+] 5 [ENTER]	2+5 7
exponents	Find 3^5.	3 [∧] 5 [ENTER]	3^5 243
multiplication	Evaluate 3(9.1 + 0.8).	3 [×] [(] 9.1 [+] .8 [)] [ENTER]	3*(9.1+.8) 29.7
roots	Find $\sqrt{14}$.	[2nd] [√] 14 [ENTER]	√14 3.741657387
opposites	Enter −3.	[(−)] 3	−3

Graphing

Before graphing, we must instruct the calculator how to set up the axes in the coordinate plane. To do this, we define a **viewing window**. The viewing window for a graph is the portion of the coordinate grid that is displayed on the **graphics screen** of the calculator. The viewing window is written as [left, right] by [bottom, top] or [Xmin, Xmax] by [Ymin, Ymax]. A viewing window of $[-10, 10]$ by $[-10, 10]$ is called the **standard viewing window** and is a good viewing window to start with to graph an equation. The standard viewing window can be easily obtained by pressing ZOOM 6. Try this. Move the arrow keys around and observe what happens. You are seeing a portion of the coordinate plane that includes the region from -10 to 10 on the *x*-axis and from -10 to 10 on the *y*-axis. Move the cursor, and you can see the coordinates of the point for the position of the cursor.

Any viewing window can be set manually by pressing the WINDOW key. The window screen will appear and display the current settings for your viewing window. Using the arrow and ENTER keys, move the cursor to edit the window settings. Xscl and Yscl refer to the *x*-scale and *y*-scale. This is the number of units between tick marks on the *x*- and *y*-axes. Xscl = 1 means that there will be a tick mark for every unit of one along the *x*-axis. The standard viewing window would appear as follows.

$$\text{Xmin} = -10$$
$$\text{Xmax} = 10$$
$$\text{Xscl} = 1$$
$$\text{Ymin} = -10$$
$$\text{Ymax} = 10$$
$$\text{Yscl} = 1$$

Programming

Programming features allow us to write and execute a series of commands to perform tasks that may be too complex or cumbersome to perform otherwise. Each program is given a name. Commands begin with a colon (:), followed by an expression or an instruction. Most of the features of the calculator are accessible from program mode.

When you press PRGM, you see three menus: EXEC, EDIT, and NEW. EXEC allows you to execute a stored program, EDIT allows you to edit or change a program, and NEW allows you to create a program. The following tips will help you as you enter and run programs.

- To begin entering a new program, press PRGM ▶ ▶ ENTER.
- After a program is entered, press 2nd QUIT to exit the program mode and return to the home screen.
- To execute a program, press PRGM. Then use the down arrow key to locate the program name and press ENTER, or press the number or letter next to the program name.
- If you wish to edit a program, press PRGM ▶ and choose the program from the menu.
- To immediately re-execute a program after it is run, press ENTER when Done appears on the screen.
- To stop a program during execution, press ON.

While a graphing calculator cannot do everything, it can make some things easier. To prepare for whatever lies ahead, you should try to learn as much as you can. The future will definitely involve technology, and using a graphing calculator is a good start toward becoming familiar with technology. Who knows? Maybe one day you will be designing the next satellite, building the next skyscraper, or helping students learn mathematics with the aid of a graphing calculator!

Investigation

Up, Up and Away!

Have you ever dreamed of breaking free of the ties of gravity and soaring among the clouds? Maybe you should try an ultralight airplane! An ultralight is a small one-person airplane that is very light, about the size and weight of a hang glider. Unlike a hang glider, an ultralight is propelled by a small engine and can be enclosed or open to the air.

The Federal Aviation Administration, or FAA, governs all types of aircraft in the United States. They set forth the following restrictions on ultralights.

Notice! All ultralight airplanes must:
- *be single seat aircrafts*
- *weigh 254 pounds or less*
- *have a top speed of 55 knots (63 miles per hour)*
- *stall at 24 knots (28 miles per hour)*
- *carry no more than 5 gallons of fuel*
- *never fly over towns or settlements*
- *never fly at night or around airports without special permission*
- *always yield the right of way to all other aircraft*

Special 2-seat exemptions, for training only, may weigh up to 496 pounds and carry 10 gallons of fuel.

Veteran fliers say that building and flying an ultralight is inexpensive when compared to other types of airplanes. A number of kits are available for you to put together yourself for between $3000 and $6000. Plan on spending 6 months to two years on construction. Flight training, while not required, is highly recommended. Instructors charge about $600 to $1200. Once the initial expenses of plane and training are completed, flying is very inexpensive. The engine of an ultralight usually burns only 2 to 3 gallons of fuel an hour. Many ultralights can be stored at home. If you need to store your plane in a hangar, budget about $30 to $90 a month.

Starting Your Investigation.........................

Each year, the Experimental Aircraft Association, or EAA, gathers in Oshkosh, Wisconsin, for their annual convention. EAA members are concerned with all types of home-built, antique, and ultralight aircraft. Suppose you and the other members of your group are on a committee chosen by the local EAA chapter to make plans to fly the group to the convention. There are many things you must consider as you make your plans. Look over the following list of considerations.

Think About . . .
- *Will you be able to fly there in one day?*
- *How much fuel will you need? Can you make it there on one tank of fuel?*
- *How should you plan your route?*
- *Are there places to store the planes while you are at the convention?*
- *Must each person fly his or her own plane, or can two-seat ultralights be used?*

*inter*NET
CONNECTION
Research For current information on ultralights, visit:
www.glencoe.com/ sec/math/prealg/ mathnet

   Work with your group to add to the list of questions to be considered. Then reread the information about ultralights on the previous page to find significant information. You may wish to look in some other sources.

You will continue working on this Investigation throughout Chapter 1. Be sure to keep your materials in your Investigation Folder.

Investigation Up, Up, and Away!

Working on the Investigation Lesson 1-1, p. 10

Working on the Investigation Lesson 1-6, p. 35

Closing the Investigation End of Chapter 1, p. 56

1 Tools for Algebra and Geometry

What You'll Learn

In this chapter, you will:

- solve problems using the four-step problem-solving plan,

- choose the appropriate method of computation,

- use variables and algebraic properties,

- graph points in a coordinate system, and

- gather and record data.

Why It's Important

Real World

Geography This map of Tallahassee, Florida, has numbers along the bottom and left side to help you pinpoint certain locations. For example, if you go over to the 8 along the bottom and then go up to the line marked 5 along the left side, you will see the marker for the state capitol. In mathematics, you can locate a point by using a coordinate system and ordered pairs. *You will learn about ordered pairs in Lesson 1-7.*

Maclay State Gardens, Tallahassee, Florida

F😊cus on...

PREREQUISITE SKILLS

To be successful in this chapter, you'll need to understand these concepts and be able to apply them. Refer to the examples to help you complete these problems.

▶ *Add and subtract decimals.*

Find the sum of 32.45 and 5.8.	Find the difference of 6 and 4.87.
$\begin{array}{r} 32.45 \\ +5.80 \\ \hline 38.25 \end{array}$ *Annex a zero and line up the decimal points. Then add.*	$\begin{array}{r} 6.00 \\ -4.87 \\ \hline 1.13 \end{array}$ *Annex two zeros and line up the decimal points. Then subtract.*
The answer is 38.25.	The answer is 1.13.

Add or subtract.

1. $5.50 + 9.92$ **2.** $4.3 + 8.61$ **3.** $8.54 + 5.34$

4. $29.3 - 7.91$ **5.** $100 - 16.99$ **6.** $34.84 - 17.69$

▶ *Multiply and divide decimals.*

Find 6.98×5.	Find $25.5 \div 1.5$.
Estimate using rounding. *Round 6.98 to 7. $7 \times 5 = 35$.*	*Estimate $26 \div 2 = 13$.*
$\begin{array}{r} 6.98 \\ \times\ \ 5 \\ \hline 34.90 \end{array}$ *Multiply as with whole numbers. Since the estimate is 35, place the decimal point after 4.*	$1.5\overline{)25.5} \rightarrow 15\overline{)255}$ $\begin{array}{r} 17 \\ \underline{15} \\ 105 \\ \underline{105} \\ 0 \end{array}$ *Multiply by 10. Divide as with whole numbers.*
The answer is 34.90.	The answer is 17.

Multiply or divide.

7. 3.25×5 **8.** 17.99×4 **9.** 0.52×0.03

10. $85.5 \div 4.5$ **11.** $40.25 \div 1.75$ **12.** $17.94 \div 2.3$

F😊cus on...

READING SKILLS

In this chapter, you will be learning about **ordered pairs** of numbers. In mathematics, ordered pairs are used in graphing. The order in which the numbers are written is important. The position of the number in the ordered pair tells you which direction to move on the coordinate grid.

Problem-Solving Strategy: Make a Plan

What You'll Learn

- To use the four-step problem-solving plan to solve problems, and

- to choose an appropriate method of computation.

Why It's Important

The four-step plan gives you an organized method for solving problems.

Real World

Application: Health and Exercise

Venus Williams is a superstar in the world of tennis. Venus learned to play tennis from her father in her home town of Compton, California. She gives back to the community by giving tennis clinics at public schools near her home in Delray Beach, Florida.

Venus spends hours exercising each week to remain in top condition. When exercising, it is important to maintain a reasonable heart rate.

Refer to the table below. If you are a 15-year-old, what heart rate should you maintain when you exercise?

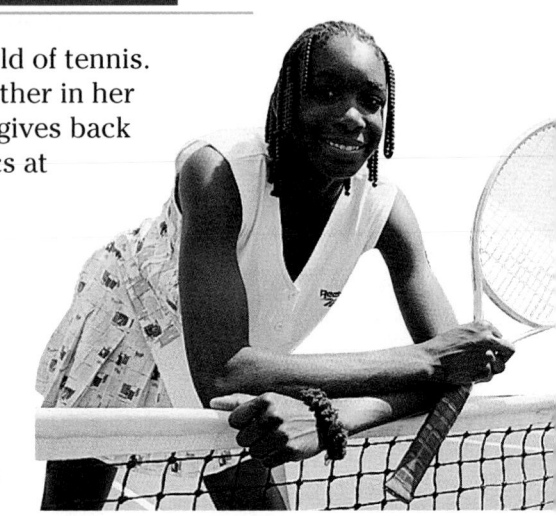

Guide to Reasonable Heart Rate (85% capacity)											
Age	20	25	30	35	40	45	50	55	60	65	70
Heart Rate (beats per minute)	174	170	166	162	157	153	149	145	140	136	132

This problem will be solved in Example 1.

Learning the Concept

You can use four steps to solve real-life, math-related problems.

1. Explore
- ▶ Read the problem carefully.
- ▶ Ask yourself questions like "What facts do I know" and "What do I need to find out?"

2. Plan
- ▶ See how the facts relate to each other.
- ▶ Make a plan for solving the problem.
- ▶ Estimate the answer.

3. Solve
- ▶ Use your plan to solve the problem.
- ▶ If your plan does not work, revise it or make a new plan.

4. Examine
- ▶ Reread the problem.
- ▶ Ask, "Is my answer close to my estimate?"
- ▶ Ask, "Does my answer make sense for the problem?
- ▶ If not, solve the problem another way.

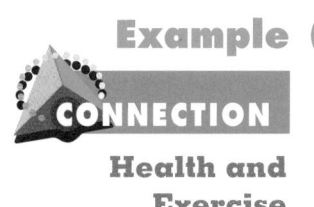

CONNECTION

Health and Exercise

Refer to the application at the beginning of the lesson. Find a reasonable heart rate for exercise for a 15-year-old.

Use the problem-solving plan to find the heart rate.

Explore   *What facts do we know and what are we trying to find out?*

The table shows the suggested heart rates for several other ages. We are trying to find the rate for a 15 year old.

Plan   *Make a plan for attacking the problem and estimate the answer.*

Since we know the heart rates for several ages, it makes sense to look for a pattern to extend to find the rate for a 15 year old. The heart rate decreases as the age increases. So the heart rate for a 15 year old should be higher than that for a 20 year old.

Solve   *Use the plan to solve the problem.*

Find the differences between consecutive ages and heart rates.

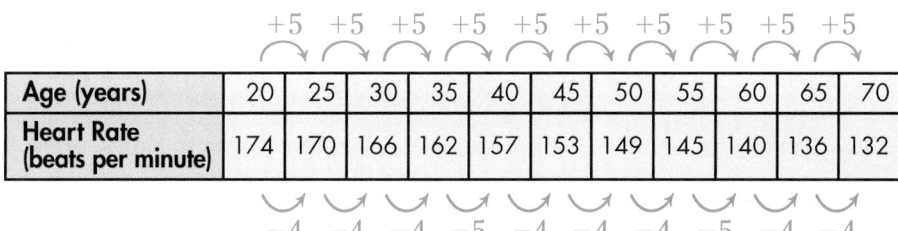

Age (years)	20	25	30	35	40	45	50	55	60	65	70
Heart Rate (beats per minute)	174	170	166	162	157	153	149	145	140	136	132

Each consecutive age increases 5 years. The pattern for the heart rates is $-4, -4, -4, -5$. Extend this pattern as follows.

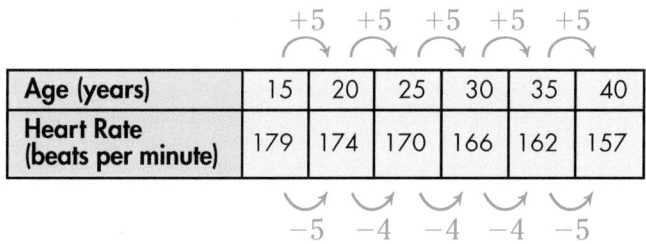

Age (years)	15	20	25	30	35	40
Heart Rate (beats per minute)	179	174	170	166	162	157

A reasonable heart rate for exercise for a 15 year old is $174 + 5$ or 179.

Examine   *Compare the answer to the problem and the estimate.*

The heart rate of 179 that we found for a 15 year old makes sense with the estimate that we made.

One of the important steps in solving problems is choosing the method of computation. The diagram below can help you decide which method is most appropriate.

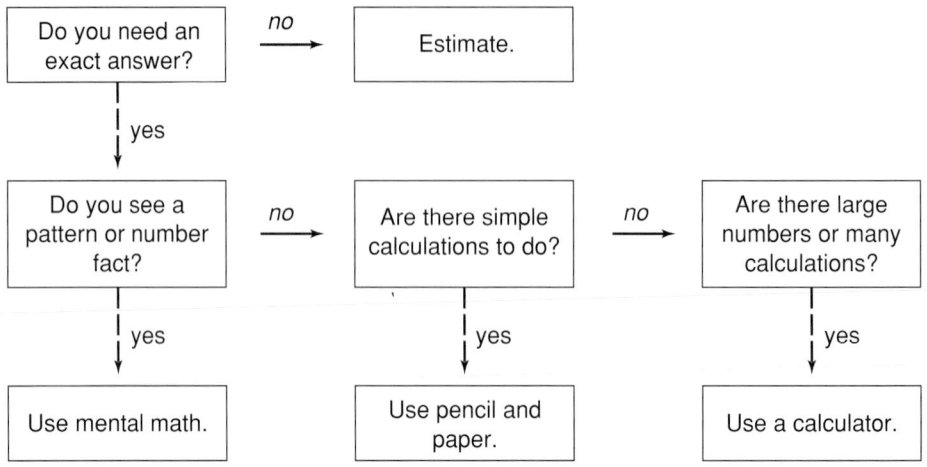

| Do you need an exact answer? | →no→ | Estimate. |

↓ yes

| Do you see a pattern or number fact? | →no→ | Are there simple calculations to do? | →no→ | Are there large numbers or many calculations? |

↓ yes   ↓ yes   ↓ yes

| Use mental math. | | Use pencil and paper. | | Use a calculator. |

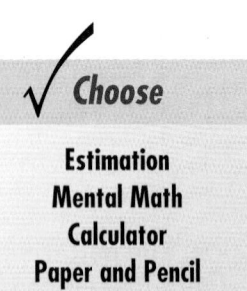

Example 2

Real World APPLICATION

Medicine

✓ **Choose**

Estimation
Mental Math
Calculator
Paper and Pencil

Modern medicine has made amazing advances in human transplants. The numbers of different types of transplants that were performed in the United States are shown at the right. Find the total number of transplants performed.

Transplant	Number
heart	2172
liver	3059
kidney	10,210
heart-lung	48
lung	535
pancreas	557

Source: U.S. Department of Health and Human Services

Explore   We are given the number of each type of transplant and wish to find the total.

Plan   Add the numbers of transplants to find the total. Use the diagram above to find the appropriate method of computation.

Since we need an exact answer, we will not solve using estimation. There is no pattern in the numbers and there are many calculations to be performed, so using a calculator may be the best choice.

Even though we will solve using a calculator, it is a good idea to estimate first so that we know whether the answer we find is reasonable. Round each number to the nearest hundred and then add mentally.

Estimate:

2172	→	2200
3059	→	3100
10,210	→	10,200
48	→	0
535	→	500
557	→	600
		16,600

Do you think the estimate is higher or lower than the actual total?

Solve *Use the plan to solve the problem.*

Use a calculator to find the total.

2172 ⊞ 3059 ⊞ 10210 ⊞ 48 ⊞ 535 ⊞ 557 ⊟ *16581*

Examine The total we found with the calculator is very close to our estimate, so the answer is reasonable.

Checking Your Understanding

Communicating Mathematics

Read and study the lesson to answer these questions.

1. **Describe** each step of the problem-solving plan in your own words.

2. **Explain** why it is important to examine your solution after you solve a problem.

3. **Demonstrate** how to use the diagram on page 8 to determine the best method of computation.

 MATH JOURNAL

4. Begin your Math Journal by writing two or three sentences that describe what you like and dislike about mathematics.

Guided Practice

5. **Sales** John Diaz sells large computer software packages to accounting corporations for Object Systems, Inc. Each month, his supervisor writes a sales plan for the number of software packages that Mr. Diaz is supposed to sell. Mr. Diaz receives a salary, plus a bonus for any software packages he sells in excess of the plan. The table at the right shows the bonus amount for different levels of sales. Determine Mr. Diaz's bonus if he sells 16 software packages more than the plan.

Sales above plan	Bonus
2	$100
4	$125
6	$150
8	$175

 a. Write the *Explore* step. What do you know and what do you need to find?

 b. Write the *Plan* step. What strategy will you use? What do you estimate the answer to be?

 c. *Solve* the problem using your plan. What is your answer?

 d. *Examine* your solution. Is it reasonable? Does it answer the question?

6. **Hygiene** A survey found that, on average, people washed their hands or face 6.8 times per day. Based on this finding, about how many times does a person wash their hands or face in one year?

 a. Which method of computation do you think is most appropriate? Justify your choice.

 b. Solve the problem using the four-step plan. Be sure to examine your solution.

 c. Do the results of the survey imply that each person actually washes his or her hands 6.8 times each day? Explain.

 Source: Lever Brothers Corporation

Lesson 1-1 *Problem-Solving Strategy* *Make a Plan* **9**

Independent Practice

Solve each problem using the four-step plan.

7. **Money Management** The Westerville North High School Band plans to travel to Miami, Florida, to perform in the Orange Bowl Parade. Ana Vera needs to save $300 for the trip. She has $265 in her bank account and earns $20 shoveling snow and $25 babysitting for the neighbors. Does Ana have enough for the trip?

8. Find the total cost of building supplies that cost $2.34, $9.28, $92.34, $501.38 and $0.34.

9. **Postal Service** The chart at the right shows the cost of mailing a first-class letter of different weights. How much would it cost to send a letter that weighs 8 ounces?

Weight (ounces)	Cost
1	$0.33
2	$0.55
3	$0.77
4	$0.99
5	$1.21

✓ **Choose**

Estimation
Mental Math
Calculator
Paper and Pencil

10. **Space Exploration** The space shuttle can carry a payload of about 65,000 pounds. A compact car weighs about 2450 pounds. About how many compact cars could be carried on the space shuttle?

11. **Geometry** Study the pattern of circles.

a. Draw the next two figures in the sequence.

b. Write a sequence of numbers for the number of pieces in each circle.

c. Write a sequence of numbers for the number of cuts in each circle.

d. What is the relationship between the number of cuts and the number of pieces in each circle?

For **Extra Practice**, see page 740.

12. When Gabriella divided 45,109.5 by 1479 the calculator showed 305. Is this a reasonable answer? Explain.

WORKING ON THE
Investigation

Up, Up, and Away!

Refer to the Investigation on pages 2–3.

Look over your research and the list of considerations that you placed in your Investigation Folder.

- Work with the members of your group to choose one of the questions to consider first.

- Explore the question. Decide what you know and what you need to know to find the answer. You may need to do some more research.
- Make a plan for solving. If possible, make an estimate.
- Execute your plan to solve. If your plan doesn't work, make another plan and solve again.
- Examine your answer. Does it come close to the estimate? Can you justify it?

Add the results of your work to your Investigation Folder.

1-2

Order of Operations

You'll Learn

■ To use the order of operations to evaluate expressions.

It's Important

Knowing the order of operations will help you find the total cost of admission for an event.

 Technology Activity

Different calculators use different methods to find the value of an expression. A mathematical **expression** is any combination of numbers and operations such as addition, subtraction, multiplication, and division. To **evaluate** an expression, you find its numerical value. The value for each of the expressions below was found on two different calculators, **A** and **B**. The answers each calculator gave are shown.

Expression	$7 + 2 \times 3$	$8 + 4 \div 2$	$12 \div 6 + 16 \div 4$	$19 - 7 + 12 \times 2 \div 8$
Calculator A	13	10	6	15
Calculator B	27	6	4.5	6

Your Turn Use a calculator to evaluate each expression above, entering the numbers and symbols in the order shown.

a. Did the calculator you used agree with the answers given by calculator A or calculator B?

b. Compare the answers you found to the ones that the rest of the class found. Did everyone agree with either calculator A or B?

c. Discuss reasons that the calculators may have given different answers. How did each calculator find the answer?

d. Which value do you think is correct for each expression? Defend your answer.

Learning the Concept

There are two methods that someone might use to find the value of an expression that contains more than one operation. Consider the expression $8 + 5 \times 2$.

Method 1
$8 + 5 \times 2 = 8 + 10$ *Multiply,*
$8 + 5 \times 2 = 18$ *then add.*

Method 2
$8 + 5 \times 2 = 13 \times 2$ *Add, then*
$8 + 5 \times 2 = 26$ *multiply.*

In order to avoid confusion, mathematicians established the **order of operations** to tell us how to find the value of an expression.

First: Do all multiplications and divisions from left to right.
Second: Do all additions and subtractions from left to right.

The order of operations guarantees that each numerical expression has a *unique* value. Using the order of operations, the correct value of the expression $8 + 5 \times 2$ is 18.

Example **Find the value of each expression.**

a. $7 \times 9 + 3 = 63 + 3$ *Multiply 7 and 9.*
             $= 66$ *Add 63 and 3.*

b. $18 + 6 - 8 \div 2 = 18 + 6 - 4$ *Divide 8 by 2.*
                 $= 24 - 4$ *Add 18 and 6.*
                 $= 20$ *Subtract 4 from 24.*

c. $144 \div 16 + 9 \div 3 + 12 = 9 + 9 \div 3 + 12$ *Divide 144 by 16.*
                          $= 9 + 3 + 12$ *Divide 9 by 3.*
                          $= 12 + 12$ *Add 9 and 3.*
                          $= 24$ *Add 12 and 12.*

TECHNO TIP

Many calculators allow you to evaluate expressions with parentheses. Consult your User's Guide to see if the calculator you use has this capability.

The order of operations can be changed by using grouping symbols like **parentheses**, ( ), and **brackets**, [ ]. The value of the expression is found by performing the operation in the grouping symbols first.

Example **Find the value of each expression.**

a. $45 - 2 \times (15 - 3)$
    $45 - 2 \times (15 - 3) = 45 - 2 \times (12)$ *Do the operations in*
                          $= 45 - 24$ *parentheses first.*
                          $= 21$ *Multiply 2 by 12.*

b. $(22 + 3) \div (9 - 4)$ *Do the operations in*
    $(22 + 3) \div (9 - 4) = 25 \div 5$ *parentheses first.*
                      $= 5$ *Divide 25 by 5.*

The order for performing operations is as follows.

Order of Operations

1. Do all operations within grouping symbols first; start with the innermost grouping symbol.
2. Next, do all multiplications and divisions from left to right.
3. Then, do all additions and subtractions from left to right.

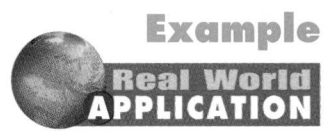

Example 3

Real World
APPLICATION

Sports

In each week of the football season, the Associated Press poll ranks football teams using votes from a group of sports reporters. A team gets 25 points for each first-place vote, 24 points for each second-place vote, 23 points for each third-place vote, and so on, to one point for each twenty-fifth place vote. The team with the highest total is ranked number one. Suppose that in one week the Miami Hurricanes received 10 first-place votes, 4 second-place votes, 2 fourth-place votes, and 1 tenth-place vote.

a. Write an expression for the number of points that the Hurricanes earned.

$$\begin{pmatrix} points\ for \\ 10 \times first\text{-}place \\ vote \end{pmatrix} + \begin{pmatrix} points\ for \\ 4 \times second\text{-}place \\ vote \end{pmatrix} + \begin{pmatrix} points\ for \\ 2 \times fourth\text{-}place \\ vote \end{pmatrix} + \begin{pmatrix} points\ for \\ 1 \times tenth\text{-}place \\ vote \end{pmatrix}$$

$$(10 \times 25) \quad + \quad (4 \times 24) \quad + \quad (2 \times 22) \quad + \quad (1 \times 16)$$

b. Find the total number of points for the Hurricanes.

$(10 \times 25) + (4 \times 24) + (2 \times 22) + (1 \times 16)$
$= 250 + 96 + 44 + 16$ *Do operations in parentheses.*
$= 406$ *Add.*

The Hurricanes earned 406 points.

There are many ways used to indicate multiplication and division in algebra.

A raised dot or parentheses can be used to indicate multiplication.

$5 \cdot 6 \xrightarrow{means} 5 \times 6$

$7(3), (7)3,$ or $(7)(3) \xrightarrow{means} 7 \times 3$

A fraction bar can be used to indicate division.

$\dfrac{37 + 38}{30 - 5} \xrightarrow{means} (37 + 38) \div (30 - 5)$

Example 4 Find the value of each expression.

a. $5(7 + 4) - 8 \cdot 4 = 5(11) - 8 \cdot 4$
$= 55 - 32$ *5(11) means 5 × 11 and*
$= 23$ *8 · 4 means 8 × 4*

b. $5[(5 + 14) - 2(7)] = 5[19 - 2(7)]$ *Do operations in innermost*
$= 5[19 - 14]$ *grouping symbols first.*
$= 5[5]$
$= 25$

c. $\dfrac{72 + 12}{35 + 7} = (72 + 12) \div (35 + 7)$ *Rewrite division using*
$= 84 \div 42$ *parentheses.*
$= 2$

Communicating Mathematics

Read and study the lesson to answer these questions.

1. **Identify** the first operation you perform when evaluating the expression $(13 + 11) \div 4$.

2. Are the expressions $9(4 + 3)$ and $9 \cdot 4 + 3$ equivalent? Explain.

3. **Make a diagram** to show the order of steps you should follow when evaluating an expression.

4. **Write** an expression involving multiplication and subtraction in which the first step when you evaluate would be to subtract.

5. Play this game with a partner.

MATERIALS
index cards
pencil

 ▶ Write each of the digits 1–9 on a card and place cards face down.
 ▶ Draw six cards from the pile. Out of the six numbers chosen, choose one to act as the answer.
 ▶ Then work with your partner to write an expression with the other five numbers so that the result is the chosen answer.
 ▶ You may use any of the operations—addition, subtraction, multiplication, and division—and add grouping symbols as needed. You may also rearrange the numbers.

Guided Practice

Name the operation that should be performed first. Then find the value of each expression.

6. $9 + 2 \cdot 5$

7. $65 - 32 \div 8$

8. $33 \div 3 + 6$

9. $17 - 6(2)$

10. $8(2 + 4)$

11. $\dfrac{20 - 8}{9 - 5}$

12. **Retail Sales** Mediaphile CDs and Books has a new pricing policy on CDs. Every single album CD is $11.99 or less. Anthony rang up four $11.99 CDs and a double-album CD that sells for $18.99.

 a. Write an expression for the total cost of the merchandise before sales tax.
 b. What was the total before sales tax was added?

Independent Practice

Find the value of each expression.

13. $15 \div 3 + 12 \div 4$

14. $3 \cdot (4 + 5) - 7$

15. $15 \div 5 \times 3$

16. $40 \cdot (6 - 2)$

17. $12(7 - 2 \times 3)$

18. $36 - 6 \cdot 5$

19. $\dfrac{18 + 66}{35 - 14}$

20. $\dfrac{16 + 8}{15 - 7}$

21. $\dfrac{2(14 - 6)}{4}$

22. $96 \div (12 \cdot 4) \div 2$

23. $72 \div (9 \times 4) \div 2$

24. $(30 \times 2) - (6 \cdot 9)$

25. $3[6(12 - 3)] - 17$

26. $7[5 + (13 - 4) \div 3]$

27. $4[3(21 - 17) + 3]$

28. $5[(12 + 5) - 3(19 - 14)]$

29. $8[(26 + 10) - 4(3 + 2)]$

30. $10[8(15 - 7) - 4 \cdot 3]$

Calculators

Copy each sentence below. Experiment with a calculator to find where to insert parentheses to make each sentence true. You may need to use the parentheses keys.

31. $71 - 17 + 4 = 50$

32. $8 - 5 \times 4 + 2 = 18$

33. $18 \div 3 + 6 + 12 = 14$

34. Only the 1, ⊞, ⊟, ⊠, ⊘, ⦗, ⦘ and ⊜ keys on a scientific calculator are working. How can a result of 75 be reached by pushing these keys less than 20 times?

Applications and Problem Solving

35. Business Phyllis Sokol leads seminars on computer technology. On an upcoming trip, she will conduct a three-day seminar in New York City followed by a two-day seminar in Washington, D.C., and then a four-day seminar in Chicago. A magazine estimates that a business traveler will spend $330 on food and lodging per day in New York City, $247 per day in Washington, D.C., and $229 per day in Chicago.

Source: *Nation's Business* magazine

a. Write an expression for the amount of money that Ms. Sokol can expect to spend on food and lodging on this trip.

b. Find the amount of money that Ms. Sokol should expect to spend on food and lodging.

c. List some other items that Ms. Sokol will have to pay for on her trip. Estimate the cost of each item and the total cost of the trip.

36. Nutrition The table below shows the number of Calories in one serving of different snack foods movie-goers love.

a. David and Marta ordered 2 boxes of chocolate mints, a bucket of popcorn, and four colas to share with friends. Find the total number of Calories in all the food.

Snack	Calories per serving
Chocolate mints (3 ounces)	360
Chocolate-covered raisins (2.3 ounces)	270
Buttered popcorn, medium bucket	1221
Cola (16 ounces)	205
Diet cola (16 ounces)	0

b. Yolanda bought two diet colas and a box of chocolate-covered raisins to share with Curtiss. How many Calories did each person consume?

37. History You can plant a bit of history in your yard. A non-profit conservation group called American Forests collects seeds from trees at historic homes, grows them into saplings, and sells them to the public. Each sapling costs $35 and $7 is added to each order for shipping and handling. How much would an order for seven saplings cost?

Mixed Review

38. Standardized Test Practice According to a magazine, a good swimmer can swim the length of a standard-sized pool in 24 or fewer strokes. If 1470 laps in a pool is equivalent to swimming the distance across the English Channel, about how many strokes would a good swimmer use to swim across the English Channel? Solve the problem using the four-step plan. (Lesson 1-1)

   A. 10,000   **B.** 20,000   **C.** 37,500   **D.** 50,000

39. Number Theory Write a sentence or two about the pattern at the right. Find the products 44 × 46, 54 × 56, and 84 × 86 by extending the pattern. (Lesson 1-1)

$$4 \times 6 = 24$$
$$14 \times 16 = 224$$
$$24 \times 26 = 624$$
$$34 \times 36 = 1224$$

For **Extra Practice**, see page 740.

1-3 Variables and Expressions

What You'll Learn

■ To evaluate expressions containing variables, and

■ to translate verbal phrases into algebraic expressions.

Why It's Important

Evaluating algebraic expressions can help you determine the amount of your paycheck.

Application: Medicine

How much do you know about your hardworking heart?

▶ An adult heart is about the size of a clenched fist.

▶ In one year, a human heart beats over 30 million times. The heart of a 70 year old has beaten over 2.5 billion times!

▶ In an average lifetime, the heart pumps 1 million barrels of blood.

One way that people maintain good health is by monitoring their blood pressure. Blood pressure is the pressure that your blood exerts on your arteries. The *systolic pressure* is the pressure when the heart contracts, and the *diastolic pressure* is the pressure when the heart is between contractions.

As people age, their blood pressure rises. You can approximate a person's normal systolic pressure by dividing his or her age by 2 and then adding 110.

Learning the Concept

Algebra, like any language, is a language of symbols. You know the symbols for division and addition, so you can write the blood-pressure relationship as *age* ÷ 2 + 110.

In arithmetic, you could write ● ÷ 2 + 110, where ● is the person's age in years. The ● serves as a placeholder.

In algebra, we use placeholders called **variables**. Variables are usually letters. The letter *x* is used very often as a variable, but it is also common to use the first letter of the value you are representing. For example, we can use *a* to represent age in the blood-pressure relationship. The relationship can then be written as *a* ÷ 2 + 110.

age ÷ 2 + 110	*Words and symbols*
● ÷ 2 + 110	*Arithmetic*
a ÷ 2 + 110	*Algebra*

a ÷ 2 + 110 is called an **algebraic expression** because it is a combination of variables, numbers, and at least one operation.

Expressions like *a* ÷ 2 + 110 can be evaluated by replacing the variables with numbers and then finding the numerical value of the expression.

If Malcolm is 18 years old, he could estimate his normal blood pressure by evaluating the expression $18 \div 2 + 110$.

$$
\begin{aligned}
a \div 2 + 110 &= 18 \div 2 + 110 & & \textit{Replace a with 18.} \\
&= 9 + 110 & & \textit{Divide first, then add.} \\
&= 119
\end{aligned}
$$

Malcolm's normal systolic blood pressure is approximately 119.

Replacing a with 18 demonstrates an important property of numbers.

Substitution Property of Equality	For all numbers a and b, if $a = b$, then a may be replaced with b.

Example **Evaluate $n + m - 34$ if $n = 18$ and $m = 49$.**

$$
\begin{aligned}
n + m - 34 &= 18 + 49 - 34 & & \textit{Replace n with 18 and m with 49.} \\
&= (18 + 49) - 34 & & \textit{Follow the order of operations.} \\
&= 67 - 34 & & \textit{18 + 49 = 67} \\
&= 33 & & \textit{67 − 34 = 33}
\end{aligned}
$$

THINK ABOUT IT

How could you rewrite $a \div 2 + 110$ using this notation?

As with numerical expression, there is special notation for multiplication and division with variables.

$$
\begin{aligned}
3d &\xrightarrow{\textit{means}} 3 \times d \\
xy &\xrightarrow{\textit{means}} x \times y \\
7st &\xrightarrow{\textit{means}} 7 \times s \times t \\
\frac{q}{4} &\xrightarrow{\textit{means}} q \div 4
\end{aligned}
$$

Example **Evaluate each expression if $x = 3$, $y = 8$, and $z = 5$.**

a.
$$
\begin{aligned}
7x - 4z &= 7(3) - 4(5) & & \textit{Replace x with 3 and z with 5.} \\
&= 21 - 20 & & \textit{Use the order of operations.} \\
&= 1
\end{aligned}
$$

b.
$$
\begin{aligned}
\frac{yz}{4} &= (8)(5) \div 4 & & \frac{yz}{4} = yz \div 4 \\
&= 40 \div 4 & & \textit{Use the order of operations.} \\
&= 10
\end{aligned}
$$

c.
$$
\begin{aligned}
3x + (z + 2y) - 12 &= 3(3) + (5 + 2 \cdot 8) - 12 \\
&= 3(3) + (5 + 16) - 12 \\
&= 9 + 21 - 12 \\
&= 18
\end{aligned}
$$

You can use variables to show patterns in expressions. For example, a babysitter may earn \$4 per hour. The chart at the right shows several possibilities for number of hours and earnings.

Hours of babysitting	Money earned
2	$4 \cdot 2$ or 8
5	$4 \cdot 5$ or 20
8	$4 \cdot 8$ or 32
h	$4 \cdot h$ or $4h$

The variable h and the algebraic expression $4h$ summarize the relationship between number of hours of babysitting and earnings.

When reading a verbal sentence and writing an algebraic expression to represent it, there are many words and phrases that suggest the operation to use. The following chart shows some common phrases and operations.

Addition	Subtraction	Multiplication	Division
plus	minus	times	divided
sum	difference	product	quotient
more than	less than	multiplied	
increased by	subtract	each	
total	decreased by	of	
in all			

Example ③ **Translate each phrase into an algebraic expression.**

a. ten more points than Marisa scored
Let p represent the points Marisa scored.
The words *more than* suggest addition.
The algebraic expression is $p + 10$ or $10 + p$.

b. three times last year's total
Let t represent last year's total.
The words *three times* suggest multiplication.
The algebraic expression is $3t$ or $3 \cdot t$.

Example ④

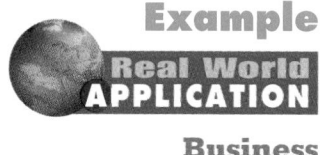

Business

Warnet Cable charges $19.95 a month for basic cable television service. Each premium channel selected costs an additional $4.95 per month. Write an expression for the cost of a month of cable service.

Let n represent the number of premium channels selected.

Since you are paying $4.95 *for each* premium channel, the total for premium channels will be $4.95 \times n$ or $4.95n$.

The word *additional* suggests addition. You will pay for basic service and the premium channels.

The algebraic expression is $19.95 + 4.95n$.

Checking Your Understanding

Communicating Mathematics

Read and study the lesson to answer these questions.

1. **Define** a variable in your own words.

2. **Write** two different expressions that are the same as $7a$.

3. If m is the number of games the Jets won, what does $m - 5$ represent?

MATERIALS

🥤 cups

counters and mat

4. Cups and counters can be used to model algebraic expressions. The cup represents the variable, and the counters are one unit each. The diagram models $2x + 3$.

   a. Use cups and counters to model the expression $4x + 2$.

   b. Model the phrase *four plus three times a number*. Then write an expression to represent the phrase.

Guided Practice

Evaluate each expression if $m = 4$, $n = 11$, $p = 2$, and $q = 5$.

5. $8 + n$ **6.** $10 - q$ **7.** $3m$

8. $24 - 4q$ **9.** $\dfrac{5m}{2}$ **10.** $n + 110 - 2p$

Translate each phrase into an algebraic expression.

11. three more than v

12. seven less than h

13. the product of a number and 6

14. twice w

15. **Space Exploration**   The force of gravity on Earth is six times greater than on the moon. As a result, objects weigh six times as much on Earth as they do on the moon.

   a. Write an expression for the weight of an object on Earth if the weight on the moon is x.

   b. A scientific instrument weighs 34 pounds on the moon. How much does the instrument weigh on Earth?

Exercises: Practicing and Applying the Concept

Independent Practice

Evaluate each expression if $a = 6$, $b = 3$, and $c = 7$.

16. $a + 22$ **17.** $2b - 4$

18. $ca - ab$ **19.** $4a - (b + c)$

20. $5c + 5b$ **21.** $\dfrac{8b}{a}$

22. $2c + 3a + 6b$ **23.** $\dfrac{6a}{b}$

24. $9a - (4b + 2c)$ **25.** $\dfrac{3(4a - 3c)}{c - 4}$

26. $12b - \dfrac{2c - 4}{a + 4}$ **27.** $\dfrac{15ab}{3c + 6}$

Translate each phrase into an algebraic expression.

28. a number divided by 8   **29.** the quotient of ninety-two and c

30. the sum of g and 8   **31.** p decreased by 5

32. eleven increased by f **33.** the difference of u and 10

34. twice a number   **35.** the quotient of eighty-eight and b

36. Gil's salary plus a $500 bonus   **37.** three times as many fouls as Sumi

38. three times a number decreased by 18

39. two more than twice the previous month's sales

Write a verbal phrase for each algebraic expression.

40. $x + 4$ **41.** $16 - b$ **42.** $7n$

43. $v \div 5$ **44.** $2(a + 1)$ **45.** $2a + 1$

Critical
Thinking

46. Write an expression for the value of a three-digit number whose hundreds, tens, and units digits are *x*, *y*, and *z*, respectively.

**Applications and
Problem Solving**

47. Meteorology You can estimate the temperature in degrees Fahrenheit by counting the number of times a cricket chirps in one minute. Count the number of chirps, divide by 4, and then add 37.

   a. Write an expression for the temperature relationship.

   b. Find the approximate temperature when a cricket chirps 124 times in a minute.

48. Statistics The *mean* is a number that is often used to represent a set of numbers. To find the mean of three numbers, find the sum of the numbers and then divide by 3.

   a. Write an expression for the mean of *a*, *b*, and *c*.

   b. Find the mean of 51, 66, and 63.

**Mixed
Review**

49. Retail Sales Jamie bought four adult tickets at $6.25 each and two children's tickets at $3.75 each. (Lesson 1-2)

   a. Write an expression for the total cost of the tickets.

   b. Find the cost.

   c. Write your own problem involving movie tickets and solve.

50. Standardized Test Practice Find the value of the expression $6 \times 8 - (9 + 12)$. (Lesson 1-2)

   A. -78

   B. 6

   C. 27

   D. 51

51. Patterns Use the four-step problem-solving plan to find the next number in the sequence. (Lesson 1-1)
1, 2, 4, 7, 11, 16, 22,   __?__

52. Art A reproduction of Vincent van Gogh's *Sunflowers* was constructed in a wheat field near Duns, Scotland. The "painting" covered a 46,000-square foot area and contained 250,000 plants and flowers. About how many flowers and plants were planted per square foot? (Lesson 1-1)

   a. Which method of computation do you think is most appropriate for solving this problem? Explain.

   b. Solve the problem.

For **Extra Practice**, see page 740.

1-3B Evaluating Expressions

An Extension of Lesson **1-3**

Graphing calculators observe the order of operations. Since you can specify the value of a variable, evaluating expressions is as simple as entering the expression with the calculator keys. The calculators also have parentheses.

Activity

Evaluate $3 + (2x - y)$ if $x = 7$ and $y = 5$ on a graphing calculator.

ENTER: 7 [STO▶] [X,T,θ,n] [ALPHA] [:]   *Enters the value of x.*

           5 [STO▶] [ALPHA] [Y] [ALPHA] [:]   *Enters the value of y.*

           3 [+] [(] 2 [X,T,θ,n] [−] [ALPHA]   *Enters and evaluates the*

           [Y] [)] [ENTER]   *expression.*

DISPLAY: *12*

If you get an error message or discover that you entered the expression incorrectly, you can use the REPLAY feature to correct your error and reevaluate without reentering the expression. Follow the steps below to use the REPLAY feature.

▶   Press [2nd] [ENTRY] to display your expression.

▶   Use the arrow keys to move to the location of the correction. Then type over, use [INS], or use [DEL] to make the correction.

▶   Press [ENTER] to reevaluate.

Your Turn

Evaluate each expression with a graphing calculator if $a = 9$, $b = 0$, $c = 7$, and $d = 3$.

   1. $5d - 1$ **2.** $b(a - c)$ **3.** $3(b + c) \div d$

   4. $4a \div d$ **5.** $9a - (4d + 2c)$ **6.** $a(b + c) - 7$

   7. How can you use the REPLAY feature to evaluate several expressions for variables of the same value?

   8. What do you observe about the display when you press the [÷] and the [×] keys?

Properties

You'll Learn

- To identify properties of addition and multiplication, and

- to use these properties to solve problems.

It's Important

Using properties can help you determine how much money you can earn each month.

MATERIALS: ⬤ counters ✺ beans, or other small objects

Hands-On Activity

Small objects like counters or beans can be used to model operations with whole numbers. The diagram at the right models the expression $5 + 3$. Counting the number of counters in the group after the additions shows that $5 + 3 = 8$.

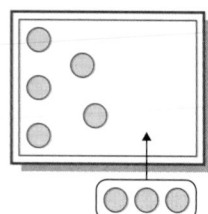

Your Turn   Use counters to model $3 + 5$.

a. What is the value of $3 + 5$ that you found using the counters?

b. What do you observe about the values of $5 + 3$ and $3 + 5$?

c. Work with a partner to write and solve other problems like $a + b$ and $b + a$. Summarize your results.

d. Investigate $3 \cdot 5$ and $5 \cdot 3$ with counters by making 3 groups of 5 counters and 5 groups of 3 counters. Then explore ab and ba. What do you observe?

Learning the Concept

The charts below and on the next page summarize some of the properties of addition and multiplication.

Commutative Properties of Addition and Multiplication	
The order in which numbers are added does not change the sum. $$5 + 3 = 3 + 5$$ For any numbers a and b, $$a + b = b + a$$	The order in which numbers are multiplied does not change the product. $$2 \cdot 4 = 4 \cdot 2$$ For any numbers a and b, $$a \cdot b = b \cdot a$$
Associative Properties of Addition and Multiplication	
The way in which addends are grouped does not change the sum. $$(2 + 4) + 6 = 2 + (4 + 6)$$ For any numbers a, b, and c, $$(a + b) + c = a + (b + c)$$	The way in which factors are grouped does not change the product. $$(6 \cdot 3) \cdot 7 = 6 \cdot (3 \cdot 7)$$ For any numbers a, b, and c, $$(a \cdot b) \cdot c = a \cdot (b \cdot c)$$

Identity Properties of Addition and Multiplication	
The sum of an addend and zero is the addend. $6 + 0 = 6$ For any number a, $a + 0 = a$.	The product of a factor and one is the factor. $6 \cdot 1 = 6$ For any number a, $a \cdot 1 = a$.
Multiplicative Property of Zero	
The product of a factor and zero is zero. $5 \cdot 0 = 0$ For any number a, $a \cdot 0 = 0$.	

These properties can be helpful when you find sums and products mentally.

Example **Find the value of each expression mentally.**

a. $17 + (13 + 15) = (17 + 13) + 15$ *Regroup addends.*
$$= 30 + 15$$
$$= 45$$

b. $2 \cdot 16 \cdot 5 = (2 \cdot 5) \cdot 16$ *Rearrange and group factors.*
$$= 10 \cdot 16 \text{ or } 160$$

Example

On a recent tour, Elton John and Billy Joel visited Knoxville, Raleigh, Richmond, and Washington, D.C. The mileages between the cities are shown in the table below. If the equipment truck traveled from each city to the next in consecutive days, how many miles did it travel?

City	Destination	Mileage
Knoxville, TN	Raleigh, NC	362
Raleigh, NC	Richmond, VA	156
Richmond, VA	Washington, D.C.	108

They would travel the total of the mileages between the cities.

Rearrange and group addends so addition is easier.

$$362 + 156 + 108 = (362 + 108) + 156$$
$$= 470 + 156$$
$$= 626$$

Connection to ⟩ Algebra

Variables represent numbers, so the properties can be used to rewrite and simplify algebraic expressions.

Example **3** Rewrite each expression using a commutative property.

  a. $n + 6$ **b.** $4 \cdot s$
  $n + 6 = 6 + n$ *Change the order.* $4 \cdot s = s \cdot 4$

Example **4** Rewrite each expression using an associative property. Then simplify.

THINK ABOUT IT

Is subtraction commutative or associative?

  a. $(m + 5) + 4$ **b.** $3(7t)$
     $(m + 5) + 4 = m + (5 + 4)$ $3(7t) = (3 \cdot 7)t$ *Regroup.*
              $= m + 9$ $= 21t$

Example **5** Evaluate xyz if $x = 2$, $y = 8$, and $z = 15$ mentally.

  $xyz = 2 \cdot 8 \cdot 15$ *Replace x with 2, y with 8, and z with 15.*
       $= 2 \cdot 15 \cdot 8$ *Use the Commutative property to rearrange factors.*
       $= (2 \cdot 15) \cdot 8$ *Group 2 and 15 to make multiplication easier.*
       $= 30 \cdot 8$
       $= 240$

Checking Your Understanding

Communicating Mathematics

Read and study the lesson to answer these questions.

1. **Compare and contrast** the associative and commutative properties.

2. **Write** mathematical sentences to illustrate each of the seven properties described in the tables on pages 22 and 23.

3. **Demonstrate** how the properties can be used to perform calculations mentally.

 MATERIALS

   counters

4. Use counters to model the expressions $(2 + 5) + 6$ and $2 + (5 + 6)$.

   a. What do you observe about the values of each expression?

   b. Model several more expressions of the form $(a + b) + c$ and $a + (b + c)$. Explain your results in terms of one of the properties of addition.

Guided Practice

Name the property shown by each statement.

5. $(3 + 8) + 7 = 3 + (8 + 7)$ 6. $8 + 9 = 9 + 8$

7. $(5 + 7) + 2 = (7 + 5) + 2$ 8. $0 + 9 = 9$

Find each sum or product mentally.

9. $12 + 9 + 8$ 10. $5 \cdot 3 \cdot 8$ 11. $4 + 13 + 26 + 5$

Rewrite each expression using an associative property. Then simplify.

12. $(y + 9) + 8$ 13. $7(6z)$ 14. $(k \cdot 8)3$

15. **Food Service** For lunch, Juana ordered an iced tea, a bowl of clam chowder, and a small Caesar salad. If the prices of the items are $1.25, $3.35, and $3.75 respectively, how much should Juana's bill be before tax and tip?

Independent Practice

Name the property shown by each statement.

16. $22 \cdot 1 = 22$

17. $19 \cdot 12 = 12 \cdot 19$

18. $6 \cdot 33 = 33 \cdot 6$

19. $3 \cdot 7 \cdot 0 = 0$

20. $8cd = 8dc$

21. $x + (2y + z) = (2y + z) + x$

22. $0 + 9 = 9 + 0$

23. $5s + 9 = 9 + 5s$

24. $(7a)(b) = 7(ab)$

25. $8t + 6 = 6 + 8t$

26. $1m = m$

27. $7 + (2 + u) = (7 + 2) + u$

Find each sum or product mentally.

28. $16 + 7 + 14$

29. $2 \cdot 9 \cdot 20$

30. $82 + 58 + 23 + 37$

31. $18 \cdot 6 \cdot 0$

32. $7 + 99 + 123$

33. $2 \cdot 13 \cdot 5$

34. $7 + 2 + 13 + 18$

35. $6 \cdot 11 \cdot 10$

36. $129 \cdot 8 \cdot 0$

Rewrite each expression using a commutative property.

37. $5 + 9$

38. $8a + 6$

39. $9 + 18w$

Rewrite each expression using an associative property. Then simplify.

40. $(y + 7) + 6$

41. $(b \cdot 6) \cdot 5$

42. $2(8f)$

43. $13 + (11 + m)$

44. $3(2z)$

45. $(p \cdot 7) \cdot 4$

46. $(u + 8) + 16$

47. $(15 + 4w) + w$

48. $0(3x)$

Critical Thinking

49. Calculator Use a calculator to find $28 \div 7$ and $7 \div 28$.

    a. Experiment with other expressions of the form $a \div b$ and $b \div a$.

    b. Write a statement about division and the commutative property.

Applications and Problem Solving

50. Sports In the game of volleyball, the net is 3 feet 3 inches tall. The bottom of the net is to be set 4 feet 8 inches from the floor.

    a. Write an expression for the distance from the floor to the top of the net.

    b. Find the distance from the floor to the top of the net. Explain how you used the associative and commutative properties.

51. Science Mr. Rex instructed students in his class to pour acid in their flasks into water in their beakers. He warned that pouring water into acid could cause spattering and burns. These actions are not commutative. Give another example of actions that are not commutative.

Mixed Review

52. Write an expression for *the product of a number and 18*. (Lesson 1-3)

53. Evaluate the expression $8xy + 6$ if $x = 0$ and $y = 2$. (Lesson 1-3)

54. Find the value of the expression $7 \cdot (9 - 3) + 4$. (Lesson 1-2)

55. Standardized Test Practice What is the value of $90 - 9 \times 2$? (Lesson 1-2)

    A. 6561   **B.** 162   **C.** 72   **D.** 9

For **Extra Practice**, see page 741.

56. Patterns Find the next number in the sequence 15, 14, 12, 9, 5, __?__. (Lesson 1-1)

The Distributive Property

What You'll Learn

■ To simplify algebraic expressions using the distributive property.

Why It's Important

You can simplify algebraic expressions using the distributive property when determining the number of lunch combinations available in the school cafeteria.

Application: Travel

Sunny beaches, exciting ski lodges, and exotic experiences in far-off lands await you! How do you get there? Call a travel agent! Elena Chávez is a travel agent for Carlson Travel. She helps people plan vacations and business trips by making arrangements for hotels, airline tickets, rental cars, and any other services her clients need.

Ms. Chávez received the brochure at the right for a travel package to Barbados. Her clients, Mr. and Mrs. Yoder, are considering Barbados for a summer vacation. If the round-trip airfare to Barbados from Columbus is $679, how much should Ms. Chávez tell the Yoders that the airfare and package trip would cost the couple? *This problem will be solved in Example 2.*

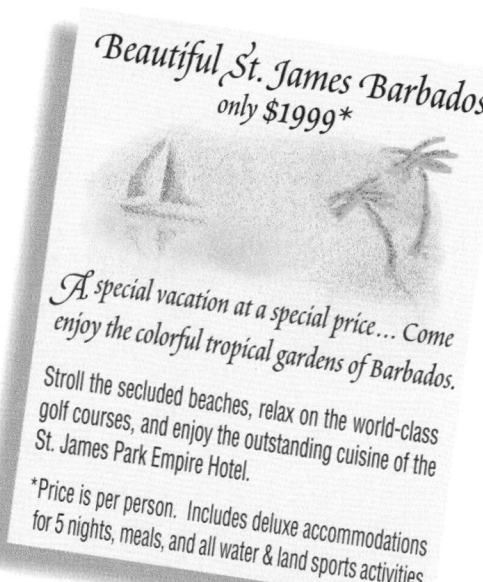

Beautiful St. James Barbados
only $1999*

A special vacation at a special price... Come enjoy the colorful tropical gardens of Barbados.

Stroll the secluded beaches, relax on the world-class golf courses, and enjoy the outstanding cuisine of the St. James Park Empire Hotel.

*Price is per person. Includes deluxe accommodations for 5 nights, meals, and all water & land sports activities.

Learning the Concept

In Lesson 1-4, you learned some of the properties of addition and multiplication with whole numbers. The **distributive property** ties addition and multiplication together.

Distributive Property	**In words:**	The sum of two addends multiplied by a number is the sum of the product of each addend and the number.
	In symbols:	For any numbers a, b, and c, $a(b + c) = ab + ac$ and $(b + c)a = ba + ca$.

The expression $a(b + c)$ is read "a times the quantity b plus c" or "a times the sum of b and c."

The distributive property allows you to write expressions in different forms.

Example **1**   **Find 14 · 12 mentally using the distributive property.**

$$14 \cdot 12 = 14(10 + 2)$$ *Replace 12 with 10 + 2.*
$$= 14 \cdot 10 + 14 \cdot 2$$ *Think of 140 + 28.*
$$= 140 + 28 \text{ or } 168$$

Example **2**

Travel

Refer to the application at the beginning of the lesson.
a. **Write two different expressions for the price of the trip for two people.**
b. **Find the cost of the trip.**

a. There are two methods of writing the total cost: find the total cost for one person and multiply by 2, or find the cost of each part of the trip for two people and then add.

Method 1			Method 2		
number of people	*times*	*total trip cost*	*airfare for two*	*plus*	*package for two*
2	×	($679 + $1999)	2($679)	+	2($1999)

Notice that if we apply the distributive property to the expression on the left, the result is the expression on the right.

b. Evaluate either expression to find the cost of the trip.
$$2(\$679 + 1999) = 2(\$2678)$$
$$= \$5356$$
Check the result by evaluating the expression 2($679) + 2($1999).

Connection to ⟩ Algebra

Refer again to the application at the beginning of the lesson. The Yoders are thinking of inviting some friends to join them on their vacation to Barbados. Ms. Chávez can use the variable *p* for the number of people.

If there are *p* people traveling, the airfare will be 679*p* dollars.
For the same number of people, the package cost will be 1999*p* dollars.

The total cost for *p* people could be represented as 679*p* + 1999*p* dollars.

THINK ABOUT IT

Are 5*x* and 5*y* like terms?

The expression 679*p* + 1999*p* has two **terms** with the same variable. These terms are called **like terms**. Some other pairs of like terms are 6*d* and 17*d*, 3*st* and 5*st*, and *x* and 4*x*. The distributive property can be helpful in simplifying expressions that have like terms. Example 3a shows how to simplify 679*p* + 1999*p*.

Example **3**   **Simplify each expression.**

a. **679*p* + 1999*p***
$$679p + 1999p = (679 + 1999)p$$ *p is a factor of both 679p and 1999p.*
$$= 2678p$$ *Use the distributive property.*

b. ***x* + 12*x***
$$x + 12x = 1x + 12x$$ *x = 1 · x by multiplicative identity*
$$x + 12x = (1 + 12)x$$ *Distributive property*
$$= 13x$$

An expression is in **simplest form** when it has no like terms and no parentheses.

Example **Write each expression in simplest form.**

a. $16k + 9 + 5k$

$$16k + 9 + 5k = 16k + 5k + 9 \qquad \textit{Commutative property of addition}$$
$$= (16 + 5)k + 9 \qquad \textit{Distributive property}$$
$$= 21k + 9$$

b. $m + 5(n + 7m)$

$$m + 5(n + 7m) = m + 5n + 5 \cdot 7m \qquad \textit{Distributive property}$$
$$= m + 5n + 35m \qquad \textit{Substitution property}$$
$$= m + 35m + 5n \qquad \textit{Commutative property of addition}$$
$$= 1m + 35m + 5n \qquad \textit{Multiplicative identity}$$
$$= (1 + 35)m + 5n \qquad \textit{Distributive property}$$
$$= 36m + 5n$$

Checking Your Understanding

Communicating Mathematics

Read and study the lesson to answer these questions.

1. **Rewrite** the expression $7(5 + 3)$ in three different ways using the distributive property.

2. **Give an example** of two like terms. Explain why they are like terms.

3. **Make a table** of expressions that are and are not in simplest form. Explain how you know whether an expression is in simplest form.

MATH JOURNAL

4. Make up a method for remembering the distributive property. Write your method and a description of how it works in your journal.

Guided Practice

Restate each expression using the distributive property. Do not simplify.

5. $5(7 + 8)$ **6.** $6(2 + 4)$ **7.** $4(x + 3)$

Simplify each expression.

8. $6 \cdot 5 + 9 \cdot 6$ **9.** $12a + a + 7$ **10.** $x + 9x$

11. $6c + c + 7$ **12.** $22c + 4(2 + 4c)$ **13.** $4ab + 6ab + 10ab$

14. Money Carlos works part time at a pizzeria to save for college. He earns $5.25 an hour plus tips.

     a. Carlos worked 4 hours on Friday and 8 hours on Saturday. Write two different expressions for his wages that weekend.

     b. Find the amount Carlos earned if his tips totaled $35.50.

Exercises: Practicing and Applying the Concept

Independent Practice

Restate each expression using the distributive property. Do not simplify.

15. $3(11 + 12)$ **16.** $(5 + 7)x$ **17.** $6t + 11t$

18. $2a + 4b$ **19.** $2r + 12s$ **20.** $6x + 6y + 1$

21. $(9 + 8)v$ **22.** $12r + 12s$ **23.** $(4x + 9y)2$

Simplify each expression.

24. $6d + d + 15$
25. $17a + 21a + 45$
26. $m + m$
27. $c + 24c + 16$
28. $q + 4 + 11 + 4q$
29. $18y + 5(7 + 3y)$
30. $14(b + 3) + 8b$
31. $30(b + 2) + 2b$
32. $3(8 + a) + 7(6 + 4a)$
33. $8(9 + 3f) + f$
34. $x + 5x + 8(x + 2)$
35. $3(x + y) + 4(2x + 3y)$

Critical Thinking

36. The distributive property given on page 26 is often called the *distributive property of multiplication over addition*. Is there a distributive property of multiplication over subtraction? Evaluate several different expressions of the form $a(b - c)$ and $ab - ac$ and explain your findings.

Applications and Problem Solving

37. **Engineering** The foundation of a building is critical to ensuring that the building is safe. Before the foundation can be poured, an engineer calculates the volume of material to be removed. The expression used to find the volume of earth to remove for an opening of the shape shown is $\dfrac{d(a + b)}{2} \times L$. Find the volume of earth to remove if $a = 52$ m, $b = 48$ m, $d = 6$ m, and $L = 40$ m.

38. **Business Operations** Most products sold in the United States carry a universal product code (UPC) like the one at the right. The last digit of the code, the check digit, is used to ensure that the numbers of the code are entered correctly. The computer uses the steps below to verify the check digit.

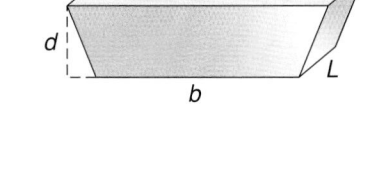

0 36800 30685 1

　　1. Add the digits in the odd-numbered positions. Multiply the total by 3.
　　2. Add the digits in the even-numbered positions. Then add to the previous total.
　　3. Subtract the result from the next highest multiple of 10. For example, a result of 57 would be subtracted from 60. The answer is the check digit.

　a. Write an expression for the first two steps of checking UPC 0 36800 30685 1. Be sure that you don't include the check digit in the calculation.

　b. Evaluate the expression you wrote for part a. Then subtract from the next highest multiple of 10. Does the check digit verify the code?

Mixed Review

39. Rewrite $(8 + 19) + 17$ using an associative property. (Lesson 1-4)

40. Name the property shown by the statement $12 \times 1 = 12$. (Lesson 1-4)

41. Translate the phrase *the quotient of a number and 9* into an algebraic expression. (Lesson 1-3)

42. Evaluate $75 - 10n$ if $n = 7$. (Lesson 1-3)

43. **Standardized Test Practice** Cindy sold 15 white chocolate hearts at $4.25 each, 36 milk chocolate hearts at $3.75 each, and 22 milk chocolate assortments at $7.45 each for her school. How much money did Cindy raise? (Lesson 1-2)

A. $1127.85   **B.** $362.65   **C.** $227.65   **D.** $163.90

44. **Geometry** Draw the next figure in the pattern. (Lesson 1-1)

For **Extra Practice**, see page 741.

45. Bandi used a calculator to find $528 \times 7 = 3696$. Without actually calculating, determine if this answer is reasonable. (Lesson 1-1)

Self Test

1. **Number Theory** The pattern at the right is known as Pascal's Triangle. Find the pattern and complete the 6th and 7th rows. (Lesson 1-1)

```
      1
    1   1
  1   2   1
 1  3   3  1
1  4   6  4  1
```

2. **Construction** Lillehammer, Norway hosted the Winter Olympic Games. Gjovick Olympic Cavern Hall, the site of the hockey competition, was carved out of Hovdeteoppen Mountain. 29,000 truckloads of rock were removed. If a truck holds about 5 cubic yards of rock, about how much rock was removed to build the hall? (Lesson 1-1)

a. Which method of computation do you think is most appropriate for this problem? Justify your choice.

b. Solve the problem using the four-step plan. Be sure to examine your solution.

Find the value of each expression. (Lesson 1-2)

3. $9(2 + 5) - 4$

4. $\dfrac{19 - 11}{2(4)}$

Evaluate each expression if $a = 3$, $b = 8$, and $c = 12$. (Lesson 1-3)

5. $\dfrac{ab}{c}$

6. $8c - (b + 3a)$

Name the property shown by each statement. (Lesson 1-4)

7. $0 + c = c + 0$

8. $19 + 7 + 1 = 19 + 1 + 7$

9. **Music** Enrique's piano teacher charges $12.50 for a lesson. If Enrique has 5 lessons in one month, how much does he owe his piano teacher? Solve mentally. (Lesson 1-5)

10. Simplify $2n + 4(n + 8n)$. (Lesson 1-5)

HANDS-ON ACTIVITY

1-5B Distributive Property

An Extension of Lesson **1-5**

MATERIALS

algebra tiles

Many algebraic principles can be modeled using geometry. You can look at the distributive property by modeling with algebra tiles of different sizes.

▶ Let $\square$ be a 1×1 square. That is, the length and width are each 1 unit. If the area of a square or a rectangle is found by multiplying the length and the width, what is the area of the square?

▶ Let $\boxed{}$ be a $1 \times x$ rectangle. The width is 1 unit, and the length is x units. What is the area of the rectangle?

Then $\boxed{}\square$ is a rectangle that has a width of 1 unit and a length of $x + 1$ units. What is the area of the rectangle?

Your Turn   Using your tiles, make rectangles with areas of $x + 2$, $2x$, $2x + 1$, and $2x + 2$ square units.

Explore   You can use geometric models to check the distributive property. Is it true that $2(x + 1) = 2x + 1$?

$2(x + 1)$ means   *Does it matter how the tiles are arranged?*

$2x + 1$ means $\boxed{}$

Therefore, $2(x + 1)$ is not the same as $2x + 1$. From the drawing above, $2(x + 1) = 2x + 2$.

 Tell whether each of the following statements is *true* or *false*. Justify your answers with algebra tiles or a drawing.

1. $2x + 3 = 6x$ **2.** $2x = x + x$ **3.** $3x + 3 = 3(x + 1)$

4. $3x + 2x = 6x$ **5.** $3x + 3 = 3(x + 3)$ **6.** $3x + 2x = x(3 + 2)$

7. You Decide   Steve says that $3(x + 2) = 3x + 2$. Linh disagrees. She thinks that $3(x + 2) = 3x + 6$. Write a paragraph with drawings to explain who you think is correct.

Variables and Equations

You'll Learn

- To identify and solve open sentences.

It's Important

You can use equations to enter information into a spreadsheet.

 Application: Entertainment

How do you like to spend a rainy Saturday afternoon? With a good book, an old movie, or how about a nice long board game? People have been playing board games for centuries and even with the advent of computer games, they still enjoy the real thing.

A board game that originated in India was one of the first games to receive a copyright in the United States. It was copyrighted in 1874.

The most famous board game was invented by Charles Darrow during the Great Depression. Darrow was an unemployed engineer who longed to be rich and spend his days on the boardwalks of Atlantic City, New Jersey. The copyright for Darrow's game was issued in 1933.

The number of years between the copyrights of the game from India and Darrow's game can be represented by $1933 - 1874 = y$.

Learning the Concept

A mathematical sentence like $1933 - 1874 = y$ is called an **equation**. Some examples of equations are $6 + 8 = 14$, $3(6) - 9 = 9$, and $6x - 2 = 11$. Equations that contain variables are **open sentences**. When the variables in an open sentence are replaced with numbers, the sentence may be true or false.

$1933 - 1874 = y$

$1933 - 1874 \stackrel{?}{=} 67$ *Replace y with 67.*
This sentence is false because $1933 - 1874 \neq 67$.

$1933 - 1874 \stackrel{?}{=} 59$ *Replace y with 59.*
This sentence is true.

A value for the variable that makes an equation true is called a **solution** of the equation. The process of finding a solution is called **solving the equation**.

The solution of $1933 - 1874 = y$ is 59.

Example **Which of the numbers 16, 24, or 32 is the solution of $66 + x = 98$?**

Replace x with each of the possible solutions to solve the equation.

$66 + 16 = 98$ *Replace x with 16.* $66 + 24 = 98$ *Replace x with 24.*
$\quad\quad 82 = 98$ $\quad\quad 90 = 98$

This sentence is false.   This sentence is false.

$66 + 32 = 98$ *Replace x with 32.*
$\quad\quad 98 = 98$

This sentence is true. The solution is 32.

THINK ABOUT IT

How could you find the solution to the equation in Example 1 without substituting each of the possible solutions?

You will learn many ways to solve equations in this course. Some equations can be solved mentally by using basic facts or arithmetic skills.

Example ②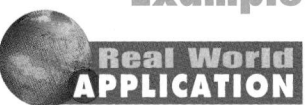

Basketball great Shaquille O'Neal is an imposing 7' 1" tall and 300 pounds. O'Neal has appeared in shoe commercials since 1992. When O'Neal began working with the shoe company, his shoe size was 19. By 1995, O'Neal's size had risen to a whopping size 22. How many shoe sizes did Shaquille O'Neal increase from 1992 to 1995?

Since the numbers are simple, this equation can be solved mentally.

$19 + s = 22$
$19 + 3 = 22$ *Replace s with 3 since this*
$\quad\quad s = 3$ *will result in a true*
$\quad\quad\quad\quad$ *sentence. The solution is 3.*

Shaquille O'Neal increased 3 shoe sizes from 1992 to 1995.

Equations involving products and quotients can also be found mentally.

Example ③ **Solve each equation mentally.**

a. $6x = 36$
$\quad 6 \cdot 6 = 36$ *Think "What number times 6 is 36?"*
$\quad\quad x = 6$

b. $\dfrac{45}{n} = 9$

$\quad \dfrac{45}{5} = 9$ *Think "What number times 9 is 45?"*

$\quad\quad n = 5$

Checking Your Understanding

Communicating Mathematics

Read and study the lesson to answer these questions.

1. **In your own words,** define an equation.

2. **Write** an equation that is always true and another equation that is always false.

3. **Find** the solution to the equation $8 + x = 12$ using the four-step problem-solving plan.

4. **Write** an open sentence and then rewrite it so that it is false.

5. You can use a graphing calculator to solve equations if the right side of the equation is zero. You need to enter the left side of the equation, the variable you wish to solve for, and a guess at the solution. Use the keystrokes below to solve $x - 5 = 0$. We guessed that the solution was 5.

   2nd catalog locate solve and press ENTER .

   X,T,θ,n – 5 , X,T,θ,n , 5) ENTER

   The solution of the equation, 5, is displayed.

   Use a graphing calculator to solve each equation below.

   a. $x - 9 = 0$ **b.** $2x = 0$ **c.** $3x - 6 = 0$

Guided Practice

Identify the solution to each equation from the list given.

6. $7 - x = 4$;   7. $y + 19 = 32$;   8. $2x + 1 = 7$;
   3, 4, 6   9, 13, 19   3, 4, 5

Solve each equation mentally.

9. $t + 5 = 10$ 10. $19 - q = 11$

11. $78 - r = 28$ 12. $\dfrac{12}{x} = 3$

13. **Food**   If you could choose only one snack food to eat for the rest of your life, what would it be? In a survey of 1000 people, 220 said hamburgers and 360 said pizza.

   a. Write an equation to find how many more people said pizza than hamburgers.

   b. Solve your equation.

Exercises: Practicing and Applying the Concept

Independent Practice

Identify the solution to each equation from the list given.

14. $12 - c = 8$; 2, 4, 7   15. $t + 33 = 72$; 29, 34, 39

16. $8 = \dfrac{16}{a}$; 1, 2, 4   17. $8 = \dfrac{g}{4}$; 32, 24, 16

18. $3x + 1 = 10$; 2, 3, 4   19. $7 = 5b + 2$; 0, 1, 2

20. $110 = 145 - t$; 35, 40, 45   21. $8 = \dfrac{48}{m}$; 3, 4, 6

Solve each equation mentally.

22. $m + 8 = 10$ 23. $129 - q = 9$ 24. $17 = k - 3$

25. $8x = 64$ 26. $\dfrac{50}{h} = 5$ 27. $63 = 7j$

28. $56 - s = 0$ 29. $99 = 9x$ 30. $\dfrac{72}{c} = 8$

31. $6x = 42$ 32. $18 = 2w$ 33. $\dfrac{21}{y} = 3$

Critical Thinking

34. Write two different open sentences that fit each description.

   a. The solution is 2.

   b. There is no solution that is a whole number. *Hint: The whole numbers are 0, 1, 2, 3, . . .*

Applications and Problem Solving

Data Update For the latest Dow Jones Industrial Average, visit:

www.glencoe.com/ sec/math/prealg/ mathnet

35. **Finance**   Stock market performance helps economists determine the health of the economy. One of the indicators of stock market performance is the Dow Jones Industrial Average. On February 21, 1995, the Dow Jones Industrial Average climbed above 4000 for the first time ever. That day, it closed at 4003.33. On February 20, 1995, the closing mark was 3973.05.

   a. Write an open sentence that describes the climb in the average.

   b. Is the solution to the open sentence 29.92, 30.28, or 33.32?

36. **Geometry**   The sum of the measures of the angles in any triangle is 180°.

   a. Write an equation for the sum of the angle measures in the triangle at the right.

   b. Solve your equation for x.

Mixed Review

37. Simplify the expression $5x + 2x + 1$. (Lesson 1-5)

38. Rewrite the expression $8(3 \cdot 6)$ using an associative property. (Lesson 1-4)

39. **Standardized Test Practice**   What is the value of $2xy - x$ if $x = 3$ and $y = 1$? (Lesson 1-3)

   A. 2   **B.** 3   **C.** 5   **D.** 6

40. Write a verbal phrase for $y - 9$. (Lesson 1-3)

41. Evaluate $8 + 9 - 6 \cdot 2$. (Lesson 1-2)

For **Extra Practice**, see page 741.

42. **Patterns**   What is the total number of rectangles, of all sizes, in the figure below? (Lesson 1-1)

WORKING ON THE

Investigation

Up, Up, and Away!

Refer to the Investigation on pages 2–3.

Look over your research and the list of considerations that you placed in your Investigation Folder at the beginning of the Investigation.

You probably listed time and fuel consumption as two of the things you need to consider as you plan your trip. You can

use the formulas *speed × time = distance* and *miles traveled ÷ miles per gallon = gallons of fuel needed* to determine your time and fuel needed.

- Use a map to determine the approximate distance to Oshkosh.
- Use the information you found to write an equation involving the speed of an ultralight and the distance you will travel.
- Then write an equation with the information about fuel. How many gallons of fuel will you need to make the trip?

Add the results of your work to your Investigation Folder.

What You'll Learn

- To use ordered pairs to locate points,

- to organize data, and

- to explore how certain quantities in expressions are related to each other.

Why It's Important

Graphing points on a coordinate plane can help you find cities on a map.

TALK ABOUT IT

MATERIALS: grid paper
Hands-On Activity

You may have played a game in which players locate their battleships on grids like the one shown here. One player tries to locate and "sink" the other player's battleships by calling out a letter-number combination like F2. If a ship is located at this point, the player says *hit*. If not, the player says *miss*. The object of the game is to sink all of your opponent's ships before he or she sinks yours.

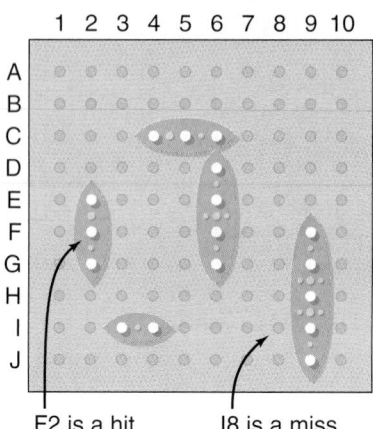

F2 is a hit.   I8 is a miss.

Your Turn   Play the game described above on grid paper.

a. Name an advantage of using a letter-number instead of a number-number combination to locate points on a grid.

b. Is the point named by C3 the same as 3C? If the locations were both named by numbers, would (3, 4) be different than (4, 3)?

c. Name a disadvantage of using a letter-number combination to locate points on a grid.

Learning the Concept

In mathematics, we can locate a point by using a **coordinate system**. The coordinate system is formed by the intersection of two number lines that meet at their zero points. This point is called the **origin**. The horizontal number line is called the **x-axis**, and the vertical number line is called the **y-axis**.

You can graph any point on a coordinate system by using an **ordered pair** of numbers. The first number in the pair is called the **x-coordinate**. The second number is called the **y-coordinate**. The coordinates are your directions to find the point.

Example **1** Graph each ordered pair.

a. **(5, 3)**
Begin at the origin. The x-coordinate is 5. This tells you to go 5 units right of the origin.

The y-coordinate is 3. This tells you to go up 3 units.

Draw a dot. You have now graphed the point whose coordinates are (5, 3).

b. **(6, 0)**
Begin at the origin again. The x-coordinate is 6, so go 6 units right of the origin. The y-coordinate is 0, so you will not go up to place the dot.

Sometimes points are named by using letters. The symbol B(3, 4) means point B has an x-coordinate of 3 and a y-coordinate of 4.

The coordinate plane allows you to locate points.

Example **2** Name the ordered pair for each point.

a. **C**
Go right on the x-axis to find the x-coordinate of point C, which is 2. Go up along the y-axis to find the y-coordinate, which is 2.

The ordered pair for point C is (2, 2).

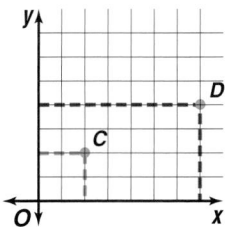

b. **D**
The x-coordinate of D is 7, and the y-coordinate is 4.

The ordered pair for point D is (7, 4).

Ordered pairs can also be used to show how the value substituted for a variable is related to the value of an expression containing the variable.

Example ③

Retailing

At Movie Madness Video Store, newly-released videos rent for $2 a day. The store paid $18 for a copy of a popular video. If x represents the number of times the video is rented in the first month, the expression $2x - 18$ gives the amount of money the store will make on this video.

a. Make a list of ordered pairs in which the x-coordinate represents the number of rentals in the first month and the y-coordinate represents the money made for 10, 15, 20, 25, or 30 rentals.

b. Then graph the ordered pairs.

a. Evaluate the expression $2x - 18$ for 10, 15, 20, 25, and 30. A table of ordered pairs that result is shown below.

x (rentals)	$2x - 18$	y (money made)	(x, y)
10	$2(10) - 18$	2	(10, 2)
15	$2(15) - 18$	12	(15, 12)
20	$2(20) - 18$	22	(20, 22)
25	$2(25) - 18$	32	(25, 32)
30	$2(30) - 18$	42	(30, 42)

b.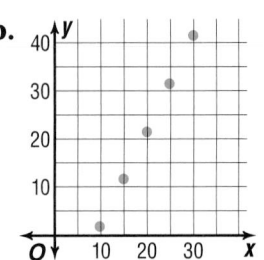

Checking Your Understanding

Communicating
Mathematics

Read and study the lesson to answer these questions.

1. **Draw** a coordinate system and label the origin, x-axis, and y-axis.

2. **Tell** three different uses of ordered pairs.

3. **Describe** a coordinate system that you have seen used in everyday life.

4. **You Decide** Takala says that the points (4, 7) and (7, 4) are the same. Jamal disagrees. Who is correct and why?

Guided
Practice

Use the map at the right to answer each question.

5. Which ordered pair indicates the location of the Tallahassee Municipal Airport?

6. Is (5, 6) the correct location for Campbell Stadium?

7. What location does (8, 5) indicate?

8. Which ordered pair indicates the location of the intersection of routes 319 and 10?

9. Is (3, 7) the correct ordered pair for the Lake Jackson Mounds State Archeological site?

Independent Practice

Use the grid at the right to name the point for each ordered pair.

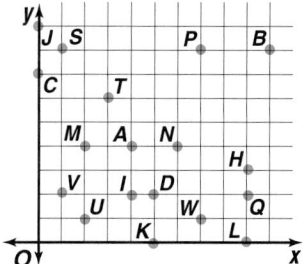

10. (5, 2)   **11.** (2, 4)

12. (1, 8)   **13.** (9, 0)

14. (4, 4)   **15.** (3, 6)

16. (9, 3)   **17.** (0, 9)

Use the grid to name the ordered pair for each point.

18. V **19.** N **20.** P

21. K **22.** W **23.** Q

24. C **25.** B **26.** U

27. a. Write an ordered pair for the coordinates of a point on the *x*-axis.

   b. Write an ordered pair for the coordinates of a point on the *y*-axis.

   c. What point lies on both the *x*- and *y*-axes?

Critical Thinking

28. Geometry   Graph (2, 1), (2, 4), and (5, 1) on a coordinate system.

   a. Connect the points with line segments. What figure is formed?

   b. Multiply each number in the set of ordered pairs by 2. Graph and connect the new ordered pairs. What figure is formed?

   c. Compare the two figures you drew. Write a sentence that tells how the figures are the same and how they are different.

Applications and Problem Solving

29. Astronomy   The table at the right contains data about the planets of our solar system. The *x*-value is the planet's mean distance from the sun in millions of miles. The *y*-value is how many years it takes each planet to complete its orbit around the Sun.

   a. Graph each ordered pair.

   b. Suppose a planet was discovered 1.5 billion miles from the sun. Use the graph to estimate how long it would take to orbit the sun.

Planet	x	y
Mercury	36.0	0.241
Venus	67.0	0.615
Earth	93.0	1.000
Mars	141.5	1.880
Jupiter	483.0	11.900
Saturn	886.0	29.500
Uranus	1782.0	84.000
Neptune	2793.0	165.000
Pluto	3670.0	248.000

30. Geology   The underground temperature of rocks varies with their depth below the surface. The temperature in degrees Celsius is estimated by the expression $35x + 20$, where *x* is the depth in kilometers.

   a. Make a list of ordered pairs in which the *x*-coordinate represents the depth and the *y*-coordinate represents the temperature for depths of 0, 2, 4, 6, and 8 kilometers.

   b. Graph the ordered pairs.

31. Traffic Control The graph below shows some information about the number of cars at the traffic light at State Street and Schrock Road at two different times of a certain weekday.

a. What information is given by the two points?

b. What do you think happened between the two times?

c. If most of the local businesses close at 5:00 P.M., copy the graph and place a point for the number of cars you think would be at the intersection at 5:15 P.M. Explain how you chose your point.

Mixed Review

32. Solve the equation $7a = 28$ mentally. (Lesson 1-6)

33. Simplify the expression $3m + 7m + 1$. (Lesson 1-5)

34. Medicine There are three-times more women enrolled in U.S. medical schools today than in the mid-70s. Write an expression for the number of women in medical schools today if x represents the number in the 1970s. (Lesson 1-3)

35. Evaluate the expression $4r$ if $r = 3$. (Lesson 1-3)

36. Standardized Test Practice Five more than twice as many students bought their lunch at the cafeteria on Tuesday than on Wednesday. Let w be the number of students who bought their lunch on Wednesday. Which expression best represents the number of students who bought their lunch on Tuesday? (Lesson 1-3)

A. $5 + 2w$

B. $5w + 2$

C. $2w - 5$

D. $5 - 2w$

For **Extra Practice**, see page 742.

37. Find the value of the expression $6 \cdot 3 \div 9 - 1$. (Lesson 1-2)

1-8

Solving Equations Using Inverse Operations

You'll Learn

- To use inverse operations to solve equations, and

- to solve problems using equations.

It's Important

You can use equations to solve many real-world applications.

Application: Transportation

When the New York City subway service began in 1904, passengers bought tickets from an attendant for five cents. Another attendant tore the ticket as you passed through the turnstile. Soon the turnstiles were made to operate on coins. When the subway fare was changed to a dime in 1948, the turnstiles could still operate on a coin. However, when the fare went to fifteen cents in July of 1953, there was a problem. That's when the New York City subway token was introduced.

The "bulls-eye" tokens used in New York subways today have brass surrounding a steel center. Since the turnstiles read tokens with a magnet, using bulls-eye tokens helps reduce the use of slugs, which are worthless pieces of metal used instead of tokens. Before the bulls-eyes, New York took in about 165,000 slugs each month. Now slugs total about 20,000 a month.

In 1995, New York subways lost about $25,000 a month to people using slugs. What was the subway fare in 1995? *This problem will be solved in Example 3.*

The New York City subway system is among the oldest and longest in the world. It has 461 stations, 230 miles of track, and most trains run 24 hours a day, 365 days a year.

Learning the Concept

You have solved some equations using mental math or patterns, but sometimes the solution is not easy to see. Another way to solve equations is to use **inverse operations**. Inverse operations "undo" each other. For example, to "undo" addition you would subtract. These can be shown by related sentences.

The sentences below show how addition and subtraction are related.

$$9 + 12 = 21 \quad \begin{cases} 9 = 21 - 12 \\ 12 = 21 - 9 \end{cases}$$

$$9 + n = 18 \quad \begin{cases} 9 = 18 - n \\ n = 18 - 9 \end{cases}$$

Lesson 1-8 Solving Equations Using Inverse Operations **41**

Multiplication and division are also inverse operations.

$$3 \cdot 5 = 15 \quad \begin{cases} 3 = 15 \div 5 \text{ or } 3 = \dfrac{15}{5} \\[2mm] 5 = 15 \div 3 \text{ or } 5 = \dfrac{15}{3} \end{cases}$$

$$12t = 48 \quad \begin{cases} 12 = 48 \div t \text{ or } 12 = \dfrac{48}{t} \\[2mm] t = 48 \div 12 \text{ or } t = \dfrac{48}{12} \end{cases}$$

You can use inverse operations to solve equations.

Example **Solve each equation by using the inverse operations.**

a. $b - 6 = 42$
$b = 42 + 6$ *Write a related addition sentence.*
$b = 48$

b. $c + 16 = 54$
$c = 54 - 16$ *Write a related subtraction sentence.*
$c = 38$ *You can check your answer by replacing c with 38 in the original equation.*

c. $\dfrac{v}{9} = 16$
$v = 16 \cdot 9$ *Write a related multiplication sentence.*
$v = 144$

d. $2.5s = 37.5$
$s = 37.5 \div 2.5$ *Write a related division sentence.*
$s = 15$ *Use a calculator.*

THINK ABOUT IT

Study each of the rewritten sentences. Why do you think each one is written with the variable alone on one side?

Using an equation to solve a problem is an important problem-solving strategy. The first step in writing an equation is to choose a variable and a quantity for the variable to represent. This is called **defining a variable**.

Example **The sum of 215 and a certain number is equal to 266. Find the number.**

Let n represent the number. Now that you have a variable, translate the words into an equation using the variable.

215	*plus*	*number*	*equals*	*266*
215	+	n	=	266

$215 + n = 266$
$n = 266 - 215$ *Write the related subtraction sentence.*
$n = 51$ The number is 51.

Check: Since 215 + 51 = 266, the solution is correct.

Example **3**

Real World
APPLICATION

Refer to the application at the beginning of the lesson. Use the four-step problem-solving plan to solve.

Explore Read the problem for the general idea. You know that about 20,000 slugs are used each month for a loss of $25,000. You need to find the subway fare. Each slug is equal to one fare. Let *f* represent the fare.

Plan Translate the words into an equation using the variable. The total amount of money lost to people using slugs is the number of slugs multiplied by the fare.

fare	*times*	*number of slugs*	*equals*	*total lost*
f	$\times$	20,000	$=$	$25,000

Solve Solve the equation for the fare.

$$f \times 20,000 = \$25,000$$
$$f = \$25,000 \div 20,000$$

Write a related division sentence.

$$25,000 \div 20,000 = 1.25$$

The fare is $1.25.

Examine The answer is close to the estimate. Since $20,000 \times \$1.25$ equals $25,000, the answer checks.

THINK ABOUT IT

Always estimate before you solve the equation. THINK: Since 25,000 is a little more than 20,000 the quotient will be a little over 1.

Checking Your Understanding

Communicating Mathematics

Read and study the lesson to answer these questions.

1. **Write** an addition sentence and then write a related subtraction sentence.

2. What is the inverse operation of gaining five yards in football?

3. **Explain** what it means to *define a variable*.

4. **Transportation** Refer to the application at the beginning of the lesson.

   a. Write an equation and find the amount of money lost each month if the New York Subway System was still taking in 165,000 slugs per month and the fare is $1.25.

   b. Check your answer using the words of the problem. Write a convincing argument for checking the answer of a problem against the words of the problem instead of the equation that you wrote.

Guided Practice

For each sentence, write a related sentence using the inverse operation.

5. $3 + 9 = 12$

6. $13 \cdot t = 39$

7. $16 - 12 = 4$

8. $44x \div 4x = 11$

Solve each equation using the inverse operation. Use a calculator when needed.

9. $8 + y = 22$ **10.** $77 = 7h$ **11.** $5.4 = \dfrac{w}{2.2}$

Translate each sentence into an equation. Then solve.

12. Twice a certain number is 6.

13. Twenty-seven more than a number is thirty-one.

Write a problem that could be solved using each equation.

14. $\$6x = \$135$ **15.** $x - 7 = 28$

16. Sports Recently, José Maria Olazábal won the $360,000 prize of the Masters Tournament in Augusta, Georgia. The following week, Hale Irwin took the prize of the Heritage Classic. Irwin won $135,000 less than Olazabal. How much was the Heritage Classic prize?

   a. Define a variable for this problem.

   b. Write an equation.

   c. Solve for the amount of the Heritage Classic prize.

   d. Check your solution against the words of the problem.

José Maria Olazábal

Exercises: Practicing and Applying the Concept

Independent Practice

For each sentence, write a related sentence using the inverse operation.

17. $14 - 3 = 11$ **18.** $23 = 17 + n$ **19.** $14 \cdot 2 = 28$

20. $\dfrac{x}{11} = 12$ **21.** $6h = 48$ **22.** $k \div 13 = 7$

Solve each equation using the inverse operation. Use a calculator when needed.

23. $6 + n = 28$ **24.** $g + 11 = 15$ **25.** $86 = c - 2$

26. $144 = 12h$ **27.** $8 = \dfrac{v}{15}$ **28.** $210 = 15w$

29. $5.56 - q = 4.73$ **30.** $3.7 = \dfrac{r}{5.4}$

31. $f + 15.98 = 19.75$ **32.** $6.8b = 34.68$

33. $1.5 + b = 2.1$ **34.** $\dfrac{p}{2} = 9$

Translate each sentence into an equation. Then solve.

35. Three times a certain number is 18.

36. Ten less than a number is 19.

37. When a number is decreased by seven, the result is 22.

38. The product of a number and eight is 88.

39. A number divided by four is 14.

40. A number divided by three is thirteen.

Write a problem that could be solved using each equation.

41. $n + 2 = 16$ **42.** $\$30n = \$90$

43. Consider the division sentence $a = \frac{5}{0}$ and the related multiplication sentence $5 = a \cdot 0$. Use these statements to explain why division by zero is not allowed.

Applications and
Problem Solving

44. Entertainment In a movie, Bret Maverick worked to raise enough money to enter a card championship. Twenty players entered the tournament, each one paying the same entrance fee. The entrance fees totaled $500,000. How much did each player pay to play?

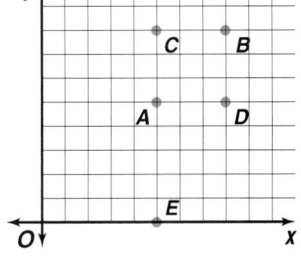

   a. Define a variable.

   b. Write an equation for the problem.

   c. Solve for the entrance fee.

   d. Check your solution against the words of the problem.

45. Sports Football on Monday night has been thrilling fans and topping the television ratings since 1969. As of the 1998 season, the San Francisco 49ers had appeared on Monday night 51 times. The Miami Dolphins had appeared the most times, with 9 more appearances than the 49ers. How many times had the Dolphins appeared?

Mixed
Review

46. Standardized Test Practice
   Which of the points on the graph at the right has the coordinates (5, 8)?
   (Lesson 1-7)

   A. *A*   **B.** *B*

   C. *C*   **D.** *D*

47. Which of the numbers 13, 16, or 19 is the solution of $77 + b = 96$? (Lesson 1-6)

48. Find $13 \cdot 13$ mentally using the distributive property. (Lesson 1-5)

49. Name the property shown by the statement $(9 + 4) + 5 = 9 + (4 + 5)$. (Lesson 1-4)

50. Evaluate $b + cd$ if $b = 2$, $c = 5$, and $d = 7$. (Lesson 1-3)

51. Translate the phrase *five times last year's total* into an algebraic expression. (Lesson 1-3)

52. Find the value of $(18 + 6) \div (8 - 2)$. (Lesson 1-2)

53. Demographics A magazine reported that American men drive an average of 16,500 miles each year. How many miles does an average American man drive in a week? (Lesson 1-1)

   a. Which method of computation do you think is most appropriate for this problem? Justify your choice.

   b. Solve the problem using the four-step plan. Be sure to examine your solution.

For **Extra Practice**, see page 742.

Source: *American Demographics* magazine

Inequalities

 Application: Health

Insomnia, the inability to get enough sleep, affects millions of Americans each year. Researchers have come up with some suggestions on how to avoid insomnia.

▶ Maintain a temperature of between 68 and 75°F in your bedroom.

▶ Stop eating at least 2 hours before bedtime.

▶ Avoid reading or watching television in bed.

▶ Go to bed and get up at the same times each day, even when you can sleep in.

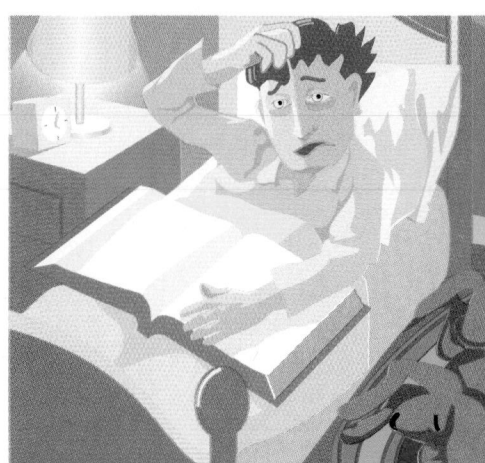

Learning the Concept

Let the variable t represent the bedroom temperature. We can write mathematical sentences to describe the temperature that physicians recommend, which is between 68°F and 75°F.

Say: t is less than 75
Write: $t < 75$

Say: t is greater than 68
Write: $t > 68$

A mathematical sentence that contains $<$ or $>$ is called an **inequality**. Inequalities, like equations, can be true, false, or open.

$18 > 13$ *This sentence is true.*
$9 < 5$ *This sentence is false.*
$t < 75$ *This sentence is open. It is neither true nor false until t is replaced with a number.*

Some inequalities contain $\leq$ or $\geq$ symbols. They are the combinations of the equals sign and the inequality symbols.

Say: n is less than or equal to 4
Write: $n \leq 4$

Say: m is greater than or equal to 8
Write: $m \geq 8$

Example **For the given value, state whether each inequality is *true* or *false*.**

a. $r - 8 > 7, r = 12$

$$12 - 8 > 7$$
$$4 > 7$$

This sentence is false.

b. $14 \leq \frac{3s}{2} + 5, s = 10$

$$14 \leq \frac{3 \times 10}{2} + 5$$
$$14 \leq 15 + 5$$
$$14 \leq 20$$

This sentence is true.

Many situations in real life can be described using inequalities. The table below shows some common phrases and corresponding inequalities.

<	>	≤	≥
• less than • fewer than	• greater than • more than • exceeds	• less than or equal to • no more than • at most	• greater than or equal to • no less than • at least

Example **A dozen eggs must weigh at least 30 ounces to be labeled "jumbo" by the U.S. Department of Agriculture. A dozen eggs that weigh less than 18 ounces are not supposed to be sold to consumers.**
a. Write an inequality for the weight of a dozen jumbo eggs.
b. Write an inequality for the weight of eggs not sold.

a. The phrase *at least* implies a greater than or equal relationship.
Let j represent the weight of a dozen jumbo eggs.
$j \geq 30$

b. Eggs that weigh less than 18 ounces per dozen are not sold.
Let u represent the weight of unsold eggs.
$u < 18$

Checking Your Understanding

Communicating Mathematics

Read and study the lesson to answer these questions.

1. **Explain** the difference between $x > 6$ and $x \geq 6$.

2. **Find** a value for which $6x > 12$ is true.

3. **Compare and contrast** equations and inequalities.

MATERIALS

index cards

4. Work with a partner.

   ▶ Write each of the digits 0–9 on a card and place cards face down.
   ▶ Then write each of the inequality symbols on a card and place these cards face down in a separate pile. (Mark the bottom of the symbol cards so you can tell which direction the symbol should face.)
   ▶ Choose one card from the digits pile.
   ▶ Then choose a card from the symbol pile.
   ▶ Choose another card from the digits pile.
   ▶ Determine if the inequality you created is true or false.
   ▶ Then try to rearrange the cards to make a true inequality.
   ▶ Shuffle the cards and draw again to create several inequalities.

State whether each inequality is *true*, *false*, or *open*.

5. $9 < 3$ **6.** $8 < 8$ **7.** $12 > 4$

State whether each inequality is *true* or *false* for the given value.

8. $n + 2 > 4, n = 8$ **9.** $15 \leq 2q, q = 9$

Translate each sentence into an inequality.

10. A number is less than 6.

11. A dozen eggs must weigh 27 ounces or more to be labeled "extra large."

12. Civics   On July 1, 1971, the 26th amendment to the U.S. Constitution was ratified. This amendment states "The right of citizens of the United States, who are 18 years of age or older, to vote shall not be denied or abridged by the United States or any state on account of age." Write an inequality showing the age of a voter.

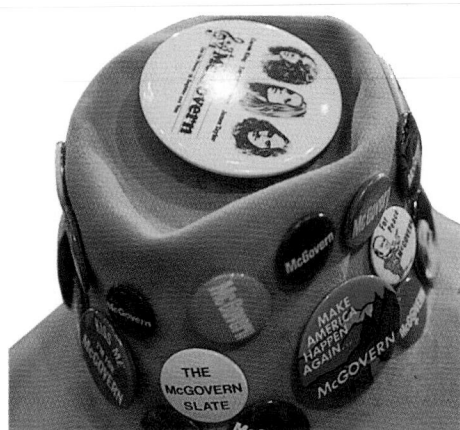

Exercises: Practicing and Applying the Concept

State whether each inequality is *true*, *false*, or *open*.

13. $7 > 6$ **14.** $40 < 28$ **15.** $2n > 4$

16. $18 > 6m$ **17.** $0 \geq 0$ **18.** $j + 7 < 34$

19. $16 \leq 16$ **20.** $17 < p$ **21.** $y - 1 < 4$

State whether each inequality is *true* or *false* for the given value.

22. $22 - w > 4, w = 11$ **23.** $24 \geq 3c, c = 6$

24. $h - 9 > 6, h = 10$ **25.** $2x - 7 \leq 7, x = 7$

26. $4x + 8x < 7, x = 1$ **27.** $5z - 16 \leq 11, z = 4$

Translate each sentence into an inequality.

28. A number exceeds 24.

29. Tertius earned $100 or more.

30. At most, 75 people applied.

31. It takes no less than three years to complete law school.

32. Julita scored at least 16 points.

33. Fewer than 15 people attended.

34. Suppose $x > 6$ and $x < 18$.

    a. Explain what the open sentence $6 < x < 18$ means.

    b. Find a value of x that makes the sentence $6 < x < 18$ true.

Applications and Problem Solving

35. Sports   On April 9, 1993, the Colorado Rockies played their first home opener in Mile High Stadium in Denver. More than 80,000 fans were in attendance. Write an inequality for the number of people who attended.

36. Finance   Some companies provide 401(k) plans that help their employees save for retirement. The Internal Revenue Service sets a limit on the maximum amount that an employee may invest in a 401(k) plan in any given year. In 1995, the limitation was changed to $9240. Write an inequality for the amount that an employee may invest in a 401(k) plan.

37. Health   Some of the benefits of quitting smoking are listed in the chart at the right.

   a. Write an inequality for the amount of time in which a person who quits smoking can expect lung cancer death risk reduced to that of a nonsmoker.

   b. Write an inequality for the amount of time it takes for a person who quits smoking to cut their risk of heart disease to half that of a smoker.

Benefits of Quitting Smoking	
Within	**Benefit**
2 weeks to 3 months	• circulation improves • walking becomes easier • lung function increases 38%
1 year	• excess risk of heart disease is half that of a smoker
10 years	• lung cancer death risk is similar to that of a nonsmoker • precancerous cells are replaced • risk of various cancers decreases

Mixed Review

38. Charity   Many airlines donate change found in seats and foreign coins that travelers no longer need to charity. One airline gives the foreign coins donated on international flights to a charity. The donations total $14,400 in an average year. How much does the airline collect for charity in a month? (Lesson 1-8)

   a. Define a variable.

   b. Write an equation for the problem.

   c. Solve for the monthly donations.

   d. Check your solution against the words of the problem.

39. Solve $7b = 77$ mentally. (Lesson 1-6)

40. Standardized Test Practice   Use the distributive property to simplify $4 \cdot 8 + 4 \cdot 2$. (Lesson 1-5)

   A. 18   **B.** 40   **C.** 72   **D.** 96

41. Rewrite the expression $8(7c)$ using an associative property. Then simplify. (Lesson 1-4)

42. Evaluate the expression $9t - 2s$ if $s = 8$ and $t = 3$. (Lesson 1-3)

43. Find the value of $\dfrac{84 - 12}{14 - 6}$. (Lesson 1-2)

For **Extra Practice,** see page 742.

44. Patterns   Find the next number in the pattern 8, 10, 14, 22, 38, __?__. (Lesson 1-1)

HANDS-ON ACTIVITY

1-10A Gathering Data

A Preview of Lesson **1-10**

Advertisers, political parties, and media such as television stations and newspapers all want to find out about public opinion. **Demographics** is the study of the characteristics of a population. **Surveys** are often used to study a population. The results of surveys are often modeled by an equation.

Your Turn

Work in small groups. Find out about the opinions of students in your school by conducting a survey. The pieces of information that you collect are called **data**.

a. Your group has been hired by a manufacturer to develop an ad campaign to convince students in your school to buy their product. Decide on a product to advertise.

b. How many people will you survey? You probably can't ask every person in your school, so you will need to ask a small group of students. A group that is used to represent a whole population is called a **sample**.

c. Write your questions.

d. Make a plan for asking students. Make sure that the group of students you will ask represents the whole school population.

e. Display your survey results in a way that is easily understood. The results of a sample survey question about television-watching habits is shown below.

How many hours of television do you watch in a week?		
Number of hours watched	**Tally**	**Frequency**
0–2	ⅢⅡ	5
3–5	Ⅲ Ⅲ ‖	12
6–8	Ⅲ Ⅲ Ⅲ Ⅲ ‖‖	24
9–11	Ⅲ Ⅲ Ⅲ Ⅲ Ⅲ Ⅲ ‖	33
12–14	Ⅲ Ⅲ Ⅲ Ⅲ Ⅲ	25
15–17	Ⅲ ‖	6

TALK ABOUT IT

1. Use the table above to find how many hours of television most of the students surveyed watch in a week.

2. Write a few sentences analyzing what you found out about the students in your school through your survey.

3. Explain how your advertising campaign can convince students to buy the product.

4. Prepare a report, with charts and graphs, that summarizes your findings.

Extension

5. Survey the cars in a parking lot. Determine what characteristics you will study. Then collect the data and analyze it.

Integration: Statistics
Gathering and Recording Data

You'll Learn

- To gather and record data in a frequency table or bar graph.

It's Important

Frequency tables are useful when you take surveys.

Application: Entertainment

"What do you think?" Opinion polls often ask that question. The results of an opinion poll are reported using **statistics**. Statistics involves collecting, analyzing, and presenting data.

A family center in Orlando, Florida conducts one of the world's largest ongoing public opinion polls. People visiting the center can gather in a 172-seat theater to participate in polls on everything from vegetables to rock videos.

The **frequency table** below shows the results of one poll on movie tastes.

What type of movie do you like most?		
	Men	**Women**
Comedy	39	47
Action/Adventure	40	18
Drama	6	20
Horror	2	2
Other	5	5
No response	8	8

The poll was conducted with 5174 adults at the center. Results are shown as if groups of 100 men and 100 women were surveyed.

Learning the Concept

Obviously, it would be very difficult to ask every man and woman in the United States which type of movie they prefer. The family center surveys smaller groups, called **samples**. The sample group is assumed to be representative of the entire population. However, since the group of people could not be chosen randomly but only from the group at the family center, the results may not be representative.

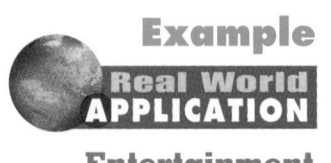

Example 1

Real World APPLICATION

Entertainment

Refer to the frequency table at the beginning of the lesson.

a. Which type of movie did most men in the group prefer?
The greatest number of men, 40, chose action/adventure. Comedy was a close second with 39 men choosing it.

b. Overall, which type of movie is most popular?
In order to determine the type of movie most preferred, find the sum of the numbers of men and women who prefer each type.

	men	+	*women*	=	*total*
Comedy	39	+	47	=	86
Action/Adventure	40	+	18	=	58
Drama	6	+	20	=	26
Horror	2	+	2	=	4
Other	5	+	5	=	10
No response	8	+	8	=	16

Comedy is the most popular type of movie overall.

Statistical information can be displayed in a **bar graph**.

Example 2

CONNECTION

Social Studies

Recently, a company conducted a poll on the women that Americans admire most. The results are shown in the frequency table at the right. Make a bar graph to display the data.

Each of the categories in a bar graph has a bar to represent it. The vertical scale shows the number of votes.

Women Americans Admire	
Woman	**Votes**
Queen Elizabeth II	28
Jacqueline Kennedy Onassis	27
Mamie Eisenhower	22
Helen Keller	18
Margaret Chase Smith	18
Claire Booth Luce	17
Mother Teresa	16
Margaret Thatcher	16

Source: The Gallup Organization

Some graphing calculators allow you to create bar graphs.

Example **3**

Real World
APPLICATION

Sports

Since the first modern Olympics in 1896, the number of events at each Summer Olympic Games has grown. The table at the right shows the number of events at recent Olympic Games. Use a graphing calculator to make a bar graph of the data.

Olympic Year	Number of Events
1968	172
1972	196
1976	199
1980	200
1984	223
1988	237
1992	257
1996	271

Begin by accessing and clearing lists L_1 and L_2 in the calculator's memory.

Enter: [STAT] [ENTER] [▲] [CLEAR] [ENTER]
[▶] [▲] [CLEAR] [ENTER]

Then enter the data. Use list L_1 for the years and L_2 for the events.

Enter: [◀] 1968 [ENTER] 1972 [ENTER] *and so on* 1996 [ENTER] [▶]
172 [ENTER] 196 [ENTER] *and so on* 271 [ENTER]

Next set the range of the data for the viewing window.

Enter: [WINDOW] 1968 [ENTER] 2000 [ENTER] 4 [ENTER] 0
[ENTER] 300 [ENTER] 25 [ENTER]

Now choose the type of graph and construct the graph.

Enter: [2nd] [STAT PLOT] [ENTER]

Use the arrow and [ENTER] keys to highlight "On", the bar graph, "L_1", and "L_2". Then press [GRAPH] to display the graph.

Checking Your Understanding

Communicating Mathematics

Read and study the lesson to answer these questions.

1. **Describe** the kind of information a frequency table shows in your own words.

2. **Explain** why a poll taken at the family center may not be representative of the entire United States population.

3. **Statistics** Find a newspaper or magazine article that contains a graph or table. Explain how you think the results were found. Do you think the display of the information is effective? Explain.

4. **Assess Yourself** When do you use statistics in your life? Do you ever make decisions based on statistics?

MATERIALS

colored candies

5. Bar graphs are excellent tools for displaying data. Examine the colors of the candies in a bag of colored candy. As you remove each candy from the bag, place it in a column to form a bar graph like the one at the right.

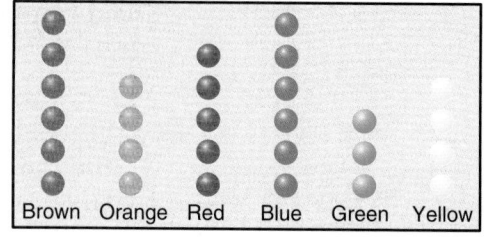

Brown Orange Red Blue Green Yellow

a. Use your data to make a bar graph.

b. How does your graph compare to the ones made by others in your class? Did each package of candy have the same color distribution and the same number of candies?

Guided Practice

The frequency table at the right contains data about the attendance at some of the major zoos in the United States.

6. Which zoo was visited by the most people?

7. Which zoo was visited least frequently?

8. How many people visited the Denver Zoo?

9. How many times were all of these zoos visited?

10. Make a bar graph of the zoo attendance figures.

Attendance of Selected Zoos	
Zoo	**Attendance (millions)**
Audubon (New Orleans)	0.9
Chicago	2.0
Cincinnati	1.3
Denver	1.3
Miami Metrozoo	0.8
National (Wash. D.C.)	3.0
San Diego	3.5
Toledo	0.8

Exercises: Practicing and Applying the Concept

Independent Practice

The table below shows the number of states with different laws about driving ages. The table shows the lowest age at which a person can receive a restricted or unrestricted driver's license.

11. What is the lowest age at which someone can obtain a driver's license in any state?

12. When can a person get a driver's license in most states? Is it a restricted or unrestricted license?

13. Are all of the states and the District of Columbia represented in the table? How can you tell?

14. Make a bar graph of the data.

15. Write a sentence to describe the driver's age laws in the United States.

16. Where in the table does your state fall?

Driver's License Laws	
Law	**States**
Restricted at 15	3
Unrestricted at 15	2
Restricted at 16	20
Unrestricted at 16	18
Restricted at 17	3
Unrestricted at 17	2
Unrestricted at 18	3

A list of the heights of the tall buildings in Seattle, Washington, is given below. Heights are given in feet.

17. Make a frequency table for the building heights. Use intervals of 100 feet.

954	740	730	722	609	605
580	574	543	514	500	498
493	487	466	456	454	409
397	389	379	371		

18. In which interval do most of the buildings fall?

19. A histogram is a special bar graph that displays the frequency of data in equal intervals. Use your frequency table to make a histogram of the data.

Critical Thinking

20. a. Find a bar graph in a newspaper or magazine.
 b. Does the graph effectively communicate the data? When would a frequency distribution be more useful?

Applications and Problem Solving

21. Social Studies The heights of the U.S. presidents are given in the table at the right.

U.S. Presidents' Heights	
Height (in.)	Frequency
63–65	1
66–68	9
69–71	12
72–74	18
75–77	1

 a. How many different presidents have there been?
 b. What range of heights describes the most presidents?
 c. How many presidents were exactly 6 feet tall? Explain your answer.
 d. Why do you think so many presidents were or are fairly tall?
 e. **Research** How do the heights of the presidents compare with the height of an average American man?

22. Family Activity Write your own survey question. You might ask family and friends about their favorite color, the number of people living in their home, or how many languages they speak. Make a frequency distribution and a bar graph of your findings.

Mixed Review

23. Family Management The Family Economics Research Group, which is a division of the U.S. Department of Agriculture, estimates that it costs a middle-income family $210,070 or more to raise a child to age 17. Write an inequality for the amount a middle income family will spend. (Lesson 1-9)

24. Solve the equation $t - 8 = 42$ using an inverse operation. (Lesson 1-8)

For **Extra Practice**, see page 743.

25. Geometry Graph the points $D(6, 3)$, $F(2, 2)$, $G(4, 5)$, and $H(5, 10)$ on a coordinate system. (Lesson 1-7)

26. Rewrite the expression $(t + 17) + 6$ using an associative property. Then simplify. (Lesson 1-4)

27. Standardized Test Practice Evaluate $6g - 12h$ if $g = 9$ and $h = 4$. (Lesson 1-3)

 A. 1   **B.** 6   **C.** 84   **D.** 102

Investigation

Up, Up, and Away!

Refer to the Investigation on pages 2–3.

Your chapter of the EAA is about to meet to finalize arrangements for the trip to Oshkosh. Your committee will be called upon to present its plan for the trip at the meeting next week. Complete your plan and make a presentation to the club.

- Suppose that your school represents the members of the club. Take a survey about the type of place they would like to stay on a trip. Would they like to camp? How about a hotel? Sometimes people rent rooms in their homes to travelers. Would that be an option? Use your results to choose a place to stay on your trip.

- Use the results of your study of the time and fuel consumption to choose the time of departure and to estimate the cost of the trip.

- Map out the route of the trip so that no Federal regulations are violated and the trip is as enjoyable as possible.

- Compile all of your results into a brochure to give to members interested in making the trip. Include your research and reasons for your choices.

PORTFOLIO ASSESSMENT

You may want to keep your work on this Investigation in your portfolio.

Vocabulary

After completing this chapter, you should be able to define each term, property, or phrase and give an example or two of each.

Algebra
algebraic expression (p. 16)
associative property of addition (p. 22)
associative property of multiplication (p. 22)
brackets (p. 12)
commutative property of addition (p. 22)
commutative property of multiplication (p. 22)
defining a variable (p. 42)
distributive property (p. 26)
equation (p. 32)
evaluate (p. 11)
evaluating an expression (p. 16)
expression (p. 11)
identity property of addition (p. 23)
identity property of multiplication (p. 23)
inequality (p. 46)
inverse operations (p. 41)
like terms (p. 27)
multiplicative property of zero (p. 23)
open sentence (p. 32)
order of operations (p. 11)
parentheses (p. 12)
simplest form (p. 28)

solution (p. 32)
solving an equation (p. 32)
substitution property of equality (p. 17)
terms (p. 27)
variable (p. 16)

Geometry
coordinate system (p. 36)
ordered pair (p. 37)
origin (p. 36)
x-axis (p. 36)
x-coordinate (p. 37)
y-axis (p. 36)
y-coordinate (p. 37)

Statistics
bar graph (p. 52)
data (p. 50)
demographics (p. 50)
frequency table (p. 51)
sample (pp. 50, 51)
statistics (p. 51)
surveys (p. 50)

Problem-Solving
method of computation (p. 8)
problem-solving plan (p. 6)

---

interNET
CONNECTION

Chapter Review For additional lesson-by-lesson review, visit:
www.glencoe.com/ sec/math/prealg/ mathnet

Understanding and Using Vocabulary

Choose the term that best completes each statement.

1. $8x + 6$ is an example of an __?__. (algebraic expression, inequality)

2. "For any numbers a and b, $a + b = b + a$" is a statement of the __?__ property of addition. (commutative, associative)

3. An expression is __?__ when it has no like terms and no parentheses. (an algebraic expression, in simplest form)

4. In (5, 6), 5 is the __?__. (x-coordinate, y-coordinate)

5. The inverse operation for subtraction is __?__. (addition, division)

6. A group used to represent a larger population is a __?__. (survey, sample)

7. $7y$ and $12y$ are __?__. (like terms, ordered pairs)

Skills and Concepts

Objectives and Examples	**Review Exercises**
Upon completing this chapter, you should be able to:	Use these exercises to review and prepare for the chapter test.

▶ **use the order of operations to evaluate expressions** (Lesson 1-2)

$$3[(7 - 3) + 8 \cdot 2] = 3[(4) + 8 \cdot 2]$$
$$= 3[4 + 16]$$
$$= 3[20]$$
$$= 60$$

Find the value of each expression.

8. $8(16 + 14)$ **9.** $3(9 - 3 \cdot 2)$

10. $14 \div 2 + 4 \cdot 3$ **11.** $12 \div 3 \cdot 4$

12. $\dfrac{14 + 4}{11 - 2}$ **13.** $\dfrac{3(2 + 6)}{4}$

14. $9(12 - 5) - 11(21 - 8 \cdot 2)$

15. $12[3(17 - 6) - 7 \cdot 4]$

▶ **evaluate expressions containing variables** (Lesson 1-3)

Evaluate $8x - 2y$ if $x = 3$ and $y = 7$.
$$8x - 2y = 8(3) - 2(7)$$
$$= 24 - 14$$
$$= 10$$

Evaluate each expression if $a = 10$, $b = 8$, $c = 6$, and $d = 12$.

16. $14 - a + b$ **17.** $2b - d$

18. $d - c$ **19.** $c + (d - b)$

20. $5a + 4b$ **21.** $\dfrac{a}{b + 2}$

22. $ab - cd$ **23.** $6(d - b)$

▶ **translate verbal phrases into algebraic expressions** (Lesson 1-3)

Forty-five fewer fans attended this football game than attended last week's game.

Let a represent last week's attendance. Then this week's attendance is $a - 45$.

Translate each phrase into an algebraic expression.

24. seven more than some number

25. the product of a number and ten

26. twice as many as last week

27. Joi's score plus 10 points extra credit

28. four more than twice last week's earnings

29. the difference of r and 12

▶ **use properties of addition and multiplication** (Lesson 1-4)

Name the property shown by the statement $a + 12 = 12 + a$.

commutative property of addition

Name the property shown by each statement.

30. $22 + 0 = 22$

31. $9 + 3t = 3t + 9$

32. $8 + (2 + y) = (8 + 2) + y$

33. $4(6 \cdot 3) = (4 \cdot 6) \cdot 3$

34. $0(b) = 0$

35. $1(m + 5) = m + 5$

Objectives and Examples

▶ **simplify expressions using the distributive property** (Lesson 1-5)

Simplify the expression $6(8 + n) - 6$.
$$6(8 + n) - 6 = 6(8) + 6(n) - 6$$
$$= 48 + 6n - 6$$
$$= 48 - 6 + 6n$$
$$= 42 + 6n$$

▶ **solve open sentences** (Lesson 1-6)

Which of the numbers 8, 12, or 16 is the solution of $84 - n = 72$?

$84 - 8 = 72$ $84 - 12 = 72$
    $76 = 72$ *false* $72 = 72$ *true*

$84 - 16 = 72$
    $68 = 72$ *false*   The solution is 12.

▶ **use ordered pairs** (Lesson 1-7)

Graph the point $(7, 3)$.

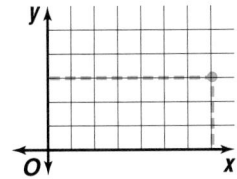

▶ **solve equations using inverse operations** (Lesson 1-8)

$\frac{h}{7} = 13$

$h = 7 \cdot 13$ *Write a related multiplication*
$h = 91$ *sentence.*

Review Exercises

Simplify each expression.

36. $7 \cdot 8 + 7 \cdot 3$ **37.** $n + 7n$

38. $7b + 12b - 10$ **39.** $16x + 2x + 8$

40. $8(5ab + 4) + 3(2ab + 2)$

41. $2(x + y) + 8(x + 3y)$

42. $10(u + 3) + 4(5 + u)$

Identify the solution to each equation from the list given.

43. $15 - h = 9; 4, 6, 7$ **44.** $7 = \frac{m}{4}; 21, 28, 35$

45. $5b = 45; 7, 9, 11$ **46.** $18 - n = 18; 18, 1, 0$

47. $12 = 9x - 15; 3, 6, 9$

48. $16 = 144 \div k; 9, 12, 16$

49. $\frac{35}{r + 2} = 5; 3, 5, 7$

Use the grid to name the point for each ordered pair.

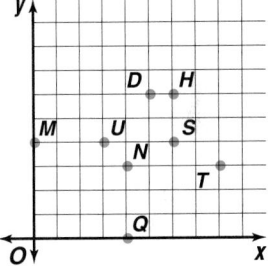

50. $(8, 3)$

51. $(4, 3)$

52. $(3, 4)$

Use the grid to name the ordered pair for each point.

53. D

54. H

55. M

Solve each equation using the inverse operation. Use a calculator when needed.

56. $72 = 8n$ **57.** $19 - g = 7$

58. $6n = 84$ **59.** $\frac{y}{4.5} = 6.3$

60. $19.9 - x = 17.6$ **61.** $9 = \frac{w}{12}$

62. The product of a number and seven is 84. Find the number.

▶ **write and solve inequalities** (Lesson 1-9)

Translate "*A number is greater than or equal to fifteen.*" into an inequality. Is the inequality *true* or *false* for the value 15?

Let n represent the number.
$n \geq 15$

$15 \geq 15$ is true.

State whether each inequality is *true* or *false* for the given value.

63. $37 - z \leq 9, z = 22$

64. $5x + 8 > 0; x = 0$

65. $7c + 9c \geq 17c, c = 8$

Translate each sentence into an inequality.

66. A number exceeds 25.

67. Fewer than 60 points were scored.

68. It takes no less than seven hours to drive to Chicago from here.

▶ **gather and record data in a frequency table or bar graph** (Lesson 1-10)

Age	Freq.
0–10	2
11–20	8
21–30	11
31–40	6

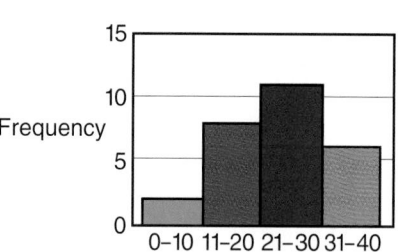

The list below shows the numbers of U.S. congressional representatives from each state for the years 1990–2000.

7	1	6	4	52	6	6	1	23	11
2	2	20	10	5	4	6	7	2	8
10	16	8	5	9	1	3	2	2	13
3	31	12	1	19	6	5	21	2	6
1	9	30	3	1	11	9	3	9	1

69. Make a frequency table of the numbers of representatives. Use intervals of 10.

70. Make a bar graph of the data.

Applications and Problem Solving

71. Space Exploration The longest flight ever made by a space shuttle ended on March 18, 1995. The astronauts aboard *Endeavour* spent 16 days, 15 hours in flight, traveling 6.9 million miles and circling Earth 262 times. About how many miles did the shuttle travel on each trip around Earth? (Lesson 1-1)

a. Which method of computation do you think is most appropriate for this problem? Justify your choice.

b. Solve the problem using the four-step plan. Be sure to examine your solution.

72. Health and Fitness One health magazine gives the equations below for finding your ideal weight, where w represents weight in pounds and h is height in inches.

males: $w = 100 + 6(h - 60)$
females: $w = 100 + 5(h - 60)$

a. Find the ideal weight of Cassie, who is 5 feet 5 inches tall. (Lesson 1-2)

b. What is Rituso's ideal weight if he is 6 feet 3 inches tall? (Lesson 1-2)

c. Rewrite each formula using the distributive property. (Lesson 1-5)

A practice test for Chapter 1 is available on page 774.

Alternative Assessment

Performance Task

Demonstrate your knowledge by giving a clear, concise solution to each problem. Be sure to include all relevant drawings and justify your answers. You may show your solutions in more than one way or investigate beyond the requirements of the problem.

A health magazine reports on issues of health and wellness. The table of Calories and grams of fat in different snack foods shown at the right accompanied an article on nutrition.

Snack	Calories	Fat grams
Chocolate wafer candy bar (4 oz)	588	33
Strawberry licorice (5 oz)	500	5
Candy bar (4 oz)	492	21
Peanut butter cups (3.2 oz)	380	22
Peanut candies (2.6 oz)	363	20
Chocolate covered raisins (2.2 oz)	320	21
Fruit flavored candy (2.6 oz)	286	2
Plain air-popped popcorn (med. bucket)	180	trace
Cola (16 oz)	205	0
Diet cola (16 oz)	<1	0

1. About how many Calories would you consume if you ate one candy item and drank a soft drink?

2. Write an expression for the number of grams of fat in two packages of peanut candies and a cola. Then find the number of Calories.

3. There are 192 Calories in a 16-ounce lemon-lime drink.
   a. Write two expressions for the number of calories in a lemon-lime drink and a package of candy.
   b. Find the number of calories in a lemon-lime drink and a chocolate wafer candy bar.

4. Explain what the entry in the Calories column for diet cola means. Could there be 6 Calories in a 16-ounce diet cola?

5. You have been asked to make a presentation on nutrition to your health class. Represent the data in the table in a different way to show the nutritious value of snack foods.

Thinking Critically

▶ Graph several points of the form (a, b) and (b, a). What do you observe? When are the graphs of (a, b) and (b, a) the same point? Write an equation that has no solution. Explain why it has no solution.

Portfolio

▶ A portfolio is representative samples of your work, collected over a period of time. Begin your portfolio by selecting an item that shows something new that you learned in this chapter.

Investigation

BON VOYAGE

You and another member of your class have been selected to participate in a student foreign exchange program. You won't be leaving the country until June, so there is plenty of time to prepare. The first thing you should do is obtain a passport.

MATERIALS

travel guides

reference books or magazines

To apply for a passport, you must:

- *complete an application in person before an official designated to accept passport applications. (A parent or legal guardian must complete the application for children less than 13 years old.)*
- *provide proof of U.S. citizenship (usually a previously-issued passport or a birth certificate).*
- *submit two identical photographs that are 2 inches by 2 inches in size. (The photographs should not be more than 6 months old and must be clear, front view, full face, with a white background.)*
- *pay $30. (The fee is $55 for passports issued to persons age 18 and older.)*

Work with your partner to prepare an estimate of your pre-flight expenses. Be sure to include costs associated with obtaining your passport. Be prepared to explain your list of pre-trip expenses to the class.

Starting the Investigation ....

Once passports have been obtained, another concern is what items to pack. There are many things to consider as you make your plans. Look over the following list of questions to consider.

Think About . . .

- *From what airport will you depart?*
- *How much luggage can you take?*
- *Are there restrictions on items you can take into the foreign country?*
- *Are there any restrictions on items you can bring or ship back?*
- *Will you have enough luggage space to pack purchases, gifts, and souvenirs?*
- *What are the costs and quality of the communication services (phone, mail) between your final destination and your home?*
- *What's the time difference between your final destination and your home?*

Brainstorm with your partner to choose a country of interest. Use this country as your destination as an exchange student. Be prepared to share the reasoning behind your choice with the class. Then work together to add to the list of questions to be considered. You may want to write to the U.S. Customs Service for more information.

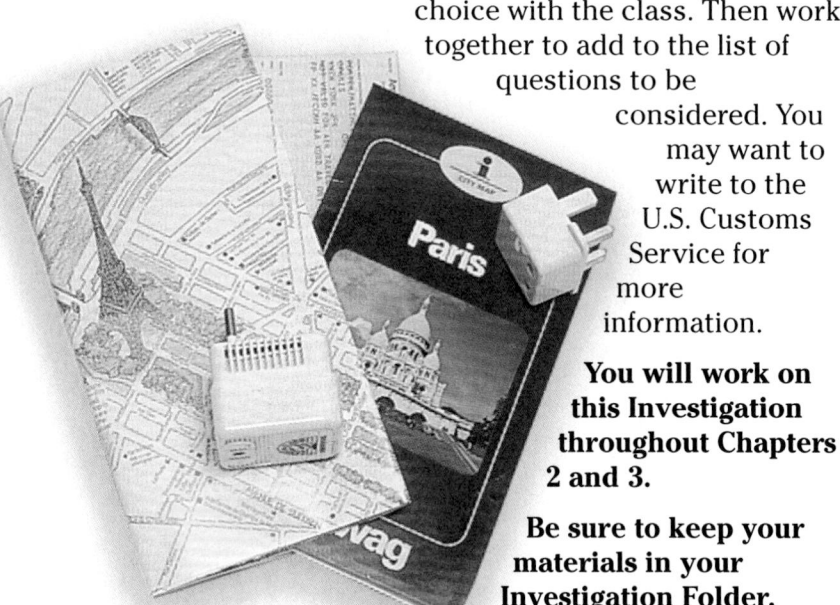

You will work on this Investigation throughout Chapters 2 and 3.

Be sure to keep your materials in your Investigation Folder.

CONNECTION

Research For current information on being an exchange student, visit the following website.

Data Collection and Comparison To share and compare your data with other students in the U.S., visit:

www.glencoe.com/ sec/math/prealg/ mathnet

Investigation: Bon Voyage!

Working on the Investigation
Lesson 2-1, p. 70

Working on the Investigation
Lesson 2-6, p. 97

Working on the Investigation
Lesson 3-2, p. 128

Closing the Investigation
End of Chapter 3, p. 160

CHAPTER 2 Exploring Integers

What You'll Learn

In this chapter, you will:

- graph on number lines and coordinate planes,

- compare and order integers,

- add, subtract, multiply, and divide integers,

- solve problems looking for a pattern, and

- solve problems with integers.

Why It's Important

Real World

Geography When reading a globe or map, you can use the longitude and latitude lines to pinpoint a location on Earth. For example, The Sydney Opera House in Sydney, Australia, is located at approximately 30°S, 150°E. In mathematics, you can locate a point precisely by using a coordinate system similar to the longitude and latitude system on a globe. *You will learn about the coordinate system in Lesson 2-2.*

The Sydney Opera House

FOCUS on...

PREREQUISITE SKILLS

To be successful in this chapter, you'll need to understand these concepts and be able to apply them. Refer to the lesson in parentheses if you need more review before beginning the chapter.

▶ **Evaluate expressions containing variables.** *(Lesson 1-3)*

Evaluate each expression if $a = 4$, $b = 10$, and $c = 8$.

1. $a + b + c$

2. $bc - ab$

3. $4c + 3b$

4. $2b - (a + c)$

▶ **Simplify expressions using the distributive property.** *(Lesson 1-5)*

Simplify each expression.

5. $6a + a$

6. $3p + p + 8$

7. $20(b + 2) + 3b$

8. $4(x + 2y) + 5x$

▶ **Use ordered pairs to locate points.** *(Lesson 1-7)*

Use the grid at the right to name the point for each ordered pair.

9. $(1, 3)$

10. $(5, 2)$

11. $(3, 4)$

12. $(0, 2)$

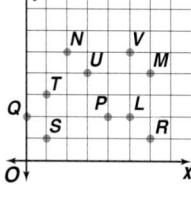

▶ **Use inverse operations to solve equations.** *(Lesson 1-8)*

Solve each equation using the inverse operation.

13. $7 + x = 22$

14. $m - 15 = 19$

15. $256 = r + 187$

16. $1566 = 27c$

FOCUS on...

READING SKILLS

You will be learning about **absolute value** in this chapter. It will be important to understand the meaning of absolute value. Absolute value is the number of units a number is from 0 on the number line. Since absolute value is a distance, it can never be less than 0.

What You'll Learn

- To graph integers on a number line, and
- to find absolute value.

Why It's Important

Integers are used to measure temperature and describe scores in sports.

Application: Geography

The shores of the Dead Sea, between Israel and Jordan, are 1312 feet *below* sea level. If you use the number 1312 to represent 1312 feet *above* sea level, what number could be used to represent 1312 feet below sea level?

Learning the Concept

Study the pattern of the following subtraction sentences.

$$5 - 1 = 4$$
$$5 - 2 = 3$$
$$5 - 3 = 2$$
$$5 - 4 = 1$$
$$5 - 5 = 0$$
$$5 - 6 = \underline{\ ?\ }$$

From the pattern, it appears that $5 - 6$ has an answer less than 0. Study the diagram below.

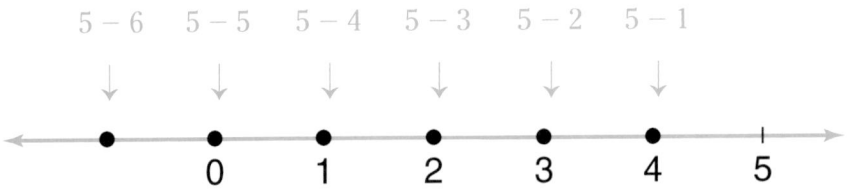

The answer for $5 - 6$ should be *one less than zero*. The number *one less than zero* is written as -1. So $5 - 6 = -1$. This is an example of a **negative** number. A negative number is a number less than zero.

You can use a negative number to write 1312 feet below sea level, the elevation of the shores of the Dead Sea. With sea level as the starting point or 0, you can express the elevation as $0 - 1312$, or -1312. So, the elevation of the Dead Sea is -1312 feet.

Negative numbers, like -1312, are members of the set of **integers**. Integers can also be represented as points on a number line.

Numbers to the left of zero are less than zero.

negative   positive

-6 -5 -4 -3 -2 -1 0 1 2 3 4 5 6

Numbers to the right of zero are greater than zero.

The numbers $-1, -2, -3,$ ... are called negative integers. The number negative 3 is written -3.

Zero is neither negative nor positive.

The numbers $1, 2, 3,$ ... are called positive integers. The number positive 4 is written $+4$ or 4.

THINK ABOUT IT

Which integer is neither positive nor negative?

This set can be written $\{\ldots, -3, -2, -1, 0, 1, 2, 3, \ldots\}$ where ... means continues indefinitely.

To graph a particular set of integers, locate the integer points on a number line. The number that corresponds to a point on the number line is called the **coordinate** of the point.

Example **a. Name the coordinates of D, E, and B.**

The coordinate of D is 4, E is -3, and B is 2.

b. Graph points F, U, and N on a number line if F has coordinate 1, U has coordinate -3, and N has coordinate 4.

To graph a number, find its location on the number line and draw a dot. Write the letter above the dot.

Looking at the number line shown below, you can see that 4 and -4 are on opposite sides of the starting point, zero. However, they are the same distance from zero. In mathematics, we say that they have the same **absolute value**, 4.

The symbol for the absolute value of a number is two vertical bars on either side of the number. $\lvert -4 \rvert = 4$ is read *The absolute value of -4 is 4.*

Absolute Value

In words: The absolute value of a number is the distance the number is from the zero point on the number line.

In symbols: $\lvert 4 \rvert = 4$ and $\lvert -4 \rvert = 4$

Example 2 **Simplify.**

a. $|9| + |-9|$

$|9| + |-9| = 9 + 9$ *The absolute value of −9 is 9.*
$= 18$ *The absolute value of 9 is 9.*

b. $|13| - |-2|$

$|13| - |-2| = 13 - 2$ *The absolute value of 13 is 13.*
$= 11$ *The absolute value of −2 is 2.*

Connection to ⟩ Algebra

Since variables represent numbers, you can use absolute value notation with algebraic expressions involving variables.

Example 3 **Evaluate the expression $|x| - 7$ if $x = -13$.**

$|x| - 7 = |-13| - 7$ *Replace x with −13.*
$= 13 - 7$ *The absolute value of −13 is 13.*
$= 6$

Checking Your Understanding

Communicating Mathematics

Read and study the lesson to answer these questions.

1. **Explain** how you would graph -7.

2. **Describe** some situations in the real world where negative numbers are used.

3. **Identify** the integers graphed on the number line at the right.

                                              −4 −3 −2 −1　0　1　2　3　4

4. **Graph** two points on a number line so that the coordinates of both points have an absolute value of 5.

MATERIALS

◯ counters
▢ integer mat

5. You can use counters to model integers. Each yellow counter represents a positive unit, and each red counter represents a negative unit.

Place three yellow counters on the mat to represent $+3$.

Place eight red counters on the mat to represent -8.

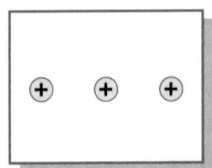

a. Model -4 using counters.   **b.** Model $+5$ using counters.

Guided Practice

6. Name the coordinate of each point graphed on the number line.

Write an integer for each situation. Then graph on a number line.

7. 5°F below zero

8. a 12-pound loss

9. 13 yards gained

10. a bank deposit of $1200

Simplify.

11. $|-5|$

12. $|8|$

13. $|13| - |-3|$

14. $|15| + |-8|$

Evaluate each expression if $d = -5$ and $r = 3$.

15. $7 + |d|$

16. $|d| - r$

17. **Scuba Diving**   Two divers are descending to a coral reef off the coast of Madagascar. One diver is 15 meters below the water's surface. The other diver is 9 meters lower.

   a. Represent the depths of the divers as integers.

   b. Graph each point on a number line.

Exercises: Practicing and Applying the Concept

Independent Practice

Graph each set of numbers on a number line.

18. $\{0, -1, 3\}$

19. $\{-2, -5, 4\}$

20. $\{-4, 1, 3, 8\}$

21. $\{-3, -5, -7, -9\}$

22. Name the coordinates of each point graphed on the number line.

```
        B       D       C       A
  +--+--+--●--+--+--●--+--+--●--+--+--+--+
    -6    -4    -2    0    2    4    6    8
```

Write an integer for each situation.

23. a salary increase of $400

24. a bank withdrawal of $50

25. 3 seconds before liftoff

26. a gain of 9 yards

27. 5° above zero

28. a loss of 6 pounds

Simplify.

29. $|-11|$

30. $-|24|$

31. $|0|$

32. $|7|$

33. $|-5| + |3|$

34. $|-8| + |-10|$

35. $|14| - |-5|$

36. $|-15| + |-12|$

37. $|0 + 12|$

38. $|16 - 2|$

39. $-|-36|$

40. $-||-4| + |-21||$

Evaluate each expression if $a = 0$, $b = 2$, and $c = -4$.

41. $|c| - 2$

42. $12 - |b|$

43. $b + a + |c|$

44. $ac + |-30|$

45. $|c| - b$

46. $|b| \cdot |c| + |a|$

Critical Thinking

Find the next three numbers in each pattern.

47. 32, 24, 16, _?_, _?_, _?_

48. $-23, -18, -13$, _?_, _?_, _?_

49. **Logical Reasoning**   Consider two numbers A and B on a number line. It is always, sometimes, or never true that the distance between A and B equals the distance between $|A|$ or $|B|$?

Applications and Problem Solving

50. Game Shows On a game show, contestants earn points for each correct response and lose points for each incorrect response. Questions are valued at 100, 200, 300, 400, 500, 600, 800, or 1000 points. Describe a situation where a contestant would have a score of -300.

51. Sports Golf scores are reported in reference to *par*. A score of $+2$ means two strokes over par. The leaders in the 1995 U.S. Open in Southampton, New York, had the scores -2, 3, 4, 1, 4, -1, -1, -5, -4, and 1 in the final round. Graph the scores on a number line.

Mixed Review

The table shows test scores for a physical science class. Use the table to answer each question. (Lesson 1-10)

Score	Test Score Tally	Frequency
19–20	IIII	5
17–18	IIII IIII	9
15–16	IIII	5
13–14	III	3
11–12	I	1
9–10	II	2

52. How many students are in the class?

53. What scores occurred most often?

54. What scores occurred least often?

55. Make a bar graph of the data.

56. Solve $16 = z - 25$. (Lesson 1-8)

57. Geometry The area of a rectangle is given by $A = \ell w$, where ℓ is the length and w is the width. Find the area of a rectangular garden if its length is 15 feet and its width is 12 feet. (Lesson 1-6)

For **Extra Practice**, see page 743.

58. Simplify $5a + 3(7 + 2a)$. (Lesson 1-5)

59. Write an algebraic expression for the phrase *three times a number decreased by 8*. (Lesson 1-3)

60. Standardized Test Practice Which expression is equivalent to $\dfrac{1 + 3 \cdot 5}{2 \cdot 2}$? (Lesson 1-2)

A. 3 **B.** $\dfrac{15}{4}$ **C.** 4 **D.** 5

WORKING ON THE

Investigation

BON VOYAGE

Refer to the Investigation on pages 62–63.

Materials: map or globe

Lines of longitude indicate degrees east or west of the prime meridian. The prime meridian is at $0°$. The prime meridian is a semicircle passing through Greenwich, England.

A city's longitude determines its time zone. The zone labeled zero is centered at the prime meridian.

- How does the way the zones are numbered differ from the integer number line?

- Compare your time zone to that of the country you chose at the beginning of this Investigation. How would you use this information when traveling abroad?
- Work with the members of your group to develop reasons why time zones exist.

Add the results of your work to your Investigation Folder.

2-1B Statistical Line Plots

An Extension of Lesson **2-1**

MATERIALS

✐ ruler

In the previous lesson, you saw how number lines helped to visualize integers. In statistics, data is also organized and presented on a number line. A picture of information on a number line is called a **line plot**.

Population figures from the census are used to determine the number of House of Representative members from each state of the United States.

Work with a partner.

▶ This chart shows the states that had a change in the number of House members after the 1990 census. Determine the change for each state. Express a decrease as a negative integer and an increase as a positive integer.

▶ Draw a number line. Since the largest and smallest changes were 7 and −3 for California and New York, respectively, use a scale of −4 to +8 and intervals of 1.

-4 -3 -2 -1 0 1 2 3 4 5 6 7 8

▶ To make the line plot, put an "x" above the number that represents the change in House members for each state. The graph below shows the first five values from the table.

State	1980	1990	Change
Arizona	5	6	+1
California	45	52	+7
Florida	19	23	+4
Georgia	10	11	+1
Illinois	22	20	−2
Iowa	6	5	
Kansas	5	4	
Kentucky	7	6	
Louisiana	8	7	
Massachusetts	11	10	
Michigan	18	16	
Montana	2	1	
New Jersey	14	13	
New York	34	31	
North Carolina	11	12	
Ohio	21	19	
Pennsylvania	23	21	
Texas	27	30	
Virginia	10	11	
Washington	8	9	
West Virginia	4	3	

New York

California

```
          ×
    ×     ×     ×       ×
+---+---+---+---+---+---+---+---+---+---+---+---+
-4 -3 -2 -1  0  1  2  3  4  5  6  7  8
```

1. What number occurs most frequently?
2. A number that is far apart from the rest of the data is called an **outlier**. Identify any numbers that appear to be outliers.
3. Data that are grouped closely together are called a **cluster**. Identify any clusters.
4. The *net change* is the change in the total number of House members. What was the net change in the number of House members?
5. What general geographic patterns do you notice from the information in the table? In which areas of the country are states gaining or losing population?

Extension

Integration: Geometry
The Coordinate System

Application: Cartography

CULTURAL CONNECTIONS

Benjamin Banneker, the grandson of a slave, was part of a team charged with designing Washington, D.C. When Pierre L'Enfant resigned and took all the plans with him, Banneker reproduced the entire set of plans from memory.

Our nation's capital has a unique system for naming its streets. The Capitol is at the center. Streets that lie east and west of the Capitol are named numerically: 1st, 2nd, 3rd, 4th, and so on. Streets that lie north and south of the Capitol are named alphabetically: A, B, C, D, . . . through W, skipping the letter J. This means that there are *two* sets each of the numerical and alphabetical street names.

Each location on the map could be indicated by a pair of numbers and letters, such as (3, D). This would represent the intersection of 3rd and D streets. However, there are *four* places where 3rd and D streets meet. The problem is resolved by separating Washington into four different sections. The sections are northeast (NE), southeast (SE), southwest (SW), and northwest (NW).

You can review the coordinate system in Lesson 1-7.

A mathematical model of the city map of Washington, D.C., could be represented by a **coordinate system**. Remember that a coordinate system is formed by the intersection of two number lines, called **axes**. The axes separate the coordinate plane into four regions called **quadrants**, like the four sections of Washington, D.C.

National Gallery of Art

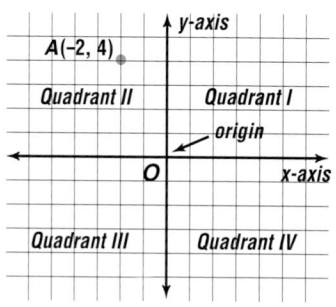

Axes is the plural of axis.

Notice that the numbers to the left of zero on the x-axis are negative as are the numbers below zero on the y-axis.

The *x-coordinate* of the ordered pair $(-2, 4)$ is -2, and the *y-coordinate* is 4.

ordered pair
$$(-2, 4)$$
x-coordinate ↑   ↑ *y-coordinate*

The dot at $(-2, 4)$ is the **graph** of point A.

The origin and the two axes do not lie in any quadrant.

Example 1

Real World APPLICATION

Cartography

In 1992, Ben Nighthorse Campbell became the first Native American to serve as a U.S. Senator.

a. **Write the ordered pair for NASA.**
The *x*-coordinate is -6, and the *y*-coordinate is -3. Thus, the ordered pair is $(-6, -3)$.

b. **What is located at (2, 3)?**
Count 2 units to the right and 3 units up. The point for the Senate Offices lies at (2, 3).

c. **What is located at the origin?**
The origin is at (0, 0). The Capitol is at the origin.

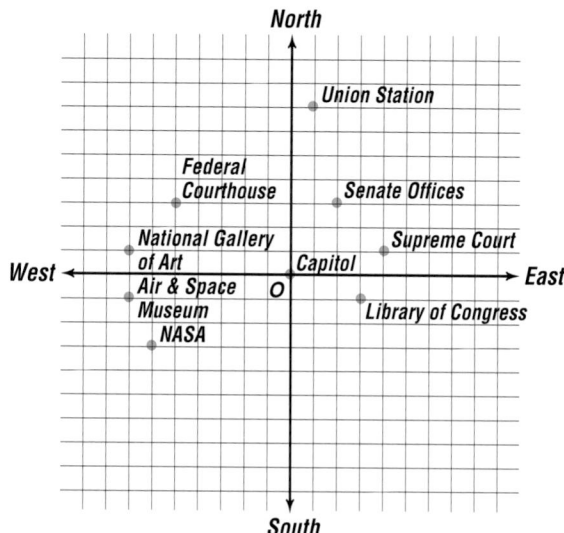

Remember that to graph a point means to place a dot at the point named by the ordered pair. This is sometimes called **plotting the point**.

Example **Graph each point on a coordinate plane. Then name the quadrant in which each point lies.**

a. D(3, −4)

Start at the origin. Move 3 units to the right. Then move 4 units down and draw a dot. Label the dot D(3, −4). Point D is located in quadrant IV.

b. E(−3, 2)

Start at the origin. Move 3 units to the left. Then move 2 units up and draw a dot. Label this dot E(−3, 2). Point E is located in quadrant II.

c. F(4, 0)

Start at the origin. Move 4 units to the right. The graph is on the x-axis. Label this dot F(4, 0). Since point F lies on the x-axis, it is not in any quadrant.

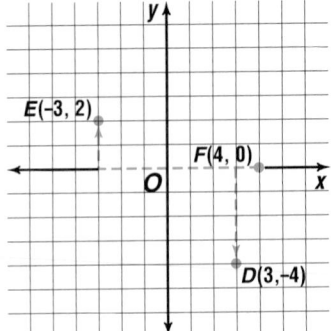

Checking Your Understanding

Communicating Mathematics

Read and study the lesson to answer these questions.

1. **Draw** a coordinate system and label the origin, the x- and y-axes, and the four quadrants.
2. **Explain** why the point (4, 8) is different from the point (8, 4).
3. **Name** two ordered pairs whose graphs are *not* located in one of the four quadrants.
4. **You Decide** Danny said that if you interchange the coordinates of any point in Quadrant I, the new point still would be in Quadrant I. Cherita disagrees. She says the new point would be in Quadrant 3. Who is right? Explain.

Guided Practice

Use the coordinate grid at the right to name the point for each ordered pair.

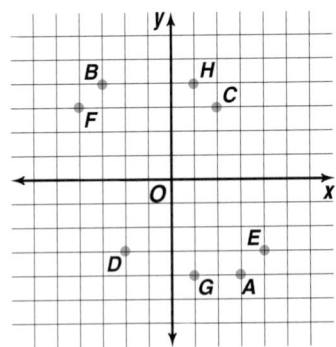

5. (−3, 4)
6. (1, 4)
7. (−2, −3)
8. (−4, 3)

On graph paper, draw coordinate axes. Then graph and label each point. Name the quadrant in which each point is located.

9. I(4, −7)
10. J(−4, 8)
11. K(0, 7)
12. L(−6, −3)

13. **Logical Reasoning** If the graph of A(x, y) satisfies the given condition, name the quadrant in which point A is located.

    a. $x > 0, y > 0$ **b.** $x < 0, y < 0$ **c.** $x < 0, y > 0$

Independent Practice

Name the ordered pair for each point graphed on the coordinate plane at the right.

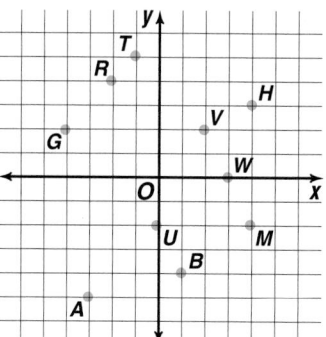

14. *R*   15. *M*
16. *A*   17. *B*
18. *G*   19. *H*
20. *U*   21. *V*
22. *W*   23. *T*

On graph paper, draw coordinate axes. Then graph and label each point. Name the quadrant in which each point is located.

24. $A(3, 5)$ 25. $K(-6, 1)$ 26. $M(2, -4)$
27. $B(-7, -4)$ 28. $R(-1, 3)$ 29. $S(3, -5)$
30. $Q(-3, -5)$ 31. $D(4, 2)$ 32. $E(7, -8)$
33. $F(-4, -7)$ 34. $G(1, 0)$ 35. $H(-7, 0)$

Graph each point. Then connect the points in alphabetical order and identify the figure.

36. $A(0, 6)$, $B(4, -6)$, $C(-6, 2)$, $D(6, 2)$, $E(-4, -6)$, $F(0, 6)$

37. $A(5, 8)$, $B(1, 13)$, $C(5, 18)$, $D(9, 13)$, $E(5, 8)$, $F(5, 6)$, $G(3, 7)$, $H(3, 5)$, $I(7, 7)$, $J(7, 5)$, $K(5, 6)$, $L(5, 3)$, $M(3, 4)$, $N(3, 2)$, $P(7, 4)$, $Q(7, 2)$, $R(5, 3)$, $S(5, 1)$

If *x* and *y* are integers, graph all ordered pairs that satisfy the given conditions.

38. $|x| < 4$ and $|y| < 3$ 39. $|x| \le 2$ and $|y| \le 1$

Critical Thinking

40. Describe the possible location in terms of quadrants or axes, for the graph of (x, y) if *x* and *y* satisfy the condition $xy = 0$.

Applications and Problem Solving

41. **Geometry** A **vertex** of a polygon is a point where two sides of the polygon intersect.

a. Identify the coordinates of the vertices in the rectangle at the right.

b. Subtract 4 from each *y*-coordinate. Graph the new ordered pairs. What figure is formed if the points are connected?

c. Multiply each *x*- and *y*-coordinate of the vertices of *PQRS* by -1. Graph the new ordered pairs. Describe the figure that is formed if the points are connected.

Sydney, Australia

42. Geography Latitude and longitude lines are used to locate places on a map or globe. Latitude is measured in degrees north or south of the equator while longitude is measured in degrees east or west of the prime meridian. When describing a location, the latitude is usually given first. For example, the location of 30°N, 90°W is New Orleans, Louisiana. Find the approximate latitude and longitude of the following locations.

a. Sydney, Australia

b. Porto Alegre, Brazil

c. Johannesburg, Republic of South Africa

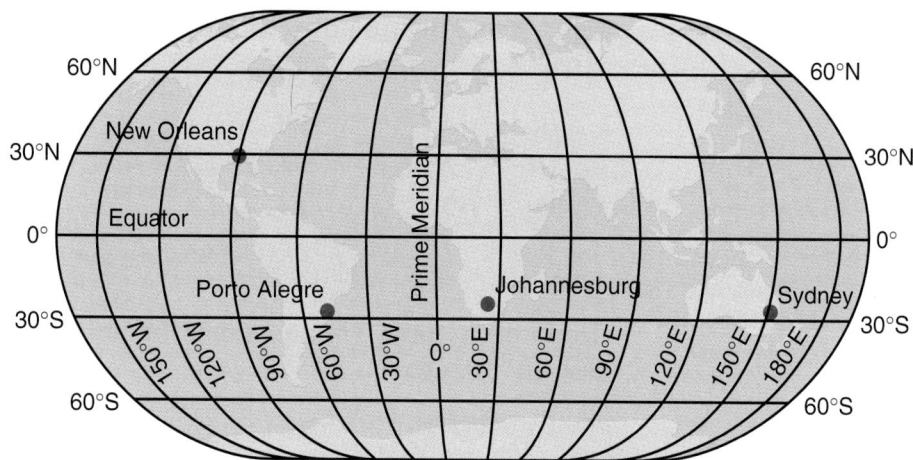

Mixed Review

43. Find the value of $|15| - |-3|$. (Lesson 2-1)

44. Graph $\{-3, 0, 2\}$ on a number line. (Lesson 2-1)

45. Recreation During the last few years, in-line skating has become a popular pastime. Today there are over 500,000 people who participate in this sport. This is more than twice as many people as there were several years ago. Which inequality, $s > 250,000$ or $s < 250,000$, describes how many skaters there were several years ago? (Lesson 1-9)

46. What property is shown by $(6 \cdot 15) \cdot 3 = 6 \cdot (15 \cdot 3)$? (Lesson 1-4)

47. Standardized Test Practice Evaluate $9a - (4b + 2c)$ if $a = 4, b = 2,$ and $c = 3$. (Lesson 1-3)

A. 12 **B.** 22 **C.** 34 **D.** 50

48. Find the value of $7[(12 + 5) - 3(19 - 14)]$. (Lesson 1-2)

49. Sports The Middletown Junior High School basketball team made 21 baskets to score 36 points. If two of the baskets were 3-point shots, how many baskets were worth 2 points and how many were worth 1 point? (Lesson 1-1)

For **Extra Practice**, see page 743.

2-2B Plotting Points

An Extension of Lesson **2-2**

You can plot points on a graphing calculator just as you do on grid paper.

Activity

Plot the ordered pairs (6, 9), (27, 13), (15, –13), (–33, 24), and (–12, –8) on a graphing calculator.

First, set the viewing window. The viewing window is the range of *x*- and *y*-values that are visible on the graphics screen. The viewing window [−47, 47] by [−31, 31], with scale factors of 10 on both axes works well for these points. This means that the scale will be from −47 to 47 on the *x*-axis and from −31 to 31 on the *y*-axis, with tick marks every 10 units.

ENTER: WINDOW (−) 47 ENTER 47 ENTER 10 ENTER (−) 31 ENTER 31 ENTER 10 GRAPH

Now draw the points.

ENTER: 2nd DRAW ▶ ENTER

Use the arrow keys to move the cursor to the first point at (6, 9) and press ENTER. As you move the cursor to the next point, at (27, 13), you will see that the first point was plotted. After you position the cursor at the next point, press ENTER again. Continue until all points are plotted. To see the last point, press CLEAR.

Your Turn

Clear the graphics screen by pressing 2nd DRAW ENTER. Then graph each set of ordered pairs on a graphing calculator.

1. (8, –3), (9, 7), (–3, 4), (–7, 4), (–6, –5)
2. (–17, 22), (8, 19), (–12, 1), (–21, –7), (16, 23)
3. (–32, 4), (26, –16), (–3, –8), (–7, –11)
4. (18, –30), (39, 17), (–33, 24), (–27, 14), (32, –21)

5. The *standard viewing window* can be set automatically by pressing ZOOM 6.
   a. What are the minimum and maximum values of the standard viewing window?
   b. What are the minimum and maximum values of the viewing window after pressing ZOOM 6 ZOOM 8 ENTER ?

6. You should choose the viewing window to best display each graph.
   a. How would you set the viewing window to display a graph of Quadrant II? Quadrant IV?
   b. If the last ordered pair in Exercise 4 was (–49, 39), how could you make sure it was visible in the viewing window?

Comparing and Ordering

You'll Learn

■ To compare and order integers.

It's Important

Knowing how to order integers can help you compare historical events.

Application: Health

Nature's way of using pollen to produce seeds in plants is remarkable. However, pollen is the enemy of millions of people who suffer from allergies. Pollen from trees or weeds can cause sneezing and eye irritation.

During allergy season, local television stations or newspapers may give daily pollen counts to warn people. As the pollen count increases, so does the discomfort level as seen moving left to right on the scale below.

Pollen Count	0–100	101–600	601–1200	over 1200
Discomfort Level	low	uncomfortable	high	very high

Learning the Concept

Just like the pollen scale, the numbers on a number line increase as you move to the right. On a number line, it is easy to determine which of two numbers is greater. -3 and 4 are graphed on the number line below.

Say: 4 is greater than -3 *The number to the right is always greater.*
Write: $4 > -3$ *The number to the left is always less.*

You might also conclude that -3 is to the left of 4.

Say: -3 is less than 4   *Remember, the symbol points*
Write: $-3 < 4$ *to the lesser number.*

Remember that any mathematical sentence containing $<$ or $>$ is called an inequality. In inequalities, numbers are compared.

Example **Use the integers graphed on the number line below for each question.**

a. **Write two inequalities involving -5 and -1.**
   Since -5 is to the left of -1, write $-5 < -1$.
   Since -1 is to the right of -5, write $-1 > -5$.

b. **State which number is greater.**
   -1 is greater since it lies to the right of -5.

c. **State which number has the greater absolute value.**
   The absolute value of -5 is greater than the absolute value of -1 because it lies farther from the origin.

Integers are used to compare numbers in many everyday applications.

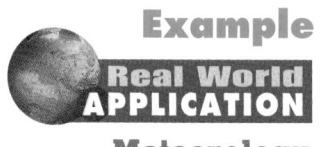

Example ② The chart at the right shows the record low temperatures for selected states. Order the temperatures from least to greatest.

Real World APPLICATION

Meteorology

Graph each integer on a number line.

Write the integers as they appear on the number line from left to right.

$-61, -37, -35, -34, -27, -27, -17,$
$-16, -2, 12$ are in order from least to greatest.

Record Low Temperatures for Selected States (°F)	
State	Temperature
Alabama	−27
California	−61
Florida	−2
Georgia	−17
Hawaii	12
Indiana	−35
Kentucky	−34
Louisiana	−16
Oklahoma	−27
West Virginia	−37

Checking Your Understanding

Communicating Mathematics

Read and study the lesson to answer these questions.

1. **Write** two inequalities that show how the highest and lowest temperatures in Example 2 are related.

2. **List** the integers graphed on the number line from least to greatest.

3. Graph two integers on a number line. Then write two inequalities that show how the integers are related.

Math Journal

4. Write a sentence that explains how to determine when one integer is less than another integer.

Guided Practice

Write an inequality using the numbers in each sentence. Use the symbols < or >.

5. $2°$ is warmer than $-5°$.

6. 5 is greater than -3.

Replace each ● with <, >, or =.

7. -19 ● -9

8. 0 ● -8

9. 4 ● $|-4|$

10. $|-3|$ ● $|8|$

11. Order the integers $\{43, 0, -9, -4, 3, -88, 234\}$ from least to greatest.

Write two inequalities for each graph.

12.

13.

14. **Sports** Patty Sheehan won the $82,500 first prize in the Rochester International LPGA tournament in June, 1995. The final round scores of the top ten finishers were $-2, +1, -5, +3, -2, -1, +2, +4, 0,$ and $+1$.

   a. Order the scores from lowest to highest.

   b. If the lowest score wins, what was the winning score for this round?

Independent Practice

Replace each ● with <, >, or =.

15. 9 ● −11 **16.** −3 ● −5 **17.** |−3| ● 3
18. −10 ● −3 **19.** −7 ● −13 **20.** −|15| ● −15
21. −4 ● 0 **22.** 0 ● −2 **23.** 0 ● |−8|
24. −33 ● 5 **25.** −439 ● −23 **26.** −10 ● −3

Write an inequality using the numbers in each sentence. Use the symbols < or >.

27. 3 m is taller than 2 m. **28.** $20 is more than $15.
29. 60 kilometers per hour is slower than 75 kilometers per hour.
30. Yesterday's high temperature was 41°F. The low temperature was −3°F.
31. Today's pollen count is 565. Yesterday's count was 344.
32. Water boils at 212°F, and it freezes at 32°F.
33. One cup of green bean casserole has 265 Calories. The same amount of fresh green beans has 66 Calories.

Write two inequalities for each graph.

34.
−4 −3 −2 −1 0 1 2

35.
−8 −7 −6 −5 −4 −3 −2 −1 0

36.
−100 −80 −60 −40 −20 0 20 40

37.
−8 −7 −6 −5 −4 −3 −2 −1 0 1 2 3

Order the integers in each set from least to greatest.

38. {7, 0, −5} **39.** {−11, 8, −3}
40. {0, −5, −8, −1} **41.** {−6, 56, 29, 1, −65}
42. {33, 9, −99, 7, 0, −4} **43.** {87, −65, −53, 48, 199}

Critical Thinking

44. Graph all integer solutions for $|x| < 4$ on a number line.

Applications and Problem Solving

45. Chemistry The chart at the right shows the freezing/melting points of water and selected elements.

   a. Graph this data on a number line.

   b. Which of these elements has the highest freezing/melting point?

   c. Which has the lowest freezing/melting point?

Element	Freezing/melting point (°C)
Water	0
Helium	−272
Lead	328
Mercury	−39
Oxygen	−219
Sodium	98
Tin	232

46. Weather The table below shows the average monthly temperatures (°F) in Fairbanks, Alaska, for the 30-year period 1961–1990. Graph these temperatures on a number line.

Jan.	Feb.	Mar.	Apr.	May	June
10	−4	11	31	49	60

July	Aug.	Sept.	Oct.	Nov.	Dec.
63	57	46	25	3	−7

a. The lowest temperature in Fairbanks was −58°F on February 1. Write an inequality to compare this with the February average.

b. The highest temperature was 93°F on July 15. Write an inequality to compare this with the July average.

Mixed Review

On graph paper, draw and label coordinate axes. Then graph and label each point. (Lesson 2-2)

47. $L(2, -8)$

48. $M(-5, -5)$

49. Space Exploration About six seconds before liftoff, the three main shuttle engines start. About 120 seconds after liftoff, the solid rocket boosters burn out. Use integers to describe these events. (Lesson 2-1)

50. State whether $15 < 3m + 7$ is *true* or *false* when $m = 4$. (Lesson 1-9)

51. Restate 6×26 using the distributive property. (Lesson 1-5)

52. Standardized Test Practice Which property is shown by the statement $(5 + 4) \cdot 0 = 0$? (Lesson 1-4)
 A. The Distributive Property
 B. The Identity Property of Multiplication
 C. The Associative Property of Multiplication
 D. The Multiplicative Property of Zero

53. Consumer Awareness The Jacksonville Middle School chorale bought 30 tickets to a musical play. The total cost of the tickets is $98. If student tickets cost $3 each and adult tickets cost $5 each, how many adults are going with the chorale? (Lesson 1-1)

For **Extra Practice**, see page 744.

2-4A Adding Integers

A Preview of Lesson **2-4**

MATERIALS

⬤ counters

▭ integer mat

You can use counters to model operations with integers. In this book, yellow counters will represent positive integers, and red counters will represent negative integers.

There are two important properties to keep in mind when modeling integers.

▶ When one positive counter is paired with one negative counter, the result is called a **zero pair**.

▶ You can add or remove zero pairs from a mat because removing or adding zero does not change the value of the counters on the mat.

Activity 1 **Find $-2 + (-4)$.**

Remember that $2 + 4$ means *combine a set of two items with a set of four items.* The expression $-2 + (-4)$ means something similar. It tells you to combine a set of two negative items with a set of four negative items.

▶ Place 2 negative counters and 4 negative counters on the mat.

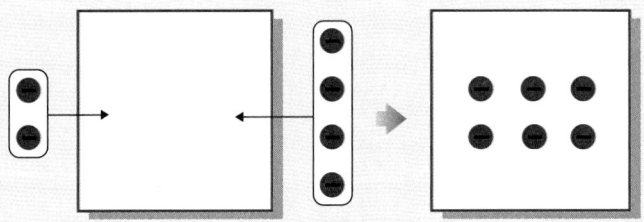

▶ Since there are 6 negative counters on the mat, the sum is -6. Therefore, $-2 + (-4) = -6$.

Activity 2 **Find $-4 + 2$.**

Remember that removing zero pairs does not change the value on the mat.

Place 4 negative counters and 2 positive counters on the mat. It is possible to remove 2 zero pairs.

▶ Since there are 2 negative counters remaining, the sum is -2. Therefore, $-4 + 2 = -2$.

Your Turn **Find each sum using counters.**

1. $3 + 2$	**2.** $3 + (-2)$	**3.** $-3 + 2$	**4.** $-3 + (-2)$
5. $1 + (-4)$	**6.** $-3 + 7$	**7.** $-4 + (-4)$	**8.** $-4 + 4$

Write About It **9.** Suppose you add a positive integer and a negative integer. Explain how you use a zero pair to find the sum.

Adding Integers

Technology Activity

You already know that the sum of two positive integers is a positive integer. What is the sign of the sum of two negative integers? The keystrokes for finding the sum of −5 and −8 on a scientific calculator are as follows. *Notice that the negative key is different than the subtraction key and that the negative is entered after the number.*

Enter: 5 ⌊(-)⌋ ⌊+⌋ 8 ⌊(-)⌋ ⌊=⌋   **Display:** -13

Your Turn Study the patterns of the sums in each column.

A	B	C	D
4 + 8 =	−3 + (−4) =	17 + (−17) =	−18 + 2 =
12 + 3 =	−4 + (−6) =	13 + (−2) =	−5 + 12 =
29 + 17 =	−13 + (−2) =	12 + (−15) =	−8 + 21 =
25 + 93 =	−7 + (−3) =	11 + (−18) =	−7 + 7 =

a. Suggest a rule for adding integers with the same signs, as in Columns A and B.

b. Suggest a rule for adding integers with different signs, as in Columns C and D.

c. Compare your rules to the rules of two of your classmates.

Learning the Concept

Another way to add integers is by using a number line.

Example **Find −5 + (−2).**

Start at zero. Move 5 units to the left. From there, move 2 more units to the left.

−5 + (−2) = −7

The results of this example and those in Columns A and B above suggest the following rule.

Adding Integers with the Same Signs	**To add integers with the same sign, add their absolute values. Give the result the same sign as the integers.**

Study Example 2 to see if you can discover the rule for adding integers that have different signs.

Example **2** Find each sum.

a. 8 + (−4)

Use counters. Put 8 positive counters on the mat. Add 4 negative counters. Remove all zero pairs.

Recall that a zero pair is one positive and one negative counter.

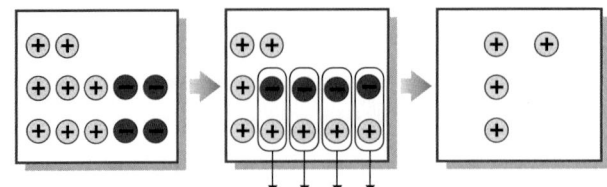

$8 + (−4) = 4$

b. −7 + 4

Use a number line. Start at zero. Move 7 units to the left. From there, move 4 units to the right.

$−7 + 4 = −3$

The results in these examples and those in Columns C and D on page 83 suggest the following rule.

Adding Integers with Different Signs	To add integers with different signs, subtract their absolute values. Give the result the same sign as the integer with the greater absolute value.

Example **3** Solve each equation.

a. $x = −9 + 6$

$x = −(|−9| − |6|)$ *Subtract absolute values. The result is*
$x = −(9 − 6)$ *negative because the integer with the*
$x = −3$ *greater absolute value, −9, is negative.*

b. $8 + (−3) + 4 = a$

$8 + 4 + (−3) = a$ *Commutative property*
$12 + (−3) = a$ *Add 8 and 4.*
$+(12 − 3) = a$ *The result is positive because $|12| > |3|$.*
$9 = a$ *Check your result with a calculator.*

Example **4**

Real World APPLICATION

Space Exploration

During a spacewalk on February 9, 1995, astronauts Bernard Harris Jr. and Michael Foale worked in the shadow of the shuttle *Discovery*. The temperature outside the cargo bay was −125°F. If the temperature inside the cargo bay was 111 degrees warmer, what was the temperature inside?

Explore   The temperature outside the cargo bay was −125°F. Inside, it was 111 degrees warmer. You need to find the temperature inside. Let *t* represent the temperature inside.

Plan   Translate the words into an equation using the variable. The temperature inside is the sum of −125 and 111.

Solve   inside temperature   is   outside temperature   plus   111°

$$t = -125 + 111$$

Solve the equation for *t*.

$$t = -125 + 111$$
$$t = -14$$

The temperature inside the cargo bay was $-14°$F.

Examine   The amount of increase was less than the absolute value of -125, so the temperature inside the cargo bay should still be negative. The answer is reasonable.

Connection to Algebra

You can use the rules for adding integers and the distributive property to combine like terms.

Example 5   **Simplify each expression.**

    a. $-15n + 22n$

$$
\begin{aligned}
-15n + 22n &= (-15 + 22)n &&\textit{Distributive property}\\
&= (7)n &&\textit{Find the sum of } -15 \textit{ and } 22.\\
&= 7n
\end{aligned}
$$

    b. $9x + (-18x) + 8x$

$$
\begin{aligned}
9x + (-18x) + 8x &= [9 + (-18) + 8]x &&\textit{Distributive property}\\
&= -1x &&\textit{Add 9, } -18, \textit{ and 8.}\\
&= -x &&-1x \textit{ can be written as } -x.
\end{aligned}
$$

Checking Your Understanding

Communicating Mathematics

Read and study the lesson to answer these questions.

1. **Explain** how to determine whether to add or subtract the absolute values to find the sum of two integers.

2. **Write** the addition sentence shown by each model.

   a.

   b.

3. **Draw a model** that shows how to find the sum of 5 and -2.

4. **Explain** how to simplify $-3a + 12a + (-14a)$.

MATH JOURNAL

5. Write, in your own words, how to add pairs of integers with the same sign and with different signs. Include an example for each situation.

Guided Practice

Write an addition sentence for each model.

6.

7.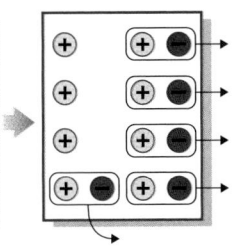

8.

9.

State whether each sum is positive or negative. Then find the sum.

10. $-6 + (-9)$

11. $13 + (-4)$

12. $-8 + 12$

13. $-3 + 7 + (-8)$

Solve each equation.

14. $r = -13 + (-5) + 7$

15. $(-83) + (-21) + (-7) = z$

Simplify each expression.

16. $-3z + (-17z)$

17. $-8a + 14a + (-12a)$

18. Golf In golf, a score of 0 is called *even par*. A score of 3 under par is written as -3. A score of 2 over par is written $+2$ or 2. In 1995, José Maria Olazábal shot 6 under par, 2 over par, even par, and even par for the four rounds of the Masters Tournament. What was his final score?

Exercises: Practicing and Applying the Concept

Independent Practice

Solve each equation.

19. $18 + (-5) = x$ **20.** $b = -8 + (-3)$ **21.** $16 + (-9) = m$

22. $v = -12 + (-4)$ **23.** $k = 9 + (-13)$ **24.** $-15 + 6 = q$

25. $g = 19 + (-7)$ **26.** $-11 + (-15) = t$ **27.** $-23 + (-43) = h$

28. $-3 + 18 = n$ **29.** $-5 + 31 = r$ **30.** $y = 6 + (-16)$

31. $m = 8 + (-17)$ **32.** $z = -12 + (-5)$ **33.** $-12 + 5 = s$

Write an addition sentence for each situation. Then find the sum.

34. Maria has $500 in the bank. She owes $700 on her car.

35. In Monday night's football game, the Jacksonville Jaguars lost 7 yards on one play. They gained 12 yards on the next.

36. The research submarine *Alvin* is at 1500 meters below the sea level. It descends another 1250 meters to the ocean floor.

37. Make Up a Problem Write a problem that can be solved using the addition sentence $-15 + 25 = c$.

Solve each equation.

38. $m = 3 + (-11) + (-5)$

39. $k = 14 + 9 + (-2)$

40. $-18 + (-23) + 10 = c$

41. $-16 + (-6) + (-5) = a$

42. $-12 + (17) + (-7) = x$

43. $y = 47 + 32 + (-16)$

Simplify each expression.

44. $6x + (-15)x$

45. $-11y + 14y$

46. $-14k + (-7)k + 15k$

47. $8m + (-23)m$

48. $14b + (-21b) + 37b$

49. $16d + (-9d) + (-27d)$

 Calculator

Use a calculator to find each sum.

50. $139 + (-316)$

51. $-249 + 7915$

52. $-3916 + (-8128)$

 Critical Thinking

53. *True* or *false*: $-n$ names a negative number.

Applications and Problem Solving

54. **Astronomy** At noon, the average temperature on the moon is 112°C. During the night, the average temperature drops 252°C. What is the average temperature on the moon's surface during the night?

55. **Pets** The chart at the right shows the numbers of dogs of different breeds registered in the American Kennel Club in a 2-year period.

Breed	Year 1 Registration	Year 2 Registration
Akita	11,574	11,383
Beagle	61,051	60,661
Chow Chow	33,824	42,670
Dachshund	48,573	50,046
Labrador Retriever	124,899	120,879
Pug	15,722	16,008

    a. Describe the change in the number of dogs of each breed registered from Year 1 to Year 2.

    b. What was the total change in the number of dogs of these breeds registered from Year 1 to Year 2?

Mixed Review

56. Order the numbers $\{0, -4, -8, -3\}$ from least to greatest. (Lesson 2-3)

57. *True* or *false*: The graph of a point whose coordinates are both negative integers would be located in Quadrant IV. (Lesson 2-2)

58. **Standardized Test Practice** Which of the points graphed below have coordinates whose absolute value is 4? (Lesson 2-1)

    A. E only   **B.** E and F   **C.** E and G   **D.** G and H

59. **Recreation** A skateboarder travels 48 meters in 16 seconds. Solve the equation $48 = r \cdot 16$ to find the speed in meters per second. (Lesson 1-8)

60. *True* or *false*: $4 + 13 = 17$ and $17 - 4 = 13$ are related sentences. (Lesson 1-8)

61. Write an equation and solve: Antonia paid $15 for a shirt on sale. It was reduced by $9. What was the regular price? (Lesson 1-8)

62. Of 19, 29, or 39, which is the solution of $3x = 87$? (Lesson 1-6)

63. Rewrite $(n + 8) + 9$ using an associative property. Then simplify. (Lesson 1-4)

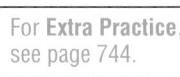 For **Extra Practice**, see page 744.

64. Write a verbal phrase for the expression $2x + 3$. (Lesson 1-3)

65. Determine whether $6 \div 2 + 5 \times 4 = 32$ is *true* or *false*. (Lesson 1-2)

2-5A Subtracting Integers

A Preview of Lesson **2-5**

MATERIALS

 counters

☐ integer mat

You can use counters to model the subtraction of integers.

Activity ① **Find −6 − (−3).**

▶ Start with 6 negative counters. Remove 3 negative counters.

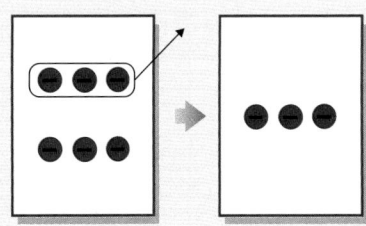

▶ Since there are 3 negative counters left, the difference is −3.
Therefore, −6 − (−3) = −3.

Activity ② **Model 4 − (−2).**

▶ Start with 4 positive counters. There are no negative counters, so you
can't remove 2 negative counters. Add 2 zero pairs to the mat.
Remember adding zero pairs does not change the value of the set.
Now remove 2 negative counters.

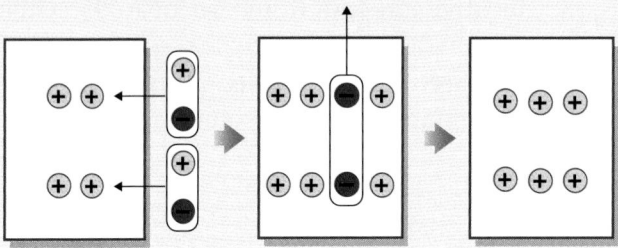

▶ Since there are 6 positive counters remaining, the difference is +6 or 6.
Therefore, 4 − (−2) = 6.

Your Turn **Find each difference using counters.**

1. 2 − (−3)   2. 2 − 3   3. −2 − (−3)   4. −2 − 3
5. −6 − 9   6. 13 − 4   7. −8 − (−12)   8. −4 − (−4)

9. Compare Exercises 5–7 above with Exercises 10–12 on page 86. How
are they alike? How are they different?

10. If you subtract a negative integer from a lesser negative integer, is
the difference positive or negative? Justify your answer with a
drawing.

Subtracting Integers

You'll Learn

■ To subtract integers.

It's Important

Knowing how to subtract integers can help you find increases and decreases in temperature.

Hands-On Activity

The diagram at the right models the expression $3 - (-3)$. Counting the number of counters on the mat after the counters are removed shows that $3 - (-3) = 6$.

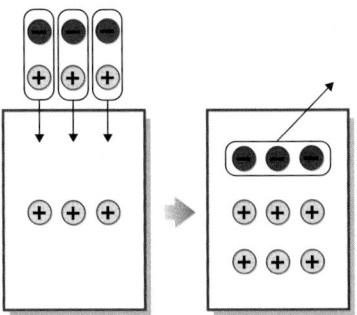

Your Turn Use counters to find $-3 - (-2)$, $4 - 6$, and $-5 - 1$.

a. Compare the solution of $-3 - (-2)$ to the solution of $-3 + 2$.

b. Compare the solution of $4 - 6$ to the solution of $4 + (-6)$.

c. Compare the solution of $-5 - 1$ to the solution of $-5 + (-1)$.

Learning the Concept

Adding and subtracting are inverse operations that "undo" each other. Similarly, when you add opposites, like 4 and -4, the sum is 0.

You can review inverse operations in Lesson 1-8.

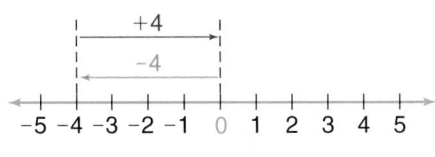

An integer and its opposite are called **additive inverses** of each other.

Additive Inverse Property	**In words:** The sum of an integer and its additive inverse is zero.
	Arithmetic **Algebra**
	$5 + (-5) = 0$ $x + (-x) = 0$

At the beginning of the lesson, you were asked to compare the result of subtracting an integer with the result of adding its additive inverse.

Subtraction	**Addition of Additive Inverse**
$-3 - (-2) = -1$	$-3 + 2 = -1$
$4 - 6 = -2$	$4 + (-6) = -2$
$-5 - 1 = -6$	$-5 + (-1) = -6$

These examples suggest a method for subtracting integers.

Subtracting Integers	**In words:** To subtract an integer, add its additive inverse.
	Arithmetic **Algebra**
	$3 - 5 = 3 + (-5)$ $a - b = a + (-b)$

Example **1** **Solve each equation.**
  a. $m = 7 - 12$
     $m = 7 + (-12)$ *To subtract 12,*
     $m = -5$ *add –12.*

  b. $6 - (-11) = r$
     $6 + 11 = r$ *To subtract –11,*
     $17 = r$ *add 11.*

  c. $y = -3 - (-8)$
     $y = -3 + 8$ *To subtract –8,*
     $y = 5$ *add 8.*

  d. $-9 - 16 = x$
     $-9 + (-16) = x$ *To subtract*
        $-25 = x$ *16, add –16.*

You can use equations with integers to represent real-life situations.

Example **2**

CONNECTION

Geography

The highest point in Africa is Kilimanjaro in Tanzania. It has an altitude of 5895 meters above sea level. The lowest point on the continent is Lake Assal in Djibouti. It is 155 meters below sea level. Find the difference between these altitudes.

Explore With sea level as the starting point, 5895 meters above sea level can be expressed as +5895 or 5895. 155 meters below sea level can be expressed as −155. Let *d* represent the difference.

Plan Translate the words into an equation using the variable.

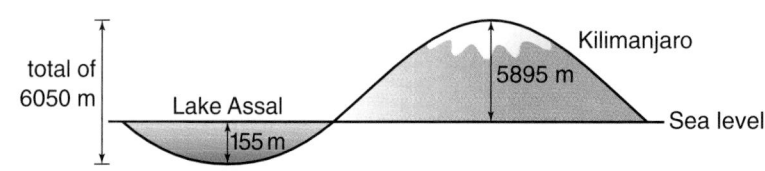

$$d \quad = \quad 5895 \quad - \quad -155$$

Solve Solve the equation for *d*.

  $d = 5895 - (-155)$ *To subtract −155*
  $d = 5895 + 155$ *add its inverse, 155.*
  $d = 6050$

The difference in altitudes is 6050 meters.

Examine Make a diagram to check the solution.

total of 6050 m

Lake Assal

Kilimanjaro

5895 m

155 m

Sea level

Connection to ▸ Algebra

You can use the rule for subtracting integers to evaluate expressions and combine like terms.

Example **3** **Evaluate $b - (-5)$ if $b = -13$.**

  $b - (-5) = -13 - (-5)$ *Replace b with −13.*
       $= -13 + 5$ *Subtract −5 by adding 5.*
       $= -8$

Example **4** **Simplify $5a - 13a$.**

  $5a - 13a = 5a + (-13)a$ *Add the additive inverse of 13a, −13a.*
       $= [5 + (-13)]a$ *Use the distributive property.*
       $= -8a$ *Add 5 and −13.*

Communicating Mathematics

Read and study the lesson to answer these questions.

1. **Explain** how subtraction of integers is related to addition of integers.

2. **Write** the subtraction sentence shown by each model.

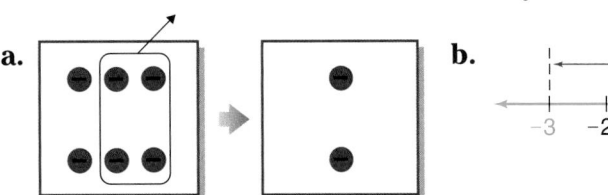

a.   b.

3. **Draw a model** that shows how to find $5 - (-2)$.

4. **Explain** how to simplify $12y - 14y$.

 MATERIALS

🎗 masking tape

5. Using masking tape, create a large number line on the floor. Then "walk out" the solution to each problem using the following rules.

▶ Stand at zero and face the positive integers.

▶ To model a positive integer, move forward.

▶ To model a negative integer, move backward.

▶ To model subtraction, do an about-face before moving.

a. $4 - 6$ b. $2 - (-3)$ c. $-3 - 1$ d. $-5 - (-4)$

Guided Practice

State the additive inverse of each of the following.

6. $+7$ 7. -8 8. $-a$ 9. $9x$

Rewrite each equation using the additive inverse. Then solve.

10. $6 - 15 = m$ 11. $-7 - 11 = x$ 12. $-16 - 7 = z$

13. $8 - (-3) = b$ 14. $-24 - (-8) = r$ 15. $-14 - (-19) = y$

Simplify.

16. $6a - 13a$ 17. $-15x - 12x$ 18. $-17y - (-3)y$

Evaluate each expression.

19. $x - 3$, if $x = -9$ 20. $y - (-8)$, if $y = 17$ 21. $a - (-7)$, if $a = -15$

22. **Meteorology**   *Windchill* factor is an estimate of the cooling effect the wind has on a person in cold weather. If the outside temperature is $10°F$ and the wind makes it feel like $-25°F$, what is the difference between the actual temperature and how cold it feels?

Research For the other windchill factors, visit: **www.glencoe.com/ sec/math/prealg/ mathnet**

Exercises: Practicing and Applying the Concept

Independent Practice

Solve each equation.

23. $y = -9 - 5$ 24. $a = -13 - 4$ 25. $r = -8 - (-5)$

26. $x = -12 - (-7)$ 27. $m = 12 - (-20)$ 28. $k = 23 - (-24)$

29. $-5 - (-14) = p$ 30. $-12 - (-16) = z$ 31. $31 - (-6) = d$

32. $-17 - 13 = a$ 33. $-21 - 8 = b$ 34. $38 - (-12) = q$

35. $x = -28 - 35$ 36. $-7 - 25 = f$ 37. $-39 - 7 = s$

Evaluate each expression.

38. $11 - n$, if $n = 5$

39. $-15 - c$, if $c = 5$

40. $h - (-17)$, if $h = -23$

41. $r - 11$, if $r = -16$

42. $8 - m$, if $m = -12$

43. $h - (-13)$, if $h = -18$

44. $5 - b$, if $b = 8$

45. $7 - (-f)$, if $f = 19$

Simplify.

46. $5b - 17b$

47. $13x - 23x$

48. $-3a - (-5a)$

49. $18p - 3p$

50. $-28d - 17d$

51. $12cd - (-12cd)$

52. $-15xy - 18xy$

53. $7a - 14a - a$

54. $-q - (-6q) - (-7q)$

Critical Thinking

55. Explain why every integer has an additive inverse. Then identify the integer that is its own additive inverse.

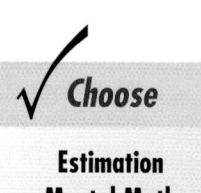

Applications and Problem Solving

56. Aviation Steve Fossett was the first person to fly a balloon solo across the Pacific Ocean. He took off from Seoul, South Korea, on February 18, 1995, and landed in Saskatchewan, Canada, three days later. When the heaters in his gondola failed, he had to endure temperatures that ranged from 10°F to −4°F. What was the change from the highest to lowest temperatures he experienced?

57. Accounting Dawn Sopher is a business information analyst. She uses spreadsheet software to keep track of accounts payable. The balance of an account can be negative or positive. Purchases are subtracted from the balance, and payments are added to the balance. Find the final balance on the spreadsheet below.

	A	B	C	D
1	Date	Purchases	Payments	Balance
2	11/1			−$12,300
3	11/8	$2500		
4	11/22		$18,345	
5	11/30	$5678		

✓ Choose

Estimation
Mental Math
Calculator
Paper and Pencil

58. Geography A region's time zone is determined by its longitude. The zone labeled zero is centered on the prime meridian. In the chart below, time zones to the east of the prime meridian are named by negative numbers, and time zones west of the prime meridian are named by positive integers. To find the time in another city, subtract the time zone of the other city from the time zone at your location.

Suppose you are in Miami and wish to know the time in Bombay. Find $5 - (-5)$, which is 10. Since 10 is positive, the time in Bombay is 10 hours *ahead* of Miami. So at 1:00 P.M. in Miami, it is 11:00 P.M in Bombay.

City	Time Zone
Bombay, India	−5
Honolulu, USA	11
Los Angeles, USA	8
Miami, USA	5
Paris, France	0
Rome, Italy	−1

a. If the time in Los Angeles is 2:00 P.M., find the time in Rome.

b. If the time in Miami is 11:00 A.M., find the time in Honolulu.

c. If the time in Honolulu is 3:00 A.M., find the time in Paris.

59. Simplify $9m + 43m + (-16m)$. (Lesson 2-4)

60. Oceanography A submarine at 1300 meters below sea level descends an additional 1150 meters. What integer represents the submarine's position with respect to sea level? (Lesson 2-4)

61. Write the inequality for *4 feet is shorter than 6 feet.* (Lesson 2-3)

62. Name the absolute value of -15. (Lesson 2-1)

63. Standardized Test Practice Evaluate $|b| - |a|$ if $a = 1$ and $b = -3$. (Lesson 2-1)

    A. -3 **B.** -2 **C.** 2 **D.** 3

64. Solve $6b = 90$ mentally. (Lesson 1-6)

65. Simplify $y + 9 + 14 + 2y$. (Lesson 1-5)

66. Rewrite $(a + 3) + 4$ using the associative property. (Lesson 1-4)

Statistics Use the information in the graph at the right to answer each question. (Lesson 1-3)

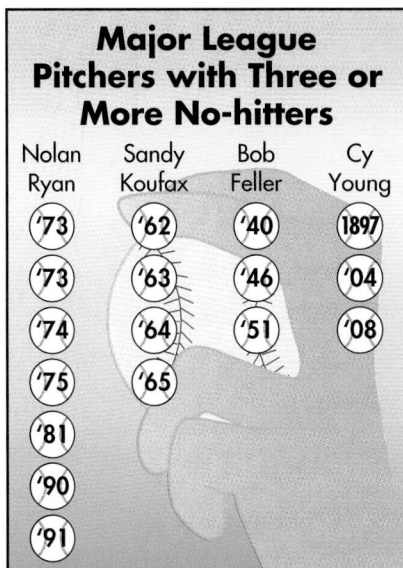

Major League Pitchers with Three or More No-hitters

Nolan Ryan	Sandy Koufax	Bob Feller	Cy Young
'73	'62	'40	1897
'73	'63	'46	'04
'74	'64	'51	'08
'75	'65		
'81			
'90			
'91			

67. If r represents the number of no-hitters that Nolan Ryan pitched, who can be represented by $r - 3$?

68. If y represents the number of no-hitters Cy Young pitched, which other pitcher can be represented by the expression y?

69. If k represents the number of no-hitters Sandy Koufax pitched, write an expression that represents the number of no-hitters Nolan Ryan pitched.

For **Extra Practice**, see page 744.

Self Test

Simplify. (Lesson 2-1)

1. $|-8|$ **2.** $|-20| + |-19|$

3. Name the coordinate of each point graphed on the number line. (Lesson 2-1)

$$-7 \quad -6 \quad -5 \quad -4 \quad -3 \quad -2 \quad -1 \quad 0 \quad 1 \quad 2 \quad 3 \quad 4 \quad 5 \quad 6 \quad 7$$

On graph paper, draw coordinate axes. Then graph and label each point. (Lesson 2-2)

4. $Q(4, 3)$ **5.** $R(-4, -3)$ **6.** $C(-3, 0)$

Replace each ● with $<$, $>$, or $=$. (Lesson 2-3)

7. -9 ● -15 **8.** -11 ● 5 **9.** $|-4|$ ● 4

Solve each equation. (Lesson 2-4, 2-5)

10. $-7 + 20 = a$ **11.** $-12 + 14 + 8 = d$ **12.** $-42 - 38 = q$ **13.** $b = -34 - (-19)$

14. Business The formula $P = I - E$ is used to find the profit (P) when income (I) and expenses(E) are known. Find P if $I = \$15,525$ and $E = \$17,000$.

Problem-Solving Strategy: Look for a Pattern

You'll Learn

■ To solve problems by looking for a pattern.

It's important

Finding a pattern will help you analyze data.

Technology Activity

Carl F. Gauss (1777–1855), a famous German mathematician, taught himself to read and to make mathematical calculations. When he was 10 years old, his teacher asked the students in his class to find the sum of the numbers from 1 to 100. The teacher thought that this problem would keep the students busy for a long time, but young Gauss wrote the answer almost immediately!

The notation "···" means continue the pattern.

Your Turn You can write this sum as $1 + 2 + 3 + \cdots + 98 + 99 + 100$. The sum of the least and greatest numbers is $1 + 100$ or 101.

▶ Find the sum of the second least and second greatest number in the set.

▶ Find the sum of the third least and third greatest number in the set.

a. What do you notice about the sums?

b. How many of these sums are in the set of numbers from 1 to 100?

Use a calculator to find 101×11, 101×12, 101×13, and 101×14.

c. What do you notice about the products?

d. Why do you think this happens?

e. How did Gauss use this pattern to amaze his teacher?

Learning the Concept

Looking for, and then extending, patterns is a good problem-solving strategy. Sometimes, solving one or more simpler problems allows you to see a pattern.

Example **1** **Use the pattern at the right to find 74×76.**

$$4 \times 6 = 24$$
$$14 \times 16 = 224$$
$$24 \times 26 = 624$$
$$34 \times 36 = 1224$$

As you look at the products of these simpler problems, observe the pattern in the factors. In each successive product, each factor is increased by 10. To find the product of 74 and 76, we will need to extend the pattern to include four more products.

Now look at the products. Each product has "24" as the last two digits. The digits before "24" follow the pattern 0, 2, 6, 12. Take a close look at this pattern and extend it four times.

$$\underbrace{0 \quad 2 \quad 6 \quad 12 \quad 20 \quad 30 \quad 42 \quad 56}_{+2 \quad +4 \quad +6 \quad +8 \quad +10 \quad +12 \quad +14}$$

Use the pattern to write the next four products.

$4 \times 6 = 24$
$14 \times 16 = 224$
$24 \times 26 = 624$
$34 \times 36 = 1224$
$44 \times 46 = 2024$
$54 \times 56 = 3024$
$64 \times 66 = 4224$ The pattern shows that $74 \times 76 = 5624$. Use a
$74 \times 76 = 5624$ calculator to verify the result.

Mathematicians have been studying patterns for centuries. Example 2 involves a pattern first made famous in a book published in 1202!

Example

INTEGRATION

Number Theory

The pattern of numbers 1, 1, 2, 3, 5, 8, . . . is called the *Fibonacci sequence*. The terms of this sequence are called *Fibonacci numbers*. List the first ten Fibonacci numbers.

What do you add to each term to get the succeeding term?

$$\overset{1+1}{} \quad \overset{1+2}{} \quad \overset{2+3}{} \quad \overset{3+5}{}$$
$$1 \quad 1 \quad 2 \quad 3 \quad 5 \quad 8$$

To continue the pattern, add the previous two terms. So, $5 + 8 = 13$, $8 + 13 = 21$, $13 + 21 = 34$, and $21 + 34 = 55$.

The first ten Fibonacci numbers are 1, 1, 2, 3, 5, 8, 13, 21, 34, and 55.

Checking Your Understanding

Communicating Mathematics

Read and study the lesson to answer these questions.

1. **Write** the sum of the numbers from 1 to 160.
2. **Explain** how $\dfrac{n(n+1)}{2}$ relates to the pattern at the beginning of the lesson.
3. **Write** a rule to find the successive terms in the pattern 63, 48, 35, 24, ...
4. **You Decide**   Julita says that the next number in the pattern below is 28. Jack thinks it is 27. Who do you think is correct? Explain.

   1   3   6   10   15   21   __?__

Guided Practice

Solve. Look for a pattern.

5. Find the next two integers in the pattern 1, 1, 2, 6, 24, __?__, __?__.
6. Use the simpler problems at the right to find 8888×5. Explain your thinking.
7. How many squares of any size are there in an 8-by-8 checkerboard?
8. **Carpentry**   If shorter boards are cut from one long board, how many cuts will it take to have 12 shorter boards if there is no waste?

$8888 \times 1 = 8888$
$8888 \times 2 = 17,776$
$8888 \times 3 = 26,664$
$8888 \times 4 = 35,552$

Solve. Use any strategy.

9. Use the pattern at the right to find 11,111 × 11,111.

$$1 \times 1 = 1$$
$$11 \times 11 = 121$$
$$111 \times 111 = 12,321$$
$$1111 \times 1111 = 1,234,321$$

10. Find the next two integers in the pattern 2, 5, 9, 14, 20, __?__, __?__.

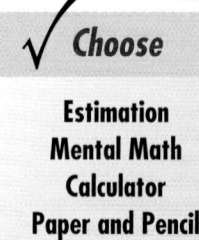

✓ Choose

Estimation
Mental Math
Calculator
Paper and Pencil

11. **Sports** A section of the sports page is torn from Saturday's *Titusville Herald* and part of the article about the Titusville High School football game is missing. Analisa reads that the team scored 34 points and made a total of 7 touchdowns and field goals. The team converted only 1 point after the touchdowns. Touchdowns score 6 points, 7 points with a point after, and field goals score 3 points. How many touchdowns were made in the game?

12. Brian travels south on his bicycle riding 8 miles per hour. One hour later, his friend Maya starts riding her bicycle from the same location. If she travels south at 10 miles per hour, how long will it take her to catch Brian?

13. A ferry service is available to take passengers and cars to South Bass Island in Lake Erie. Michelle has part of a ferry schedule. She wishes to take the ferry to the island, but she cannot leave until 1:30 P.M. Assuming the ferry departs at regular intervals, what is the earliest time Michelle can catch it?

South Bass Island Ferry Schedule

Departures	Arrivals
8:45 A.M.	9:
9:33 A.M.	1
10:21 A.M.	
11:09 A.M.	

14. Two hot dogs and one hamburger cost $3.39. Three hot dogs and three hamburgers cost $6.57. What is the cost of one hot dog?

15. **Geometry** The green line segments in the figures at the right are called the *diagonals* of the polygon. The square has two diagonals, and the pentagon has five diagonals.

 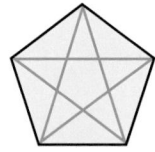

a. Draw the next two figures in the pattern.

b. Write a sequence to represent the number of diagonals in successive figures. How many diagonals does a 9-sided polygon have?

16. What is the sum of the first 50 odd numbers?

17. **Geometry** Find the number of line segments determined by nine points on a line.

18. A college student sent home the message at the right. If each letter stands for one digit 0–9, how much money did she ask for?

$$\begin{array}{r} \text{SEND} \\ +\text{MORE} \\ \hline \text{MONEY} \end{array}$$

19. **Patterns** What is the next number in the pattern 1, 5, 14, 30, . . . ?

20. **Family Activity** Survey your family members about their hair and eye color. Do you see a pattern? What color hair and eyes do you think your children will have? Explain.

Critical Thinking

21. The product of a 2-digit number and 101 contains the digits of the 2-digit number repeated twice. For example, $42 \times 101 = 4242$. What should you multiply a three-digit number by in order to have the product repeat the digits?

Mixed Review

22. **Standardized Test Practice**   The temperature on Venus can rise to $480°C$, while on Jupiter the temperature can reach about $-148°C$. What is the difference between these two temperatures? (Lesson 2-5)

A. $628°C$ **B.** $332°C$ **C.** $-332°C$ **D.** $-628°C$

23. Solve the equation $h = 8 + (-15) + 13$. (Lesson 2-4)

24. State whether $3x + 2x - 9 \geq 17$ is *true* or *false* if $x = 6$. (Lesson 1-9)

For **Extra Practice**, see page 744.

25. Simplify $5y + 3(7 + 2y)$. (Lesson 1-5)

26. Find the value of $2[5(4 + 6) - 3]$. (Lesson 1-2)

WORKING ON THE

Investigation

BON VOYAGE

Therefore, the chilling effect of cold increases as the speed of the wind increases.

The *heat index* is a measure of how high temperatures accompanied by high humidity affect the body's ability to cool itself. Increased humidity makes abnormally high temperatures "feel" even hotter to the average person.

Refer to the Investigation on pages 62–63.

Depending on which season it is in the country you chose at the beginning of the Investigation, you may want to consider the *windchill factor* or the *heat index* when you are deciding what clothes to pack. If the country chosen is south of the equator, it will be winter when you visit that country. You may need to prepare for windchill factors. It may be summer when you visit and you need to be aware of the heat index.

The *windchill factor* is an estimate of the cooling effect the wind has on a person in cold weather. Increased wind velocity causes the body to lose heat more rapidly.

- Determine which season it will be in June in the country you chose at the beginning of the Investigation. Then find a windchill table (winter) or a heat index table (summer).
- Explain why you would use the heat index table if the country is between the Tropic of Cancer and the Tropic of Capricorn.
- Find the average daily temperature for the country for a month in the appropriate season (winter or summer). Then find the average wind velocity (winter) or average relative humidity (summer) for the country.
- Look for a pattern in the table you are using. Use the pattern to estimate an extreme measure that is not listed in the table. For example, if your table for windchill goes to 0 degrees and to a wind of 40 mph, estimate the windchill for 0 degrees and a 50-mph wind.
- Write some general safety guidelines with respect to windchill factors or heat index.

Add the results of your work to your Investigation Folder.

2-7A Multiplying Integers

A Preview of Lesson **2-7**

MATERIALS

 counters

integer mat

You can use counters to model multiplying integers. Remember that 2×4 means *two sets of four items*. Using models, 2×4 means **put in** *two sets of four* **positive** *counters*.

Activity 1 Find $2 \times (-4)$.

▶ Put in two sets of four negative counters.

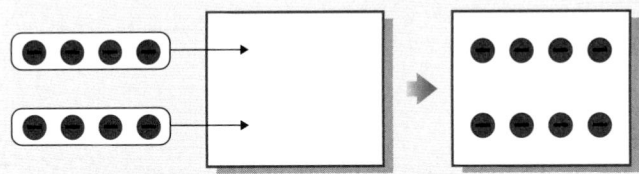

Therefore, $2 \times (-4) = -8$.

Activity 2 Find $-2 \times (-4)$.

▶ Since one meaning of the negative sign is *the opposite of*, $-2 \times (-4)$ means to **remove** *two sets of four* **negative** *counters*. How can you take out two sets? First, put in as many zero pairs as you need. Then, remove 2 sets of 4 negative counters.

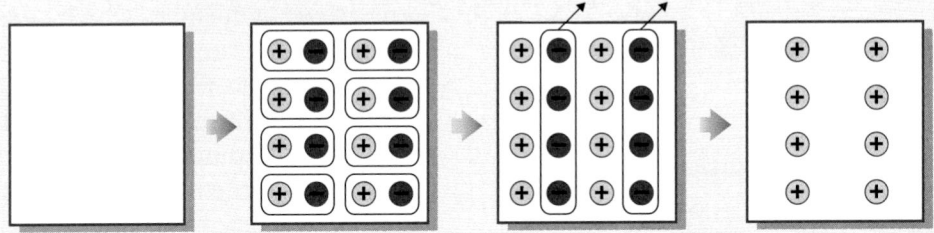

Therefore, $-2 \times (-4) = 8$.

Your Turn **Find each product using counters.**

1. $2 \times (-3)$ 2. 2×3
3. $-2 \times (-3)$ 4. -2×3
5. -3×2 6. $-3 \times (-6)$
7. -4×4 8. $-4 \times (-4)$

TALK ABOUT IT

9. What does -2×4 mean? Write your explanation in paragraph form. Model this operation.

10. How are the expressions $2 \times (-3)$ and -3×2 alike? How do they differ?

2-7 Multiplying Integers

What You'll Learn
- To multiply integers.

Why It's Important
Multiplying integers can help you estimate how far you travel, if you know the speed and the time.

Application: Oceanography

Robert Ballard and his team from Woods Hole Oceanographic Institution are scientists who have discovered such famous shipwrecks as the *Titanic* and the *Lusitania*. In 1989, they discovered the sunken remains of the German battleship *Bismarck*, which sank in 1942 during World War II.

The *Bismarck* sank in 16,000 feet of water in the North Atlantic. The team used the research submarine, *Alvin*, to locate the hulk. It took *Alvin* over 2 hours to reach the *Bismarck* on the ocean floor. Its rate of descent was about 100 feet per minute. At this rate, how far would *Alvin* descend in 3 minutes?

Learning the Concept

The submarine descends 100 feet per minute, or it moves -100 feet. In 3 minutes, it will move $3 \times (-100)$ feet. You can find this product by using a number line or by observing a pattern.

Method 1: Number Line

$$
\begin{array}{c}
\text{-100} \mid \text{-100} \mid \text{-100} \\
\end{array}
$$

-300 -200 -100   0   100

Method 2: Patterns

$$3 \cdot 200 = 600$$
$$\Big) -300$$
$$3 \cdot 100 = 300$$
$$\Big) -300$$
$$3 \cdot 0 = 0$$
$$\Big) -300$$
$$3 \cdot (-100) = -300$$

So, $3 \times (-100) = -300$. In three minutes, Alvin would descend 300 feet.

If $3 \times (-100) = -300$, then $-100 \times 3 = -300$ by the commutative property of multiplication. Notice that in each sentence one integer is positive, one integer is negative, and the product is negative. The pattern above and other similar examples suggest the following rule.

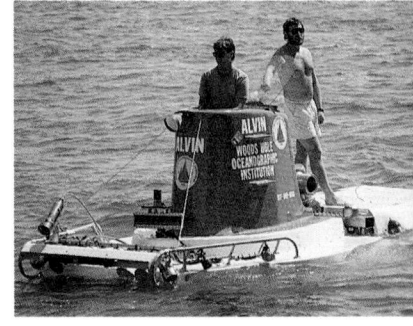

Multiplying Integers with Different Signs

The product of two integers with different signs is negative.

Example **1** Solve each equation.

    a. $m = -5(18)$

        The two factors have different signs. The product is negative.

        $m = -5(18)$
        $m = -90$

    b. $6(-12) = n$

        The two factors have different signs. The product is negative.

        $6(-12) = n$
        $-72 = n$

You already know that the product of two positive integers is positive.

Both factors are positive. $3 \cdot 7 = 21$ The product is positive.

How would you find $-3(-2)$? You can solve this problem by again observing a pattern or by using counters.

Method 1:   **Patterns**

$-3(2) = -6$
$\big)\ +3$
$-3(1) = -3$
$\big)\ +3$
$-3(0) = 0$
$\big)\ +3$
$-3(-1) = 3$
$\big)\ +3$
$-3(-2) = 6$

Method 2:   **Counters**

$-3(-2)$ means that you will remove 3 sets of 2 negative counters.

So, $-3(-2) = 6$. This suggests the following rule.

> **Multiplying Integers with the Same Sign**
> The product of two integers with the same sign is positive.

Example **2** Solve each equation.

    a. $m = -8(-15)$

        The two factors have the same sign. The product is positive.

        $m = -8(-15)$
        $m = 120$

    b. $-7(-5)(-4) = q$

        $[-7(-5)](-4) = q$ *Associative property*
           $35(-4) = q$ *$-7(-5) = 35$*
            $-140 = q$

Many real-life situations can be modeled using products of integers.

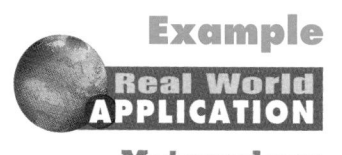

Example **3**

Real World
APPLICATION

Meteorology

The temperature between the ground and 11 kilometers above the ground drops about 7°C for each kilometer of increase in altitude. Suppose the temperature 2 kilometers above the ground is 0°C. What would the ground temperature be?

Explore   The temperature 2 km above ground is 0°C, and the temperature drops 7°C for each km of increase in altitude. You need to find the ground temperature. Let t represent the ground temperature, -2 represent 2 km lower, and -7 represent a drop of 7°C.

Plan   Write an equation.

Solve

| ground temperature = | temperature at 2 km | + (drop of 7° per km)(2 km lower) |

$$t = 0 + (-7)(-2)$$
$$t = 0 + 14$$
$$t = 14$$

The ground temperature is 14°C.

Examine   Since the altitude is decreasing, the temperature rises. The answer is reasonable.

Connection to ⟩ Algebra

You can use the rules for multiplying integers to evaluate algebraic expressions and to simplify expressions.

Example **4**   **Evaluate $2ab$ if $a = -3$ and $b = -2$.**

$2ab = 2(-3)(-2)$ *Replace a with −3 and b with −2.*
$\quad = 2[(-3)(-2)]$ *Associative property of multiplication*
$\quad = 2(6)$ *The product of −3 and −2 is positive.*
$\quad = 12$

Example **5**   **Find the product of –3 and $8m$.**

$-3(8m) = (-3 \cdot 8)m$ *Associative property of multiplication*
$\quad\quad\quad = -24m$

The product is $-24m$.

Checking Your Understanding

Communicating Mathematics

Read and study the lesson to answer these questions.

1. **Explain** why the solution of $-3(-4) = p$ must be 12. Use a pattern in your explanation.

2. **Explain** how to multiply 12 and $-4y$.

Lesson 2-7   Multiplying Integers   **101**

3. Write the multiplication sentence shown by each model.

a.

b.

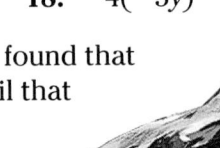

4. Draw a model that shows the solution for $q = 5(-2)$.

Guided Practice

State whether each product is positive or negative. Then find the product.

5. $-8 \cdot 7$ **6.** $5(-9)$ **7.** $14 \cdot 37$

8. $-35(-57)$ **9.** $(-9)(2)(-3)$ **10.** $(-3)(-5)(-7)$

Solve each equation.

11. $t = -9 \cdot 3$ **12.** $-7(-3) = u$ **13.** $8(-10)(4) = v$

Evaluate each expression.

14. $4m$, if $m = -12$ **15.** $-3gh$, if $g = -4$, $h = -11$

Multiply.

16. $-5 \cdot 3w$ **17.** $8(-5x)$ **18.** $-4(-3y)$

19. Geology   In December of 1994, geologists found that the Bering Glacier had come to a stop. Until that time, the glacier had been rushing forward at a rate of up to 300 feet per day. Before the surge began, the glacier was retreating at a rate of about 2 feet per day. If the retreat resumes at the old rate, what integer represents how far the glacier will have advanced after 28 days?

Exercises: Practicing and Applying the Concept

Independent Practice

Solve each equation.

20. $m = -6(-8)$ **21.** $x = -7(-4)$ **22.** $a = -3(12)$

23. $r = 13(-2)$ **24.** $-5 \cdot 7 = b$ **25.** $-15(-5) = y$

26. $-4(-5)(6) = z$ **27.** $9(-13)(5) = q$ **28.** $6(-5)(-8) = g$

29. $h = 2 \cdot 7 \cdot 12$ **30.** $-7(-7)(3) = p$ **31.** $11(-4)(7) = s$

Evaluate each expression.

32. $-5n$, if $n = -7$ **33.** st, if $s = 7$, $t = 8$

34. $12ab$, if $a = -3$, $b = -7$ **35.** $-6b$, if $b = 8$

36. $-15c$, if $c = -1$ **37.** $12x$, if $x = -3$

Find each product.

38. $8(-5x)$ **39.** $-4 \cdot 3b$ **40.** $12(-5n)$

41. $(-12)(-4b)$ **42.** $(7y)(5z)$ **43.** $(-10r)(5s)$

44. $2ab \cdot 6 \cdot (-2)$ **45.** $-3(-a)(-b)$ **46.** $7(-m)(k)$

Critical Thinking

47. Write a rule that will help you determine the sign of the product if you are multiplying two or more integers.

Applications and Problem Solving

48. **Geometry** On a sheet of graph paper, copy the figure at the right.

   a. Name the ordered pairs for the coordinates of the vertices of $\triangle PQR$.

   b. Multiply each coordinate in the ordered pairs by -1. Label the new ordered pairs as P', Q', and R'.

   c. Now graph $\triangle P'Q'R'$.

   d. How do the two triangles compare?

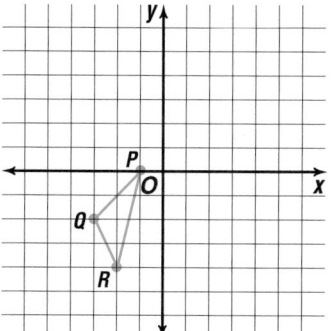

49. **Weather** In the great flood of 1993, the Mississippi River was so high that it caused the Illinois River to flow backward. If the Illinois River flowed at the rate of -1500 feet per hour, how far would the water travel in 24 hours?

Mixed Review

50. **Patterns** Look at the pattern of numbers below.

        1   2   4   7

   Is the next number found by multiplying by 2? Explain. (Lesson 2-6)

51. **Weather** In Spearfish, South Dakota, on January 22, 1943, the temperature was $-4°F$ at 7:30 A.M. At 7:32 A.M., it was $45°F$. How much did the temperature rise in two minutes? (Lesson 2-5)

52. State whether the sum $-10 + 9 + 8$ is positive or negative. Then find the sum. (Lesson 2-4)

53. Write an inequality that compares yesterday's high temperature of $65°F$ with yesterday's low temperature of $34°F$. (Lesson 2-3)

54. Graph the numbers $\{-1, 1, 4\}$ on a number line. (Lesson 2-1)

55. **Standardized Test Practice** The attendance at Shea Stadium was 35,892 on Friday. This was 3285 less than the attendance Saturday. Solve the equation $35,892 = a - 3285$ to find Saturday's attendance. (Lesson 1-8)

   A. 39,267   **B.** 39,177   **C.** 32,697   **D.** 32,607

56. Solve $\dfrac{18}{a} = 9$ mentally. (Lesson 1-6)

57. What property is shown by the statement $9a + b = b + 9a$? (Lesson 1-4)

For **Extra Practice**, see page 745.

58. Find the value of $21 \div 7 + 4 \cdot 11$. (Lesson 1-2)

2-8 Dividing Integers

What You'll Learn

■ To divide integers.

Why It's Important

You can divide integers to find the average amount of decrease in prices over time.

 Hands-On Activity

Remember that you used counters in Lesson 2-7 to understand the rules for multiplying integers. You can also use counters to model division. The diagram at the right shows 12 negative counters separated into 4 equal groups. Counting the number of counters in each group shows that $-12 \div 4 = -3$.

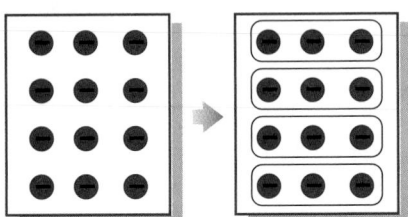

Your Turn   Use counters to solve $-10 \div 5 = t$.

TALK ABOUT IT   **a.** How is $-12 \div 4 = -3$ related to $-3 \times 4 = -12$?

b. What do you think is the quotient of 12 and -4?

Learning the Concept

You could also find $-12 \div 4$ by looking for a pattern.

$$4 \div 4 = 1$$
$$) -1$$
$$0 \div 4 = 0$$
$$) -1$$
$$-4 \div 4 = -1$$
$$) -1$$
$$-8 \div 4 = -2$$
$$) -1$$
$$-12 \div 4 = -3$$

THINK ABOUT IT

What is the next division sentence in the pattern?

In the example above, the dividend is negative and the divisor is positive. Let's try a case where the signs are reversed.

Solve $f = 12 \div (-4)$.

$$f = 12 \div (-4)$$
$$= -3$$

So, the solution of $f = 12 \div (-4)$ is -3.

These and other similar examples suggest the following rule.

Dividing Integers with Different Signs	The quotient of two integers with different signs is negative.

Example **1** **Solve each equation.**

a. $-34 \div 2 = x$

The dividend and the divisor have different signs. The quotient is negative.

$$-34 \div 2 = x$$
$$-17 = x$$

b. $y = \dfrac{42}{-3}$

The dividend and the divisor have different signs. The quotient is negative.

$$y = \dfrac{42}{-3}$$
$$y = -14$$

You already know that the quotient of two positive integers is positive.

$$21 \div 3 = 7$$

Both the dividend and the divisor are positive. The quotient is positive.

How would you solve $z = -27 \div (-9)$? You can solve this problem by again looking for a pattern.

$$18 \div (-9) = -2$$
$$9 \div (-9) = -1 \qquad \Big) +1$$
$$0 \div (-9) = 0 \qquad \Big) +1$$
$$-9 \div (-9) = 1 \qquad \Big) +1$$
$$-18 \div (-9) = 2 \qquad \Big) +1$$
$$-27 \div (-9) = 3 \qquad \Big) +1$$

So, the solution to $z = -27 \div (-9)$ is 3. This pattern suggests the following rule.

Dividing Integers with the Same Sign — **The quotient of two integers with the same sign is positive.**

Example **2** **Solve each equation.**

a. $m = -40 \div (-8)$

The dividend and the divisor have the same sign. The quotient is positive.

$$m = -40 \div (-8)$$
$$m = 5$$

b. $\dfrac{90}{15} = r$

The dividend and the divisor have the same sign. The quotient is positive.

$$\dfrac{90}{15} = r$$
$$6 = r$$

Connection to Algebra

You can find the value of expressions involving division by using the same rules as division with integers.

Example ③ Evaluate $cd \div (-3)$, if $c = -5$ and $d = -9$.

$$cd \div (-3) = -5(-9) \div (-3)$$ *Replace c with –5 and d with –9.*
$$= 45 \div (-3)$$ *The product of –5 and –9 is positive.*
$$= -15$$ *The quotient of 45 and –3 is negative.*

Example ④

Real World APPLICATION

Meteorology

People in Duluth, Minnesota, find ways to enjoy their lengthy cold spells. Energy providers use *degree days* to estimate the energy that will be needed for heating in these cold days. The degree-day number indicates how far the day's average temperature was above or below 65°F. The formula $d = 65 - \dfrac{h + \ell}{2}$ can be used to find degree days. In the formula, *d* represents degree days, *h* represents the high temperature of the day, and ℓ represents the low temperature of the day. If *d* is positive, it represents heating degree days. If *d* is negative, it represents cooling degree days. If the high temperature in Duluth was $-2°F$ and the low temperature was $-16°F$, find the degree days.

$$d = 65 - \frac{h + \ell}{2}$$
$$= 65 - \frac{-2 + (-16)}{2}$$
$$= 65 - \frac{-18}{2}$$
$$= 65 - (-9)$$
$$= 74$$

The day had 74 heating degree days.

Checking Your Understanding

Communicating Mathematics

Read and study the lesson to answer these questions.

1. **Explain** why the solution of $q = -10 \div 2$ must be -5. Use a pattern.
2. **Draw a model** that shows the solution for $15 \div 3 = r$.
3. **Write** the division sentence shown at the right.

MATH JOURNAL

4. **Assess Yourself** In your own words, describe a situation in your life where you use negative numbers.

State whether each quotient is positive or negative. Then find the quotient.

5. $-56 \div 8$

6. $63 \div 9$

7. $42 \div (-7)$

8. $\frac{91}{7}$

9. $\frac{105}{-7}$

10. $\frac{222}{-37}$

Solve each equation.

11. $d = \frac{-240}{-6}$

12. $-96 \div 24 = k$

13. $\frac{-105}{-15} = m$

14. $n = \frac{-804}{67}$

Evaluate each expression.

15. $\frac{x}{-5}$, if $x = -65$

16. $\frac{y}{5}$, if $y = -50$

17. $\frac{42}{z}$, if $z = -14$

18. $\frac{b}{-7}$, if $b = -98$

19. **Sports** The Green Bay Packers football team was penalized the same amount 4 times during the third quarter. The total of the 4 penalties was 60 yards. If -60 represents a loss of 60 yards, write a division sentence to represent this situation. Then express the number of yards of each penalty as an integer.

Exercises: Practicing and Applying the Concept

Divide.

20. $56 \div (-7)$

21. $48 \div (-12)$

22. $-72 \div 8$

23. $-36 \div (-6)$

24. $84 \div (-12)$

25. $-52 \div (-4)$

26. $64 \div 4$

27. $-555 \div 5$

28. $\frac{24}{6}$

29. $\frac{27}{-3}$

30. $\frac{-570}{19}$

31. $\frac{-804}{67}$

32. Divide -54 by 9.

33. Divide 63 by -7.

34. Find the quotient of 121 and 11.

35. Find the quotient of -28 and -7.

Solve each equation.

36. $m = \frac{91}{-7}$

37. $r = \frac{76}{-4}$

38. $\frac{-63}{-21} = a$

39. $\frac{-90}{18} = z$

40. $\frac{45}{-15} = k$

41. $\frac{-119}{-17} = x$

42. $p = \frac{-143}{13}$

43. $b = \frac{-144}{16}$

44. $t = \frac{-175}{-25}$

Evaluate each expression.

45. $\frac{x}{6}$, if $x = -102$

46. $\frac{105}{m}$, if $m = -21$

47. $\frac{y}{-9}$, if $y = -108$

48. $\frac{112}{z}$, if $z = -7$

49. $\frac{126}{a}$, if $a = -18$

50. $\frac{-144}{p}$, if $p = 9$

51. $\frac{v}{q}$, if $v = -24$, and $q = 8$

52. $\frac{n \times (-n)}{m}$, if $m = 27$, and $n = -9$

53. Find values for x, y, and z, so that all of the following statements are true.

▸ $y > x$, $z < y$, and $x < 0$.

▸ $x \div z = -z$

▸ z is divisible by 2 and 3.

▸ $x \div y = z$

Applications and Problem Solving

54. Aerospace To simulate space travel, NASA's Lewis Research Center in Cleveland, Ohio, uses a 430-foot shaft. The air is pumped out of the shaft to create a near-vacuum for free-fall experiments. If the free fall of an object in the shaft takes 5 seconds to travel the -430 feet, on average how far does the object travel in each second?

55. Economics At the end of 1997, the United States Federal Government had a debt of 5502 billion dollars. By the end of 1998, the debt had grown to more than 5614 billion. The term *per capita* means per person. If there were about 270,000,000 U.S. citizens in 1998, how much did the per capita debt grow from 1997 to 1998? (*Hint*: Find the amount of change in the debt, then divide by the number of citizens.)

Mixed Review

56. Solve $y = 7(-11)$. (Lesson 2-7)

57. Education At Clay Battelle Junior High School, the bell rings at 8:05, 8:51, 8:55, 9:41, and 9:45 each weekday morning. When do the next three bells ring? (Lesson 2-6)

The frequency table at the right contains data from a survey on message service. (Lesson 1-10)

Who do you trust more with a message?	Frequency
Postal Service	649
e-mail	304
Both Postal Service and e-mail equally	20
Neither Postal Service nor e-mail/Don't know	41

Source: ICR Survey Research Group

58. What message service was chosen most often?

59. How many people responded that they didn't trust the Postal Service or e-mail, or didn't know which one they trusted more?

60. Make a bar graph of the data in the table.

61. How many people gave information for this survey?

62. Standardized Test Practice Silvia Vidal has a job as a veterinary assistant. The paycheck for her first week was $280 before taxes were deducted. If Ms. Vidal worked 35 hours that week, how much did she earn per hour? (Lesson 1-8)

   A. $6   **B.** $7   **C.** $8   **D.** $9

63. Solve $r + 11 = 17$ mentally. (Lesson 1-6)

64. Name the property shown by $(9 + 7) + 4 = 4 + (9 + 7)$. (Lesson 1-4)

For **Extra Practice**, see page 745.

Find the value of each expression. (Lesson 1-2)

65. $12 \div 3 + 12 \div 4$ **66.** $(25 \cdot 3) + (15 \cdot 3)$ **67.** $144 \div 16 \cdot 9 \div 3$

Vocabulary

After completing this chapter, you should be able to define each term, property, or phrase and give an example or two of each.

Algebra

absolute value (p. 67)

adding integers with different signs (p. 84)

adding integers with the same sign (p. 83)

additive inverses (p. 89)

dividing integers with different signs (p. 104)

dividing integers with the same signs (p. 105)

integers (p. 66)

multiplying integers with different signs (p. 99)

multiplying integers with the same sign (p. 100)

negative (p. 66)

subtracting integers (p. 89)

zero pair (p. 82)

Geometry

axes (p. 73)

coordinate (p. 67)

coordinate system (p. 73)

graph (p. 73)

plotting a point (p. 73)

quadrants (p. 73)

vertex (p. 75)

Problem Solving

look for a pattern (p. 94)

Statistics

cluster (p. 71)

line plot (p. 71)

outlier (p. 71)

*inter*NET
CONNECTION
Chapter Review For additional lesson-by-lesson review, visit:
www.glencoe.com/ sec/math/prealg/ mathnet

Understanding and Using Vocabulary

Choose the letter of the term that best matches each statement or phrase.

1. an integer less than zero

2. the four regions separated by the axes on a coordinate plane

3. the number that corresponds to a point on the number line

4. an integer and its opposite

5. positive and negative whole numbers

6. a dot placed at the point named by the ordered pair

7. the distance a number is from the zero point on the number line

a. absolute value

b. additive inverses

c. coordinate

d. graph

e. integers

f. negative integer

g. quadrants

Skills and Concepts

Objectives and Examples	**Review Exercises**

Upon completing this chapter, you should be able to:

Use these exercises to review and prepare for the chapter test.

▶ **graph integers on a number line** (Lesson 2-1)

Graph points *A*, *B*, and *C* on a number line if *A* has coordinate −2, *B* has coordinate 3, and *C* has coordinate −1.

Graph each set of numbers on a number line.

8. {−3, 6, −4, −6, 11, −1}

9. {22, 74, −30, 82, −10, −12}

10. {112, 98, −21, −41, 67}

11. {−1, −16, −4, −6, 4}

▶ **find absolute value** (Lesson 2-1)

$|-3| = 3$

$|4| - |-9| = 4 - 9 = -5$

Simplify.

12. $|-31|$	**13.** $|24|$
14. $-|18|$	**15.** $-|40|$
16. $-|6-2|$	**17.** $|-8| + |-21|$
18. $|17-29|$	**19.** $||7| - |-19||$

▶ **graph points in all quadrants of the coordinate plane** (Lesson 2-2)

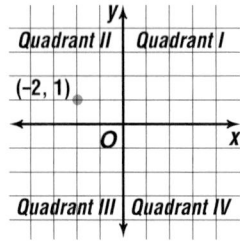

On graph paper, draw coordinate axes. Then graph and label each point. Name the quadrant in which each point is located.

20. $A(4, 3)$	**21.** $R(-1, -1)$
22. $W(-5, 3)$	**23.** $G(-6, 5)$
24. $M(-4, -1)$	**25.** $L(3, -2)$

▶ **compare integers** (Lesson 2-3)

Compare −4 and −6.

Since −6 is to the left of −4 on the number line, −4 > −6.

Replace each ● with <, >, or = .

26. 9 ● −9	**27.** −3 ● −3
28. 6 ● 13	**29.** −12 ● −21
30. −2 ● 0	**31.** −43 ● 43

Write an inequality using the numbers in each sentence. Use the symbols < or >.

32. The population of Detroit, Michigan, is 1,027,974. The population of Ann Arbor, Michigan, is 109,592.

33. The winning team in Super Bowl XXVIII won by 17 points. In Super Bowl XXVII, the winning team won by 35 points.

Objectives and Examples	Review Exercises

▶ **order integers** (Lesson 2-3)

Order the integers $\{-3, 5, -1, 8, -6\}$ from least to greatest

$-6, -3, -1, 5, 8$ are in order from least to greatest.

Order the integers in each set from least to greatest.

34. $\{8, 0, -3, 6, 5\}$

35. $\{-16, 3, -4, 2\}$

36. $\{-75, 21, 4, -34, -88\}$

37. $\{210, -300, 124, 9, -33\}$

▶ **add integers** (Lesson 2-4)

$d = 4 + (-5)$
$d = -1$

$r = -4 + (-5)$
$r = -9$

$m = -4 + 5 + (-7)$
$m = 1 + (-7)$
$m = -6$

Solve each equation.

38. $a = -4 + (-5)$ 39. $8 + (-11) = s$

40. $-12 + 4 = t$ 41. $k = -15 + 9$

42. $g = 18 + (-25)$ 43. $h = -13 + (-2)$

44. $-6 + (-18) + 40 = q$

45. $a = 7 + (-3) + (-14)$

▶ **subtract integers** (Lesson 2-5)

$r = 5 - 12$
$r = 5 + (-12)$
$r = -7$

$-8p - (-11p) = -8p + 11p$
$\qquad\qquad = 3p$

Solve each equation.

46. $g = -8 - 7$ 47. $-33 - (-3) = s$

48. $-17 - 8 = w$ 49. $t = 15 - (-6)$

Simplify.

50. $2a - 5a$ 51. $-12b - (-5b)$

52. $14m - (-12m)$ 53. $-15r - (-21r)$

▶ **multiply integers** (Lesson 2-7)

$m = -7 \cdot (-4)$
$m = +(7 \cdot 4)$
$m = 28$

$-4a(3b) = (-4 \cdot 3)(a \cdot b)$
$\qquad\quad = -12ab$

Solve each equation.

54. $z = -9 \cdot 4$ 55. $(-4)(2) = k$

56. $(-11)(-15) = y$ 57. $p = -5 \cdot (-13)$

Find each product.

58. $(-2)(8z)(3)$ 59. $6 \cdot (-6c) \cdot (-d)$

60. $-15x(5y)$ 61. $(-25y) \cdot (-6) \cdot (-2z)$

Objectives and Examples	Review Exercises

▶ **divide integers** (Lesson 2-8)

$u = -105 \div 15$
$u = -(105 \div 15)$
$u = -7$

Evaluate $t \div 18$, if $t = -108$.

$t \div 18 = -108 \div 18$
$\quad\quad = -6$

Solve each equation.

62. $k = 88 \div -4$ **63.** $-14 \div (-2) = h$

64. $99 \div -11 = n$ **65.** $q = -45 \div (-3)$

Evaluate each expression.

66. $d \div 8$, if $d = 120$

67. $64 \div m$, if $m = -4$

68. $-700 \div k$, if $k = -10$

69. $z \div -12$, if $z = -228$

Applications and Problem Solving

70. Physics Jalisa did an experiment to see how the mass of an object attached to a spring affected the distance a spring stretched. The data are shown at the right. Graph the ordered pairs (mass, distance) in the coordinate plane. (Lesson 2-2)

Stretching of a Spring	
Mass	Distance
100 g	3 cm
200 g	6 cm
300 g	9 cm
400 g	12 cm
500g	15 cm

71. Weather A combination of wind and cold makes you feel colder than the actual temperature. According to the National Weather Service, a temperature of 20 degrees Fahrenheit with a wind of 30 miles per hour makes you feel like it is 38 degrees colder than it is. How cold does it feel when it is 20 degrees with a wind of 30 mph? (Lesson 2-4)

72. Patterns Find the next two integers in the pattern 1, -2, 3, -4, __?__, __?__. (Lesson 2-6)

74. Physics When an object that is traveling increases its speed it is *accelerating*, when it slows, it is *decelerating*. Acceleration and deceleration are described by positive and negative numbers respectively. Suppose a car is decelerating at a rate of 1.2 meters per second per second. This is represented by -1.2 m/s^2.

  a. How much will a car's speed change if it decelerates at a rate of -1.2 m/s^2 for 11 seconds? (Lesson 2-7)

  b. If a motorcycle slows down its rate of speed by 3.6 meters per second in 12 seconds, what is its rate of deceleration? (Lesson 2-8)

A practice test for Chapter 2 is available on page 775.

73. Patterns Use the pattern at the right to find $1089 \cdot 7$. (Lesson 2-6)

$1089 \cdot 1 = 1089$
$1089 \cdot 2 = 2178$
$1089 \cdot 3 = 3267$
$1089 \cdot 4 = 4356$

Temperature changes occur hourly, daily, weekly, and so on. For this project, you will need to be aware of this change and record it as follows.

1. Pick three cities in the world and record their temperature at the same time each day for a period of five days.

2. Make a table showing the time of day, name of the city, and the temperature for each day.

3. Describe the change in temperature for each city from day 1 to day 2, day 2 to day 3, day 3 to day 4, and day 4 to day 5.

4. On a coordinate graph, plot the daily temperature change for each city. The x-coordinate should be the first day of the series. The y-coordinate should be the change in temperature.

5. What was the total change in temperature for each city from day 1 to day 5?

6. Find the average change in temperature for each city. (Step 1: Add the change in temperature from day to day. Step 2: Divide that number by the number of temperature changes.)

Thinking Critically

▶ The point (a, b) is located in quadrant IV. What are all the possible values for a and b?

▶ Describe all the points located in each quadrant using + (for positive numbers) and − (for numbers) for the values of the x-coordinate and the y-coordinate. (*Hint*: You should have four ordered pairs as your answer.)

 Portfolio

▶ Select an item from this chapter that you feel shows your best work and place it in your portfolio.

Section One: Multiple Choice

There are ten multiple-choice questions in this section. After working each problem, write the letter of the correct answer on your paper.

1. Which is equivalent to $3 \times 8 - 6 \div 2$?

   A. 3
   B. 9
   C. 13
   D. 21

2. You have two more brothers than sisters. If you have b brothers, which sentence could be used to find s, the number of sisters you have?

   A. $b = s - 2$
   B. $b - 2 = s$
   C. $s = b + 2$
   D. $s = 2b$

3. The perimeter of a rectangle is $2(\ell + w)$, where ℓ is the length and w is the width. What is the perimeter of a rectangle with length 10 cm and width 8 cm?

   A. 36 cm
   B. 28 cm
   C. 20 cm
   D. 18 cm

4. If a diver descends at a rate of 5 meters per minute, at what depth will she be after 12 minutes?

   A. $+60$ m
   B. -17 m
   C. -50 m
   D. -60 m

5. If $36 - b = 20$, what is the value of b?

   A. 56
   B. 26
   C. 16
   D. 6

6. $7 \times (10 + 5) =$

   A. $(7 + 10) + 5$
   B. $(7 + 10) \times 5$
   C. $(7 \times 10) + (7 \times 5)$
   D. $(7 + 10) \times (7 + 5)$

Use the following graph for exercises 7 and 8.

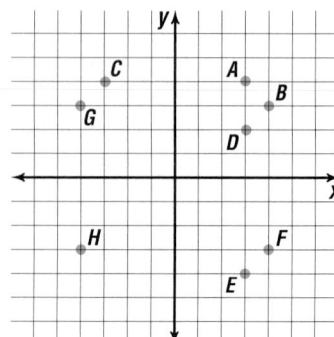

7. Which letter represents the ordered pair $(4, 3)$?

   A. A
   B. B
   C. C
   D. D

8. Which letter represents the ordered pair $(-4, 3)$?

   A. E
   B. F
   C. G
   D. H

9. The frequency table below contains students' test scores on a 25-point test.

Score	18	19	20	21	22	23	24	25
Number of Students	2	1	5	4	5	6	4	1

   How many students had a score greater than 20?

   A. 4
   B. 5
   C. 20
   D. 25

10. If $a = 5$ and $b = 20$, what is the value of $2a + b$?

   A. 50
   B. 45
   C. 40
   D. 30

Section Two: Free Response

This section contains nine questions for which you will provide short answers. Write your answer on your paper.

11. Rewrite $(x + 2) + 12$ using an associative property. Then simplify.

12. Many of the roller coasters at Cedar Point in Sandusky, Ohio, require the rider to be taller than 48 inches. Write an inequality for the height requirement.

13. A bolt of fabric contains about 18 yards of fabric. You want to make T-shirts for your friends. You need 1.875 yards for each T-shirt. About how many T-shirts can you make from one bolt of fabric?

14. During a cold spell in January, 1994, the low temperatures were $-22°F$, $-20°F$, and $-18°F$. The average is the sum of these temperatures divided by 3. What was the average low temperature for those three days?

15. Translate *sixteen less than a number is twelve* into an equation and solve.

16. During one of the Chicago Bears' offensive drives, they gained 12 yards on one play. On the next play, the quarterback was sacked for a loss of 6 yards. Write and solve an addition sentence representing the net yards for these two plays.

17. Write a verbal phrase for $6u + 7$.

18. The Branson Middle School football team is 2 yards from their opponent's goal line. On the next two plays, their offense loses 17 yards and then gains 11 yards. How many yards are they from the goal line now?

19. The temperature at 6:00 A.M. was $-5°F$. What was the temperature at 8:00 A.M. if it had risen 7 degrees?

Test-Taking Tip You can prepare for taking standardized tests by working through practice tests like this one. The more you work with questions in a format similar to the actual test, the better you become in the art of test taking.

Section Three: Open-Ended

This section contains four open-ended problems. Demonstrate your knowledge by giving a clear, concise solution to each problem. Your score on these problems will depend on how well you do the following.
- Explain your reasoning.
- Show your understanding of the mathematics in an organized manner.
- Use charts, graphs, and diagrams in your explanation.
- Show the solution in more than one way or relate it to other situations.
- Investigate beyond the requirements of the problem.

20. Show that $7ac + 14ab + 28a = 7a(c + 2b + 4)$. Use values for a, b, c, and d to justify your answer.

21. You are at a party with 9 other guests. If everyone is to shake hands with every other guest, how many handshakes will occur?

22. If y is an integer, what is the least value that $y \cdot y$ could have? Explain your answer.

23. Rivera High School will graduate 268 seniors on June 1. The ceremony will be held in the school gymnasium. The gymnasium holds 1400 people in addition to the graduates. How many tickets should be offered to each graduate for family and friends?

Test Practice For additional test practice questions, visit:
www.glencoe.com/sec/math/prealg/mathnet

CHAPTER 3

Solving One-Step Equations and Inequalities

What You'll Learn

In this chapter, you will:

- solve problems by eliminating possibilities,

- solve equations and inequalities using addition, subtraction, multiplication and division, and

- use formulas to solve real-world and geometry problems.

Why It's Important

Real World

Architecture The floor plan of this house not only shows you the location of the various rooms, but it also shows the dimensions of each room. You can use mathematical formulas to find the area and perimeter of each room in this floor plan. For example, the dimensions of the family room are 15 feet by 21.3 feet. The formula for the area of a rectangle tells you to multiply the two dimensions to find the area of the room. *You will learn about area and perimeter in Lesson 3-5.*

Focus on...

To be successful in this chapter, you'll need to understand these concepts and be able to apply them. Refer to the lesson in parentheses if you need more review before beginning the chapter.

▶ *Evaluate expressions containing variables.* *(Lesson 1-3)*

Evaluate each expression if $a = 2$, $b = -3$, and $c = 5$.
1. $a + b + c$ **2.** $2a + 1$ **3.** $3b + c$ **4.** $ab - c$

▶ **Find the additive inverse of an integer.** *(Lesson 2-5)*

State the additive inverse of each of the following.
5. $+4$ **6.** -3 **7.** $+15$ **8.** -1

▶ **Multiply integers.** *(Lesson 2-7)*

Find the product.
9. $6(8)$ **10.** $-3 \cdot 5$ **11.** $-6(-25)$ **12.** $4(-9)$

▶ **Divide integers.** *(Lesson 2-8)*

Divide.
13. $64 \div (-32)$ **14.** $-15 \div 3$ **15.** $-12 \div (-3)$ **16.** $-6 \div (-6)$

▶ **Write an inequality from a verbal statement.** *(Lesson 1-9)*

Translate each sentence into an inequality.
17. A number is less than 12.
18. The cost is at least $50.

Focus on...

In this chapter, you'll be learning about **equations** and **inequalities**. The word equation contains the word "equal." An equation is a mathematical sentence in which the value of one side is equal to the value of the other side. The word inequality contains the prefix "in" which can mean "not." An inequality is a mathematical sentence in which the value of one side is not necessarily equal to the value of the other side.

Problem-Solving Strategy: Eliminate Possibilities

You'll Learn

■ To solve problems by eliminating possibilities.

It's Important

Knowing how to eliminate possibilities can help you take standardized tests.

Application: Education

Was there ever a question on a multiple-choice test for which you weren't quite sure of the answer? How did you go about choosing an answer? Eliminating some of the possible answers may help you choose the correct answer without having to guess.

Learning the Concept

This problem-solving strategy, **eliminating possibilities**, is helpful in solving many types of problems.

Example

Education

1 **The question below appeared on a multiple choice test. Use the strategy of eliminating possibilities to choose the correct answer.**

For what values of y will $2y - 4$ be equal to $2y + 6$?
a. all negative numbers
b. 0
c. all positive numbers
d. no value

Explore You know that y can represent any number. You want to find the values of y, if any, for which $2y - 4 = 2y + 6$.

Plan Look at the choices of values of y to see which ones are impossible and eliminate those.

Solve Choose a test value of y for each of the possible answers. Then test to see if $2y - 4 = 2y + 6$ for that value. If $2y - 4 \neq 2y + 6$ for that value, then we can eliminate that answer choice.

Choice	Test value for y	$2y - 4$	$2y + 6$	Is $2y - 4 = 2y + 6$?
a	-3	$2(-3) - 4$ $= -6 - 4$ or -10	$2(-3) + 6$ $= -6 + 6$ or 0	no
b	0	$2(0) - 4$ $= 0 - 4$ or -4	$2(0) + 6$ $= 0 + 6$ or 6	no
c	2	$2(2) - 4$ $= 4 - 4$ or 0	$2(2) + 6$ $= 4 + 6$ or 10	no

We can eliminate a, b, and c as possible answers. Therefore, d must be the correct choice.

Examine   Study the equation $2y - 4 = 2y + 6$. Notice that $2y$ appears on both sides. Since $2y$ will have the same value for any given value of y, the equation will only be true when $-4 = 6$. But $-4 \neq 6$, so there is no value of y that makes the two expressions equal.

The process of elimination is also useful in solving some logic problems. Making a chart will help you organize the facts. The process in Example 2 is called **matrix logic**.

Example Shalana, Ken, Rolon, and Erica each take their pets to Dr. McDermott for veterinary care. They each have a different pet: a dog, a cat, a gerbil, or a parrot. Use the following information to match each person with his or her pet.

▶ Rolon lives next door to the person with the gerbil.
▶ Erica and Ken frequently care for the dog when its owner is out of town.
▶ Shalana cannot have a dog or cat because of allergies.
▶ Erica is teaching her pet how to talk.

Make a chart to organize the information. Mark an X whenever you eliminate a possibility and draw a circle when you find a match.

	Shalana	Ken	Rolon	Erica
Dog				
Cat				
Gerbil				
Parrot				

Since Rolon cannot live next door to himself, the first clue tells you that he does not own the gerbil. The second clue tells us that neither Ken nor Erica owns the dog. And the third clue tells us that Shalana does not own the cat or the dog.

	Shalana	Ken	Rolon	Erica
Dog	X	X		X
Cat	X			
Gerbil			X	
Parrot				

So Rolon must own the dog. Draw a circle to indicate the match. Then mark X's for the other parts of Rolon's column.

(continued on the next page)

	Shalana	Ken	Rolon	Erica
Dog	X	X	◯	X
Cat	X		X	
Gerbil			X	
Parrot			X	

Since the only pet that could possibly talk is a parrot, the fourth clue tells us that Erica must own the parrot. Draw a circle to indicate the match. Then mark Xs for each of the other choices of pets for Erica and owners for the parrot.

Complete the table to determine the answers.

	Shalana	Ken	Rolon	Erica
Dog	X	X	◯	X
Cat	X	◯	X	X
Gerbil	◯	X	X	X
Parrot	X	X	X	◯

Shalana owns the gerbil, Ken owns the cat, Rolon owns the dog, and Erica owns the parrot.

Checking Your Understanding

Communicating Mathematics

Read and study the lesson to answer these questions.

1. **Explain** why eliminating possibilities can be helpful in solving problems.

2. **Explain** how eliminating possibilities is different from guessing and checking to find the answer.

MATERIALS

25 square tiles (5 each of 5 different colors)

3. Arrange the tiles in a 5-by-5 square so that no row, column, or diagonal has more than one of the same color.

   a. Compare your arrangement with others in your class.

   b. Does the order of the first row make a difference to your final arrangement?

   c. Is there more than one way to arrange the tiles?

Guided Practice

Solve by eliminating possibilities.

4. Matt, Stacey, and Keisha each ate either cheese pizza, a hot dog, or a hamburger for dinner. Use the clues to find each person's dinner.

   ▶ Keisha did not have any meat for dinner.

   ▶ Stacey did not have a hot dog.

5. Use the following clues to eliminate all but the correct number.

4235	8765	4223	6087	6444	5163
9635	7123	6668	5228	8224	9296

   ▶ It is an odd number that is divisible by 3.

   ▶ It is greater than 6000 and less than 7000.

   ▶ The sum of the digits is 21.

6. If a number is increased by 5 and the result is multiplied by 8, the product is 168. What is the original number?

    a. 42   **b.** 26   **c.** 16   **d.** 128

Exercises: Practicing and Applying the Concept

Independent Practice

Solve using any strategy.

7. Study the list of numbers below. Using the clues, find the mystery number.

    ▶ It is not divisible by 5.

    ▶ It is even.

    ▶ It is divisible by 3.

    ▶ It has 2 digits.

    ▶ The sum of its digits is 6.

6	124	76	25	7
13	252	321	60	9
24	120	15	27	8
43	561	18	78	5

8. How many packages of hot dogs balance with the ketchup?

    ketchup + mustard = 4 packages   ketchup = mustard + 2 packages

9. Statistics Trax Music Store took a survey of 100 people. Sixty-three said they like rock, 49 said they like rap, and 24 said they like jazz. Twenty-two people like rock and rap but don't like jazz, 2 like jazz and rap but not rock, and 11 like rock and jazz but not rap. If 3 people like all three types of music, how many people do not like rock, rap, or jazz?

10. If 9 less than the product of a number and -4 is greater than 7, which of the following could be that number?

    a. 5   **b.** 4   **c.** -3 **d.** -5

11. What is the least positive number that you can divide by 7 and get a remainder of 4, divide by 8 and get a remainder of 5, and divide by 9 and get a remainder of 6?

12. Business A concession stand at the football game sold 500 candy bars for a total of $255.00. A small candy bar sells for 40¢ and a large candy bar sells for 65¢. How many of the small candy bars were sold?

    a. 320   **b.** 280   **c.** 220   **d.** 77

13. Consumerism According to a magazine, most Americans choose to buy shoes that cost $70 and last two years instead of shoes that cost $200 and last seven years. Is this good money management? Explain.

    Source: *Money magazine*

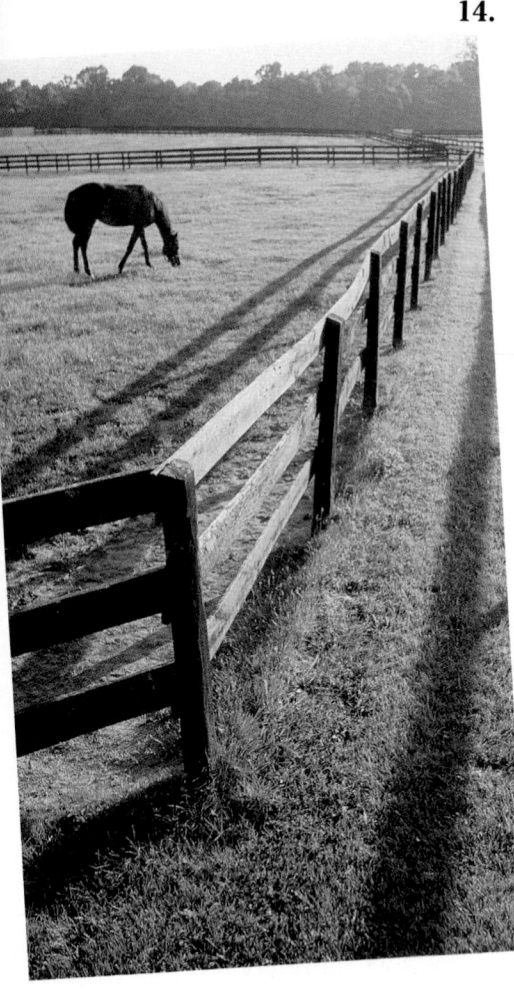

14. Horse Riding Five members of the Riding Club at Kingston High School went for a two-day riding trip to Manchester Lodge. They rode Sommie, Samantha, Friar, Claude, and Blaze on the first day. The stable staff wanted the club members to ride the same horses on the second day, but they could not remember who was on which horse. Use the clues below to match the rider with the horse.

▸ Blaze is the fastest horse and always the first horse on the trail.

▸ Friar is the most patient and gentle horse.

▸ Sommie is old so she always has the lightest rider, and on the trail she liked to be near the rear.

▸ Samantha and Claude are spirited and always start off together with the most experienced riders.

▸ Thomas was nervous because this was his first riding trip.

▸ Amber and Cynthia were the first two riders back to the stable.

▸ Amber taught Cynthia how to ride, but her horse had trouble keeping up with Cynthia's horse.

▸ Vernita was afraid that she was not tall enough to handle the horses she saw in the stable, so she rode behind the experienced riders.

▸ Carlos rode a male horse. He started out with Amber, but stopped on the trail to check his horse's hooves. He was one of the most experienced riders, so the rest of the group knew he would catch up.

Critical Thinking

15. Three students are blindfolded and stand in a single-file line. Kieko takes three hats from a box containing three blue hats and two green hats and places one on each of the students. She tells the students how many hats of each color were in the box and removes the blindfolds. The student in the back looks at the hats on the two students in front of him and says "I don't know what color hat I am wearing." The student in the middle hears him, looks at the hat on the student in front of her and says the same thing. The student in front says "I *do* know what color hat I am wearing." What color hat is he wearing? Explain how he knows.

Mixed Review

16. Solve $x = \dfrac{-120}{-15}$. (Lesson 2-8)

17. Patterns Find the next number in the pattern 0, 1, 3, 6, 10, 15, . . . (Lesson 2-6)

18. Standardized Test Practice The lowest temperature last week was $-9°F$. The lowest temperature this week was $15°$ higher. What was the lowest temperature this week? (Lesson 2-4)

A. $-24°F$ **B.** $-6°F$ **C.** $6°F$ **D.** $24°F$

For **Extra Practice**, see page 745.

19. Simplify $21k + 3(k + 1)$. (Lesson 1-5)

20. Evaluate $35 - c - b$ if $b = 13$ and $c = 7$. (Lesson 1-3)

HANDS-ON ACTIVITY

3-2A Solving Equations with Cups and Counters

A Preview of Lesson **3-2**

MATERIALS

⬜ cups and counters

▱ equation mat

You can use cups and counters as a model for solving equations.

Activity ①

An equation mat allows you to model an equation using cups and counters. The equation mat at the right represents the equation $x + 4 = 6$.

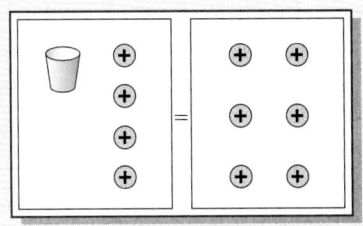

$x + 4 = 6$

To solve $x + 4 = 6$, find the value of x that makes the equation true. Think of this as finding the number of counters in the cup. To do that, get the cup by itself on one side of the mat.

$x + 4 - 4 = 6 - 4$

If you remove four counters from each side of the mat, then you can see that the cup is equal to two counters. So, $x = 2$. Since $2 + 4 = 6$, our solution is correct.

$x = 2$

Activity ②

LOOK BACK◄

You can review adding integers with counters in Lesson 2-4A.

Some equations involve subtraction. The mat at the right represents $7 = y - 5$. To get the cup by itself, remove 5 negative counters from each side. Since there are no negative counters on the left side, add 5 positive counters to each side to make 5 zero pairs on the right side.

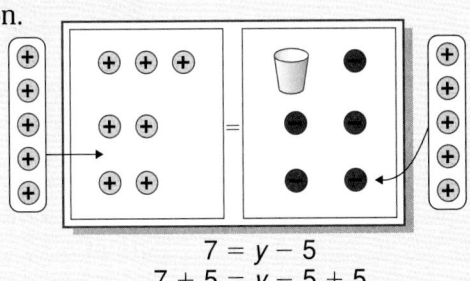

$7 = y - 5$
$7 + 5 = y - 5 + 5$

Now remove the zero pairs. The remaining cup and counters show that $y = 12$. Check by substituting 12 for y in the original equation.

$12 = y$

Your Turn   Use cups and counters to model and solve each equation.

1. $3 + x = 7$ 2. $n + 6 = 9$ 3. $y + 15 = 18$
4. $x + 6 = 10$ 5. $t + 2 = 9$ 6. $4 + w = 5$
7. $x - 2 = 10$ 8. $n - 4 = 9$ 9. $6 = q - 5$

Math Lab 3-2A Solving Equations with Cups and Counters **123**

Solving Equations by Adding or Subtracting

You'll Learn

■ To solve equations by using the addition and subtraction properties of equality.

It's Important

You can add and subtract equations to find increases and decreases in prices.

Technology Activity

Graphing calculators are computers, so you can run programs. The program at the right allows you to enter an equation of the form $x + a = b$ or $x - a = b$ and guess the solution. Using this program and eliminating possibilities, you can solve equations.

Your Turn Use the program to choose the solution of each equation from the possible values. Record the solutions.

$x + 5 = 9$	$x = 1, 3, 14, 4$
$x - 9 = 6$	$x = 3, 6, 15, 18$
$x + 7 = 8$	$x = 1, 7, 8, 15$
$x - 2 = 14$	$x = 2, 12, 14, 16$
$x + 10 = 15$	$x = 2, 5, 10, 20$
$x - 8 = 8$	$x = 0, 16, 1, 64$

PROGRAM: ONESTEP
: Disp "PRESS 1: X + A = B"
: Input "PRESS 2: X − A = B", N
: Input "ENTER A", A
: Input "ENTER B", B
: Lbl 1
: Input "GUESS X", X
: If N = 1 and X + A = B or N = 2 and X − A = B
: Then
: Disp "CORRECT"
: Stop
: Else
: Disp "TRY AGAIN"
: Goto 1

---

a. What do you observe about the solutions of equations like $x + a = b$?

b. What do you observe about the solutions of equations like $x - a = b$?

c. Solve $x + 3 = 12$ and $x - 5 = 15$. Use the program to verify the solutions.

Learning the Concept

Consider $x + 10 = 14$. One way to solve is to mentally find the number that makes the mathematical sentence true. Since $4 + 10 = 14$, the solution is 4.

Not all equations are easy to solve mentally. You can subtract to isolate the variable. If you subtract 10 from the left side of the equation, you must also subtract 10 from the right side to keep the equation balanced.

$$x + 10 = 14$$
$$x + 10 - 10 = 14 - 10 \quad \textit{Subtract 10 from each side.}$$
$$x + 0 = 4 \quad \textit{10 − 10 = 0 and 14 − 10 = 4.}$$
$$x = 4$$

These are **equivalent equations** because they all have the same solution, 4. We used the **subtraction property of equality** to solve this equation.

Subtraction Property of Equality	**In words:**	If you subtract the same number from each side of an equation, the two sides remain equal.
	In symbols:	For any numbers a, b, and c, if $a = b$, then $a - c = b - c$.

Example 1 Solve $r + 56 = -5$.

$$r + 56 = -5$$
$$r + 56 - 56 = -5 - 56 \quad \textit{Subtract 56 from each side.}$$
$$r = -61 \qquad\quad \textit{Check your solution by replacing r with −61.}$$

Some equations can be solved by using the **addition property of equality**.

Addition Property of Equality	**In words:** If you add the same number to each side of an equation, the two sides remain equal.
	In symbols: For any numbers a, b, and c, if $a = b$, then $a + c = b + c$.

Example 2 Solve $y - 9 = -17$.

$$y - 9 = -17$$
$$y - 9 + 9 = -17 + 9 \quad \textit{Add 9 to each side.}$$
$$y = -8 \qquad\quad \textit{Check your solution.}$$

You can often use an equation to represent a situation in real life.

Example 3

Real World APPLICATION

Entertainment

The only one of the top ten money-making films to win an award for best picture earned $317 million at the box office. This is $83 million less than the all-time top money maker. How much did the all-time top money maker earn?

Write an equation. Use r for the top money maker.
$$317 = r - 83$$
$$317 + 83 = r - 83 + 83 \quad \textit{Add 83 to each side.}$$
$$400 = r \qquad\qquad \textit{Check your solution.}$$

The all-time top money maker earned $400 million at the box office.

LOOK BACK

You can review additive inverses in Lesson 2-5.

Recall that to subtract an integer, you can add its inverse.

subtraction   addition

additive inverses

$$8 - (-2) = 10 \qquad\qquad 8 + 2 = 10$$

same result

You can also use the additive inverse to solve equations involving addition.

Example **Solve $c + 8 = 12$.**

Method 1: Subtracting	**Method 2: Adding Inverse**
$c + 8 = 12$	$c + 8 = 12$
$c + 8 - 8 = 12 - 8$	$c + 8 + (-8) = 12 + (-8)$
$c = 4$	$c + 0 = 4$
	$c = 4$

Check: $c + 8 = 12$
$4 + 8 \stackrel{?}{=} 12$ *Replace c with 4.*
$12 = 12$ ✓ The solution is 4.

Solutions to equations can be represented on a *number line*. Each solution is a **coordinate** of a point. The coordinate tells its distance and direction from the 0-point on the line. The dot marking the point is called the **graph** of the number.

The solution in the example above, 4, is represented by placing a dot above 4 on a number line.

Example **Solve $y - 24 = 16$. Graph the solution on a number line.**

$y - 24 = 16$
$y - 24 + 24 = 16 + 24$ *Add 24 to each side.*
$y = 40$ The solution is 40.

Graph the solution, 40, on a number line.

Checking Your Understanding

Communicating Mathematics

Read and study the lesson to answer these questions.

1. **Explain** when you use the addition property of equality to solve an equation.

2. **Tell** what property you would use to solve $x + 34 = -6$.

3. **Write** two equations that are equivalent. Then write two equations that are not equivalent.

4. **Write** an equation in the form $x - a = b$ where the solution is -12.

5. **You Decide** Ellen and Tertius solved $x - (-25) = 56$ in two different ways. Who is correct? Explain.

Ellen	**Tertius**
$x - (-25) = 56$	$x - (-25) = 56$
$x + 25 = 56$	$x - (-25) + (-25) = 56 + (-25)$
$x + 25 - 25 = 56 - 25$	$x = 56 - 25$
$x = 31$	$x = 31$

MATERIALS
cups and counters

6. Use cups and counters to model the solution to $9 = w + (-5)$.

Solve each equation and check your solution. Then graph the solution on a number line.

7. $w + 8 = -17$ **8.** $25 = -4 + y$ **9.** $m - (-3) = 40$

10. $y + (-7) = 9$ **11.** $k - 36 = -37$ **12.** $h - 8 = -22$

13. Entertainment   Movie A earned $322 million in revenues while its sequel, Movie B, earned $264 million. How much more did Movie A earn?

Exercises: Practicing and Applying the Concept

Solve each equation and check your solution. Then graph the solution on a number line.

14. $y + 7 = 21$ **15.** $k - 6 = 32$

16. $19 = g - 5$ **17.** $-24 = h - 22$

18. $x + 49 = 13$ **19.** $f + 34 = 2$

20. $d - (-3) = 2$ **21.** $a - (-26) = 5$

22. $12 + p = -14$ **23.** $x + 18 = 14$

24. $59 = s + 95$ **25.** $x - 27 = 63$

26. $-7 = z - (-12)$ **27.** $34 + r = 84$

28. $-234 = x + 183$ **29.** $y - 94 = 562$

30. $-591 = m - (-112)$ **31.** $846 + t = -538$

Choose the equation whose solution is graphed.

32.

```
←――+――+――+――+――+――+――+――+――+――+――+――+――●――→
  -6 -5 -4 -3 -2 -1  0  1  2  3  4  5  6
```
a. $x + 5 = -11$ **b.** $x - 5 = 11$ **c.** $x + 5 = 11$ **d.** $x - 5 = -11$

33.
```
←――+――+――+――●――+――+――+――+――+――+――+――+――+――→
  -8 -7 -6 -5 -4 -3 -2 -1  0  1  2  3  4
```
a. $-8 = t - 4$ **b.** $-8 = t + 4$ **c.** $8 = t - 4$ **d.** $8 = t + 4$

Solve each equation. Check each solution.

34. $(f + 5) + (-2) = 6$ **35.** $[y + (-3)] + 2 = 4$

36. $16 = [n - (-2)] + (-3)$ **37.** $-10 = [b + (-4)] + 2$

38. Write an equation whose solution is represented by the number line.

```
←――+――●――+――+――+――+――+――+――+――+――+――+――→
    -6 -5 -4 -3 -2 -1  0  1  2  3  4  5
```

39. Advertising   Morris has been the official mascot for a cat food company for 26 years. Actually, there have been two cats named Morris. Morris I died unexpectedly at the age of 19. If he was the official mascot for 14 years, how long has Morris II been the official mascot?

40. Meteorology   Weather fronts usually move across the United States from west to east and can bring with them astounding changes in temperatures.

  a. On January 23–24, 1916, the temperature in Browning, Montana, dropped 100° in 24 hours. If the temperature started at 44°F, how cold did it get?

  b. In 1892, at Fort Assinaborn, Montana, the temperature rose 43° in 15 minutes. At 2:00 A.M., the temperature was −5°F. If that is when the temperature began to rise, what was the temperature at 2:15 A.M.?

41. **Illustration**   In 1995, Charles Schulz, the creator of a comic strip, was 70 years old. He created the comic in 1950. How old was he when he started drawing the comic?

Mixed Review

42. **Travel**   On a trip to Florida, a bus traveled 600 miles in 10 hours. At that rate, how far could the bus travel in 4 hours? (Lesson 3-1)
   a. 2400 miles   **b.** 400 miles   **c.** 240 miles   **d.** 150 miles

43. Solve $y = (-9)(18)$. (Lesson 2-7)

44. Solve $7 - 25 = h$. (Lesson 2-5)

45. Order the integers in the set $\{3, -1, 15, -3\}$ from least to greatest. (Lesson 2-3)

46. **Standardized Test Practice**   Which is the value of the expression $12 - |-3|$? (Lesson 2-1)
   A. 6   **B.** 9   **C.** 15   **D.** 36

47. **Exercise**   A 150-pound person burns 384 Calories an hour jogging at a rate of 6 miles per hour. That is 120 Calories more than the same person burns playing volleyball for one hour. How many Calories does a 150-pound person burn playing volleyball for one hour? (Lesson 1-8)

   a. Define a variable for this problem.

   b. Write an equation.

   c. Solve for the number of Calories burned in an hour of volleyball.

For **Extra Practice**, see page 745.

48. Find the value of $7 + 8 \div 2 \cdot 6 - 1$. (Lesson 1-2)

WORKING ON THE Investigation

BON VOYAGE

Refer to the Investigation on pages 62–63.

Look over the list of considerations that you placed in your Investigation Folder at the beginning of Chapter 2.

You probably listed weather as one thing to consider before packing clothes. You don't want to pack shorts and sandals if temperatures are going to be in the 20s—or do you?

Unless you selected Great Britain, Australia, New Zealand, or Ireland, as your destination, your hosts will refer to temperatures measured in *degrees Celsius*.

Like the Fahrenheit scale, the Celsius scale has two fixed points: the boiling point of water and the freezing point of water (or the melting point of ice).

On the Celsius scale, the boiling point of water was set at 100 and the freezing point of water was set at 0 so that the scale would have 100 equal divisions (called degrees Celsius) between them.

If you've grown up using the Fahrenheit scale, you can probably translate temperatures into "feel". You know, for example, that 85°F feels very warm.

- Work with your partner to choose at least four reference Fahrenheit temperatures that you "feel."
- Find a thermometer that has both scales on it. Use it to estimate the equivalent Celsius temperature for each Fahrenheit temperature you selected.
- If you haven't already done so, obtain environmental data for the country you and your partner selected at the beginning of the investigation. Then provide general guidelines for your wardrobe.

Add your results to your Investigation Folder.

3-3

Solving Equations by Multiplying or Dividing

You'll Learn

- To solve equations by using the multiplication and division properties of equality.

It's Important

You can multiply and divide equations to find how long you can expect a car trip to take.

Application: Toys

Popular toys are often created by individuals who have a simple idea. Richard James came up with an idea for a toy in 1943 when he watched a spring fall from his desk to the floor. His toy is a wiry coil, that with a nudge, appears to walk down the stairs. After more than fifty years, the toy is still a "walking" success.

Millions of these toys have been sold over the past 50 years. They are still made using the same machines designed years ago. One such toy is produced every 8 seconds, 24 hours a day. How many of these toys are made each day?

Learning the Concept

To find out how many of these toys are made in a day, translate what we know into an equation.

First find the number of seconds in a day. There are 60 seconds in a minute, 60 minutes in an hour, and 24 hours in a day.

$$\left(\frac{60 \text{ s}}{\text{min}}\right)\left(\frac{60 \text{ min}}{\text{hr}}\right)\left(\frac{24 \text{ hr}}{\text{day}}\right) = 86{,}400 \ \frac{\text{s}}{\text{day}}$$

Let d represent the number of toys made in a day. If each toy takes 8 seconds to make, we can write an equation to represent the daily production of toys.

seconds per toy	*times*	*toys made in a day*	*equals*	*seconds in a day*
8	×	d	=	86,400

To solve this equation, undo the multiplication by dividing each side by the same number.

$8d = 86{,}400$

$\dfrac{8d}{8} = \dfrac{86{,}400}{8}$ *Divide each side by 8 to undo the multiplication $8 \cdot d$.*

$d = 10{,}800$

Check: $8d = 86{,}400$

         $8(10{,}800) \stackrel{?}{=} 86{,}400$ *Replace d with 10,800.*

            $86{,}400 = 86{,}400$ ✓   The solution is 10,800.

Therefore, 10,800 of these toys are made in a day.

Lesson 3-3   *Solving Equations by Multiplying or Dividing*   **129**

| Division Property of Equality | **In words:** | If you divide each side of an equation by the same nonzero number, the two sides remain equal. |
| | **In symbols:** | For any numbers a, b, and c, where $c \neq 0$, if $a = b$, then $\dfrac{a}{c} = \dfrac{b}{c}$. |

Example **1** **Solve** $-68 = -4m$. **Check the solution and graph it on a number line.**

$$-68 = -4m$$

$$\dfrac{-68}{-4} = \dfrac{-4m}{-4} \quad \textit{Divide each side by} -4.$$

$$17 = m \quad -68 \div (-4) = 17$$

Check: $\quad -68 = -4m$

$$-68 \overset{?}{=} -4(17) \quad \textit{Replace m with 17.}$$

$$-68 = -68 \ \checkmark$$

The solution is 17.

Just as division can be used to solve equations in some situations, multiplication can also be used to solve equations. This makes use of the multiplication property of equality.

| Multiplication Property of Equality | **In words:** | If you multiply each side of an equation by the same number, the two sides remain equal. |
| | **In symbols:** | For any numbers a, b, and c, if $a = b$, then $a \cdot c = b \cdot c$. |

Example **2** **Solve** $\dfrac{d}{3} = -3$. **Check the solution and graph it on a number line.**

$$\dfrac{d}{3} = -3$$

$$\dfrac{d}{3}(3) = -3(3) \quad \textit{Multiply each side by 3 to undo the division in } \dfrac{d}{3}.$$

$$d = -9$$

Check: $\quad \dfrac{d}{3} = -3$

$$\dfrac{-9}{3} \overset{?}{=} -3 \quad \textit{Replace d with} -9.$$

$$-3 = -3 \ \checkmark$$

The solution is -9.

Example **3**

Sports

In one baseball season, the fans at the Baltimore Orioles' Camden Yards bought 232,618 bags of peanuts at ball games. There were 80 games in the park that season.

a. On average, how many bags of peanuts were sold at each game?

b. How many bags of peanuts should the concession manager order for the next season if there will be 86 games in the park that year?

a. Let p represent the number of bags of peanuts sold at each game. Then the equation $80p = 232{,}618$ will allow us to find the average number of bags sold at each game.

$$80p = 232{,}618$$
$$\frac{80p}{80} = \frac{232{,}618}{80} \qquad \textit{Divide each side by 80.}$$
$$p = \frac{232{,}618}{80} \qquad \textit{Estimate: } 240{,}000 \div 80 = 3000$$

Use a calculator to find p.

232618 ÷ 80 = *2907.725*

On average, about 2908 bags of peanuts were sold at each game.

b. You can estimate to find out how many bags of peanuts the manager should order. Round to 3000 bags per game and 90 games in the season. So, the manager should order 3000×90 or 270,000 bags of peanuts for a season with 86 games.

Checking Your Understanding

Communicating Mathematics

Read and study the lesson to answer these questions.

1. **In your own words, explain** how you would solve $\frac{x}{4} = -3$.

2. **Compare and contrast** solving equations using the multiplication property of equality and solving equations using the division property of equality.

3. **Write** an equation in the form $ax = c$ where the solution is -5.

4. In an equation of the form $\frac{x}{a} = b$, explain why a cannot be 0.

Guided Practice

Solve each equation and check your solution. Then graph the solution on a number line.

5. $5x = 45$ 6. $8y = -64$ 7. $-\frac{p}{3} = 4$

8. $\frac{a}{-5} = 8$ 9. $-3g = -51$ 10. $-2 = \frac{p}{6}$

Lesson 3-3   *Solving Equations by Multiplying or Dividing*   **131**

Data Update For the latest exchange rates, visit:
www.glencoe.com/sec/math/prealg/mathnet

11. **Traveling** Paloma's social studies class was planning an imaginary trip to Kenya. Paloma had to find out about the rate of exchange for money in Kenya. She learned that money in Kenya is based on shillings and that the current exchange rate was one U.S. dollar for 40 shillings. If one night at a hotel cost 960 shillings, how much would that be in U.S. dollars?

Exercises: Practicing and Applying the Concept

Independent Practice

Solve each equation and check your solution. Then graph the solution on a number line.

12. $8x = 72$

13. $-5y = 95$

14. $\dfrac{k}{-2} = 7$

15. $-\dfrac{h}{7} = 20$

16. $86 = 2v$

17. $-21 = -\dfrac{g}{8}$

18. $\dfrac{a}{45} = -3$

19. $\dfrac{b}{-9} = -4$

20. $672 = -21t$

21. $-56 = -7p$

22. $-116 = -4u$

23. $-\dfrac{y}{11} = 132$

24. $\dfrac{f}{-34} = -14$

25. $\dfrac{y}{8} = 117$

26. $17r = -357$

27. $-18p = 306$

28. $-144 = 8x$

29. $-384 = -3m$

30. $\dfrac{b}{46} = 216$

31. $-71 = \dfrac{x}{31}$

32. $-171 = \dfrac{x}{-12}$

33. $584 = -\dfrac{s}{23}$

34. $\dfrac{p}{47} = 123$

35. $-67w = -5561$

Graphing Calculator

36. You can check your solutions to equations using a graphing calculator. Enter one side of the equation as an expression using your solution and see if the value is the same as the other side of the equation.

For example, if $\dfrac{x}{5} = -2$, then $x = -10$.

Enter: (−) 10 STO▸ X,T,θ,n ENTER   *Enters the value of x.*

X,T,θ,n ÷ 5 ENTER   *-2*   *Enters equation and evaluates it.*

The solution checks.

Solve each equation. Then check your solution using a graphing calculator.

a. $0.013x = 0.546$ b. $\dfrac{x}{1200} = -0.12$ c. $-16.7x = 367.4$

Critical Thinking

37. Write a word problem that can be solved using the equation $4x = 13$.

Applications and Problem Solving

38. **Manufacturing** Refer to the beginning of the lesson. One toy is made from 80 feet of wire. If a standard spool of wire contains 4000 feet, how many toys can be made from one spool?

39. Biology The mola is an interesting fish because it appears to have no body. Weighing more than two tons, it looks like a large head with two perpendicular fins. Female molas can have 300 million eggs. This is 6 to 15 times the number in most species of fish.

    a. Write an equation to represent 6 times the number of eggs is 300 million eggs.

    b. Write an equation to represent 15 times the number of eggs is 300 million eggs.

    c. Write a sentence that states the range of eggs produced by most species of fish.

40. Finance Refer to Exercise 11. On the trip to Kenya, the students in Paloma's class wanted to take a safari to the Masai Mara. They estimated that the cost of the safari would be $300 a person.

    a. Research to find out how many shillings are exchanged for a U.S. dollar today.

    b. Write an equation to find out how much the safari would cost in shillings.

    c. Solve the equation.

41. Geometry The perimeter of any square is 4 times the length of one of its sides. If the perimeter of a square is 72 centimeters, what is the length of each side of the square?

Perimeter = 72 cm x cm

x cm

Mixed Review

42. Solve $121 = k - 34$ and check your solution. Then graph the solution on a number line. (Lesson 3-2)

43. Standardized Test Practice If a number is multiplied by 4 and the result is subtracted from 52, the difference is -12. What is the original number? (Lesson 3-1)

    A. -3 **B.** 0   **C.** 12   **D.** 16

44. Find the product of -7 and -15. (Lesson 2-7)

45. Simplify $|-5| + |12|$. (Lesson 2-1)

46. State whether the inequality $7 > 2$ is *true*, *false*, or *open*. (Lesson 1-9)

47. Solve $7n = 70$ mentally. (Lesson 1-6)

48. Careers Curt Harris received a raise. He now makes $150 less than twice his old salary. Write an expression for Mr. Harris's current salary if his old salary was s. (Lesson 1-3)

49. Shopping Is $7 enough money to buy a loaf of bread for $0.98, a pound of hamburger for $2.29, and a pound of roast turkey for $3.29? (Lesson 1-1)

    a. Which method of computation do you think is most appropriate? Explain.

    b. Solve the problem using the four-step plan.

For **Extra Practice**, see page 746.

3-4 Using Formulas

You'll Learn

■ To solve problems by using formulas.

It's Important

Using the formula for finding the area of a circle can help you determine the area affected by an earthquake.

Application: Travel

You slide behind the wheel of a red convertible. The engine roars as you squeal away down the road. With the top down, the sound of the whipping wind drowns out the world until suddenly the engine stops.

"Just how far back was that gas station? Wish I would have figured the gas mileage on this thing before I left home."

You can use a formula to compute gas mileage.

Learning the Concept

A **formula** shows the relationship among certain quantities. The formula below can be used to find the miles per gallon achieved by a car.

number of miles driven	*divided by*	*number of gallons of gas*	*equals*	*miles per gallon*
m	$\div$	g	$=$	mpg

Example 1 **You drove 192.8 miles before your convertible ran out of gas. The gas tank holds 11 gallons and was full when you left. What gas mileage does the car get?**

$m \div g = \text{mpg}$
$192.8 \div 11 = \text{mpg}$ *Replace m with 192.8 and g with 11.*
$192.8 \boxed{\div} 11 \boxed{=} \mathit{17.527272}$ *Estimate: 200 ÷ 10 = 20*

The car gets about 17.5 miles per gallon.

Connection to ⟩ Geometry

There are formulas associated with almost every field of study. Example 2 uses a formula from geometry.

INTEGRATION
Geometry

Example 2 **A pentagon is a polygon with five sides. In a regular pentagon, all of the sides are the same length, and all of the angles have the same measure. The area, *A*, of a regular pentagon can be found using the formula $A = \frac{5}{2}sa$, where *s* is the length of a side of the pentagon and *a* is the length of the apothem. Find the area of the pentagon shown at the right.**

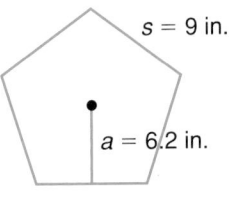

$s = 9$ in.

$a = 6.2$ in.

$$A = \frac{5}{2}sa$$

$$A = \frac{5}{2}(9)(6.2)$$ *Replace s with 9 and a with 6.2.*

$$A = 139.5$$

The area of the pentagon is 139.5 square inches.

Checking Your Understanding

Communicating Mathematics

MATERIALS

✎ ruler

Read and study the lesson to answer these questions.

1. **Explain** how to use a formula.
2. **Write** a formula that you have used in your life.
3. **Draw** polygons with the numbers of sides in the table below. Pick a vertex and draw all the diagonals from this vertex. Count the number of triangles formed. Copy and complete the table.

Number of sides	3	4	6	8	10
Number of triangles					

Can you find a formula that will predict how many triangles will be formed in a polygon with 100 sides? 1250 sides? x sides?

Guided Practice

Solve by replacing the variables with the given values.

4. $d = rt$, if $d = 315$, $r = 45$ 5. $s = l - d$, if $s = \$35$, $d = \$10$

Solve. Use the correct formula.

6. **Travel**   Find the miles per gallon for a car that travels 271.7 miles on 9.6 gallons of fuel.
7. **Transportation**   The formula $d = rt$ relates distance, d, rate, r, and time, t, traveled. Find the distance you travel if you drive at 55 miles per hour for 3 hours.
8. **Sports**   In league bowling, each team has an equal chance of winning because team members are given a handicap based on their averages. They use a formula to find the handicap score. The formula is as follows.

   handicap score(s) = game score(g) + handicap(h) or $s = g + h$

   Monique has a game score of 145 and a handicap of 35. Dennis has a game score of 153 and a handicap of 30. Who has the higher handicap score?

Exercises: Practicing and Applying the Concept

Independent Practice

Solve by replacing the variables with the given values.

9. $d = rt$, if $r = 120$ and $t = 3$ 10. $F = \frac{9}{5}C + 32$, if $C = 30$

11. $P = 5s$, if $P = 65$ 12. $h = 69 + 2.2F$, if $F = 14$

13. $d = 2r$, if $r = 1.7$ 14. $A = \frac{b + d}{2}$, if $b = 11$ and $d = 13$

Solve. Use the correct formula.

15. **Ballooning**   What is the speed in miles per hour of a balloon that travels 56 miles in 4 hours?

16. **Transportation**   A car can travel an average of 464 miles on one tank of gas. If the tank holds 16 gallons, how many miles per gallon does it get?

17. **Travel**   The Flynn family plans to drive 600 miles from Harrisburg, Pennsylvania to Chicago, Illinois for their summer vacation. The speed limit on the highways they plan to use is 55 miles per hour. How long will the trip take if they don't exceed the speed limit?

18. **Air Travel**   How long does it take an Air Force jet fighter to fly 5200 miles at 650 miles per hour?

Translate each sentence into a formula.

19. The sale price of an item, s, is equal to the list price, ℓ, minus the discount, d.

20. **Geometry**   In a circle, the diameter, d, is twice the length of the radius, r.

21. **Geometry**   The area, A, of a triangle is equal to the product of the length of the base, b, and the height, h, divided by two.

    a. What is the area of a triangle with a base of 12 inches and a height of 5 inches?

    b. What is the length of the base of a triangle with an area of 24 square centimeters and a height of 8 centimeters?

Critical Thinking

22. Measure the height and head circumference of five people. Use these measures to write a formula to predict a person's height from the size of his or her head.

Applications and Problem Solving

23. **Jobs**   Dorinda Pinel found a job about 8 miles from home. For a few weeks, she rode the bus to and from work. But the connections were poor, and it took an hour to get home in the afternoon. She thought that she could get home earlier if she rode her bicycle to and from work. Use a calculator to answer the following questions.

    a. If she rode at an average speed of 10 miles an hour, how long would it take for her to get home?

    b. The trip to work was uphill most of the way. It took Dorinda 54 minutes, or 0.9 hours, to get to work. How fast was she riding?

24. **Physical Therapy**   Physical therapists work to help people regain strength after surgery or an accident. The formula they use to determine a person's arm strength, S, is as follows.

$$S = \frac{D + P}{\frac{W}{10} + H - 60}$$

Find the arm strength of a person with the following information.

D(dips on a parallel bar)   = 4 dips
P(pull-ups)   = 8 pull-ups
W(weight in pounds)   = 120 pounds
H(height in inches)   = 60 inches

25. **Physical Science** Acceleration is the rate at which velocity is changing with respect to time. To find the acceleration, first find the change in velocity by subtracting the starting velocity, *s*, from the final velocity, *f*. Then divide by the time, *t*, it took to make the change.

   a. Write the formula for acceleration, *a*.

   b. A motorcycle goes from 2 m/s to 14 m/s in 6 seconds. Find its acceleration.

   c. Find the acceleration of a skateboard that goes from 6 m/s to 0 m/s in 3 seconds.

   d. When something slows down, it *decelerates*. Describe the value of the acceleration when an object decelerates.

Mixed Review

26. Solve $17p = 119$ and check your solution. Then graph the solution on a number line. (Lesson 3-3)

27. Find the quotient of -81 and 3. (Lesson 2-8)

28. **Standardized Test Practice** Solve $n = 8 + (-21)$. (Lesson 2-4)

   A. -29 **B.** -13 **C.** 13   **D.** 29

29. **Sports** There are 25 players on the Lady Scot soccer team. There are 11 players on the field at one time. Write an equation for the number of players on the bench. Then solve. (Lesson 1-8)

30. Name the property shown by $(2 + 6) + 9 = 2 + (6 + 9)$. (Lesson 1-4)

For **Extra Practice**, see page 746.

Self Test

1. **Entertainment** David Ogden Stiers, Mel Gibson, Russell Means, Linda Hunt, and Irene Bedard all lent their voices to characters in a popular film. The characters they played were Governor John Ratcliffe, Captain John Smith, Pocahontas, Grandmother Willow, and Powhatan. Use the following information to match each person with his or her character. (Lesson 3-1)

   - Irene Bedard is part Inupiat and was proud to play the part of a Native American heroine.

   - The character of Grandmother Willow is a tree.

   - Governor John Ratcliffe and the character played by Mel Gibson came to America from England. The character played by Russell Means is Native American.

   - Powhatan is Pocahontas' father and chief of the Powhatan tribe.

Solve each equation and check your solution. Then graph the solution on a number line.
(Lessons 3-2 and 3-3)

2. $18 + m = -57$ 3. $v - 11 = -5$ 4. $67 = h + 38$ 5. $d - (-16) = 9$

6. $4h = -52$ 7. $\frac{x}{5} = 20$ 8. $\frac{m}{11} = -10$ 9. $-2b = -13$

10. **Electronics** Ohm's Law states that the relationship between the current in amperes, *I*, the voltage in volts, *V*, and the resistance in ohms, *R*, in an electrical circuit is $I = \frac{V}{R}$. Current is usually stated in amperes. An automobile has a 12-volt system. What is the current in an automobile system that has a resistance of 3 ohms? (Lesson 3-4)

3-5A Area and Perimeter

A Preview of Lesson **3-5**

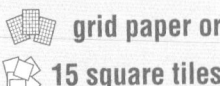
The distance around a geometric figure is called its **perimeter**. The measure of the surface enclosed by a geometric figure is its **area**. You can investigate the relationship between perimeter and area by drawing figures on grid paper or making models with square tiles.

Activity

The figures below show several figures with an area of 4 square units.

It appears that for an area of 4 square units, the greatest possible perimeter is 10 units.

Your Turn

Repeat the activity for areas of 1, 2, 3, 5, and 6 square units. Find the greatest possible perimeter for each area. Record your results in a table.

For each given area, find the greatest possible perimeter. Draw a figure to show the greatest possible perimeter.

1. 8 square units

2. 10 square units

3. 15 square units

TALK ABOUT IT

4. A figure has an area of 50 square units. Predict the greatest possible perimeter.

5. Explain why the two figures at the right have the same perimeter. Draw another figure with the same perimeter.

6. Suppose a figure has an area of x square units. Write an expression for the greatest possible perimeter.

3-5

Integration: Geometry
Area and Perimeter

Hands-On Activity

What You'll Learn

■ To find the perimeter and area of rectangles and squares.

Why It's Important

Finding the areas of the walls of a room can help you estimate the amount of paint needed to paint that room.

Below are all the different rectangles that have sides with measures that are whole numbers and a perimeter of 12 units. Model each with tiles or by drawing on grid paper.

A

B

C

On a paper, copy and complete the table at the right using the information from figures A, B, and C.

Rectangle	Length of Rectangle	Width of Rectangle	Perimeter of Rectangle
A	5	1	12
B	4		12
C			12

Your Turn Use grid paper or tiles to make as many different rectangles as you can with a perimeter of 16 units. Make and complete a table similar to the one above.

a. What observations can you make about the tables?

b. Work with a partner to write a rule on how to find the perimeter of a rectangle.

c. Will your rule also work for squares?

Learning the Concept

One method of finding the perimeter of a rectangle is to add the measures of the four sides. Let *P* represent the measure of the perimeter of a rectangle. Let ℓ represent the measure of the length and *w* the measure of the width.

$P = \ell + w + \ell + w$ *Definition of perimeter*
$P = \ell + \ell + w + w$ *Commutative property*
$P = 2\ell + 2w$ *of addition*
$P = 2(\ell + w)$ *Distributive property*

	In words:	If a rectangle has a length of ℓ units and a width of *w* units, then the perimeter is twice the sum of the length and width.
Perimeter of a Rectangle	**In symbols:**	$P = 2(\ell + w)$

$2(\ell + w)$ is read two times the quantity of ℓ plus w.

Example ① **Find the perimeter of each rectangle.**

a.

9 in.

4 in.

b.

8 m

8 m

$P = 2(\ell + w)$
$P = 2(9 + 4)$ $\ell = 9, w = 4$
$P = 2(13)$ *Add 9 and 4.*
$P = 26$ inches

The perimeter is 26 inches.

$P = 2(\ell + w)$
$P = 2(8 + 8)$ $\ell = 8, w = 8$
$P = 2(16)$
$P = 32$ m

The perimeter is 32 meters.

Another measurement of a rectangle is its **area**. By counting the number of 1 unit-by-1 unit squares, you can find the area of the rectangle shown at the right. The area of the rectangle at the right is 12 square units.

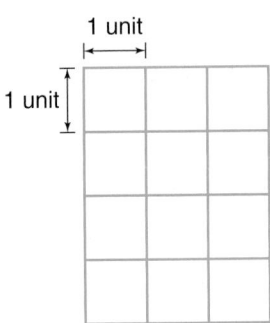

1 unit

1 unit

Look back at the models of rectangles that you investigated at the beginning of this lesson. Add another column called Area of Rectangle to your table. Count the squares in each rectangle to fill in the column. Do you notice a pattern? The area can also be found by multiplying the measures of the length and the width.

| **Area of a Rectangle** | **In words:** | If a rectangle has a length of ℓ units and a width of w units, then the area is $\ell \cdot w$ square units. |
| | **In symbols:** | $A = \ell w$ |

Example ② **Find the area of a rectangle with length 15 meters and width of 8 meters.**

First draw a picture of the rectangle and label the length and width.

$A = \ell \cdot w$ *formula for area of a rectangle*
$A = 15 \cdot 8$ *Replace ℓ with 15 and w with 8.*
$A = 120$

15 m

8m

The area is 120 square meters.

Many real-life problems involve finding the area or perimeter of a rectangle.

Example ③

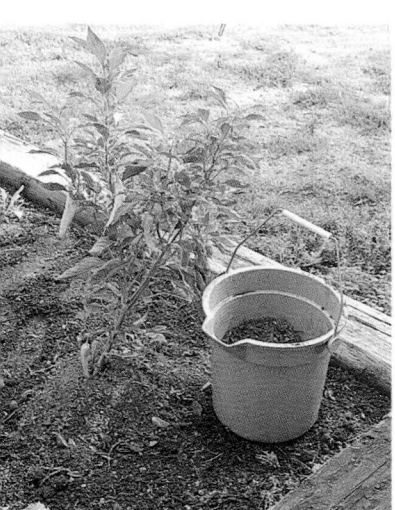

Sumi has just come home from the plant nursery with 20 outdoor plants. The man at the plant nursery told her that each of these particular types of plants needs 24 square feet of space to grow, so that it won't be choked off by its neighbors.

a. How many square feet does Sumi need in a garden plot so all the plants have enough room?

square feet per plant	times	number of plants	equals	square feet
24	·	20	=	480

b. Sumi has decided to use old railroad ties to enclose a rectangular space along the side of her house for the plants. The space is 16 feet wide, so how long should the garden be?

$A = \ell \cdot w$ *formula for area of a rectangle*

$480 = \ell \cdot 16$ *Replace A with 480 and w with 16.*

$\dfrac{480}{16} = \dfrac{\ell \cdot 16}{16}$ *Divide each side by 16.*

$30 = \ell$ The length should be 30 feet.

Example ④

Real World APPLICATION

Real Estate

Real estate agents help their clients choose a home that meets their needs. One of the ways that real estate agents compare homes is by their total square footage of living space. Find the square footage of each room of the home whose floor plan is shown at the right.

Room	$\ell \times w$	Area
master bedroom	12.6×15.2	191.52
family room	15×21.3	319.5
dining room	12×9.2	110.4
living room	12.5×12.3	153.75
den/bedroom	10×12.2	122
bedroom 2	11.6×10.3	119.48

Checking Your Understanding

Communicating Mathematics

Read and study the lesson to answer these questions.

1. **Restate** the definition of perimeter in your own words.

2. **Compare and contrast** the perimeter and area of a rectangle

3. **Draw and label** a rectangle that has a perimeter of 18 inches.

4. **Explain** how to find the perimeter and area of a rectangle that is 9 cm long and 5 cm wide.

Find the perimeter and area of each rectangle.

5.

4 cm

3 cm

6.

30 m

30 m

7. rectangle with length of 15 feet and a width of 6 feet

Find the missing dimension of each rectangle.

8. 12 in.

Perimeter = 32 in. w

9.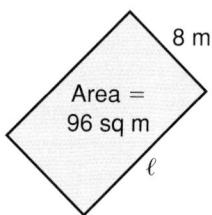

8 m

Area = 96 sq m

ℓ

10. Community Project   A neighborhood uses an empty city lot for a community vegetable garden. Each participant is allotted a space of 18 feet by 90 feet. What is the perimeter and area of each plot?

Exercises: Practicing and Applying the Concept

Find the perimeter and area of each rectangle.

11. 11 ft

24 ft

12. 9.1"

9.1"

13.

15 km

2 km

14.

8 cm

8 cm

15.

5.8 m

1.7 m

16. 180"

500"

17.

0.9 m

0.9 m

18. 21ft

1ft

19. 3.1 cm

3.1 cm

20. a rectangle that is 3.8 meters long and 1.1 meters wide

21. a square that is 4.8 millimeters on each side

22. an $8\frac{1}{2}$-inch by 11-inch rectangle

Find the missing dimension of each rectangle.

	Length	Width	Perimeter	Area
23.	7 m		24 m	35 sq m
24.		15 cm	66 cm	270 sq cm
25.	16 yd		54 yd	176 sq yd
26.	11 km		70 km	264 sq km
27.		12 ft	102 ft	468 sq ft
28.		91 yd	216 yd	1547 sq yd

Find the area of the blue part of each rectangle.

29. **30.**

Critical Thinking

31. Copy and complete the following table that compares changes in dimension, perimeter, and area of a square.

Dimension of square	Change in each from original	Perimeter	Change in perimeter from original	Area	Change in area from original
2 by 2	—	8	—	4	—
4 by 4	multiplied by 2		multiplied by 2		multiplied by 4
6 by 6					
8 by 8					

a. Show a relationship between the length of a side of a square and its perimeter using a graph.

b. Show a relationship between the length of a side of a square and its area using a graph.

c. The change in each square from the original is called the *scale factor*. Suppose the scale factor is *x*. What are the new perimeter and area?

Applications and Problem Solving

32. Landscaping Tachiyuki Mizumoto wishes to fertilize his lawn. Fertilizer comes in bags that cover 5000 square feet. His yard is 110 by 210 feet. His house occupies a rectangle of 30 by 45 feet and the driveway is 18 by 50 feet.

a. What is the area he must fertilize?

b. How many bags should Mr. Mizumoto buy?

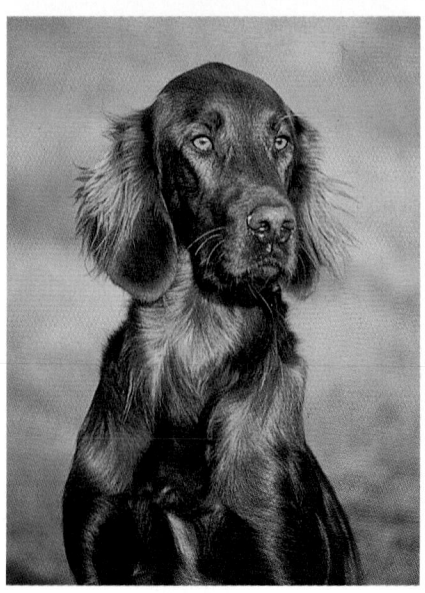

33. Kennels Liam raises Irish setters at his home in the country. He wants to enclose an area for a run in which his dogs can exercise.

   a. If the dimensions of the run are 60 by 15 feet, how much fencing would he need?

   b. What is the area enclosed?

34. Construction Jackie Rockford wishes to add a 15-by-40-foot rectangular deck to her house. However, she needs to allow for two square wells for the large trees in the deck area. The wells are 5-by-5 feet each. How much area must Jackie allow for decking materials?

35. Family Activity Suppose you are going to add a deck to the back of your home or have a garden in a community plot. Design two different plans and give the area and perimeter of each option.

Mixed Review

36. Travel Ana Perez travels 371 miles in 7 hours on a bus trip to a concert. What is the average rate of speed for the bus? (Lesson 3-4)

37. Solve $b + (-14) = 6$ and check your solution. Then, graph the solution on a number line. (Lesson 3-2)

For **Extra Practice**, see page 746.

38. Evaluate $-6ab$, if $a = -3$ and $b = -5$. (Lesson 2-7)

39. Cooking A casserole made with ground beef has 375 Calories per serving. The same recipe made with ground turkey has 210 Calories per serving. Write an inequality using the numbers and $<$ or $>$. (Lesson 2-3)

40. Solve $121 = 11x$ mentally. (Lesson 1-6)

41. Simplify $6d + 14(d + 2)$. (Lesson 1-5)

42. Standardized Test Practice Evaluate $8 + 9 \cdot (7 - 4)$. (Lesson 1-2)

   A. 68   **B.** 67   **C.** 51   **D.** 35

From the → FUNNY PAPERS

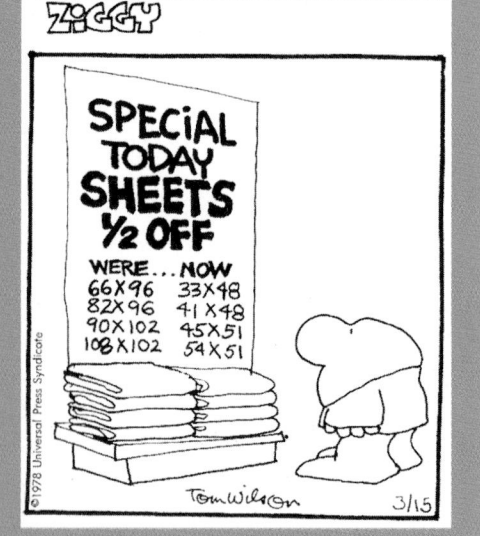

1. What does "half off" usually mean at a sale?
2. What is actually half off in this sale?
3. Draw a diagram of one of the sheets. Is the sheet half of the size that it was when it was not on sale? Explain.

3-5B Area and Coordinates

An Extension of Lesson **3-5**

You have learned about rectangles and squares, but many figures have irregular shapes. Coordinates can be used to find the area of these shapes. Aerial surveyors use this technique to find areas of plots of land. Use the graphing calculator program below to find the area of the figure at the right.

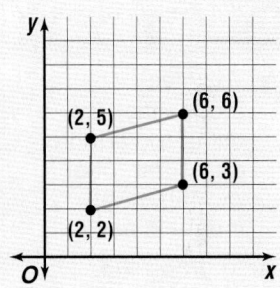

Activity

See page 1 for information on programming the graphing calculator.

Before you run the program, choose one vertex as the first and list the ordered pairs for the vertices in counterclockwise order. Repeat the first ordered pair at the end of the list.

(2, 2)
(6, 3)
(6, 6)
(2, 5)
(2, 2)

The program will ask for the number of vertices in the figure. For this figure, press 4 and ENTER.

When asked for the *x*- and *y*-coordinates, enter the lists in counterclockwise order. Use braces, { }, before and after the lists.

The program will display area in square units.

```
PROGRAM:AREA
: ClrList L1, L2
: 0 → D
: 0 → U
: 0 → X
: Input "VERTICES", N
: Disp "X COORDINATES"
: Input L1
: Disp "Y COORDINATES"
: Input L2
: For (X, 1, N, 1)
: L1(X)*L2(X+1)+D → D
: L2(X)*L1(X+1)+U → U
: End
: (D−U)/2 → A
: Disp "AREA ="
: Disp A
: Stop
```

Your Turn

Draw a coordinate plane and graph each figure whose vertices are listed. Then use the graphing calculator program to find the area of each figure.

1. (2, 3), (4, 1), (6, 8)
2. (−4, 2), (5, 2), (5, 5), (−4, 5)
3. (0, 4), (4, 0), (−4, 0), (0, −4)
4. (0, 0), (8, 0), (9, 5), (6, 8), (−1, 4)

Extension

5. Graph a rectangle or square on a coordinate grid and determine the coordinates of its vertices. Use the formula and the graphing calculator program to find its area. How do the methods compare?

Solving Inequalities by Adding or Subtracting

- To solve inequalities by using the addition and subtraction properties of inequalities, and

- to graph the solution set.

It's Important

Inequalities can be used to determine how many hours you must work to earn a certain amount of money.

MATERIALS: ⚖ balance 🔲 blocks ▭ paper cups

Hands-On Activity

Small objects and a balance can be used to represent mathematical inequalities.

The diagram at the right shows $9 > 5$ since 9 objects are heavier than 5. The balance scale models an *inequality* because the two sides are not equal.

Let's see if the addition and subtraction properties of equality apply to inequalities.

First, set up your balance like the diagram. Remove 4 blocks from the left side. Remove 4 blocks from the right side. The scale returns to the position it was originally. This means that subtracting the same amount from each side does not change the inequality.

Your Turn Work with a partner. Model the inequality $2 < 5$ and draw a picture of your model. Add 6 blocks to the side with 2. Draw a picture of what happens to the balance. Next add 6 blocks to the side with 5. Draw a picture of the result. Write down your observations.

Use a paper cup and blocks with a balance to model the inequality $x + 3 > 6$.

  a. How many blocks have to be in the paper cup to model this inequality?

  b. Is this the only answer?

  c. Remove the 3 blocks on the side with the variable and draw a picture of what happened.

  d. Next, remove 3 blocks from the other side.

  e. What does this tell you about the variable? Is there more than one answer to this inequality?

Learning the Concept

To solve an inequality that involves addition, you can use subtraction just as you did to solve equations. Likewise, you can use addition to solve an inequality that involves subtraction.

<table>
<tr><td rowspan="2">**Addition and Subtraction Properties of Inequalities**</td><td>**In words:**</td><td>Adding or subtracting the same number from each side of an inequality does not change the truth of the inequality.</td></tr>
<tr><td>**In symbols:**</td><td>For all numbers a, b, and c:
1. If $a > b$, then $a + c > b + c$ and $a - c > b - c$.
2. If $a < b$, then $a + c < b + c$ and $a - c < b - c$.</td></tr>
</table>

The rules are similar for $a \geq b$ and $a \leq b$.

Example **1** **Solve $m + 4 < 9$. Check the solution.**

$$m + 4 < 9$$
$$m + 4 - 4 < 9 - 4 \quad \textit{Subtract 4 from each side.}$$
$$m < 5$$

Check: Try -3, a number less than 5.

$$-3 + 4 \overset{?}{<} 9 \quad \textit{Replace m with -3.}$$
$$1 < 9 \checkmark \quad \textit{This statement is true, so it checks.}$$

Any number less than 5 will make the statement true. Therefore, the solution is $m < 5$, all numbers less than 5.

Like solutions to equations, solutions to inequalities can be graphed. The solutions to $m < 5$ are graphed on the number line below.

The arrow represents all numbers to the left of (less than) 5. *The point, 5, is not included. Thus, the circle is open.*

$$-4 \; -3 \; -2 \; -1 \quad 0 \quad 1 \quad 2 \quad 3 \quad 4 \quad 5 \quad 6 \quad 7$$

Example **2** **Solve $-13 \geq x - 8$ and check the solution. Graph the solution.**

$$-13 \geq x - 8$$
$$-13 + 8 \geq x - 8 + 8 \quad \textit{Add 8 to each side.}$$
$$-5 \geq x$$

Check: Try -5 and -7, numbers less than or equal to -5.

$$-13 \overset{?}{\geq} -5 - 8 \qquad\qquad\qquad -13 \overset{?}{\geq} -7 - 8$$
$$-13 \geq -13 \checkmark \quad \textit{Both are true, so they check.} \quad -13 \geq -15 \checkmark$$

The solution is $-5 \geq x$. This means that the inequality is true for all numbers less than or equal to -5. The solution may also be written as $x \leq -5$. The graph of the solution set is shown below.

The point, -5, is included. Thus, the circle is solid.

$$-8 \; -7 \; -6 \; -5 \; -4 \; -3 \; -2 \; -1 \quad 0 \quad 1$$

Example 3

Real World APPLICATION

Hobbies

Christopher has $30 to buy new baseball cards for his collection. He has chosen a Greg Maddux rookie card that costs $8. What is the most he can spend on other baseball cards?

Explore   We need to find out how much money he can spend on other new cards.

Plan   Translate the words into an equation. Let s represent the amount he can spend on the other cards.

Cost of Maddux card	*plus*	*cost of other cards*	*must be less than or equal to*	*$30*
8	+	s	$\leq$	30

Solve
$$8 + s \leq 30$$
$$8 - 8 + s \leq 30 - 8 \quad \textit{Subtract 8 from each side.}$$
$$s \leq 22$$

Christopher can spend no more than $22 on other baseball cards.

Examine   Suppose Christopher spent $20, a number less than $22, on other baseball cards. Then he would have spent $20 + $8 or $28 in all. $28 \leq $30, so the answer checks.

Checking Your Understanding

Communicating Mathematics

Read and study the lesson to answer these questions.

1. **Describe** how you would explain to a friend how to solve the inequality $m - 6 < 3$.

2. **Restate** in your own words the addition property of inequality.

3. **Compare and contrast** the solutions of $n + 5 = -3$ and $n + 5 > -3$.

4. **You Decide** Kelli translated the sentence *"Rachel scored no less than 8 points."* as $p < 8$. Derrick translated it as $p \geq 8$. Who is correct and why?

MATH JOURNAL

5. **Assess Yourself** Measure your height and a friend's height. Write an inequality that compares your heights. Suppose that each of you were to grow 3 more inches. Now write an inequality that compares your new heights.

Guided Practice

Write an inequality for each solution set graphed below.

6.

      -3 -2 -1 0 1 2 3 4 5

7.   -6 -5 -4 -3 -2 -1 0 1

Write an inequality for each sentence. Then draw its graph.

8. Roberto spent more than $7 for lunch.

9. Liz drove less than 75 miles today.

Solve each inequality and check your solution. Then graph the solution on a number line.

10. $z + (-5) > -3$

11. $m + 4 < -9$

12. $9 < x - 7$

13. $-6 > y - 2$

14. $k + (-1) \geq -6$

15. $18 + a \leq -13$

16. Money Miguel spent more than \$6 at a fast food restaurant. If he had \$25 to start, what was his remaining pocket money?

Exercises: Practicing and Applying the Concept

Independent Practice

Write an inequality for each solution set graphed below.

17.

18.

19.

20.

Write an inequality for each sentence. Then draw its graph.

21. With the windchill factor, the temperature is equal to or less than 14 degrees.

22. Each of the NFL team rosters must have no more than 45 members at the beginning of the season.

23. Maria Vasquez is less than 5 ft. 4 in. (64 in.) tall.

24. The U.S. Constitution is more than 200 years old.

Solve each inequality and check your solution. Then graph the solution on a number line.

25. $m - 6 < 13$

26. $r \leq -12 - 8$

27. $y + 7 > 13$

28. $k + 9 \geq 21$

29. $-4 + x > 23$

30. $14 \geq a + -2$

31. $-7 + x < -3$

32. $-13 + y \geq 8$

33. $m + (-2) > -11$

34. $f + (-5) \geq 14$

35. $19 \geq z + (-9)$

36. $-41 < m - 12$

37. $73 + k < 47$

38. $33 < m - (-6)$

39. $-31 \geq x + (-5)$

40. $22 < n - (-16)$

41. $-30 \leq c + (-5)$

42. $56 > w + 72$

Critical Thinking

43. Graph the solutions for the compound inequality $y < -1$ or $y > 3$. (*Hint:* An **or** in a sentence means that either part is true.)

44. Graph the solutions for the compound inequality $x > -5$ and $x \leq 4$. (*Hint:* An **and** in a sentence means that both parts are true.)

Applications and Problem Solving

45. Shopping Martin is saving money to buy a new mountain bike. Bikes that he likes start at \$375. If he already has saved \$285, what is the least amount he must still save?

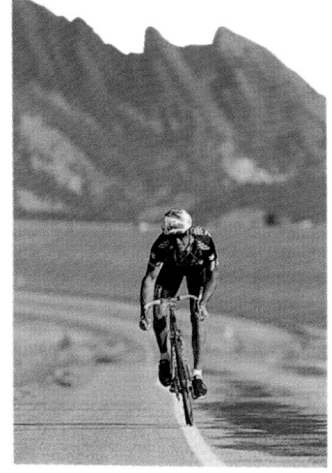

46. Grades On three extra credit problems on a math test, students are given up to 4 points each, provided that their scores do not exceed 100 (points). What is the minimum raw score students could have and still score 100 with the extra credit?

47. Taxes Residents of New York and Connecticut pay the highest taxes in the country. They must work 3 hours 9 minutes (189 minutes) of each 8-hour day to earn enough to pay their federal, state, and local taxes. The inequality $x + 52 \leq 189$ compares the time in these states with x, where x is the minimum time required in Alaska. What is the time needed in Alaska to pay off taxes?

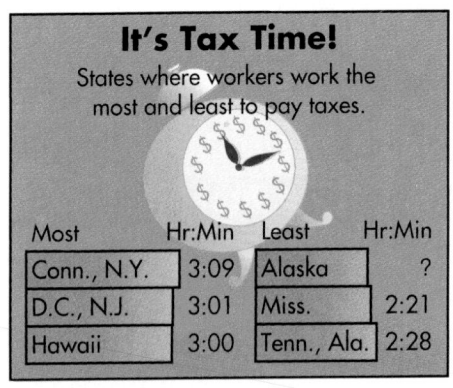

It's Tax Time!
States where workers work the most and least to pay taxes.

Most	Hr:Min	Least	Hr:Min
Conn., N.Y.	3:09	Alaska	?
D.C., N.J.	3:01	Miss.	2:21
Hawaii	3:00	Tenn., Ala.	2:28

Source: Tax Foundation

Mixed Review

48. Landscaping Cherita has a rectangular garden that is 96 square feet. If the garden is 8 feet long, what is its width? (Lesson 3-5)

49. Solve $f = t - h$, if $t = 125$ and $h = 25$ by replacing the variables with the given values. (Lesson 3-4)

50. Standardized Test Practice Jennifer made a $150 down payment for soccer camp. Her unpaid balance was $300. What is the total fee for the soccer camp? (Lesson 3-2)

   A. $150

   B. $300

   C. $450

   D. $600

51. Simplify $10a - 12a$. (Lesson 2-5)

52. Find the sum of -12 and 7. (Lesson 2-4)

53. Communication The table below shows the numbers of millions of different types of greeting cards sent in the United States each year. (Lesson 1-10)

   a. How many more cards are sent for Mother's Day than are sent for Father's Day?

   b. Make a bar graph of the data.

54. Solve $\dfrac{36}{b} = -4$. (Lesson 1-6)

For **Extra Practice**, see page 747.

55. Name the property shown by $8 \cdot 1 = 8$. (Lesson 1-4)

Occasion	Estimated number sent (millions)
Christmas	2700
Valentine's Day	1000
Easter	160
Mother's Day	150
Father's Day	101
Thanksgiving	40
Halloween	35
St. Patrick's Day	19
Jewish New Year	12
Hanukkah	11

3-7

Solving Inequalities by Multiplying or Dividing

You'll Learn

- To solve inequalities by using the multiplication and division properties of inequalities.

It's Important

Knowing how to use multiplication and division properties of inequalities can help you with recycling issues.

Application: Recycling

Environmental protection is an ever-growing concern for the world. When we pollute the environment, we are polluting *our* living space. Recycling materials is a good way to keep our habitat cleaner.

Recycling Resources

Source: Bureau of Mines

By weight, the reuse of steel and paper lead the way. The recycling of steel, by weight, is greater than that of paper.

Learning the Concept

The comparison of paper to steel recycling provides a way to look at inequalities. Rounding the weights, 35 million tons of paper being recycled is less than 64 million tons of steel being recycled. In symbols, $35 < 64$.

It is estimated that people use four times as much paper and steel as they recycle. Does the inequality between paper and steel still keep its order for the amount used?

$$35 \times 4 \overset{?}{<} 64 \times 4$$
$$140 < 256 \checkmark \qquad \textit{The inequality is still true.}$$

Another way people can help the environment is to use as little of these products as possible. What if people cut the amount of paper and steel that they used in half? Would this affect the inequality?

$$140 < 256 \qquad \textit{The original amount of tons of paper and}$$
$$\qquad\qquad\qquad \textit{steel used.}$$
$$\frac{140}{2} \overset{?}{<} \frac{256}{2} \qquad \textit{Each amount is divided by 2.}$$
$$70 < 128 \checkmark \qquad \textit{The inequality is still true.}$$

These examples suggest the properties on the next page.

Lesson 3-7 *Solving Inequalities by Multiplying or Dividing* **151**

| **Multiplication and Division Properties of Inequalities** | **In words:** | When you multiply or divide each side of a true inequality by a *positive* integer, the result remains true. |
| | **In symbols:** | For all integers a, b, and c, where $c > 0$, if $a > b$, then $a \cdot c > b \cdot c$ and $\frac{a}{c} > \frac{b}{c}$. |

The rule is similar for $a < b$, $a \geq b$, and $a \leq b$.

Next, we need to consider if an inequality remains true if we multiply or divide each side by a negative number. Consider the inequality $6 < 10$.

Multiply each side by -1.

$$6 < 10$$
$$6\,(-1) \overset{?}{<} 10\,(-1)$$
$$-6 < -10 \quad \textbf{False}$$

Divide each side by -2.

$$6 < 10$$
$$\frac{6}{-2} < \frac{10}{-2}$$
$$-3 < -5 \quad \textbf{False}$$

The inequalities $-6 < -10$ and $-3 < -5$ are both false. However, notice that both inequalities would be true if you reverse the order symbol. That is, change $<$ to $>$.

$$-6 > -10 \quad \text{and} \quad -3 > -5 \quad \text{are both true.}$$

These examples suggest the following properties.

| **Multiplication and Division Properties of Inequalities** | **In words:** | When you multiply or divide each side of an inequality by a *negative* integer, you must *reverse the order symbol*. |
| | **In symbols:** | For all integers a, b, and c, where $c < 0$, if $a > b$, then $a \cdot c < b \cdot c$ and $\frac{a}{c} < \frac{b}{c}$. |

The rule is similar for $a < b$, $a \geq b$, and $a \leq b$.

Example ① Solve each inequality and check the solution. Then graph the solution on a number line.

a. $4x < -28$

$$4x < -28$$
$$\frac{4x}{4} < \frac{-28}{4} \qquad \textit{Divide each side by 4.}$$
$$x < -7$$

Check: Try -10, a number less than -7.
$$4(-10) \overset{?}{<} -28$$
$$-40 < -28 \quad \textit{✔ This is true.}$$

The solution is $x < -7$, all numbers less than -7. The graph is shown below.

$$\xleftarrow{\quad} \overset{}{\underset{-9 \;\; -8 \;\; -7 \;\; -6 \;\; -5 \;\; -4 \;\; -3 \;\; -2 \;\; -1 \;\; 0}{\rule{6cm}{0.4pt}}}$$

b. $\frac{m}{-3} \leq 4$

$$\frac{m}{-3} \leq 4$$

$$\frac{m}{-3} \times (-3) \geq 4 \times (-3)$$ *Multiply each side by -3 and reverse the order symbol.*

$$m \geq -12$$

Check: Try 0, a number greater than -12.

$$\frac{m}{-3} \leq 4$$

$$\frac{0}{-3} \overset{?}{\leq} 4$$ *Replace m with 0.*

$$0 \leq 4 \; ✔$$

The solution is all numbers greater than or equal to -12. The graph is shown below.

$$\xleftarrow{\hspace{1cm}}\overset{\textstyle -15\,-14\,-13\,-12\,-11\,-10\,-9\,-8}{\rule{0pt}{0pt}}\xrightarrow{\hspace{1cm}}$$

Inequalities are often used to solve problems involving money.

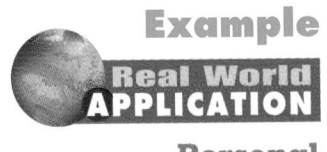

Example ② **Real World APPLICATION**

Personal Finance

Carlissa earns \$5.00 per hour busing tables at The Italian Eatery Restaurant. How many hours must she work to earn at least \$110 each week?

Let x represent the hours required to earn at least \$110.

$$5x \geq 110$$

$$\frac{5x}{5} \geq \frac{110}{5}$$ *Divide each side by 5.*

$$x \geq 22$$

Carlissa must work at least 22 hours to earn at least \$110 a week.

Checking Your Understanding

Communicating Mathematics

Read and study the lesson to answer these questions.

1. **Write** an inequality that is always true.

2. **Write** an inequality that would be solved using division where the solution is $x < 4$.

3. **Compare and contrast** solving an inequality involving multiplication and solving an inequality involving division.

MATH JOURNAL

4. **Write** a note to a classmate about how to solve the inequalities $5x < 35$ and $\frac{y}{3} \geq 12$.

State the number needed to multiply or divide each side by to solve each inequality. Then tell whether the order symbol should be reversed.

5. $-15 \leq 5a$ **6.** $\frac{y}{4} > -20$ **7.** $-3m < -33$

Solve each inequality and check your solution. Then graph the solution on a number line.

8. $-3y \geq -18$ **9.** $-8 \leq -4x$ **10.** $\frac{z}{-2} > -2$

11. $\frac{p}{-3} > -35$ **12.** $13y \geq -26$ **13.** $6 > \frac{a}{-7}$

14. Demographics According to *Condé Nast Traveler,* at least two-fifths of single Americans consider a sport-oriented vacation to be their ideal honeymoon. If at least 29 million unmarried Americans would choose a sports vacation as a honeymoon, how many unmarried Americans are there?

Exercises: Practicing and Applying the Concept

Solve each inequality and check your solution. Then graph the solution on a number line.

15. $-4y < -36$ **16.** $-3m \geq -24$ **17.** $-72 > -4x$

18. $-6s \geq -78$ **19.** $-70 \leq 5p$ **20.** $9x \geq -72$

21. $\frac{x}{3} < -7$ **22.** $-8 > \frac{y}{4}$ **23.** $\frac{m}{-2} \leq -7$

24. $\frac{p}{-4} < 20$ **25.** $-7 \leq \frac{z}{14}$ **26.** $-13a \geq -273$

27. $-378 < 18r$ **28.** $\frac{a}{14} \leq -17$ **29.** $-14 \leq \frac{s}{-23}$

30. $\frac{g}{-21} > 33$ **31.** $-4 > \frac{x}{3}$ **32.** $-25a \leq 400$

33. The product of an integer and -4 is greater than -24. Find the greatest integer that meets this condition.

34. Statistics The Boston Marathon had more than 2,600,000 spectators along its 26-mile route. What was the average number of spectators per mile?

35. Paper Recycling A giant roll of paper towels made from recycled paper can be cut into 80,000 small rolls for retail sale. If the smallest carton for shipment to a retail store holds 100 small rolls, what is the maximum number of cartons required to package and ship the rolls of towels made from one giant roll of towels?

36. Pets For those who like a well-dressed pet, NuBreed Pet Hats in Batavia, Illinois, makes baseball, cowboy, beach, and other hats for dogs and cats. The hats cost $10–20 in retail stores.

   a. Write an inequality for the lowest possible retail cost of 15 pet hats.

   b. Write an inequality for the greatest possible retail cost of 15 pet hats.

37. Lifestyles According to a study by *Men's Fitness* magazine, the average American spends six months of his or her life waiting at stop lights. More than twelve times that amount of time is spent eating. How long does the average American spend eating in a lifetime?

38. Health Americans bought almost 59,000 *miles* of dental floss in one year.

   a. There are 5280 feet in a mile. How many feet of dental floss were purchased in that year?

   b. If there were about 248,710,000 Americans in that year, how much dental floss did the average American buy?

   c. Dentists recommend that people floss their teeth once each day. Based on your answer to part b, do you think most Americans follow this recommendation? Explain.

Mixed Review

39. Solve $m + (-5) < 23$. (Lesson 3-6)

40. Pets Find the perimeter and area of a square dog kennel that is 6.5 feet on each side. (Lesson 3-5)

41. Standardized Test Practice If $x + 5 = \frac{1}{3}(3x - 5)$, what is the value of x? (Lesson 3-1)

   A. 5

   B. any real number

   C. 3

   D. no real number

42. Find the quotient of -126 and 3. (Lesson 2-8)

43. Find the product $8 \cdot (-7y)$. (Lesson 2-7)

44. Find the value of $-|-4|$. (Lesson 2-1)

45. Grades When finding a grade point average, an A is worth 4 points, a B is worth 3 points, a C is worth 2 points, a D is 1 point, and an F is 0. After totaling the points, the sum is divided by the number of grades. (Lesson 1-2)

   a. Write an expression for the grade point average of a student who received 3 As, 2 Bs, 1 C, and 1 D.

   b. Find the grade point average.

For **Extra Practice**, see page 747.

Lesson 3-7 *Solving Inequalities by Multiplying or Dividing* **155**

3-8 Applying Equations and Inequalities

You'll Learn

- To solve verbal problems by translating them into equations and inequalities, and

- to write verbal problems for equations and inequalities.

It's Important

Translating verbal problems into equations and inequalities can help you to determine sports averages.

Application: World Records

What takes 400 pounds of butter, 500 volunteers, and 3 blocks of tables? The World's Biggest Breakfast! Springfield, Massachusetts, and Battle Creek, Michigan, compete every year to serve the world's largest pancake breakfast. 66,344 people were served 132,688 pancakes in two hours in Springfield. Each person was served the same number of pancakes. How many pancakes were served to each person?

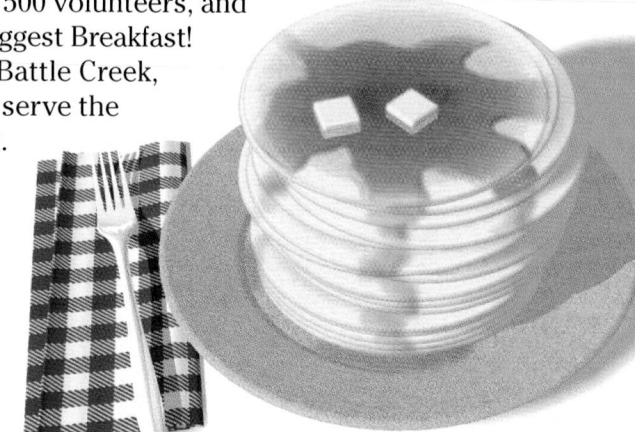

Learning the Concept

We can use the four-step plan to solve this problem.

Explore We know how many pancakes were served to how many people. We want to find out how many pancakes each person had.

Plan Translate the words into an equation. Let p represent the number of pancakes served per person.

number of people served	times	number of pancakes per person	equal	total number of pancakes
66,344	$\times$	p	$=$	132,688

Solve $66,344p = 132,688$

$\dfrac{66,344p}{66,344} = \dfrac{132,688}{66,344}$ *Estimate: $\frac{130,000}{65,000} = 2$.*

$p = 2$ *Use a calculator.*

Examine If each of the 66,344 people was served 2 pancakes, $66,344 \times 2$ or 132,688 pancakes were served. The answer checks.

Each person was served 2 pancakes.

Example 1

Real World APPLICATION

Sports

Special Olympics athlete, Andy Leonard, is one of the few American athletes who can lift up to four times his own body weight. The most weight he has lifted is 385 pounds. What is his body weight, rounded to the nearest whole number?

Andy Leonard

Explore   Mr. Leonard's maximum lift is 385 pounds, which is up to four times his weight.

Plan   First, estimate the solution by rounding.

$$\frac{400}{4} = 100 \text{ pounds}$$

Translate the words into an inequality. Let w represent Mr. Leonard's body weight.

lifting weight	is less than or equal to	4 times body weight
385	$\leq$	$4w$

Solve $385 \leq 4w$

$\dfrac{385}{4} \leq \dfrac{4w}{4}$ *Divide each side by 4.*

$96.25 \leq w$

Examine   Compare to the estimate. The answer appears to be reasonable. Therefore, Mr. Leonard's weight is about 96 pounds or less.

Many types of situations can be represented using an equation.

Example 2

Real World APPLICATION

Entertainment

Write a realistic problem that could be solved using the equation $25x = 325$.

There is an endless number of possible situations that this equation could describe. One situation is as follows.

The pep band went on a trip to an amusement park. Admission was $25 per person. If the total price of admission was $325, how many people went to the park?

Checking Your Understanding

Communicating Mathematics

Read and study the lesson to answer these questions.

1. **Explain** the advantages in using the four step problem-solving plan on word problems using equations or inequalities.

2. **List** as many word phrases as you can think of for which you would use the symbol $\leq$.

3. **You Decide**   Over 2100 pounds of pancake mix were used to make the largest breakfast in 1994. Rosa solved the inequality $5b > 2100$ to find the number of 5-pound bags of mix that were used. Kiele used the inequality $5b < 2100$. Who is correct and why?

Define a variable and translate each sentence into an equation or inequality. Then solve.

4. An integer increased by 12 is 28.

5. Four times some number is less than 96.

6. **Cycling** Dwain took his bike to the Leshnock Bike Shop to have the gear-shift mechanism repaired. At the shop, Dwain was told the total bill for parts and labor would be at most $63. The cost of the parts was $39. How much could Dwain expect to pay for labor?

   a. What is being asked?

   b. What does the phrase *at most* mean?

   c. Write an equation or inequality that describes the problem.

   d. Solve the problem.

7. Write a realistic problem that could be solved using the equation $169 + x = 187$.

Exercises: Practicing and Applying the Concept

Define a variable and translate each sentence into an equation or inequality. Then solve.

8. The bill for labor at $13 per hour was at least $52.

9. The quotient when dividing by -3 is greater than 5.

10. The 10-km race time of 88 minutes was twice as long as the winner's time.

11. A savings account increased by $50 is now more than $400.

12. Seven times a number is less than 84.

13. The house sold for $85,000. That is at least four times the original purchase price.

Write a realistic problem that could be solved using each equation or inequality.

14. $9 + 3x = 24$ 15. $b + 9 \leq 16$ 16. $30m > 300$

17. Write an inequality to represent the numbers whose absolute values are less than 4. Find all the integer solutions to the inequality.

Define a variable and write an equation or inequality. Then solve.

18. Brisa delivers pizza for Mama's Pizzeria. Her average tip is $1.50 for each pizza that she delivers. How many pizzas must she deliver to earn at least $20 in tips?

    a. What is being asked?

    b. What does the phrase *at least $20* mean?

    c. Write an equation or inequality that describes the problem. Solve.

    d. Can your answer be anything other than a whole number? Explain.

19. **Sports** The University of Connecticut Huskies women's basketball team was the undefeated NCAA Champion for the 1994–95 season. The team played 35 games, counting both regular and post-season play. Rebecca Lobo, one of Connecticut's star players, scored almost 600 points for the season. What was her average per game?

20. **Transportation** A minivan is rated for maximum carrying capacity of 900 pounds.

   a. If the luggage weighs 100 pounds, what is the maximum weight allowable for passengers?

   b. What is the maximum average weight allowable for 5 passengers?

21. **Postage** In 1971, first-class postage stamps cost 8 cents for the first ounce. In 1995, stamps for the same first-class mail cost 32 cents for the first ounce. How many times more did stamps cost in 1995 than in 1971?

22. **Air Travel** In 1903, the Wright brothers' first flight was at 30 mph. A Boeing 747 cruises at 600 mph. How many times faster is the flight speed of a 747?

23. **Geometry** The volume of a rectangular solid is found by multiplying its length, width, and depth. If the volume of the solid shown at the right is 440 cubic centimeters, what is its height?

height = ?
length = 8 cm
width = 5 cm

24. **Agriculture** Between 1987 and 1992, the number of female farmers increased 14,386 to 145,156. What was the number of female farmers in 1987?

25. **Sports** Danika is on her high school golf team. Her personal goal is to shoot 80. Her practice scores during the week were within 6 strokes of her desired score. What could be some of her scores?

26. **Tournaments** In the NBA playoffs, a team advances from the first round if they win the "best of five." Best of five means that they win more than half of the five games played between the teams. The semifinal, conference final, and final rounds are all best of seven.

   a. What is the minimum number of games an NBA Champion would have to win?

   b. How many games could an NBA Champion lose? Express your answer as an inequality.

NBA playoffs

Eastern Conference   Western Conference

First Round   Conference Conference Conference Conference First Round
Best of Five   semifinals   final   final   semifinals Best of Five
               Best of seven Best of seven Best of seven Best of seven

Orlando   San Antonio
Boston   Denver
Charlotte   Seattle
Chicago   LA Lakers

Indiana   NBA   Phoenix
Atlanta   FINALS   Portland
New York   Best of seven Utah
Cleveland   Houston

Mixed Review

27. Solve $-3h < -39$ and check your solution. Then graph the solution on a number line. (Lesson 3-7)

28. **Standardized Test Practice** What is the length of a rectangle whose area is 16 square feet and whose width is 2 square feet? (Lesson 3-5)

   A. 8 ft

   B. 14 ft

   C. 32 ft

   D. 36 ft

29. Simplify $-12y + 5y$. (Lesson 2-4)

30. Name the absolute value of -9. (Lesson 2-1)

For **Extra Practice**, see page 747.

31. Solve $\frac{t}{3} = 8$ mentally. (Lesson 1-6)

BON VOYAGE

Refer to the Investigation on pages 62–63.

You are ready to finalize arrangements for your trip. You and your partner will be called upon to present your plan for the trip at the parents' meeting next week. Complete your research and make your presentation to the parents.

- Find the most current foreign exchange rate for the currency used in the country you chose at the beginning of the investigation. In other words, how much is *one* franc (or one krone, one peso, one rupee or one yen) worth in American dollars? These can be found in *The Wall Street Journal* and in the business section of other major newspapers.
- Explain how to find the dollar value of the foreign currency. Provide both an exact formula (that could be done on a calculator) and a simple way to estimate the value. You should be able to determine, for example, what a pair of jeans would cost in the United States if they cost 45,280 lire in Rome, Italy.

- Furnish the time zone data you found when working on the Investigation in Lesson 2-1. Work with your family to determine a convenient time to make phone calls.
- Supply environmental data for the country. Then provide the data you found when working on the Investigation in Lesson 3-2.
- Provide a short refresher course on metric units of length, weight, and liquid volume.
- Review any important business and social customs. If you can also explain the rationale behind them, it may be easier for you to remember. For example, in some Middle Eastern countries, it is considered unclean to eat or drink with your left hand!

You may wish to organize all of the information on the country you studied in a travel scrapbook to be displayed in your classroom.

PORTFOLIO ASSESSMENT

You may want to keep your work on this Investigation in your portfolio.

Vocabulary

After completing this chapter, you should be able to define each term, property, or phrase and give an example or two of each.

interNET
C O N N E C T I O N
Chapter Review For additional lesson-by-lesson review, visit:
www.glencoe.com/ sec/math/prealg/ mathnet

Algebra
addition property of equality (p. 125)
addition property of inequality (p. 147)
division properties of inequality (p. 152)
division property of equality (p. 130)
equivalent equations (p. 124)
formula (p. 134)
multiplication properties of inequality (p. 152)
multiplication property of equality (p. 130)
subtraction property of equality (p. 124)
subtraction property of inequality (p. 147)

Geometry
area (p. 138)
area of a rectangle (p. 140)
coordinate (p. 126)
graph (p. 126)
perimeter (p. 138)
perimeter of a rectangle (p. 139)

Problem Solving
eliminating possibilities (p. 118)
matrix logic (p. 119)

Understanding and Using Vocabulary

Determine whether each statement is *true* or *false*.

1. When two numbers that are additive inverses of each other are added, the sum is zero.

2. The formula *rate · time = distance* can be used to find the rate when only the distance is known.

3. The coordinates of the points on a number line are greater as you move from right to left.

4. The area of a geometric figure is the measure of the surface it encloses.

5. A solid dot on a number line indicates that the point is included in the solution set.

6. The perimeter of a geometric figure is the distance around it.

7. The division property of inequality says that if you divide both sides of an inequality by the same number, the order of the inequality symbol must be reversed.

Skills and Concepts

Objectives and Examples	Review Exercises

Upon completing this chapter, you should be able to:

Use these exercises to review and prepare for the chapter test.

▶ **solve equations using the addition and subtraction properties of equality.**
(Lesson 3-2)

Solve $g - (-19) = 23$.

$$g - (-19) = 23$$ *Rewrite as addition*
$$g + 19 = 23$$ *sentence.*
$$g + 19 - 19 = 23 - 19$$ *Subtract 19 from*
$$g = 4$$ *each side.*

Solve each equation and check your solution. Then graph the solution on a number line.

8. $f + 31 = -5$ **9.** $r + (-11) = 40$

10. $12 = k - 7$ **11.** $-9 = r + (-15)$

12. $16 = h - (-4)$ **13.** $7 = q - 12$

14. $m - 3 = -10$ **15.** $z - (-51) = 36$

▶ **solve equations using the multiplication and division properties of equality.** (Lesson 3-3)

Solve $\dfrac{r}{-7} = 15$.

$$\frac{r}{-7} = 15$$

$$(-7) \cdot \frac{r}{-7} = 15 \cdot (-7)$$ *Multiply each side*
$$r = -105$$ *by (−7).*

Solve each equation and check your solution. Then graph the solution on a number line.

16. $\dfrac{c}{6} = -29$ **17.** $\dfrac{p}{-3} = -42$

18. $840 = -28r$ **19.** $13b = -299$

20. $14x = 224$ **21.** $-\dfrac{y}{10} = -33$

22. $\dfrac{g}{-9} = 21$ **23.** $25d = -1775$

▶ **solve problems using formulas.** (Lesson 3-4)

The Herrs flew 360 miles to Tallahassee in 2 hours. What was the speed of the plane? Use $d = rt$.

$$d = rt$$
$$360 = r(2)$$ *Replace d with 360 and t with 2.*
$$\frac{360}{2} = \frac{r(2)}{2}$$ *Divide each side by 2.*
$$180 = r$$

Solve by replacing the variables with the given values.

24. $A = \ell \cdot w, A = 105, \ell = 7$

25. $I = p \cdot r \cdot t, p = 200, r = 0.06, t = 2$

26. $A = \dfrac{b \cdot h}{2}, A = 72, b = 6$

27. $C = 2\pi r, \pi = 3.14, r = 9$

Objectives and Examples	Review Exercises

▶ **find the perimeter and area of rectangles and squares.** (Lesson 3-5)

Find the area and perimeter of a rectangle with length 11 cm and width 7 cm.

7 cm

11 cm

$P = 2(\ell + w)$ $A = \ell \cdot w$
$= 2(11 + 7)$ $= 11 \cdot 7$
$= 2(18)$ or 36 cm $= 77$ sq cm

Find the perimeter and area of each rectangle.

28. 4 cm 6 cm

29. 6 in. 12 in.

30. a rectangle that is 4 meters wide and 12 meters long

31. a 15 foot-by-5 foot rectangle

32. a square with side 22 yards

▶ **solve inequalities by using the addition and subtraction properties and graph the solution set of the inequalities.** (Lesson 3-6)

Solve $-6 < n - 6$. Then graph the solution.

$-6 < n - 6$
$-6 + 6 < n - 6 + 6$ *Add 6 to each side.*
$0 < n$

-2 -1 0 1 2 3 4 5 6

Solve each inequality and check your solution. Then graph the solution on a number line.

33. $x - 2 < -14$ 34. $12 + a > -9$

35. $-7 + b < -5$ 36. $4 < y - 23$

37. $f + (-8) > -12$ 38. $w + 9 \le -5$

▶ **solve inequalities by using the multiplication and division properties.** (Lesson 3-7)

Solve $-7s > 70$. Then graph the solution.

$-7s > 70$
$\dfrac{-7s}{-7} < \dfrac{70}{-7}$ *Divide each side by -7 and*
$s < -10$ *reverse the order symbol.*

-14 -12 -10 -8 -6 -4 -2 0

Solve each inequality and check your solution. Then graph the solution on a number line.

39. $\dfrac{m}{-15} < 9$ 40. $-72 \ge -4k$

41. $\dfrac{t}{12} < 11$ 42. $144 < 9k$

43. $88 > -8p$ 44. $\dfrac{s}{-22} < -7$

45. $285 \le 19f$ 46. $18 \le \dfrac{r}{6}$

Objectives and Examples

▶ solve verbal problems by translating them into equations and inequalities. (Lesson 3-8)

After spending $17 to get into the amusement park, Marcus had $21 left. How much money did he have before he bought his ticket?

Let x be the amount of money Marcus had at the beginning.

money at start	minus	ticket price	equals	money left
x	$-$	17	$=$	21

$$x - 17 = 21$$
$$x - 17 + 17 = 21 + 17$$
$$x = 38$$

Marcus had $38 before buying his ticket.

Review Exercises

Define a variable and write an equation or inequality. Then solve.

47. Three friends went to dinner to celebrate one of them getting a job promotion. The total check was less than $36. If the cost was shared evenly, how much did each person pay for dinner?

48. Four times a number is greater than 76. What is the number?

49. A Girl Scout troop must sell more than 800 boxes of cookies to earn a trip to camp. If troop #332 has sold 576 boxes, how many boxes must they still sell to win a trip?

Applications and Problem Solving

50. Four runners represented South, Kennedy, Lincoln, and Anthony High Schools in an 880-meter race. The runner from Lincoln was faster than the runner from Kennedy. The runner from South beat the runner from Anthony, but lost to the runner from Kennedy. Which school did the runner who came in last represent? (Lesson 3-1)

    a. Lincoln   b. Kennedy
    c. South   d. Anthony

51. **Statistics** For many years, immigrants to the United States arrived at Ellis Island in New York. The flow of immigrants peaked in 1907 with more than 1 million. During that year, 37,807 people arrived from Germany. Find the average number of Germans who arrived daily at Ellis Island in 1907. (Lesson 3-3)

52. **Physics** A 110-car train goes from a velocity of 1.2 miles per hour to 6.8 miles per hour in 4 minutes. Use the formula $a = \dfrac{f - s}{t}$, where a is acceleration, f is final velocity, s is starting velocity, and t is time it took to make the change in velocity to find the acceleration of the train. (Lesson 3-4)

A practice test for Chapter 3 is available on page 776.

Demonstrate your knowledge by giving a clear, concise solution to each problem. Be sure to include all relevant drawings and justify your answers. You may show your solutions in more than one way or investigate beyond the requirements of the problems.

	Oatmeal Creme Pies	Peanut Butter Cookies	Chocolate-Chip Cookies
Calories per serving	170		90
Grams of Fat	8 g		
Sodium	190 mg	130 mg	80 mg

Define a variable, write an equation or inequality, and then solve. Then copy and complete the table.

1. Chocolate-chip cookies have 100 Calories less than peanut butter cookies.

2. Oatmeal Creme Pies have half as much fat as peanut butter cookies.

3. Chocolate-chip cookies have 20 times more sodium than grams of fat.

4. Write a realistic problem involving the numbers given in the chart.

5. Examine the nutrition label below. Could you eat six servings of this product and still consume less than 5 grams of sugar? Explain.

Nutrition Facts	Amount/serving	%DV	Amount/serving	%DV
Serv. Size 43 g	Total fat 12 g	19%	Total Carb. 23 g	8%
Serv. per package 1	Sat. fat 2.5 g	13%	Fiber 1 g	5%
Calories 220	Cholest. 5 mg	1%	Sugars < 1 g	
Fat Cal. 110	Sodium 340 mg	14%	Protein 6 g	
* Percent Daily Values (DV) are based on a 2000 calorie diet.	Vitamin A 0% • Vitamin C 0% • Calcium 6% • Iron 10%			

Thinking Critically

▶ A rectangle is 7 inches by w inches. Suppose you increase the width by 1 inch.

    a. How do the perimeter and area change?

    b. If the length was 10 inches instead of 7 inches, would the change in the perimeter be different? Explain.

    c. If the length was 10 inches instead of 7 inches, would the change in the area be different? Explain.

 Portfolio

▶ Place your favorite word problem from this chapter in your portfolio and attach a note explaining why it is your favorite.

Investigation

TAKING STOCK

Good news: You have just inherited $10,000!

Bad news: You can't claim it until your 21st birthday.

More good news: By investing wisely, you'll have even more money in a few years!

There are several kinds of investments. You are probably most familiar with savings accounts, stocks, and bonds. Before making any kind of investment, you should learn as much as possible about investing in general.

MATERIALS

 newspaper

> **Some important investment terms are:**
> - *Capital Gain* – profit made by buying something at one price and selling it at a higher price
> - *Dividend* – a payment made by a corporation to its stockholders
> - *Market Price* – the price per share that a stock may be purchased for or sold for at a particular time
> - *Security* – a document specifying the right to property in the form of stocks or bonds
> - *Share* – unit used for owning stock
> - *Stock Exchange* – place where the buying and selling of stocks takes place

Upon the advice of a relative who is a financial counselor, you decide to invest your inheritance in the stock market.

When you purchase stock, you become part owner of a corporation. The stockholders of a company share in the profits and losses of the firm. If a company's profits are high, stockholders may receive a dividend.

How to Read a Newspaper Stock Report

Numbers in this column show the current annual dividend paid per share.

These columns show, in dollars, the highest and lowest prices paid in the last 52 weeks, excluding yesterday's trading.

Bold face type denotes something special about the stock. In this case, it shows stock that rose 3% or more.

This column lists the name of the stock or an abbreviation for it. You can find the full name for an abbreviation at the library or from a broker.

| 52-Week | | Stock | Div | Sales | | |
High	Low			100s	Cls	Chg
19 1/4	5 1/8	Corprp		177	5 5/8	
26 3/4	**12 3/8**	**CntwCrd**	**.32**	7171	21	+7/8
19 5/8	16	CoutsPr	1.08f	102	19 1/8	+1/8
21 1/2	14	CosCm		3266	18 3/8	+1/8
39 1/2	25 3/4	Crane	.75	480	37 1/8	+1/8
17 3/4	14 3/8	CrwfdA	.58	27	16 1/8	+1/8
17 3/4	14 3/8	CrwfdB	.54	47	16 1/4	-3/8
29 1/4	14 5/8	CrayRs		947	23 7/8	+3/8
18 1/2	14 3/4	Credicp n		764	17 1/4	+1/4
35 7/8	25	CrescRE	2.20f	552	33 7/8	
61	36 3/4	Crestar	1.80	661	59 1/8	+1/8
9 1/4	6 1/2	CrimMa	.94f	215	8 7/8	+1/8
20	12	CrmpKno	.54	2238	12 3/4	-1/4
11 7/8	**9 1/2**	**CrosTim**	**1.00e**	**316**	**10 1/4**	**+3/8**
14 1/8	6 1/2	CrwnAm	.80	1357	7 5/8	
50 5/8	33 1/2	CwnCork		5613	41 1/8	+3/8
18	11 1/2	CwnCr	.12	119	12	
21 5/8	17	CwnPac	2.04	657	18 1/4	-1/8
24	23 1/4	CullgW		2216	24	
48 5/8	34	CumEng	1.00	1400	37 5/8	+1/8
27 3/4	10 3/4	CySem		38153	12 1/2	-1/2
32 1/8	24 1/4	Cyprus	.80a	2596	26 5/8	-1/4
64 1/4	31 3/4	Cytec		379	61 3/4	+1/2
14	10 1/8	CzechFd	.10e	172	13 1/2	

This column lists the number of *round lots* (group of 100s) traded during the day. *Odd lots* (groups with less than 100) are not included.

This is the *closing price*, the price per share in the last sale of the day.

This indicates that the closing price was one-half dollar or 50 cents less than on the preceding business day.

Before you decide in which stocks to invest your money, you should familiarize yourself with how a stock exchange reports its daily activity. These reports use several columns of numbers, symbols, and abbreviations.

Starting the Investigation ·····················

A list of investments, such as stocks, that a person or a company owns is called a **portfolio**. Stockbrokers recommend selecting stock from several companies so that if the price of one company's stock drops, it will not cause a large decrease in the total value of your portfolio.

Create your portfolio by "spending" no more than $10,000 on at least six different stocks. You will need to track your holdings and periodically analyze their net worth. You may change stock holdings by selling and buying other stocks at any time during this Investigation. Your goal is to have the largest capital gain possible by the end of this Investigation.

Data Update For current information on the stock market, visit the following website.

Data Collection and Comparison To share and compare your data with other students in the U.S., visit:

www.glencoe.com/ sec/math/prealg/ mathnet

Investigation Taking Stock

Working on the Investigation Lesson 4-1, p. 174

Working on the Investigation Lesson 4-8, p. 209

Working on the Investigation Lesson 5-3, p. 238

Closing the Investigation End of Chapter 5, p. 264

4 Exploring Factors and Fractions

What You'll Learn

In this chapter, you will:

- use powers and exponents,

- solve problems by drawing a diagram,

- find the greatest common factor of two or more numbers or monomials,

- simplify and compare fractions, and

- multiply and divide monomials.

Why It's Important

Real World

Food In a recent summer, Americans ate about 606 million pounds of potato chips. This graph shows the fraction of the total chips eaten in each of seven regions. Fractions can help you to compare parts of a whole. If you want to know whether the Southeast or the Southwest ate more chips, you need to write the fractions with the same denominator. Then you can easily compare the fractions. *You will learn about the least common denominator in Lesson 4-7.*

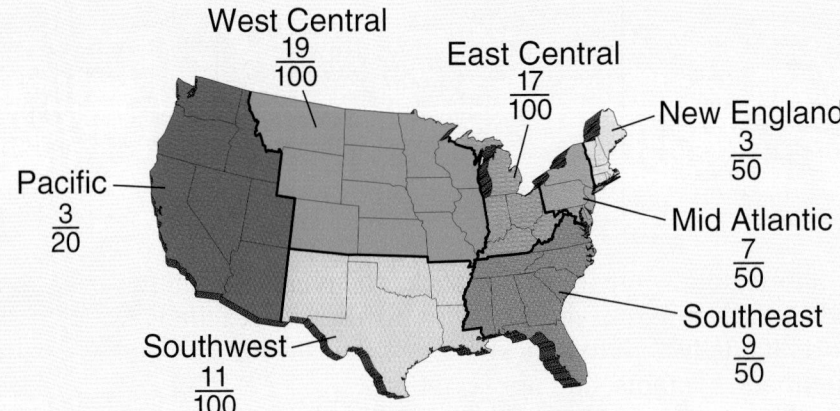

West Central $\frac{19}{100}$

East Central $\frac{17}{100}$

New England $\frac{3}{50}$

Pacific $\frac{3}{20}$

Mid Atlantic $\frac{7}{50}$

Southeast $\frac{9}{50}$

Southwest $\frac{11}{100}$

Focus on...

PREREQUISITE SKILLS

To be successful in this chapter, you'll need to understand these concepts and be able to apply them. Refer to the lesson in parentheses if you need more review before beginning the chapter.

▶ **Find the area of rectangles.** *(Lesson 3-5)*

Find the area of each rectangle.

1.
6 ft
9 ft

2.
10.2 m
22.5 m

3.
1 in.
5 in.

▶ **Recognize and extend a pattern.** *(Lesson 1-1)*

Find the next three terms in each sequence.
4. 10, 100, 1000, 10,000,...   5. 625, 125, 25, 5,...

6. 1, 0.1, 0.01, 0.001,...   7. 216, 36, 6, 1,...

▶ **Use the order of operations to evaluate expressions.** *(Lesson 1-2)*

Find the value of each expression.
8. $24 \div 6 + 3 \times 5$ 9. $2[5(10 - 1)] - 25$
10. $3[(6 - 2)] \div 4$ 11. $54 - 8 \cdot 5$

▶ **Evaluate expressions containing variables.** *(Lesson 1-3)*

Evaluate each expression if $a = 4$, $b = 5$, and $c = 2$.
12. $b - 2$ 13. $3a + c$ 14. $ab - bc$ 15. $2b - (a + c)$

Focus on...

READING SKILLS

In this chapter, you'll be learning about **factors** and **fractions**. The word fraction can mean "a piece broken off." Some fractions indicate parts of a whole, so you might think of them as pieces broken off of a whole. Factors are numbers that divide evenly into another number. Factors are important in the study of fractions because you often need to write two or more fractions with the same denominator. Factors can help you to do this.

Factors and Monomials

What You'll Learn

- To determine whether one number is a factor of another.

Why It's Important

You can use factors to determine which years are leap years.

 MATERIALS: grid paper

Hands-On Activity

Although it may seem like they are not at all related, algebra and geometry are so interconnected that you can learn about algebra by exploring geometric figures.

Your Turn Two different rectangular arrangements of 12 squares are shown at the right. There are six possible rectangular arrangements of 12 squares. Use grid paper to draw the remaining four arrangements.

 TALK ABOUT IT

a. You know that $3 \times 4 = 12$. List all of the pairs of whole numbers that have a product of 12.

b. Compare your list of whole number pairs to the lengths of the sides of the rectangles that you drew. What do you observe?

c. Write a sentence about the relationship between rectangular arrangements of squares and pairs of whole number factors of a number.

d. Can you arrange 14 squares in a rectangle so that one side of the rectangle is 2 units long? 3 units long? Did you know whether or not it could be done before you tried?

Learning the Concept

The **factors** of a whole number divide that number with a remainder of 0. For example, 2 is a factor of 12 because $12 \div 2 = 6$, and 5 is not a factor of 12 because $12 \div 5 = 2$ with a remainder of 2.

Another way of saying that 6 is a factor of 12 is to say that 12 is **divisible** by 6.

Sometimes you can test for divisibility mentally. These rules will help you determine whether a number is divisible by 2, 3, 5, 6, or 10.

A number is divisible by:

▶ 2 if the ones digit is divisible by 2.

▶ 3 if the sum of its digits is divisible by 3.

▶ 5 if the ones digit is 0 or 5.

▶ 6 if the number is divisible by 2 and 3.

▶ 10 if the ones digit is 0.

A number that is divisible by 2 is called ***even***. *An* ***odd*** *number is not divisible by 2.*

Example 1 Determine whether 146 is divisible by 2, 3, 5, 6, or 10.

Number	Divisible?	Reason
2	yes	The ones digit is 6 and 6 ÷ 2 = 3.
3	no	The sum of the digits is 1 + 4 + 6 or 11. 11 is not divisible by 3.
5	no	The ones digit of 146 is 6, not 0 or 5.
6	no	146 is divisible by 2 but not by 3.
10	no	The ones digit of 146 is not 0.

Use a calculator to verify the divisibility of each number.
So, 146 is divisible by 2, but not 3, 5, 6, or 10.

THINK ABOUT IT

Explain why you can say that a number is divisible by 10 if it is divisible by 2 and 5.

As the rules for divisibility suggest, a pattern can be found in many numbers that share factors. Looking for a pattern can help you determine other rules for divisibility.

Example 2

PROBLEM SOLVING

Look for a Pattern

Look for a pattern in the digits of numbers to write a rule for determining if a number is divisible by 4.

Explore We know a rule to determine if a number is divisible by 2. We need to find a rule for numbers divisible by 4.

Plan Use a calculator to determine many numbers that are divisible by 4. Then look for a pattern among those numbers.

Solve Since 2 is a factor of 4, any number divisible by 4 must also be divisible by 2. So, let's look at some even numbers.

Number	Last Two Digits	Divisible by 4?
16	16	yes
28	28	yes
112	12	yes
148	48	yes
254	54	no
284	84	yes
1016	16	yes
2026	26	no

In all of the numbers that are divisible by 4, the last two digits form a number that is divisible by 4. For example in 112, 12 is divisible by 4. In 254, 54 is not divisible by 4, so neither is 254 divisible by 4.

A possible rule for determining whether a number is divisible by 4 is that if the last two digits of a number are divisible by 4, then the number is divisible by 4.

Examine According to the rule, 524, 816, and 1248 are divisible by 4. Use a calculator to check.

Being able to determine whether one number is divisible by another is a useful skill that can be applied in everyday tasks.

In the Gregorian Calendar, each "common year" is 365 days long. Years that are divisible by 4, called "leap years" are 366 days long. Also, years ending in "00" that are divisible by 400 are leap years and those that are not are common. Determine whether 1998, 2000, 2066, and 2116 are leap years.

Use the rule for divisibility by 4 that we discovered in Example 2.

Year	Divisible by 4?	Leap Year?
1998	no, 98 is not divisible by 4	no
2066	no, 66 is not divisible by 4	no
2116	yes, 16 is divisible by 4	yes

Since 2000 ends in "00", it must be divisible by 400 to be a leap year. The quotient $2000 \div 400$ is 5, so 2000 is a leap year.

The years 2000 and 2116 are leap years, but 1998 and 2066 are not.

Connection to ▷ Algebra

We know that 5 and 12 are factors of 60 because $5 \cdot 12 = 60$. Consider the algebraic expression $5x$. Since this notation means $5 \cdot x$, it follows that 5 and x are factors of $5x$.

An expression like $5x$ is called a **monomial**. A monomial is an integer, a variable, or a product of integers or variables. Other monomials are z, 16, mn, and $6ab$. Expressions like $4x - 2$ and $x + 3$ are not monomials since they involve addition or subtraction.

Example ④ Determine whether each expression is a monomial.
a. $5(x + y)$ b. $-25xyz$

This expression is not a monomial because it involves addition.

This expression is a monomial because it is the product of integers and variables.

Checking Your Understanding

Communicating Mathematics

Read and study the lesson to answer these questions.

1. **Explain** why the diagram at the right demonstrates that 8 is a factor of 40.

2. **Determine** whether 6 is a factor of 78. Explain.
3. Is $7n$ a monomial? Justify your answer.
4. What number is a factor of every nonzero number?

5. You know that the factors of a number can be modeled using rectangles. Arrange 36 small cubes in a shape like a rectangular box in as many ways as you can. Write a multiplication sentence for each model you created.

Using divisibility rules, state whether each number is divisible by 2, 3, 5, 6, or 10.

6. 39   **7.** 118   **8.** 876   **9.** 5010

Determine whether each expression is a monomial. Explain why or why not.

10. 24   **11.** $-3m$ **12.** $4y + 2$ **13.** $5xyz$

14. Music The 78 members of the Brookhaven High School Marching Band will be performing in the Independence Day parade on Saturday. Due to the width of the street they will be marching down, the parade director has recommended that all bands march in rows of 6. Can the whole Brookhaven band be arranged in rows of 6?

Exercises: Practicing and Applying the Concept

Using divisibility rules, state whether each number is divisible by 2, 3, 5, 6, or 10.

15. 111   **16.** 11,222   **17.** 5050   **18.** 4444

19. 10,505   **20.** 24,640   **21.** 117   **22.** 330

23. 49   **24.** 10,523   **25.** 434   **26.** 378

Determine whether each expression is a monomial. Explain why or why not.

27. -234 **28.** $3c - 3$ **29.** $-4b$ **30.** $6(xy)$

31. $4(3x - 1)$ **32.** v **33.** $q + r$ **34.** $166h$

35. Replace the _?_ in 82_?_ so that the number is divisible by 4.

✓ Choose

Estimation
Mental Math
Calculator
Paper and Pencil

Use mental math, paper and pencil, or a calculator to find at least one number that satisfies each condition.

36. is divisible by 2, 3, and 5

37. has four digits and is divisible by 3 and 5, but is not divisible by 10

38. has 3 digits and is not divisible by 2, 3, 5, or 10

39. has four equal digits and is divisible by 5

40. a. What is the greatest three-digit number that is not divisible by 2, 3, 5, or 10?

    b. What is the least three-digit number that is not divisible by 2, 3, 5, or 10?

41. Social Studies Chinese New Year, the most important holiday of the Chinese year, comes in late January or early February. The Chinese calendar follows an annual cycle, which repeats every twelve years. For example, 1996 was the Year of the Rat, so $1996 + 12$ or 2008 and $1996 - 12$ or 1984 are also the Year of the Rat.

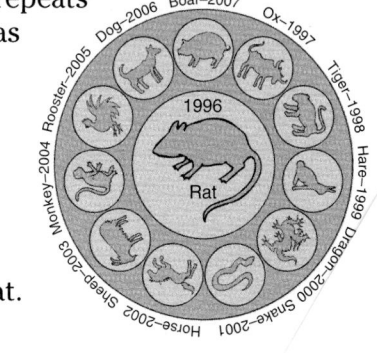

    a. 1998 is the Year of the Tiger. How old is the youngest person born in a Year of the Tiger?

    b. List five years that are Years of the Rat.

42. **Civics** Each state has a star on the U.S. flag and the stripes represent the thirteen British colonies that became the United States.

    a. How do you think the 48 stars were arranged before Alaska and Hawaii joined the Union?

    b. Suggest an arrangement for the stars if another state were to be added. Could it be rectangular?

43. **Number Theory** A perfect number is equal to the sum of its factors, except itself. There are two perfect numbers less than 30. The third perfect number is 496, since $496 = 1 + 2 + 4 + 8 + 16 + 31 + 62 + 124 + 248$.

    a. Find a perfect number between 1 and 10.

    b. Find a perfect number between 20 and 30.

Mixed Review

44. **Geography** Colorado is less than twice as large as Arkansas. Colorado is 103,730 square miles. How large is Arkansas? (Lesson 3-8)

45. Solve $89 = \dfrac{x}{-3}$. (Lesson 3-3)

46. **Standardized Test Practice** Solve $f = -64 \div -4$. (Lesson 2-8)

    A. -256 **B.** -16 **C.** 16   **D.** 256

47. Solve $t = -6 + (-14) + 12$. (Lesson 2-4)

48. Solve $5b = 95$ mentally. (Lesson 1-6)

For **Extra Practice**, see page 748.

49. Evaluate $ac - 3b$ if $a = 7$, $b = 3$, and $c = 6$. (Lesson 1-3)

50. Find the value of $3 \cdot (3 + 5) \div 2$. (Lesson 1-2)

WORKING ON THE

Investigation

Refer to the Investigation on pages 166–167.

One of the factors to consider when investing in stock is the trends in the prices. You may see something that looks like the diagram at the right in the financial pages of a newspaper. This diagram 'charts' the average stock price of 30 large industrial companies whose stocks are traded on the New York Stock Exchange.

You can also chart an individual stock. When you chart a stock's activity for a day, the top of the vertical bar represents the highest price at which the stock traded, while the bottom of the bar represents the lowest price at which

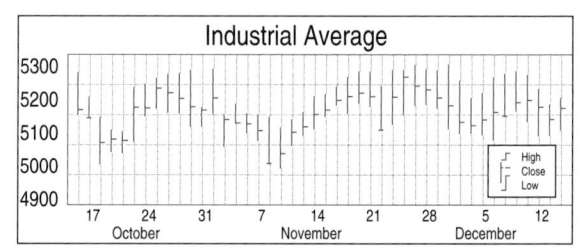

the stock traded. The horizontal bar indicates the closing price.

- Go to the library and use the last three months' editions of a local or national newspaper to chart the price of one of the stocks you have chosen. If the paper you are using doesn't provide the daily highs and lows, prepare a line graph of the closing prices.
- Write a brief analysis of your stocks' recent performance.

Add the results of your work to your Investigation Folder.

4-2 Powers and Exponents

Application: Games

In a certain board game, players move pieces around a board determined by tossing two dice.

Since each die has six faces, we know that in rolling one die we can come up with six different results. How many different results are there with two dice? Make a table of the possible results.

First Die

		1	2	3	4	5	6
	1	(1, 1)	(2, 1)	(3, 1)	(4, 1)	(5, 1)	(6, 1)
	2	(1, 2)	(2, 2)	(3, 2)	(4, 2)	(5, 2)	(6, 2)
Second Die	**3**	(1, 3)	(2, 3)	(3, 3)	(4, 3)	(5, 3)	(6, 3)
	4	(1, 4)	(2, 4)	(3, 4)	(4, 4)	(5, 4)	(6, 4)
	5	(1, 5)	(2, 5)	(3, 5)	(4, 5)	(5, 5)	(6, 5)
	6	(1, 6)	(2, 6)	(3, 6)	(4, 6)	(5, 6)	(6, 6)

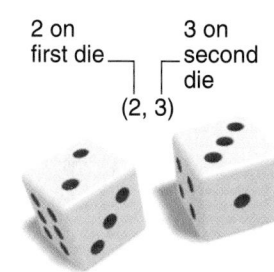

2 on first die ─ ─ 3 on second die
(2, 3)

Six choices on the first die for each of the six choices on the second die gives us 6×6 or 36 possibilities.

Learning the Concept

The expression 6×6 can be written in a shorter way using exponents. An **exponent** tells how many times a number, called the **base**, is used as a factor. Numbers that are expressed using exponents are called **powers**.

The expression 6×6 can be written as 6^2.

$$base \rightarrow \underline{6}^2 \leftarrow exponent$$
$$power$$

The powers 7^2, 6^3, 10^4 and $(-2)^5$ are read as follows.

7^2	is seven to the second power or seven squared.
6^3	is six to the third power or six cubed.
10^4	is ten to the fourth power.
$(-2)^5$	is negative two to the fifth power.

Any number, except 0, raised to the zero power, like 5^0, is defined to be 1.

The value of any number to the first power is the number. For example, $6^1 = 6$.

Example **Write each multiplication expression using exponents.**

a. $3 \cdot 3 \cdot 3 \cdot 3$

The base is 3. It appears as a factor 4 times. The exponent is 4.

$3 \cdot 3 \cdot 3 \cdot 3 = 3^4$

b. $y \cdot y \cdot y \cdot y \cdot y$

The base is y. It appears as a factor 5 times. The exponent is 5.

$y \cdot y \cdot y \cdot y \cdot y = y^5$

c. 5

The base is 5. It
appears once.
The exponent is 1.

$5 = 5^1$

d. $(-2)(-2)(-2)(-2)$

The base is -2. It
appears as a factor 4
times. The exponent is 4.

$(-2)(-2)(-2)(-2) = (-2)^4$

Example ② Write each power as a multiplication expression.

a. 5^4

The base is 5. The
exponent 4 means
5 is a factor 4 times.

$5^4 = 5 \cdot 5 \cdot 5 \cdot 5$

b. b^2

The base is b. The
exponent 2 means
b is a factor 2 times.

$b^2 = b \cdot b$

The number 12,496 is in **standard form**. You can use exponents to
express a number in **expanded form**.

Example ③ Express 12,496 in expanded form.

To express a number in expanded form, use place value to write the
value of each digit in the number.

$$12{,}496 = 10{,}000 + 2000 + 400 + 90 + 6$$
$$= (1 \times 10{,}000) + (2 \times 1000) + (4 \times 100) + (9 \times 10) + (6 \times 1)$$

Then write the multiples of 10 using exponents.
$$12{,}496 = (1 \times 10^4) + (2 \times 10^3) + (4 \times 10^2) + (9 \times 10^1) + (6 \times 10^0)$$
Recall that $10^0 = 1$.

Since powers are forms of multiplication, they need to be included in the
rules for order of operations.

**Order of
Operations**

1. Do all operations within grouping symbols first; start with the
   innermost grouping symbols.
2. *Evaluate all powers in order from left to right.*
3. Next do all multiplications and divisions in order from left to right.
4. Then do all additions and subtractions in order from left to right.

Connection to ＞ Algebra

Follow the order of operations as you evaluate algebraic expressions.

Example ④ Evaluate $3a + b^3$ if $a = 2$ and $b = 5$.

$$\begin{aligned} 3a + b^3 &= 3(2) + 5^3 && \textit{Replace a with 2 and b with 5.}\\ &= 3(2) + 125 && \textit{Evaluate the power.}\\ &= 6 + 125 && \textit{Find the product of 3 and 2.}\\ &= 131 && \textit{Find the sum.} \end{aligned}$$

Many scientific formulas involve evaluating expressions containing
exponents.

Example ⑤

Real World
APPLICATION

Space Flight

TECHNO
TIP

You can evaluate expressions involving exponents by using a calculator. Enter the base first, press the y^x key, and then enter the exponent.

At liftoff, the space shuttle *Discovery* has a constant acceleration, a, of 16.4 ft/s². The initial velocity, v, is 1341 ft/s. Use the formula $d = vt + \frac{1}{2}at^2$, where d represents distance and t represents time in seconds, to find the distance that the shuttle has traveled after 30 seconds.

$$d = vt + \frac{1}{2}at^2$$

$$= (1341)(30) + \frac{1}{2}(16.4)(30)^2 \quad \textit{Replace v with 1341, t with 30, and a with 16.4.}$$

Estimate: *40,000 + 8 · 900 is about 47,000.*

Evaluate this expression by using a calculator. Many calculators have a special key labeled $\boxed{x^2}$ for finding squares.

Enter: 1341 $\boxed{\times}$ 30 $\boxed{+}$ .5 $\boxed{\times}$ 16.4 $\boxed{\times}$ 30 $\boxed{x^2}$ $\boxed{=}$ 47610

The shuttle has traveled 47,610 feet after 30 seconds.

Checking Your Understanding

Communicating Mathematics

Read and study the lesson to answer these questions.

1. **Explain** how to find the value of the expression *3 squared*.
2. **Write** 10^5 as a product of the same factor. What is the value of 10^5?
3. Regardless of the value of n, what can you say about 1^n? Explain.
4. **You Decide** Nichelle says that for every value of n, $n^2 > n$. Cardida disagrees. Who is correct? Explain.

MATH **J**OURNAL

5. **Assess Yourself** Why do you think that exponents were invented?

Guided Practice

Write each multiplication expression using exponents.

6. $10 \cdot 10 \cdot 10$ 7. $m \cdot m \cdot m$ 8. $(3)(3)(3)$

Write each power as a multiplication expression.

9. 2^3 10. $(-8)^4$ 11. 10^6

Evaluate each expression if $x = 5$, $y = 9$, and $z = -3$.

12. x^4 13. $y^2 + z^3$

14. Write 598 in expanded form.

15. **Geometry** The edges of the cube shown at the right are 5 inches long.

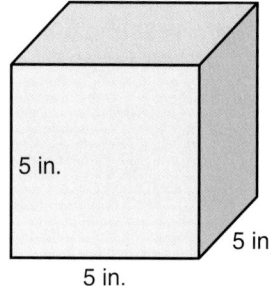

5 in.
5 in.
5 in.

   a. The surface area of a cube is the sum of the areas of the faces. Use exponents to write an expression for the surface area of the cube.
   b. The *volume* of a cube is the amount of space that it occupies. The volume can be found by multiplying the length, width, and height of the cube. Write an expression for the volume of the cube.
   c. If you double the length of each edge of the cube, what is the effect on the surface area and volume?

Independent Practice

Write each multiplication expression using exponents.

16. $(5)(5)(5)$

17. $n \cdot n \cdot n \cdot n$

18. $1 \cdot 1 \cdot 1 \cdot 1 \cdot 1$

19. 14

20. $(p \cdot p)(p \cdot p)$

21. $\underbrace{3 \cdot 3 \cdot \cdots \cdot 3 \cdot 3}_{15 \; factors}$

22. $6 \cdot 6 \cdot 6 \cdot 6 \cdot 6$

23. $b \times b \times b \times b$

24. $(-5)(-5)$

Write each power as a multiplication expression of the same factor.

25. 12^2

26. y^6

27. $(-7)^3$

28. $(-f)^2$

29. 6^{20}

30. $(x + 1)^2$

31. 1^{55}

32. q^3

33. $(-1)^4$

Evaluate each expression if $a = 3$, $b = 9$, and $c = -2$.

34. $a^4 + b$

35. $c^2 + ab$

36. $b^0 - 10$

37. $2b^2$

38. $10a^5 + c^2$

39. $a - b^2$

Write each number in expanded form.

40. 56

41. 149

42. 2053

43. Write $(2 \times 10^4) + (3 \times 10^3) + (4 \times 10^2) + (5 \times 10^0)$ in standard form.

44. Patterns Study the pattern at the right. Extend the pattern to find the value of each expression.

$$
\begin{aligned}
10^0 \cdot 10^2 &= 1 \cdot 100 &= 100 && \text{or } 10^2 \\
10^1 \cdot 10^2 &= 10 \cdot 100 &= 1000 && \text{or } 10^3 \\
10^2 \cdot 10^2 &= 100 \cdot 100 &= 10{,}000 && \text{or } 10^4 \\
10^3 \cdot 10^2 &= 1000 \cdot 100 &= 100{,}000 && \text{or } 10^5
\end{aligned}
$$

a. $10^5 \cdot 10^2$

b. $10^{10} \cdot 10^2$

c. $10^x \cdot 10^2$

Critical Thinking

45. Patterns Study the pattern of exponents at the right.

$$
\begin{aligned}
2^4 &= 16 \\
2^3 &= 8 \\
2^2 &= 4 \\
2^1 &= 2 \\
2^0 &= 1
\end{aligned}
\quad \Big\} \div 2
$$

a. If you continue the pattern, what is the next exponent for 2?

b. What is the next value in the pattern?

c. Use a calculator to verify your results.

d. Make a conjecture about the value of 3^{-1}.

Applications and Problem Solving

46. Electricity Niagara Falls and Hoover Dam can produce electric energy with little pollution. As the power goes to its point of use, energy is lost to the resistance in the wires. The amount of power lost can be found by using the formula $P = I^2 R$, where P is power in watts, I is current in amps, and R is resistance in ohms.

a. The resistance of the wire leading from the source of power to a home is 2 ohms. If an electric stove causes a current of 41 amps to flow through the wire, write an equation for the power lost.

b. Find the power lost from the wire powering the stove.

47. Geometry The second and third powers have special names related to geometry. Since the area of a square whose side is s units long is $s \cdot s$ or s^2, s^2 is often read as "s squared." A third power like s^3 is often read as "s cubed." Can you explain why?

48. Entertainment   In a movie, Albert Einstein tries to make his niece Catherine Boyd fall in love with auto mechanic Ed Walters. In one scene, Ed asks Catherine how many stars she thinks are in the sky. Her answer is $10^{12} + 1$. Write this number in standard form.

49. Literature   There is a story of a knight who slew a dragon that was destroying a small kingdom. The knight asked the king to give him reward money for a month following this pattern. The first day of the month, he would get 1 cent, the second day, 2 cents, the third day, 4 cents, and so on, continuing to double the amount for 30 days.

  a. Express his reward on each of the first three days as a power of 2.

  b. What is his reward on the fourth day? Express it as a power of 2.

  c. How much would the knight be getting from the king on the 30th day? Express it as a power of 2.

  d. Write your answer to part c in standard form.

  e. Write an expression using a sum and exponents for the amount of money the knight would receive for the whole month.

  f. The king agreed to the knight's request because he thought it would be a cheap reward. Was it? Explain.

Mixed Review

50. State whether 945 is divisible by 2, 3, 5, 6, or 10. (Lesson 4-1)

51. Solve and graph $-7x < 84$. (Lesson 3-7)

52. Geometry   Find the perimeter and area of a rectangle that is 4 meters wide and 12 meters long. (Lesson 3-5)

53. Meteorology   A maximum and minimum thermometer records both the high and low temperatures of the day. If the difference between one day's high and low was 49 degrees and the high was 67°F, what was the low temperature? (Lesson 3-2)

For Extra Practice, see page 748.

54. Solve $m = \dfrac{126}{-21}$. (Lesson 2-8)

55. Simplify $18p - 26p$. (Lesson 2-5)

56. Standardized Test Practice   Which number is a solution of $r - 12 = 15$? (Lesson 1-6)

  A. 27   **B.** 23   **C.** 7   **D.** 3

57. Write an expression to represent the phrase *four more than the Bears scored in the last game*. (Lesson 1-3)

From the → FUNNY PAPERS

1. Explain why the comic is funny.
2. Explain the differences in the formulas on the chalkboard.
3. Einstein's famous formula $E = mc^2$ relates energy, E, mass, m, and the speed of light, c. The speed of light is about 300,000 km/s. Use a calculator to find the value of c^2.

4-2B Evaluating Expressions with Exponents

An Extension of Lesson **4-2**

You learned how to evaluate expressions with a graphing calculator in Lesson 1-3B. You can also use a graphing calculator to evaluate expressions involving exponents.

Storing the values of the variables allows you to evaluate complicated expressions accurately. Also, it will save time when you want to evaluate several expressions for the same values.

Activity 1 Evaluate $\dfrac{x}{(y-2)^2}$ on a graphing calculator if $x = 24$ and $y = 4$.

ENTER: 24 [STO▸] [X,T,θ,n] [ALPHA] [:]   *Enters the value of x.*
4 [STO▸] [ALPHA] [Y] [ALPHA] [:]   *Enters the value of y.*
[X,T,θ,n] [÷] [(] [ALPHA]   *Enters and evaluates*
[Y] [−] 2 [)] [x^2] [ENTER] *6*   *the expression.*

Expressions involving exponents other than 2 can be evaluated using the [∧] key.

Activity 2 Use a graphing calculator to evaluate $3ab^5$ if $a = -4$ and $b = 2$.

ENTER: [(−)] 4 [STO▸] [ALPHA] [A] [ALPHA] [:] *Enters the value of a.*
2 [STO▸] [ALPHA] [B] [ALPHA] [:] *Enters the value of b.*
3 [ALPHA] [A] [ALPHA] [B] *Enters and evaluates*
[∧] 5 [ENTER] *-384* *the expression.*

Notice that the calculator was able to evaluate the expression without entering multiplication symbols between 3, a, and b^5.

Your Turn Use a graphing calculator to evaluate each expression if $a = 9$, $b = -1$, $c = 7$, and $d = 3$.

1. $7d^2 - 1$ 2. $\dfrac{32}{(a-c)^3}$ 3. $3(b+c) \div d^2$

4. $12b^4 \div d$ 5. $(2c-4d)^{10}$ 6. $11 - a^2(b+c)$

7. Evaluate $12 - 4d^2$ and $(12 - 4d)^2$ if $d = 3$. Are the values the same? Explain.

Problem-Solving Strategy: Draw a Diagram

Application: Entertainment

The game of Totolospi was invented by the Hopi Indians. You play the game with three cane dice, a counting board inscribed in stone, and a counter for each player. Each cane die can land round side up, (R), or flat side up, (F). When two players play Totolospi, each player places a counter on the nearest circle. The moves in the game are determined by tossing three cane dice.

▶ Toss three round sides up, (RRR), and advance 2 lines.

▶ Toss three flat sides up, (FFF), and advance 1 line.

▶ No advance with any other combination.

The first player to reach the opposite side wins.

If you are playing Totolospi, how many combinations of tosses will allow an advance?

Learning the Concept

When we are able to draw a diagram for a given situation, we often are able to understand it better. Drawing a diagram is a powerful problem-solving strategy.

Explore There are 3 dice being tossed. We need to know how many combinations of round and flat side up can be tossed.

Plan We can draw a diagram showing how each of the dice lands to determine the number of possible combinations.

Solve

First Die	**Second Die**	**Third Die**

```
                R ——————— R  (RRR)  ← advance 2
        R  <                F  (RRF)
                F ——————— R  (RFR)
                                 F  (RFF)
                                             no advance
                R ——————— R  (FRR)
        F  <                F  (FRF)
                F ——————— R  (FFR)
                                 F  (FFF)  ← advance 1
```

There are 8 possible outcomes. Two of them, RRR and FFF, allow you to advance. So 8 − 2 or 6 tosses allow no advance.

Examine Each die can land two different ways. So there should be 2 · 2 · 2 or 8 possible tosses.

Communicating Mathematics

Read and study the lesson to answer these questions.

1. **Explain** why a diagram is a good strategy for problem solving.

2. Are diagrams helpful in solving *every* type of problem?

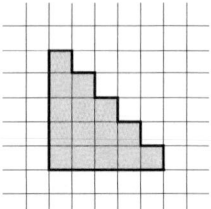

MATERIALS

grid paper

3. The diagram at the right shows a model of a five-step staircase built with concrete blocks. Use grid paper to draw a staircase with seven steps. How many blocks would be required to build a staircase with seven steps?

Guided Practice

4. **Draw** a diagram to determine how many flights an airline would have to schedule if it wants to provide nonstop flights between Miami, Ft. Lauderdale, Tallahassee, Tampa, Orlando, and Daytona.

Tallahassee

Daytona

Tampa

Orlando

Ft. Lauderdale

Miami

5. **Sports** The Centerville High School volleyball team is participating in the state championship. Eight teams will participate in a single-elimination tournament; that is, only winning teams will continue to play. Draw a diagram to determine the number of games that will be played.

Practicing and Applying the Concept

Independent Practice

Solve. Use any strategy.

6. The West Peoria Baseball League has 7 teams and the East Peoria League has 8 teams. Every team from the West League must play every team from the East League at least once. What is the least number of games that must be played?

7. Replace each ● with an operation symbol to make the equation true. Add grouping symbols if necessary.

$$4 ● 3 ● 6 ● 3 = 45$$

8. During the first minute of a game of tug-of-war, the Red Team pulled the Blue Team forward 2 feet. During the second minute, the Blue Team pulled the Red Team forward 1 foot. During the third minute, the Red Team pulled the Blue Team forward 2 feet and during the fourth minute, the Blue Team pulled the Red Team forward 1 foot. If this pattern continues, how long will it take the Red Team to pull the Blue Team forward a total of 10 feet to win the game?

9. Meg is conditioning for the start of soccer season. She does seven minutes of stretching, followed by an 18-minute run. Then she cools down with 8 minutes of walking and 7 minutes of leg lifts. Finally, she finishes with 4 minutes of stretching. If Meg finishes her workout at 2:07 P.M., when did she begin?

10. History Jorge has to write a paper and complete a family tree for history class. If part of the assignment was to find the names and birthdates of the five generations that preceded him, direct relatives only, no step-parents, how many people will Jorge have to investigate?

11. Magic squares are arrays of numbers in which the sum of the numbers on each row, column, and diagonal are the same. According to legend, the most famous magic square, the *lo-shu*, was discovered on the back of a divine tortoise by China's Emperor Yu in about 2200 B.C. Copy and complete the magic square pictured at the right.

4	15		14
		12	
	3		
5		8	

12. Geometry A diagonal is a line segment that can be drawn from one vertex of a polygon to another vertex that does not share a side. The diagram at the right shows the two diagonals of a rectangle.

   a. A pentagon has five sides. How many diagonals does it have?

   b. How many diagonals can be drawn in a hexagon? (*Hint:* A hexagon has six sides.)

   c. Use a pattern to predict how many diagonals a heptagon, 7 sides, has. Check your answer by drawing a diagram.

   d. In your own words, write a description of the pattern.

13. Charlie traded in his old car which averaged 22 miles per gallon. The EPA sticker on the new car stated that it should average 37 miles per gallon. If Charlie drives about 12,000 miles per year and gasoline is about $1.10 per gallon, how much should he expect to save on gasoline in the first year of owning the new car?

For **Extra Practice**, see page 748.

14. At Grandma's Bakery, muffins are sold in boxes of 4, 6, or 13. You can buy 8 muffins by choosing two boxes of 4 muffins, but you can't buy 9 muffins with any combination of boxes. Find all of the numbers of muffins less than 25 that you can't buy.

Critical Thinking

15. City code requires that a party house provide 9 square feet for each person on a dance floor. If the owners of the Westman Hotel want to provide a square dance floor that is large enough for 100 people, how long should each side be?

Mixed Review

16. Write $x \cdot x \cdot x \cdot x \cdot x$ using exponents. (Lesson 4-2)

17. Standardized Test Practice Solve $x - 8 \le -17$. (Lesson 3-6)

   A. $x \le -9$ **B.** $x \ge -9$ **C.** $x \le 9$ **D.** $x \ge 9$

18. Patterns What is the next number in the pattern 3, 3, 6, 18, ... ? (Lesson 2-6)

19. Find the sum $5 + (-8) + 7$. (Lesson 2-4)

20. Name the property shown by $(3 \cdot 2) \cdot 8 = 3 \cdot (2 \cdot 8)$. (Lesson 1-4)

Prime Factorization

MATERIALS: grid paper
Hands-On Activity

As you learned in Lesson 4-3, drawing a diagram can be very helpful as you solve problems.

Your Turn   The grid at the right shows one way of drawing a rectangle consisting of 10 squares. Use grid paper to draw as many different rectangular arrangements of 1, 2, 3, 4, 5, 6, 7, 8, 9, and 10 squares as possible.

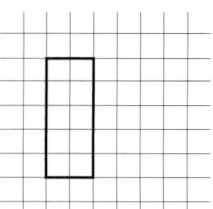

a. Which numbers of squares could only be arranged in either one row or one column?

b. Which numbers of squares could be arranged in a way other than one row or column?

c. Choose one of the numbers that could only be arranged in one row or one column. Write as many multiplication sentences with the number as the product and two whole numbers as factors as you can. What do you observe?

Learning the Concept

In the activity above, the numbers of squares that can be arranged in only one row or one column are prime numbers. A **prime number** is a whole number greater than one that has *exactly* two factors, 1 and itself.

Numbers of squares like 4, 6, 8, 9, and 10 that can be arranged in rectangles other than a row or a column are called composite numbers. A **composite number** is a whole number greater than one that has more than two factors. A composite number can always be expressed as a product of two or more primes.

▶ The numbers 0 and 1 are considered *neither* prime *nor* composite.
▶ Every number is a factor of 0, since any number multiplied by zero is zero.
▶ The number 1 has only one factor, itself.
▶ Every whole number greater than 1 is either prime or composite.

Sometimes a number can be factored in several ways. Three different ways to find all of the prime factors of 24 using diagrams called **factor trees** are shown below.

$$24 = 2 \cdot 12$$
$$= 2 \cdot 3 \cdot 4$$
$$= 2 \cdot 3 \cdot 2 \cdot 2$$

$$24 = 3 \cdot 8$$
$$= 3 \cdot 2 \cdot 4$$
$$= 3 \cdot 2 \cdot 2 \cdot 2$$

$$24 = 4 \cdot 6$$
$$= 2 \cdot 2 \cdot 2 \cdot 3$$

The factors are in a different order, but the result is the same. The order in which you factor is not important. The factoring process ends when all of the factors are prime. When a positive integer (other than 1) is expressed as a product of factors that are all prime, the expression is called the **prime factorization**.

Example **Use a factor tree to factor 64 completely.**

Begin with any two whole number factors of 64.

$$64 = 8 \cdot 8$$
$$= 4 \cdot 2 \cdot 4 \cdot 2 \quad \textit{Factor each 8.}$$
$$= 2 \cdot 2 \cdot 2 \cdot 2 \cdot 2 \cdot 2 \quad \textit{Factor each 4.}$$

The factorization is complete when all the factors are prime. The prime factorization of 64 is $2 \cdot 2 \cdot 2 \cdot 2 \cdot 2 \cdot 2$ or 2^6.

The prime factorization of a negative integer (other than -1) contains a factor of -1, and the rest of the factors are prime.

Example **Use a factor tree to factor -72 completely.**

$$-72 = -1 \cdot 72 \quad \textit{A negative integer may be expressed as the product of -1 and a whole number.}$$
$$= -1 \cdot 8 \cdot 9 \quad \textit{Begin with any two whole number factors of 72.}$$
$$= -1 \cdot 2 \cdot 4 \cdot 3 \cdot 3 \quad \textit{Factor 8 and 9.}$$
$$= -1 \cdot 2 \cdot 2 \cdot 2 \cdot 3 \cdot 3 \quad \textit{Factor 4.}$$

You are finished factoring when all of the factors, other than -1, are prime. The factorization of -72 is $-1 \cdot 2 \cdot 2 \cdot 2 \cdot 3 \cdot 3$. You can write this product as $-1 \cdot 2^3 \cdot 3^2$.

You can also use a strategy called the *cake method* to find a prime factorization. The cake method uses division to find factors. For example, the prime factorization of 210 is shown below.

▶ Begin with the smallest prime that is a factor.
▶ Then divide the quotient by the smallest possible prime factor.
▶ Repeat until the quotient is prime.

$$\begin{array}{r} 7 \\ 5\overline{)35} \\ 3\overline{)105} \\ 2\overline{)210} \end{array}$$

The prime factorization of 210 is $2 \cdot 3 \cdot 5 \cdot 7$. *Use a tree diagram to check the result.*

Example **3**

Prime numbers are the key to coding and decoding information using the Rivest-Shamirs-Adleman (RSA) cryptoalgorithm. Two prime numbers p and q are chosen and the key to the code is $n = pq$. Find p and q if $n = 1073$.

The greater the prime numbers, the more difficult it is to break the code.

Some possibilities can be eliminated by using the divisibility rules you learned in Lesson 4-1.

1073 is not divisible by 2 since the ones digit, 3, is not divisible by 2.
1073 is not divisible by 3 since $1 + 0 + 7 + 3$ or 11 is not divisible by 3.
1073 is not divisible by 5 since the ones digit, 3, is not 0 or 5.

Now use a calculator and the guess-and-check problem-solving strategy to continue checking primes.

x	$1073 \div x$	factor of 1073?
7	153.286	no
11	97.5455	no
13	82.5385	no
17	63.1176	no
19	56.4737	no
23	46.6522	no
29	37	yes

If $n = 1073$, then p and q are 29 and 37.

Connection to > Algebra

A monomial can also be written in factored form as a product of prime numbers, -1, and variables with no exponent greater than 1.

Example **4**

Factor each monomial completely.
a. $28x^2y$

$$28x^2y = 2 \cdot 14 \cdot x^2 \cdot y$$
$$= 2 \cdot 2 \cdot 7 \cdot x \cdot x \cdot y$$

b. $64ab^3$

$$64ab^3 = 8 \cdot 8 \cdot a \cdot b^3$$
$$= 2 \cdot 4 \cdot 2 \cdot 4 \cdot a \cdot b \cdot b \cdot b$$
$$= 2 \cdot 2 \cdot 2 \cdot 2 \cdot 2 \cdot 2 \cdot a \cdot b \cdot b \cdot b$$

Checking Your Understanding

Communicating Mathematics

Read and study the lesson to answer these questions.

1. **Explain** the differences between prime and composite numbers.

2. **Find** the prime factorization of 84 by using both a factor tree and the cake method. Are the results the same? Explain.

3. **You Decide**   Mansi and Wes each factored -124. Whose factorization is complete and why?

Mansi

$-124 = -1 \cdot 124$

$= -1 \cdot 4 \cdot 31$

$= -1 \cdot 2 \cdot 2 \cdot 31$

Wes

$$\begin{array}{r} 31 \\ 4\overline{)124} \\ -1\overline{)-124} \end{array}$$

$-124 = -1 \cdot 4 \cdot 31$

Guided Practice

Determine whether each number is *prime* or *composite*.

4. 17

5. 9

6. 27

Factor each number or monomial completely.

7. 38

8. 66

9. 56

10. $8x^2y^3$

11. $-42abc$

12. $25(xy)^2$

13. **Entertainment**   In a recent movie about a boy, Damon is called "the Mathemagician" because he can perform complicated calculations mentally. In a contest, he is asked how many factors the number 3067 has. Damon answers "Come on, guys. There are no factors of 3067. The number is prime." Is Damon correct? Explain.

Exercises: Practicing and Applying the Concept

Independent Practice

Determine whether each number is prime or composite.

14. 35

15. 19

16. 55

17. 101

18. 2

Factor each number or monomial completely.

19. 51

20. -63

21. 41

22. -95

23. 110

24. -333

25. 81

26. 59

27. 13

28. 1024

29. $42xy^2$

30. $38mnp$

31. $21xy^3$

32. $560x^4y^2$

33. $28f^2g$

34. $275st^3$

35. $210mn^3$

36. $-8a^3b^2$

37. $75m^2k$

38. $-400a^2b^3$

Critical Thinking

39. Find the least number that gives you a remainder of 1 when you divide it by 2 or by 3 or by 5 or by 7.

40. February the third is a *prime day* because the month and day (2/3) are represented by prime numbers. How many prime days are there in a leap year?

Applications and Problem Solving

41. Cryptography Use the formula in Example 3 to find *p* and *q* if *n* = 1643 for an RSA code.

42. Number Theory Mathematicians have many theories that are unproved. One theory is that there is an infinite number of *twin primes*. Twin primes are prime numbers that differ by 2, like 3 and 5. List all the twin primes that are less than 100.

43. History Chinese mathematician Sun-Tsŭ lived in the 1st century A.D. He studied *Chinese remainders*. This involves the remainders left when a number is divided by different primes.

   a. Find the least positive integer having the remainders 2, 3, 2 when divided by 3, 5, and 7 respectively.

   b. Is the least integer described in **part a** prime?

44. Family Activity The movie *Sneakers* involves a supposed invention that would break codes. Find the movie in a library or video store and watch it to see what the invention did.

Mixed Review

45. Probability If you toss a coin 4 times, how many different combinations of outcomes are possible? (Lesson 4-3)

46. Write $(-8)^7$ as the product of the same factor. (Lesson 4-2)

47. Geometry Find the perimeter and area of the rectangle at the right. (Lesson 3-5)

9 inches

17 inches

48. Solve $\frac{x}{-7} = -28$. (Lesson 3-3)

49. Replace ● in 6 ● −6 with <, >, or = to make a true sentence. (Lesson 2-3)

50. Rewrite $3 \cdot (12 - 4)$ using the distributive property. (Lesson 1-5)

51. Standardized Test Practice Which equation is equivalent to $6 + (x + 7) = 25$? (Lesson 1-4)

   A. $6x + 7 = 25$ **B.** $6x + 42 = 25$

   C. $x + 13 = 25$ **D.** $42 + x = 25$

52. Demographics The average American earns $1,235,720 and pays $178,364 in taxes in his or her lifetime. The amount of money that is left in a paycheck after taxes is called take-home pay. What is the average American's lifetime take-home pay? (Lesson 1-1)

   a. Which method of computation do you think is most appropriate? Justify your choice.

   b. Solve the problem using the four-step plan. Be sure to examine your solution.

For **Extra Practice**, see page 749.

4-4B Factor Patterns

An Extension of Lesson **4-4**

You already know that prime numbers have exactly two factors. Which numbers have exactly three factors? or four factors? In this Math Lab, you will investigate factor patterns.

The chart below shows the numbers 2 through 18 and their factors arranged according to the number of factors.

Exactly 2 Factors	Exactly 3 Factors	Exactly 4 Factors	Exactly 5 Factors	Exactly 6 Factors
2: 1, 2 **3:** 1, 3 **5:** 1, 5 **7:** 1, 7 **11:** 1, 11 **13:** 1, 13 **17:** 1, 17	**4:** 1, 2, 4 **9:** 1, 3, 9	**6:** 1, 2, 3, 6 **8:** 1, 2, 4, 8 **10:** 1, 2, 5, 10 **14:** 1, 2, 7, 14 **15:** 1, 3, 5, 15	**16:** 1, 2, 4, 8, 16	**12:** 1, 2, 3, 4, 6, 12 **18:** 1, 2, 3, 6, 9, 18

Your Turn Copy the chart above. Then use the graphing calculator program below to find the factors of the numbers 19 through 50. Place each number in the correct column of the chart.

Begin entering the graphing calculator program by pressing [PRGM] [◄] [ENTER]. Then type in the program title, FACTOR. Consult your User's Guide if you need help finding the keys for the functions. When you finish entering the program, press [2nd] [QUIT] to exit the programming mode. To run the program, press [PRGM], the number next to the title FACTOR, and [ENTER].

> **PROGRAM: FACTOR**
> : Input "ENTER NUMBER ", N
> : For (D, 1, N)
> : If iPart (N/D) = (N/D)
> : Disp D
> : End

1. Predict a number from 50 to 150 that can be placed in each column. Check your prediction by using the graphing calculator program.

2. Write a paragraph that describes the pattern in each column.

4-5 Greatest Common Factor (GCF)

You'll Learn

- To find the greatest common factor of two or more integers or monomials.

It's Important

You can use greatest common factors when determining the amount of lumber needed for a project.

Application: History

Very little of the mathematics of ancient China remains known to us today. The Chinese recorded information on bamboo, which did not last very long. Also, many of the books were destroyed during the reign of Emperor Shï Huang-ti. But the *Jiuzhang Suanshu*, or *The Nine Chapters on the Mathematical Art* survived to tell us a great deal about the mathematical accomplishments of the culture.

One of the chapters in the *Jiuzhang Suanshu* contains information on field measurement. In it, a procedure for finding the **greatest common factor (GCF)** of two numbers is described.

In Exercise 40, you will use the Chinese method to find the GCF.

Learning the Concept

The greatest of the factors of two or more numbers is called the greatest common factor. There are many methods for finding the greatest common factor. One method is to simply list the factors of each number and identify the greatest of the factors common to the numbers.

For example, consider finding the GCF of 78 and 91.

Factors of 78: 1, 2, 3, 6, 13, 26, 39, 78
Factors of 91: 1, 7, 13, 91

The common factors of 78 and 91, shown in blue, are 1 and 13. The GCF of 78 and 91 is 13.

Another method for finding the greatest common factor of two or more numbers is to find the prime factorization of the numbers and then find the product of their common factors.

Example **Use prime factorization to find the GCF of each set of numbers.**
a. 315 and 135

First find the prime factorization of each number.

$$315 = 3 \cdot 105 \qquad\qquad 135 = 5 \cdot 27$$
$$= 3 \cdot 5 \cdot 21 \qquad\qquad = 5 \cdot 3 \cdot 9$$
$$= 3 \cdot 5 \cdot 3 \cdot 7 \qquad\qquad = 5 \cdot 3 \cdot 3 \cdot 3$$

Then find the common factors.

$315 = 3 \cdot 5 \cdot 3 \cdot 7$ or $3 \cdot 3 \cdot 5 \cdot 7$ *The loops indicate each common factor.*

$135 = 5 \cdot 3 \cdot 3 \cdot 3$ or $3 \cdot 3 \cdot 5 \cdot 3$ *They are 3, 3, and 5.*

The greatest common factor of 315 and 135 is $3 \cdot 3 \cdot 5$ or 45.

b. 66, 90, and 150

Express each number as a product of prime factors.

$66 = 2 \cdot 3 \cdot 11$

$90 = 2 \cdot 3 \cdot 3 \cdot 5$ *The common factors are 2 and 3.*

$150 = 2 \cdot 3 \cdot 5 \cdot 5$

The GCF is $2 \cdot 3$ or 6.

Some real-life problems can be solved using the GCF.

Example **2**

Real World APPLICATION

Hobbies

Shameka is covering the surface of an end table with equal-sized ceramic tiles. The table is 30 inches long and 24 inches wide.
a. What is the largest square tile that Shameka can use and not have to cut any tiles?
b. How many tiles will Shameka need?

a. The size of the tile is the greatest common factor of 30 and 24.

$30 = 2 \cdot 3 \cdot 5$

$24 = 2 \cdot 2 \cdot 2 \cdot 3$

The GCF of 30 and 24 is $2 \cdot 3$ or 6. So Shameka should use tiles that are 6-inch squares.

b. $30 \div 6 = 5$ and $24 \div 6 = 4$

So Shameka will need $5 \cdot 4$ or 20 tiles to cover the table.

Connection to > Algebra

The product of the common prime factors of two or more monomials is their GCF.

Example **3**   **Find the greatest common factor of $36x^3y$ and $56xy^2$.**

$36x^3y = 2 \cdot 2 \cdot 3 \cdot 3 \cdot x \cdot x \cdot x \cdot y$

$56xy^2 = 2 \cdot 2 \cdot 2 \cdot 7 \cdot x \cdot y \cdot y$

The GCF of $36x^3y$ and $56xy^2$ is $2 \cdot 2 \cdot x \cdot y$ or $4xy$.

Communicating Mathematics

Read and study the lesson to answer these questions.

1. **Explain** how to find the greatest common factor of two or more numbers.

2. **Find** the common factors of 21 and 45 by using factor trees.

3. Name two numbers whose GCF is 15.

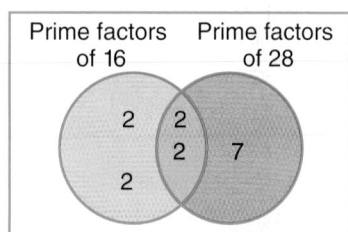

4. In 1880, English mathematician John Venn developed the use of diagrams to show the relationships between collections of objects. The diagram at the right shows the prime factors of 16 and 28. The common factors are in both circles, so the GCF is $2 \cdot 2$ or 4. Find the greatest common factor of 36 and 48 by making a Venn diagram.

Guided Practice

Find the GCF of each set of numbers or monomials.

5. 6, 8

6. 12, 8

7. 1, 20

8. $12x, 40x^2$

9. $-5ab, 6b^2$

10. $15a^2b^2, 27a^3b^3$

11. **Interior Design** Mrs. Garcia has picked out two different fabrics to use to make square pillows for her living room. One fabric comes in a width of 48 inches, and the other comes in a width of 60 inches. How long should each side of the squares for the pillows be if all the pillows are the same size and there is no fabric wasted?

Exercises: Practicing and Applying the Concept

Independent Practice

Find the GCF of each set of numbers or monomials.

12. 16, 56

13. 24, 40

14. 6, 8, 12

15. 12, 24, 36

16. 20, 21, 25

17. 108, 144

18. $14n, 42n^2$

19. $40x^2, 16x$

20. $24a^2, -60a$

21. $14b, -56b^2$

22. $33y^2, 44y$

23. $-18, 45mn$

24. $32mn^2, 16n, 12n^3$

25. $18a, 30ab, 42b$

26. $15v^2, 36w^2, 70vw$

27. $24x^3, 36y^3, 18xy$

28. $15r^2, 35s^2, 70rs$

29. $18ab, 6a^2, 42a^2b$

Two numbers are *relatively prime* if their only common factor is 1. Determine whether the numbers in each pair are relatively prime. Write *yes* or *no*.

30. 8 and 9   **31.** 11 and 13   **32.** 21 and 14

33. 25 and 30   **34.** 21 and 23   **35.** 9 and 12

Critical Thinking

36. Can the GCF of any set of numbers be greater than any one of the numbers? Explain.

Applications and Problem Solving

37. Patterns What is the GCF of all the numbers in the sequence 15, 30, 45, 60, 75, . . . ?

38. History Eratosthenes was a Greek mathematician who lived during the third century B.C. One of his contributions to the field was the *sieve of Eratosthenes*. Follow the steps below to complete a sieve.

MATERIALS

colored pencils or crayons

▶ Write the numbers 2 to 50 in a list.

▶ Circle 2, the first prime number. Then cross out every second number after 2.

▶ Using a different color, circle 3, the second prime number. Then cross out every third number after 3.

▶ Continue using different colors to circle prime numbers and cross out their multiples until all the numbers in the list have been circled or crossed out.

Determine the greatest common factor of 30 and 42 using the sieve. Explain how you found the GCF.

39. Carpentry Elena and her father are making shelves to store sports equipment and garden supplies in the garage. They would like to make the best use of a 48 in. by 72 in. piece of plywood. How many shelves measuring 12 in. by 16 in. could be cut from the plywood if there were no waste?

40. History Follow the steps outlined below to find the GCF of 42 and 86 by using the Chinese method mentioned at the beginning of the lesson.

▶ Subtract the lesser number, *a*, from the greater number, *b*.

▶ If the result is a factor of both numbers, it is the greatest common factor. If the result is not a factor of both numbers, subtract the result from *a* or subtract *a* from the result so that the difference is a positive number.

▶ Continue subtracting and checking the results until you find a number that is a factor of both numbers.

Mixed Review

41. Find the prime factorization of 3080. (Lesson 4-4)

42. Solve and graph $\frac{r}{4} \geq -14$. (Lesson 3-7)

43. Evaluate $|a| - |b| \cdot |c|$ if $a = -16$, $b = 2$, and $c = 3$. (Lesson 2-1)

44. Sports The table at the right shows the average cost of equipping an athlete for different sports. Make a bar graph of the costs. (Lesson 1-10)

Sport	Cost
Bowling	$100
Golf	$400
Racquetball	$60
Skiing	$500
Tennis	$75

45. Solve $c - 15 = 120$ using the inverse operation. (Lesson 1-8)

46. Business Book Distributors adds a $1.50 shipping and handling charge to the total price of every order. If the cost of a book is c, write an expression for the total cost. (Lesson 1-3)

47. Standardized Test Practice Manufacturers often deliver small items in a quantity called a great gross. There are 12 items in a dozen, 12 dozens in a gross, and 12 gross in a great gross. How many items are in a great gross? (Lesson 1-2)

A. 36   **B.** 144   **C.** 288   **D.** 1728

48. Automobiles Mechanics recommend changing the motor oil in your car every 3000 miles in order to keep your car running its best. If you just changed the oil and the odometer reads 56,893 miles, at what odometer reading should you change the oil next? (Lesson 1-1)

a. Which method of computation do you think is most appropriate? Justify your choice.

For **Extra Practice**, see page 749.

b. Solve the problem using the four-step plan. Be sure to examine your solution.

Self Test

Using divisibility rules, state whether each number is divisible by 2, 3, 5, 6, or 10. (Lesson 4-1)

1. 117   **2.** 1002   **3.** 57

Write each product using exponents. (Lesson 4-2)

4. $9 \cdot 9 \cdot 9 \cdot 9 \cdot 9$ **5.** $2(2 \cdot 2)$ **6.** $\underbrace{k \cdot k \cdot k \cdot \ldots \cdot k \cdot k \cdot k}_{28 \text{ factors}}$

7. How many cubes are in the twenty-fifth building in the sequence at the right? (Lesson 4-3)

Building 1   Building 2   Building 3

Factor each number or monomial completely. (Lesson 4-4)

8. 80   **9.** -26 **10.** $42xy^2$

Find the GCF of each set of numbers or monomials. (Lesson 4-5)

11. 42, 56   **12.** 9, 15, 24   **13.** $-18, 45xy$

HANDS-ON ACTIVITY

4-6A Equivalent Fractions

A Preview of Lesson **4-6**

MATERIALS

 ruler

It is often helpful to use a model to help you understand a concept. In this lab, you will use models to help you understand fractions.

Activity ① **Work with a partner to represent $\frac{1}{8}$ using a model.**

Use a ruler to draw a rectangle and separate the rectangle into eight equal parts. Shade one part. Since one part in eight is shaded, $\frac{1}{8}$ of the rectangle is shaded.

Your Turn   Draw a model to represent $\frac{1}{6}$.

1. How would you represent $\frac{2}{6}$ using a model?

2. If you shaded five parts of a rectangle with eight equal parts, what fraction is represented?

Activity ② **Compare $\frac{2}{8}$ and $\frac{3}{12}$ using models.**

Draw two identical rectangles and separate one rectangle into eight equal parts. Separate the other into twelve equal parts. Shade two parts of the rectangle that has eight sections and shade three parts of the rectangle that has twelve sections.

3. What do you notice about the shaded regions in the two rectangles in Activity 2?

4. Do the fractions $\frac{2}{8}$ and $\frac{3}{12}$ name the same number? Explain.

5. Are there other fractions that name the same number? If so, draw a diagram to show an example.

Extension   **Use models to determine whether each pair of fractions represents the same number.**

6. $\frac{1}{2}, \frac{5}{10}$ 7. $\frac{1}{3}, \frac{3}{9}$ 8. $\frac{1}{7}, \frac{2}{12}$

Math Lab 4-6A   *Equivalent Fractions*   **195**

Simplifying Fractions

You'll Learn

■ To simplify fractions.

It's Important

Simplifying fractions makes it easy to compare fractions.

Application: Forestry

When the National Arbor Day Foundation, the world's largest tree-planting advocacy group, needs a farmer to speak on their behalf, they turn to Chuck Leavell.

Fifteen years ago, Mr. Leavell was a rock 'n' roller who had no experience with tree farming. His life changed when his wife's grandparents left her a 1200-acre farm in Twiggs County, Georgia. "The more I learned, the more I fell in love with it," said Leavell. Now the Leavells divide their time between their two loves, music and forestry. Their farm has now grown to 1500 acres. You can compare the original size of their farm to its current size by using a ratio.

Chuck Leavell

Learning the Concept

A **ratio** is a comparison of two numbers by division. A ratio can be expressed in several ways. The expressions below all represent the same ratio.

$$2 \text{ to } 3 \qquad 2 : 3 \qquad \frac{2}{3} \qquad 2 \div 3$$

A ratio is most often written as a fraction in **simplest form**. A fraction is in simplest form when the GCF of the numerator and the denominator is 1.

Example ① **Real World APPLICATION**

Forestry

Refer to the application at the beginning of the lesson. Write the ratio of the amount of land that Mrs. Leavell inherited to the size of the farm today in simplest form.

Mrs. Leavell inherited 1200 acres, and the farm is 1500 acres today. So the ratio is $\frac{1200}{1500}$.

First find the GCF of the numerator and denominator.

$1200 = 2 \cdot 2 \cdot 2 \cdot 2 \cdot 3 \cdot 5 \cdot 5$ *The GCF of 1200 and*
$1500 = 2 \cdot 2 \cdot 3 \cdot 5 \cdot 5 \cdot 5$ *1500 is $2 \cdot 2 \cdot 3 \cdot 5 \cdot 5$ or 300.*

Now divide the numerator and the denominator by 300 to write the fraction in simplest form.

$$\frac{1200}{1500} \xlongequal{\div 300, \div 300} \frac{4}{5}$$

Since the GCF of 4 and 5 is 1, the fraction $\frac{4}{5}$ is in simplest form.

The division in Example 1 can be represented in a different way.

$$\frac{1200}{1500} = \frac{\overset{1}{\cancel{2}} \cdot \overset{1}{\cancel{2}} \cdot 2 \cdot 2 \cdot \overset{1}{\cancel{3}} \cdot \overset{1}{\cancel{5}} \cdot \overset{1}{\cancel{5}} \cdot 5}{\underset{1}{\cancel{2}} \cdot \underset{1}{\cancel{2}} \cdot \underset{1}{\cancel{3}} \cdot \underset{1}{\cancel{5}} \cdot \underset{1}{\cancel{5}} \cdot 5} = \frac{2 \cdot 2}{5} \text{ or } \frac{4}{5}$$
The slashes indicate that the numerator and denominator are both divided by $2 \cdot 2 \cdot 3 \cdot 5 \cdot 5$, the GCF.

Example 2 Write $\frac{12}{40}$ in simplest form.

$$\frac{12}{40} = \frac{\overset{1}{\cancel{2}} \cdot \overset{1}{\cancel{2}} \cdot 3}{\underset{1}{\cancel{2}} \cdot \underset{1}{\cancel{2}} \cdot 2 \cdot 5}$$
Divide both the numerator and denominator by $2 \cdot 2$.

$$= \frac{3}{10}$$

Connection to > Algebra

A fraction with variables in the numerator or denominator is called an **algebraic fraction**. Algebraic fractions can also be written in simplest form.

Example 3 Simplify $\frac{16x^2y^3}{12xy^4}$. Assume that x and y are not equal to zero.

THINK ABOUT IT
What is the GCF of $16x^2y^3$ and $12xy^4$?

$$\frac{16x^2y^3}{12xy^4} = \frac{\overset{1}{\cancel{2}} \cdot \overset{1}{\cancel{2}} \cdot 2 \cdot 2 \cdot \overset{1}{\cancel{x}} \cdot x \cdot \overset{1}{\cancel{y}} \cdot \overset{1}{\cancel{y}} \cdot \overset{1}{\cancel{y}}}{\underset{1}{\cancel{2}} \cdot \underset{1}{\cancel{2}} \cdot 3 \cdot \underset{1}{\cancel{x}} \cdot \underset{1}{\cancel{y}} \cdot \underset{1}{\cancel{y}} \cdot \underset{1}{\cancel{y}} \cdot y}$$
Divide both numerator and denominator by $2 \cdot 2 \cdot x \cdot y \cdot y \cdot y$.

$$= \frac{4x}{3y}$$

Checking Your Understanding

Communicating Mathematics

Read and study the lesson to answer these questions.

1. **Explain** what is meant by expressing a fraction in simplest form.

2. **Express** the ratio 12:14 as a fraction in simplest form.

3. **Give examples** of fractions in simplest form and fractions that are not in simplest form.

4. Why is it important to be able to simplify fractions?

5. The fractions $\frac{3}{4}$ and $\frac{6}{8}$ are graphed on the number lines at the right. Notice that the fractions have the same graph. Thus, $\frac{3}{4}$ and $\frac{6}{8}$ are equivalent. Use number lines to determine whether $\frac{3}{9}$ and $\frac{1}{3}$ are equivalent.

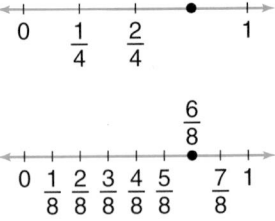

Guided Practice

Express each ratio as a fraction. Then if the fraction is not in simplest form, write it in simplest form.

6. 6 to 10   7. $1 \div 7$ 8. $20 : 100$ 9. eleven to fifteen

Write each fraction in simplest form. If the fraction is already in *simplest* form, write simplified.

10. $\frac{10}{37}$ 11. $\frac{15}{21}$ 12. $\frac{51}{60}$ 13. $\frac{18}{44}$

14. $\frac{x}{x^3}$ 15. $\frac{11t}{121t^2}$ 16. $\frac{8z^2}{16z}$ 17. $\frac{8ab}{15cd}$

18. **Economics** How does Uncle Sam spend a dollar? The graph at the right shows how each dollar spent by the Federal Government is used.

    a. Write the ratio of the amount spent on housing assistance to total spending as a fraction in simplest form.

    b. Write the ratio of the amount spent on the national debt to total spending as a fraction in simplest form.

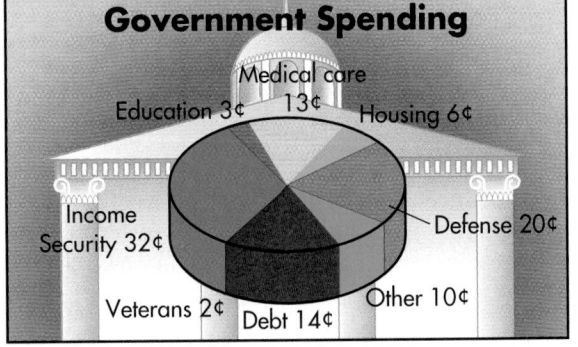

Source: The Tax Foundation

Exercises: Practicing and Applying the Concept

Independent Practice

Write each fraction in simplest form. If the fraction is already in simplest form, write *simplified*.

19. $\frac{2}{14}$ 20. $\frac{17}{20}$ 21. $\frac{34}{38}$ 22. $\frac{20}{53}$ 23. $\frac{17}{51}$

24. $\frac{9}{15}$ 25. $\frac{25}{40}$ 26. $\frac{30}{51}$ 27. $\frac{30}{37}$ 28. $\frac{16}{64}$

29. $\frac{11}{13}$ 30. $\frac{8}{27}$ 31. $\frac{124}{222}$ 32. $\frac{12m}{15m}$ 33. $\frac{12x}{15y}$

34. $\frac{40a}{42a}$ 35. $\frac{82t}{14t}$ 36. $\frac{12cd}{19ef}$ 37. $\frac{xyz^3}{x^2y}$ 38. $\frac{30x^2}{51xy}$

39. $\frac{40p^2q}{52pq^2}$ 40. $\frac{17k^2z}{51z}$ 41. $\frac{31gh}{14}$ 42. $\frac{52rst}{26rst}$ 43. $\frac{765}{2023}$

Critical Thinking

44. Suppose $(1 \times 10^a) + (2 \times 10^b) + (3 \times 10^c) + (4 \times 10^d) = 24{,}130$, and a, b, c and d are all positive integers. Find the value of $\dfrac{a + b + c + d}{16}$.

Applications and Problem Solving

45. Transportation According to a demographics magazine, the average American worker spends 22 minutes traveling to work. What fraction of the day is this?

46. Demographics The United States is aging. Not only is the country getting older, but the number of older Americans increases each year. There are 2,500,000 Americans over the age of 85 and 30,000 of them are 100 or older.

  a. What is the ratio of the number of Americans age 100 or over to the number of Americans over age 85?

  b. Write the ratio in simplest terms.

  c. Explain what the ratio in part b means.

47. Medicine The University of California recommends using a soft gel pack to relieve muscle pain. To make a soft gel pack, fill a plastic bag with 2 ounces of rubbing alcohol and 6 ounces of water. Seal the bag and put it in a freezer. Apply to the injured area.

  a. What is the ratio of rubbing alcohol to water for the gel pack?

  b. Find a fraction in simplest form to represent the ratio.

  c. Find other amounts of rubbing alcohol and water that could be used to make a gel pack.

Mixed Review

48. Find the GCF of $42x^2y$ and $38xy^2$. (Lesson 4-5)

49. Solve $v = \ell \cdot w \cdot h$, by replacing the variables with the values $\ell = 6$, $w = 3$, and $h = 6$. (Lesson 3-4)

50. Standardized Test Practice Solve $180 = 15r$. (Lesson 3-3)
  A. 12   **B.** 165   **C.** 195   **D.** 2700

51. Find the product of -4 and $-17d$. (Lesson 2-7)

52. History Ohio was declared a state in 1803. Hawaii became a state in 1959. Write a mathematical statement comparing the ages of these states in 1996. (Lesson 2-3)

For **Extra Practice**, see page 749.

53. Solve $\dfrac{144}{x} = 12$ mentally. (Lesson 1-6)

54. Evaluate $324 \div (2 \times 3 + 6)$. (Lesson 1-2)

4-7 Using the Least Common Multiple (LCM)

You'll Learn

- To find the least common multiple of two or more numbers and,

- to compare fractions.

It's Important

Using the least common multiple will help you determine when the arrival of two flights will coincide.

Application: Astronomy

Some clear night, go to a park or out in the country and spend some time looking at the night sky. In ideal conditions, you will be able to see about 2500 stars, although there are many more than this. Some nights you will even be able to see some planets.

The chart at the right shows the number of Earth years it takes for some of the planets to revolve around the Sun. In 1982, the planets Jupiter and Saturn were in conjunction; that is, they appeared very close to each other in the sky. How often do these planets line up in these positions? *This problem will be solved in Example 1.*

Planet	Revolution Time (in Earth Years)
Earth	1
Jupiter	12
Saturn	30
Uranus	84

Learning the Concept

A **multiple** of a number is a product of that number and any whole number. *Recall that the whole numbers are 0, 1, 2, 3, . . .*

multiples of 2: 0, 2, 4, 6, 8, 10, 12, 14, 16, 18, 20, . . .
multiples of 5: 0, 5, 10, 15, 20, 25, 30, 35, 40, . . .

Multiples that are shared by two or more numbers are called **common multiples**. Some of the common multiples of 2 and 5, shown in blue, are 0, 10, and 20.

The least of the nonzero common multiples of two or more numbers is called the **least common multiple (LCM)** of the numbers. The least common multiple of 2 and 5 is 10.

Example ① Refer to the application at the beginning of the lesson.
  a. How often will Jupiter and Saturn appear close in the sky?
  b. When will this happen next?
  c. How often do Saturn, Jupiter, and Uranus coincide?

Astronomy

  a. According to the table, Jupiter revolves around the Sun every 12 Earth years and Saturn every 30 years. The time required for the planets to coincide in this way again will be a multiple of 12 and 30.

  multiples of 12: 0, 12, 24, 36, 48, 60, . . .
  multiples of 30: 0, 30, 60, 90, 120, . . .

  The least common multiple of 12 and 30 is 60. Jupiter and Saturn coincide every 60 years.

b. The last time Jupiter and Saturn coincided was 1982. The next time will be 1982 + 60 or 2042.

c. Find the LCM of 12, 30, and 84 to find the number of years for Saturn, Jupiter, and Uranus to coincide in the same way.

A second method for finding an LCM is to find the prime factorization of the numbers. A common multiple of the numbers contains *all* of the prime factors of each number. The LCM contains *each* factor the greatest number of times it appears for any of the numbers.

$$12 = 2 \cdot 2 \cdot 3$$
$$30 = 2 \cdot 3 \cdot 5$$
$$84 = 2 \cdot 2 \cdot 3 \cdot 7$$

Find the common factors. Then multiply all of the factors, using the common factors only once.

$$2 \cdot 2 \cdot 3 \cdot 5 \cdot 7 = 420$$

Jupiter, Saturn, and Uranus will coincide in this way every 420 years.

Connection to Algebra

The LCM of two or more monomials is found in the same way as the LCM of two or more numbers.

Example 2   **Find the LCM of $12a^2b$ and $8ac$.**

$$12a^2b = 2 \cdot 2 \cdot 3 \cdot a \cdot a \cdot b$$
$$8ac \;\;\;= 2 \cdot 2 \cdot 2 \cdot a \cdot c$$

The LCM is $2 \cdot 2 \cdot 2 \cdot 3 \cdot a \cdot a \cdot b \cdot c$ or $24a^2bc$.

One way to compare fractions is to write them with the *same* denominator. While any common denominator could be used, the least common multiple of the denominators, or the **least common denominator (LCD)** is usually the most convenient.

Example Which is greater, $\frac{5}{6}$ or $\frac{7}{9}$?

First, find the LCM of 6 and 9.

$6 = 2 \cdot 3$
$9 = 3 \cdot 3$

The LCM is $2 \cdot 3 \cdot 3$ or 18. Find equivalent fractions with 18 as the denominator.

$\frac{5}{6} = \frac{\bullet}{18} \quad \rightarrow \quad \frac{5}{6} = \frac{15}{18}$ ($\times 3$)

$\frac{7}{9} = \frac{\bullet}{18} \quad \rightarrow \quad \frac{7}{9} = \frac{14}{18}$ ($\times 2$)

Since $\frac{15}{18} > \frac{14}{18}$, $\frac{5}{6} > \frac{7}{9}$.

Connection to Algebra

You can also find the LCD for algebraic fractions with different denominators.

Example Find the LCD for $\frac{5}{6k^2}$ and $\frac{3}{8km}$.

$6k^2 = 2 \cdot 3 \cdot k \cdot k$
$8km = 2 \cdot 2 \cdot 2 \cdot k \cdot m$

The LCM of $6k^2$ and $8km$ is $2 \cdot 2 \cdot 2 \cdot 3 \cdot k \cdot k \cdot m$ or $24k^2m$.

Thus, the LCD of $\frac{5}{6k^2}$ and $\frac{3}{8km}$ is $24k^2m$.

Checking Your Understanding

Communicating Mathematics

Read and study the lesson to answer these questions.

1. **Explain** why the definition of the least common multiple says that it is the least *nonzero* multiple common to the numbers involved.

2. **Demonstrate** how to find the LCM of two numbers by using the prime factorizations.

3. **Compare and contrast** the LCM and the LCD.

4. The Venn diagram at the right shows the prime factorizations of 12 and 16. Using each number shown as a factor will give you the LCM of the numbers. The LCM of 12 and 16 is $2 \cdot 2 \cdot 2 \cdot 2 \cdot 3$ or 48. Use a Venn diagram to find the LCM of 12, 28, and 48.

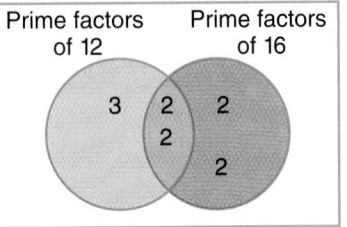

Guided Practice

Find the least common multiple (LCM) of each set of numbers or algebraic expressions.

5. 10, 14
6. 7, 9
7. $20a$, $12a$
8. 12, 16, 24
9. 45, 30, 35
10. a, b

Find the least common denominator (LCD) for each pair of fractions.

11. $\frac{4}{5}, \frac{1}{2}$ **12.** $\frac{3}{4}, \frac{1}{8}$ **13.** $\frac{3a}{25}, \frac{17a}{20}$

Replace each ● with <, >, or = to make a true statement.

14. $\frac{5}{12}$ ● $\frac{1}{2}$ **15.** $\frac{5}{7}$ ● $\frac{7}{9}$ **16.** $\frac{14}{15}$ ● $\frac{9}{10}$

17. Food Americans ate about 606 million pounds of potato chips one summer. That is about the weight of six ships! The graph at the right shows the regions where most of the chips were eaten.

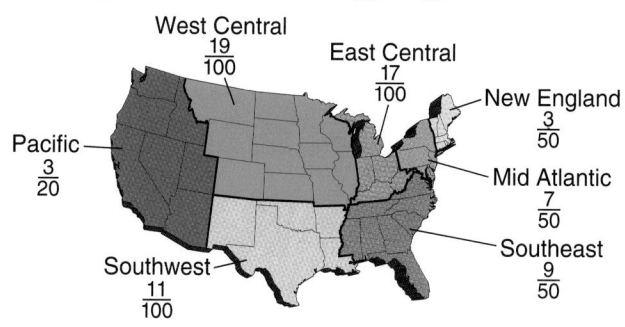

West Central $\frac{19}{100}$

East Central $\frac{17}{100}$

New England $\frac{3}{50}$

Pacific $\frac{3}{20}$

Mid Atlantic $\frac{7}{50}$

Southeast $\frac{9}{50}$

Southwest $\frac{11}{100}$

   a. Were more chips eaten in the Pacific or in the East Central region?

   b. In which region were the most chips eaten?

   c. In which region were the fewest chips eaten?

Exercises: Practicing and Applying the Concept

Independent Practice

Find the least common multiple (LCM) of each set of numbers or algebraic expressions.

18. 20, 12   **19.** 2, 9   **20.** 15, 75

21. 18, 32   **22.** 24, 32   **23.** 21, 28

24. $6c, 8cd$ **25.** $7y, 12y$ **26.** $20cd, 50d$

27. 10, 20, 40   **28.** 7, 21, 84   **29.** 9, 12, 15

30. $3b, 5c, 2$ **31.** $3t, 5t^2, 7$ **32.** $3a^2, 6a^3, 9a^4$

Find the least common denominator (LCD) for each set of fractions.

33. $\frac{1}{4}, \frac{3}{16}$ **34.** $\frac{5}{6}, \frac{3}{8}$ **35.** $\frac{3}{5}, \frac{4}{7}$ **36.** $\frac{2}{9}, \frac{5}{12}$

37. $\frac{5}{11}, \frac{11}{20}$ **38.** $\frac{1}{2}, \frac{1}{3}, \frac{2}{5}$ **39.** $\frac{6c}{5b}, \frac{7c}{25b}$ **40.** $\frac{7}{8a}, \frac{2}{15a}$

Replace each ● with <, >, or = to make a true statement.

41. $\frac{5}{6}$ ● $\frac{9}{11}$ **42.** $\frac{3}{5}$ ● $\frac{4}{7}$ **43.** $\frac{35}{51}$ ● $\frac{12}{17}$ **44.** $\frac{10}{12}$ ● $\frac{9}{11}$

45. $\frac{15}{32}$ ● $\frac{1}{2}$ **46.** $\frac{21}{100}$ ● $\frac{1}{5}$ **47.** $\frac{8}{9}$ ● $\frac{19}{21}$ **48.** $\frac{7}{9}$ ● $\frac{8}{10}$

49. Use $\frac{3}{8}, \frac{5}{12}$, and < to write a true statement.

Critical Thinking

50. Two numbers have a LCM of $2^2 \cdot 3 \cdot 5^2$. Their GCF is $2 \cdot 5$. If one of the numbers is $2 \cdot 3 \cdot 5$, what is the other number?

Applications and Problem Solving

51. Veterinary Medicine According to the Humane Society, caring for a dog costs an average of $7815 over 12 years or $\frac{\$7815}{12}$ a year. Caring for a cat costs $4723 over 15 years or $\frac{\$4723}{15}$ a year. Which is more costly, caring for a cat or a dog? Explain.

Carol Moseley Braun

52. Politics In the United States, presidents are elected every four years. Senators are elected every six years. In 1992, a presidential election year, Carol Moseley Braun of Illinois became the first African-American woman elected to the U.S. Senate. If she continues to run and win each time her term expires, when is the next presidential election year in which Senator Braun will run?

53. Technology A research firm studied the amount of technology that American teens use. The table at the right shows the fractions of the teens surveyed who had different technology at home.

    a. Did more teens say that they had cable TV or a video game player?

    b. Which technology did most teens use?

    c. Which technology was used the least?

Technology	Users
Cable TV	$\frac{2}{3}$
Cellular phone	$\frac{7}{20}$
Computer	$\frac{23}{50}$
On-line services	$\frac{17}{100}$
Video game player	$\frac{77}{100}$
VCR	$\frac{24}{25}$

Source: Chilton Research

Mixed Review

54. Write $\frac{30}{36}$ in simplest form. (Lesson 4-6)

55. If the quotient of x and 8 is greater than 72, find x. (Lesson 3-8)

56. Standardized Test Practice The daily cost of renting a car is $30.00 plus $0.25 per mile driven. This is represented by the equation $c = \$30 + 0.25m$, where c is the total cost and m is the number of miles driven. What is the cost of renting the car for one day and driving 35 miles? (Lesson 3-4)

    A. $10.75

    B. $35.50

    C. $38.75

    D. $55

57. Solve $14 = \frac{a}{-7}$. (Lesson 3-3)

58. Evaluate $a \cdot b$ if $a = -3$ and $b = 14$. (Lesson 2-7)

The table below shows the sizes of several national parks in acres. (Lesson 1-10)

59. Which state has the most parks listed?

60. What is the difference between the size of the largest and smallest parks?

For **Extra Practice**, see page 749.

61. Make a bar graph of the data.

62. Solve $5g = 65$ mentally. (Lesson 1-6)

Park	Size (acres)
Acadia, ME	41,951
Arches, UT	73,374
Badlands, SD	242,756
Big Bend, TX	801,163
Biscayne, FL	172,925
Canyonlands, UT	337,570
Capital Reef, UT	241,904
Carlsbad Caverns, NM	46,766

Source: World Almanac

4-8

Multiplying and Dividing Monomials

You'll Learn

■ To multiply and divide monomials.

It's Important

Knowing how to multiply monomials will help you determine the increase of a financial investment.

Technology Activity

In dealing with large numbers and small numbers, calculators can simplify and speed our computation.

Your Turn Copy the tables below. Then use a calculator to find each product and quotient and complete the table.

Factors	Product	Product written as a power
$10^1 \cdot 10^1$		
$10^1 \cdot 10^2$		
$10^1 \cdot 10^3$		
$10^1 \cdot 10^4$		
$10^1 \cdot 10^5$		

Division	Quotient	Quotient written as a power
$10^5 \div 10^1$		
$10^4 \div 10^1$		
$10^3 \div 10^1$		
$10^2 \div 10^1$		
$10^1 \div 10^1$		

a. Compare the exponents of the factors to the exponent in the products. What do you observe?

b. Compare the exponents of the division expressions to the exponents in the quotients. What do you observe?

c. Write a rule for determining the exponent of the product when you multiply powers. Test your rule by multiplying $2^2 \cdot 2^4$ using a calculator.

d. Write a rule for determining the exponent in the quotient when you divide powers. Test your rule by dividing 8^6 by 8^4 on a calculator.

Learning the Concept

The pattern you observed in the products table above leads to the following rule for multiplying powers that have the same base.

Product of Powers

In words: You can multiply powers *that have the same base* by adding their exponents.

In symbols: For any number a and positive integers m and n, $a^m \cdot a^n = a^{m+n}$.

Example ❶   Find $8^4 \cdot 8^5$.

$8^4 \cdot 8^5 = 8^{4+5}$ or 8^9

Check: $8^4 \cdot 8^5 = (8 \cdot 8 \cdot 8 \cdot 8)(8 \cdot 8 \cdot 8 \cdot 8 \cdot 8)$
$= 8 \cdot 8 \cdot 8 \cdot 8 \cdot 8 \cdot 8 \cdot 8 \cdot 8 \cdot 8$ or 8^9

Connection to Algebra

Some monomials can also be multiplied using the rule for the product of powers.

Example ❷   Find each product.

   a. $n^2 \cdot n^4$

   $n^2 \cdot n^4 = n^{2+4}$ or n^6

   b. $(-6x^2)(5x^3)$

   $(-6x^2)(5x^3) = (-6 \cdot 5)(x^2 \cdot x^3)$ *Commutative and associative*
                                            properties
               $= (-30)(x^{2+3})$ *Product of powers*
               $= -30x^5$

You can also write a rule for finding quotients of powers. Study the pattern of division sentences below. Each dividend, divisor, and quotient has been replaced with a power of 2.

$2^1 = 2$		
$2^2 = 4$		
$2^3 = 8$		
$2^4 = 16$		
$2^5 = 32$		
$2^6 = 64$		
$2^7 = 128$		
$2^8 = 256$		

$16 \div 2 = 8$ $64 \div 8 = 8$ $256 \div 8 = 32$
$\downarrow \quad \downarrow \quad \downarrow$ $\downarrow \quad \downarrow \quad \downarrow$ $\downarrow \quad \downarrow \quad \downarrow$
$2^4 \div 2^1 = 2^3$ $2^6 \div 2^3 = 2^3$ $2^8 \div 2^3 = 2^5$

Look at the exponents only. Do you see a pattern?

Quotient of Powers	**In words:**	You can divide powers *that have the same base* by subtracting their exponents.
	In symbols:	For any nonzero number a and whole numbers m and n, $\dfrac{a^m}{a^n} = a^{m-n}$.

Example ❸   Find each quotient.

   a. $\dfrac{9^7}{9^5}$ **b.** $\dfrac{c^8}{c^8}$

   $\dfrac{9^7}{9^5} = 9^{7-5}$ or 9^2 $\dfrac{c^8}{c^8} = c^{8-8}$

                                   $= c^0$ or 1

   Check: $\dfrac{9^7}{9^5} = \dfrac{\overset{1}{9} \cdot \overset{1}{9} \cdot \overset{1}{9} \cdot \overset{1}{9} \cdot \overset{1}{9} \cdot 9 \cdot 9}{\underset{1}{9} \cdot \underset{1}{9} \cdot \underset{1}{9} \cdot \underset{1}{9} \cdot \underset{1}{9}}$ **Check:** $\dfrac{c^8}{c^8} = \dfrac{\overset{1}{c} \cdot \overset{1}{c} \cdot \overset{1}{c} \cdot \overset{1}{c} \cdot \overset{1}{c} \cdot \overset{1}{c} \cdot \overset{1}{c} \cdot \overset{1}{c}}{\underset{1}{c} \cdot \underset{1}{c} \cdot \underset{1}{c} \cdot \underset{1}{c} \cdot \underset{1}{c} \cdot \underset{1}{c} \cdot \underset{1}{c} \cdot \underset{1}{c}}$

                    $= 9 \cdot 9$ $= 1$
                    $= 9^2$

Many applications involve calculations with very large or very small numbers. The quotient of powers rule often makes these calculations less complicated.

Example 4

Real World **APPLICATION**

Seismology

The intensity of an earthquake is usually measured on the Richter scale. On this scale the number designation given to an earthquake represents the intensity as a power of 10. For example, an earthquake measuring 8 on the Richter scale has an intensity of 10^7. In April, 1992, an earthquake measuring 5 on the Richter scale shook the Netherlands. Twelve days later, northern California experienced an earthquake measuring 7. How many times more intense was the California earthquake?

The Netherlands earthquake had an intensity of 10^4, and the California earthquake had an intensity of 10^6. Divide 10^6 by 10^4 to find out how many times more intense the California earthquake was.

$$\frac{10^6}{10^4} = 10^{6-4} \quad \textit{Quotient of powers}$$
$$= 10^2 \text{ or } 100$$

The California earthquake was 100 times more intense.

Checking Your Understanding

Communicating Mathematics

Read and study the lesson to answer these questions.

1. **Explain** how to find the product $6^6 \cdot 6^3$.

2. **Justify** the product of powers rule by multiplying 2^2 and 2^5 using the definition of exponents.

3. **Make up** a division problem whose solution is 4^4.

4. **Describe**, in your own words, the relationship between the rules for multiplying powers and dividing powers.

Guided Practice

✓ **Choose**

Estimation
Mental Math
Calculator
Paper and Pencil

Find each product or quotient. Express your answer in exponential form.

5. $10^4 \cdot 10^3$　　6. $\frac{2^5}{2^2}$　　7. $3^6 \cdot 3^4$　　8. $\frac{8^4}{8^3}$

9. $8^2 \cdot 8^3 \cdot 8$　　10. $m \cdot m^6$　　11. $\frac{y^{11}}{y^9}$　　12. $\frac{ab^4}{b^2}$

13. **Seismology** On June 6, 1994, an earthquake measuring 6.8 on the Richter scale struck southwest Colombia. Three days later an earthquake measuring 8.2 hit La Paz, Bolivia. *Approximately* how many times more intense was the earthquake in Bolivia?

Independent Practice

Find each product or quotient. Express your answer using exponents.

14. $10^5 \cdot 10^5$ **15.** $w \cdot w^5$ **16.** $3^3 \cdot 3^2$ **17.** $b^5 \cdot b^2$

18. $\dfrac{5^5}{5^2}$ **19.** $\dfrac{10^{10}}{10^3}$ **20.** $\dfrac{z^3}{z}$ **21.** $\dfrac{a^{10}}{a^6}$

22. $\dfrac{(-2)^6}{(-2)^5}$ **23.** $(5x^3)(4x^4)$ **24.** $\dfrac{(-x)^4}{(-x)^3}$ **25.** $\dfrac{f^{20}}{f^8}$

26. $(3x^4)(-5x^2)$ **27.** $3a^2 \cdot 4a^3$ **28.** $(-10x^3)(2x^2)$

29. $a^2b \cdot ab^3$ **30.** $x^4(x^3y^2)$ **31.** $t^3t^2 \div t^4$

32. $\dfrac{y^{100}}{y^{100}}$ **33.** $k^3m^2 \div km$ **34.** $(-3y^3z)(7y^4)$

35. $w^5xy^2 \div wx$ **36.** $7ab(a^{16}c)$ **37.** $15n^9q^2 \div 5nq^2$

Find each missing exponent.

38. $(5^\bullet)(5^3) = 5^{11}$ **39.** $x(x^2)(x^6) = x^\bullet$ **40.** $\dfrac{t^\bullet}{t} = t^{14}$

41. $3^5 \cdot 3^\bullet = 3^8$ **42.** $\dfrac{12^5}{12^\bullet} = 1$ **43.** $\dfrac{2^5}{4^2} = 2^\bullet$

Critical Thinking

44. If $10^n \div 10^m = 1$, what can you conclude about n and m?

Applications and Problem Solving

45. Chemistry The pH of a solution is a measure of its acidity. A low pH indicates an acid and a high pH indicates a base. Neutral water has a pH of 7; this is the dividing line between acids and bases. Each one-unit increase or decrease in the pH means that the intensity of the acid or base is changed by a factor of ten. For example, a pH of 4 is ten times more acidic than a pH of 5.

    a. Acid rain is an environmental problem related to the burning of fossil fuels. Suppose the pH of a lake is 5 due to acid rain. How much more acidic is the lake than neutral water?

    b. Human blood usually has a pH of about 7.4. Is it an acid or a base?

46. Biology When bacteria reproduce, they split so that one cell becomes two. The number of cells after t time periods is 2^t.

    a. *E. coli* reproduce very quickly, about every 15 minutes. If there are 100 *e. coli* in a dish now, how many will there be in 30 minutes?

    b. How many more *e. coli* are there in a population after 3 hours than there were after 1 hour?

Mixed Review

47. Replace $\bullet$ with $<$ or $>$ to make $\dfrac{5}{8} \bullet \dfrac{8}{11}$ a true inequality. (Lesson 4-7)

48. Find the prime factorization of $16ab^3$. (Lesson 4-4)

49. Use divisibility rules to determine whether 298 is divisible by 2, 3, 5, 6, or 10. (Lesson 4-1)

50. **Art** A photograph measuring 9" by 11" is matted and placed in a 10" by 12" frame. Find the area of the matting. (Lesson 3-5)

51. Solve $5 + n = -2$ and check your solution. Graph the solution on a number line. (Lesson 3-2)

52. **Standardized Test Practice** Solve $s = 40 - (-17)$. (Lesson 2-5)

   A. -57 **B.** -23 **C.** 23 **D.** 57

53. In which quadrant is the graph of $(-3, 4)$ located? (Lesson 2-2)

54. Is $4x < 16$ *true*, *false*, or *open*? (Lesson 1-9)

55. **Music** A survey of 6500 Top-10 songs released between 1900 and 1975 showed that "I" is the most frequent first word of a song. "My" was the second most popular first word. "I" was the first word of four less than twice as many songs as the word "my." Write an expression for the number of songs that begin with "my" if x songs started with "I." (Lesson 1-3)

For **Extra Practice**, see page 750.

WORKING ON THE

Investigation

Refer to the Investigation on pages 166–167.

You can use a **spreadsheet** to find the value of your portfolio on any given day. Using a spreadsheet, it is possible to project results, make calculations, and print almost anything that can be arranged in a table.

The basic unit of a spreadsheet is called a **cell**. A cell may contain numbers, words (called labels), or a formula. Each cell is named by the column and row that describes its location. The cell C2 is the box at the intersection of column C and row 2. *Cell C2 in the spreadsheet at the right contains the number 20.*

As you know, stock prices are listed as mixed numbers. You can enter a mixed number in a spreadsheet as an expression. For example, the price of Company M's stock was listed as $48\frac{3}{8}$. It can be entered in cell B2 of the spreadsheet as $48 + \left(\frac{3}{8}\right)$ or 48.375.

When cells are related, a formula can be used to generate this relationship. If you type B2*C2 into cell D2, the cell will not show the formula, but the result of the multiplication of the contents of B2 and C2.

	A	B	C	D
1	Company	Price	Shares	Value
2	M	48.375	20	
3	N	70.375	10	
4	O	50	100	
5	P	48	40	
6	Q	109.625	10	
7	R	32.75	10	
8	Total Value			

- Write spreadsheet formulas for cells D3, D4, D5, D6, D7, and D8.

- Create your own spreadsheet to find the value of your portfolio. Update the value using share prices in a local or national newspaper each week. Also document any changes in your portfolio such as selling old stock holdings and buying new ones.

Add the results of your work to your Investigation Folder.

Negative Exponents

Technology Activity

You can often discover important concepts by looking for a pattern.

Your Turn Copy the table at the right. Then use a calculator to complete the second column.

a. Describe how successive expressions differ.

b. Describe how successive values differ.

c. If the pattern continues, what is the next expression?

d. If the pattern continues, what is the next value?

e. Use a calculator to verify that the expression you wrote for part c is equal to the value you wrote for part d.

Expression	Value
2^6	64
2^5	32
2^4	
2^3	
2^2	
2^1	
2^0	

Learning the Concept

Extending the pattern above suggests that $2^{-1} = \frac{1}{2}$. You can use the quotient of powers rule and the definition of a power to simplify the expression $\frac{x^3}{x^6}$ and write a general rule about negative powers.

Method 1: Quotient of Powers Rule

$$\frac{x^3}{x^6} = x^{3-6}$$

$$= x^{-3}$$

Method 2: Definition of Powers

$$\frac{x^3}{x^6} = \frac{\overset{1}{\cancel{x}} \cdot \overset{1}{\cancel{x}} \cdot \overset{1}{\cancel{x}}}{\underset{1 \quad 1 \quad 1}{\cancel{x} \cdot \cancel{x} \cdot \cancel{x} \cdot x \cdot x \cdot x}}$$

$$= \frac{1}{x \cdot x \cdot x}$$

$$= \frac{1}{x^3}$$

Since $\frac{x^3}{x^6}$ cannot have two different values, you can conclude that x^{-3} is equal to $\frac{1}{x^3}$. This and other examples suggest the following definition.

Negative Exponents	For any nonzero number a and any integer n, $a^{-n} = \frac{1}{a^n}$.

Example 1 **Write each expression using positive exponents.**

a. 10^{-5}

$10^{-5} = \dfrac{1}{10^5}$ *Definition of negative exponents*

b. ab^{-3}

$ab^{-3} = a \cdot b^{-3}$

$= a \cdot \dfrac{1}{b^3}$ *Definition of negative exponents*

$= \dfrac{a}{b^3}$

Negative exponents are often used in science when dealing with very small numbers.

Example 2

Red blood cells are circular-shaped cells that carry oxygen through your bloodstream. A red blood cell is about 7.75×10^{-7} meters across.
a. **Write the width of a red blood cell using positive exponents.**
b. **Write the width of a red blood cell as a decimal.**

a. $7.75 \times 10^{-7} = 7.75 \times \dfrac{1}{10^7}$ or $\dfrac{7.75}{10^7}$ meters

b. Use a calculator to write the number as a decimal.

7.75 $\boxed{\div}$ 10 $\boxed{y^x}$ 7 $\boxed{=}$ *0.000000775*

You can use the prime factorization of a number to write a fraction as an expression with negative exponents.

Example 3 **Write each fraction as an expression using negative exponents.**

a. $\dfrac{1}{8}$

$\dfrac{1}{8} = \dfrac{1}{2^3}$

$= 2^{-3}$ *Definition of negative exponents*

b. $\dfrac{x}{y^2}$

$\dfrac{x}{y^2} = x \cdot \dfrac{1}{y^2}$

$= x \cdot y^{-2}$ or xy^{-2}

Connection to > Algebra

Expressions involving variables can also contain negative exponents.

Example **4** **Evaluate the expression $4x^{-3}$ if $x = 3$.**

$$4x^{-3} = 4(3)^{-3} \quad \textit{Replace x with 3.}$$
$$= 4\left(\frac{1}{3^3}\right) \quad \textit{Definition of negative exponents}$$
$$= 4\left(\frac{1}{27}\right) \quad \textit{Find } 3^3.$$
$$= \frac{4}{27} \quad \textit{Multiply.}$$

Checking Your Understanding

Communicating Mathematics

Read and study the lesson to answer these questions.

1. **Express** x^{-3} with a positive exponent.

2. **Write** a convincing argument that $2^0 = 1$ using the fact that $2^4 = 16$, $2^3 = 8$, $2^2 = 4$, and $2^1 = 2$.

3. Write in your own words the relationship between the rules for multiplying powers and dividing powers.

Guided Practice

Write each expression using positive exponents.

4. 3^{-3} 5. 15^{-5} 6. a^{-12}

Write each fraction as an expression using negative exponents.

7. $\dfrac{1}{10^7}$

8. $\dfrac{1}{4^3}$

9. $\dfrac{1}{9}$

Evaluate each expression.

10. 2^x if $x = -2$

11. $3n^{-2}$ if $n = 4$

12. **Biology** Deoxyribonucleic acid, or DNA, contains the genetic code of an organism. The length of a DNA strand is about 10^{-7} meters.

   a. Write the length of a DNA strand using positive exponents.

   b. Write the length of a DNA strand as a decimal.

Practicing and Applying the Concept

Independent Practice

Write each expression using positive exponents.

13. 4^{-1} 14. 5^{-2} 15. $(-2)^{-5}$ 16. x^{-1}

17. $s^{-2}t^{-1}$ 18. mn^{-2} 19. $2(xy)^{-2}$ 20. $\dfrac{1}{2^{-2}}$

Write each fraction as an expression using negative exponents.

21. $\dfrac{1}{5^3}$ **22.** $\dfrac{1}{a}$ **23.** $\dfrac{1}{16}$ **24.** $\dfrac{1}{x^6}$

25. $\dfrac{2}{3^2}$ **26.** $\dfrac{a}{b^6}$ **27.** $\dfrac{f}{g^3}$ **28.** $\dfrac{3m}{n^2}$

Evaluate each expression.

29. 3^n if $n = -3$ **30.** $4x^{-3}y^2$ if $x = 2$ and $y = 6$

31. $6t^{-2}$ if $t = 3$ **32.** $(2b)^{-3}$ if $b = -2$

Find each product or quotient. Express using positive exponents.

33. $(a^6)(a^{-3})$ **34.** $(b^{-10})(b^5)$ **35.** $\dfrac{c^5}{c^{-2}}$ **36.** $\dfrac{d^2}{d^3}$

Critical Thinking

37. Simplify $\dfrac{a^b}{a^{a-b}}$.

Applications and Problem Solving

38. Physical Science Electromagnetic waves are used to transmit television and radio signals. The diagram at the right shows the electromagnetic wave spectrum.

  a. Write the length of a light wave using a positive exponent.

  b. What is the range of lengths of infrared waves written in decimal form?

39. Astronomy Any two objects in space have an attraction that can be calculated by using a formula written by English scientist Henry Cavendish. The formula uses the universal gravitational constant, which is 6.67×10^{-11} Nm²/kg². (N is newtons.)

  a. Write the universal gravitational constant by using positive exponents.

  b. Write the constant as a decimal.

40. Physics An object has energy if it is able to produce a change in itself or its surroundings. For example, people use energy to run or to digest food. Scientists use the joule as the unit in which they measure energy. A joule is $1 \text{ kg} \cdot \text{m}^2 \cdot \text{s}^{-2}$. Write this unit as a fraction using positive exponents.

41. Find the product of $(10x)$ and $(-5x^3)$. (Lesson 4-8)

42. Sports Calida made 8 of 14 free throws in her last basketball game. How would you describe her success as a fraction in simplest form? (Lesson 4-6)

43. Write the product $7 \cdot 7 \cdot 7 \cdot 7 \cdot 7$ using exponents. (Lesson 4-2)

44. Solve $V = \frac{4}{3}\pi r^3$ for V if $\pi = 3.14$ and $r = 5$. (Lesson 3-4)

45. Personal Finance When Chandra and Scott got married, they put their savings into one account. Chandra had $2230 and together they had $4750. Write and solve an equation to find the amount of money Scott had. (Lesson 3-2)

46. Standardized Test Practice The graph below shows the number of *U.S. Teens* newsletters sold during the last four months of 2000. How many more newsletters were sold in November than October? (Lesson 1-10)

A. 4000 **B.** 6000 **C.** 8000 **D.** 18,000

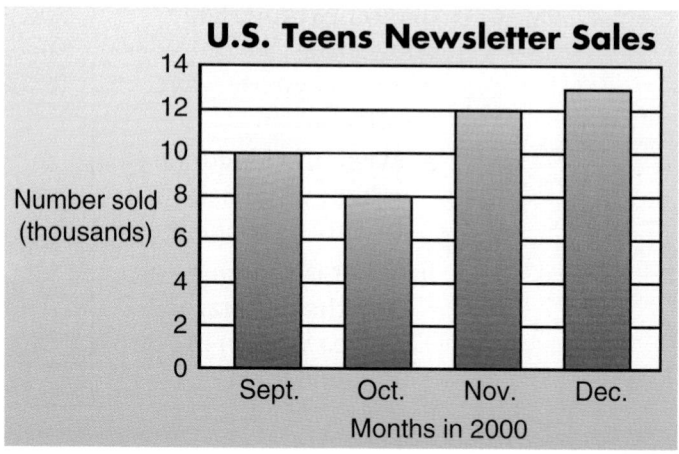

47. Write and solve an equation for *eight less than a number is three*. (Lesson 1-8)

48. Exercise According to one medical study, you can add 21 minutes to your life for every mile that you walk. If you walked 2 miles each day for a year, how much time would you add to your life? (Lesson 1-1)

a. Which method of computation do you think is most appropriate? Justify your choice.

b. Solve the problem using the four-step plan. Be sure to examine your solution.

For **Extra Practice**, see page 750.

Vocabulary

After completing this chapter, you should be able to define each term, property, or phrase and give an example or two of each.

Algebra

algebraic fraction (p. 197)

base (p. 175)

common multiples (p. 200)

composite number (p. 184)

divisible (p. 170)

expanded form (p. 176)

exponent (p. 175)

factor trees (p. 185)

factors (p. 170)

greatest common factor (GCF) (p. 190)

least common denominator (LCD) (p. 201)

least common multiple (LCM) (p. 200)

monomial (p. 172)

multiple (p. 200)

negative exponents (p. 210)

order of operations (p. 176)

powers (p. 175)

prime factorization (p. 185)

prime number (p. 184)

product of powers (p. 205)

quotient of powers (p. 206)

ratio (p. 196)

simplest form (p. 196)

standard form (p. 176)

Problem Solving

draw a diagram (p. 181)

guess and check (p. 186)

look for a pattern (p. 171)

▣*inter*NET
C O N N E C T I O N
Chapter Review For additional lesson-by-lesson review, visit:
www.glencoe.com/ sec/math/prealg/ mathnet

Understanding and Using Vocabulary

Complete each statement.

1. The greatest integer that is a factor of each of two or more integers is called their _____.

2. Numbers that are expressed using exponents are called _____.

3. The _____ of a whole number divide that number with a remainder of zero.

4. A _____ is a whole number greater than one that has exactly two factors, 1 and itself.

5. The product of a number and any whole number is called a _____ of the original number.

6. _____ is writing numbers using place values and exponents.

7. The least common multiple of the denominators of two or more fractions is called the _____.

8. A _____ is an integer, a variable, or a product of integers or variables.

9. A comparison of two numbers by division is called a ____.

Skills and Concepts

Objectives and Examples	Review Exercises
After completing this chapter, you should be able to:	Use these exercises to review and prepare for the chapter test.

▶ **determine whether one number is a factor of another.** (Lesson 4-1)

Is 342 divisible by 2, 3, 5, 6, or 10?

It is divisible by 2 since its last digit is 2.

It is divisible by 3 because $3 + 4 + 2$ or 9 is divisible by 3.

The last digit is not 5 or 0, so it is not divisible by 5.

It is divisible by 6 since it is divisible by 2 and 3.

It is not divisible by 10 since its last digit is not 0.

Using divisibility rules, state whether each number is divisible by 2, 3, 5, 6, or 10.

10. 863   **11.** 635

12. 4200   **13.** 14,577

14. 40, 180   **15.** 582

16. 5103   **17.** 1962

▶ **use powers and exponents in expressions and equations.** (Lesson 4-2)

Write a^5 as a product of the same factor.

$a^5 = a \cdot a \cdot a \cdot a \cdot a$

Write $3 \cdot 3 \cdot 3 \cdot 3$ using exponents.

$3 \cdot 3 \cdot 3 \cdot 3 = 3^4$

Write each multiplication expression using exponents.

18. $6 \cdot 6 \cdot 6 \cdot 6 \cdot 6$

19. $(c + 2)(c + 2)(c + 2)$

20. $(-4)(-4)(-4)(-4)$

21. $(1 \cdot 1) \cdot 1$

Write each power as a multiplication expression.

22. $(-16)^3$ **23.** k^9

24. $(-w)^4$ **25.** 32^{11}

▶ **write the prime factorization of composite numbers.** (Lesson 4-4)

Write the prime factorization of 648.

$648 = \quad 9 \quad \cdot \quad 72$

$\quad = 3 \cdot 3 \cdot 8 \cdot 9$

$\quad = 3 \cdot 3 \cdot 2 \cdot 4 \cdot 3 \cdot 3$

$\quad = 3 \cdot 3 \cdot 2 \cdot 2 \cdot 2 \cdot 3 \cdot 3$

Factor each number or monomial completely.

26. 120   **27.** -114

28. $630x^3$ **29.** $-825x^2y$

30. 2805   **31.** $-1827jk^3$

32. $66g^5h^2$ **33.** $550mnq^2$

Objectives and Examples

▶ **find the greatest common factor of two or more integers or monomials.** (Lesson 4-5)

Find the GCF of $42x^2$ and $110xy$.

$42x^2 = 2 \cdot 3 \cdot 7 \cdot x \cdot x$
$110xy = 2 \cdot 5 \cdot 11 \cdot x \cdot y$
The GCF is $2 \cdot x$ or $2x$.

▶ **simplify fractions using the GCF.** (Lesson 4-6)

Write $\dfrac{6}{45}$ in simplest form.

$$\overset{\div\ 3}{\frac{6}{45} = \frac{2}{15}} \quad \textit{The GCF of 6 and 45 is 3.}$$
$$\underset{\div\ 3}{}$$

Simplify $\dfrac{15ac^2}{24ab}$.

$$\frac{15ac^2}{24ab} = \frac{\overset{1}{3} \cdot 5 \cdot \overset{1}{a} \cdot c \cdot c}{2 \cdot 2 \cdot 2 \cdot \underset{1}{3} \cdot \underset{1}{a} \cdot b}$$
$$= \frac{5c^2}{8b}$$

▶ **find the least common multiple of two or more numbers.** (Lesson 4-7)

Find the LCM of 18 and 15.

$18 = 2 \cdot 3 \cdot 3$
$15 = 3 \cdot 5$
The LCM is $2 \cdot 3 \cdot 3 \cdot 5 = 90$

▶ **compare fractions.** (Lesson 4-7)

Which is greater, $\dfrac{3}{4}$ or $\dfrac{7}{12}$?

$$\overset{\times\ 3}{\frac{3}{4} = \frac{9}{12}} \quad \textit{The LCM of 4 and 12 is 12.}$$
$$\underset{\times\ 3}{}$$

$\dfrac{9}{12} > \dfrac{7}{12}$, so $\dfrac{3}{4} > \dfrac{7}{12}$.

Review Exercises

Find the GCF of each set of numbers or monomials.

34. 70, 66
35. 1092, 325
36. $100x$, $84xy$, $-76x^2$
37. $210a^3b^2$, $875a^2b^3$
38. $112j^5k$, $-144j^2$
39. $40x$, $25y$
40. $115wx$, $224wz$
41. $441ac$, $223bd$

Write each fraction in simplest form. If the fraction is already in simplest form, write *simplified*.

42. $\dfrac{72}{88}$
43. $\dfrac{133}{140}$
44. $\dfrac{225}{315}$
45. $\dfrac{48}{66}$
46. $\dfrac{23a}{32b}$
47. $\dfrac{500x}{1000x}$
48. $-\dfrac{125mn}{625m}$
49. $\dfrac{68b^2}{105c}$

Find the least common multiple of each set of numbers or algebraic expressions.

50. $11ab$, $6b$
51. 12, 4, 5
52. $7x^3$, $49xy$
53. $10j$, $6jk$, 3
54. $5f^4g$, $13fg^2$
55. 12, 15, 18
56. $72x^3$, $64x^5$
57. $6ab$, $18c$, $36d$

Replace each ● with <, >, or = to make each statement true.

58. $\dfrac{6}{7}$ ● $\dfrac{5}{14}$
59. $\dfrac{9}{16}$ ● $\dfrac{8}{15}$
60. $\dfrac{9}{100}$ ● $\dfrac{1}{9}$
61. $\dfrac{1}{11}$ ● $\dfrac{2}{23}$

Objectives and Examples	Review Exercises

▶ **multiply and divide monomials.** (Lesson 4-8)

Find each product or quotient.

$$g^2 \cdot g^7 = g^{2+7}$$
$$= g^9$$

$$\frac{(-3)^5}{(-3)^2} = (-3)^{5-2}$$
$$= (-3)^3$$

Find each product or quotient. Express your answer in exponential form.

62. $12^4 \cdot 12^7$ **63.** $\dfrac{d^9}{d^4}$

64. $n^5 \cdot n^{14}$ **65.** $\dfrac{(-4)^5}{(-4)}$

66. $(6w^3)(10w^4)$ **67.** $\dfrac{6x^{40}}{3x^{18}}$

68. $\dfrac{30c^2d}{15c}$ **69.** $22ab^3 \cdot 5a^9$

▶ **use negative exponents in expressions.** (Lesson 4-9)

Write $-2x^3y^{-2}$ **using positive exponents.**

$$-2x^3y^{-2} = -2x^3 \cdot y^{-2}$$
$$= -2x^3 \cdot \frac{1}{y^2}$$
$$= \frac{-2x^3}{y^2}$$

Write each expression using positive exponents.

70. $(-5)^{-3}$ **71.** $4c^{-2}d$

72. $\dfrac{1}{6^{-18}}$ **73.** $4x(yz)^{-2}$

74. $\left(\dfrac{2}{x}\right)^{-4}$ **75.** $\dfrac{3}{(f)^{-9}}$

Applications and Problem Solving

76. Business The Carpet Experts cleaning team can clean carpet in a room that is 10 feet by 10 feet in 20 minutes. Draw a diagram to find how long it would take them to do a walk-in closet that is 5 feet by 5 feet. (Lesson 4-3)

77. Sports The City Monday-Night Softball League has 7 teams. The director is making a schedule where every team plays every other team once. How many games must be scheduled? (Lesson 4-3)

78. Number Theory Numbers that have a greatest common factor of 1 are said to be **relatively prime**. Find the least two composite numbers that are relatively prime. (Lesson 4-5)

79. Literature A newspaper polled Americans on the types of books that they buy. The graph below shows the results of the poll. Which type of book is more popular, travel or home improvement? (Lesson 4-7)

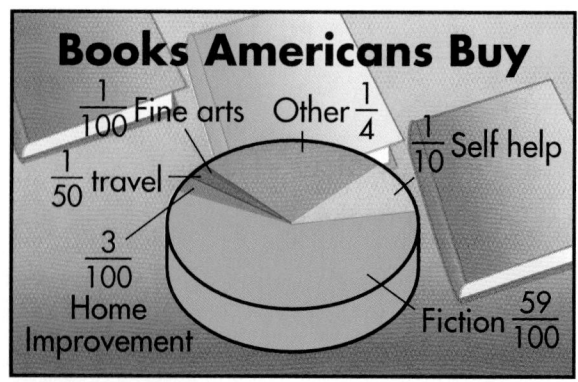

Source: USA-Today

A practice test for Chapter 4 is available on page 777 .

Performance Task

Demonstrate your knowledge by giving a clear, concise solution to each problem. Be sure to include all relevant drawings and justify your answers. You may show your solutions in more than one way or investigate beyond the requirements of the problem.

A planned community housing project is providing a gardening area for its residents. The area will be laid out in individual garden plots as shown.

1. Explain how to find the prime factorization of a number.
2. Find the prime factorizations of 30 and 24.
3. Find the common factors of 30 and 24.
4. The garden area will be fenced. The fence posts in front are to be equally spaced along the entire length with posts at the corners of each garden plot. Would a spacing of 3 feet between posts work? How do you know?
5. What is the longest spacing between posts that can be used? What is this distance called?
6. List the multiples of the widths of each size lot.
7. If the back fence of the overall gardening area is to be straight, give at least two possible widths, *w*, of the area.
8. What is the shortest possible width of the gardening area? Explain.
9. If the fence posts are to be equally spaced along the width of the garden area with posts at the corners of each garden plot, what spacing would you recommend? Why?

Thinking Critically

▶ Find the GCF and LCM of 45 and 72.

▶ Explain why you think the greatest common factor is called that when it is less than the numbers given.

▶ Explain why you think the least common multiple is called that when it is greater than the numbers given.

 Portfolio

Select an item from your work in this chapter that shows your creativity and place it in your portfolio.

Section One: Multiple Choice

There are eight multiple-choice questions in this section. After working each problem, write the letter of the correct answer on your paper.

1. The family sizes of students in a mathematics class are recorded in the following table.

Family Size	2	3	4	5	6	7	8
Number of Students	1	4	3	6	5	2	3

   How many students have fewer than 5 members in their family?

   A. 6
   B. 8
   C. 9
   D. 14

2. A group of divers needs to descend to a depth 3 times their present depth of -25 meters. At what depth do they need to be?

   A. 75 meters
   B. 28 meters
   C. -25 meters
   D. -75 meters

3. Simplify $|-12| - |-3|$.

   A. -9
   B. -15
   C. 9
   D. 15

4. An airplane descended 250 feet in 5 minutes. What was its average rate of descent?

   A. 50 ft/min
   B. 250 ft/min
   C. 1250 ft/min
   D. Not here

5. Which of the following is equivalent to 4^{-3}?

   A. $3 \cdot 3 \cdot 3 \cdot 3$
   B. $4 \cdot 4 \cdot 4$
   C. $\dfrac{1}{3 \cdot 3 \cdot 3 \cdot 3}$
   D. $\dfrac{1}{4 \cdot 4 \cdot 4}$

6. Which fraction is in simplest form?

   A. $\dfrac{12a^2}{15ab}$
   B. $\dfrac{15a}{50}$
   C. $\dfrac{12d^2}{25ab}$
   D. $\dfrac{5b^2c}{7bh^2}$

7. Which set of numbers is in order from least to greatest?

   A. $\{-5, -3, 1, 14\}$
   B. $\{-1, -6, 8, 11\}$
   C. $\{15, -8, 1\}$
   D. $\{10, 0, -5\}$

8. The distance traveled is given by the formula $d = rt$, where r is the rate of speed and t is the time spent traveling. How long would it take to ride a bike 9 miles at the rate of 6 miles per hour?

   A. 15 min
   B. 1.5 hours
   C. 54 hours
   D. 54 miles

Section Two: Free Response

This section contains eight questions for which you will provide short answers. Write your answer on your paper.

9. The temperature when you left for school this morning was 10°F. The wind speed was 20 mph, which produced a windchill of −24°F. What is the difference between the actual temperature and the windchill temperature?

10. Before she paid the electric bill, Rocio's checking account had a balance of $131. Her new account balance is $109. How much was her electric bill?

11. Juwan's yard is 10 yards wide and 20 yards long. His dad wants to build a rectangular swimming pool that has an area of 100 square feet. What are the possible dimensions of the pool?

12. Simplify $12a + 8b - 6a$.

13. A college basketball court is 94 feet long by 50 feet wide. What is the area of the court?

14. Nituna must earn 250 points to win a prize. She has already earned 176 points. Write a sentence to find p, the number of points she still needs to earn.

15. Carrie is cutting a round pizza. What is the greatest number of pieces she can get from 1 pizza with 5 straight cuts?

16. In high school football, a team can score 6 points for a touchdown, 7 points for a touchdown and an extra point, 8 points for a touchdown and a two-point conversion, 3 points for a field goal, or 2 points for a safety. Can a team have a score of 23 points? Explain.

Test-Taking Tip You can solve many problems without much calculating if you understand the basic mathematical concepts. Always look carefully at what is asked, and think of possible shortcuts for solving the problem.

Section Three: Open-Ended

This section contains two open-ended problems. Demonstrate your knowledge by giving a clear, concise solution to each problem. Your score on these problems will depend on how well you do the following.

- Explain your reasoning.
- Show your understanding of the mathematics in an organized manner.
- Use charts, graphs, and diagrams in your explanation.
- Show the solution in more than one way or relate it to other situations.
- Investigate beyond the requirements of the problem.

17. Drina, Susan, Latisha, and Joshua are in the orchestra at school. Each student plays one of the following instruments: drums, violin, trumpet, or trombone. If each student plays a different instrument, use the following information to match each student with an instrument.
   — Drina and Joshua both play brass instruments.
   — The trumpet player and Joshua also play in the jazz band.
   — Susan lost one of her sticks last week.

18. Determine whether 435 is divisible by 2, 3, 5, 6, or 10. Justify your answers.

Test Practice For additional test practice questions, visit:
www.glencoe.com/sec/math/prealg/mathnet

5 Rationals: Patterns in Addition and Subtraction

What You'll Learn

In this chapter, you will:

- estimate and find sums and differences of rational numbers,

- solve equations and inequalities involving rational numbers,

- solve problems using logical reasoning, and

- find terms of an arithmetic sequence.

Why It's Important

Real World

Transportation The table shows the number of passengers for each type of public transportation for 1997 and 1998. You can subtract decimals to find the difference in the number of passengers from 1997 to 1998 for each type of transportation. For example, you can subtract 358.9 from 378.4 to find the increase in passengers for commuter rail transportation. *You will learn about decimal subtraction in Lesson 5-3.*

Type of Transportation	1997 (billions)	1998 (billions)
Commuter rail	358.9	378.4
Heavy rail	2398.5	2530.1
Light rail	220.5	234.4

Source: American Public Transit Association

Focus on...

PREREQUISITE SKILLS

To be successful in this chapter, you'll need to understand these concepts and be able to apply them. Refer to the lesson in parentheses if you need more review before beginning the chapter.

▶ **Simplify fractions using the GCF.** *(Lesson 4-6)*

Write each fraction in simplest form. If the fraction is already in simplest form, write *simplified*.

1. $\frac{12}{18}$ **2.** $\frac{19}{57}$ **3.** $\frac{30}{73}$ **4.** $\frac{46}{115}$

▶ **Add integers.** *(Lesson 2-4)*

Find each sum.
5. $-3 + (-8)$ **6.** $9 + (-4)$ **7.** $12 + (-20)$ **8.** $-3 + 1$

▶ **Subtract integers.** *(Lesson 2-5)*

Find each difference.
9. $15 - 12$ **10.** $23 - 32$ **11.** $-9 - (-3)$ **12.** $-1 - (-10)$

▶ **Find the least common denominator (LCD) for a pair of fractions.** *(Lesson 4-7)*

Find the least common denominator (LCD) for each pair of fractions.

13. $\frac{1}{2}, \frac{3}{4}$ **14.** $\frac{2}{5}, \frac{7}{8}$ **15.** $\frac{5}{6}, \frac{4}{9}$ **16.** $\frac{7}{15}, \frac{4}{21}$

Focus on...

READING SKILLS

In this chapter, you'll be learning about **rational numbers.** The word rational contains the word "ratio." You can think of a rational number as a number that can be written as a ratio, such as $\frac{1}{2}$. Some people like to think of rational numbers as fractions or decimals. Recall that a decimal like 0.3 can be written as $\frac{3}{10}$. Whole numbers and integers can also be written as ratios. For example, -5 can also be written as $\frac{-5}{1}$.

5-1 Rational Numbers

You'll Learn

- To identify and compare rational numbers, and

- to rename decimals as fractions.

It's Important

Comparing rational numbers can help you determine which sports team has a better record.

Application: Movies

Are you frightened by dinosaurs in movies? Or amazed by the flight of space ships in movies? Then you have Jeff Olson to thank. Mr. Olson has created miniature models for many blockbuster movies. Movie makers often use models to simulate situations that would be difficult or impossible to do in real life.

To create models, Mr. Olson uses a ratio called a **scale**. A scale is a comparison of the size of a model to the size of the real object.

$$\text{scale} = \frac{size\ of\ model}{size\ of\ real\ object}$$

The scale tells how much smaller or larger the model has to be than the real object. Suppose Mr. Olson was building a 15-inch model of a Bigfoot that would be 120 inches tall in real life. The scale would be $\frac{15}{120}$ or $\frac{1}{8}$. That means absolutely everything on the model is $\frac{1}{8}$ the size of the same item on the original Bigfoot. Numbers such as $\frac{1}{8}$ that can be written as fractions are called **rational numbers**.

Learning the Concept

Numbers can be organized into sets. One set is the **whole numbers**.

Whole Numbers
This set includes 0, 1, 2, 3, . . . It also includes any number that can be written as a whole number, such as $\frac{5}{5}$ or $\frac{9}{1}$.

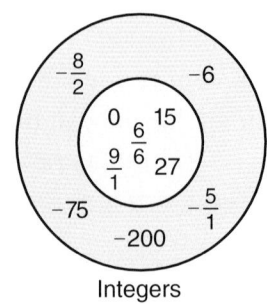
Whole Numbers

Another set of numbers is the **integers**.

Integers
This set includes . . . , −2, −1, 0, 1, 2, . . . Notice that all whole numbers are included in the set of integers.

Integers

Refer to the application at the beginning of the lesson. The model of Bigfoot had a scale of $\frac{1}{8}$. Both 1 and 8 are integers, but $\frac{1}{8}$ is not. The number system is extended to include numbers like $\frac{1}{8}$. Numbers like $\frac{1}{8}$ are rational numbers.

Definition of a Rational Number	Any number that can be expressed in the form $\frac{a}{b}$, where a and b are integers and $b \neq 0$, is called a rational number.

Rational Numbers

This set includes common fractions, such as $\frac{1}{4}$. It also includes mixed numbers, decimals, integers, and whole numbers because all these numbers can be written in the form $\frac{a}{b}$.

Rational Numbers

You can review simplifying fractions in Lesson 4-6.

Some decimals are rational numbers. Decimals either terminate or they go on forever. The decimal 3.74 is an example of a **terminating decimal**. Every terminating decimal can be written as a fraction with a denominator of 10, 100, 1000, and so on. For example, $0.75 = \frac{75}{100}$ or $\frac{3}{4}$, and $1.3 = 1\frac{3}{10}$ or $\frac{13}{10}$. So terminating decimals are always rational numbers.

Example 1

CONNECTION

Biology

The average adult's fingernails grow 0.125 inch each month. Write 0.125 as a fraction.

$$0.125 = \frac{125}{1000}$$

$$= \frac{1}{8} \quad \text{Simplify. The GCF of 125 and 1000 is 125.}$$

The period of a repeating decimal are the digits that repeat. For $6.\overline{23}$ and $8.4\overline{613}$, the periods are 23 and 613 respectively.

Decimals like 0.4444444444 . . . are called **repeating decimals**. Because it is inconvenient to write all of these digits, you can use the **bar notation** $0.\overline{4}$ to indicate that the 4 repeats forever. Here are some other examples.

6.23232323232323 . . . = $6.\overline{23}$ The digits 23 repeat.

8.4613613613613 . . . = $8.4\overline{613}$ The digits 613 repeat.

Repeating decimals can always be written as fractions. For example, $0.\overline{4}$ is equivalent to $\frac{4}{9}$. So, repeating decimals are always rational numbers.

Connection to Algebra

Example 2 shows how to rename repeating decimals as fractions.

Example Express each repeating decimal as a fraction or mixed number in simplest form.

a. $0.\overline{5}$

Let $N = 0.555\ldots$ Then $10N = 5.555\ldots$ *Multiply each side by 10 because one digit repeats.*

Subtract N from $10N$ to eliminate the repeating part.

$$10N = 5.555\ldots \quad \textit{Recall } 10N - N = 9N.$$
$$-\ 1N = 0.555\ldots$$
$$\overline{9N = 5}$$
$$N = \frac{5}{9} \qquad \textit{Divide each side by 9.}$$

Therefore $0.\overline{5} = \frac{5}{9}$.

b. $3.363636\ldots$

Let $N = 3.363636\ldots$ Then $100N = 336.3636\ldots$ *Multiply each side by 100 because two digits repeat.*

$$100N = 336.3636\ldots$$
$$-\ \ \ N = \ \ \ 3.3636\ldots$$
$$\overline{99N = 333}$$
$$N = \frac{333}{99} \text{ or } 3\frac{4}{11}$$

So $3.363636\ldots = 3\frac{4}{11}$.

THINK ABOUT IT

What multiplier should you use to express $0.\overline{718}$ as a fraction?

Decimals that do not terminate and do not repeat such as $4.252627\ldots$, cannot be written as fractions. So they are not rational numbers.

Thus, the set of rational numbers includes whole numbers, integers, fractions, terminating decimals, and repeating decimals.

Example Name the set(s) of numbers to which each number belongs.

In this example and in exercises 10-13 and 30-41, choose from these sets:
- *whole numbers*
- *integers*
- *rationals*

a. -7 -7 is an integer and a rational number.

b. $-3\frac{2}{5}$ Since $-3\frac{2}{5}$ can be written as $-\frac{17}{5}$, it is a rational number. It is not a whole number nor is it an integer.

c. $0.545556\ldots$ This is a non-terminating, non-repeating decimal. So it is not rational.

d. 1.469 1.469 is a terminating decimal. It is a rational number.

Like integers, rational numbers can be graphed on a number line. Examples of rational numbers graphed on a number line are shown below.

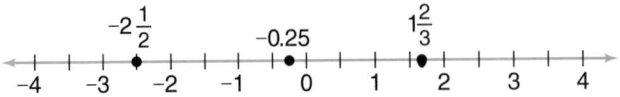

You can use a number line to compare rational numbers.

Example Use a number line to choose the greater number from each pair.

a. $-\frac{2}{3}, -\frac{1}{2}$

b. $0.3, \frac{1}{3}$

$$0.3 = \frac{3}{10}$$

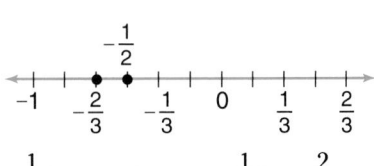

$-\frac{1}{2}$ is greater or $-\frac{1}{2} > -\frac{2}{3}$.

$\frac{1}{3}$ is greater or $\frac{1}{3} > 0.3$.

Checking Your Understanding

Communicating Mathematics

Read and study the lesson to answer these questions.

1. **Explain** what a rational number is.

2. Are all integers also rational numbers? Explain.

3. Use $\frac{2}{3}, \frac{4}{5}$, and $<$ to write a true statement.

4. **Explain** why 0.13 and $0.\overline{13}$ are rational numbers.

 MATERIALS
✎ ruler

5. Use a ruler to draw a number line. Mark tick marks that are 1 centimeter apart from -8 to 8.

a. Mark $3\frac{1}{5}, -1\frac{1}{3}, 0.7, -5.6, 4\frac{1}{3}, 4.3,$ and -2.9 on the number line.

b. Which of the points indicates the greatest number?

c. Which number in the list is the least?

Guided Practice

Express each decimal as a fraction or mixed number in simplest form.

6. 0.8   7. 0.05   8. -9.64 9. $0.\overline{2}$

Name the set(s) of numbers to which each number belongs.

10. -3 11. $-8\frac{3}{4}$ 12. 15   13. $0.414243\ldots$

CONNECTION

Research For the latest on models, visit:
www.glencoe.com/
sec/math/prealg/
mathnet

Replace each ● with $<$, $>$, or $=$ to make each sentence true. Use a number line if necessary.

14. $\frac{3}{7}$ ● $-\frac{3}{7}$ 15. $\frac{3}{4}$ ● 0.75   16. -3.78 ● -3.88

17. **Hobbies** A model airplane has a wingspan of 27 inches. The full-sized aircraft has a wingspan of 36 feet or 432 inches.

a. What is the scale of the model plane?

b. Is the scale a whole number, an integer, or a rational number? Explain.

Exercises: Practicing and Applying the Concept

Independent Practice

Express each decimal as a fraction or mixed number in simplest form.

18. 0.4   19. -0.7 20. 0.23   21. 0.57

22. 0.48   23. 0.03   24. $0.\overline{8}$ 25. $-0.333\ldots$

26. -0.51 27. $0.\overline{25}$ 28. 0.375   29. $2.\overline{34}$

Lesson 5-1 *Rational Numbers* **227**

Name the set(s) of numbers to which each number belongs.

30. -14

31. $-\frac{6}{7}$

32. $-3\frac{7}{8}$

33. $\frac{35}{7}$

34. 0.12

35. $-3.3738\ldots$

36. -0.17

37. 3.11

38. -10

39. $29\frac{1}{2}$

40. -32.0

41. $0.1234\ldots$

Replace each ● with >, <, or = to make a true sentence. Use a number line if necessary.

42. $\frac{3}{4}$ ● $-\frac{3}{8}$

43. $-6\frac{1}{2}$ ● $-6.\overline{5}$

44. 9.9 ● $9.\overline{8}$

45. $\frac{3}{5}$ ● 0.6

46. $\frac{1}{7}$ ● $0.222\ldots$

47. $\frac{13}{26}$ ● $-\frac{1}{2}$

48. $0.3\overline{4}$ ● $0.\overline{34}$

49. $\frac{1}{11}$ ● 0.1

50. $-1\frac{1}{3}$ ● $-1.\overline{3}$

51. Which number is the greatest, $\frac{2}{7}$, $\frac{2}{9}$, or $\frac{2}{11}$?

52. Which number is the greatest, $-\frac{3}{5}$, $-\frac{5}{6}$, or $-\frac{6}{7}$?

Critical Thinking

53. Does $\frac{6}{2.4}$ name a rational number? Explain.

54. A machinist made a steel peg 2.37 inches in diameter for a $2\frac{3}{8}$-inch diameter hole. Will the peg fit? How do you know?

Applications and Problem Solving

55. Measurement A gauge measured the thickness of a piece of metal as 0.025 inch. What fraction of an inch is this?

56. Manufacturing A garbage bag has a thickness of 0.8 mils. This is 0.0008 inch. What fraction of an inch is this?

57. Engineering Ingrid Proctor-Fridia works for the United States Department of Defense checking to see that airplane parts conform to the blueprints. Blueprints are drawings of objects that show their scale.

   a. The wingspan of an airplane on a blueprint is 26.9 centimeters. If the wingspan of the actual airplane is 8.07 meters, or 807 centimeters, what is the scale of the blueprints?

   b. Is the scale a rational number? Explain.

Ingrid Proctor-Fridia

Mixed Review

58. Physical Science The wavelength of visible red light is 7.50×10^{-7} meters. (Lesson 4-9)

   a. Write the size of the wavelength using positive exponents.

   b. Write the size of the wavelength using decimals.

59. Find the least common multiple of $4x^2$, $3x$, and 5. (Lesson 4-7)

60. Is 945 divisible by 2, 3, 5, 6, or 10? (Lesson 4-1)

61. Solve $-4b > 72$. (Lesson 3-7)

62. Solve $\frac{y}{12} = -7$. (Lesson 3-3)

For **Extra Practice**, see page 750.

63. Solve $q = \frac{323}{-17}$. (Lesson 2-8)

64. Is $17 > 40$ a true, false, or open sentence? (Lesson 1-9)

65. Standardized Test Practice A carton of orange juice contains 64 ounces. If the nutritional information lists the serving size as 8 ounces, how many servings are in the carton? (Lesson 1-3)

   A. 6 servings   **B.** 8 servings   **C.** 10 servings   **D.** 12 servings

5-2

Estimating Sums and Differences

You'll Learn

- To estimate sums and differences of decimals and fractions.

It's Important

Estimating with decimals will help you determine whether you have enough money to purchase the items you want from a store.

Application: Business

The exercise craze of the 1980s changed the world of fitness *and* the world of fashion. Once seen only in the locker room, athletic shoes are now a part of the everyday wardrobe. The graph at the right shows the athletic shoe sales made to women, men, and children. What were the total sales of athletic shoes?

Source: Sporting Goods Manufacturers Association

Learning the Concept

Estimation is often used to provide a quick and easy answer when a precise answer is not necessary. It is also an excellent way to quickly see if your answer is reasonable or not.

To solve the problem about shoe sales, you can use *rounding* to estimate the answer. Round each number to a convenient place-value position. Often the greatest place-value position is used. Then complete the operation.

To find the total sales, round each of the sales figures to the greatest place value that they all share and then add.

$$\begin{array}{r} 153.2 \\ 115.9 \\ +\ 96.0 \end{array} \quad \rightarrow \quad \begin{array}{r} 150 \\ 120 \\ +100 \\ \hline 370 \end{array} \quad \textit{Round to the nearest ten.}$$

There were approximately 370 million pairs of athletic shoes sold.

Example ① **Estimate using rounding.**
 a. **8.890 + 15.98**   b. **132.62 − 45.81**

$8.890 + 15.98 \rightarrow 9 + 16 = 25$ $132.62 - 45.81 \rightarrow 130 - 50 = 80$
$8.890 + 15.98$ is about 25. $132.62 - 45.81$ is about 80.

Before you use a calculator to compute an answer, it is a good idea to estimate the answer. This helps make sure that you entered the numbers correctly.

Lesson 5-2 *Estimating Sums and Differences* **229**

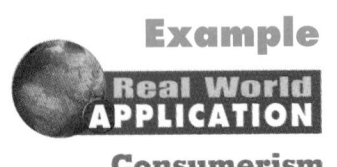

Example **2**

Suppose you go to Miller Cycles to buy equipment for in-line skating. You purchase a helmet for $29.89, a set of elbow pads for $19.26, knee pads for $21.67, and some energy bars for $6.03. The cashier tells you that the total is $82.88. Is this a reasonable cost?

Round each item to the nearest dollar amount.

$$
\begin{array}{lll}
29.89 & \rightarrow & 30 \\
19.26 & \rightarrow & 19 \\
21.67 & \rightarrow & 22 \\
+\ 6.03 & \rightarrow & +\ 6 \\
\hline
& & 77
\end{array}
$$

Obviously the amount charged is not correct. Probably the cashier accidentally charged you twice for the energy bars. The correct total is $76.85, which is very close to the estimate.

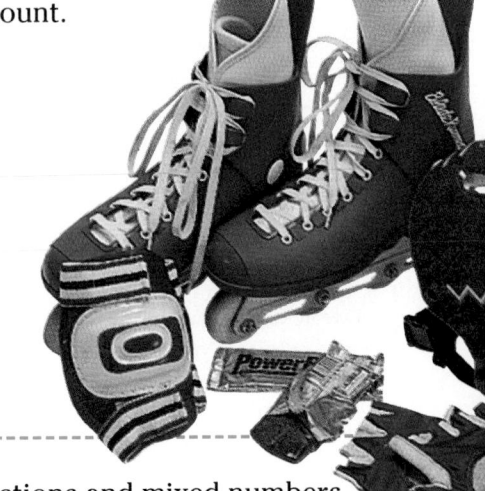

You can estimate sums and differences of fractions and mixed numbers by rounding also. To estimate the sum or difference of mixed numbers, round each mixed number to the nearest whole number. To estimate the sum or difference of proper fractions, round each fraction to 0, $\frac{1}{2}$, or 1.

Example **3** **Estimate each sum or difference.**

a. $4\frac{1}{12} + 16\frac{19}{32}$

$$4\frac{1}{12} + 16\frac{19}{32} \quad \rightarrow \quad 4 + 17 = 21$$

The sum of $4\frac{1}{12}$ and $16\frac{19}{32}$ is about 21.

b. $\frac{7}{8} + \frac{7}{16}$

$$\frac{7}{8} + \frac{7}{16} \quad \rightarrow \quad 1 + \frac{1}{2} = 1\frac{1}{2}$$

The sum of $\frac{7}{8}$ and $\frac{7}{16}$ is about $1\frac{1}{2}$.

c. $\frac{4}{5} - \frac{3}{8}$

$$\frac{4}{5} - \frac{3}{8} \quad \rightarrow \quad 1 - \frac{1}{2} = \frac{1}{2}$$

$\frac{4}{5} - \frac{3}{8}$ is about $\frac{1}{2}$.

Checking Your Understanding

Read and study the lesson to answer these questions.

1. **Give** two examples of numbers that round to a nearest whole number. One should round up; the other should not.

2. **State** two reasons for using estimation.

3. **You Decide**   Pascual estimates that $569 - 345 = 300$. Nida estimates the difference as 220. Which is the better estimate? Explain.

MATH JOURNAL

4. **Assess Yourself** Write about a situation or an example in your life in which only an estimate is necessary, not an exact answer.

Guided Practice

Round to the nearest whole number.

5. 6.19

6. 12.641

7. 803.487

8. $6\frac{3}{4}$

9. $11\frac{1}{6}$

10. $33\frac{7}{24}$

Round each fraction to 0, $\frac{1}{2}$, or 1.

11. $\frac{10}{11}$

12. $\frac{1}{9}$

13. $\frac{15}{32}$

Choose the best estimate.

14. $4\frac{1}{8} - 2\frac{9}{10}$ **a.** almost 1   **b.** more than 1   **c.** more than $1\frac{1}{2}$

15. $5\frac{2}{3} + 2\frac{1}{11}$ **a.** almost 8   **b.** more than 8   **c.** less than $7\frac{1}{2}$

Estimate each sum or difference.

16. $34.32 + 19.51$

17. $159.7 - 124.8$

18. $6\frac{7}{10} - 3\frac{1}{6}$

19. $12\frac{1}{9} - 2\frac{9}{14}$

20. $1\frac{7}{12} + \frac{8}{19}$

21. $23\frac{8}{9} + 5\frac{4}{25}$

22. **Conservation** The chart at the right shows the amount of water used for average and quick showers with standard and low-flow showerheads. About how much water could be saved by changing from an average shower with a standard showerhead to a quick shower with a low-flow showerhead?

Rub-A-Dub-Dub

Gallons used by one shower daily per year:

Standard head

Average shower[1] 💧💧💧💧💧💧💧💧💧💧💧💧💧 26,718

Quick shower[2] 💧💧💧💧💧💧💧 13,140

Low-flow head

Average shower[1] 💧💧💧💧💧💧 11,132

Quick shower[2] 💧💧💧 5,475   💧 = 2,000

1-National average 12.2 minutes, 2-Recommended 6.0 minutes

Source: Opinion Research Corp. for Teledyne Water Pik

Exercises: Practicing and Applying the Concept

Independent Practice

Estimate each sum or difference.

23. $12.5 + 44.8$

24. $\$20.00 - \$15.34$

25. $8.6 + 11.9$

26. $32 - 29.75$

27. $\frac{8}{9} + \frac{1}{5}$

28. $\frac{11}{15} + \frac{7}{8}$

29. $6\frac{9}{10} + \frac{2}{5}$

30. $\frac{17}{20} + 8\frac{3}{4}$

31. $\frac{4}{5} - \frac{1}{10}$

32. $\frac{7}{9} - \frac{13}{18}$

33. $18\frac{1}{8} - 12\frac{1}{2}$

34. $11\frac{89}{100} - 4\frac{1}{7}$

35. $125.8 - 22.4$

36. $16.432 + 11.910$

37. $32\frac{8}{10} - 4\frac{3}{8}$

38. $\frac{65}{131} + \frac{3}{35}$

39. $22\frac{2}{95} + 28\frac{3}{75}$

40. $145\frac{4}{5} - 121\frac{2}{15}$

The DECA club at Union High School runs a store in the school lobby as a fundraiser. Use their price list at the right to estimate each purchase price or change amount to the nearest dollar.

Pirate Corner Store Price List	
felt-tip pen	$1.39
one inch binder	$2.89
dividers	$0.89
loose-leaf paper	$1.29
9 x 12 envelopes	$0.93
liquid correction fluid	$1.59
package of highliters	$2.79
calculator	$13.89
composition books	
1 theme	$2.69
2 theme	$3.99
5 theme	$4.19
index tabs	$1.53
3 x 5 index cards	$0.79
ruler	$0.89
tape	$1.69
can of soda or juice	$0.79
ice cream bar	$1.29
candy bars	$0.69
bag of popcorn	$1.74

41. price of binder, loose-leaf paper, index cards

42. price of loose-leaf paper, dividers, liquid correction fluid

43. change from $10 for purchase of 5-theme composition book, and package of hi-liters

44. change from $20 for purchase of a calculator and one-inch binder

45. Is $8.15 the correct change from $10.00 for a ruler, ice cream bar, and a candy bar?

46. What supplies would you purchase for $10.00, allowing yourself a candy bar? Let your goal be to get back less than $0.50 change.

Critical Thinking

47. Make up a problem using estimation in which the answer is $12\frac{1}{2}$.

Applications and Problem Solving

48. Recycling A magazine estimated the amounts of steel, paper, and aluminum available for recycling. Which material accounted for the greatest amount of recycling?

Source: *National Geographic*

49. Retail Sales Jaime bought a sweatshirt with his high school name and mascot for $41.29 and a T-shirt with the track team logo for $18.25. Estimate his cost to the nearest dollar.

50. Fundraising The freshman class needs to earn $500 for their annual trip to an amusement park. They made $95.60 at a car wash, $41.75 at a cookie and bake sale, $150.95 at a band concert, and $125.16 selling candy bars.

    a. Estimate their total earnings.

    b. Estimate, if necessary, the amount needed to reach their goal.

51. Business A soft drink company reported that $\frac{3}{10}$ of the sales of their soft drinks were made through fountains and restaurants in 1994. Another $\frac{2}{5}$ of the sales were through supermarkets and other retail stores. Did the sales through these channels account for more or less than half of the total soft drink sales for the company in 1994?

Mixed Review

52. Express -0.88 as a fraction in simplest form. (Lesson 5-1)

53. Food Tortilla chips account for $\frac{26}{100}$ of the snack chip retail sales. Write $\frac{26}{100}$ as a fraction in simplest form. (Lesson 4-6)

54. Standardized Test Practice Which graph shows the solution of $-5 \leq x + 6$? (Lesson 3-6)

  A.
```
  ←+—+—+—+—+—●——+—+—+→
   -4 -3 -2 -1  0  1  2  3  4
```

  B.
```
  ←+—+—+—+—+—+—●—+—+—+→
   -4 -3 -2 -1  0  1  2  3  4
```

  C.
```
  ←+—●—+—+—+—+—+—+—+→
   -12  -10   -8   -6   -4
```

  D.
```
  ←+—●—+—+—+—+—+—+—+→
   -12  -10   -8   -6   -4
```

55. Divide $48 \div (-3)$. (Lesson 2-8)

56. In which quadrant is the graph of $(-6, 0)$ located? (Lesson 2-2)

57. Is the inequality $4 < 3$ *true, false,* or *open*? (Lesson 1-9)

58. Solve $12n = 48$ mentally. (Lesson 1-6)

59. Entertainment When Bob and Neva went to the movies, the tickets were $6.25 apiece. They also bought two small boxes of popcorn for $2.25 each. (Lesson 1-5)

    a. Write two different expressions for the amount of money Bob and Neva spent.

    b. Find the amount of money they spent.

For **Extra Practice**, see page 751.

60. Find the value of $(11 + 9) \div (3 - 1)$. (Lesson 1-2)

Adding and Subtracting Decimals

What You'll Learn

■ To add and subtract decimals.

Why It's Important

Balancing a checkbook involves adding and subtracting decimals.

Technology Activity

You have experience with adding and subtracting integers. Use a calculator to develop rules for adding and subtracting with decimals.

Your Turn Copy the table below. Use a calculator to find each sum or difference and complete the table.

Column A		Column B	
Sum or Difference	Answer	Sum or Difference	Answer
18 + 66		1.8 + 6.6	
56 + (−78)		5.6 + (−7.8)	
44 − 15		4.4 − 1.5	
220 − (−104)		2.20 − (−1.04)	

a. Compare the sums and differences in Column A to the sums and differences in Column B.

b. Compare the answers in Column A to the answers in Column B.

c. Describe how you could find the sum of 6.8 and 9.3 if you know that 68 + 93 = 161.

Learning the Concept

Adding and subtracting rational numbers follow the same principles as addition and subtraction of integers.

Properties of Addition	Examples
Commutative Property For any rational numbers a and b, $a + b = b + a$.	$-7 + 9 = 9 + (-7)$ $-8.1 + 9.6 = 9.6 + (-8.1)$
Associative Property For any rational numbers a, b, and c, $(a + b) + c = a + (b + c)$.	$(-3.5 + 5.2) + 3.1 = -3.5 + (5.2 + 3.1)$
Identity Property For every rational number a, $a + 0 = a$ and $0 + a = a$.	$8 + 0 = 8$ $0 + -5.2 = -5.2$
Inverse Property For every rational number a, $a + (-a) = 0$.	$6 + (-6) = 0$ $15.8 + (-15.8) = 0$

You will use these properties and the rules you learned for adding and subtracting integers to add and subtract rational numbers. Estimating will help you determine the reasonableness of your answer.

Example **Solve each equation**

a. $m = 13.2 + 11.7$ *Estimate: 13 + 12 = 25*

$$\begin{array}{r} 13.2 \\ +11.7 \\ \hline 24.9 \end{array}$$

$m = 24.9$ *Compare with the estimate.*

b. $g = 231 - 126.7$ *Estimate: 230 − 130 = 100*

$$\begin{array}{r} 231.0 \\ -126.7 \\ \hline 104.3 \end{array}$$

$g = 104.3$ *Compare with the estimate.*

c. $d = 8.3 + (-12.5)$ *Estimate: 8 − 13 = −5*

$d = 8.3 - 12.5$ *The addition can be written as subtraction.*
$d = -|12.5 - 8.3|$ *The difference will be negative since*
$|-12.5| > |8.3|$.

$$\begin{array}{r} 12.5 \\ - 8.3 \\ \hline 4.2 \end{array}$$

$d = -4.2$ *Compare with the estimate.*

TECHNO TIP

You can save time by entering a sum involving a negative as a subtraction problem. For example, find 8.3 + (−12.5) by entering 8.3 − 12.5.

Connection to > Algebra

Simplifying expressions may involve adding and subtracting decimals.

Example **Simplify $4.7x + 6.3x - 13.7x$.**

$$\begin{aligned} 4.7x + 6.3x - 13.7x &= (4.7 + 6.3 - 13.7)x & \textit{Distributive property} \\ &= (11.0 - 13.7)x & \textit{Associative property} \\ &= -2.7x \end{aligned}$$

You use sums and differences of rational numbers everyday when you deal with money.

Example 3

Personal Finance

Kiana uses a spreadsheet program to keep track of her money. The printout at the right shows her expenses and budgeted amounts for the month of January. Find the amount of money that Kiana is under or over budget for the month.

Item	Budget	Actual
Car payment	159.14	159.14
Car insurance	195.63	195.63
Fuel	35.00	37.32
Car repairs	150.00	135.67
Entertainment	60.00	52.44
Savings	50.00	50.00

(continued on the next page)

Find the sums of the budget and the actual amounts. Use a calculator.

budget = 159.14 + 195.63 + 35.00 + 150.00 + 60.00 + 50.00
        = 649.77
actual = 159.14 + 195.63 + 37.32 + 135.67 + 52.44 + 50.00
        = 630.20

The actual total is less than the budgeted total, so Kiana is under budget by 649.77 − 630.20 or $19.57.

Checking Your Understanding

Communicating Mathematics

Read and study the lesson to answer these questions.

1. Using the expression $12.3 + (8.3 + 2.5)$, illustrate the commutative property of addition and the associative property of addition.

2. **State** the additive inverse of 7.398.

3. **You Decide** Mariana and Leslie used different methods to solve $-3.8 - (-2.9)$. Who is correct? Explain.

Mariana	**Leslie**
$-3.8 - (-2.9) = -0.9$	$-3.8 + 2.9 = -0.9$

Guided Practice

State where the decimal point should be placed in each sum or difference.

4. $4.6 + 5.9 = 105$

5. $7.97 - 4.29 = 368$

6. $3.16 + 4.2 = 736$

7. $-8.47 + 3.56 = -491$

8. $8.17 + 9.123 = 17293$

9. $70.3 + 7.03 = 7733$

Solve each equation.

10. $a = 41.3 + 0.28$

11. $b = -9.6 + 3.2$

12. $34.2 - 43.0 = c$

13. $t = 81.9 - 38$

Simplify each expression.

14. $4.7x + 2x$

15. $5.3m - 1.4m - 8m$

16. **Transportation** Many people think that trains are a part of the past. But railroads are still a big part of the transportation industry in the United States. Use the table below to find the change in the number of passengers on each type of public transportation listed from 1997 to 1998.

Type of Transportation	1997 (billions)	1998 (billions)
Commuter rail	358.9	378.4
Heavy rail	2398.5	2530.1
Light rail	220.5	234.4

Source: *American Public Transit Association*

Independent Practice

Solve each equation.

17. $a = 4.3 + 9.8$

18. $x = 31.92 + 14.2$

19. $15.3 - 13.8 = m$

20. $85.3 - 37.07 = r$

21. $72.47 - 9.039 = b$

22. $c = 0.735 - 0.3879$

23. $-7.5 + 9.8 = x$

24. $7.4 + (-3.9) = y$

25. $z = -13.9 + (-12.5)$

26. $k = -3.91 - (-0.6)$

27. $a = -34.1 + (-17.63)$

28. $y = -18.12 - (-7.3)$

Simplify each expression.

29. $5.3m + 7m$

30. $47.9w - 31.8w$

31. $0.3y + 4.1y + 2.5y$

32. $27y - 4.7y - 13.8y$

33. $3.5x - 5 + 8x$

34. $6.93 + (3.1 + 4.07)m$

Evaluate each expression if $a = 5.3$, $b = 8.07$, $x = 21.33$, and $y = 0.7$.

35. $b + x$

36. $a - y$

37. $x - (b + y)$

38. $b + x + y$

39. $a + b - y$

40. $(x - b) + a$

Critical Thinking

41. Cassie has 45 feet of chicken wire for a fence. She wants to build a triangular pen that is 11.2 feet on one side and 9.35 feet on a second side. How much wire will she have left for the third side? Can she use all of it? Why or why not?

Applications and Problem Solving

42. Sports The physical size of the top 10 men and women tennis players has changed in 20 years. How much more do the men and women, respectively, weigh now than 20 years ago?

Average weight of top ten players

Men: 161.7 (1974), 169.9 (1994)

Women: 127.2 (1974), 135.7 (1994)

Source: *USA TODAY* Research

43. Cycling Lance Armstrong of the United States won the Tour de France, the world's most prestigious bicycle race, in 1999. At the beginning of one of his workouts, the odometer on his bicycle reads 201.9 kilometers. If Lance rides 176.6 kilometers that day, what will the odometer read at the end of the workout?

44. Recycling It is estimated that for every 13.2 million tons of glass used per year in the U.S., 2.6 million tons are recycled. How many millions of tons are simply discarded?

Lance Armstrong

Lesson 5-3 *Adding and Subtracting Decimals* **237**

45. Travel   The Department of Transportation keeps tabs on the quality of service by the airlines. Recently, the average number of pieces of luggage lost per 10,000 passengers on major airlines was 54.4 bags. The worst month was January with 78.7 bags lost per 10,000 travelers.

  a. What is the difference between the worst month and the average?

  b. Why do you think January is the worst month for lost baggage?

46. Geometry   Find the perimeter of the triangle shown at the right.

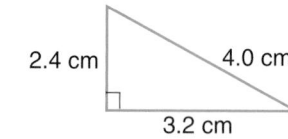
2.4 cm   4.0 cm
3.2 cm

Mixed Review

47. Personal Finance   When packing for a trip, Deandra bought toothpaste for $1.89, suntan lotion for $4.39, hair spray for $1.27, and soap for $2.04. She estimated that the $10 she had would be enough to pay the bill. Do you agree? Explain. (Lesson 5-2)

48. Write $p \cdot p \cdot p \cdot p$ using exponents. (Lesson 4-2)

49. Solve $19 < 17 + a$. (Lesson 3-6)

50. Standardized Test Practice   What is the value of x in the equation $-3 + x = 7$? (Lesson 3-2)

  A. -10 **B.** -4 **C.** 4   **D.** 10

51. Solve $b = -121 \div 11$. (Lesson 2-8)

52. Replace ● with $<$, $>$, or $=$ in $|-10|$ ● 9 to make a true sentence. (Lesson 2-3)

For **Extra Practice**, see page 751.

53. Find the value of $|8| + |-4|$. (Lesson 2-1)

54. Evaluate $2m - n$, if $m = 8$ and $n = 5$. (Lesson 1-3)

WORKING ON THE

Investigation

Refer to the Investigation on pages 166–167.

To get an overall feel of how the stock market is performing, you can consult one of the market indexes. An **index** is an analysis of a selected group of individual stocks such as communications, pharmaceuticals, technology, or utilities.

There are two types of indexes. An index that is calculated by adding the price per share of all the stocks in the index and dividing by the number of stocks in the index is called a **price-weighted index**. The *Dow Jones Industrial Average* is an example of a price-weighted index.

If an index tracks the total market value, it is called a **value-weighted index**. The total market value is the product of the price per share and the total number of shares outstanding. To evaluate this type of index, the total market value is divided by the total market value on the date when the index was first started. The quotient is then multiplied by 100. The *Standard and Poor's 500* is an example of a value-weighted index.

Refer to your work with a spreadsheet on page 209 of Lesson 4-8.

- Using the stocks in your portfolio, create a spreadsheet to calculate your own price-weighted stock market index.
- Update your index each day for a five-day period and note any changes.
- Show the changes in your index by drawing a line graph of the five-day period.

Add the results of your work to your Investigation Folder.

Adding and Subtracting Like Fractions

You'll Learn

■ To add and subtract fractions with like denominators.

It's Important

Adding and subtracting fractions with like denominators will help you determine the fraction of each dollar of your allowance you spend.

MATERIALS: colored pencils ✏ ruler

Hands-On Activity

As you learned in Chapter 4, a fraction can be represented using a model like the one at the right. Different colors of shading can be used to show addition of fractions. This model represents $\frac{1}{6} + \frac{2}{6}$. Since 3 of the 6 sections are shaded, the sum is $\frac{3}{6}$ or $\frac{1}{2}$.

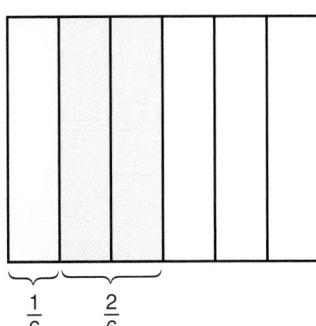

$\underbrace{}_{\frac{1}{6}}$ $\underbrace{}_{\frac{2}{6}}$

Your Turn Draw and shade a model that represents the sum of $\frac{1}{8}$ and $\frac{5}{8}$.

a. How many sections of your model are shaded? How does this compare to the sum of the numerators?

b. How many sections are there in your rectangle? How does this compare to the denominator of the addends?

c. What fraction represents the shaded sections of your rectangle?

d. Write a rule for finding the sum of fractions with the same denominator.

Learning the Concept

Adding fractions with like denominators can be described by the following rule.

Adding Like Fractions	**In words:**	To add fractions with like denominators, add the numerators and write the sum over the same denominator.
	In symbols:	For fractions $\frac{a}{c}$ and $\frac{b}{c}$, where $c \neq 0$, $\frac{a}{c} + \frac{b}{c} = \frac{a+b}{c}$.

When the sum of two fractions is greater than one, the sum is usually written as a mixed number in simplest form. A **mixed number** indicates the sum of a whole number and a fraction.

Example **1** Solve each equation.

a. $s = \frac{3}{4} + \frac{3}{4}$ *Estimate: 1 + 1 = 2*

$s = \frac{6}{4}$ *Since the denominators are the same, add the numerators.*

$s = 1\frac{2}{4}$ *Rename $\frac{6}{4}$ as a mixed number.*

$s = 1\frac{1}{2}$ *Write the mixed number in simplest form.*

b. $y = 3\frac{2}{5} + 4\frac{3}{5}$ *Estimate: 3 + 5 = 8*

$y = (3 + 4) + \left(\frac{2}{5} + \frac{3}{5}\right)$ *Use the Associative and Commutative properties to add the whole numbers and fractions separately.*

$y = 7 + \frac{5}{5}$ *Add the numerators of the fractions.*

$y = 7 + 1$ or 8 *Simplify.*

The rule for subtracting fractions with like denominators is similar to the rule for addition.

Subtracting Like Fractions	**In words:**	To subtract fractions with like denominators, subtract the numerators and write the difference over the same denominator.
	In symbols:	For fractions $\frac{a}{c}$ and $\frac{b}{c}$, where $c \neq 0$, $\frac{a}{c} - \frac{b}{c} = \frac{a-b}{c}$.

Example **2** Solve $y = \frac{7}{15} - \frac{28}{15}$. Write the solution in simplest form.

$y = \frac{7}{15} - \frac{28}{15}$ *Since the denominators are the same, subtract the numerators.*

$y = \frac{-21}{15}$

$y = -1\frac{6}{15}$ or $-1\frac{2}{5}$ *Rename as a mixed number and simplify.*

Connection to > Algebra

You can use the same rules for adding and subtracting like fractions when you evaluate algebraic expressions.

Example **3** Evaluate $m + n$ if $m = 3\frac{1}{6}$ and $n = -1\frac{5}{6}$.

$m + n = 3\frac{1}{6} + \left(-1\frac{5}{6}\right)$ *Replace m with $3\frac{1}{6}$ and n with $-1\frac{5}{6}$.*

$= 3\frac{1}{6} - 1\frac{5}{6}$

$= 2\frac{7}{6} - 1\frac{5}{6}$ *Rename $3\frac{1}{6}$ as $2\frac{7}{6}$.*

$= (2 - 1) + \left(\frac{7}{6} - \frac{5}{6}\right)$

$= 1\frac{2}{6}$ or $1\frac{1}{3}$

Example **4**

Real World APPLICATION

Entertainment

How do you like to spend time in the great outdoors? The results of a survey on that subject are shown in the graph at the right.

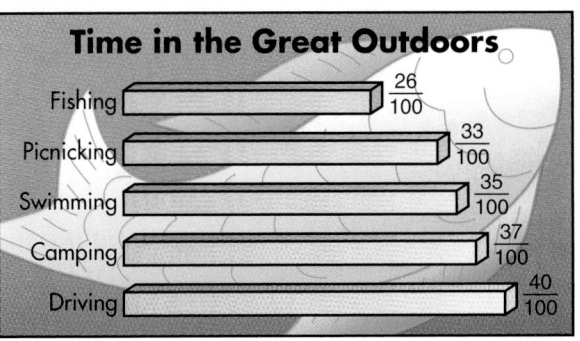

Time in the Great Outdoors

Fishing $\frac{26}{100}$

Picnicking $\frac{33}{100}$

Swimming $\frac{35}{100}$

Camping $\frac{37}{100}$

Driving $\frac{40}{100}$

Source: Roper Starch Worldwide

a. What fraction of the people said that they enjoyed driving or swimming?

$$\frac{40}{100} + \frac{35}{100} = \frac{40 + 35}{100}$$
$$= \frac{75}{100} \text{ or } \frac{3}{4}$$

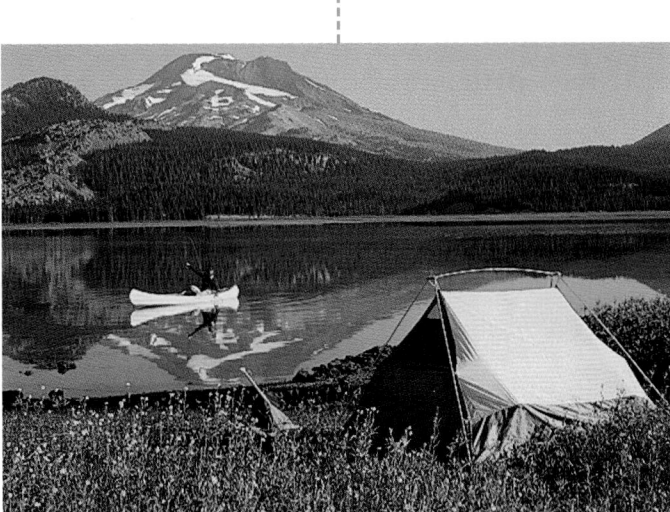

b. What fraction of the people chose camping or picnicking as favorite activities?

$$\frac{37}{100} + \frac{33}{100} = \frac{37 + 33}{100}$$
$$= \frac{70}{100} \text{ or } \frac{7}{10}$$

Checking Your Understanding

Communicating Mathematics

Read and study the lesson to answer these questions.

1. What number properties allow you to write $4\frac{2}{7} + 8\frac{3}{7}$ as $(4 + 8) + \left(\frac{2}{7} + \frac{3}{7}\right)$?

2. **Draw** and shade a model or use objects like paper plates to represent the sum of $\frac{3}{5}$ and $\frac{1}{5}$.

3. **Express** $\frac{33}{7}$ as a mixed number.

MATH JOURNAL

4. List several examples of like fractions and unlike fractions. Then in your own words state a simple rule for adding and subtracting like fractions.

Guided Practice

Solve each equation. Write the solution in simplest form.

5. $\frac{4}{7} + \frac{5}{7} = m$

6. $1\frac{2}{19} + \frac{17}{19} = a$

7. $\frac{19}{24} - \frac{11}{24} = r$

8. $-\frac{19}{30} + \frac{7}{30} = z$

Evaluate each expression if $x = \frac{3}{8}$, $y = \frac{7}{8}$, and $z = \frac{5}{8}$. Write in simplest form.

9. $x + z$

10. $z - y$

Lesson 5-4 Adding and Subtracting Like Fractions **241**

11. Home Maintenance The graph at the right shows the fraction of each dollar an owner spends on different expenses for a house.

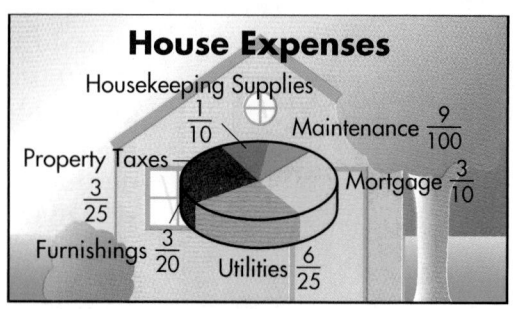

House Expenses

Housekeeping Supplies $\frac{1}{10}$

Maintenance $\frac{9}{100}$

Property Taxes $\frac{3}{25}$

Mortgage $\frac{3}{10}$

Furnishings $\frac{3}{20}$ Utilities $\frac{6}{25}$

Source: Bureau of Labor Statistics

a. What fraction of each dollar is spent on utilities and property taxes?

b. How much more is spent on the mortgage than is spent on paying for housekeeping supplies?

Exercises: Practicing and Applying the Concept

Independent Practice

Solve each equation. Write the solution in simplest form.

12. $\frac{5}{7} + \frac{1}{7} = b$

13. $m = \frac{11}{15} - \frac{8}{15}$

14. $a = \frac{19}{31} - \frac{8}{31}$

15. $\frac{19}{27} - \frac{7}{27} = r$

16. $1\frac{13}{16} - 1\frac{5}{16} = x$

17. $\frac{13}{18} + \frac{11}{18} = a$

18. $m = 1\frac{9}{16} + \frac{15}{16}$

19. $1\frac{16}{21} - \frac{9}{21} = y$

20. $p = -\frac{19}{27} + \left(-\frac{7}{27}\right)$

21. $1\frac{5}{12} + \frac{11}{12} = z$

22. $\frac{9}{20} - \frac{17}{20} = c$

23. $1\frac{19}{41} + \left(-\frac{19}{41}\right) = w$

Evaluate each expression if $a = \frac{5}{12}$, $b = \frac{7}{12}$, and $c = \frac{1}{12}$. Write in simplest form.

24. $a + b$

25. $b - c$

26. $a + c$

27. $b + c$

28. $c - a$

29. $a - b$

Simplify each expression.

30. $4\frac{1}{3}x + \frac{2}{3}x - 3\frac{1}{3}x$

31. $-4\frac{1}{4}r + \left(-2\frac{3}{4}\right)r + 5r$

32. $5\frac{1}{7}m + \left(-3\frac{3}{7}\right)m + 1\frac{2}{7}m$

33. $6\frac{3}{5}a - 3\frac{4}{5}a - 2\frac{2}{5}a$

34. $5\frac{5}{6}y + 3\frac{5}{6}y - 2\frac{1}{6}y$

35. $3\frac{5}{9}b - 2\frac{4}{9}b - \frac{7}{9}b$

Critical Thinking

36. Write $\frac{x + y}{8}$ as the sum of two fractions.

37. Write $\frac{2x - 1}{5}$ as the difference of two fractions.

38. Find the sum of $\frac{1}{20} + \frac{2}{20} + \frac{3}{20} + \cdots + \frac{17}{20} + \frac{18}{20} + \frac{19}{20}$. Describe any pattern or shortcut you used to find the sum.

Applications and Problem Solving

39. Carpentry A carpenter is installing a countertop in a new kitchen. The countertop will be between two walls that are $54\frac{5}{8}$ inches apart. If the piece of countertop material is five feet long, how much will have to be cut off before the countertop is installed?

40. **Home Projects** The Stadlers added an exercise room to the rear of their house. They installed $\frac{3}{8}$-inch thick paneling over a layer of dry wall $\frac{5}{8}$ inch thick. How thick are the wall coverings?

41. **Sewing** Nyoko is making a linen suit. The portion of the pattern envelope that shows the amounts of fabric needed for different sizes of skirts and jackets is shown at the right. If Nyoko is making a size 6 from 45-inch fabric, how much fabric should she buy for the jacket and skirt?

SIZE	(6	8	10)
JACKET			
45"	$2\frac{5}{8}$	$2\frac{3}{4}$	$2\frac{3}{4}$
60"	2	2	2
SKIRT			
45"	$\frac{7}{8}$	$1\frac{1}{8}$	$1\frac{1}{8}$
60"	$\frac{7}{8}$	$\frac{7}{8}$	$\frac{7}{8}$

Mixed Review

42. **Shopping** Cheryl has $38.78 in her wallet. If she buys a purse for $25.59, how much will she have left? (Lesson 5-3)

43. **Economics** The Valdosta Recreation Center needs $17 million for renovations. The table at the right shows the contributions that have been made so far. (Lesson 4-6)

Contributor	Amount
City of Valdosta	$2.5 million
First One Federal Bank	$0.5 million
Montgomery County	$0.5 million
State of Georgia	$1.3 million
Recreation Center Fundraisers	$3 million
Sale of tax credits	$6.2 million

 a. Write the ratio of the amount of money the city contributed to the total amount of money needed.

 b. Write the ratio of the amount of money Montgomery County is contributing to the total amount of money received so far.

44. Solve $\frac{d}{6} < -14$. (Lesson 3-7)

45. Solve $c - 12 \geq -25$. (Lesson 3-6)

46. **Standardized Test Practice** Find the missing dimension in the rectangle below. (Lesson 3-5)

Perimeter = 42 meters   6 meters

x meters

 A. 36 meters   **B.** 30 meters   **C.** 15 meters   **D.** 6 meters

47. Simplify $-8a + (-17a) - (-3a)$. (Lesson 2-4)

48. Name the property shown by $1z = z$. (Lesson 1-4)

For **Extra Practice**, see page 751.

Adding and Subtracting Unlike Fractions

You'll Learn

- To add and subtract fractions with unlike denominators.

It's Important

You'll add and subtract fractions with unlike denominators when you analyze graphs.

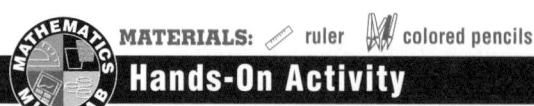 **MATERIALS:** ✏ ruler colored pencils

Hands-On Activity

You can add or subtract fractions with unlike denominators by changing them into fractions with like denominators.

Your Turn Make a model that represents the sum of $\frac{1}{4}$ and $\frac{1}{2}$. First shade $\frac{1}{4}$ of a rectangle like the one at the right. Then shade $\frac{1}{2}$ using a second color to represent $\frac{1}{4} + \frac{1}{2}$.

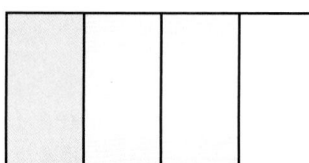

TALK ABOUT IT

a. What fraction of the model is shaded?

b. What is the least common denominator of $\frac{1}{4}$ and $\frac{1}{2}$?

c. Write an addition sentence equivalent to $\frac{1}{4} + \frac{1}{2}$ by using the least common denominator. Then find the sum.

Learning the Concept

The activity leads us to the following rule for adding and subtracting unlike fractions.

Adding and Subtracting Unlike Fractions	To find the sum or difference of two fractions with unlike denominators, rename the fractions with a common denominator. Then add or subtract and simplify.

One way to rename unlike fractions before you add or subtract is to use the LCD.

Example **Solve each equation. Write the solution in simplest form.**

a. $m = \frac{7}{9} + \frac{11}{12}$ *Estimate: 1 + 1 = 2*

$m = \frac{7}{9} \cdot \frac{4}{4} + \frac{11}{12} \cdot \frac{3}{3}$ *The LCD is 3 · 3 · 2 · 2 or 36.*

$m = \frac{28}{36} + \frac{33}{36}$ *Rename each fraction with the LCD.*

$m = \frac{61}{36}$ or $1\frac{25}{36}$

b. $\frac{1}{4} - \frac{2}{3} = a$ *Estimate:* $\frac{1}{2} - 1 = -\frac{1}{2}$

$\frac{3}{12} - \frac{8}{12} = a$ *The LCD is 12.* $\frac{1}{4} = \frac{3}{12}, \frac{2}{3} = \frac{8}{12}$

$-\frac{5}{12} = a$

To add and subtract mixed numbers with unlike denominators, rename using the LCD.

Example **2** **Solve each equation. Write the solution in simplest form.**

a. $n = 4\frac{1}{6} + 7\frac{11}{18}$ *Estimate:* $4 + 8 = 12$

$n = 4\frac{3}{18} + 7\frac{11}{18}$ *Use the LCD to rename* $\frac{1}{6}$ *as* $\frac{3}{18}$.

$n = 11\frac{14}{18}$ or $11\frac{7}{9}$

b. $7\frac{2}{3} - 9\frac{1}{12} = y$ *Estimate:* $8 - 9 = -1$

$7\frac{8}{12} - 9\frac{1}{12} = y$ *The LCD is 12. Rename the fractions.*

$7\frac{8}{12} - 8\frac{13}{12} = y$ *Rename* $9\frac{1}{12}$ *as* $8\frac{13}{12}$.

$-1\frac{5}{12} = y$

Stock market prices are expressed as mixed numbers.

Example **3**

Real World
APPLICATION

Business

News that sales of a certain software were brisk after its August, 2000, debut sent stocks of technological companies up. On Monday, September 11, 2000, one company's stock rose $\$1\frac{3}{8}$ to close at $\$45\frac{1}{4}$. What was the opening price for the company that day?

Let p represent the opening price. Then write an equation.

$p = 45\frac{1}{4} - 1\frac{3}{8}$

$p = 45\frac{2}{8} - 1\frac{3}{8}$ *The LCD is 8. Rename* $\frac{1}{4}$ *as* $\frac{2}{8}$.

$p = 44\frac{10}{8} - 1\frac{3}{8}$ *Rename* $45\frac{2}{8}$ *as* $44\frac{10}{8}$.

$p = 43\frac{7}{8}$

The opening price was $\$43\frac{7}{8}$.

Checking Your Understanding

Communicating Mathematics

Read and study the lesson to answer these questions.

1. What is the first step in adding or subtracting fractions with unlike denominators?

2. **Explain** why you cannot add or subtract fractions without first finding a common denominator.

3. **Make up a problem** in which you would add $12\frac{1}{2}$ and $18\frac{3}{4}$.

4. **You Decide** Hank found the sum of $\frac{7}{9}$ and $\frac{11}{12}$ to be $\frac{18}{12}$ or $1\frac{1}{2}$. Keri found the sum to be $1\frac{25}{36}$. Who is correct? Explain.

5. **Draw** a model or use objects such as paper plates to represent the difference $\frac{3}{4} - \frac{1}{2}$.

Guided Practice

Solve each equation. Write the solution in simplest form.

6. $\frac{1}{3} + \frac{5}{6} = h$ 7. $m = \frac{3}{4} - \frac{5}{8}$ 8. $t = \frac{3}{5} + \frac{3}{10}$

9. $n = \frac{3}{8} - \frac{1}{2}$ 10. $\frac{7}{12} - \frac{2}{3} = g$ 11. $\frac{3}{7} + \frac{5}{14} = w$

12. **Travel** A research firm studied the factors that travelers consider when choosing an airline.

     a. What fraction of those asked said they look at cost or safety record?

     b. What is the difference between the fraction of people who said they considered time of arrival or departure and the number who chose based on the size or type of aircraft?

Source: Chilton Research

Exercises: Practicing and Applying the Concept

Independent Practice

Solve each equation. Write the solution in simplest form.

13. $x = \frac{1}{6} + \frac{7}{18}$ 14. $m = \frac{4}{7} + \frac{9}{14}$ 15. $\frac{5}{12} - \frac{1}{2} = d$

16. $y = \frac{6}{7} - \frac{5}{21}$ 17. $r = \frac{9}{26} + \frac{3}{13}$ 18. $b = \frac{11}{12} + \frac{3}{4}$

19. $p = 4\frac{1}{3} - 2\frac{1}{2}$ 20. $k = \frac{7}{8} + \frac{3}{16}$ 21. $5\frac{1}{3} - 2\frac{1}{6} = m$

22. $\frac{5}{7} - \frac{10}{21} = a$ 23. $9\frac{3}{4} - 5\frac{1}{2} = d$ 24. $x = 7\frac{1}{3} - 3\frac{1}{2}$

25. $y = 12\frac{3}{7} + 4\frac{1}{21}$ 26. $19\frac{3}{8} - 4\frac{3}{4} = s$ 27. $8\frac{9}{10} + 1\frac{1}{6} = r$

28. $b = \frac{1}{3} + \frac{2}{35}$ 29. $\frac{7}{10} - \frac{8}{9} = c$ 30. $\frac{64}{143} - \frac{21}{208} = a$

Evaluate each expression if $x = \frac{5}{8}$, $y = -\frac{3}{4}$, and $z = 2\frac{7}{12}$. Write in simplest form.

31. $x + z$ 32. $z + y$ 33. $y - x$

34. $x + y + z$ 35. $z - x + y$ 36. $z + x - y$

Critical Thinking

37. Studies of unit fractions have been found in Greek papyrus dating back to 500 and 800 A.D. A *unit fraction* is a fraction that has a numerator of 1, such as $\frac{1}{5}$, $\frac{1}{7}$, or $\frac{1}{4}$. Greeks wrote all fractions as the sum of unit fractions. Express $\frac{2}{9}$ as the sum of two different unit fractions.

Applications and Problem Solving

38. **Construction** The outside walls of a new home have $\frac{5}{8}$-inch drywall, $5\frac{1}{2}$ inches of insulation, $\frac{3}{4}$-inch outside wall plywood sheathing, and $\frac{7}{8}$-inch siding. How thick is the wall?

39. **Nutrition** A survey found that $\frac{1}{6}$ of American households bought bottled water in 2000. Only $\frac{1}{17}$ of the households bought bottled water in 1993. What fraction of the population bought bottled water in 2000 that did not in 1993?

Mixed Review

40. Simplify $5\frac{1}{3}b + 3\frac{2}{3}b - 2\frac{1}{3}b$. (Lesson 5-4)

41. Express $s^{-2}t^3$ with positive exponents. (Lesson 4-9)

42. **Boating** The U.S. Coast Guard recommends using the formula $p = \frac{\ell w}{15}$, where ℓ is the length of the boat and w is the width, to determine the number of people, p, who can safely occupy a pleasure boat. If a Daysailer is 18 feet long and 5 feet wide, how many people can occupy it? (Lesson 3-4)

43. Solve $b = -40 \div 8$. (Lesson 2-8)

44. **Standardized Test Practice** Find $-8(3)(-2)$. (Lesson 2-7)
   A. -48 **B.** -22 **C.** 22 **D.** 48

45. Solve $y = 9 - (-4)$. (Lesson 2-5)

46. Order the integers in the set $\{9, -3, -1, 12, -11\}$ from least to greatest. (Lesson 2-3)

47. Simplify $3x + 15x$. (Lesson 1-5)

For **Extra Practice**, see page 751.

Self Test

Express each decimal as a fraction or mixed number in simplest form. (Lesson 5-1)

1. $-0.1666\ldots$ 2. 2.98 3. $1.\overline{4}$

Estimate each sum or difference. (Lesson 5-2)

4. $\$50.00 - \$37.52$ 5. $\frac{8}{9} + \frac{1}{15}$ 6. $\frac{9}{11} - \frac{5}{12}$

Solve each equation. (Lesson 5-3)

7. $r = 5.4 + 9.12$ 8. $90.6 - 33.1 = k$ 9. $m = 17.4 - (-13.2)$

Solve each equation. Write the solution in simplest form. (Lesson 5-4)

10. $\frac{8}{10} + \frac{3}{10} = a$ 11. $b = 4\frac{6}{7} + \frac{2}{7}$ 12. $8\frac{5}{35} - 2\frac{8}{35} = d$

13. **Publishing** The length of a page in a yearbook is 10 inches. The top margin is $\frac{1}{2}$ inch, and the bottom margin is $\frac{3}{4}$ inch. What is the length of the page inside the margins? (Lesson 5-5)

5-6 Solving Equations

What You'll Learn

■ To solve equations with rational numbers.

Why It's Important

You solve equations with rational numbers when you determine patterns and changes in the weather.

Application: Meteorology

Hurricanes can be devastating to life and property. The graph at the right compares the statistics for the five most costly hurricanes to strike the U.S. mainland. The two most costly hurricanes were Andrew and Hugo. Together, the damage they caused totaled $38.96 billion. How much did Hurricane Andrew cost?

Source: National Hurricane Center

Learning the Concept

FYI

The earliest date for a hurricane to strike the U.S. was Hurricane Allison on June 5, 1995. The previous early hurricane record holder was Hurricane Alma on June 9, 1966, which like Allison, struck the Florida panhandle.

To find the cost of Hurricane Andrew, let c represent its cost and write an equation.

$$c \qquad + \qquad 8.49 \qquad = \qquad 38.96$$

$$c + 8.49 = 38.96$$
$$c + 8.49 - 8.49 = 38.96 - 8.49 \qquad \textit{Subtract 8.49}$$
$$c = 30.47 \qquad \textit{from each side.}$$

Hurricane Andrew cost $30.47 billion.

You can solve rational number equations using the same skills you used to solve equations involving integers.

Example **Solve each equation. Check the solution.**

    a. $b + \dfrac{3}{5} = \dfrac{3}{2}$

$$b + \frac{3}{5} = \frac{3}{2}$$

$$b + \frac{3}{5} - \frac{3}{5} = \frac{3}{2} - \frac{3}{5} \qquad \textit{Subtract } \frac{3}{5} \textit{ from each side.}$$

$$b = \frac{15}{10} - \frac{6}{10} \qquad \textit{The LCD is 10.}$$

$$b = \frac{9}{10}$$

Check: $b + \dfrac{3}{5} = \dfrac{3}{2}$

$$\dfrac{9}{10} + \dfrac{3}{5} \overset{?}{=} \dfrac{3}{2}$$

$$\dfrac{9}{10} + \dfrac{6}{10} \overset{?}{=} \dfrac{3}{2}$$

$$\dfrac{15}{10} \overset{?}{=} \dfrac{3}{2} \quad \text{✓} \quad \text{The solution is } \dfrac{9}{10}.$$

b. $a - 3.2 = 2.7$

$$a - 3.2 = 2.7$$
$$a - 3.2 + 3.2 = 2.7 + 3.2 \quad \textit{Add 3.2 to}$$
$$a = 5.9 \qquad\qquad \textit{each side.}$$

Check: $\quad a - 3.2 = 2.7$

$$5.9 - 3.2 \overset{?}{=} 2.7$$
$$2.7 = 2.7 \quad \text{✓} \quad \text{The solution is 5.9.}$$

You can solve equations that represent real-world situations involving rational numbers.

Example 2

Real World APPLICATION

Meteorology

Meteorologists use barometers to measure the air pressure in the atmosphere. Changes in the pressure indicate that there will be a change in the weather. As a hurricane approaches, the barometric pressure at 7:00 P.M. was 28.65 inches. It had dropped 0.37 inches from the previous reading at noon. What was the noon reading?

Let p represent the barometric pressure at noon.

pressure at noon   dropped   0.37 inches   to   28.65

$$p \qquad\qquad - \qquad 0.37 \quad = \quad 28.65$$

$$p - 0.37 = 28.65$$
$$p - 0.37 + 0.37 = 28.65 + 0.37 \quad \textit{Add 0.37 to each side.}$$
$$p = 29.02$$

The reading was 29.02 inches at noon.

Checking Your Understanding

Communicating Mathematics

Read and study the lesson to answer these questions.

1. What property of equality would you use to solve the equation $x - 5.3 = 6.8$?

2. **Explain** how to solve $n - 8.9 = 6.8$.

3. **Tell** how you know that 6.9 is not a solution of $n + 8.5 = 14.7$.

Guided Practice

Solve each equation. Check your solution.

4. $\dfrac{7}{6} = m + \dfrac{5}{12}$ 5. $\dfrac{2}{3} + r = \dfrac{3}{5}$ 6. $7\dfrac{1}{2} = x - 5\dfrac{2}{3}$

7. $m - 4.1 = -9.38$ 8. $y + 7.2 = 21.9$ 9. $x - 1.5 = 1.75$

10. $a - 1\dfrac{1}{3} = 4\dfrac{1}{6}$ 11. $-13.7 = b - 5$ 12. $y + 3.17 = -3.17$

13. **Driver's Education** The odometer on the driver's education car read 26,375.4 miles at the start of Rachel's first in-car class. After the class, the odometer read 26,397.3 miles. How far did Rachel drive?

Independent Practice

Solve each equation. Check your solution.

14. $y + 3.5 = 14.9$

15. $r - 8.5 = -2.1$

16. $a - \dfrac{3}{5} = \dfrac{5}{6}$

17. $b - 1\dfrac{1}{2} = 14\dfrac{1}{4}$

18. $a + 7.1 = 4.7$

19. $b - 5.3 = 8.1$

20. $y + 1\dfrac{1}{3} = 3\dfrac{1}{18}$

21. $m + \dfrac{7}{12} = -\dfrac{5}{18}$

22. $x + \dfrac{5}{8} = 7\dfrac{1}{2}$

23. $k - \dfrac{3}{8} = 1\dfrac{3}{5}$

24. $n + 1.4 = 0.72$

25. $d - (-31.4) = 28.6$

26. $w - 0.04 = 1.2$

27. $d + (-7.03) = 0.98$

28. $5\dfrac{3}{10} + z = 2\dfrac{2}{3}$

29. $v - 4\dfrac{7}{9} = 8\dfrac{1}{6}$

30. $501.1 = y - 9.32$

31. $\dfrac{1}{312} + t = \dfrac{5}{78}$

Critical Thinking

32. Find two rational numbers, a and b, such that $a + b = ab$.

Applications and Problem Solving

33. Oil Production   Recently, the top two oil-producing states were Texas and Alaska. Together they produced 1372.2 million barrels of oil. Alaska produced 684.0 million barrels. How many barrels of oil were produced in Texas?

34. Geometry   The sum of the measures of the angles in a triangle is 180°. If two of the angles measure 75.5° and 60.3°, what is the measure of the third angle?

35. Airports   The graph at the right gives information on the world's busiest cargo airports. Memphis tops the list at 2.36 million metric tons. O'Hare is the world's second busiest passenger airport. What is the difference in cargo handling between Memphis and O'Hare?

Move 'em Out!

Busiest airports in million metric tons of freight:

Memphis	2.36		NY-JFK	1.60
Los Angeles	1.86		Frankfurt	1.46
Miami	1.79		Chicago-O'Hare	1.44
Hong Kong	1.65		Seoul	1.42
Tokyo	1.63		Louisville	1.39

Source: Airports Council International

36. Trains   As a train begins to roll, the cars are "jerked" into motion, and the movement spreads like a ripple towards the back of the train. Slack is built into the couplings so that the engine does not have to move every single car all at once. If the slack built into each coupling is 3 inches or $\dfrac{1}{4}$ foot, how many feet of slack is there between two freight cars and the engine?

Mixed Review

37. Solve $y = \dfrac{7}{8} + 4\dfrac{1}{24}$. Write the solution in simplest form. (Lesson 5-5)

38. Factor $420ab^3$ completely. (Lesson 4-4)

39. Solve $\dfrac{d}{6} \geq -14$. (Lesson 3-7)

40. Geometry   Find the area of a 4" by 6" rectangle. (Lesson 3-5)

41. Solve $x + 5 = 23$. (Lesson 3-2)

42. Standardized Test Practice   What is the value of $-4mn$ if $m = 2$ and $n = -3$? (Lesson 2-7)

   A. -24 **B.** -6 **C.** -4 **D.** 24

For **Extra Practice**, see page 752.

43. Solve $\dfrac{90}{x} = 6$ mentally. (Lesson 1-6)

5-7

Solving Inequalities

Application: Sports

"So let's root, root, root, for the home team. . . ." Commentators and athletes often talk about the advantage of being the home team. Being the home team with a *new* home field may be even better.

When a baseball team has a winning average of 0.600, this means they have won 0.600 or $\frac{3}{5}$ of their games. The table at the right lists the winning rates for teams in the first season in new homes. In 1995, the Colorado Rockies opened in a new park. As of May 22nd, their winning average at home was 0.600. How much would the Rockies have to raise their winning average to have the best new field record?

Home Sweet (New) Home	
Jacobs Field (1994 Indians)	0.686
SkyDome (1989 Blue Jays)	0.618
Comiskey Park (1991 White Sox)	0.568
Camden Yards (1992 Orioles)	0.531
The Ballpark (1994 Rangers)	0.492
Metrodome (1982 Twins)	0.457

Source: Major League Baseball teams

Learning the Concept

The amount that the Colorado team needs to raise their average should be an inequality since anything greater than the present record will give them the best new field record.

current average   *plus*   *increase*   *is greater than*   *current record*

$$0.600 \quad + \quad x \quad > \quad 0.686$$
$$0.600 - 0.600 + x > 0.686 - 0.600 \qquad \textit{Subtract 0.600}$$
$$x > 0.086 \qquad\qquad \textit{from each side.}$$

The Rockies would have to increase their winning record by more than 0.086 to establish a new home-field advantage record.

You can solve rational number inequalities using the same skills you used to solve inequalities involving integers.

Example **Solve each inequality. Graph each solution on a number line.**

a. $a + \frac{3}{8} > 2$

$$a + \frac{3}{8} - \frac{3}{8} > 2 - \frac{3}{8} \qquad \textit{Subtract } \frac{3}{8} \textit{ from each side.}$$

$$a > \frac{16}{8} - \frac{3}{8} \qquad \textit{Rename 2 as a fraction with a}$$
$$\textit{denominator of 8.}$$

$$a > \frac{13}{8} \textit{ or } 1\frac{5}{8}$$

(continued on the next page)

Check: Try 2, a number greater than $1\frac{5}{8}$.

$$a + \frac{3}{8} > 2$$

$$2 + \frac{3}{8} \overset{?}{>} 2$$

$$2\frac{3}{8} > 2 \quad \text{✓ It checks.}$$

The solution is $a > 1\frac{5}{8}$, all numbers greater than $1\frac{5}{8}$.

b. $14.92 + r \geq 7.65$

$$14.92 + r \geq 7.65$$
$$14.92 - 14.92 + r \geq 7.65 - 14.92 \quad \textit{Subtract 14.92 from each side.}$$
$$r \geq -7.27$$

Check: Try -7, a number greater than -7.27.
$$14.92 + r \geq 7.65$$
$$14.92 + (-7) \overset{?}{>} 7.65$$
$$7.92 \geq 7.65 \quad \textit{✓ It checks.}$$

The solution is $r \geq -7.27$, all numbers greater than or equal to -7.27.

THINK ABOUT IT

Are $1\frac{5}{8}$ and -7.27 solutions to the inequalities in Examples 1a and 1b, respectively? Explain.

You can translate a phrase into an inequality involving rational numbers.

Example ❷

Real World APPLICATION

Health

They say "If the shoe fits, wear it." But people often wear shoes that don't fit properly. Podiatrists recommend that a shoe be at least $\frac{3}{8}$ inch but no more than $\frac{1}{2}$ inch longer than your foot. If a shoe is $9\frac{1}{2}$ inches long, how long should a wearer's foot be?

Explore   We know the length of the shoe. We need to find the length of the wearer's foot.

Plan   This situation can be represented using two inequalities. Let f represent the length of the foot.

Solve   Write the inequalities.

foot length   plus $\frac{3}{8}$ inch   no more than $9\frac{1}{2}$ inches

$$f \qquad + \qquad \frac{3}{8} \qquad \leq \qquad 9\frac{1}{2}$$

$$f + \frac{3}{8} \leq 9\frac{1}{2}$$

$$f + \frac{3}{8} - \frac{3}{8} \leq 9\frac{1}{2} - \frac{3}{8}$$

$$f \leq 9\frac{4}{8} - \frac{3}{8}$$

$$f \leq 9\frac{1}{8}$$

$$\text{foot length} \quad \text{plus} \quad \tfrac{1}{2} \text{ inch} \quad \text{is no more than} \quad 9\tfrac{1}{2} \text{ inches}$$

$$f \quad + \quad \tfrac{1}{2} \quad \geq \quad 9\tfrac{1}{2}$$

$$f + \frac{1}{2} \geq 9\frac{1}{2}$$

$$f + \frac{1}{2} - \frac{1}{2} \geq 9\frac{1}{2} - \frac{1}{2}$$

$$f \geq 9$$

The foot should be between 9 and $9\frac{1}{8}$ inches long.

Examine Since $9\frac{1}{8} + \frac{3}{8} = 9\frac{1}{2}$ and $9 + \frac{1}{2} = 9\frac{1}{2}$, the answer is reasonable.

Checking Your Understanding

Communicating Mathematics

Read and study the lesson to answer each question.

1. **Compare and contrast** solving an inequality involving rational numbers to solving an inequality involving integers.

2. **Write a problem** that could be solved using the inequality $65.50 + x > 89.75$.

3. **Write** the inequality whose solution is graphed below.

Guided Practice

Solve each inequality and check your solution. Graph the solution on a number line.

4. $m - 2\frac{3}{4} \geq \frac{7}{12}$ 5. $x + \frac{5}{6} \leq 3\frac{3}{5}$ 6. $\frac{5}{9} < 1\frac{1}{2} + y$

7. $x - 1.4 \leq 7.9$ 8. $n - 3.7 \geq -7.2$ 9. $y + 5.2 < 7.12$

10. **Computers** The number of electronic mailboxes on private networks was 2.5 million in 1990. It is predicted that more than 36 million will exist in 1996. How many mailboxes will have been created between 1990 and 1996?

Exercises: Practicing and Applying the Concept

Independent Practice

Solve each inequality and check your solution. Graph the solution on a number line.

11. $r + 7.5 \leq 13.2$ 12. $x + 4.2 \geq -7.3$ 13. $y + \frac{3}{4} < \frac{7}{12}$

14. $a - 2\frac{2}{3} \geq 8\frac{5}{6}$ 15. $m - 3.1 < 7.4$ 16. $b - 8.9 > -7.2$

17. $d - 1\frac{2}{3} < 1\frac{1}{6}$ 18. $q - \frac{5}{12} \leq \frac{7}{18}$ 19. $r + \frac{7}{8} \leq 2\frac{3}{4}$

20. $z - 1\frac{1}{2} \geq 4\frac{5}{9}$ 21. $\frac{3}{4} \leq -1\frac{1}{2} + w$ 22. $13.5 \leq 18.3 + k$

23. $h - 9.76 < 6.2$ 24. $7\frac{6}{11} + f > 2\frac{14}{15}$ 25. $u - 9.03 \leq 0.8$

26. $n - 289.90 < 479.21$ 27. $\frac{17}{18} + t > \frac{107}{108}$ 28. $0.027 + k \leq 0.00013$

Critical Thinking

29. If $\frac{1}{x} > x$, what can you say about the value of x?

Applications and Problem Solving

30. Hobbies Susan plans to spend no more than $30 on model airplanes and supplies. If she buys a model for $19.95, how much more can she spend on supplies? Express your solution as an inequality.

31. Figure Skating Michelle Quan has scores of 9.4, 9.3, 9.8, and 8.9 in a figure skating competition. She has one more event to skate. Michelle will win the competition if her total score is greater than 47.1. What must Michelle's final event be rated in order for her to win?

32. Family Activity Imagine that you have won $1000.00 in a radio contest. Choose one item that you would purchase for each person in your family including yourself. Research the prices in an advertisement or catalog. Would you have enough money? Write an inequality to show the amount of money you would have left over to put in savings.

33. Employment According to the Census Bureau, a high school graduate earns less than half what a college graduate earns in a lifetime.

  a. What do you think this statement means?

  b. According to the statement, the average high school graduate could earn one-tenth of what an average college graduate does. Do you think that is true? Explain.

34. Geometry The length of each side of a triangle is always less than the sum of the lengths of the other two sides. Find s for the triangle shown at the right.

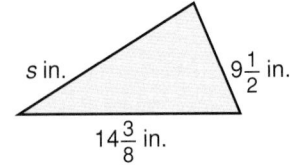

s in. $9\frac{1}{2}$ in. $14\frac{3}{8}$ in.

Mixed Review

35. Solve $x + 3\frac{1}{2} = 7\frac{1}{4}$. (Lesson 5-6)

36. Write $\frac{3}{a^2}$ as an expression using negative exponents. (Lesson 4-9)

37. Find the GCF of 46 and 72. (Lesson 4-5)

38. Geometry Find the perimeter and area of the rectangle shown at the right. (Lesson 3-5)

8 ft

15 ft

39. Simplify $2a(-3b)(5c)$. (Lesson 2-7)

40. Standardized Test Practice For which value is the inequality $16 - x < 9$ true? (Lesson 1-9)

  A. 0   **B.** 6   **C.** 7   **D.** 10

41. Driving The table at the right shows the cost of driving a car in different cities in cents per mile. The costs include insurance, depreciation of the car, license fees, taxes, fuel, oil, tires, and maintenance. (Lesson 1-10)

  a. Which of the listed cities is the most expensive for driving?

  b. What is the difference between the cost of driving in the most expensive city and in the least expensive city listed?

City	Cost (cents per mile)
Bismarck, ND	36.3
Boston, MA	49.8
Burlington, VT	36.4
Hartford, CT	48.0
Los Angeles, CA	55.8
Nashville, TN	37.1
Philadelphia, PA	49.0
Providence, RI	48.5
Raleigh, NC	37.1
Sioux Falls, SD	35.8

Source: Runzheimer International

For **Extra Practice**, see page 752.

Problem-Solving Strategy: Using Logical Reasoning

You'll Learn

- To use deductive and inductive reasoning.

Application: Comics

Peppermint Patty has used **inductive reasoning**. Inductive reasoning *makes* a rule after seeing several examples. Patty made a conclusion based on what happened in the past.

Example **What is the next term in the sequence 1, 2, 4, 8, 16, . . . ?**

Notice that every number is twice the previous number. For example, 4 is twice 2 and 8 is twice 4. The next term is probably twice 16, or 32.

Suppose Patty gets back her algebra paper and finds that this time $x = 9$ and $y = 15$. Then her method of inductive reasoning did not work. Since inductive reasoning is based on past evidence only, it may sometimes fail.

The Greek letter π, called pi, represents the number 3.14159. . . . If the question on Patty's paper had been "What is the value of π?", she would know that the answer was 3.14159. . . . This is an example of **deductive reasoning**. Deductive reasoning *uses* a rule to make a conclusion.

Example **All octagons have eight sides. A stop sign is shaped like an octagon. How many sides does a stop sign have?**

Traffic

You do not need to count the sides of a stop sign. Using deductive reasoning, you can conclude that all stop signs have eight sides.

Checking Your Understanding

Communicating Mathematics

Read and study the lesson to answer these questions.

1. **Compare and contrast** deductive reasoning and inductive reasoning.

2. Which type of reasoning are you using when you look for a pattern?

3. The law requires that a person be born in the United States in order to serve as president. What type of reasoning are you using when you say that President Clinton was born in the United States? Explain.

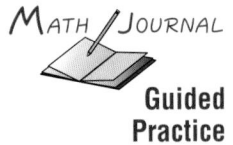

MATH JOURNAL

4. Make up your own examples of inductive and deductive reasoning.

Guided Practice

State whether each is an example of *inductive* or *deductive* reasoning. Explain your answer.

5. Every year for the past five years it has rained during spring break. Jamal says that it will rain during spring break this year.

6. If a student earns an "A" for each nine weeks in a class, then he or she will not have to take the final exam in that class. Stacey earned an A each nine weeks in Pre-Algebra, so she will not have to take the final.

7. The citizens of San Juan Capistrano celebrate when the swallows leave and return. For many years, the swallows have left on October 13 and returned on March 19. The citizens are predicting the swallows will return this year on March 19.

Exercises: Practicing and Applying the Concept

Independent Practice

8. Is the statement *integers that are even are always divisible by 2* an example of inductive or deductive reasoning? Explain.

9. Find the next two numbers in the pattern 3, 6, 9, 12, 15, _?_, _?_.

10. The product of a number and itself is 196. Find the number.

Choose

Estimation
Mental Math
Calculator
Paper and Pencil

11. **Patterns** Write the equation that you think should come next in the pattern at the right. Check your answer with a calculator.

$$1^2 = 1$$
$$11^2 = 121$$
$$111^2 = 12,321$$

12. Insert one set of parentheses in $4 \times 5 - 2 + 7 = 19$ to make the equation true.

13. **School** If you do not pass your math test, then your parents will ground you for the weekend.

   a. Suppose you do not pass your math test. What will your parents do? What type of reasoning did you use?

   b. Suppose you do pass your math test. Does that mean you will definitely not be grounded? Why or why not?

14. At Chicken Little's, you can order Chicken Fingers in boxes of 6, 9, or 20. If you order a box of 6 and a box of 9, you can get 15 fingers. Since no combination of 6, 9, and 20 adds up to 13, you cannot order 13 fingers. What is the greatest number of fingers you *cannot* order?

15. **Patterns** Study the pattern at the right.

   a. If you were to continue the pattern, how would you write 11×10?

   b. What is the actual value of 11×10?

   c. What does this tell you about inductive reasoning? Could you use inductive reasoning to prove something?

$$11 \times 1 = 11$$
$$11 \times 2 = 22$$
$$11 \times 3 = 33$$
$$11 \times 4 = 44$$

16. At Champion High School, if you take an art class you do not have to take a music class. Anastasia is taking a technical drawing course, so she does not have to take a music course. Is this an example of inductive or deductive reasoning?

17. What color piece is needed to complete the quilt at the right?

Critical Thinking

18. A set of dominoes has 28 rectangular pieces with two numbers of dots on each one. Each number, 0 through 6, is paired with every other number, including itself, on exactly one domino. The grid at the right was made by arranging dominoes and then recording the numbers. Four dominoes are shown. Copy the grid and use deductive reasoning to draw the positions of the remaining dominoes.

1	0	2	0	0	5	4	1
1	1	5	3	6	2	4	2
3	3	1	0	3	5	3	4
0	6	6	4	6	5	1	1
0	4	0	2	5	4	2	6
1	2	3	2	6	4	5	2
3	5	5	0	3	4	6	6

Mixed Review

19. Solve $z - 4.71 \leq -3.8$. (Lesson 5-7)

20. Solve $a - 9\frac{5}{6} = 2\frac{3}{24}$. (Lesson 5-6)

21. Find the GCF of 78 and 35. (Lesson 4-5)

22. Standardized Test Practice The formula that relates men's shoe size, s, and foot length, f, in inches is $s = 3f - 21$. What is the shoe size for a man whose foot is 11 inches long? (Lesson 3-4)

  A. 10   **B.** 11   **C.** 12   **D.** 13

23. Find $\left|-6\right| + \left|4\right|$. (Lesson 2-1)

For **Extra Practice**, see page 752.

From the FUNNY PAPERS

1. What type of reasoning did the caveman use? Explain.

2. Is the conclusion the caveman reached correct? Give an example to justify your answer.

"Water boils down to nothing . . . snow boils down to nothing . . . ice boils down to nothing . . . everything boils down to nothing."

Integration: Discrete Mathematics
Arithmetic Sequences

You'll Learn

■ To find the terms of arithmetic sequences, and

■ to represent a sequence algebraically.

It's Important

You can use sequences to find the cost of a long-distance telephone call.

Technology Activity

Have you ever seen a flash of lightning and heard the roll of thunder and wondered how far away it was? We can generate a table to estimate the distance to a lightning bolt for which you heard the thunder. Sound travels approximately 1100 feet or $\frac{1}{5}$ of a mile per second. For every second that passes, the soundwaves of the thunder travel an additional 1100 feet.

Time (seconds)	0	1	2	3	4	5	6
Distance (feet)	0	1100	2200				
Distance (miles)	0	$\frac{1}{5}$	$\frac{2}{5}$				

Your Turn   Copy the table above. Use a calculator and a pattern to complete the table.

a. How far away is the flash of lightning if the thunderclap takes 4 seconds to reach the observer?

b. What pattern do you see in the table?

c. Write a rule for finding the next numbers in the patterns.

Learning the Concept

A branch of mathematics called **discrete mathematics** deals with topics like logic and statistics. One topic of discrete mathematics is **sequences**. A sequence is a list of numbers in a certain order, such as 0, 1, 2, 3, or 2, 4, 6, 8. Each number is called a **term** of the sequence. A sequence like the distances for thunder, where the difference between any two consecutive terms is the same, is called an **arithmetic sequence**. The difference is called the **common difference**.

Arithmetic Sequence	An arithmetic sequence is a sequence in which the difference between any two consecutive terms is the same.

Example **1** **State whether each sequence is arithmetic. Then write the next three terms of each sequence.**

a. 1.25, 1.45, 1.65, 1.85, 2.05

Find the difference between consecutive terms in the sequence.

$$1.25 \quad 1.45 \quad 1.65 \quad 1.85 \quad 2.05$$
$$+0.20 \quad +0.20 \quad +0.20 \quad +0.20$$

Since the difference between any two consecutive terms is the same, the sequence is arithmetic.

Continue the sequence to find the next three terms.

$$\ldots 1.85 \quad 2.05 \quad 2.25 \quad 2.45 \quad 2.65$$
$$+0.20 \quad +0.20 \quad +0.20 \quad +0.20$$

The next three terms of the sequence are 2.25, 2.45, and 2.65.

b. 1, 3, 7, 13, 21, . . .

Since there is no common difference, the sequence is not arithmetic.

$$1 \quad 3 \quad 7 \quad 13 \quad 21$$
$$+2 \quad +4 \quad +6 \quad +8$$

Although the sequence is not arithmetic, it is still a sequence. Notice that the differences are in a pattern of 2, 4, 6, 8. So, the next three differences will be 10, 12, 14. The next three terms are 21 + 10 or 31, 31 + 12 or 43, and 43 + 14 or 57.

c. 13, 8, 3, −2, −7, . . .

$$13 \quad 8 \quad 3 \quad -2 \quad -7 \quad \mathbf{-12} \quad \mathbf{-17} \quad \mathbf{-22}$$
$$-5 \quad -5 \quad -5 \quad -5 \quad -5 \quad -5 \quad -5$$

Since there is a common difference of −5, the sequence is arithmetic. The next three terms are −12, −17, and −22.

Connection to Algebra

If we know any term of an arithmetic sequence and the common difference, we can list the terms of the sequence. Consider the following sequence.

3, 7, 11, 15, 19, 23, . . .

We can write an expression that represents any term in the sequence. The first term is 3; call the first term a. The common difference between terms is 4; call the common difference d. Study this pattern.

1st term	a	3
2nd term	$a + d$	$3 + 4 = 7$
3rd term	$a + d + d$ or $a + 2d$	$3 + 2 \cdot (4) = 11$
4th term	$a + d + d + d$ or $a + 3d$	$3 + 3 \cdot (4) = 15$
⋮	⋮	
nth term	$a + (n - 1)d$	

Lesson 5-9 **INTEGRATION** *Discrete Mathematics* *Arithmetic Sequences* **259**

Because *n* represents any term, you can use the expression $a + (n - 1)d$ to find any term in the sequence.

Example **2**

Real World
APPLICATION

Postage

In 1999, the first class postage rates were raised to 33 cents for the first ounce and 22 cents for each additional ounce. A chart showing the postage for weights up to 6 ounces is shown below. What was the cost for a 10-ounce letter?

Weight (ounces)	1	2	3	4	5	6
Postage (cents)	33	55	77	99	121	143

Use the expression $a + (n - 1)d$ to find the 10th term of the postage sequence.

The first term, *a*, is 33. The common difference, *d*, is 22. Since we are looking for the tenth term, *n* is 10.

$$a + (n - 1)d = 33 + (10 - 1)(22) \quad \text{Substitute 33 for a, 10 for n, and 22 for d.}$$
$$= 33 + 9(22) \quad \text{Use the order of operations.}$$
$$= 33 + 198$$
$$= 231 \quad \text{The 10th term is 231.}$$

A 10-ounce letter would cost $2.31.

THINK ABOUT IT

Which would be easier to find the cost of a 16-ounce letter: extending the table to 16 ounces or using the formula?

Checking Your Understanding

Communicating Mathematics

Read and study the lesson to answer these questions.

1. **Give** an example of an arithmetic sequence.

2. **Tell** the relationship between consecutive terms of an arithmetic sequence.

3. **Explain** how to determine if a sequence is arithmetic.

4. Given the expression $a + (n - 1)d$, record in your Math Journal the meaning of the variables *a*, *n*, and *d*. Illustrate the meaning with a numerical example.

Guided Practice

State whether each sequence is arithmetic. Then write the next three terms of each sequence.

5. 2, 5, 8, 11, 14, . . .

6. 5, 9, 13, 17, 21, . . .

7. 17, 16, 14, 11, 7, . . .

8. $\frac{3}{2}, 2, \frac{5}{2}, 3, \ldots$

9. 3, 5, 8, 12, 17, . . .

10. 15, 25, 40, 60, 85, . . .

11. **On-Line Service** An Internet service provider charged $9.95 and $2.95 per hour after the first five hours to use their service. Let *n* represent the number of hours after five.

   a. Write the sequence of prices for *n* = 0 to 5.

   b. How much would it cost if you used the service 9 hours more than the allotted five hours in one month?

Independent Practice

State whether each sequence is arithmetic. Then write the next three terms of each sequence.

12. 0.75, 1.5, 2.25, 3, . . .

13. 4.5, 4.0, 3.5, 3.0, . . .

14. 0, 1100, 2200, 3300, . . .

15. 1, 4, 9, 16, . . .

16. 1, 1, 1, 1, 1, . . .

17. 1, 2, 4, 8, 16, . . .

18. 7, 10, 13, 16, . . .

19. 91, 82, 73, 64, . . .

20. 83, 77, 71, 65, . . .

21. 19, 15, 11, 7, . . .

22. 0.1, 0.3, 0.9, 2.7, . . .

23. 5.47, 6.49, 7.51, 8.53, . . .

24. 10, 12, 15, 19, 24, . . .

25. 11, 14, 19, 26, 35, . . .

26. $1, \frac{1}{2}, \frac{1}{3}, \frac{1}{4}, \ldots$

27. $\frac{3}{4}, \frac{4}{5}, \frac{5}{6}, \frac{6}{7}, \ldots$

28. Write the first six terms in an arithmetic sequence with a common difference of -3. The first term is 35.

29. Given the sequence 7, 18, 29, 40, . . . , name the first term and the common difference.

30. Find the 25th even integer.

31. The first term of an arithmetic sequence is 100, and the common difference is 25. Write the first six terms of the sequence.

Critical Thinking

32. Draw the next figure in the sequence below.

Applications and Problem Solving

33. Meteorology   Refer to the application at the beginning of the lesson. Thunder can be heard from up to 10 miles away.

   a. How many seconds would it take for the thunder to travel 10 miles?

   b. Write a formula for finding the distance, *d*, in miles from lightning for which the thunder is heard *t* seconds after the lightning is seen.

34. Geometry   Russian mathematician Sonya Kovalevsky (1850–1891) studied number sequences. One of the sequences she studied is represented by the shaded squares below.

 + + + +

$\frac{1}{2}$ + $\frac{1}{4}$ + $\frac{1}{8}$ + $\frac{1}{16}$ + $\frac{1}{32}$

   a. Write the first ten numbers of the sequence represented by the model.

   b. Is the sequence arithmetic? Explain.

   c. What do you think the sum of the first ten terms is close to? Do not actually add.

35. **Physics**   While flying his hot-air balloon at 4000 feet, Rey dropped a sandbag in order to climb higher. A free-falling object increases its velocity by 32 feet per second every second. Use a sequence to determine how fast the sandbag will be falling at the end of 10 seconds.

36. **Meteorology**   Television weather reporters frequently give the daily barometric pressure, such as 29.56 inches. This means that atmospheric pressure alone will hold up a column of mercury to a height of 29.56 inches. Centuries ago, scientists experimented with other materials such as water and discovered that atmospheric pressure would hold up a column of water more than thirty feet high.

Mercury (inches)	25	26	27	28	29	30	31	32
Water (feet)	27.2	28.33	29.46	30.59				

a. Copy and complete the table of heights above.

b. What is a in the sequence of water heights shown? What is the common difference?

c. How many feet of water are held by the pressure that holds up thirty inches of mercury?

Mixed Review

For **Extra Practice**, see page 752.

37. **Business**   The long-distance telephone company that Lathan uses charges $0.10 per minute for a long distance call. Lathan talked for one hour, so he knows his bill will be $6.00. Is this an example of inductive or deductive reasoning? (Lesson 5-8)

38. Write $\frac{x}{125}$ as an expression with negative exponents. (Lesson 4-9)

39. Find the missing exponent in $(x^{\bullet})(x^9) = x^{15}$. (Lesson 4-8)

40. **Standardized Test Practice**   In which quadrant does the graph of $(-17, -4)$ lie? (Lesson 2-2)
    A. I   **B.** II   **C.** III   **D.** IV

41. **Statistics**   Have you ever heard of "sleeping like a baby?" How about "sleeping like a koala!" A koala sleeps more than any other animal. The table at the right shows the average number of hours that different animals sleep each day. Make a bar graph of the data. (Lesson 1-10)

Animal	Average hours of sleep
Koala	22
Sloth	20
Armadillo	19
Opossum	19
Lemur	16
Hamster	14
Squirrel	14
Cat	13
Pig	13
Spiny anteater	12

42. Write a related sentence for $35 = 16 + m$ using an inverse operation. (Lesson 1-8)

5-9B Fibonacci Sequence

An Extension of Lesson **5-9**

As you have learned, there are sequences that are not arithmetic. One special sequence is called the **Fibonacci sequence**.

The Fibonacci sequence is not arithmetic, but its terms are found by adding. The sequence begins with 1, 1, and each term after the second is the sum of the previous two terms of the sequence.

$$1, \quad 1, \quad \overset{1+1}{2,} \quad \overset{1+2}{3,} \quad \overset{2+3}{5,} \quad \overset{3+5}{8,} \ldots$$

Your Turn List the first ten terms of the Fibonacci sequence.

There are many patterns in the Fibonacci sequence. Here are some of the patterns.

▶ Every third term is divisible by 2.
▶ Every fourth term is divisible by 3.
▶ Every fifth term is divisible by 5.
▶ Every sixth term is divisible by 8.

TALK ABOUT IT

1. What number do you think may be a factor of every seventh term? Use a calculator to check your conjecture.

2. There is only one perfect square and one perfect cube, except for 1, in the Fibonacci sequence. Name them.

3. If you square the Fibonacci numbers and then add adjacent squares, a new sequence is formed. Describe the sequence.

Extension

4. Investigate how the Fibonacci numbers are related to a pineapple, a pine cone, and a sunflower.

Investigation

TAKING STOCK

Refer to the Investigation on pages 166–167.

Throughout this investigation, the terms *capital gain*, *closing price*, and *share* were often used.

What do these terms mean?

What do these words mean in terms of the work you did in this investigation?

Plan and give an oral presentation to describe what you know about these and any other investment terms. Your presentation should include the following.

- Examples from your portfolio to help you describe and define the investment terms you used most often in this investigation.

- A table showing how you initially invested your $10,000 and the source of the prices.

- A stock chart or line graph showing a price-weighted index of your portfolio's performance over the course of this investigation.

- An explanation of the analysis of your portfolio's performance.

- A comparison of your portfolio's performance to how the *Standard and Poor's (S&P) 500* performed over the same time period.

- A statement of what your shares would be worth if you sold all of them today.

- An explanation of what you would do differently if you had to do this investigation over again.

Extension

Schedule an interview with a financial analyst or stock broker. Make a list of questions to ask him or her.

- Be sure to ask about the requirements and duties of their job.

- Include questions about investment strategies for teenagers and how investment strategies differ depending on the age of the investor.

PORTFOLIO ASSESSMENT

You may want to keep your work on this Investigation in your portfolio.

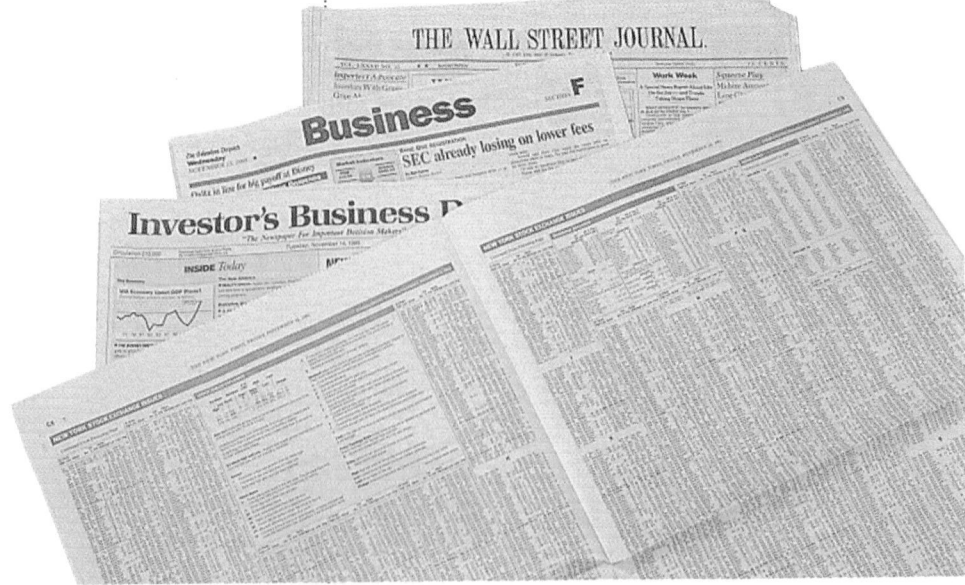

Vocabulary

After completing this chapter, you should be able to define each term, property, or phrase and give an example or two of each.

Algebra
adding like fractions (p. 239)
adding unlike fractions (p. 244)
associative property of addition
 (p. 234)
bar notation (p. 225)
commutative property of addition
 (p. 234)
identity property of addition
 (p. 234)
integers (p. 224)
inverse property of addition
 (p. 234)
mixed number (p. 239)
rational numbers (p. 224)
repeating decimals (p. 225)
subtracting like fractions (p. 240)
subtracting unlike fractions
 (p. 244)
terminating decimal (p. 225)
whole numbers (p. 224)

Geometry
scale (p. 224)

Problem Solving
deductive reasoning (p. 255)
inductive reasoning (p. 255)

Discrete Mathematics
arithmetic sequence (p. 258)
common difference (p. 258)
discrete mathematics (p. 258)
Fibonacci sequence (p. 263)
sequence (p. 258)
term (p. 258)

Chapter Review For additional lesson-by-lesson review, visit:
www.glencoe.com/ sec/math/prealg/ mathnet

Understanding and Using Vocabulary

Choose the letter of the term that best matches each statement.

1. This is any number that can be expressed in the form $\frac{a}{b}$, where a and b are integers, and $b \neq 0$.

2. This type of reasoning creates a rule based on consistent examples.

3. A comparison of the size of a model to the size of the real object is called a __?__.

4. This property justifies the statement $3 + 5 = 5 + 3$.

5. This is any number which is positive, negative, or zero and whose absolute value is a whole number.

6. This type of reasoning uses a given rule to make a conclusion or a decision.

7. This property justifies the statement $4 + 0 = 4$.

A. Identity property of addition
B. rational number
C. integers
D. Commutative property of addition
E. whole numbers
F. deductive reasoning
G. Inverse property of addition
H. inductive reasoning
I. scale
J. Associative property of addition

Skills and Concepts

Objectives and Examples

Upon completing this chapter, you should be able to:

▶ **identify and compare rational numbers**
(Lesson 5-1)

0 is a rational number, an integer, and a whole number.

0.1234567 . . . is a non-terminating, non-repeating decimal. So it is not rational.

▶ **rename decimals as equivalent fractions.**
(Lesson 5-1)

Let $N = 0.888$ . . . .   Then $10N = 8.888$ . . .

$$\begin{array}{r} 10N = 8.8888 \ldots \\ -N = 0.8888 \ldots \\ \hline 9N = 8 \\ N = \dfrac{8}{9} \end{array}$$

$0.\overline{8} = \dfrac{8}{9}$

▶ **estimate sums and differences of decimals and fractions.** (Lesson 5-2)

$48.89 + 91.21 \quad \rightarrow \quad 50 + 90 = 140$

$6\dfrac{9}{10} - 2\dfrac{3}{8} \quad \rightarrow \quad 7 - 2\dfrac{1}{2} = 4\dfrac{1}{2}$

▶ **add and subtract decimals.** (Lesson 5-3)

$f = 4.56 - 2.358$

$$\begin{array}{r} 4.560 \\ -2.358 \\ \hline 2.202 \end{array}$$

$f = 2.202$

Review Exercises

Use these exercises to review and prepare for the chapter test.

Name the set(s) of numbers to which each number belongs.

8. $\dfrac{9}{10}$ 9. -4

10. 18 11. $0.122333 \ldots$

Express each decimal as a fraction or mixed number in simplest form.

12. 3.25 13. -9.45

14. 0.40 15. -0.13

16. $0.\overline{72}$ 17. $4.\overline{24}$

18. $0.0\overline{6}$ 19. 8.89

Estimate each sum or difference.

20. $53.6 + 41.2$ 21. $4.99 + 3.29$

22. $325.44 - 249.25$ 23. $50.00 - 39.89$

24. $\dfrac{11}{12} + \dfrac{6}{10}$ 25. $\dfrac{24}{25} + \dfrac{1}{11}$

26. $18\dfrac{1}{10} - 3\dfrac{1}{9}$ 27. $24\dfrac{4}{7} - 22\dfrac{1}{6}$

Solve each equation.

28. $t = 8.5 + 42.25$ 29. $9.43 - (-1.8) = p$

30. $j = -7.43 + 5.34$ 31. $m = 17.19 - 24.87$

32. $y = 13.983 + 4.52$ 33. $-8.52 - 9.43 = d$

Objectives and Examples	Review Exercises

▶ **add and subtract fractions with like denominators.** (Lesson 5-4)

$n = 3\frac{3}{5} + 4\frac{4}{5}$

$n = (3 + 4) + \left(\frac{3}{5} + \frac{4}{5}\right)$

$n = 7 + \frac{7}{5}$

$n = 7 + 1\frac{2}{5}$ or $8\frac{2}{5}$

Solve each equation. Write the solution in simplest form.

34. $w = \frac{4}{7} + \frac{6}{7}$

35. $t = 5\frac{2}{9} + 8\frac{5}{9}$

36. $7\frac{4}{7} + \left(-2\frac{5}{7}\right) = s$

37. $\frac{7}{8} - \frac{5}{8} = b$

38. $k = 5\frac{1}{5} - 4\frac{4}{5}$

39. $\frac{4}{15} - \frac{13}{15} = y$

▶ **add and subtract fractions with unlike denominators.** (Lesson 5-5)

$r = \frac{2}{5} + \frac{7}{8}$

$r = \frac{2 \cdot 8}{5 \cdot 8} + \frac{7 \cdot 5}{8 \cdot 5}$ *The LCD is 40.*

$r = \frac{16}{40} + \frac{35}{40}$

$r = \frac{51}{40}$ or $1\frac{11}{40}$

Solve each equation. Write the solution in simplest form.

40. $\frac{3}{7} + \frac{11}{14} = q$

41. $s = \frac{7}{15} - \frac{16}{30}$

42. $6\frac{4}{5} + 1\frac{3}{4} = h$

43. $4\frac{1}{6} - \left(-2\frac{3}{4}\right) = m$

44. $1\frac{2}{5} + \frac{1}{3} = a$

45. $9\frac{5}{12} - 4\frac{7}{18} = h$

▶ **solve equations with rational numbers.** (Lesson 5-6)

$y + 5.7 = 3.1$
$y + 5.7 - 5.7 = 3.1 - 5.7$
$y = -2.6$

Solve each equation. Check your solution.

46. $v - 4.72 = 7.52$ **47.** $s + (-13.5) = -22.3$

48. $x + \frac{3}{4} = -1\frac{2}{5}$ **49.** $b - \frac{5}{8} = 1\frac{3}{16}$

50. $z - (-5.8) = 1.36$ **51.** $k - \frac{4}{9} = 1\frac{7}{18}$

▶ **solve inequalities with rational numbers.** (Lesson 5-7)

$p - 4\frac{9}{10} > -8\frac{6}{10}$

$p - 4\frac{9}{10} + 4\frac{9}{10} > -8\frac{6}{10} + 4\frac{9}{10}$

$p > -3\frac{7}{10}$

Solve each inequality and check your solution. Graph the solution on a number line.

52. $w + \frac{7}{12} \geq \frac{5}{18}$ **53.** $f - 3\frac{1}{4} \leq \frac{7}{8}$

54. $31.6 < -5.26 + g$ **55.** $q - (-6.7) > 12$

56. $w - 4.32 \leq 1.234$ **57.** $m + 7.17 > 1.019$

58. $\frac{7}{6} + c \geq 3\frac{5}{24}$ **59.** $a - \left(\frac{-3}{10}\right) \leq 1\frac{1}{5}$

Objectives and Examples	Review Exercises

▶ **find the terms of arithmetic sequences and represent a sequence algebraically.**
(Lesson 5-9)

2   4   8   16
  +2   +4   +8

This is not an arithmetic sequence because the difference between any two consecutive terms is not the same.

The next three terms are 16 + 16 or 32, 32 + 32 or 64, and 64 + 64 or 128.

State whether each sequence is arithmetic. Then write the next three terms of each sequence.

60. 75, 90, 105, . . .

61. 45, 37, 29, 21, . . .

62. 0.0625, 0.125, 0.25, . . .

63. $\frac{10}{11}, \frac{8}{9}, \frac{6}{7}$

64. 67, 58, 49, 40, . . .

Applications and Problem Solving

65. Hobbies   Tedrick is building a model yacht that is 35 centimeters long. The actual yacht is 12 meters long. (Lesson 5-1)

  a. What is the scale of the model?

  b. Is the scale a whole number, an integer, or a rational number? Explain.

67. Carpentry   Currito is making a porch swing for his house. He bought a 4-foot piece of poplar for the seat. If the finished seat on the pattern is to be $41\frac{3}{8}$ inches long, how much should Currito cut off? (Lesson 5-5)

69. Physical Science   The Italian scientist Galileo discovered that there was a relationship between the time of the back-and-forth swing of a pendulum and its length. The table at the right shows the time for a swing and the length of several pendulums. (Lesson 5-9)

  a. How long do you think a pendulum with a swing of 5 seconds is?

  b. Describe the sequence of lengths. Is it an arithmetic sequence? Explain.

66. Politics   In the 1992 presidential election, 186.7 million people were eligible to vote. Only 113.9 million people actually voted. How many people were eligible to vote in 1992 but did not? (Lesson 5-3)

68. Numbers that can be divided by both 2 and 3 are always divisible by 6. 18 is divisible by 2 and 3, so it is divisible by 6. Is this inductive or deductive reasoning? Explain. (Lesson 5-8)

Time of swing (seconds)	Pendulum length (units)
1	1
2	4
3	9
4	16

A practice test for Chapter 5 is available on page 778.

Demonstrate your knowledge by giving a clear, concise solution to each problem. Be sure to include all relevant drawings and justify your answers. You may show your solutions in more than one way or investigate beyond the requirements of the problem.

1. A house is 36 feet from front to back. On the blueprints, the drawing of the house is 18 inches from front to back.

   a. What is the scale factor of the blueprints to the house?

   b. What type of number is the scale factor? Explain.

2. The windows on the side of the house are $2\frac{7}{8}$ feet from the ground and $7\frac{3}{4}$ feet from the roof. The house is 15 feet from the ground to the roof.

   a. Approximately how tall are the windows?

   b. Exactly how tall are the windows on the side of the house?

3. The window on the front of the house is 6.72 feet wide, and there are 4.14 feet on either side of the window. What is the width of the front section of the house?

4. The house covers 1023.5 square feet. The house and the yard together cover 1 acre, or 43,560 square feet. How large is the yard?

5. The law in Genoa Township states that an individual may erect a fence as long as it is no closer than 3 inches from the property line. Mr. Owens says that the fence he is erecting 5 inches from the property line is legal. Did Mr. Owens use inductive or deductive reasoning? Explain.

6. A series of bushes is to be planted along the foundation of the house. Devise a plan for the planting that uses an arithmetic sequence.

Thinking Critically

▶ When you estimate an area in order to purchase wallpaper, it is wise to overestimate the area you need to cover. Give an example of a situation when it is better to estimate low.

▶ Do you come closer to the actual answer when you round to the nearest tenth or to the nearest whole number? Explain.

 Portfolio

Review the items in your portfolio. Make a list or table of contents of the items, noting why each item was chosen and place it in your portfolio.

Investigation

ROLLER COASTER MATH

MATERIALS

✐ ruler

⌒ protractor

△ compass

▦ graph paper

"Please remain seated and keep your hands and feet inside the ride at all times. Enjoy your ride."

You are a member of a group of three engineers that has been hired to design and build a new roller coaster ride. You will oversee every detail of the design of the ride from the length of its track to how riders will wait in line and board the cars. You'll even name the ride. The amusement park where the roller coaster will be built wants you to design it so that it establishes at least one new record.

Some of the existing world records are:

Longest

- *1.42 miles—*The Ultimate *at Lightwater Valley Theme Park in Ripon, Great Britain*

Tallest

- *222 feet above its footings—*Pepsi Max-The Big One *at Blackpool Pleasure Beach, Great Britain*

Fastest

- *85 mph—*Desperado *at Buffalo Bill's, Jean, NV*

Greatest Drop

- *225 feet—*Steel Phantom *at Kennywood Amusement Park, West Mifflin, PA*

Coasters are classified as either steel or wooden thus allowing for separate records. Some records depend on whether passengers remain seated or are standing during the ride.

Starting the Investigation ....

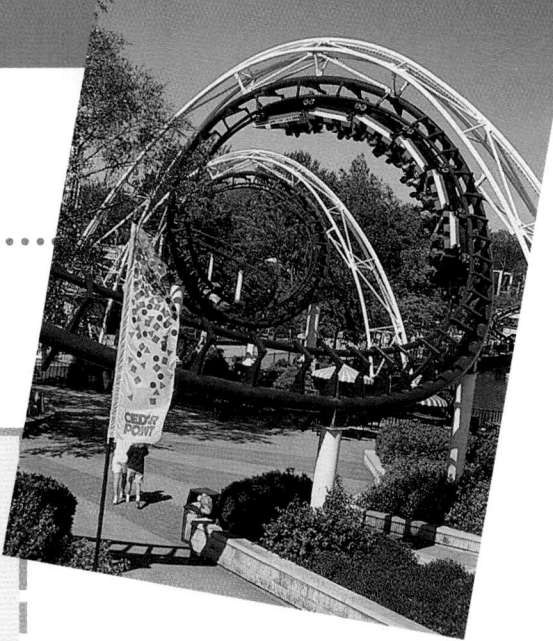

To start the design, you will want to draw a detailed model of the path your roller coaster will take. Before you begin drawing, familiarize yourself with some roller coaster terms.

A few important roller coaster definitions are:

- **Banked Turn**—*Turn in which the tracks are tilted to allow trains to turn at high speed.*
- **Car**—*The part of a coaster train that carries between two and eight passengers.*
- **Chain Lift**—*The rolling chain that carries the train to the crest of a hill.*
- **Circuit**—*A completed journey on a coaster track.*
- **First Drop**—*Usually the highest and fastest drop on a coaster. It most often follows immediately after the chain lift.*
- **Gully Coaster**—*One that uses the natural terrain to give the added feeling of speed by keeping the track close to the ground.*
- **Lift Hill**—*The hill that the train is carried up by the chain lift.*
- **Loading Platform**—*The part of the station where passengers board the coaster trains.*
- **Speed Dip**—*A small hill taken at high speeds usually lifting riders off their seats.*
- **Train**—*A series of cars hooked together to travel the circuit of the coaster track.*

inter NET
CONNECTION

Research For current information on roller coasters, visit:

www.glencoe.com/ sec/math/prealg/ mathnet

Since the design of a roller coaster can be very complex, you will need to include a scale drawing from several different angles including an overhead view. Be sure to include a sketch of the loading platform. Also include the number of cars in each train and how many passengers each car will carry.

Brainstorm with your group to decide what record you want to break with your design. You may wish to look in some other sources for other records and for guidance on how to design your coaster.

You will work on this investigation throughout Chapters 6 and 7. Be sure to keep your materials in your Investigation Folder.

Investigation
Roller Coaster Math

Working on the Investigation
Lesson 6-1, p. 279

Working on the Investigation
Lesson 6-4, p. 293

Working on the Investigation
Lesson 7-2, p. 337

Closing the Investigation
End of Chapter 7, p. 362

6 Rationals: Patterns in Multiplication and Division

What You'll Learn

In this chapter, you will:

- write fractions as decimals,

- estimate and find products and quotients of rational numbers,

- find the mean, median, and mode of a set of data,

- identify and extend geometric sequences, and

- write numbers in scientific notation.

Why It's Important

Technology A computer virus is a destructive program that is placed on a person's computer without their knowledge. Once inside the system, a virus manipulates or destroys the programs and data. In the graph, Total Number of Computer Viruses, the number of viruses for each year from 1990 through 1998 is shown. You can use estimation to find out how many times more viruses there were in 1998 than in 1991. For example, suppose there were 907 viruses in 1991. 16,000 ÷ 907 is about 16,000 ÷ 1000 = 16. There were about 16 times more viruses in 1998 than in 1991. *You will learn about estimating quotients in Lesson 6-2.*

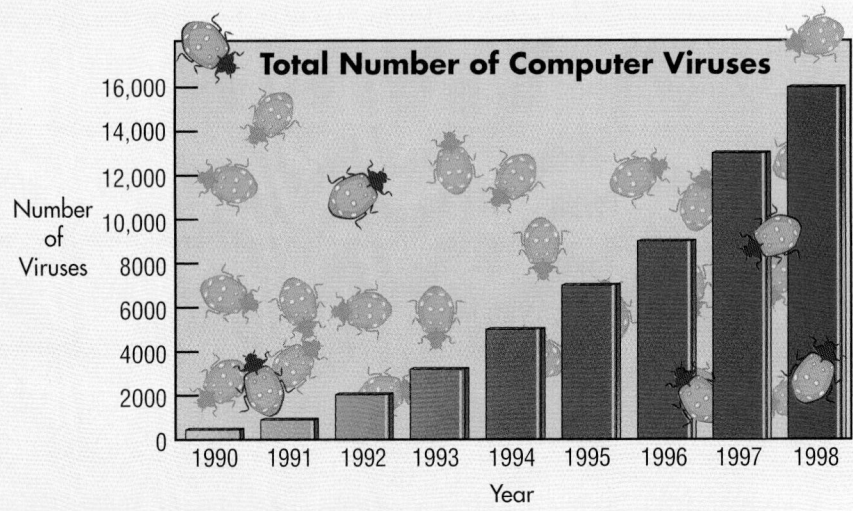

Total Number of Computer Viruses

Focus on...

PREREQUISITE SKILLS

To be successful in this chapter, you'll need to understand these concepts and be able to apply them. Refer to the lesson in parentheses if you need more review before beginning the chapter.

▶ *Rename decimals as fractions. (Lesson 5-1)*

Express each decimal as a fraction or mixed number in simplest form.

1. 0.6 **2.** $0.333...$ **3.** $5.87...$ **4.** $3.\overline{36}$

▶ *Evaluate expressions containing variables. (Lesson 4-2)*

Evaluate each expression if $x = 6$, $y = 2$, and $z = 5$.

5. $y + 3$ **6.** $4z^2 - 1$ **7.** $xz + xy$ **8.** $3x + (y^3 - z)$

▶ *Solve inequalities by using the multiplication and division properties of inequalities and graph the solution set. (Lesson 3-7)*

Solve each inequality and check your solution. Then graph the solution on a number line.

9. $-2y > 10$ **10.** $\frac{a}{6} \geq -5$ **11.** $-5x \leq -15$ **12.** $14 < \frac{r}{-1}$

▶ *Use negative exponents in expressions. (Lesson 4-9)*

Write each fraction as an expression using negative exponents.

13. $\frac{1}{2}$ **14.** $\frac{1}{10}$ **15.** $\frac{1}{3^4}$ **16.** $\frac{1}{10^5}$

Focus on...

READING SKILLS

In this chapter, you'll be learning more about **rational numbers**. You will multiply and divide fractions and decimals. In the word **decimal**, the prefix "deci" means tenth part. Decimals can be written as fractions with denominators that are powers of 10, like 10, 100, 1000, and so on. For example, 0.134 can also be written as $\frac{134}{1000}$.

6-1 Writing Fractions as Decimals

You'll Learn

■ To write fractions as terminating or repeating decimals, and

■ to compare rational numbers.

It's Important

You can calculate a batting average by knowing how to express a fraction as a decimal.

Application: Food

"You scream, I scream, we all scream for ice cream!" What is your favorite flavor? The International Ice Cream Association polled Americans about their favorite flavors. Some of the top finishers and the approximate fractions of the votes they received are shown in the graph below. If $\frac{3}{76}$ of those surveyed chose chocolate chip as their favorite, did more people choose cookies-and-cream or chocolate chip?

This problem will be solved in Example 3.

Learning the Concept

You can review the definition of rational numbers in Lesson 5-1.

In the case above, it is convenient to write rational numbers as decimals instead of fractions because decimals are easier to compare. Any fraction can be written as a decimal.

Consider the fraction $\frac{3}{5}$. The fraction $\frac{3}{5}$ indicates $3 \div 5$.

Method 1
Divide using paper and pencil.

$$\frac{3}{5} \rightarrow \begin{array}{r} 0.6 \\ 5\overline{)3.0} \\ \underline{-3\,0} \\ 0 \end{array}$$

Method 2
Divide using a calculator.

3 ÷ 5 = .6

The fraction $\frac{3}{5}$ can be written as the decimal 0.6. This means that $\frac{3}{5}$ and 0.6 are **equivalent** rational numbers. Remember that a decimal like 0.6 is called a *terminating decimal* because the division ends, or terminates, when the remainder is zero.

Example **Write $2\frac{3}{4}$ as a decimal.**

$$2\frac{3}{4} = 2 + \frac{3}{4}$$

Method 1
Use paper and pencil.

$$\frac{3}{4} \rightarrow \begin{array}{r} 0.75 \\ 4\overline{)3.00} \\ \underline{-2\,8} \\ 20 \\ \underline{-20} \\ 0 \end{array}$$ *Consider only $\frac{3}{4}$.*

$2 + 0.75 = 2.75$

The mixed number $2\frac{3}{4}$ equals the decimal 2.75.

Method 2
Use a calculator.

2 + 3 ÷ 4 = 2.75

Be sure your calculator follows the order of operations.

Not all fractions can be written as terminating decimals. Consider the fraction $\frac{2}{3}$. How can $\frac{2}{3}$ be expressed as a decimal?

Use paper and pencil.

$$\frac{2}{3} \rightarrow \begin{array}{r} 0.666 \\ 3\overline{)2.000} \\ \underline{-1\,8} \\ 20 \\ \underline{-18} \\ 20 \end{array}$$

Use a calculator.

2 ÷ 3 = .6666666

When you divide using paper and pencil, the remainder after each step is 2. If you continue dividing, the pattern repeats. The calculator display fills with 6s. Remember, a decimal like 0.6666666 . . . , or $0.\overline{6}$, is called a *repeating decimal*.

Some calculators round answers and others truncate answers. **Truncate** means to cut off at a certain place-value position, ignoring the digits that follow.

Example **Write $\frac{5}{11}$ as a decimal.**

Use paper and pencil.

$$\frac{5}{11} \rightarrow \begin{array}{r} 0.4545 \\ 11\overline{)5.0000} \\ -4\ 4 \\ \hline 60 \\ -55 \\ \hline 50 \\ -44 \\ \hline 60 \\ -55 \\ \hline \end{array}$$

$\frac{5}{11} = 0.\overline{45}$

Use a calculator.

5 ÷ 11 = .4545455
This calculator rounds.

5 ÷ 11 = .4545454
This calculator truncates.

Some calculators round and others truncate. Find the decimal value of $\frac{5}{11}$ on your calculator to determine whether it rounds or truncates.

Sometimes you need to compare numbers, as in the application at the beginning of the lesson. You can use what you have learned about writing fractions as decimals to compare fractions.

Example **Refer to the application at the beginning of the lesson. Which is more popular, chocolate chip or cookies-and-cream ice cream?**

Write the fractions as decimals and then compare the decimals.

chocolate chip

$\frac{3}{76} \rightarrow$ 0.0394736842

cookies-and-cream

$\frac{2}{55} \rightarrow$ 0.036363636

In thousandths place, 9 > 6.

Note that on a number line, 0.039 . . . is to the right of 0.036 . . . .

0.039

0.020   0.030   0.040

0.036

Since 0.039 . . . > 0.036 . . . , you know that $\frac{3}{76} > \frac{2}{55}$.

More people chose chocolate chip than chose cookies-and-cream.

Another way to compare two rational numbers is to write them as equivalent fractions with like denominators.

Example **Replace each ● with <, >, or = to make a true sentence.**

a. $\frac{7}{8}$ ● $\frac{5}{6}$

$\frac{21}{24}$ ● $\frac{20}{24}$ *The LCM is 24. $\frac{7}{8} = \frac{21}{24}$ and $\frac{5}{6} = \frac{20}{24}$*

$\frac{21}{24} > \frac{20}{24}$ *Since 21 > 20, $\frac{21}{24} > \frac{20}{24}$.*

Therefore, $\frac{7}{8} > \frac{5}{6}$.

b. $-\frac{3}{5}$ ● -0.75

-0.6 ● -0.75 *Write $-\frac{3}{5}$ as a decimal.*

$-0.6 > -0.75$ *In tenths place, $-6 > -7$.*

Therefore, $-\frac{3}{5} > -0.75$.

Verify this answer using a number line.

Checking Your Understanding

Read and study the lesson to answer these questions.

1. **Explain** why we use division to change a fraction to a decimal.

2. **Describe** the difference between terminating and repeating decimals. Give an example of each.

3. **Explain** the difference between 0.6 and $0.\overline{6}$. Which is greater?

4. **You Decide**   Juanita says that $\frac{2}{3} > \frac{3}{5}$. Drew disagrees. Who is correct and why?

 MATERIALS
 grid paper

5. The grid at the right represents the number 0.45. Represent 0.5, $\frac{1}{5}$, $\frac{1}{3}$, and 0.25 by shading 10-by-10 grids.

    a. Which number is greatest?

    b. Which number is least?

    c. Order the numbers from least to greatest.

Guided Practice

Write each fraction as a decimal. Use a bar to show a repeating decimal.

6. $\frac{5}{9}$ 7. $2\frac{2}{3}$ 8. $-3\frac{1}{8}$ 9. $\frac{1}{8}$

State the greater number for each pair.

10. $\frac{1}{2}, \frac{1}{3}$ 11. $-\frac{3}{4}, -\frac{7}{8}$ 12. $\frac{3}{8}, \frac{10}{33}$ 13. $-\frac{3}{4}, -0.\overline{75}$

14. **Business**   On Monday, June 12, 1995, the price of a hamburger chain's stock closed at $16\frac{5}{8}$. The previous Friday, the stock had closed at $16\frac{3}{4}$. On which day did the stock close at a higher price?

Exercises: Practicing and Applying the Concept

Independent Practice

Write each fraction as a decimal. Use a bar to show a repeating decimal.

15. $\frac{7}{10}$ 16. $-\frac{1}{3}$ 17. $\frac{1}{2}$ 18. $-1\frac{1}{4}$

19. $\frac{14}{20}$ 20. $-\frac{2}{9}$ 21. $\frac{5}{16}$ 22. $\frac{21}{25}$

23. $-3\frac{4}{11}$ 24. $\frac{23}{45}$ 25. $-4\frac{5}{16}$ 26. $\frac{31}{40}$

Replace each ● with >, <, or = to make a true sentence.

27. $5\frac{1}{5}$ ● 5.18

28. 7.56 ● $7\frac{12}{25}$

29. −3.45 ● $3\frac{2}{3}$

30. $-1\frac{1}{20}$ ● 1.01

31. $\frac{7}{8}$ ● $\frac{8}{9}$

32. $\frac{2}{5}$ ● $\frac{1}{3}$

33. $3\frac{4}{7}$ ● $3\frac{5}{8}$

34. $2\frac{5}{8}$ ● $2\frac{2}{3}$

35. $-\frac{11}{14}$ ● $-\frac{13}{16}$

36. $\frac{9}{11}$ ● $\frac{19}{23}$

37. $1\frac{4}{5}$ ● 1.857

38. $-5\frac{1}{12}$ ● −5.09

Calculators

Several different calculators' displays for 2 ÷ 3 are shown below. What would the display for 7 ÷ 9 be for each calculator?

39. .6666666

40. 0.6666666

41. 0.6666667

42. .67

Critical Thinking

43. Find the decimal equivalents for $\frac{1}{3}, \frac{1}{4}, \frac{7}{10}, \frac{4}{5}, \frac{5}{12}, \frac{4}{7},$ and $\frac{3}{25}$.

    a. Which are terminating decimals? Which are repeating decimals?

    b. Find the prime factorization of each denominator.

    c. What prime factors are present in the denominator of a fraction whose equivalent decimal is terminating? Write about what you observe.

44. Find a terminating and a repeating decimal between $\frac{1}{6}$ and $\frac{8}{9}$. Explain how you found them.

45. Zoology Monarch butterflies migrate up to 2000 miles from Canada and the northern United States to Florida, California, and Mexico. The fastest monarch butterfly can fly $\frac{1}{3}$ mile in one minute. Write $\frac{1}{3}$ as a decimal rounded to the nearest hundredth.

Applications and Problem Solving

46. Physics For best performance, 0.91 to 0.93 of a rocket's total mass should be propellant. Aerospace engineers call the following ratio the *mass fraction (MF)* and usually express it as a decimal.

$$MF = \frac{\text{mass of propellant}}{\text{total mass}}$$

If the total mass of a rocket is 6500 pounds and the mass of the propellant is 6000 pounds, would this ratio fall into the range for best performance? Explain.

47. Business The cost in dollars for one share of stock is represented by a mixed number. Find the cost of one share of the following stock for the stock market values shown below. (Round to the nearest cent.)

```
Kine 82 3/8     Sipse 46 7/8     Grap 36 3/4
```

48. Demographics Do you believe in love at first sight? In a survey, $\frac{2}{5}$ of those asked said "It happens," $\frac{7}{25}$ said "It could happen," and $\frac{8}{25}$ said "No way." What did most people respond? **Source:** Roper Starch Worldwide

Mixed Review

49. Write the next three terms of the sequence 88, 82, 76, 70, . . . (Lesson 5-9)

50. Solve $t - 3\frac{1}{4} = \frac{7}{2}$. (Lesson 5-6)

51. Find the product of $b^5 \cdot b^6$. (Lesson 4-8)

52. Find a three-digit number that is divisible by 2, 3 and 6. (Lesson 4-1)

53. Standardized Test Practice The Home Economics Class made a quilt measuring 4 feet by 6 feet to donate for a raffle. How much binding should they buy to trim the outer edges? (Lesson 3-5)

   A. 2 feet   **B.** 10 feet   **C.** 20 feet   **D.** 24 feet

54. Order the integers {0, −7, 15, −6, 20} from least to greatest. (Lesson 2-3)

55. Sports The Cleveland Indians beat the California Angels by a score of 10 to 3. Write an inequality that would tell you how many more runs the Angels would have needed to win the game. (Lesson 1-9)

For **Extra Practice**, see page 753.

56. Solve $13 + y = 25$ by using the inverse operation. (Lesson 1-8)

57. Solve $3t = 81$ mentally. (Lesson 1-6)

WORKING ON THE
Investigation

Refer to the Investigation on pages 270–271.

Begin working on the investigation by doing a group activity to determine the steepness of your lift hill.

• Build a scale model of your lift hill using a yardstick to represent the length of the track on the downhill side.

• While one member of your group holds the yardstick at the angle matching your design, another member measures how high the end of the yardstick is from the floor. Record the ratio of the height to the length of the yardstick. For example, if the height is 30 inches, you would write $\frac{30}{36}$ or $\frac{5}{6}$. Write the ratio as a decimal.

• Next, use the same method to find the ratio of an angle that is not as steep as yours and another angle that is steeper than yours.

• Write a short summary explaining the relationship between steepness and the decimal values of the ratios.

Add the results of your work to your Investigation Folder.

Estimating Products and Quotients

Technology Activity

In this activity, you and a partner will try to score as many points as possible by estimating decimal products. Products receive points as shown in the table below.

Your Turn   Pick two numbers from the chart below and use your calculator to find the product. You may use each number only once.

Numbers:	4.6	0.02	12.1	30.5	11.5	0.6	31.8	0.05
	19.8	0.115	10.6	0.43	0.08	2.9	0.96	18.8
	24.2	9.7	1.7	5.64				

Add the points for your products to find your score. You can do this activity more than once to see if you can increase your team score.

 a. What strategy did you use to choose the numbers for each product?
b. How could you increase your score?

Product is between:	Points Scored
0 and 0.1	1
0.1 and 1	2
1 and 10	3
10 and 100	2
100 and 1000	1

Learning the Concept

In the activity above, it is useful to know how to estimate the product of the decimals before actually multiplying them. It is also important to estimate products and quotients when using a calculator. We can make mistakes when entering numbers, and estimating will tell us if our result is reasonable.

Using rounding and **compatible numbers**, it is possible to estimate the products and quotients of rational numbers. Compatible numbers are rounded so it is easy to compute with them mentally.

$47.5 \div 5.23 \rightarrow 48 \div 6$ or 8   *Even though 5.23 rounds to 5, 6 is a compatible number because 48 is divisible by 6.*

$8 \times 98.24 \rightarrow 8 \times 100$ or 800.

Example **Use rounding or compatible numbers to estimate 67.45 ÷ 7.6.**

$67.45 \div 7.6 \rightarrow 70 \div 7$ *Round 67.45 to 70 and round 7.6 to 7 since 70 and 7 are compatible numbers.*

Think: 70 ÷ 7 = 10

67.45 ÷ 7.6 is about 10.

The strategy of using compatible numbers works for fractions as well as decimals.

Example **Use rounding or compatible numbers to estimate $\frac{5}{14} \times 30$.**

$\frac{5}{14} \times 30 \rightarrow \frac{1}{3} \times 30$ *Replace $\frac{5}{14}$ with $\frac{1}{3}$ because $\frac{5}{14}$ is close to $\frac{5}{15}$ and $\frac{5}{15} = \frac{1}{3}$.*

Think: $\frac{1}{3} \times 30 = 10$

$\frac{5}{14} \times 30$ is about 10.

Being able to estimate products and quotients of rational numbers is very useful in everyday life.

Example

Employment

Kelly found a part-time job working at the Discus Record Store for $5.85 an hour. The manager told Kelly that she will probably work about $7\frac{1}{2}$ hours each week. She will be paid every two weeks. Estimate Kelly's pay before taxes are deducted.

hourly rate ×	*number of hours per week* ×	*number of weeks per pay period*
$5.85 ×	$7\frac{1}{2}$ ×	2

Think: $5.85 rounds to $6 and $7\frac{1}{2}$ hours rounds to 8 hours.

$\$5.85 \times 7\frac{1}{2} \rightarrow \$6 \times 8$ or $48 a week

For two weeks, multiply by 2.
$48 × 2 → $50 × 2 = $100

Kelly will earn about $100 every two weeks.

THINK ABOUT IT

Will Kelly's actual pay be more or less than the estimate?

Communicating Mathematics

Read and study the lesson to answer these questions.

1. **Explain** why estimation is more important in mathematics since calculator use has become more widespread.

2. **Give an example** of two compatible numbers. Explain why they are compatible.

3. Choose the correct phrase: The product of a whole number and a fraction less than one is always (less than, greater than, equal to) the whole number.

 MATH JOURNAL

4. **Assess Yourself** Describe a time in your daily life when estimation is useful.

Guided Practice

Choose the best estimate for each product or quotient.

5. $\left(1\frac{1}{8}\right)\left(3\frac{1}{4}\right)$
   a. almost 3
   b. less than 1
   c. more than 3
   d. none of these

6. $49 \times \frac{2}{5}$
   a. about 20
   b. more than 25
   c. less than 10
   d. none of these

7. $\$24.99 \div 4$
   a. more than $8
   b. almost $6
   c. about $5
   d. none of these

Rewrite the problem using rounding and compatible numbers. Then estimate each product or quotient.

8. $40 \times \frac{2}{9}$

9. $17.8 \div 3.2$

10. $\frac{5}{12} \times 44$

11. $8.4 \div 0.95$

12. $75.4 \div 9.8$

13. $20\frac{1}{9} \times \frac{1}{4}$

14. **Geometry** The formula for the perimeter of a rectangle is $P = 2(\ell + w)$, where ℓ is the length and w is the width. Estimate the perimeter of the rectangle at the right. Is it greater than or less than 20 meters?

3.1 m

5.8 m

Independent Practice

Estimate each product or quotient.

15. 15.93×9.8

16. $2.94 \cdot 1.8$

17. $8.1 \div 2.2$

18. $47.6 \div 7.8$

19. $15.2 \div 2.7$

20. $(84.2)(3.9)$

21. $\frac{1}{4} \cdot 9$

22. $\left(\frac{1}{3}\right)(14)$

23. $\frac{3}{8} \times 17$

24. $\frac{9}{19} \times 120$

25. $\frac{11}{20} \times 41$

26. $75 \div 6\frac{7}{8}$

27. $\frac{31}{40} \times 200$

28. $146 \div 13\frac{1}{15}$

29. $43.8 \div 9.2$

30. Estimate $\frac{5}{6}$ times 10.

31. Estimate the quotient of 27.26 and 2.6.

Calculators

32. Use a calculator to find the following quotients.

    a. $80 \div 0.5$ **b.** $10 \div 0.1$ **c.** $4 \div 0.25$

        $6 \div 0.5$ $2 \div 0.1$ $20 \div 0.25$

        $1.5 \div 0.5$ $6.5 \div 0.1$ $1.1 \div 0.25$

    d. What patterns do you observe?

    Use the patterns to estimate each quotient.

    e. $14 \div 0.48$ **f.** $5 \div 0.12$ **g.** $12 \div 0.23$

    h. $2\frac{1}{5} \div 0.25$ **i.** $21.2 \div 0.085$ **j.** $3\frac{1}{4} \div 0.51$

Critical Thinking

33. Explain how you would know if an estimated answer is greater than or less than the actual answer.

Applications and Problem Solving

34. Decorating Jennifer Green ran out of wallpaper before she finished wallpapering her room. The remaining area is 12 feet by $10\frac{1}{2}$ feet and has one window, which is $3\frac{1}{2}$ feet by $4\frac{3}{4}$ feet.

    a. Estimate the amount (square feet) of wall paper Jennifer needs to cover the wall.

    b. If the wallpaper comes on rolls that cover 51 square feet, how many rolls should she purchase?

35. Population In 1994, Heather Whitestone became the first hearing-impaired Miss America. In the United States, $\frac{3}{50}$ of the people are hearing impaired or deaf. If 1990 census showed that the population of the United States was 248,709,873 at that time, estimate how many people were hearing impaired.

36. Stock Market Stock prices are stated as mixed numbers. Suppose a share of stock is listed as $107\frac{1}{4}$. The cost of one share of stock is $107.25. (The fraction indicates a part of a dollar.)

    a. Estimate the cost of 37 shares of stock listed at $107\frac{1}{4}$ a share.

    b. Suppose you had $1000 to invest in the stock. About how many shares could you buy?

37. Biology A dolphin can swim at a speed of 37 miles per hour. A human can swim about one-eighth that speed. About how fast can a human swim?

Heather Whitestone

Mixed Review

38. Write $\frac{3}{8}$ as a decimal. (Lesson 6-1)

39. Solve $\frac{6}{7} + \frac{5}{14} = s$. (Lesson 5-5)

40. Food A foot-long submarine sandwich was cut into 6 equal pieces. There are 2 pieces left. Write the ratio of pieces eaten to total pieces as a fraction in simplest form. (Lesson 4-6)

For **Extra Practice**, see page 753.

41. Solve $-14 + c = 13$. (Lesson 3-2)

42. Standardized Test Practice Find $-13 + 6$. (Lesson 2-4)

    A. -19 **B.** -7 **C.** 7   **D.** 19

43. Simplify $|-6| - |-4|$. (Lesson 2-1)

44. Solve $\frac{r}{9} = 15$ by using the inverse operation. (Lesson 1-8)

6-3 Multiplying Fractions

You'll Learn

■ To multiply fractions.

It's Important

Knowing how to multiply fractions can help you make half of a recipe.

MATERIALS: grid paper colored pencils
Hands-On Activity

You can use grid paper to make an **area model** for multiplication of fractions. The diagram at the right models the product $\frac{2}{3} \cdot \frac{5}{6}$. Notice that the rectangle has three rows to represent thirds. Two rows are shaded yellow to represent $\frac{2}{3}$. The rectangle has six columns to represent sixths. Five columns are shaded blue to represent $\frac{5}{6}$. Notice that the overlapping area, $\frac{10}{18}$ of the rectangle, is shaded green. The product of $\frac{2}{3}$ and $\frac{5}{6}$ is $\frac{10}{18}$, or $\frac{5}{9}$.

Your Turn Use grid paper to make an area model of $\frac{3}{5} \cdot \frac{1}{4}$.

a. How did you represent $\frac{3}{5}$ in your model?

b. How did you represent $\frac{1}{4}$?

c. What is the value of $\frac{3}{5} \cdot \frac{1}{4}$ that you found using the grid paper?

d. Write a description of how an area model shows multiplication of fractions.

Learning the Concept

An area model shows what happens when you multiply fractions. You will get the same results if you multiply fractions as shown below.

$$\frac{2}{3} \cdot \frac{5}{6} = \frac{2 \cdot 5}{3 \cdot 6} \qquad \textit{Multiply the numerators and multiply the denominators.}$$

$$= \frac{10}{18} \text{ or } \frac{5}{9} \qquad \textit{Simplify by dividing numerator and denominator by 2.}$$

Multiplying Fractions	**In words:**	To multiply fractions, multiply the numerators and multiply the denominators.
	In symbols:	For fractions $\frac{a}{b}$ and $\frac{c}{d}$, where $b \neq 0$ and $d \neq 0$, $\frac{a}{b} \cdot \frac{c}{d} = \frac{ac}{bd}$.

If the fractions have common factors in the numerators and denominators, you can simplify before you multiply. Compare the two methods used in Example 1.

Example **Solve** $x = \frac{5}{12} \cdot \frac{2}{9}$.

Method 1: Use the rule.

$x = \frac{5}{12} \cdot \frac{2}{9}$

$= \frac{5 \cdot 2}{12 \cdot 9}$ *Multiply.*

$= \frac{10}{108}$ or $\frac{5}{54}$

Method 2: Use the GCF.

$x = \frac{5}{12} \cdot \frac{2}{9}$

$= \frac{5 \cdot \overset{1}{\cancel{2}}}{\underset{6}{\cancel{12}} \cdot 9}$ *The GCF of 2 and 12 is 2. Divide 2 and 12 by 2.*

$= \frac{5 \cdot 1}{6 \cdot 9}$ or $\frac{5}{54}$

You can use the multiplication skills that you developed with integers and fractions to multiply negative fractions.

Example **Solve each equation.**

a. $3\frac{3}{8}\left(-5\frac{1}{3}\right) = x$ *Estimate: 3(−5) = −15.*

$\frac{27}{8}\left(-\frac{16}{3}\right) = x$ *Rename $3\frac{3}{8}$ as $\frac{27}{8}$ and $-5\frac{1}{3}$ as $-\frac{16}{3}$.*

$\frac{\overset{9}{\cancel{27}}}{\underset{1}{\cancel{8}}}\left(\frac{-\overset{-2}{\cancel{16}}}{\underset{1}{\cancel{3}}}\right) = x$ *Simplify.*

$\frac{-18}{1} = x$

$-18 = x$ *Compare with the estimate.*

b. $a = \left(-1\frac{4}{9}\right)\left(-2\frac{3}{5}\right)$ *Estimate: (−1)(−3) = 3.*

$a = \left(\frac{-13}{9}\right)\left(\frac{-13}{5}\right)$ *Rename $-1\frac{4}{9}$ as $-\frac{13}{9}$ and $-2\frac{3}{5}$ as $-\frac{13}{5}$.*

$a = \frac{169}{45}$ or $3\frac{34}{45}$ *The product of two rational numbers with the same sign is positive.*

Example

Cooking

Manuel Mendes is the chief cook at the U.S. Military Academy at West Point, New York. He and his 96 staff members work to feed the 4500 cadets three times each day. Their sloppy joe recipe calls for $7\frac{1}{2}$ pounds of ketchup and serves 100 people. How many pounds of ketchup are needed for one meal of sloppy joes?

The recipe serves 100 people, so the staff will need to purchase enough ingredients for 4500 ÷ 100 or 45 batches of sloppy joes.

$45 \cdot 7\frac{1}{2} = \frac{45}{1} \cdot \frac{15}{2}$ *Rename 45 as $\frac{45}{1}$ and $7\frac{1}{2}$ as $\frac{15}{2}$.*

$= \frac{45 \cdot 15}{1 \cdot 2}$

$= \frac{675}{2}$ or $337\frac{1}{2}$ *Write as a mixed number.*

$337\frac{1}{2}$ pounds of ketchup are needed.

Long a staple in Latin countries, salsa has now passed ketchup in the race for most popular condiment in America.

You can evaluate algebraic expressions involving multiplication of fractions in the same way that you evaluated expressions using integers.

Example ④ Evaluate the expression ab^2 if $a = 1\frac{1}{2}$ and $b = \frac{2}{3}$.

$ab^2 = 1\frac{1}{2} \cdot \left(\frac{2}{3}\right)^2$ *Substitute $1\frac{1}{2}$ for a and $\frac{2}{3}$ for b.*

$= 1\frac{1}{2} \cdot \frac{2}{3} \cdot \frac{2}{3}$ *Rewrite the problem without exponents.*

$= \frac{3}{2} \cdot \frac{4}{9}$ *Write $1\frac{1}{2}$ as $\frac{3}{2}$.*

$= \frac{12}{18}$ or $\frac{2}{3}$ *Multiply and write in simplest form.*

Checking Your Understanding

Communicating Mathematics

Read and study the lesson to answer these questions.

1. **Explain** how the model at the right shows the product of $\frac{3}{4}$ and $\frac{1}{2}$. What is the product?

2. **You Decide**   Jamal and Penny multiplied the fractions $\frac{18}{21}$ and $\frac{3}{14}$ in the following ways. Who is correct and why?

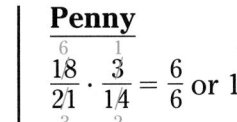

Jamal	Penny
$\overset{9}{\underset{7}{18}} \cdot \overset{1}{\underset{7}{3}} = \frac{9}{49}$ i.e. $\frac{\overset{9}{18}}{\underset{7}{21}} \cdot \frac{\overset{1}{3}}{\underset{7}{14}} = \frac{9}{49}$	$\frac{\overset{6}{18}}{\underset{3}{21}} \cdot \frac{\overset{1}{3}}{\underset{2}{14}} = \frac{6}{6}$ or 1

3. **Identify** which point shown on the number line below could be the product of the numbers graphed at A and B. Explain your reasoning.

MATERIALS

🖊 ruler
📇 construction paper

4. Fold a piece of $8\frac{1}{2}$" by 11" construction paper in half and then in half again as shown. Cut out a square from the open end of the folded paper. Unfold the paper. Then fold up the edges to form a box with no lid. Use your ruler to measure the length, width, and height of your box to the nearest sixteenth of an inch.

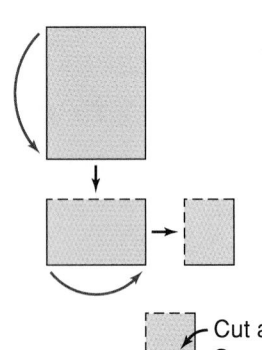

Cut a Square.

    a. Find the total area of the bottom and four sides of your box.

    b. How does the area of your box compare with your neighbor's? Why do you think they are the same or different?

Guided Practice

Choose the correct product.

5. $\frac{1}{6} \cdot \left(-\frac{3}{5}\right)$ **a.** $\frac{2}{5}$ **b.** $-\frac{2}{5}$ **c.** $\frac{1}{10}$ **d.** $-\frac{1}{10}$

6. $\left(\frac{1}{2}\right)^2$ **a.** $\frac{1}{2}$ **b.** $\frac{1}{4}$ **c.** 1   **d.** 2

Solve each equation. Write each solution in simplest form.

7. $\frac{3}{4} \cdot \frac{2}{3} = a$

8. $b = \frac{1}{2}\left(-\frac{5}{6}\right)$

9. $c = \frac{8}{12} \cdot \frac{4}{6}$

10. $x = \left(3\frac{1}{2}\right)4$

11. $m = \frac{8}{15}(-45)$

12. $\left(-\frac{4}{5}\right)^2 = p$

Evaluate each expression if $a = \frac{2}{3}$, $x = \frac{3}{5}$, and $y = \frac{5}{8}$.

13. xy

14. $2a$

15. x^2

16. ay

17. **Economics** The U.S. Congress is debating whether to replace paper one-dollar bills with coins in order to save money.

   a. It costs 8¢ to produce a one-dollar coin. The cost of producing a paper dollar is $\frac{19}{40}$ of that. How much does it cost to produce a paper dollar?

   b. An average paper dollar lasts $1\frac{1}{4}$ years. A coin lasts about $22\frac{1}{2}$ times longer. How long does a coin last?

   c. Do you think it would save money to use coins instead of paper dollars? Explain.

Exercises: Practicing and Applying the Concept

Independent Practice

Solve each equation. Write each solution in simplest form.

18. $\frac{3}{4} \cdot \left(-\frac{1}{3}\right) = b$

19. $\frac{1}{2} \cdot \frac{2}{7} = t$

20. $k = -\frac{5}{6}\left(-\frac{2}{5}\right)$

21. $d = -4\left(\frac{3}{8}\right)$

22. $(-7)\left(-2\frac{1}{3}\right) = h$

23. $c = 1\frac{4}{5} \cdot \left(-2\frac{1}{2}\right)$

24. $v = 2\frac{5}{6} \cdot 3\frac{1}{3}$

25. $s = \left(2\frac{1}{4}\right)\left(-\frac{4}{3}\right)$

26. $\left(-9\frac{3}{5}\right)\left(\frac{5}{12}\right) = y$

27. $\left(-3\frac{1}{5}\right)\left(7\frac{1}{2}\right) = w$

28. $m = (9)\left(-1\frac{5}{6}\right)$

29. $p = \left(-\frac{3}{5}\right)^2$

30. $r = \left(\frac{7}{11}\right)^2$

31. $\left(-\frac{9}{10}\right)^2 \cdot 8 = f$

32. $\left(\frac{5}{7}\right)^2 = n$

33. What is the product of $\frac{4}{9}$ and $\frac{3}{5}$?

34. What is $\frac{5}{8}$ of 36?

Evaluate each expression if $a = -\frac{1}{3}$, $b = \frac{3}{4}$, $x = 1\frac{2}{5}$, and $y = -3\frac{1}{6}$.

35. ay

36. $2a^2$

37. $b(x + a)$

38. $3a - 5x$

39. $a^2(b + 2)$

40. $-a(a - b)$

41. Use a fraction calculator and the numbers 2, 3, 4, 5, 6, or 7 to make two fractions with the greatest product. Use each digit only once. Explain your answer.

Critical Thinking

42. **Number Theory** Every composite number can be written as a product of primes.

   a. Can you write $\frac{42}{45}$ as the product of fractions with prime numerators and denominators? You may use as many fractions as you need.

   b. Is there more than one way to write the product? Explain.

43. Health If you usually burn after $\frac{1}{2}$ hour exposure to sun, you can lengthen the time you are in the sun to three hours by applying SPF number 6. How long could a person who burns after $\frac{1}{4}$ hour stay in the sun using SPF 15 lotion?

44. Cooking The West Point sloppy joe recipe mentioned in Example 3 uses $3\frac{1}{4}$ ounces of cider vinegar and $1\frac{2}{3}$ ounces of garlic powder. How much of these ingredients would be needed to make sloppy joes for 4500 people?

45. Measurement People who design printed products use a system of measurement that is different from the customary and metric systems. The measurements used in this system are the *point* and the *pica*. One point equals $\frac{1}{72}$ inch, and 12 points make one pica.
 a. How many picas are in an inch?
 b. The margin of a book is $\frac{3}{4}$ inches. How many picas is this?
 c. If a particular kind of type has an average measure of $2\frac{3}{4}$ characters per pica, about how many characters would fit across a $5\frac{1}{2}$-inch line?

46. Statistics At Westside High School, $\frac{3}{5}$ of the students are female. The dance preferences of those students are shown in the graph at the right. What part of the students are female and interested in each kind of dance?

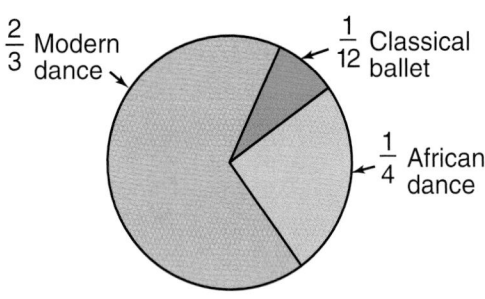

47. Estimate $\frac{13}{28} \times 30$. (Lesson 6-2)

48. Consumerism At the movies, it costs $6.50 for a ticket, $1.25 for a drink, and $2.50 for popcorn. Will $10 be enough to purchase one of each? Explain. (Lesson 5-2)

49. Find the prime factorization of 72. (Lesson 4-4)

50. Solve $f + (-3) > 8$. (Lesson 3-6)

51. Find the product $-4(-x)(-5y)$. (Lesson 2-7)

52. Standardized Test Practice Supplementary angles are angles whose measures have a sum of 180°. One angle measures 57°. Solve the equation $y + 57 = 180$ to find the measure of a supplementary angle. (Lesson 1-8)
 A. 3.15°   **B.** 123°   **C.** 130°   **D.** 237°

53. Find the value of $18 \div 3 + 18 \div 2$. (Lesson 1-2)

For **Extra Practice,** see page 753.

54. When Kim divided 1033.5 by 26 the calculator showed 397.5. Is this a reasonable answer? Explain. (Lesson 1-1)

6-4 Dizviding Fractions

What You'll Learn

- To divide fractions using multiplicative inverses.

Why It's Important

Knowing how to divide fractions will help you know how much food to order for a group.

MATERIALS: colored pencils

Hands-On Activity

One way to think of $6 \div 2$ is to say "How many 2s are there in 6?" This way of thinking and an area model can help you divide fractions.

The drawing at the right shows a model for $\frac{3}{4} \div \frac{1}{2}$. To find the quotient, first draw a rectangle to represent one unit. Divide it into four equal parts and shade three to represent $\frac{3}{4}$. Then determine how many halves are contained in the shaded area. There is one half, and half of a second. So, $\frac{3}{4} \div \frac{1}{2} = 1\frac{1}{2}$.

Your Turn

▶ Model $2\frac{2}{3} \div \frac{4}{9}$. Draw three rectangles and divide into thirds. Then shade $2\frac{2}{3}$.

▶ Add additional marks to show ninths in each rectangle.

▶ Mark as many groups of $\frac{4}{9}$ as you can.

TALK ABOUT IT

a. How many groups of $\frac{4}{9}$ did you find?

b. What is the quotient of $2\frac{2}{3} \div \frac{4}{9}$?

Learning the Concept

THINK ABOUT IT

Are $\frac{3}{4}$ and $-\frac{4}{3}$ multiplicative inverses?

You can use the same skills to divide fractions that you used to divide integers. For example, dividing 40 by 2 is the same as multiplying 40 by $\frac{1}{2}$. In each case, the solution is 20. In other words, dividing by 2 is the same as multiplying by $\frac{1}{2}$.

Notice that $2 \times \frac{1}{2} = 1$. Two numbers whose product is 1 are **multiplicative inverses**, or **reciprocals**, of each other. So, 2 and $\frac{1}{2}$ are multiplicative inverses. In the same way, $-\frac{3}{2}$ and $-\frac{2}{3}$ are multiplicative inverses because $-\frac{3}{2} \cdot \left(-\frac{2}{3}\right) = 1$.

Inverse Property of Multiplication	**In words:**	The product of a number and its multiplicative inverse is 1.
	In symbols:	For every nonzero number $\frac{a}{b}$, where $a, b \neq 0$, there is exactly one number $\frac{b}{a}$ such that $\frac{a}{b} \cdot \frac{b}{a} = 1$.

Remember that dividing by 2 was the same as multiplying by $\frac{1}{2}$, its multiplicative inverse. This is true for any rational number.

Division with Fractions	**In words:** To divide by a fraction, multiply by its multiplicative inverse.
	In symbols: For fractions $\frac{a}{b}$ and $\frac{c}{d}$ where $b, c, d \neq 0$, $\frac{a}{b} \div \frac{c}{d} = \frac{a}{b} \cdot \frac{d}{c}.$

Example **1**

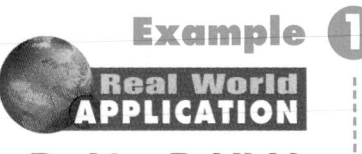

Desktop Publishing

Robert G. Fernandez, a vice-president at a financial corporation in Austin, Texas, wrote an article on money management in a magazine. The magazine is printed on $8\frac{1}{2}$-by-11 inch pages, and the articles are printed in three columns. If the margins are $\frac{3}{8}$ inch on the left and right side of the page and there is $\frac{1}{4}$ inch of space between the columns, how wide should the columns on his computer be set?

Explore You know the total width of the page, the width of the margins, and the amount of space between each column. There are three equal columns. You need to know how wide to make each column.

Plan Make a diagram and mark the widths you know. Subtract the known widths from the total width. Then divide the remaining width by 3.

Solve $8\frac{1}{2} - \frac{3}{8} - \frac{3}{8} - \frac{1}{4} - \frac{1}{4} = 7\frac{1}{4}$

$7\frac{1}{4} \div 3 = \frac{29}{4} \div \frac{3}{1}$ *Rename $7\frac{1}{4}$ as $\frac{29}{4}$ and 3 as $\frac{3}{1}$.*

$= \frac{29}{4} \times \frac{1}{3}$ *Multiply by the multiplicative inverse*

$= \frac{29}{12}$ or $2\frac{5}{12}$ *of 3, $\frac{1}{3}$.*

Each column should be $2\frac{5}{12}$ inches wide.

Examine Add the widths of all the margins and the three columns. The sum should be $8\frac{1}{2}$.

$\frac{3}{8} + 2\frac{5}{12} + \frac{1}{4} + 2\frac{5}{12} + \frac{1}{4} + 2\frac{5}{12} + \frac{3}{8} \stackrel{?}{=} 8\frac{1}{2}$

$8\frac{1}{2} = 8\frac{1}{2}$ ✔   *It checks.*

You can use the division skills that you developed with integers and fractions to divide negative fractions.

Example 2 Solve $p = -3\frac{3}{5} \div \frac{6}{7}$.

$p = -3\frac{3}{5} \div \frac{6}{7}$ *Estimate: $-4 \div 1 = -4$*

$= -\frac{18}{5} \div \frac{6}{7}$ *Rename $-3\frac{3}{5}$ as $-\frac{18}{5}$.*

$= -\frac{18}{5} \times \frac{7}{6}$ *Multiply by the multiplicative inverse of $\frac{6}{7}$, $\frac{7}{6}$.*

$= -\frac{21}{5}$ or $-4\frac{1}{5}$ *Rename as a mixed number in simplest form. Compare to the estimate.*

Connection to ⟩ Algebra

Evaluating algebraic expressions involving division of fractions uses the same procedures as evaluating algebraic expressions with integers.

Example 3 Evaluate $a^2 \div x$ if $a = \frac{1}{3}$, and $x = 1\frac{4}{5}$.

$a^2 \div x = \left(\frac{1}{3}\right)^2 \div 1\frac{4}{5}$ *Replace a with $\frac{1}{3}$ and x with $1\frac{4}{5}$.*

$= \left(\frac{1}{3}\right)\left(\frac{1}{3}\right) \div 1\frac{4}{5}$ *Evaluate powers first.*

$= \frac{1}{9} \div 1\frac{4}{5}$

$= \frac{1}{9} \div \frac{9}{5}$ *Rename $1\frac{4}{5}$ as $\frac{9}{5}$.*

$= \frac{1}{9} \times \frac{5}{9}$ *Multiply by the multiplicative inverse of $\frac{9}{5}$, $\frac{5}{9}$.*

$= \frac{5}{81}$

Checking Your Understanding

Communicating Mathematics

Read and study the lesson to answer these questions.

1. In your own words, what are reciprocals? Give an example.

2. How and when do you use a reciprocal in division of fractions?

3. **Write** an equation that you could use the multiplicative inverse of $\frac{3}{8}$ to solve.

4. **Give a counter example** to the statement that every rational number has a reciprocal.

5. A number is both multiplied and divided by the same rational number n, where $0 < n < 1$. Which is greater, the product or the quotient? Explain your reasoning.

Guided Practice

State whether each pair of numbers are multiplicative inverses. Write *yes* or *no*.

6. $7, \frac{1}{7}$ 7. $-\frac{2}{5}, -1\frac{1}{2}$ 8. $5\frac{2}{3}, \frac{17}{3}$ 9. $1.4, \frac{5}{7}$

Write each division expression as a multiplication expression. Then find its value.

10. $\frac{1}{2} \div \frac{6}{7}$

11. $-\frac{3}{4} \div \frac{3}{4}$

12. $\frac{7}{9} \div \frac{2}{3}$

13. $5 \div \left(-1\frac{1}{3}\right)$

14. $2\frac{3}{5} \div 3\frac{6}{7}$

15. $-8 \div \left(-22\frac{4}{5}\right)$

Evaluate each expression.

16. $x \div y$, if $x = 1\frac{1}{2}$ and $y = \frac{1}{2}$

17. $c \div d + e$, if $c = \frac{2}{3}$, $d = 1\frac{1}{2}$, and $e = \frac{1}{6}$

18. Carpentry How many boards, each 2 feet 8 inches long, can be cut from a board 16 feet long?

Exercises: Practicing and Applying the Concept

Independent Practice

Name the multiplicative inverse for each rational number.

19. $-\frac{7}{3}$

20. 8

21. $1\frac{3}{5}$

22. $-3\frac{2}{7}$

23. 1.5

24. -0.4

25. $\frac{x}{y}$

26. $\frac{m}{n}$

Estimate the solution to each equation. Then solve. Write the solution in simplest form.

27. $r = -\frac{3}{5} \div \frac{5}{9}$

28. $u = -1\frac{1}{9} \div \frac{2}{3}$

29. $-8 \div \frac{4}{5} = t$

30. $6 \div \frac{1}{3} = m$

31. $2\frac{1}{4} \div \left(-1\frac{1}{2}\right) = h$

32. $-2\frac{3}{7} \div \left(-4\frac{4}{7}\right) = d$

33. $12 \div \frac{4}{9} = c$

34. $-10 \div \frac{3}{8} = f$

35. $z = 24 \div \frac{7}{10}$

36. $q = -2 \div \left(-\frac{1}{3}\right)$

37. $s = -3\frac{1}{4} \div 2\frac{1}{6}$

38. $7\frac{1}{2} \div 1\frac{1}{5} = n$

39. $m = -\frac{16}{7} \div \left(-\frac{12}{35}\right)$

40. $a = \frac{21}{30} \div \frac{7}{15}$

41. $12\frac{1}{4} \div \left(-\frac{14}{3}\right) = j$

Evaluate each expression.

42. $a \div b$, if $a = \frac{2}{3}$ and $b = 1\frac{1}{3}$

43. $r \div s$, if $r = -\frac{8}{9}$ and $s = \frac{7}{18}$

44. $a^2 \div b^2$, if $a = -\frac{3}{4}$ and $b = 1\frac{1}{3}$

45. $m + n \div p$, if $m = \frac{2}{3}$, $n = 1\frac{1}{3}$, and $p = \frac{1}{9}$

Critical Thinking

46. a. Divide the fraction $\frac{3}{4}$ by $\frac{1}{2}, \frac{1}{4}, \frac{1}{8}$, and $\frac{1}{12}$.

b. What happens to the quotient as the divisor changes?

c. Make a conjecture about what happens when you divide $\frac{3}{4}$ by fractions that increase in size. Then give examples that support your conjecture.

Applications and Problem Solving

47. Business Bertha Thangelane of Soweto, South Africa used the political changes in South Africa to start her business. She is the co-owner and managing director of RNB Creations (Pty) Ltd., which makes school uniforms. If she can make an elementary school child's uniform with $2\frac{1}{3}$ yards of fabric, how many uniforms can she make with a 50-yard bolt of fabric?

48. Sequences What is the quotient if the eighth term of the sequence 1, $\frac{1}{2}$, $\frac{1}{4}$, $\frac{1}{8}$, . . . is divided by the ninth term? Explain your method and strategy for solving this problem.

49. Food According to the U.S. Department of Agriculture, the average young American woman drinks $1\frac{1}{2}$ cans of cola each day. At this rate, how long would a 12-pack of cola last?

Mixed Review

50. Solve $\frac{3}{5} \cdot \frac{2}{7} = a$. Write the solution in simplest form. (Lesson 6-3)

51. Which number is greater, $\frac{2}{5}$ or 0.25? (Lesson 6-1)

52. Health Teri lost $1\frac{3}{4}$ pounds the first week of her diet. The following week she lost $3\frac{1}{4}$ pounds. How much did Teri lose? (Lesson 5-4)

53. Find the greatest common factor (GCF) of 30 and 12. (Lesson 4-5)

54. Standardized Test Practice Which number is divisible by 3? (Lesson 4-1)

    A. 1451   **B.** 1592   **C.** 1811   **D.** 2001

55. Geometry Name the quadrant in which $(-6, 18)$ is located. (Lesson 2-2)

For **Extra Practice**, see page 754.

56. State whether $9 > 6$ is *true*, *false*, or *open*. (Lesson 1-9)

57. Simplify $h + 12h + 23$. (Lesson 1-5)

WORKING ON THE

Investigation

Refer to the Investigation on pages 270–271.

Look over the scale drawings that you placed in your Investigation Folder at the beginning of Chapter 6. Make any necessary revisions to them based on what you've learned so far.

- Using the scale drawings, determine the speed of your train for each section of the circuit you

designed. A typical speed down the first drop is about 60 miles an hour (88 feet per second). The length of each section should be determined by the group and will, for the most part, be dictated by the design, that is, uphill, downhill, straight, or curved.

- Determine the time it will take for your coaster train to travel each section of the circuit you designed. This can be found by dividing the length of the section of track by the speed you think the train will be traveling over that section.

- Find the total time it will take for the coaster train to complete the circuit you designed.

Add the results of your work to your Investigation Folder.

6-5A Multiplying and Dividing Decimals

A Preview of Lesson **6-5**

MATERIALS

 grid paper

 colored pencils

In the activities below, you will explore the multiplication and division of decimals using grid paper decimal models. The whole decimal model represents 1, and each row or column represents 0.1.

Activity ① Work with a partner.

▶ Use a 10-by-10 decimal model to find 0.3 × 0.8.

▶ Color 3 rows of the model yellow to represent 0.3.

▶ Color 8 columns of the model blue to represent 0.8.

TALK ABOUT IT

1. How many squares are green? What is the product of 0.3 and 0.8?

2. How many decimal places are there in both factors? How many are in the product? What is the relationship between the total number of decimal places in the factors and the number of decimal places in the product?

3. Use a model and a calculator to find the product 0.4 × 0.2. Explain your results.

Activity ② Use decimal models to find 1.5 ÷ 0.3.

▶ Outline one complete model and 5 columns in the second model to represent 1.5.

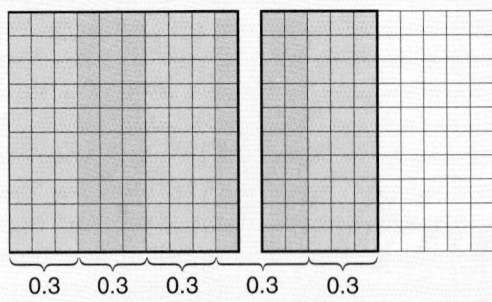

0.3   0.3   0.3   0.3   0.3

▶ Shade the first three columns yellow to represent 0.3.

▶ Shade the next three columns blue to represent another 0.3.

▶ Continue shading groups of three columns in alternating colors until you have filled the outline representing 1.5.

TALK ABOUT IT

4. How many groups of three tenths do you have?

5. What is the quotient of 1.5 ÷ 0.3?

6. Use decimal models to divide 2.4 by 0.4.

Multiplying and Dividing Decimals

6-5

You'll Learn

■ To multiply and divide decimals.

It's Important

You will divide decimals when comparing the costs of different-sized items.

Technology Activity

You can multiply decimals in the same way that you multiply whole numbers. Look for a pattern in the products below to help you write a rule for placing the decimal point in the product.

Your Turn   Use a calculator to find each product.

$15.6 \times 38 =$

$15.6 \times 3.8 =$

$15.6 \times 0.38 =$

$15.6 \times 0.038 =$

a. Describe the pattern in the factors of the problems above.

b. Describe the pattern in the products.

c. What do you think 15.6×0.0038 equals? Explain your answer.

d. Write a rule for placing the decimal point in a product of two decimals.

Learning the Concept

When you multiply decimals, it is helpful to use estimation to verify the placement of the decimal point.

Example   **Solve each equation.**

a. $c = (3.9)(8.2)$ *Estimate: $4 \times 8 = 32$*

$$
\begin{array}{rl}
3.9 & \leftarrow \quad \textit{one decimal place} \\
\times\, 8.2 & \leftarrow \quad \textit{one decimal place} \\
\hline
7\,8 & \\
312\,0 & \\
\hline
31.98 & \leftarrow \quad \textit{two decimal places}
\end{array}
$$

$c = 31.98$ *Compare to the estimate.*

b. $x = (-6.302)(0.81)$ *Estimate: $-6 \times 1 = -6$*

$$
\begin{array}{rl}
-6.302 & \leftarrow \quad \textit{three decimal places} \\
\times\, 0.81 & \leftarrow \quad \textit{two decimal places} \\
\hline
6302 & \\
504160 & \\
\hline
-5.10462 & \leftarrow \quad \textit{five decimal places}
\end{array}
$$

$x = -5.10462$ *Compare to the estimate.*

All the properties that were true for multiplication of integers are also true for multiplication of rationals. The properties of rationals are summarized in the following chart.

Properties of Multiplication	Examples
Commutative Property For all rational numbers x and y, $x \cdot y = y \cdot x$.	$-5(-7) = -7(-5)$ $0.6 \cdot 1.5 = 1.5 \cdot 0.6$ $\dfrac{5}{6} \cdot \dfrac{8}{9} = \dfrac{8}{9} \cdot \dfrac{5}{6}$
Associative Property For all rational numbers x, y, and z, $(x \cdot y) \cdot z = x \cdot (y \cdot z)$.	$(-6 \cdot 5) \cdot 8 = -6 \cdot (5 \cdot 8)$ $(0.3 \cdot 4) \cdot 5.7 = 0.3 \cdot (4 \cdot 5.7)$ $\left(-\dfrac{1}{3} \cdot \dfrac{9}{10}\right) \cdot \dfrac{2}{5} = -\dfrac{1}{3} \cdot \left(\dfrac{9}{10} \cdot \dfrac{2}{5}\right)$
Identity Property For every rational number x, $x \cdot 1 = x$ and $1 \cdot x = x$.	$-8 \cdot 1 = -8$ $1 \cdot \dfrac{1}{4} = \dfrac{1}{4}$ $10.3 \cdot 1 = 10.3$
Inverse Property For every rational number $\dfrac{x}{y}$, where $x, y \neq 0$, there is a unique number $\dfrac{y}{x}$ such that $\dfrac{x}{y} \cdot \dfrac{y}{x} = 1$.	$6 \cdot \dfrac{1}{6} = 1$ $-\dfrac{5}{12}\left(-\dfrac{12}{5}\right) = 1$ $0.7 \cdot \dfrac{1}{0.7} = 1$

To divide by a decimal, think of the division as a fraction and write the divisor, or denominator, as a whole number. For example, consider $50 \div 0.26$.

Think of it as $\dfrac{50}{0.26}$.

$$50 \div 0.26 \;\rightarrow\; \frac{50}{0.26} \times \frac{100}{100} = \frac{5000}{26}$$

Since $\dfrac{100}{100} = 1$, the value of the fraction is unchanged. This becomes a division problem involving whole numbers.

$$\frac{50}{0.26} \;\rightarrow\; 0.26\overline{)50.00} \;\rightarrow\; 26\overline{)5000.00}$$

$$
\begin{array}{r}
192.30 \\
26\overline{)5000.00} \\
\underline{26} \\
240 \\
\underline{234} \\
60 \\
\underline{52} \\
80 \\
\underline{78} \\
20
\end{array}
$$

To the nearest tenth, the quotient is 192.3.

Example **Solve $y = 42 \div (-0.8)$.** *Estimate: $42 \div -1 = -42$.*

$0.8\overline{)42.0}$ *Multiply 0.8 and 42 by 10 to get a whole number in the divisor.*

$$
\begin{array}{r}
52.5 \\
8\overline{)420}
\end{array}
$$

$y = -52.5$ *Compare to the estimate.*

Decimals are used every day when people work with money.

Example 3

Recycling

In Columbus, Ohio, art teachers can make use of re:Art, a warehouse that recycles materials for use by artists. Membership in re:Art costs the Columbus Public Schools $500 a year, a cost of about $3.49 for each art teacher. How many art teachers are there in the school system?

$500 \div \$3.49 = ?$ *Estimate: 500 ÷ 4 = 125*

$$
\begin{array}{r}
143.2 \\
3.49\overline{)500.000} \\
-349 \\
\hline
1510 \\
-1396 \\
\hline
1140 \\
-1047 \\
\hline
930 \\
-698 \\
\hline
232
\end{array}
$$

There are 143 or 144 art teachers in the system.
Check by comparing your answer to the estimate.

Connection to Algebra

Evaluate expressions involving division of decimals in the same way you evaluate other algebraic expressions.

Example 4   Evaluate $\dfrac{6m}{n}$ if $m = 0.5$ and $n = -3.2$.

$\dfrac{6m}{n} = \dfrac{6(0.5)}{-3.2}$ *Replace m with 0.5 and n with −3.2.*

$\phantom{\dfrac{6m}{n}} = \dfrac{3}{-3.2}$ or -0.9375.

The value of the expression is -0.9375.

Checking Your Understanding

Communicating Mathematics

1. **Illustrate** or explain the rule for placement of the decimal point in $(2.54)(0.067)$.

2. What is the first step you should take before you actually begin to multiply or divide decimals?

3. **Decide** where to place a decimal point in each factor so that the product is correct. $\boxed{493 \times 17 = 8.381}$

MATERIALS

grid paper
colored pencils

4. In Lesson 6-5A, you used decimal models to represent multiplication and division problems. Work with a partner to represent these multiplication and division problems.

    a. $(1.2)(3)$ **b.** $(1.2)(0.3)$ **c.** $4.8 \div 4$ **d.** $4.8 \div 0.4$

Guided Practice

State where the decimal point should be placed in each product or quotient.

    5. $(7.2)(0.2) = 144$ **6.** $(0.06)(3) = 18$

    7. $0.63 \div 0.9 = 7$ **8.** $8.4 \div 0.4 = 21$

Solve each equation.

9. $x = (-0.2)(-3.1)$

10. $y = (1.2)(-0.05)$

11. $s = 27 \div (-0.3)$

12. $t = 0.4 \div 2$

Evaluate each expression if $a = 15.7$, $b = 0.4$, and $c = 1.6$.

13. ac

14. $5c \div b$

Name the property shown by each statement.

15. $\left(-\dfrac{5}{6}\right)\left(-\dfrac{6}{5}\right) = 1$

16. $(-4 \cdot 5) \cdot \dfrac{1}{4} = [5 \cdot (-4)] \cdot \dfrac{1}{4}$

17. Recycling Sun Shares, a recycling company, is paid $0.65 a pound for aluminum. How many pounds of aluminum would they have to collect a week to meet their weekly phone costs of $45.50?

Exercises: Practicing and Applying the Concept

Independent Practice

Solve each equation.

18. $g = (-6.5)(0.13)$

19. $k = (0.47)(3.01)$

20. $-14.9(-0.56) = n$

21. $7.45(-0.75) = t$

22. $x = (0.001)(7.09)$

23. $b = (1.03)(-6.4)$

24. $r = 14.4 \div (0.16)$

25. $q = 0.384 \div 1.2$

26. $-85 \div (-1.7) = k$

27. $-0.51 \div 0.03 = g$

28. $s = -15.3 \div (-9)$

29. $h = 2.92 \div 0.002$

30. Find the product of 7.5 and 0.03.

31. What is the quotient of 52.64 and 9.4?

Evaluate each expression.

32. $3y^2$ if $y = 0.4$

33. ab if $a = -1.5$, $b = 10$

34. $\dfrac{5t}{s}$ if $t = 6.2$, $s = 2.5$

35. $n^2 \div q$ if $n = 2.2$, $q = 4$

36. $12y \div z$ if $y = 9.8$, $z = 6.4$

37. $(x + 4)(x - 2)$ if $x = 1.8$

Name the property shown by each statement.

38. $\left(2 \cdot \dfrac{5}{2}\right) \cdot \dfrac{3}{4} = \dfrac{3}{4}\left(2 \cdot \dfrac{5}{2}\right)$

39. $-\dfrac{8}{9} \cdot 1 = -\dfrac{8}{9}$

40. $0.6\left(\dfrac{1}{0.6}\right) = 1$

41. $(-6.5 \cdot 9.3) \cdot 2.1 = -6.5 \cdot (9.3 \cdot 2.1)$

42. $-9.3 = 1 \cdot -9.3$

43. $(0.6 \cdot 3.2) \cdot 0.3 = (3.2 \cdot 0.6) \cdot 0.3$

Calculators

Estimate. Then compute with a calculator. Round quotients to the nearest tenth if necessary.

44. $(8.01)(3.33) = w$

45. $(-0.56)(4.59) = r$

46. $-8.022(-0.03) = y$

47. $c = 90.5 \div (-8.9)$

48. $f = 93.702 \div 2.4$

49. $z = -0.36 \div (-2.1)$

Critical Thinking

Applications and Problem Solving

50. a. Name two decimals whose sum is 2.4 and whose product is 1.44.
   b. Name two decimals whose sum is 2.7 and whose quotient is 2.

51. Consumer Economics A 13.5 ounce box of muesli cereal costs $2.45, and a 19.5 ounce box of frosted corn flakes costs $3.59. Which costs less per ounce?

52. Economics When Lilia's softball team visited Mexico, $1 in American money could be exchanged for 2702.7 pesos. If Lilia exchanged $25 in American money, how many pesos did she receive?

53. Sports American consumers spent $0.8 billion dollars for golf equipment in a recent year. That same year they spent four times that much on athletic shoes. How much did consumers spend on athletic shoes?

Mixed Review

54. Write $\frac{5}{11} \div \frac{10}{11}$ as a multiplication expression. Then find its value. (Lesson 6-4)

55. Solve $3\frac{4}{9} \cdot \frac{3}{4} = h$. (Lesson 6-3)

56. Solve $b - 1.6 \le 4.3$. (Lesson 5-7)

57. Name the set(s) of numbers to which -10 belongs. (Lesson 5-1)

58. Solve $8 = -\frac{b}{11}$. (Lesson 3-3)

59. Banking Account #8234620 had an opening balance of $3245. Three checks cleared on Tuesday, leaving a balance of $2931. What was the total of the checks? (Lesson 3-2)

For **Extra Practice**, see page 754.

60. Meteorology The *heat index* is an estimate of the warming effect that humidity has on a person in hot weather. If the outside temperature is 86°F and the humidity makes it feel like 93°F, how much warmer does the humidity make it feel? (Lesson 2-5)

61. Name the property shown by $3 + a = a + 3$. (Lesson 1-4)

62. Standardized Test Practice Find the value of $18 \div (18 - 6 \cdot 2)$. (Lesson 1-2)

    A. -11 **B.** -10 **C.** 3   **D.** 6

Self Test

Express each fraction as a decimal. Use a bar to show a repeating decimal. (Lesson 6-1)

1. $\frac{7}{8}$ **2.** $-1\frac{2}{3}$ **3.** $-\frac{8}{11}$

4. Business On Tuesday, a company's stock fell $2\frac{1}{8}$ points from the previous day's closing price. If the closing price on Monday was $75.50, what was the closing price, in dollars, on Tuesday? (Lesson 6-1)

Estimate each product or quotient. (Lesson 6-2)

5. $\frac{2}{3} \times 25$ **6.** 3.85×6 **7.** $6.2 \div 2.1$

Solve each equation. Write each solution in simplest form. (Lessons 6-3, 6-4, and 6-5)

8. $x = \frac{1}{4}\left(-\frac{3}{8}\right)$ **9.** $y = -9 \div \frac{3}{8}$ **10.** $w = (0.5)(1.46)$

6-6A Mean, Median, and Mode

A Preview of Lesson **6-6**

MATERIALS

small boxes of raisins

🖩 calculator

Measures of central tendency are numbers that represent a set of data. Researchers use these numbers when analyzing data.

Your Turn

Work in small groups.

▶ Get a small box of raisins for your group from your teacher.

▶ Estimate the number of raisins you think will be in the box.

▶ Count the number of raisins and record the number.

▶ Record the data for each group.

▶ Use a calculator to add the data. Then divide the sum by the number of groups. This number is called the **mean** of the data.

▶ List the data in order and mark the middle number. If there is no middle number in the set of data, circle the middle two numbers and find their mean. This number is called the **median**.

▶ If a number appears more often than the others in the data, this number is called the **mode**.

TALK ABOUT IT

1. If you add the number 5 to your group's set of data, predict which measure of central tendency will be affected the most.

2. Find the mean, median, and mode of the new set of data.

3. Which measure of central tendency will be most representative of the data before adding 5? after adding 5?

4. If you add the number 400 to your group's set of data, predict which measure of central tendency will be affected the most.

5. Find the mean, median, and mode of the new set of data.

6. Which measure of central tendency will be most representative of the data after adding 400?

Extension

7. Suppose you count the number of raisins in a large box. If you add this number to your set, will it affect the mean? the median? the mode?

8. Choose your own numbers to add to your set of data. Study the effects on the mean, median, and mode.

6-6

 You'll Learn

- To use the mean, median, and mode as measures of central tendancy.

 It's Important

When analyzing test scores you will use mean, median, and mode.

Integration: Statistics
Measures of Central Tendency

 Application: Meteorology

"It's hot fun in the summertime today, folks! The high will reach a scorching 92 degrees, way above the normal of 84 for this date."

You have heard meteorologists report the normal high temperature for a given date. The normal high temperature is determined by averaging the high temperatures for a given date over several years. The high temperatures on March 1 for twelve consecutive years in a certain city are recorded below. What is the normal high temperature for the date?

1990	1991	1992	1993	1994	1995	1996	1997	1998	1999	2000	2001
59°F	50°F	48°F	43°F	40°F	46°F	50°F	53°F	58°F	61°F	64°F	73°F

Learning the Concept

To analyze sets of data, researchers often try to find a number that can represent the whole set. These numbers or pieces of data are **measures of central tendency**. Three that we will study are the **mean**, the **median**, and the **mode**.

The mean is what people usually are talking about when they say average. It is the arithmetic average of the data. For the temperatures above, the mean is

$$\frac{59 + 50 + 48 + 43 + 40 + 46 + 50 + 53 + 58 + 61 + 64 + 73}{12} = \frac{645}{12} \text{ or } 53.75.$$

The mean is about 53.8°.

Notice that the mean may not be a member of the set of data.

Definition of Mean	The mean of a set of data is the sum of the data divided by the number of pieces of data.

The median is the middle number when the data are in order. Consider two cases.

Case 1: An Odd Number of Data

Write the set of data in order from least to greatest. Consider the set of data below.

8   12   22   22   25   34   35   36   55
                         ↑

The middle number is 25.

Case 2: An Even Number of Data

If the number of data is even, there are two middle numbers. In this case, the median is the mean of the two numbers. Look back at the set of temperatures.

40   43   46   48   50   50   53   58   59   61   64   73
                               ↑   ↑

$$\frac{50 + 53}{2} = \frac{103}{2} \text{ or } 51.5$$

The median is 51.5°.

If there is an even number of data in a set, the median may not be a member of the data set.

Definition of Median	The median is the number in the middle when the data are arranged in order. When there are two middle numbers, the median is their mean.

The number or item in a data set that appears most often is called the mode. There can be one mode, more than one mode, or no mode in a data set.

In the temperatures above, the temperature 50° appears more times than any other (two times), so 50° is the mode.

The mode, if there is one, is always a member of the set of data. If there is no one piece of data that appears more often than the rest, the set of data has no mode.

Definition of Mode	The mode of a set of data is the number or item that appears most often.

FYI

In 1985, Wilma Mankiller was elected chief of the Cherokee Nation. As chief, she placed a high priority on the economic growth of the nation.

Wilma Mankiller

The table below shows the populations of several American Indian tribes as of the 1990 U.S. census. Organize the data and then find the mean, median, and mode.

Tribe	Population (thousands)
Apache	50
Cherokee	308
Chippewa	104
Choctaw	82
Creek	44
Iroquois	49
Lumbee	48
Navajo	219
Pueblo	53
Sioux	103

Median List the populations in order from greatest to least. The median is the middle number. There are 10 populations listed, so the average of the 5th and 6th populations is the median.

44 48 49 50 53 82 103 104 219 308

The median population is $\frac{53 + 82}{2}$ or 67.5 thousand.

Mode Since no population occurs more than the rest, there is no mode.

Mean To find the mean, add the populations and divide by 10.

$$\frac{50 + 308 + 104 + 82 + 44 + 49 + 48 + 219 + 53 + 103}{10}$$

$$= \frac{1060}{10} \text{ or } 106$$

The mean population is 106 thousand.

You can use the mean, median, and mode to compare two different sets of data.

A research group monitors the popularity of television programs and assigns a rating to each one. The ratings for programs airing at 8:00 P.M. on two networks for each day of the week of July 3, 1995 are shown at the right.

Network A	Network B
7.2	6.1
5.5	5.7
7.1	4.1
8.4	9.6
5.6	5.6
5.1	5.3
9.9	12.5

a. Find the mean, median, and mode of the ratings for each network.

Begin by putting the data in order.

Network A	5.1	5.5	5.6	7.1	7.2	8.4	9.9
Network B	4.1	5.3	5.6	5.7	6.1	9.6	12.5

Median Because there are 7 pieces of data in each set, the medians are the 4th piece of data.

(continued on the next page)

Network A The median is 7.1.

Network B The median is 5.7.

Mode None of the pieces of data occur more than once in either set of data. Thus, neither set has a mode.

Mean The means are the sums of the data divided by 7.

Network A $\dfrac{5.1 + 5.5 + 5.6 + 7.1 + 7.2 + 8.4 + 9.9}{7} = \dfrac{48.8}{7}$ or about 7.0

Network B $\dfrac{4.1 + 5.3 + 5.6 + 5.7 + 6.1 + 9.6 + 12.5}{7} = \dfrac{48.9}{7}$ or about 7.0

b. Compare the averages to determine whether Network A or Network B programs rated better for the week.

The median rating for Network A programs was higher than for Network B programs. The means for the two networks were about the same. It appears that the popularity of Network A programs was more consistent. Network B had more programs that were rated very high or very low.

Checking Your Understanding

Communicating Mathematics

Read and study the lesson to answer these questions.

1. Which "average" do you think best represents the data in Example 2? Explain your answer.

2. **Write** a set of data that has a mean of 7. Is the median 7 also?

MATERIALS
roll of 50 pennies

3. Complete this activity with a group of three or four students. Arrange the 50 pennies in a roll in order from oldest to newest.

   a. Find the mean, median, and mode for the year the coins were minted.

   b. Based on your sample of pennies, what can you say about all pennies in circulation?

   c. Compare your results with the rest of the class and check your conclusion in **part b**.

Guided Practice

List the data in each set from least to greatest. Then find the mean, median, and mode. When necessary, round to the nearest tenth.

4. 4, 5, 7, 3, 9, 11, 23, 37

5. 11, 45, 62, 12, 47, 8, 12, 35, 33

6. 25.98, 30.00, 45.36, 25.00, 45.36

7. 105, 116, 125, 78, 78

8. **Health** If 50 males have a resting heart rate of 59 beats a minute, 30 males have 72, 15 males have 84, and 5 males have 93, what is the mean resting heart rate for all 100 males represented?

Independent Practice

Find the mean, median, and mode for each set of data. When necessary, round to the nearest tenth.

9. 78, 45, 32, 64, 22, 63, 45
10. 22, 63, 53, 26, 41, 44, 12, 26
11. 3.6, 7.2, 9.0, 5.2, 7.2, 6.5, 3.6
12. 0.6, 0.7, 1.4, 0.9, 0.2, 0.7, 0.7, 1.4
13. 100, 113, 94, 86, 113, 120, 105
14. 2.3, 3.6, 4.1, 3.6, 2.9, 3.0
15. 0.4, 1.6, 0.8, 0.9, 0.7, 1.1
16. 3.4, 5.6, 1.2, 4.8, 8.8, 4.5, 3.4, 7.6

Use the following spreadsheet to answer exercises 17–25.

	A	B	C	D	E	F
		Test Scores				
	Student	**Test 1**	**Test 2**	**Test 3**	**Test 4**	**Test 5**
1	Student	Test 1	Test 2	Test 3	Test 4	Test 5
2	Colleen	93	86	84	92	85
3	Lakesha	83	85	93	78	86
4	Bryan	88	92	73	78	86
5	Phongsava	90	88	83	76	90
6	Raul	74	90	86	89	88

17. What is the mean of all the test scores?
18. What is the median of all the test scores?
19. What is the mode of all the test scores?
20. **a.** What is Colleen's mean on the five tests?

    b. How does Colleen's mean affect the class mean?
21. **a.** What is Colleen's median on the first five tests?

    b. How does Colleen's median affect the class median?
22. **a.** What is Phongsava's mean on the five tests?

    b. How does Phongsava's mean affect the class mean?
23. **a.** Find Phongsava's mode on the first five tests.

    b. How does Phongsava's mode affect the class mode?
24. If test 4 is removed, how will this change the mean?
25. The lowest average grade that receives a B on a report card is 86. If Lakesha wanted a B after the sixth test, what grade does she need?

Critical Thinking

26. Several years ago, the mayor of Glasshouse City quit and moved to the city of Stonethrowers. An elated citizen of Glasshouse City was heard to say that the move raised the average IQ in both cities. Suppose that the average referred to is the mean. Explain how the statement could be true.

27. Construct a set of data with at least seven different items in which the mean, median, and mode are the same number.

28. **Employment**   The table at the right shows the earnings for women for every $100 earned by a man in the same occupation for two years.

Occupation	Year 1	Year 2
Nurses	99.5	104.7
Teachers	88.6	90.3
Police	91.2	94.2
Food service	102.5	105.6
Packagers	91.6	94.6
Secretaries	76.7	91.6
Orderlies	81.0	96.0
Postal clerks	93.4	94.6

   a. Find the mean, median, and mode of the earnings for each year.

   b. Compare the averages for the two years. What might you conclude about the changes in women's employment situations from Year 1 to Year 2? Explain.

29. **Family Activity**   For one hour while you are watching television or listening to the radio with your family, record the length of each commercial break. Find the mean, median, and mode. Compare your results with three other people in your class. Find the mean, median, and mode for the four sets of data. Write about your findings.

30. **Business**   Revenues of the top ten money-making businesses that are owned by women are listed in the chart below.

   a. What are the mean, median, and mode revenues for the ten companies?

   b. Which measure of central tendency would complete this statement correctly?

   "There are more companies below the _____ than above it."

Company Rank	Revenues*
1	$1820
2	$1800
3	$1400
4	$1000
5	$815
6	$800
7	$789
8	$465
9	$450
10	$380

*Revenues are in millions of dollars.

Mixed Review

31. Solve $w = (-3.2)(4.057)$. (Lesson 6-5)

32. Write $\dfrac{36a^3b}{52ab^2}$ in simplest form. (Lesson 4-6)

33. Solve $-3y \geq 72$. (Lesson 3-7)

34. **Standardized Test Practice**   A submarine at 2200 feet below sea level descends another 1325 feet. How far below sea level is the submarine? (Lesson 2-4)

   A. 3525   **B.** 1325   **C.** 875   **D.** -1105

35. Name the property shown by $3 \cdot 0 = 0$. (Lesson 1-4)

36. **Nutrition**   You can find the number of calories from fat in a food item by multiplying the number of grams of fat by nine. (Lesson 1-3)

   a. Write an expression for the relationship.

   b. Find the number of calories from fat in a snack with 5 grams of fat.

For **Extra Practice**, see page 754.

6-6B Finding Mean and Median

An Extension of Lesson **6-6**

A graphing calculator is a powerful computer. You can use a graphing calculator to find the mean and median of a set of data.

Activity

Find the mean and median of the bicycle helmet prices (in dollars) shown below.

51 40 58 60 30 45 66 40 87 65 41 60 40 35 47 49 54 50 52 47

Begin by clearing list L1.

Enter: | STAT | 4 | 2nd | L1 | ENTER |

Then enter the prices into list L1.

Enter: | STAT | ENTER | 51 | ENTER | 40 | ENTER | ... 52 | ENTER | 47 | ENTER |

Now compute the values.

Enter: | STAT | ▶ | ENTER | ENTER |

The calculator displays many statistical values at one time. The first value, $\bar{x}$, is the mean. The calculator found the mean to be 50.85 or $50.85.

Use the ▼ key to scroll through the list to locate "Med". The value displayed is the median. The median is 49.5 or $49.50.

Your Turn

Clear list L1 and find the mean and median of each data set with a graphing calculator. Round to the nearest hundredth.

1. 6.4, 5.6, 7.3, 1.2, 5.7, 8.9, 3.9, 5.8, 6.0, 4.7
2. 123, 423, 190, 289, 99, 178, 156, 217, 217
3. $-23, -13, -16, -21, -15, -34, -4, -6, -11, -15, -22$
4. $-8.4, 2.2, 7.3, -5.3, 6.7, -4.3, 5.1, 1.3, -1.1, -3.2, 2.2,$
   $2.9, -2.8, 1.4$

5. **Geography** The table at the right shows the populations per square mile in thousands for different regions as of 1994. Find the mean and median of the data.

Region	Population
N. America	30.7
Latin America	68.7
Europe	133.9
Asia	192.2
Oceania	16.2
Antarctica	0

TALK ABOUT IT

6. Look back at the medians that you found. When is the median a member of the data set?

7. When does the median best represent the data? When does the mean best represent the data? Explain.

Solving Equations and Inequalities

You'll Learn

- To solve equations and inequalities containing rational numbers.

It's Important

Inequalities can be used to find the minimum number of signatures needed for a petition.

Application: Music

Whether it's rock'n'roll, jazz, rap, or classical, the art of making music is the same. About 2600 years ago, Pythagoras discovered the mathematical principles behind the musical scale. Musical sounds are made by vibrations. He discovered the relationships between the vibrations and the notes of the scale. If *n* represents the vibrations for *do*, or middle C, the vibrations for the other notes going up the scale are given below.

middle C   1 octave up

do	re	mi	fa	so	la	ti	do
n	$\frac{9}{8}n$	$\frac{5}{4}n$	$\frac{4}{3}n$	$\frac{3}{2}n$	$\frac{5}{3}n$	$\frac{15}{8}n$	$\frac{2}{1}n$

A guitar string vibrates 352 times per second to produce fa. Find the number of vibrations per second needed to produce middle C.

Learning the Concept

What you learned about solving equations with integers can be used to solve equations with rational numbers.

Let *n* represent vibrations for middle C.

vibrations for fa → $\frac{4}{3}n = 352$ ← *guitar string vibrations for fa*

$$\frac{3}{4} \cdot \frac{4}{3}n = \frac{3}{4}(352)$$ *Multiply each side by the multiplicative inverse of $\frac{4}{3}$, $\frac{3}{4}$.*

$$n = 264$$

The guitar vibrates 264 times a second to produce middle C.

Example **1** **Solve each equation. Check your solution.**

a. $6.5y = -63.7$

$$\frac{6.5}{6.5}y = \frac{-63.7}{6.5} \quad \textit{Divide each side by 6.5.}$$

$$y = -9.8$$

Check: $6.5y = -63.7$

$$(6.5)(-9.8) \stackrel{?}{=} -63.7 \quad \textit{Replace y with }-9.8.$$

$$-63.7 = -63.7 \quad \textit{✔}$$

b. $-\dfrac{y}{5.3} = 8.9$

$$(-5.3)\left(-\frac{y}{5.3}\right) = (-5.3)(8.9) \quad \textit{Multiply each side by }-5.3.$$

$$y = -47.17 \quad \textit{Check by estimation: }-5 \times 9 = -45$$

c. $\dfrac{7}{9}k = -\dfrac{5}{12}$

$$\frac{9}{7} \cdot \frac{7}{9}k = \frac{9}{7} \cdot \left(-\frac{5}{12}\right) \quad \textit{Multiply each side}$$

$$k = -\frac{15}{28} \quad \textit{by }\frac{9}{7}.\textit{ Why?}$$

Check: $\dfrac{7}{9}k = -\dfrac{5}{12}$

$$\frac{7}{9}\left(-\frac{15}{28}\right) \stackrel{?}{=} -\frac{5}{12}$$

$$-\frac{5}{12} = -\frac{5}{12} \quad \textit{✔}$$

You can use the same skills you developed to solve inequalities with integers to solve inequalities with rational numbers.

Example **2** **Solve each inequality.**

a. $\dfrac{t}{4.3} \geq 5$

You can review graphing inequalities in Lesson 3-6.

$$(4.3)\frac{t}{4.3} \geq (4.3)5 \quad \textit{Multiply each side by 4.3.}$$

$$t \geq 21.5 \quad \textit{Any number greater than or equal to 21.5 is}$$
$$\textit{a solution.}$$

$$\overset{\hspace{6.5cm}}{\underset{17\quad18\quad19\quad20\quad21\quad22\quad23\quad24\quad25\quad26\quad27\quad28}{\longleftrightarrow}}$$

Check: Try 25, a number greater than 21.5.

$$\frac{25}{4.3} \stackrel{?}{\geq} 5$$

$$5.8 \geq 5 \quad \textit{✔}$$

b. $-\dfrac{2}{3}b > -9\dfrac{1}{2}$

$$\left(-\frac{3}{2}\right)\left(-\frac{2}{3}\right)b < \left(-\frac{3}{2}\right)\left(-\frac{19}{2}\right) \quad \textit{Multiply each side by }-\frac{3}{2}\textit{ and reverse}$$
$$\textit{the order symbol.}$$

$$b < \frac{57}{4} \textit{ or } 14\frac{1}{4} \quad \textit{Any number less than } 14\frac{1}{4} \textit{ is a solution.}$$

Example **3**

Real World APPLICATION

Theater

A replica of the Globe Theater, where Shakespeare's plays were originally performed, is currently being constructed in London, England. Supporters of the project have raised almost $13 million, which is over four-fifths of the budget. How much is the project budget?

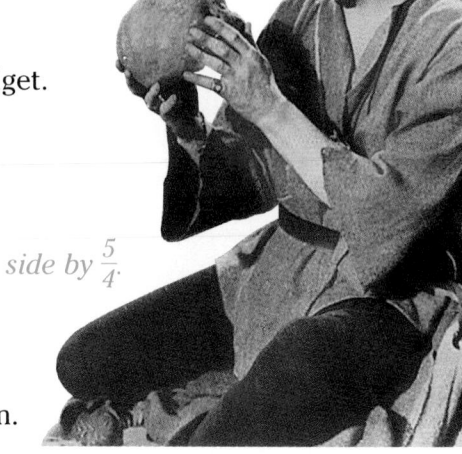

Write an equation. Let b represent the budget.

$\underbrace{\$13\ million}_{13} \quad \underbrace{is\ over}_{>} \quad \underbrace{four\text{-}fifths\ of\ the\ budget}_{\frac{4}{5}b}$

$$\frac{5}{4}(13) > \frac{5}{4}\left(\frac{4}{5}b\right) \qquad \textit{Multiply each side by } \frac{5}{4}.$$

$$\frac{65}{4} > b$$

The budget is less than $\frac{65}{4}$ or \$16.25 million.

Checking Your Understanding

Communicating Mathematics

Read and study the lesson to answer these questions.

1. **Explain** how you would solve the equation $0.3x = 4.5$.

2. In the application at the beginning of the lesson, the equation was solved by multiplying each side by the multiplicative inverse, or reciprocal, of $\frac{4}{3}$.

   a. How does this help you solve the equation?

   b. Describe the property of equality that allows you to do this.

3. **State** two ways to solve the equation $\frac{1}{4}x = 28$.

4. **Write** an inequality that has the solution graphed below.

Guided Practice

Explain how to solve each equation or inequality. Then solve. Graph the solutions on a number line.

5. $0.6n = -9$ 6. $4.1p = 16.4$ 7. $-2.7 < 0.3f$

8. $-\frac{2}{3}x > 6$ 9. $\frac{5}{6}y = \frac{3}{5}$ 10. $\dfrac{r}{-4.6} \geq 0.01$

11. **Geometry** You are asked to draw a rectangle with one side measuring 2.5 cm and with an area of 15 cm². How long should you make the other side?

> x cm
>
> Area = 15 cm²   2.5 cm

Independent Practice

Solve each equation or inequality. Check your solution.

12. $7y < 3.5$

13. $-3y = 1.5$

14. $5a > -7\frac{1}{2}$

15. $\frac{1}{4}x = 4\frac{1}{8}$

16. $-\frac{1}{8}r = \frac{1}{4}$

17. $\frac{x}{3.2} < -4.5$

18. $\frac{7}{8}r \le 1.4$

19. $-15 = 2\frac{2}{5}x$

20. $-6.7b = 14.07$

21. $\frac{p}{7.1} = -0.5$

22. $\frac{1}{3}d = -0.36$

23. $-1.3z \ge 1.69$

24. $\frac{3}{4}y > 1\frac{4}{5}$

25. $-7\frac{1}{2}x = 5\frac{1}{4}$

26. $-2h > 4.6$

27. $-0.05x = -0.95$

28. $\frac{x}{-2.5} \le 3.2$

29. $7x < 1\frac{4}{10}$

Solve each equation or inequality and graph the solution on a number line.

30. $-\frac{1}{4}c = 3.4$

31. $\frac{4}{5}y < \frac{3}{8}$

32. $-12.6p \ge 28.98$

Critical Thinking

33. Write an inequality involving multiplication or division, using fractions, whose solution is $p < \frac{1}{3}$.

Applications and Problem Solving

34. Geometry The formula for the area of a triangle is $A = \frac{1}{2}bh$. A triangle has an area greater than 18 square inches. If the base of the triangle is 2.5 inches, what is the height of the triangle?

35. Business The Rose Department Store chain was closing one of their stores so they had to sell most of the inventory in the store. In the final markdown, they marked all items $\frac{1}{3}$ off the ticketed price. How much would a shirt that was ticketed at 24.99 cost?

36. Food The National Restaurant Association polled a group of Americans about their food preferences. Less than $\frac{1}{3}$ of those polled had tried Indian or Thai food.

    a. If 332 people had tried Indian or Thai food, how large a group was polled?

    b. How large do you think the group polled was? Explain your answer.

For **Extra Practice**, see page 755.

Mixed Review

37. Education If you scored 93, 86, 84, 92, 85, and 93 on your first six pre-algebra tests, what is your mean test score? (Lesson 6-6)

38. Write $\frac{28}{45}$ as a decimal. (Lesson 6-1)

39. Solve $\frac{6}{7} + \frac{5}{14} = s$. (Lesson 5-5)

40. Standardized Test Practice Find the LCM of $15c$ and $12cd$. (Lesson 4-7)

    A. $60c$ **B.** $60cd$ **C.** $180c$ **D.** $180cd$

41. Solve $a = (-12)(3)$. (Lesson 2-7)

42. Name the property shown by $5(6 \cdot 3) = (5 \cdot 6)3$. (Lesson 1-4)

Integration: Discrete Mathematics
Geometric Sequences

Application: Genealogy

In 1974, Alex Haley finished his book, *Roots*, which traced his family's history. His story began in 1767 with the African Kunta Kinte. Alex Haley's great-great-great-great-great-grandfather, Kunta Kinte, was brought to America to be a slave. With each preceding generation that Alex Haley researched, he found more and more ancestors.

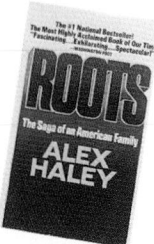

Learning the Concept

You can use a pattern to find the number of ancestors Alex Haley had back to Kunta Kinte's generation. The number of ancestors doubles as shown below.

OUR FAMILY TREE

Generation 1   *Alex Haley*   1 $\big)\times 2$
Generation 2   *Parents*   2 $\big)\times 2$
Generation 3   *Grandparents*   4 $\big)\times 2$
Generation 4   8 $\big)\times 2$
Generation 5   16 $\big)\times 2$
Generation 6   32 $\big)\times 2$
Generation 7   64 $\big)\times 2$
Generation 8   128 $\big)\times 2$

The list of numbers 1, 2, 4, 8, 16, 32, 64, 128 forms a **geometric sequence**. Each term in a geometric sequence increases or decreases by a common *factor*, called the **common ratio**. The number of ancestors doubles in each generation so the common ratio for this sequence is 2.

Geometric Sequence	A geometric sequence is a sequence in which the ratio between any two successive terms is the same.

Example State whether each sequence is geometric. If so, state the common ratio and list the next three terms.

a. 3, 15, 75, 375, . . .

3 15 75 375   *Notice $\frac{15}{3} = 5$, $\frac{75}{15} = 5$, and $\frac{375}{75} = 5$.*
$\times 5 \ \times 5 \ \times 5$

The sequence has a common ratio so it is a geometric sequence. The common ratio is 5. The next three terms are 375 × 5 or 1875, 1875 × 5 or 9375, and 9375 × 5 or 46,875.

b. $-4, -2, -\frac{2}{3}, -\frac{1}{6}, \ldots$

$$-4 \quad -2 \quad -\frac{2}{3} \quad -\frac{1}{6}$$
$$\times \frac{1}{2} \quad \times \frac{1}{3} \quad \times \frac{1}{4}$$

Since there is no common ratio, the sequence is *not* geometric.

c. $-4, 1, -\frac{1}{4}, \frac{1}{16}, \ldots$

$$-4 \quad 1 \quad -\frac{1}{4} \quad \frac{1}{16}$$
$$\times \left(-\frac{1}{4}\right) \quad \times \left(-\frac{1}{4}\right) \quad \times \left(-\frac{1}{4}\right)$$

The sequence is geometric. The common ratio is $-\frac{1}{4}$.

$$\frac{1}{16} \times \left(-\frac{1}{4}\right) = -\frac{1}{64} \qquad -\frac{1}{64} \times \left(-\frac{1}{4}\right) = \frac{1}{256} \qquad \frac{1}{256} \times \left(-\frac{1}{4}\right) = -\frac{1}{1024}$$

The next three terms are $-\frac{1}{64}, \frac{1}{256}$, and $-\frac{1}{1024}$.

Connection to > Algebra

If we know the first term of a geometric sequence and the common ratio, we can find any other term of the sequence. Consider the sequence 3, 12, 48, 192, ...

Here's how to write an expression that represents a term in the sequence. The first term is 3; call the first term a. The common ratio is 4; call the common ratio r.

Term	1st	2nd	3rd	4th	*n*th
Sequence	3	12	48	192	
Factors of term	3	$3 \cdot 4$	$3 \cdot 4 \cdot 4$	$3 \cdot 4 \cdot 4 \cdot 4$	$\underbrace{3 \cdot 4 \cdot 4 \cdot 4 \cdot \ldots \cdot 4}_{n-1}$
Term using exponents	3	$3 \cdot 4^1$	$3 \cdot 4^2$	$3 \cdot 4^3$	$3 \cdot 4^{n-1}$
Variables	a	$a \cdot r^1$	$a \cdot r^2$	$a \cdot r^3$	$a \cdot r^{n-1}$

↑
The exponent is one less than the number of the term.

Because n represents any term, you can use this expression to find any term in the sequence.

Example ② **Use the expression $a \cdot r^{n-1}$ to find the eighth term in the sequence 4, 12, 36, 108, . . .**

The first term is 4, so $a = 4$. The common ratio is 3, so $r = 3$, and $n = 8$.
$a \cdot r^{n-1} = 4 \cdot 3^{8-1}$ or $4 \cdot 3^7$

$4 \boxed{\times} 3 \boxed{y^x} 7 \boxed{=} \; \mathit{8748}$

The eighth term is 8748.

Some real-life problems can be solved using geometric sequences.

Example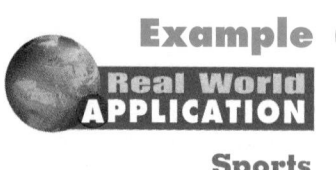

Real World APPLICATION

Sports

The NCAA Women's Basketball tournament is a single elimination tournament. After each game, the winning team progresses to the next level of competition and the loser is out of the tournament. How many teams are in the first round of play if there are 6 rounds in the tournament?

Explore   With each round, half of the teams are out of the competition. So, this is a geometric sequence. If you start with the final game and work backward, each round had twice as many teams as the one after it. The first term is 2, the 2 teams who played in the final game of the tournament. The common ratio is 2 because there are twice as many games in each previous round. The number of terms is 6 because there were 6 rounds.

Plan   Find out how many teams played in each round by starting with 2 and multiplying by 2 until you find the 6*th* term.

Solve $2 \times 2 \times 2 \times 2 \times 2 \times 2 = 2^6$ or 64

There were 64 teams.

Examine   Another strategy is to use the expression $a \cdot r^{n-1}$
$a = 2, r = 2$, and $n = 6$.

$2 \cdot 2^{(6-1)} = 2 \cdot 2^5$ or 64

The solution checks.

Checking Your Understanding

Communicating Mathematics

Read and study the lesson to answer these questions.

1. **Describe** the common ratio in a geometric sequence.

2. You studied arithmetic sequences in Lesson 5-9. Compare and contrast geometric and arithmetic sequences.

3. **Give an example** of a geometric sequence with a first term of $\frac{1}{4}$.

4. **Find** the number of ancestors Alex Haley has in the generation before Kunta Kinte.

5. You can model a geometric sequence by folding a piece of paper and counting the number of sections.

   a. Start with an $8\frac{1}{2}$"-by-11" piece of paper. There are 0 folds and 1 section. The first term of the sequence is 1.

   b. Fold the paper in half. There is 1 fold and 2 sections. The second term of the sequence is 2. Continue to fold the paper in half. Keep a record of the number of folds and the number of sections.

   c. How many sections result from the 5th fold?

   d. Write the sequence. Does the expression $a \cdot r^{(n-1)}$ represent the terms in the sequence? Explain.

Guided Practice

State whether each sequence is a geometric sequence. If so, state the common ratio and list the next three terms.

   6. $1, 3, 9, 27, \ldots$ **7.** $2, 4, 6, 8, \ldots$ **8.** $1, -3, 5, -7, \ldots$

   9. $81, 27, 9, 3, \ldots$ **10.** $10, 14, 18, 22, \ldots$ **11.** $\frac{1}{2}, 1, 2, 4, \ldots$

   12. Find the fourth term of a geometric sequence if $a = -5$ and $r = -2$.

   13. Genealogy Find the number of ancestors in your family 11 generations ago.

Exercises: Practicing and Applying the Concept

Independent Practice

State whether each sequence is a geometric sequence. If so, state the common ratio and list the next three terms.

   14. $-5, 1, -\frac{1}{5}, \frac{1}{25}, \ldots$ **15.** $24, 12, 6, 3, \ldots$ **16.** $9, 3, -3, -9, \ldots$

   17. $144, 12, 1, \frac{1}{12}, \ldots$ **18.** $-14, 7, -1, -\frac{1}{7}, \ldots$ **19.** $\frac{1}{2}, \frac{1}{4}, \frac{1}{8}, \frac{1}{16}, \ldots$

   20. $-8, 16, -32, 64, \ldots$ **21.** $7, -14, 28, -56, \ldots$ **22.** $25, 5, 1, \frac{1}{5}, \ldots$

   23. $2\frac{1}{2}, \frac{1}{2}, -1\frac{1}{2}, \ldots$ **24.** $-\frac{1}{4}, -\frac{1}{2}, -1, -2, \ldots$ **25.** $1, 1, 2, 3, 5, \ldots$

   26. $18, -6, 2, -\frac{2}{3}, \ldots$ **27.** $24, 6, \frac{3}{2}, \frac{3}{8}, \ldots$ **28.** $5, 7, 9, 11, 13, \ldots$

   29. Write the first five terms in a geometric sequence with a common ratio of -2. The first term is 2.

   30. Write the first four terms of a geometric sequence if $a = -6$ and $r = \frac{2}{3}$.

   31. Write the first three terms of a geometric sequence if $a = 80$ and $r = \frac{5}{4}$.

   32. Use the expression $ar^{(n-1)}$ to find the ninth term in the geometric sequence $-14, -\frac{7}{2}, -\frac{7}{8}, -\frac{7}{32}, \ldots$.

33. Use your calculator to find the first eight terms in the geometric sequence defined by $1 \cdot 7^n$, where n represents the number of the term. Determine the pattern that develops in the ones digits of each term. Use the pattern to find the ones digit of the 100th term.

34. Write the first five terms in a geometric sequence with a common ratio that is greater than 1. Then write the first five terms in a geometric sequence with a common ratio that is between 0 and 1.

    a. What happens to the size of the terms in a geometric sequence if the common ratio is greater than 1?

    b. What happens to the size of the terms in a geometric sequence if the common ratio is less than 1?

Applications and Problem Solving

35. Geography Between 1980 and 1990, the population of Florida increased from approximately 10 million to 13 million.

    a. Find the common ratio between these numbers.

    b. If the population grows in a geometric pattern over each of the next four decades, what would the population of Florida be in 2030?

36. Investment The face value of a high return bond increases to $1\frac{1}{10}$ of its previous value each year. If the starting value is $1250, what is the face value after 3 years?

37. Patterns Jacob told three friends a secret. Each of these friends told three more friends, who each told three more friends. How many people in all knew about the secret?

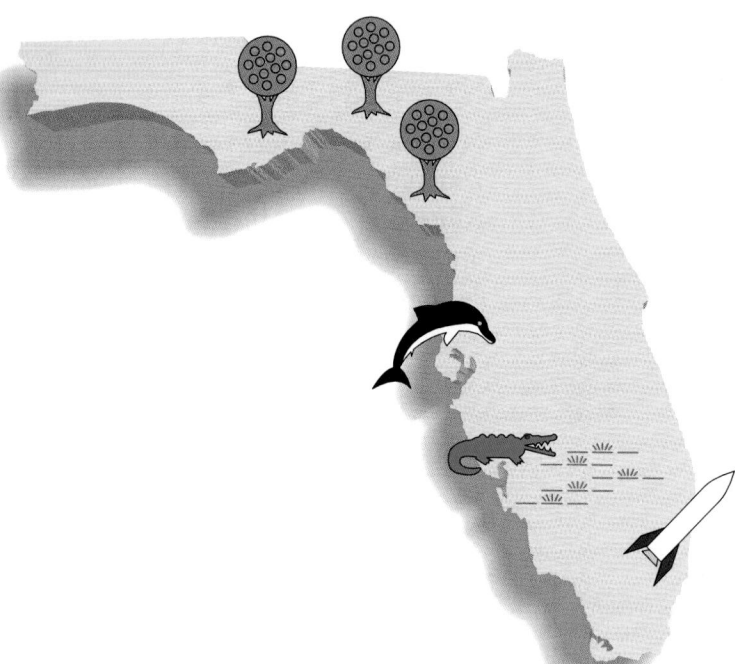

38. Games A popular television game show holds a tournament each year. Three people participate in each game of the competition and only the winners proceed to the next round. How many players participate if there are five rounds?

39. Solve $\frac{5}{9}k = 15$. (Lesson 6-7)

40. Solve $d = 119 - 9.8$. (Lesson 5-3)

41. Find the quotient $\frac{3^3}{3^7}$. (Lesson 4-9)

42. Geometry The area of a square is the product of its length and width. Use exponents to write an expression for the area of a square that has sides b units long. (Lesson 4-2)

43. Standardized Test Practice Solve $t = \frac{120}{24}$. (Lesson 2-8)

    A. 0.2   **B.** 5   **C.** 6   **D.** 2880

44. Simplify $5x + 3x + 1$. (Lesson 1-5)

45. Find the value of $6 \cdot 5 + 12 \div 3$. (Lesson 1-2)

For **Extra Practice**, see page 755.

46. Patterns Find the next number in the pattern 8, 9, 11, 14, 18, 23, . . . (Lesson 1-1)

Scientific Notation

You'll Learn

- To write numbers in scientific notation.

It's Important

Scientific notation allows you to express the distances between planets.

Application: Geology

Dr. William Rose, a geologist, specializes in the study of volcanoes and their effect on the atmosphere. He and his team studied the Augustine volcano about 280 kilometers from Anchorage, Alaska. The Augustine volcano erupted twice in 1986, once on March 27 and again on April 4.

Even though these eruptions were relatively small, on March 27th and 28th, the volcano produced 75,000,000 metric tons of fallout per day. The individual particles in the fallout were very small. The diameter of a particle was 0.000022 meter.

Learning the Concept

It's easy to make mistakes when computing with small numbers such as 0.000022 and large numbers such as 75,000,000. People who deal regularly with such numbers use **scientific notation**.

Scientific Notation	Numbers expressed in scientific notation are written as the product of a factor and a power of 10. The factor must be greater than or equal to 1 and less than 10.

To write a large positive or negative number in scientific notation, move the decimal point to the right of the left-most digit, and multiply this number by a power of ten.

To find the power of ten, count the number of places you moved the decimal point.

Example Write each number in scientific notation.

a. 687,000

$687,000 = 6.87 \times 10^5$ *Move the decimal point 5 places to the left. Multiply by 10^5.*

b. 50,000,000

$50,000,000 = 5 \times 10^7$ *Move the decimal point 7 places to the left. Multiply by 10^7.*

You can review negative exponents in Lesson 4-9.

To write a small positive number in scientific notation, move the decimal point to the right of the first nonzero digit, and multiply this number by a power of ten. Remember that $10^{-n} = \frac{1}{10^n}$. Count the number of places you moved the decimal point and use the negative of that number as the exponent of ten.

Example Write 0.000003576 in scientific notation.

$$0.000003576 = 3.576 \times 10^{-6}$$

Move the decimal point 6 places to the right. Multiply by 10^{-6}.

Remember that $10^{-6} = \dfrac{1}{1,000,000}$.

Example 3 Write each number in standard form.

a. 8.495×10^5

$$8.495 \times 10^5 = 8.495 \times 100,000$$

Move the decimal point 5 places to the right.

$$= 849,500 \qquad 8.49500$$

b. 3.08×10^{-4}

$$3.08 \times 10^{-4} = 3.08 \times \left(\dfrac{1}{10}\right)^4$$

Remember that $10^{-4} = \left(\dfrac{1}{10}\right)^4$.

$$= 3.08 \times \dfrac{1}{10,000}$$

$$= 3.08 \times 0.0001$$

Move the decimal point 4 places to the left.

$$= 0.000308 \qquad 00003.08$$

TECHNO TIP

Use the [EE] key to enter numbers in scientific notation in a calculator. Enter the decimal portion, press [EE], then enter the exponent.

You can compare and order numbers in scientific notation. First compare the exponents. With positive numbers, any number with a greater exponent is greater. If the exponents are the same, compare the numbers.

Example 4 The diameter of Venus is 1.208×10^4 km, the diameter of Earth is 1.276×10^4 km, and the diameter of Mars is 6.79×10^3 km. Rank these planets by diameter, from least to greatest.

Science

$1.208 \times 10^4 < 1.276 \times 10^4$
The exponent is the same; $1.208 < 1.276$.
So Venus is smaller than Earth.

$6.79 \times 10^3 < 1.208 \times 10^4 \qquad 10^3 < 10^4$
So Mars is smaller than Venus.

The rank from least to greatest is Mars, Venus, Earth.

Checking Your Understanding

Communicating Mathematics

Read and study the lesson to answer these questions.

1. In your own words, explain how to write these numbers in scientific notation.

   a. 634,930   **b.** 0.0000407

2. **Explain** the relationship between a number in standard form and the sign of the exponent when the number is written in scientific notation.

MATH JOURNAL

3. **Assess Yourself** What career are you interested in pursuing after high school? How might you use scientific notation in that career?

Guided
Practice

State whether each number is in scientific notation. Write *yes* or *no*. If *no*, explain why not.

4. 23.45×10^5

5. 5.689×10^{-3}

6. 0.23×10^8

State where the decimal point should be placed in order to express each number in scientific notation. State the power of ten by which you should multiply. Then write the number in scientific notation.

7. 8490

8. 0.000045

9. 847.9

Write each number in standard form.

10. 5.2×10^5

11. 6.1×10^4

12. 9.34×10^{-2}

13. **Physics** The speed of light is 3.00×10^5 kilometers per second. If a light-year is the distance light travels in one Earth year, how many kilometers are in one light-year? (One Earth year = 365.25 days. Remember to convert days to seconds)

Exercises: Practicing and Applying the Concept

Independent
Practice

Write each number in scientific notation.

14. 5,894,000

15. 0.0059

16. 269,000

17. 0.00015

18. 80,000,000

19. 0.0000000498

20. the possible hands in a 5-card game, 2,598,960

21. the population of Africa in 1994, 701,000,000

22. the diameter of a spider's thread, 0.001 inch

Write each number in standard form.

23. 5.6×10^{-6}

24. 1.399×10^2

25. 9.001×10^{-3}

26. 7.8×10^{11}

27. 5.985×10^{-5}

28. 4.054×10^1

29. the height of the Sears Tower, 1.454×10^3 feet

30. the diameter of a uranium atom, 3.5×10^{-8}

31. the volume of a drop of liquid, 5×10^{-5} liter

32. Write $(4 \times 10^4) + (8 \times 10^3) + (3 \times 10^2) + (9 \times 10^1) + (6 \times 10^0)$ in standard notation.

33. Write $(7 \times 10^8) + (8 \times 10^4) + (3 \times 10^0)$ in standard notation.

34. Write $(6 \times 10^0) + (4 \times 10^{-3}) + (3 \times 10^{-5})$ in standard notation.

Critical
Thinking

35. Convert the numbers in each expression to scientific notation. Then evaluate each expression. Express your answer in scientific notation and in decimal notation.

a. $\dfrac{20,000}{0.01}$

b. $\dfrac{(420,000)(0.015)}{0.025}$

c. $\dfrac{(0.078)(8.5)}{0.16\,(250,000)}$

d. $\dfrac{(520)(0.008)(0.0072)}{2.4\,(130)(50,000)}$

36. In standard form, $3.14 \times 10^{-4} = 0.000314$, and $3.14 \times 10^4 = 31,400$. What is 3.14×10^0 in standard form?

37. Science Everything around you is made up of atoms. Atoms are extremely small. An atom is about two millionths of an inch in diameter. Write this number in standard form and then in scientific notation.

Data Update For the latest volcano eruption rates, visit:

www.glencoe.com/ sec/math/prealg/ mathnet

38. Geology This graph below gives the maximum eruption rate in cubic meters per second of seven volcanoes in the last century.

Eruption Rates

Mount St. Helens, 1980 $\quad 2.0 \times 10^4$

Ngauruhoe, 1975 $\quad 2.0 \times 10^3$

Helka, 1970 $\quad 4.0 \times 10^3$

Agung, 1963 $\quad 3.0 \times 10^4$

Bezymianny, 1956 $\quad 2.0 \times 10^5$

Helka, 1947 $\quad 2.0 \times 10^4$

Santa Maria, 1902 $\quad 4.0 \times 10^4$

Source: University of Alaska

a. Which volcano had the greatest eruption rate?

b. Which had the least eruption rate?

c. Use standard form to rank the volcanoes in order from greatest to least eruption rate.

Mixed Review

39. State whether $9, -3, 1, -\frac{1}{3}, \ldots$ is a geometric sequence. If so, state the common ratio and list the next three terms. (Lesson 6-8)

40. Standardized Test Practice Brett's parents do not allow him to work more than 10 hours during any school week. Brett worked $4\frac{1}{4}$ hours on Monday and $2\frac{2}{3}$ hours on Wednesday. Which expression gives n, the number of additional hours he can work this week? (Lesson 5-7)

   A. $n \leq 3\frac{1}{12}$

   B. $n > 4\frac{1}{12}$

   C. $n \leq 6\frac{3}{4}$

   D. $n \leq 6\frac{11}{12}$

41. Solve $-3\frac{1}{5} - \left(-\frac{2}{5}\right) = a$. (Lesson 5-4)

42. Evaluate $\frac{b}{7}$, if $b = -91$. (Lesson 2-8)

For **Extra Practice,** see page 755.

43. Sports Translate *three more than the Bears' score* into an algebraic expression. (Lesson 1-3)

Vocabulary

After completing this chapter, you should be able to define each term, property, or phrase, and give an example or two of each.

Algebra
common ratio (p. 312)
compatible numbers (p. 280)
equivalent (p. 275)
geometric sequence (p. 312)
inverse property of multiplication (p. 289)
multiplicative inverses (p. 289)
reciprocal (p. 289)
repeating decimal (p. 275)
scientific notation (p. 317)
terminating decimal (p. 275)
truncate (p. 275)

Statistics
mean (p. 301)
measures of central tendency (p. 301)
median (p. 301)
mode (p. 301)

Geometry
area model (p. 284)

Problem Solving
work backward (p. 314)

Understanding and Using Vocabulary

Determine whether each statement is *true* or *false*. If the statement is false, rewrite it to make it true.

1. Two numbers whose product is 1 are multiplicative inverses, or reciprocals, of each other.

2. The mode is the number in the middle when the data are arranged in order.

3. Each term in a common ratio increases or decreases by a geometric sequence.

4. Mean is what people are usually talking about when they say average.

5. A decimal is called a terminating decimal because the division ends when the divisor is zero.

6. Equivalent numbers are rounded so that they fit together.

7. Numbers written in scientific notation are expressed as the product of a factor and a power of 10.

8. A calculator that truncates will display .66666666 when you divide 2 by 3. A calculator that rounds will display .66666667 when you divide 2 by 3.

9. It is possible to have a set of numerical data with no mean.

10. The inverse property of multiplication says that every number, including zero, has a multiplicative inverse. Zero is its own multiplicative inverse.

Skills and Concepts

Objectives and Examples

Upon completing this chapter, you should be able to:

▶ **write fractions as terminating or repeating decimals.** (Lesson 6-1)

$$\frac{2}{5} \rightarrow 5\overline{)2.0}^{\,0.4}$$

$$\frac{2}{5} = 0.4$$

$$\frac{4}{9} \rightarrow 4\;\boxed{\div}\;9\;\boxed{=}\;.44444$$

$$\frac{4}{9} = 0.\overline{4}$$

▶ **estimate products and quotients of rational numbers.** (Lesson 6-2)

$$14\frac{1}{5} \times 4\frac{8}{9} \rightarrow 14 \times 5 = 70$$

$$84.73 \div 4.930 \rightarrow 85 \div 5 = 17$$

▶ **multiply fractions.** (Lesson 6-3)

$$a = \frac{4}{3} \cdot \frac{5}{12}$$

$$a = \frac{\overset{1}{4}}{3} \cdot \frac{5}{\underset{3}{12}}$$

$$a = \frac{5}{9}$$

▶ **divide fractions using multiplicative inverses.** (Lesson 6-4)

Evaluate $c \div d$ if $c = \frac{1}{3}$ and $d = \frac{7}{10}$.

$$c \div d = \frac{1}{3} \div \frac{7}{10}$$

$$= \frac{1}{3} \cdot \frac{10}{7} \quad \textit{Multiply by the inverse.}$$

$$= \frac{10}{21}$$

Review Exercises

Use these exercises to review and prepare for the chapter test.

Write each fraction as a decimal. Use a bar to show a repeating decimal.

11. $\frac{5}{8}$ **12.** $\frac{6}{9}$ **13.** $\frac{17}{40}$

14. $\frac{2}{25}$ **15.** $\frac{4}{15}$ **16.** $\frac{7}{12}$

Estimate each product or quotient.

17. $\frac{9}{10} \cdot 21$ **18.** $8.36 \div 16.53$

19. $9 \div 1\frac{1}{5}$ **20.** 18.17×6.19

Solve each equation. Write each solution in simplest form.

21. $-\frac{5}{9} \cdot \frac{8}{25} = n$

22. $6\frac{2}{3} \cdot \frac{1}{2} = t$

23. $w = -3\left(\frac{6}{15}\right)$

24. $-7\left(\frac{8}{21}\right) = d$

25. $g = \left(\frac{4}{5}\right)^3$

26. $q = \frac{2}{5} \cdot \frac{7}{8}$

Evaluate each expression.

27. $a \div b$, if $a = \frac{1}{4}$ and $b = \frac{2}{6}$

28. $bc \div f$, if $b = -\frac{8}{15}$, $c = 3$, and $f = 1\frac{1}{2}$

29. $\frac{x}{y}$, if $x = \frac{8}{12}$ and $y = -6$

30. $\frac{gh}{8}$, if $g = 3\frac{2}{7}$ and $h = -5\frac{1}{3}$

Objectives and Examples	Review Exercises

▶ **multiply and divide decimals.** (Lesson 6-5)

Solve $w = -5.7(3.8)$.

$$\begin{array}{rl} -5.7 & \leftarrow \quad \textit{one decimal place} \\ \underline{\times\ 3.8} & \leftarrow \quad \textit{one decimal place} \\ 456 & \\ \underline{1710} & \\ -21.66 & \leftarrow \quad \textit{two decimal places} \end{array}$$

Solve each equation.

31. $4 \times 7.07 = y$ **32.** $k = -1.25 \times 12$

33. $b = 62.9 \div 1000$ **34.** $2.65 \cdot 3.46 = c$

35. $0.7 \div 2.4 = d$ **36.** $-25.9 \div 2.8 = s$

37. $n = 5.3 \cdot (-9)$ **38.** $f = 1.3 \div 100$

▶ **find the mean, median, and mode of a set of data.** (Lesson 6-6)

Find the mean, median, and mode of the test scores 90, 88, 95, 73, and 92.

mean $\quad \dfrac{90 + 88 + 95 + 73 + 92}{5} = \dfrac{438}{5}$ or 87.6

median $\quad 73 \quad 88 \quad 90 \quad 92 \quad 95$
$\qquad\qquad\qquad \uparrow$ *The median is 90.*

mode $\quad$ Since no scores occur more than once, there is no mode.

Find the mean, median, and mode for each set of data. When necessary, round to the nearest tenth.

39. 21, 25, 16, 18, 18, 26, 21, 17, 19, 25

40. 2.6, 3.1, 6.8, 4.9, 5.7, 3.4, 4.3

41. 0, 9, 14, 22, 26, 14

42. 5.4, 4.6, 6.2, 2.7, 8.0

▶ **solve equations and inequalities containing rational numbers.** (Lesson 6-7)

Solve $-4.7x < 14.1$.

$$-4.7x < 14.1$$

$$\dfrac{-4.7x}{-4.7} > \dfrac{14.1}{-4.7} \qquad \textit{Reverse the order symbol.}$$

$$x > -3$$

Solve each equation or inequality. Check your solution.

43. $8.67k = 78.03$ **44.** $6.3a \leq 52.92$

45. $-\dfrac{1}{4}r = 3$ **46.** $4.5 < -\dfrac{5}{8}f$

47. $-25 \geq 6.25n$ **48.** $3x = 55.8$

49. $8x = 42.4$ **50.** $2.6v > 19.76$

▶ **recognize and extend geometric sequences.** (Lesson 6-8)

State whether 768, 192, 48, . . . is a geometric sequence. If so, state the common ratio and list the next three terms.

$768, \quad 192, \quad 48, \ldots$
$\qquad \times\dfrac{1}{4} \quad \times\dfrac{1}{4}$

The sequence is geometric. The common ratio is $\dfrac{1}{4}$. The next three terms are $48 \times \dfrac{1}{4}$ or 12, $12 \times \dfrac{1}{4}$ or 3, and $3 \times \dfrac{1}{4}$ or $\dfrac{3}{4}$.

State whether each sequence is a geometric sequence. If so, state the common ratio and list the next three terms.

51. 6.6, 5.7, 4.9, . . .

52. 3, 9, 27, . . .

53. 0.5, 2, 3.5, 5, . . .

54. $\dfrac{1}{4}, \dfrac{1}{2}, 1, 2, \ldots$

55. 17, 13, 18, 14, 19, . . .

Objectives and Examples	Review Exercises

▶ **write numbers in standard form and scientific notation.** (Lesson 6-9)

Write 58,970,000,000 in scientific notation.

$58,970,000,000 = 5.897 \times 10^{10}$

Write 6.892×10^{-2} in standard form.

$$6.892 \times 10^{-2} = 6.892 \times \left(\frac{1}{10}\right)^2$$
$$= 6.892 \times 0.01$$
$$= 0.06892$$

Write each number in scientific notation.

56. 13,490,000   **57.** 0.00000674

58. 0.00032   **59.** 5810

Write each number in standard form.

60. 4.24×10^2

61. 5.72×10^4

62. 3.347×10^{-1}

63. 2.02×10^8

Applications and Problem Solving

64. Sports Of the medals that the United States' 1992 Summer Olympic Team earned, 0.34 were gold. $\frac{44}{112}$ of the medals earned by the Unified Team were gold. Which team had a greater portion of their medals as gold? (Lesson 6-1)

66. Population Canada is divided into ten provinces. The table at the right shows the population in thousands of each province in 1991. (Lesson 6-6)

  a. Find the mean population for the provinces.

  b. Find the median population for the provinces.

  c. Find the mode population for the provinces.

67. Astronomy The first new moon discovered by *Voyager* revolves around Uranus. Called Puck, this moon makes one orbit around Uranus every 18 hours. How many times has Puck circled Uranus in 273.6 hours? (Lesson 6-7)

65. Economics The sales tax for many cities is $5\frac{1}{2}$%. This means that for every dollar you spend, you pay $5\frac{1}{2}$¢ in tax. If you purchase an item for $24, how much will you owe in tax? (Lesson 6-3)

Province	Population (thousands)
Alberta	2546
British Columbia	3282
Manitoba	1092
New Brunswick	724
Newfoundland	568
Nova Scotia	900
Ontario	10,085
Prince Edward Island	130
Quebec	6896
Saskatchewan	989

68. Forestry The U.S. National Park System covers a total of 80,663,217.42 acres. Write this number in scientific notation. (Lesson 6-9)

A practice test for Chapter 6 is available on page 779.

Alternative Assessment

Demonstrate your knowledge by giving a clear, concise solution to each problem. Be sure to include all relevant drawings and justify your answers. You may show your solutions in more than one way or investigate beyond the requirements of the problem.

1. How much does one share of Kine stock cost in dollars?

2. How much would it cost if you bought 6 shares of Sunny stock?

3. If you spent $951 buying shares of stock in Mickeys, how many shares would you have purchased?

4. Find the mean of the stock prices listed.

Stock	Price per share
Mickey's	$39\frac{5}{8}$
Big Bob's	19
Sunny	$55\frac{3}{8}$
Kine	$92\frac{3}{4}$

5. Find the median stock price.

6. If you bought 32 shares of Big Bob's at $17\frac{3}{4}$ per share and sold it all at the price listed, how much money would you make?

7. Find a list of stock prices in a local newspaper. Choose ten shares of at least three stocks to "sell". Record the prices and the number of shares of each stock that you chose in a table. Then look at a paper from one week ago and record the price of each stock then.

   a. If you bought the shares last week and sold this week, how much did you make or lose?

   b. Find the gain or loss on each stock. Then find the mean, median, and mode of the numbers. If you were bragging about your financial accomplishments, which number would you tell? If you were complaining about the stock market which number would you tell? Explain.

► Why is scientific notation important?

► One meter is 1×10^2 centimeters. The approximate size of an atom is 1×10^{-10} centimeters. How many meters is that? Write the number of meters in scientific notation and standard form.

 Portfolio

Select some of your work from this chapter that shows how you used a calculator or a computer and place it in your portfolio.

Section One: Multiple Choice

There are ten multiple-choice questions in this section. After working each problem, write the letter of the correct answer on your paper.

1. Carl bought a bike that cost $532.16. He paid for the bike in equal payments for 12 months. Estimate: the amount of each payment was between

   A. $10 and $20.
   B. $25 and $35.
   C. $40 and $50.
   D. $55 and $65.

2. The Sanchez's added a 12-foot-by-15-foot rectangular deck to the back of their house. How much wood do they need for the railing? (The deck is attached to the house on one of the long sides.)

   A. 39 feet
   B. 27 feet
   C. 42 feet
   D. 90 feet

3. Gladys bought grocery items for the following prices: $1.39, $2.89, $0.58, and $1.19. The best estimate of the total cost is

   A. $9.
   B. $6.
   C. $5.
   D. $4.

4. If $m = 12$ and $p = 10$, what is the value of $m + mp$?

   A. 120
   B. 132
   C. 240
   D. 1440

5. If $3.2y = 80$, what is the value of y?

   A. 25
   B. 2.5
   C. 256
   D. 76.8

6. Which statement is *not* true?

   A. $\frac{1}{2} > \frac{3}{8}$
   B. $\frac{4}{5} < \frac{5}{6}$
   C. $\frac{8}{11} < \frac{7}{13}$
   D. $\frac{1}{3} < \frac{10}{13}$

7. Brent knows that to have an average of more than 85 in biology, his two exam scores must total more than 170. If his first grade was 82, which inequality could he use to find g, his second grade needed for an average greater than 85?

   A. $82 + g > 170$
   B. $82 > 170 + g$
   C. $g + 82 < 170$
   D. $\frac{82 + g}{2} > 170$

8. The mileage reading on the Garcia's car was 256.8 before they left on vacation. When the family returned, the reading was 739.4. How many miles did the Garcia family travel on their vacation?

   A. 996.2 mi
   B. 583.5 mi
   C. 483.6 mi
   D. 482.6 mi

9. Which is equivalent to $\frac{a^2}{a^4}$?

   A. a^{-2}
   B. a^2
   C. a^6
   D. a^8

10. The labor charge of repairing a car is $42.50 per hour. If it takes 2.5 hours to repair the car, what will be the charge for labor?

   A. $812.50
   B. $106.25
   C. $13.00
   D. none of these

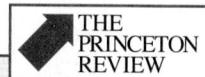

Section Two: Free Response

This section contains eight questions for which you will provide short answers. Write your answer on your paper.

11. Solve $\dfrac{a}{-5} < 1$ and graph on a number line.

12. Express 0.36 as a fraction.

13. Find the GCF of $60x^2y$, $24x^3y^2$, and $84xy^4$.

14. Write the first five terms of a geometric sequence that has a common ratio of -2.

15. Carlos used $2\dfrac{5}{8}$ yards of material to make a shirt. He began with $3\dfrac{1}{4}$ yards of material. How much material did he have left?

16. The difference between the average high temperature in January and July in New York City is 37°F. The average high in July is 85°F. What is the average high in January?

17. The dimensions of Adita's yard are 25 feet long by 40 feet wide. Express the ratio of the length to the width in simplest form.

18. Ms. Clinton calculated that it would take $2\dfrac{5}{8}$ gallons of paint to paint one classroom at her school. She is going to paint 5 classrooms. How many gallons of paint will she need?

Test-Taking Tip You may be able to eliminate all or most of the answer choices by estimating. Also, look to see which answer choices are not reasonable for the information given in the problem.

When questions involve variables, check your solutions by substituting values for the variables.

Section Three: Open-Ended

This section contains three open-ended problems. Demonstrate your knowledge by giving a clear, concise solution to each problem. Your score on these problems will depend on how well you do the following.
- Explain your reasoning
- Show your understanding of the mathematics in an organized manner.
- Use charts, graphs, and diagrams in your explanation.
- Show the solution in more than one way or relate it to other situations.
- Investigate beyond the requirements of the problem.

19. Give an example of either inductive or deductive reasoning. Explain why your example is either inductive or deductive reasoning.

20. The high temperatures for a week in April were 56°F, 58°F, 60°F, 63°F, 58°F, 62°F, and 70°F. What were the mean and the median temperatures? Explain the difference between the two temperatures.

21. Give an example of a decimal that neither terminates nor repeats.

Test Practice For additional test practice questions, visit:
www.glencoe.com/sec/math/prealg/ mathnet

CHAPTER 7

Solving Equations and Inequalities

What You'll Learn

In this chapter, you will:

- solve problems by working backward,

- write and solve multi-step equations and inequalities,

- find the circumferences of circles, and

- convert measurements within the metric system.

Why It's Important

Zoos The San Diego Zoo is known throughout the world for its leadership in animal research. But did you know that it is not one of the largest zoos in the United States? The table shows the number of acres in the top ten largest zoos. The San Diego Zoo has some growing to do to top the list. If its size were tripled, it would still be 104 acres smaller than the third largest zoo!

You can use the information about the sizes of the zoos to find the number of acres in the San Diego Zoo. First, you would choose a variable to represent the number of acres in the zoo. Then you would write and solve a two-step equation to find the San Diego Zoo size. *You will learn to write and solve two-step equations in Lesson 7-3.*

Major U.S. Public Zoological Parks

Zoo	Acres
San Diego Wild Animal Park	2200
Minnesota Zoo (Apple Valley)	500
Columbus Zoo and Aquarium (Powell, OH)	404
Miami Metrozoo	300
Bronx Zoo (N.Y.C.)	265
Brookfield Zoo (Chicago area)	216
Milwaukee County Zoo	194
Riverbanks Zoo & Garden (Columbia, SC)	170
Cleveland Metroparks Zoo	165
National Zoo (Washington, D.C.)	163

Source: *World Almanac*

Focus on...

PREREQUISITE SKILLS

To be successful in this chapter, you'll need to understand these concepts and be able to apply them. Refer to the lesson in parentheses if you need more review before beginning the chapter.

▶ *Solve verbal problems by translating them into equations and inequalities.* (Lesson 3-8)

Define a variable and translate each sentence into an equation. Then solve.

1. The amount of money in a savings account increased by $100 is $450. What was the amount of money in the account?
2. The product of 22 and a number is -44. What is the number?
3. The quotient of a number and -16 is less than 4. What is the number?
4. A car was traveling at least 20 miles per hour over the speed limit of 70 miles per hour. What was the speed of the car?

▶ *Solve inequalities by using the addition and subtraction properties of inequalities.* (Lesson 3-6)

Solve each inequality. Graph the solution on a number line.
5. $c - 3 > 5$ **6.** $a + (-8) \le -1$ **7.** $13 < y + 4$ **8.** $-4 \ge x - (-3)$

▶ *Solve inequalities by using the multiplication and division properties of inequalities.* (Lesson 3-7)

Solve each inequality. Graph the solution on a number line.
9. $4b < -20$ **10.** $\dfrac{x}{-6} > -1$ **11.** $15r \le -30$ **12.** $-z \le 1$

Focus on...

READING SKILLS

In Chapter 3, you learned to write and solve one-step equations and inequalities. In this chapter, you learn to write and solve more complicated equations and inequalities. Recall that addition and subtraction are **inverse operations,** because they "undo" each other. Multiplication and division are also inverse operations. Solving more complicated equations and inequalities is easy if you work backward and use inverse operations.

Problem-Solving Strategy: Work Backward

Technology Activity

A certain computer software simulates a factory assembly line. Machines rotate, punch holes, and paint a vertical stripe on geometric-shaped objects so that they match a given design.

Your Turn Copy and complete the chart.

Operation	Begin with a designed square.	Rotate 45° clockwise.	Remove thin line.	Rotate 45° clockwise.	Remove thick line.
Result					

TALK ABOUT IT

a. What process is involved in completing the chart above?

b. Look at the two squares at the right. What would you do to the first square to get the square on the far right as the end result from the software?

Learning the Concept

In most problems, a set of conditions or facts is given and an end result must be found. However, the software shows the result and asks for something that happened earlier.

The strategy of **working backward** can be used to solve problems like this. To use this strategy, start with the end result and *undo* each step.

Marita put $10 of her paycheck in savings. Then she spent one-half of what was left on clothes. She paid $26 for a haircut and then spent one-half of what was left on a concert ticket. When she got home, she had $17 left. What was the amount of Marita's paycheck?

Explore   You know all the amounts Marita spent and what she had when she got home. You want to find the amount of her paycheck.

Plan   Since this problem gives the end result and asks for something that happened earlier, start with the result and work backward. *Undo* each step.

Solve   When Marita got home, she had \$17. ⟶ 17

Undo the half that she spent on a concert ticket. ⟶ × 2
                                                       ——
                                                       34

Undo the \$26 she spent on a haircut. ⟶ + 26
                                          ——
                                          60

Undo the half she spent for clothes. ⟶ × 2
                                          ——
                                          120

Undo the \$10 she put in savings. ⟶ + 10
                                      ——
                                      130

The amount of Marita's paycheck was \$130.

Examine   Assume that Marita started with \$130. After putting \$10 in savings, she had \$120. She spent half of \$120, so she had 120 ÷ 2 or \$60. Then she spent \$26, so she had 60 − 26 or \$34. Finally, she spent half of \$34, so she had 34 ÷ 2 or \$17. In the end, Marita would have \$17, so the answer is correct.

Checking Your Understanding

Communicating Mathematics

Read and study the lesson to answer these questions.

1. **Describe** the circumstances under which you could use the strategy of working backward.

2. **Explain** how to solve a problem by working backward.

3. **Write** a problem that can be solved by working backward. Trade problems with a classmate and solve his or her problem.

4. Create a design that could be made using the software described at the beginning of the lesson. Record the procedure as you make it. Trade designs with a friend and reproduce your friend's design.

Guided Practice

Solve by working backward.

5. Mr. Fuentes uses half of a can of evaporated milk to make pumpkin pie. The can and the milk that is left weigh 9 ounces. If the can and the milk weighed 15 ounces before it was opened, how much does the can weigh?

6. A certain number is divided by 5, and then 1 is subtracted from the result. The final answer is 32. Find the number.

Exercises: Practicing and Applying the Concept

Independent Practice

Solve. Use any strategy.

7. Mr. and Mrs. Gentry each own an equal number of shares of a stock. Mr. Gentry sells one-third of his shares for \$2700. What was the total value of Mr. and Mrs. Gentry's stock before the sale?

8. A certain number is multiplied by 3 and then 5 is added to the result. The final answer is 41. Find the number.

9. Chris has $1.50. Half of the money he had when he left home this morning was spent on lunch. He lent Bill a dollar after lunch. How much money did Chris start with?

10. Forty-four students took the bus to the St. Louis Science Center. Each student paid $1.25. How much did the driver collect?

11. On Monday, Fumiko told a joke to 3 of her friends. On Tuesday, each of those friends told the joke to 3 other friends. On Wednesday, each person who heard the joke on Tuesday told 3 other people. If this pattern continues, how many people will hear the joke on Saturday?

12. An ice sculpture is melting at the rate of half its weight every hour. After 8 hours, it weighs $\frac{5}{16}$ of a pound. How much did it weigh in the beginning?

13. In Canada and other countries, dates are abbreviated as "date.month.year." For example, September 19, 1991 would be abbreviated as 19.9.91 and September 5, 1995 would be 5.9.95. Dates like these, where the pattern of numbers is the same when read forward and backward, are called *palindromic dates*. What are the two palindromic dates in the 1900s that are closest together?

Critical Thinking

14. At the end of the second round of a game, Bart's and Carlos' scores were double the score each had at the end of the first round. The total number of points they gained were subtracted from Al's first round score. After the third round, Al's and Carlos' scores were double their second round scores. The total number of points they gained were subtracted from Bart's second round score. After the fourth round, Al's and Bart's third round scores doubled. The total number of points they gained were subtracted from Carlos' third round score. Now each of them has 8 points. How many points did each of them have after the first round?

Mixed Review

15. Write 6.789×10^{-7} in standard form. (Lesson 6-9)

16. **Entertainment** The table at the right shows the attendance in millions of people of the largest fairs in the United States. How many more people attended the New Mexico State Fair than attended the Los Angeles County Fair? (Lesson 5-3)

For **Extra Practice**, see page 756.

17. Name the sets of numbers to which $\frac{42}{7}$ belongs—whole numbers, integers, or rationals. (Lesson 5-1)

18. Evaluate $ab \div (-7)$ if $a = 49$ and $b = -2$. (Lesson 2-8)

19. **Standardized Test Practice** Find the value of $[4(3 + 8) \div (4 - 2)] + 9$. (Lesson 1-2)

    A. 21   **B.** 25   **C.** 30   **D.** 31

Fair	Attendance (millions)
State Fair of Texas, Dallas	3.15
State Fair of Oklahoma, Oklahoma City	1.79
	1.68
New Mexico State Fair, Albuquerque	1.60
Minnesota State Fair, St. Paul	1.57
Houston (Texas) Livestock Show	1.42
Western Washington Fair, Puyallup	1.40
Los Angeles County Fair, Pomona	1.11
De Mar (California) Fair	1.08
Colorado State Fair, Pueblo	1.03
Tulsa (Oklahoma) State Fair	

Source: *Top Ten of Everything*

MATERIALS

cups

counters

equation mat

In this activity, you will extend what you know about solving one-step equations to model and solve two-step equations.

Recall from Chapter 3 that you can use cups and counters to model one-step equations. You can model two-step equations also. A two-step equation contains two operations. For example, the equation $2x + 3 = 7$ involves multiplication and addition.

Activity

Model and solve the equation $2x + 3 = 7$.

▶ Place 2 cups and 3 positive counters on the left side of the mat. Then place 7 positive counters on the right side of the mat.

$2x + 3 = 7$

▶ To determine how many counters are in each cup, the goal is to isolate the cups on one side of the mat. Since the two sides of the mat represent equal quantities, you can remove an equal number of counters from each side without changing the value of the equation.

$2x + 3 - 3 = 7 - 3$ $2x = 4$

▶ Now the equation is $2x = 4$. Match an equal number of counters with each cup. Therefore, each cup must contain 2 counters, so $x = 2$.

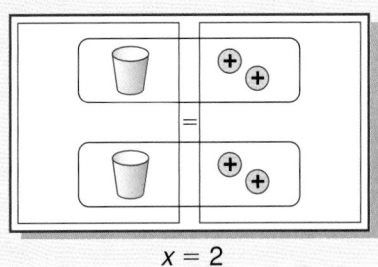

$x = 2$

Your Turn

Model each equation and solve.

1. $2x + 2 = -4$ **2.** $3x + 3 = 12$ **3.** $2x - 2 = 6$

4. $2x - 2 = -4$ **5.** $4 + 2x = 8$ **6.** $3x - (-2) = -4$

7. Explain how this activity models the work-backward strategy presented in Lesson 7-1.

8. Explain how you can use models to solve $-2x + 1 = -5$.

Solving Two-Step Equations

You'll Learn

- To solve equations that involve more than one operation.

It's Important

Using two-step equations can help you to change temperatures from Fahrenheit to Celsius.

Application: Consumerism

"At Carpet World, we will not be undersold! Get 18 square yards of the best stain-resistant carpet installed for just $305.89! That *includes* the $54 installation charge. Hurry in! At this price, the carpet will not last!"

The Jackson family is comparing prices at different carpet stores. Most carpet stores price carpet by the square yard. What price should the Jacksons use to compare the price of one square yard of carpet at Carpet World to the prices at other stores? *You will solve this problem in Exercise 36.*

Learning the Concept

To solve this problem, you will use the equation $54 + 18p = 305.89$. Some equations, like this one, contain more than one operation. To solve an equation with more than one operation, use the work-backward strategy and undo each operation.

Example **Solve each equation. Check your solution.**

a. $\frac{c}{4} - 19 = 17$

$\frac{c}{4} - 19 + 19 = 17 + 19$ *Add to undo the subtraction.*

$\frac{c}{4} = 36$

$4 \cdot \frac{c}{4} = 4 \cdot 36$ *Multiply to undo the division.*

$c = 144$

Check: $\frac{c}{4} - 19 = 17$

$\frac{144}{4} - 19 \overset{?}{=} 17$ *Replace c with 144.*

$36 - 19 \overset{?}{=} 17$ *Do the division first.*

$17 = 17$ ✔   The solution is 144.

b. $-3n + 8 = -7$

$$-3n + 8 - 8 = -7 - 8 \quad \textit{Subtract 8 from each side.}$$
$$-3n = -15$$
$$\frac{-3n}{-3} = \frac{-15}{-3} \quad \textit{Divide each side by } -3.$$
$$n = 5$$

Check: $-3n + 8 = -7$
$$-3(5) + 8 \stackrel{?}{=} -7$$
$$-15 + 8 \stackrel{?}{=} -7$$
$$-7 = -7 \quad \checkmark \quad \text{The solution is 5.}$$

In Chapter 3, you learned that a formula is an algebraic expression that can be used to show the relationship among certain quantities. Some formulas involve more than one operation. You can use the work-backward strategy to solve problems involving formulas.

Example ❷

INTEGRATION

Geometry

The area A of a trapezoid can be found by multiplying the height h and one-half the sum of the lengths of the bases b_1 and b_2. The formula is $A = \frac{1}{2} \cdot h(b_1 + b_2)$. The area of the trapezoid at the right is 32 square inches. The trapezoid is 4 inches high, and the length of one of the bases is 9 inches. What is the length of the other base?

base 1 (b_1)

height (h)

base 2 (b_2)

$$A = \frac{1}{2} \cdot h(b_1 + b_2)$$
$$32 = \frac{1}{2} \cdot 4(b_1 + 9) \quad \textit{Replace A with 32, h with 4, and } b_2 \textit{ with 9.}$$
$$32 = 2(b_1 + 9)$$
$$\frac{32}{2} = \frac{2(b_1 + 9)}{2} \quad \textit{Because } 2(b_1 + 9) \textit{ means } 2 \cdot (b_1 + 9),$$
$$16 = b_1 + 9 \quad \textit{undo the multiplication first.}$$
$$16 - 9 = b_1 + 9 - 9 \quad \textit{Subtract to undo the addition.}$$
$$7 = b_1 \quad \textit{Check the solution.}$$

The second base is 7 inches long.

Checking Your Understanding

Communicating Mathematics

Read and study the lesson to answer these questions.

1. How is the order of operations used in solving two-step equations?
2. **You Decide** Kelsey says that to solve the equation $3x + 6 = 24$, the first step should be to subtract 6 from each side. Dana says the first step should be to divide each term by 3.
   a. Which method would you use and why?
   b. Whose method would be easier to use to solve the equation $2x + 3 = 5$? Explain.
3. **Write** a problem that could be solved using the equation $3x - 5 = 15$.
4. This diagram models the equation $4 + 2x = 8$.
   a. What is the solution of $4 + 2x = 8$?
   b. Model the equation $2x - 3 = 7$. Then solve the equation.

MATERIALS

🗏 cups
🔘 counters
⊟ equation mat

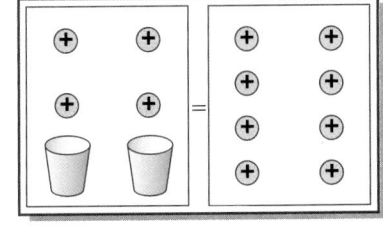

Solve each equation. Check your solutions.

5. $2n - 5 = 21$ **6.** $3 + \dfrac{t}{2} = 35$ **7.** $7.5r + 2 = -28$

8. $3(x + 5) = 9$ **9.** $-2.5(a + 2a) = 22.5$ **10.** $\dfrac{a + 5}{2} = 10.5$

11. Postage Judi mailed some photographs to her friend Misae. She paid $1.43 to send them first-class. The Post Office charges 33 cents for the first ounce and 22 cents for each additional ounce. Use the equation $33 + 22w = 143$ to find how much Judi's package weighs.

FOR ALL YOUR
PHOTO AND VIDEO NEEDS
Enlargements Slide Developing Cameras Video Albums

Exercises: Practicing and Applying the Concept

Solve each equation. Check your solution.

12. $85 = 4d + 5$ **13.** $2r - 7 = 1$ **14.** $4 - 2b = -8$

15. $-8 - t = -25$ **16.** $-4y + 3 = 19$ **17.** $1.8 = 0.6 - y$

18. $19 = 7 + \dfrac{b}{7}$ **19.** $-12 + \dfrac{j}{4} = 9$ **20.** $-3 = -31 + \dfrac{c}{6}$

21. $8 = \dfrac{h}{-3} + 19$ **22.** $\dfrac{-4x}{3} = 24$ **23.** $13 + \dfrac{-p}{3} = -4$

24. $\dfrac{3}{4}n - 3 = 9$ **25.** $\dfrac{y}{3} + 6 = -45$ **26.** $\dfrac{c}{-4} - 8 = -42$

27. $\dfrac{n - 10}{5} = 2.5$ **28.** $16 = \dfrac{-8 + s}{-7}$ **29.** $\dfrac{-d - (-5)}{7} = 14$

30. $\dfrac{6 + c}{-13} = -3$ **31.** $\dfrac{n}{2} + (-3) = 5$ **32.** $4.7 = 1.2 - 7m$

33. Make Up a Problem Write a two-step equation using the numbers 2, 3, and 6, in which the solution is $-\dfrac{1}{6}$.

34. Medicine Dr. Bell recommended that Marcie take eight tablets on the first day and then 4 tablets each day until the prescription was used. The prescription contained 28 tablets. Use the equation $8 + 4d = 28$ to find how many days Marcie will be taking pills after the first day.

35. Chemistry Chemical equations are like algebraic equations; they must be balanced on either side of the arrow. The symbol $3H_2O$ represents 3 molecules made up of 2 hydrogen and 1 oxygen atoms. Find the numbers that belong in the boxes in order for the equation $\Box P + \Box O_2 \rightarrow P_4O_{10}$ to be balanced.

36. Consumerism Refer to the application at the beginning of the lesson. How much is the carpet per square yard?

37. Academics Swedish scientist Alfred Nobel left a major part of his fortune to establish annual prizes to be awarded to those who have achieved the greatest common good in the fields of physics, chemistry, physiology or medicine, literature, peace, and economic science. People from the United States have won Nobel Prizes 213 times. That is 33 more than twice as many as the people from the United Kingdom, which is second in producing prize winners. How many people from the United Kingdom have won Nobel Prizes?

In 1950, Ralph Johnson Bunche became the first African-American to win the Nobel Peace Prize. Mr. Bunche worked at the U.S. State Department and the United Nations.

Mixed Review

38. **Standardized Test Practice** Before Shantal left on her trip, she decided to plan carefully how much money she should set aside to cover the cost of the vacation. First, the plane ticket would cost $156. Shantal decided she would use three-fourths of the remaining money in her budget for a hotel, rental car, and meals. She added $30 for souvenirs and she wanted to have $100 left over for emergencies. How much money should Shantal start with in her vacation budget? (Lesson 7-1)

    A. $676   **B.** $553   **C.** $403   **D.** $286

39. Replace ● with $<$, $>$, or $=$ to make 16.43 ● $16\frac{3}{7}$ a true sentence. (Lesson 6-1)

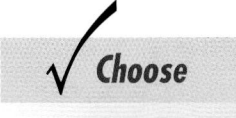

✓ Choose

Estimation
Mental Math
Calculator
Paper and Pencil

40. Solve $c - \frac{3}{5} > 14\frac{2}{9}$. (Lesson 5-7)

41. Write $y \cdot y \cdot y$ as a product using exponents. (Lesson 4-2)

42. Simplify $|18| - |-4|$. (Lesson 2-1)

43. Is $x - (-17) > 29$ *true*, *false*, or *open* if $x = 11$? (Lesson 1-9)

44. **Food** A survey found that the average American family spent $1664 eating out in 1994. Based on this finding, about how much did the average family spend on eating out in a month? (Lesson 1-1)

    a. Which method of computation do you think is most appropriate? Justify your choice.

    b. Solve the problem using the four-step plan. Be sure to examine your solution.

For **Extra Practice**, see page 756.

WORKING ON THE
Investigation

Refer to the Investigation on pages 270–271.

Look at your design of the loading platform that you placed in your Investigation Folder at the beginning of Chapter 6.

Determine how long it would take to load a coaster train by doing a whole-class experiment.

- Simulate the unloading and loading of different numbers of passengers in different numbers and sizes of cars. For example, determine how long it would take to unload and load 24 passengers in six cars or 20 passengers in ten cars. Record the time it takes from when the train stops to the time it starts moving again.

- Using the total ride time you calculated at the end of Lesson 6-4, determine how many trains you can safely operate on the track at the same time using your loading time and total ride time.

Add the results of your work to your Investigation Folder.

Writing Two-Step Equations

You'll Learn

- To solve verbal problems by writing and solving equations.

It's Important

Many real-world situations can be modeled using two-step equations.

Application: Baseball

"It has been an interesting year for the Red Sox. Here at the All-Star Game, they stand at 10 games above 0.500. So 68 games into the season, they are atop the eastern division of the American League."

How many games had the Red Sox won before the All-Star Game?

Learning the Concept

If a baseball team is at 0.500, that means that they have won $\frac{500}{1000}$ or $\frac{1}{2}$ of their games. If a team wins more than half of their games, then their status is given as a number of games "over 0.500." At 10 over 0.500, the Red Sox have won 10 more games than they have lost. Let w represent the number of games the Red Sox have won. Then they have lost $w - 10$ games. Write an equation.

68 equals the number of wins plus the number of losses

$$68 \quad = \quad w \quad + \quad (w - 10)$$

Work backward to solve the equation.

$$68 = w + (w - 10)$$
$$68 = 2w - 10 \qquad \textit{Simplify.}$$
$$68 + 10 = 2w - 10 + 10 \qquad \textit{Add 10 to each side.}$$
$$78 = 2w$$
$$\frac{78}{2} = \frac{2w}{2} \qquad \textit{Divide each side by 2.}$$
$$39 = w$$

The Boston Red Sox had won 39 games before the All-Star Game.

Example **Five more than twice some number is 19. Find the number.**

First define a variable. Let n represent the number.

five more than twice some number is 19

$$5 \quad + \quad 2n \quad = 19$$
$$5 - 5 + 2n = 19 - 5 \qquad \textit{Subtract 5 from each side.}$$
$$2n = 14$$
$$\frac{2n}{2} = \frac{14}{2} \qquad \textit{Divide each side by 2.}$$
$$n = 7 \qquad \text{The number is 7.}$$

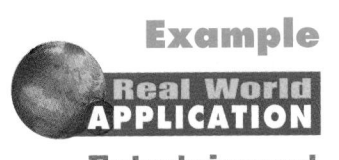
"Play it Sam." is a well known line from the classic film *Casablanca.* Rick Blaine, played by Humphrey Bogart, spoke this line to Sam, played by Dooley Wilson. Humphrey Bogart earned $36,667 for his role in the movie. This is $1667 more than ten times the salary Dooley Wilson earned. How much did Mr. Wilson earn for *Casablanca?*

Explore You know Mr. Bogart's salary. You are looking for Mr. Wilson's salary.

Plan Let *w* represent Mr. Wilson's salary. Write an equation.

Solve 36,667 is 1667 more than ten times Mr. Wilson's salary

$$36,667 = 1667 \quad + \quad 10w$$
$$36,667 - 1667 = 1667 - 1667 + 10w \quad \textit{Subtract 1667 from each side.}$$
$$35,000 = 10w$$
$$\frac{35,000}{10} = \frac{10w}{10} \quad\quad \textit{Divide each side by 10.}$$
$$3500 = w$$

Examine Since $10(3500) + 1667 = 36,667$, the answer is reasonable. Mr. Wilson's salary for *Casablanca* was $3500.

Checking Your Understanding

Communicating Mathematics

Read and study the lesson to answer these questions.

1. **List** the steps you would use to solve a verbal problem.

2. **a. Pick** any number. Double it and then add 30. Divide the sum by 2. Subtract the original number. What do you get?

   b. Repeat the exercise with two other numbers.

   c. Write an algebraic equation to explain your answers.

Guided Practice

Define a variable and write an equation for each situation. Then solve.

3. If 17 is decreased by twice a number, the result is 5. Find the number.

4. Three times a number plus twice the number plus one is 6. What is the number?

5. **Pets** According to a survey by the Pet Food Institute, about 2682 thousand dogs in the United States can "speak." This is 354 thousand fewer than four times the number of dogs that can "sing."

   a. Let *d* represent the number of dogs that can "sing." Write an equation for this situation.

   b. How many dogs can "sing"?

Exercises: Practicing and Applying the Concept

Independent Practice

Match each sentence with the equation it represents.

6. Four plus three times a number is 18.

7. Eighteen equals seven plus twice some number.

8. Four times a number minus 15 is 92.

9. Ten minus three times a number is forty-five.

A.	$4 + 3x = 18$
B.	$4n - 15 = 92$
C.	$18 = 7 + 2n$
D.	$18 = 7n + 2$
E.	$10 - 3x = 45$
F.	$15 - 4n = 92$
G.	$10x - 3 = 45$

Define a variable and write an equation for each situation. Then solve.

10. Three times a number less four is 17. What is the number?

11. The sum of a number and 6, divided by 7 is 5. Find the number.

12. Twenty more than twice a number is -30. What is the number?

13. The quotient of a number and -4, less 8 is -42. What is the number?

14. The difference between twice a number and 9 is 16. Find the number.

15. The product of 5 and a number is 10 more than 145. Find the number.

16. Write a real-life problem that could be solved using $3x + 2 = 17$.

17. **Geometry** The perimeter of the triangle at the right is 27 yards. What are the lengths of the sides of the triangle?

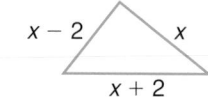

Critical Thinking

18. Numbers that are in order and differ by 1, such as 2, 3, 4, are *consecutive numbers*. If you begin with an even integer and count by two, you are counting *consecutive even integers*. Write an expression for the sum of two consecutive even integers where $2x$ is the lesser integer.

Applications and Problem Solving

19. **Fencing** Wanda used 130 feet of fence to enclose a rectangular flower garden. She also used the 50-foot wall of her house as one side of the garden. What was the width of the garden?

   a. Write an equation that represents this situation.

   b. Solve the equation to find the width of the garden.

20. **Running** In cross-country, a team's score is the sum of the place numbers of the first five finishers on the team. The co-captains finished first and fourth in a meet. The next three finishers on the team placed in consecutive order. The team score was 35.

   a. Write an equation that represents this situation.

   b. In what places did the other three members finish?

For **Extra Practice**, see page 756.

Mixed Review

21. Solve $8 + \frac{2}{5}n = 28$. (Lesson 7-2)

22. Estimate $188.76 + 14.31 - 106.5$. (Lesson 5-2)

23. **Standardized Test Practice** Solve $y - 19 = 53$. (Lesson 3-2)

   A. 19   **B.** 34   **C.** 72   **D.** 1007

24. **Money** What is the least positive number of coins that is *impossible* to give as change for a dollar? (Lesson 3-1)

From the → FUNNY PAPERS

1. Explain why the comic is funny.

2. Use the information in the comic to write an equation. Let *y* represent the number of years ago.

3. What is the value of *y*?

AN INTEGRAL NUMBER OF YEARS AGO, THE INTEGER BEING SEVEN GREATER THAN FOUR TIMES THE LEAST COMMON MULTIPLE OF 4 AND 5, OUR FATHERS...

7-4

Integration: Geometry
Circles and Circumference

What You'll Learn

■ To find the circumference of a circle.

Why It's Important

You can use circumference to find how far a bicyclist travels.

MATERIALS: 4 circular objects ∿ string ✏ metric ruler

Hands-on Activity

A **circle** is the set of all points in a plane that are the same distance from a given point in the plane. The given point is called the **center**. The distance from the center to any point on the circle is called the **radius (r)**. The distance across the circle through its center is its **diameter (d)**. The **circumference (C)** of a circle is the the distance around the circle.

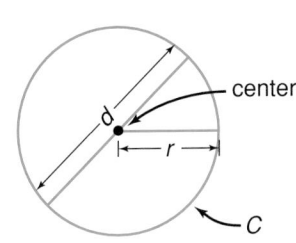

You can measure the circumference of a circle with a metric ruler.

THINK ABOUT IT

What is the relationship between the diameter and the radius of a circle?

▶ Use a metric ruler to measure the diameter of a circular object. Record your findings in a table like the one below.

▶ Wrap a string around the circular object once. Mark the string where it meets itself.

▶ Lay the string out straight and measure its length with your ruler. Record your finding. This is the circumference of the circular object.

▶ Use a calculator to divide the circumference by the diameter. Round to the nearest hundredth and record your finding.

▶ Repeat this activity with three other circular objects of various sizes.

	Object	Diameter	Circumference	Circumference / Diameter
#1				
#2				
#3				
#4				

TALK ABOUT IT

a. What is the mean of the quotients in the last column?

b. Press the $\boxed{\pi}$ key on your calculator. How does your mean compare to this value?

Learning the Concept

The relationship you discovered in the activity above is true for all circles. The circumference of a circle is always 3.1415926… times the diameter. The Greek letter **π (pi)** stands for this number. Although π is not a rational number, the rational numbers 3.14 and $\frac{22}{7}$ are two generally accepted approximations for π.

<table>
<tr><td>Circumference</td><td>In words: The circumference of a circle is equal to its diameter times π, or 2 times its radius times π.
In symbols: $C = \pi d$ or $C = 2\pi r$</td></tr>
</table>

Example **Find the circumference of each circle described below.**

a. The diameter is 8.25 cm. Use a calculator.

$C = \pi d$

$C = \pi \cdot 8.25$

| π | $\times$ | 8.25 | $=$ | 25.91813939 |

The circumference is about 25.9 cm.

b. Use $\frac{22}{7}$ for π.

$C = 2\pi r$

$C \approx \frac{2}{1} \cdot \frac{\overset{}{22}}{\underset{1}{7}} \cdot \frac{\overset{4}{28}}{3}$

$C \approx \frac{176}{3}$ or $58\frac{2}{3}$

The circumference is about $58\frac{2}{3}$ in.

($9\frac{1}{3}$ in.)

Connection to Algebra

Sometimes you'll know the circumference of a circle and need to find its radius or diameter. In this case, you can write an equation and solve it.

Example 2 **Real World APPLICATION** **Geography**

Earth's circumference is approximately 25,000 miles. If you could dig a tunnel to the center of the Earth, how long would the tunnel be?

Explore You know the circumference of Earth. You want to know the distance to the center, or the radius of Earth.

Plan Replace C with 25,000 and solve $C = 2\pi r$ for r.

Solve $25{,}000 = 2\pi r$

$\dfrac{25{,}000}{2\pi} = \dfrac{2\pi r}{2\pi}$ *Divide each side by 2π.*

$\dfrac{25{,}000}{2\pi} = r$ **Estimate:** $24{,}000 \div 6 = 4000$

| 25,000 | $\div$ | (| 2 | $\times$ | π |) | $=$ | 3978.8736 |

The distance to the center of Earth is about 3979 miles.

Examine The radius is about 4000 and π is about 3. Estimate the circumference by finding their product.

$2 \cdot 3 \cdot 4000 = 24{,}000$

The answer is reasonable.

Checking Your Understanding

Communicating Mathematics

Read and study the lesson to answer these questions.

1. **Explain** how to find the radius of a circle if you know its diameter.

2. **Name** two rational numbers that are approximations for π.

3. **Calculate** the circumference of a circle with a radius of 6 inches using the pi key on a calculator and each of the rational approximations.

4. A *chord* of a circle is a segment whose endpoints both lie on the circle. What is the longest chord you can draw?

5. Write a note to someone at home that explains how to estimate the circumference of a circle.

Guided Practice

Find the circumference of each circle described below.

6.
3 ft

7.
1.8 cm

8. The diameter is $5\frac{1}{4}$ inches.

9. The radius is 1.75 meters.

10. **Recreation**   A 10-speed bicycle tire has a diameter of 27 inches. Find the distance the bicycle will travel in 10 rotations of the tire.

Exercises: Practicing and Applying the Concept

Independent Practice

Find the circumference of each circle.

11.
4 cm

12.
7 m

13.
$5\frac{1}{2}$ ft

14.
9 in.

15.
14 mm

16.
34 ft

✓ **Choose**

Estimation
Mental Math
Calculator
Paper and Pencil

17. The diameter is 18.8 m.

18. The radius is 0.5 meters.

19. The radius is 1.3 yd.

20. The diameter is $13\frac{1}{2}$ in.

21. The diameter is $2\frac{1}{3}$ ft.

22. The radius is $4\frac{1}{2}$ km.

Match each circle described in the column on the left with its corresponding measurement in the column on the right.

23. $C = 628$ cm

24. $r = 30$ cm

25. $2r = 28$ cm

26. $C = 47.728$ cm

A. $d = 15.2$ cm

B. $C = 87.92$ cm

C. $r = 100$ cm

D. $C = 188.4$ cm

27. The circumference of a circle is 4.082 meters. Find its radius.

28. The circumference of a circle is π units. What is the radius of this circle?

29. The radius of a circle is $\frac{7}{22}$ units. Find its circumference.

30. A *central angle* of a circle is an angle whose vertex is the center of a circle. A 90° angle intersects an arc of the circle that has a length that is $\frac{1}{4}$ the circumference. Find the length of the arc intersected by a 145° central angle of a circle that has a diameter of 8 feet.

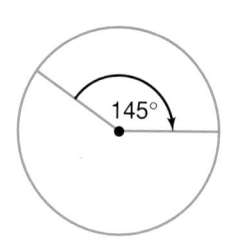

31. A central angle separates a circle into a *major arc* and a *minor arc*. (A 90° central angle intersects a minor arc of 90° and a major arc of 270°.) An *inscribed angle* has its vertex on the circle and sides that contain chords. (The measure of an inscribed angle equals one-half the measure of its intercepted arc.) Find the measures of the following angles and arcs.

  a. Minor arc *PQ*

  b. Angle 1

  c. Major arc *QP*

  d. Angle 2

  e. Arc *TR*

  f. Arc *RQ*

  g. Arc *TS*

  h. Arc *SR*

**Applications and
Problem Solving**

32. Entertainment The first Ferris wheel, designed by George Ferris, was built in 1893 in Paris, France. The diameter of that Ferris wheel was 76 meters. What distance would you travel in one revolution if you rode on the first Ferris wheel?

33. Geometry Each side of a square is 7 cm long. A circle has a diameter that is 7 cm long.

  a. Which is longer, the perimeter of the square or the circumference of the circle?

  b. Justify your answer to part a without doing any computation.

34. Research Mathematicians have worked for centuries to find an accurate value for π. Investigate the methods that have been used. Who was the first to calculate a value of π? How many decimal places are known now?

35. Standardized Test Practice The Louvre Museum, home to the *Mona Lisa*, is one of the most-visited museums in the world. The National Gallery of Art in Washington, D.C. is the most-visited art museum in the United States with 7.5 million visitors each year. This is 0.1 million more than twice what the second most visited museum, the Metropolitan Museum of Art, in New York, receives. How many visitors go to the Metropolitan Museum each year? (Lesson 7-3)

  A. 0.37 million   **B.** 3.7 million   **C.** 37 million   **D.** 370 million

36. Solve $\frac{x}{-1.7} \geq 6.8$. (Lesson 6-7)

37. Factor $220pq^2$ completely. (Lesson 4-4)

38. Solve $-4 + m \geq -3$ and graph the solution. (Lesson 3-6)

39. Replace ● with $<$, $>$, or $=$ to make $|-13|$ ● $|7|$ a true sentence. (Lesson 2-3)

40. Simplify $18z + 7(2 + 3z)$. (Lesson 1-5)

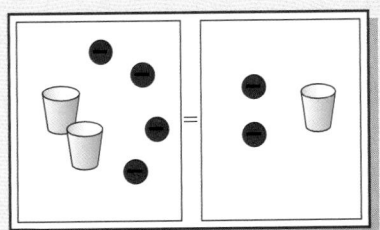

HANDS-ON ACTIVITY

7-5A Equations with Variables on Each Side

A Preview of Lesson **7-5**

In this activity, you will use cups and counters to model and solve equations with variables on each side.

Activity

MATERIALS

- cups
- counters
- equation mat

Model and solve the equation $2x - 4 = x - 2$.

▶ Place 2 cups and 4 negative counters on the left side of the mat. Then place 1 cup and 2 negative counters on the right side of the mat.

▶ Since the two sides of the mat represent equal quantities, you can remove an equal number of cups or counters from each side without changing the value of the equation.

▶ Now the equation is $x - 2 = 0$. Since there are no counters on the right, add 2 zero pairs. Then you can remove 2 negative counters from each side.

Therefore, each cup must contain 2 counters and $x = 2$.

Your Turn **Model and solve each equation.**

1. $2x + 3 = x - 5$ 2. $2x + 3 = x + 1$ 3. $3x - 2 = x + 6$

4. $3x - 7 = x + 1$ 5. $6 + x = 5x + 2$ 6. $x - 1 = 3x + 7$

7. Does it matter whether you remove cups or counters first? Explain.

8. Explain how you could use models to solve $-2x + 3 = -x - 5$.

Math Lab 7-5A   *Equations with Variables on Each Side*   **345**

Solving Equations with Variables on Each Side

- To solve equations with the variable on each side.

You can use equations to determine when two city populations will be the same.

Technology Activity

Most countries use the Celsius scale to measure temperature. In the United States, we use the Fahrenheit scale most of the time. There is one temperature which is the same on both scales. You can use the graphing calculator program at the right to find that temperature. In this program, *C* represents Celsius temperature, and *F* represents Fahrenheit temperature.

```
PROGRAM: TEMPS
: Lbl 1
: Input "ENTER F",
  F
: (5/9)*(F−32) → C
: If F≠C
: Then
: Disp "C=", C,
  "TRY AGAIN"
: Goto 1
: End
: Disp "CORRECT,
  F=C=", C
: Stop
```

Your Turn

▶ Enter the program into the calculator's memory.

▶ Run the program.

Enter a guess at the Fahrenheit temperature that has the same value on the Celsius scale. Continue guessing and checking until you determine the temperature.

a. What temperature is the same on both scales?

b. Explain what the formula in the third line of the program does.

Learning the Concept

The formula, $F = \frac{9}{5}C + 32$, is used for finding the Fahrenheit temperature when a Celsius temperature is known. When the temperature is the same on both scales, $F = C$. Replacing *F* with *C* results in the equation $C = \frac{9}{5}C + 32$.

In $C = \frac{9}{5}C + 32$, the variable is on each side of the equal sign. Use the properties of equality to eliminate the variable from one side. Then solve the equation.

$$C = \frac{9}{5}C + 32$$

$$C - \frac{9}{5}C = \frac{9}{5}C - \frac{9}{5}C + 32 \quad \textit{Subtract } \frac{9}{5}C \textit{ from each side.}$$

$$\frac{5}{5}C - \frac{9}{5}C = 32 \quad \textit{Rename C as } \frac{5}{5}C. \textit{ Why?}$$

$$-\frac{4}{5}C = 32$$

$$\left(-\frac{5}{4}\right)\left(-\frac{4}{5}\right)C = \left(-\frac{5}{4}\right) \cdot \frac{\overset{8}{\cancel{32}}}{1} \quad \textit{Multiply each side by } -\frac{5}{4}. \textit{ Why?}$$

$$C = -40$$

Check: $C = \frac{9}{5}C + 32$

$-40 \stackrel{?}{=} \frac{9}{5}(-40) + 32$ *Replace C with −40.*

$-40 \stackrel{?}{=} -72 + 32$

$-40 = -40$ ✔

The temperature is the same on the Celsius and Fahrenheit scales at $-40°$.

Example **1** **Solve $7n + 12 = 2.5n - 2$.**

$7n + 12 = 2.5n - 2$

$7n - 2.5n + 12 = 2.5n - 2.5n - 2$ *Subtract 2.5n from each side.*

$4.5n + 12 = -2$

$4.5n + 12 - 12 = -2 - 12$ *Subtract 12 from each side.*

$4.5n = -14$

$\frac{4.5n}{4.5} = \frac{-14}{4.5}$ *Divide each side by 4.5.*

14 (−) ÷ 4.5 = -3.11111

$n = -3.\overline{1}$

Sometimes a rectangle is described in terms of only one of its dimension. To find the dimensions, you will have to solve an equation that contains grouping symbols. When solving equations of this type, first use the distributive property to remove the grouping symbols.

Example **2**

Geometry

The perimeter of a rectangle is 74 inches. Find the dimensions if the length is 7 inches greater than twice the width.

Let w represent the width in inches. Then $2w + 7$ represents the length in inches. Recall that the formula for perimeter can be represented as $2w + 2\ell = P$.

$2w + 7$

$2w + 2\ell = P$

$2w + 2(2w + 7) = 74$ *Replace ℓ with $2w + 7$ and P with 74.*

$2w + 4w + 14 = 74$ *Use the distributive property.*

$6w + 14 = 74$ *Simplify.*

$6w + 14 - 14 = 74 - 14$ *Subtract 14 from each side.*

$6w = 60$

$\frac{6w}{6} = \frac{60}{6}$ *Divide each side by 6.*

$w = 10$

The width is 10 inches. Now evaluate $2w + 7$ to find the length.

$2w + 7 = 2(10) + 7$ or 27

The length is 27 inches. *Check to see if this length and width results in the correct perimeter.*

Some equations may have *no* solution. The solution set is the **null** or **empty set**. It is shown by the symbol { } or $\varnothing$. Other equations may always be true and have every number in their solution set.

Example **3**   Solve each equation.

a. $2x + 5 = 2x - 3$

$$2x + 5 = 2x - 3$$
$$2x - 2x + 5 = 2x - 2x - 3 \quad \textit{Subtract 2x from each side.}$$
$$5 = 3$$

This sentence is *never* true. The solution is $\varnothing$.

b. $3(x + 1) - 5 = 3x - 2$

$$3(x + 1) - 5 = 3x - 2$$
$$3(x + 1) - 3 = 3x \qquad \textit{Add 2 to each side.}$$
$$3x + 3 - 3 = 3x \qquad \textit{Distributive property}$$
$$3x = 3x \qquad \textit{3 − 3 = 0}$$
$$\frac{3x}{3} = \frac{3x}{3} \qquad \textit{Divide each side by 3.}$$
$$x = x$$

This sentence is *always* true. The solution set is all numbers.

Checking Your Understanding

Communicating Mathematics

Read and study the lesson to answer these questions.

1. **Describe** the steps you would take to solve $2p = 15 + 7p$.

2. **Name** the property of equality that allows you to subtract the same term from each side of an equation.

MATERIALS

 cups and counters

⊟ equation mat

3. The diagram at the right models the equation $2x - 1 = x - 4$.

   a. Solve $2x - 1 = x - 4$.

   b. Solve $3x - 1 = x + 3$ using cups and counters.

   c. Solve $x + 5 = x + 7$ using cups and counters.

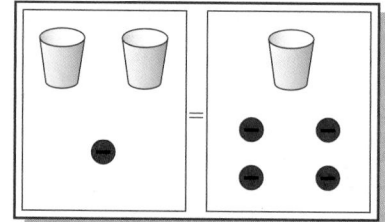

Guided Practice

Name the first two steps you should take to solve each equation. Then solve. Check your solution.

4. $12k + 15 = 35 + 2k$
5. $3x + 2 = 4x - 1$
6. $3b + 8 = -10b + b$
7. $3(a + 22) = 12a + 30$
8. $3k + 10 = 2k - 21$
9. $n + 4 = -n + 10$

10. **Geometry** The perimeter of a rectangle is 32 feet. Find the dimensions if the length is 4 feet longer than three times the width.

Exercises: Practicing and Applying the Concept

Independent Practice

Solve each equation. Check your solution.

11. $6n - 42 = 4n$
12. $8 - 3g = -2 + 2g$
13. $3(k + 2) = 12$
14. $\frac{4}{7}y - 8 = \frac{2}{7}y + 10$
15. $6 - 8x = 20x + 20$
16. $13 - t = -t + 7$
17. $3n + 7 = 7n - 13$
18. $7b - 3 = -b + 4$
19. $2 + 7(d + 1) = 9 + 7d$
20. $2a - 1 = 3.5a - 3$
21. $5m + 4 = 7(m + 1) - 2m$
22. $12x - 24 = -14x + 28$

23. $2(f - 3) + 5 = 3(f - 1)$ 　　　**24.** $4[z + 3(z - 1)] = 36$

25. $-3(4b - 10) = \frac{1}{2}(-24b + 60)$ 　　**26.** $\frac{3}{4}a + 16 = 2 - \frac{1}{8}a$

27. $\frac{d}{0.4} = 2d + 1.24$ 　　　　　**28.** $\frac{a - 6}{12} = \frac{a - 2}{4}$

Find the dimensions of each rectangle. The perimeter is given.

29. $P = 460$ m 　　　**30.** $P = 440$ yd 　　　**31.** $P = 110$ ft

w | [rectangle: $w + 30$]

w | [rectangle: $3w - 60$]

w | [rectangle: $2w - 20$]

Define a variable and write an equation for each situation. Then solve.

32. Twice a number is 220 less than six times the number. What is the number?

33. Fourteen less than three times a number equals the number. What is the number?

Critical Thinking

34. An apple costs the same as 2 oranges. Together, an orange and a banana cost 10¢ more than an apple. Two oranges cost 15¢ more than a banana. What is the cost for one of each fruit?

Applications and Problem Solving

35. Census　The table at the right shows the 1990 populations and the average rates of change in population in the 1980s for Shreveport, Louisiana and Orlando, Florida. Suppose the population of each city continued to increase or decrease at these rates.

City	Shreveport, LA	Orlando, FL
1990 Population	199,000	165,000
Annual Population Change in 1980s	−700	+3700

a. Write an expression for the population of Shreveport after x years.

b. Write an expression for the population of Orlando after x years.

c. Use the expressions in parts a and b to write an equation to find the number of years until the populations are the same.

d. In how many years would the population of the two cities be the same?

36. Cooking　Water boils at 212°F at sea level. However, as the altitude increases, the air pressure drops. The lower air pressure causes water to boil at a lower temperature. For every 550 feet above sea level, the boiling point is lowered 1°F. The boiling point rises 1°F for every 550 feet below sea level.

a. Write a formula for finding the boiling point b of water at any altitude A above sea level.

b. Denver is 5280 feet above sea level. If you were boiling an egg in Denver, how hot would the water be?

c. Find the temperature at which water would boil in Death Valley, which is 282 feet *below* sea level.

37. Sales Shoe World offers Veronica a temporary job during her spring break. The manager gives her a choice as to how she wants to be paid, but she must decide before she starts working.

Plan 1 $2 an hour plus 10¢ for every dollar's worth of shoes she sells.
Plan 2 $3 an hour plus 5¢ for every dollar's worth of shoes she sells.

   a. How much would Veronica's sales need to be in one hour to earn the same amount in either plan?

   b. If you were in Veronica's place, which method of payment would you choose? Explain your reasoning.

38. An empty bucket is put under two faucets. If one faucet is turned on alone, the bucket fills in 6 minutes. If the other faucet is turned on alone, the bucket fills in 4 minutes. If both are turned on, how many seconds will it take to fill the bucket?

Mixed Review

39. Solve $3x + 3 = 99$. (Lesson 7-2)

40. Standardized Test Practice Solve $15 \div \frac{3}{7} = n$. (Lesson 6-4)

   A. 0.71   **B.** 6.42   **C.** 35   **D.** 105

41. Write $\frac{36}{126}$ in simplest form. (Lesson 4-5)

42. Solve $A = \ell w$ for w if $A = 65$ and $\ell = 13$. (Lesson 3-4)

For **Extra Practice**, see page 757.

43. Solve $n - 10 = 42$ by using the inverse operation. (Lesson 1-8)

44. Solve $7m = 56$ mentally. (Lesson 1-6)

Self Test

1. Banking Dorinda is balancing her checkbook. Since the bank statement was mailed to her, she has written checks for $13.85, $19.72, and $59.66 and made a deposit in the amount of $497.92. If the balance she has written in her check register is $973.88, how much should the bank statement say is in her account? (Lesson 7-1)

Solve each equation. (Lesson 7-2)

2. $12 - z = 28$ **3.** $9 - 4z = 57$ **4.** $\dfrac{d - 5}{7} = 14$

Define a variable and write an equation for each situation. Then solve. (Lesson 7-3)

5. Four times a number less twelve is 18. What is the number?

6. The product of a number and 6 is eight greater than 10.

7. The circumference of a circle is 6.398 yards. Find its radius. (Lesson 7-4)

8. The diameter of a circle is 16.2 inches. What is the circumference of the circle? (Lesson 7-4)

9. Sports A soccer field is 75 yards shorter than 3 times its width. If the perimeter is 370 yards, find the dimensions of the soccer field. (Lesson 7-5)

7-6 Solving Multi-Step Inequalities

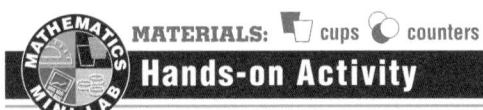
MATERIALS: cups ⊙ counters
Hands-on Activity

The diagram at the right models the inequality $4 + 2x > x + (-2)$. Notice that the side that is lower has the heavier weight. To solve an inequality, use the same process used for equations.

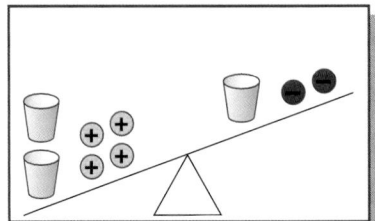

▶ Draw a diagram of the unbalanced scale on a sheet of paper and model the diagram shown above.

▶ Remove one cup from each side. *Perform operations on each side.*

▶ Place four zero pairs on the right side.

▶ Remove four positive counters from each side.

a. State the inequality after removing one cup from each side.

b. Why did you have to place four zero pairs on the right side?

c. What is the solution to the inequality?

d. How could you check the solution?

Learning the Concept

Inequalities can be solved by applying the methods for solving equations. Remember that if each side of an inequality is multiplied or divided by a negative number, the sign of the inequality should be reversed.

Example **1** Solve each inequality and check the solution. Graph the solution on a number line.

a. $7x + 29 > 15$

$$7x + 29 > 15$$
$$7x + 29 - 29 > 15 - 29 \quad \textit{Subtract 29 from each side.}$$
$$7x > -14$$
$$\frac{7x}{7} > \frac{-14}{7} \quad \textit{Divide each side by 7.}$$
$$x > -2$$

Any number greater than -2 is a solution.

Check: Try 0, a number greater than -2.

$$7(0) + 29 \overset{?}{>} 15 \quad \textit{Replace x with 0.}$$
$$0 + 29 \overset{?}{>} 15$$
$$29 > 15 \quad ✔$$

$$\underset{-5\ -4\ -3\ -2\ -1\ \ 0\ \ 1\ \ 2\ \ 3\ \ 4}{\longleftrightarrow}$$

b. $13 - 6n \leq 49 - 2n$

THINK ABOUT IT

Why was the order symbol reversed?

$$13 - 6n \leq 49 - 2n$$
$$13 - 6n + 2n \leq 49 - 2n + 2n \quad \textit{Add 2n to each side.}$$
$$13 - 4n \leq 49$$
$$13 - 13 - 4n \leq 49 - 13 \quad \textit{Subtract 13 from each side.}$$
$$-4n \leq 36$$
$$\frac{-4n}{-4} \geq \frac{36}{-4} \quad \textit{Divide each side by } -4 \textit{ and change } \leq \textit{ to } \geq.$$
$$n \geq -9$$

Any number greater than or equal to -9 is a solution.

Check: Try -5, a number greater than -9.
$$13 - 6n \leq 49 - 2n$$
$$13 - 6(-5) \overset{?}{\leq} 49 - 2(-5) \quad \textit{Replace n with } -5.$$
$$13 + 30 \overset{?}{\leq} 49 + 10$$
$$43 \leq 59 \quad \checkmark$$

When solving inequalities that contain grouping symbols, remember to first use the distributive property to remove the grouping symbols.

Example 2

Real World APPLICATION

Tourism

Route 1A along the Maine coast is called one of the most scenic highways in America. Every year, thousands of people go there to see the ocean and the charming New England villages. Suppose you rented a car in Portland for $139.95 plus $0.25 per mile for each mile over 100 miles. If you had budgeted up to $200 for car rental in your vacation budget, how many miles could you drive and stay within the budget?

Let m represent the number of miles driven.

139.95 plus	0.25 per mile over 100	is no more than	200
$139.95 +$	$0.25(m - 100)$	$\leq$	200

$$139.95 + 0.25(m - 100) \leq 200$$
$$139.95 + 0.25m - 25 \leq 200 \quad \textit{Simplify.}$$
$$114.95 + 0.25m \leq 200$$
$$114.95 - 114.95 + 0.25m \leq 200 - 114.95 \quad \textit{Subtract 114.95 from each side.}$$
$$0.25m \leq 85.05$$
$$\frac{0.25m}{0.25} \leq \frac{85.05}{0.25} \quad \textit{Divide each side by 0.25.}$$
$$m \leq 340.2 \quad \textit{Use a calculator.}$$

You could drive up to 340.2 miles and stay within your budget.

Communicating Mathematics

Read and study the lesson to answer these questions.

1. **Explain** how to check the solution for Example 2.

2. **You Decide** Tinesha says that to solve the inequality $2(2y + 3) > y + 1$, the first step should be to divide each side by 2. Kiki says the first step should be to subtract 3 from each side. Who is correct, and why?

MATERIALS

cups and counters

3. **Draw a model** that you could use to solve the inequality $2x + 3 < 3x - 7$.

4. Write an inequality for the model at the right. Then use cups and counters to solve the inequality.

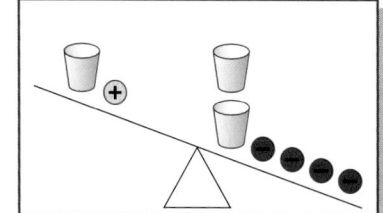

Guided Practice

Solve each inequality and check your solution. Graph the solution on a number line.

5. $3x + 2 \leq 23$ 6. $18 - 2v < 16$ 7. $2k + 7 > 13 - k$

8. $y + 1 \geq 5y + 5$ 9. $-3(m - 2) > 12$ 10. $-5 \leq \frac{x}{4} - 7$

11. **Consumer Awareness** Heather has \$90 that relatives gave her for her birthday. She needs to buy a new pair of shoes and would also like to buy two CDs. The shoes she likes cost \$65. Use the inequality $2d + 65 \leq 90$ to determine how much she can spend on each CD.

Exercises: Practicing and Applying the Concept

Independent Practice

Solve each inequality and check your solution. Graph the solution on a number line.

12. $2x + 9 > 25$ 13. $-2u + 3 \leq 11$ 14. $16 < 18 - 2n$

15. $2x - 5 < 2x - 9$ 16. $1.2z - 2 > 7 + 0.9z$ 17. $3(j - 1) \geq -12$

18. $\frac{m}{3} - 7 > 11$ 19. $36 + \frac{k}{5} \geq 51$ 20. $\frac{2x}{9} - 2 < -4$

21. $0.47 < \frac{t}{-9} + 0.6$ 22. $-4.4 > \frac{b}{-5} - 4.8$

23. $12 - \frac{5z}{4} < 37$ 24. $\frac{n - 11}{2} \leq -6$

25. $1.3x + 6.7 \geq 3.1x - 1.4$ 26. $-5x + 3 < 3x + 23$

27. $-5(k + 4) \geq 3(k - 4)$ 28. $8c - (c - 5) > c + 17$

29. $\frac{c + 8}{4} < \frac{5 - c}{9}$ 30. $\frac{2(n + 1)}{7} \geq \frac{n + 4}{5}$

Critical Thinking

31. The sum of three times a number and 5 lies between -10 and 8. Solve the *compound inequality* $-10 < 3x + 5 < 8$ to find the solution(s). (*Hint*: Any operation must be done to every part of the inequality.)

32. **Geometry**   The area of trapezoid *ABCD* is at least 320 square meters. The height of the trapezoid is 16 meters and the length of one of the bases is 15 meters. Use the inequality $16\left(\dfrac{15 + b}{2}\right) \geq 320$ to find the length of the other base, *b*.

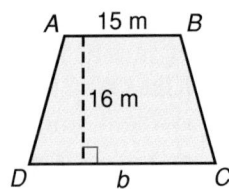

33. **Statistics**   Odina's scores on the first three of four 100-point tests were 89, 92, and 82. Use the inequality $\dfrac{89 + 92 + 82 + s}{4} \geq 90$ to determine what scores she could receive on the fourth test and have an average of at least 90 for all the tests.

Mixed Review

34. Solve $-75 - m = 25 + 4m$. (Lesson 7-5)

35. **Patterns**   Duncan received a chain letter that said he should copy the letter and send it to six friends in order to have good luck. Each of Duncan's friends sent copies to six other friends, who each sent copies to six other friends. How many people have received the chain letter after Duncan? (Lesson 6-8)

36. **Standardized Test Practice**   Gastone is training for a bicycle race. The chart shows the number of miles he bicycled during the last four days. Which is the best estimate of the total number of miles he bicycled during this 4-day period? (Lesson 5-2)

Day	Miles
Monday	9.2
Tuesday	13.3
Wednesday	7.9
Thursday	11.5

   A. 41 miles

   B. 42 miles

   C. 43 miles

   D. 44 miles

37. Find the missing exponent $\dfrac{t^{17}}{t^{\bullet}} = t^{9}$. (Lesson 4-8)

38. Find the value of $17 + (-13) + 8$. (Lesson 2-4)

39. Translate *3 more days of summer vacation than Northridge High School* into an algebraic expression. (Lesson 1-3)

7-7 Writing Inequalities

You'll Learn

■ To solve verbal problems by writing and solving inequalities.

It's Important

You can use inequalities to find the amount of postage needed to mail a letter.

Application: Backpacking

Ahh, the great outdoors. Nearly 10 million Americans hit the trail to go backpacking each year. State and national parks are among the most popular getaways. A fitness magazine says that three times the weight of your backpack and its contents should be less than your body weight in order for you to avoid an injury. If you weigh 120 pounds and your empty backpack weighs 15 pounds, how much should the contents of your backpack weigh?
This problem will be solved in Example 2.

Learning the Concept

LOOK BACK

You can review writing simple inequalities in Lesson 1-9.

Verbal problems with phrases like *greater than* or *less than* describe inequalities. Other phrases that suggest inequalities are *at least*, *at most*, and *between*. Inequalities that correspond to each statement are below.

x is *at least* 3.	$x \geq 3$	*At least 3 means 3 or greater.*
y is *at most* 9.	$y \leq 9$	*At most 9 means 9 or less.*
z is *between* 2 and 4.	$2 < z < 4$	*Between 2 and 4 means greater than 2 and less than 4.*

Example ❶ **Three times a number increased by eight is at least twenty-five. What is the number?**

Let *x* represent the number. Write an inequality.

three times a number   *increased by*   *eight*   *is at least*   *twenty-five*

$$3x \qquad\qquad + \qquad 8 \quad \geq \qquad 25$$

$3x + 8 \geq 25$

$3x + 8 - 8 \geq 25 - 8$ *Subtract 8 from each side.*

$3x \geq 17$

$\dfrac{3x}{3} \geq \dfrac{17}{3}$ *Divide each side by 3.*

$x \geq \dfrac{17}{3} \text{ or } 5\dfrac{2}{3}$

In many real-life problems, you will use a multi-step inequality to solve.

Example ❷

Real World APPLICATION

Backpacking

Refer to the application at the beginning of the lesson. How much should the contents of your backpack weigh?

Explore   You want to find what the backpack contents should weigh.

Plan   Let *c* represent the weight of the backpack contents.

(continued on the next page)

$$\text{Solve} \quad \underbrace{\text{three times}}_{3} \left(\underbrace{\substack{\text{weight of} \\ \text{backpack}}}_{15} \underbrace{\text{plus}}_{+} \underbrace{\substack{\text{weight of} \\ \text{contents}}}_{c} \right) \underbrace{\text{is less than}}_{<} \underbrace{\text{your weight}}_{120}$$

$$3(15 + c) < 120$$

$$\frac{3(15 + c)}{3} < \frac{120}{3} \qquad \textit{Divide each side by 3.}$$

$$15 + c < 40$$

$$15 - 15 + c < 40 - 15 \qquad \textit{Subtract 15 from each side.}$$

$$c < 25$$

The backpack contents should weigh less than 25 pounds.

Examine Suppose the contents of your backpack weigh 20 pounds, a number less than 25. The backpack and its contents would weigh 20 + 15 or 35 pounds. 3(35) = 105 and 105 < 120, so this would be a safe weight to carry.

Checking Your Understanding

Communicating Mathematics

Read and study the lesson to answer these questions.

1. **a.** **Write** an inequality for *six times a number decreased by four is at most 41.*

   b. **Explain** how you would know if $4\frac{1}{2}$ is a member of the solution set to the inequality you wrote for part a.

2. **Write** an inequality for *a number is between −7 and −9.*

3. **Assess Yourself**   Think of a time when you needed to use an inequality. Write an inequality about the situation.

M_{ATH} J_{OURNAL}

Guided Practice

Define a variable and write an inequality. Then solve.

4. Four less than a number is at most 10. What is the number?

5. Twice a number decreased by 9 is greater than 11. Find the number.

6. You earn $2.00 for every magazine subscription you sell plus a salary of $10 a week. How many subscriptions do you need to sell each week to earn at least $40 a week?

7. You buy some candy bars at 55¢ each and one newspaper for 35¢. How many candy bars can you buy if you only have $2?

8. **Sports**   Gao Min represents China in springboard diving. Suppose Gao scored 68.2, 68.9, 67.5, and 71.7 for her first four dives and has one more dive remaining. If the diver in first place has a score of 345.4, what must Gao receive on her fifth dive to overtake her?

Exercises: Practicing and Applying the Concept

Independent Practice

Define a variable and write an inequality. Then solve.

9. Four times a number increased by four is at least 16. Find the number.

10. Nine less than six times a number is at most 33. What is the number?

Write a problem that could be solved using each inequality.

11. $f + 115 \geq 260$

12. $\dfrac{87 + 92 + 89 + 97 + s}{5} \geq 90$

13. You can hike along the Appalachian Trail at 3 miles per hour. You will stop for one hour for lunch. You want to walk at least 18 miles. How many hours should you expect to spend on the trail?

14. Twelve times a number decreased by $\frac{1}{20}$ of the number is less than 3250. What is the number?

15. The sum of an integer and the next greater integer is at least 35. Find the least pair of such integers.

16. The sum of an odd integer and the next greater odd integer is more than 136. Find the least pair of such integers.

17. Three times a number t exceeds six by at least five. What are the possible values of t?

18. A number is less than the sum of twice its opposite and five. Find the number.

19. If 12.15 times an integer is increased by 9.348, the result is between 39 and 75. Find the integer.

Critical Thinking

20. A *compound inequality* is formed when two inequalities are connected by *and* or *or*. A compound inequality containing *and* is true only if *both* inequalities are true.
    a. Write an inequality equivalent to $2x - 5 < 3$ and $x + 5 > 4$ without using *and*.
    b. **Geometry** An obtuse angle is one that measures more than $90°$ but less than $180°$. Write an inequality to represent the possible measures of angle A if you know that angle A is obtuse.

Applications and Problem Solving

21. **Geometry** The sum of the lengths of any two sides of a triangle is greater than the length of the third side. Triangle ABC has side lengths of 22 inches and 28 inches. Write a compound inequality to describe the length of the third side of triangle ABC.

22. **Agriculture** Two companies are working on a system that uses satellites to help dispense the correct amount of fertilizer on crops. The basic equipment costs about $7000. Annual fees for signal reception will be about $1200.

   a. How much additional revenue would the technology have to generate for a farmer in ten years for the investment to have been a good one?

   b. One farm expert estimated that the system would produce an additional $100 per acre each year on a certain crop. How large a farm should a farmer have in order to expect to make money over a five-year period using the system?

Mixed Review

23. Solve $\frac{2}{5}x - 1 > -3$. (Lesson 7-6)

24. Solve $\frac{20}{27} - \frac{11}{27} = b$. (Lesson 5-4)

25. **Standardized Test Practice** Which expression is equivalent to 2^{-5}? (Lesson 4-9)
    A. -32 **B.** -10 **C.** $-\frac{1}{32}$ **D.** $\frac{1}{32}$

For **Extra Practice**, see page 757.

26. Solve $6 \leq \frac{n}{-7}$. (Lesson 3-7)

27. Find the product of -14 and $9b$. (Lesson 2-7)

7-8

Integration: Measurement
Using the Metric System

What You'll Learn

- To convert measurements within the metric system.

Why It's Important

Knowing how to convert metric measurements can help you find the better grocery value.

 Application: Travel

You're finally here! Your dream vacation abroad. Noticing that the gas gauge is low, you pull into a gas station to buy some fuel. As you finish filling the tank, you notice that the pump stopped at 41.6!

How can the gas tank in your little rental car hold 41.6 gallons? It doesn't! The tank holds 41.6 *liters*. Most countries other than the United States use the **metric system of measurement**. They use liters to measure capacity instead of gallons.

Learning the Concept

CAREER CHOICES

The metric system was developed so that scientists in different countries could communicate. For more information on science-related careers, contact:

National Science Teachers Association
1840 Wilson Blvd.
Arlington, VA 22201

The metric system was developed by a group of French scientists in 1795. The **meter (m)** was defined as $\frac{1}{10,000,000}$ of the distance between the North Pole and the Equator. All units of length in the metric system are defined in terms of the meter. A prefix is added to indicate the decimal place-value position of the measurement. The chart below shows the relationships between the prefixes and the decimal place-value positions.

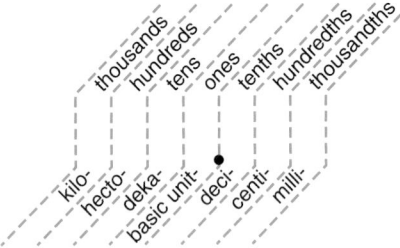

Note that each place value is 10 times the place value to its right.

Note that the value of each metric prefix is 10 times the value of the prefix to its right.

You can review multiplying and dividing by powers of ten in Lesson 6-9.

Converting units within the metric system follows the same procedure as multiplying or dividing by powers of ten.

Example Complete each sentence.

a. 4.8 km = _?_ m

Kilometers are larger than meters. Converting to smaller units means there will be more units. Multiply by 1000.

4.8 × 1000 = 4800
4.8 km = 4800 m

b. 9 mm = _?_ cm

Millimeters are smaller than centimeters. Converting to larger units means there will be fewer units. Divide by 10.

9 ÷ 10 = 0.9
9 mm = 0.9 cm

The application at the beginning of the lesson involved units of *capacity*. Capacity is the amount of liquid or dry substance a container can hold. The basic unit of capacity in the metric system is the **liter (L)**. A liter and milliliter are related in a manner similar to meter and millimeter.

$$1 \text{ L} = 1000 \text{ mL}$$

Example

Consumer Awareness

A two-liter bottle of cola sells for 99¢. A six-pack of the same cola sells for the same price. If each can contains 354 mL, which is the better value?

Explore　Both packages sell for the same price. You want to determine which package gives you more cola for your money.

Plan　To find the total capacity of the six-pack, multiply 354 times 6. To compare capacities, change from milliliters to liters. A milliliter is less than a liter, so divide by 1000.

capacity of cans in liters $= \dfrac{(354 \text{ mL/can})(6 \text{ cans})}{1000 \text{mL/L}}$

Estimate: 350 × 6 ÷ 1000 = 2.1

Solve　354 ☒ 6 ÷ 1000 ═ 2.124　The six-pack holds 2.124 L.

The six-pack is the better value since you get 2.124 liters of cola for the same price as the 2-liter bottle.

Examine　The capacity we found with the calculator is very close to our estimate, so the answer is reasonable.

The *mass* of an object is the amount of matter that it contains. The basic unit of mass in the metric system is the **kilogram (kg)**. Kilogram, gram, and milligram are related in a manner similar to kilometer, meter, and millimeter.

$$1 \text{ kg} = 1000 \text{ g} \qquad 1 \text{g} = 1000 \text{ mg}$$

Example Complete each sentence.

a. 3.2 kg = _?_ g

Kilograms are larger than grams. Larger to smaller means more units. Multiply by 1000.

3.2 × 1000 = 3200
3.2 kg = 3200 g

b. 200 mg = _?_ g

Milligrams are smaller than grams. Smaller to larger means fewer units. Divide by 1000.

200 ÷ 1000 = 0.2
200 mg = 0.2 g

Communicating Mathematics

Read and study the lesson to answer these questions.

1. Write a formula for converting a measure in meters to an equivalent measure in centimeters. Then write a second formula for converting a measure in centimeters to an equivalent measure in meters.

Math Journal

2. **Show** how many times longer a kilometer is than a millimeter by using the diagram on page 358.

3. Which concept in this chapter was the most difficult for you? Why was it challenging?

Guided Practice

State which metric unit you would probably use to measure each item.

4. width of a computer screen   5. juice in a small glass

6. thickness of a coin   7. a load of lumber

Complete each sentence.

8. 0.035 m = _?_ mm   9. 40 mL = _?_ L   10. 3 kg = _?_ g

11. **Running**   Last week, Ramon ran 8 laps around a 400-meter track on Monday, 4 laps on Tuesday, 8 laps on Wednesday, 4 laps on Thursday, and 8 laps on Friday. How many kilometers did he run last week?

Exercises: Practicing and Applying the Concept

Independent Practice

Write which metric unit you would probably use to measure each item.

12. distance between two cities   13. height of a 15-year-old female

14. weight of an aspirin   15. water in a washing machine

Complete each sentence.

16. 3 m = _?_ cm   17. 400 m = _?_ km   18. 30 mm = _?_ cm

19. 9400 mL = _?_ L   20. 5 L = _?_ mL   21. 600 cm = _?_ m

22. 5000 mg = _?_ g   23. 8 kg = _?_ g   24. 67 g = _?_ mg

25. How many millimeters are in a meter?

26. How many centimeters are in a kilometer?

27. How many grams are in 0.316 kilograms?

28. **Biology**   The thickness of a leg of an ant is 0.035 centimeters. How many millimeters is this?

29. A bottle of ketchup holds 0.946 liters. How many milliliters is this?

30. How many milligrams are in 0.019 kilograms?

31. How many centimeters are in 6.032 kilometers?

32. The hair on your scalp grows about 13 millimeters per month. How long would it take for hair to grow 0.5 meter?

Critical Thinking

33. Four volumes of International Recipes are in order on a shelf. The total pages of each volume are 5 cm thick. Each cover is 5 mm thick. A bookworm started eating at page 1 of Volume I and stopped eating at the last page of Volume IV. Describe the amount the bookworm ate in terms of distance.

Applications and Problem Solving

✓ **Choose**

Estimation
Mental Math
Calculator
Paper and Pencil

34. **Ecology** Every metric ton of recycled office paper saves about 19 trees. Glencoe Publishing's office in Westerville, Ohio, recycled 16,427 kilograms of paper in 1994. How many trees did this save?

35. **Geometry** The perimeter of a triangle is 1.6 meters. The length of the shortest side is 40 centimeters. Another side is 150 millimeters longer than the shortest side. Find the lengths of the other sides.

36. **Refreshments** To mix a punch, Lavonna starts with 3 liters of lemon-lime soda and adds 1893 milliliters of cranberry juice.

    a. Estimate how many liters are in the punch bowl when the punch is completed.

    b. Use your estimate to find how many 250-mL glasses of punch can be served.

37. **Photography** Most cameras use 35-millimeter film. This means that the film is 35 millimeters wide. Write the width of the film in meters.

38. **Physical Science** Electromagnetic waves, like light and radar waves, are measured in angstroms and microns. An angstrom is 10^{-10} meters, and a micron is 10^{-6} meters.

    a. Write the length of an angstrom and a micron in millimeters.

    b. The visible portion of the electromagnetic spectrum, including white light, has waves that are approximately 3900 to 7700 angstroms long. Write the lengths of these waves in millimeters.

    c. The waves used for radio and television transmission are 1 to 10 meters long. Write these lengths in angstroms.

39. **Family Activity** Find several measurements in your home or neighborhood and express them as both metric units and customary units. Use each of the eight units listed below at least once.

1m ≈ 1yd (width of doorway)	1ft ≈ 30 cm (length of adult shoe)
1cm ≈ 0.5 in. (width of a dime)	1kg ≈ 2.2 lb (a textbook)
1km ≈ 0.6 mi (length of city block)	1L ≈ 1qt (small pitcher)
1in. ≈ 25 mm or 2.5 cm	1oz ≈ 27 g (small box of raisins)

Mixed Review

40. The sum of three integers, x, $x + 1$, and $x + 2$, is at least 216. Find the three smallest such integers. (Lesson 7-7)

41. Solve $7b - 39 > 2(b - 2)$. (Lesson 7-6)

42. Solve $2z - 500 = 100 - z$. (Lesson 7-5)

43. Evaluate $\dfrac{5x}{y}$ if $x = 31.74$ and $y = 9.2$. (Lesson 6-5)

44. **Standardized Test Practice** Edgar has a board that is 17 feet long. He wishes to cut it into three equal length pieces to use as bookshelves. How long should each piece be? (Lesson 6-4)

    A. 5 ft

    B. $5\frac{1}{3}$ ft

    C. $5\frac{1}{2}$ ft

    D. $5\frac{2}{3}$ ft

For **Extra Practice**, see page 758.

45. Solve $\left(3\frac{1}{4}\right)\left(-6\frac{2}{5}\right) = r$. (Lesson 6-3)

46. Solve $15x = 45$. (Lesson 3-3)

Investigation

ROLLER COASTER MATH

Refer to the Investigation on pages 270–271.

You are ready to begin construction of your new roller coaster ride. The amusement park's information director is preparing a press release to announce when the new ride will open.

Using information from your design and the data from your class experiments, write a report for the information director. Your report should include the following.

- The name of your roller coaster and an explanation of why you chose it.

- Any records and track data such as height, speed, and so on.

- Ride data such as how long a ride lasts and how many passengers will be carried per hour.

- Any special-interest items like providing entertainment for people waiting in line.

Plan and give an oral presentation to describe what you know about roller coasters and designing roller coasters. Your presentation could include the following.

- A brief history of roller coasters.

- Examples from your portfolio to help you describe and define the roller coaster terms you used most often in this investigation.

- A table showing how you determined the most efficient number of passengers and trains for your design.

- An explanation of what you would do differently if you had to do this Investigation over again.

Extension

Build a scale model of your roller coaster using wooden craft sticks or other building materials.

PORTFOLIO ASSESSMENT

You may want to keep your work on this Investigation in your portfolio.

Vocabulary

After completing this chapter, you should be able to define each term, property, or phrase and give an example or two of each.

Algebra

empty set (p. 347)

null set (p. 347)

Geometry

center (p. 341)

circle (p. 341)

circumference (p. 341)

diameter (p. 341)

pi, π (p. 341)

radius (p. 341)

Measurement

gram (p. 359)

liter (p. 359)

meter (p. 358)

metric system (p. 358)

Problem Solving

working backward (p. 330)

_inter_NET CONNECTION

Chapter Review For additional lesson-by-lesson review, visit:
www.glencoe.com/ sec/math/prealg/ mathnet

Understanding and Using Vocabulary

Determine whether each statement is _true_ or _false_.

1. The basic unit of capacity in the metric system is the gram.
2. The distance around a circle is its circumference.
3. In the metric system, the meter is the basic unit used for measuring distance.
4. A radius of a circle is twice as long as its diameter.
5. The null set, or the empty set, contains no elements.
6. A segment from any point on a circle to the center of the circle is a radius.
7. A formula for the circumference of a circle is $C = \pi d$.

Skills and Concepts

| Objectives and Examples | Review Exercises |

Upon completing this chapter, you should be able to:

Use these exercises to review and prepare for the chapter test.

▶ **solve equations that involve more than one operation.** (Lesson 7-2)

$$3x + 6 = 27$$
$$3x + 6 - 6 = 27 - 6 \quad \text{Subtract 6 from each side.}$$
$$3x = 21$$
$$\frac{3x}{3} = \frac{21}{3} \quad \text{Divide each side by 3.}$$
$$x = 7$$

Solve each equation. Check your solution.

8. $54 = 7x - 9$
9. $\frac{g}{3} + 10 = 14$
10. $12 = 7 - m$
11. $\frac{x}{4} - 17 = -2$
12. $24 - y = -9$
13. $4d - 8 = -88$
14. $-2f - 37 = -11$
15. $2.9 = 3.1 - t$

▶ **solve verbal problems by writing and solving equations.** (Lesson 7-3)

One more than twice a number is 17. What is the number?

one   more than   twice a number   is   17

$$1 \quad + \quad 2n \quad = 17$$
$$1 + 2n = 17$$
$$1 - 1 + 2n = 17 - 1$$
$$2n = 16$$
$$\frac{2n}{2} = \frac{16}{2}$$
$$n = 8$$

Define a variable and write an equation for each situation. Then solve.

16. The sum of six times a number and 4 is 52. What is the number?
17. Twice a number less 7 is 19. Find the number.
18. Six hundred is 15 more than 15 times a number. What is the number?
19. Vashawn earns $50 a week less than twice his old salary. If he earns $450 a week now, what was his old salary?

▶ **find the circumference of a circle.** (Lesson 7-4)

Find the circumference of a circle with a diameter 12.3 cm long.

$$C = \pi d$$
$$C = \pi(12.3)$$
$$C \approx 38.6 \text{ cm}$$

12.3 cm

Find the circumference of each circle shown or described below.

20.

$14\frac{2}{3}$ km

21. The radius is $3\frac{1}{2}$ inches.
22. The diameter is 18.45 centimeters.
23. The radius is 4.9 meters.

| **Objectives and Examples** | **Review Exercises** |

▶ **solve equations with the variable on each side.** (Lesson 7-5)

$$3t - 1 = t + 8$$
$$3t - t - 1 = t - t + 8 \quad \textit{Subtract t from each side.}$$
$$2t - 1 = 8$$
$$2t - 1 + 1 = 8 + 1 \quad \textit{Add 1 to each side.}$$
$$2t = 9$$
$$\frac{2t}{2} = \frac{9}{2} \quad \textit{Divide each side by 2.}$$
$$t = \frac{9}{2} \text{ or } 4\frac{1}{2}$$

Solve each equation. Check your solution.

24. $6n - 8 = 2n$

25. $4 + 2w = 8w + 16$

26. $4m + 11 = 6 - m$

27. $5(3 - k) = 20$

28. $5(q + 2) = 2q + 1$

29. $\frac{2}{7}h - 10 = h - 20$

30. $\frac{x - 5}{6} = \frac{x - 11}{2}$

▶ **solve inequalities that involve more than one operation.** (Lesson 7-6)

$$5b - 10 < 7b + 4$$
$$5b - 5b - 10 < 7b - 5b + 4 \quad \textit{Subtract 5b from each side.}$$
$$-10 < 2b + 4$$
$$-10 - 4 < 2b + 4 - 4 \quad \textit{Subtract 4 from each side.}$$
$$-14 < 2b$$
$$\frac{-14}{2} < \frac{2b}{2} \quad \textit{Divide each side by 2.}$$
$$-7 < b$$

Solve each inequality and check your solution. Graph the solution on a number line.

31. $5n + 4 > 34$

32. $-7m - 12 < 9$

33. $\frac{r}{7} - 11 \leq 3$

34. $16f > 13f + 45$

35. $0.52c + 14.7 > 2.48c$

36. $\frac{a - 5}{6} \leq \frac{a + 9}{8}$

▶ **solve verbal problems by writing and solving inequalities.** (Lesson 7-7)

Six times a number less 12 is at least twenty-two. What is the number?

| *six times a number* | *less* | *12* | *is at least* | *twenty-two* |
| $6x$ | $-$ | 12 | $\geq$ | 22 |

$$6x - 12 \geq 22$$
$$6x - 12 + 12 \geq 22 + 12 \quad \textit{Add 12 to each side.}$$
$$6x \geq 34$$
$$\frac{6x}{6} \geq \frac{34}{6} \quad \textit{Divide each side by 6.}$$
$$x \geq \frac{34}{6} \text{ or } 5\frac{2}{3}$$

Define a variable and write an inequality for each situation. Then solve.

37. Eight times a number increased by two is at least 18. What is the number?

38. Twelve less than three times a number is at most 27. Find the number.

39. The sum of an integer and the next greater integer is more than 47. What is the integer?

40. Four times a number s exceeds 6 by at least 8. What are the possible values of s?

Objectives and Examples	Review Exercises

▶ **convert measurements within the metric system.** (Lesson 7-8)

4.5 km = _?_ m

Kilometers are larger than meters, so there will be more units. There are 1000 meters in a kilometer.

$4.5 \times 1000 = 4500$
4.5 km = 4500 m

Complete each sentence.

41. 6 mm = _?_ m
42. 8.3 g = _?_ kg
43. 43 L = _?_ mL
44. 560 mg = _?_ g
45. 3 km = _?_ cm
46. 40 mL = _?_ L
47. 88 m = _?_ cm

Applications and Problem Solving

48. **Patterns** "One potato, two potato, . . ." says Alejándro as the children decide who is "it." At the beginning, six children were standing in a circle and each round of the rhyme eliminated the eighth child. Alejándro counts clockwise and each round begins with the person next to the one who was eliminated. Let A represent Alejándro and B through F the rest of the children clockwise around the circle. If E is declared "it," where did Alejándro begin the first rhyme? (Lesson 7-1)

49. **Health** Doctors recommend that you not exceed a safe heart rate when you exercise. The formula $r = \dfrac{4(220 - a)}{5}$, gives the desirable heart rate, r, in beats per minute for a person a years old. Find the maximum safe heart rate for a 44-year old. (Lesson 7-2)

50. **Music** Before valves were invented in the late 19th century, musicians used to adjust the scale of their brass instruments by adding a crook. A crook is a metal piece of tubing inserted between the mouthpiece and the main tubing of the instrument. If a circular crook has a diameter of 3.4 inches, how long was the piece of metal used to make it? (Lesson 7-4)

51. **Aviation** The pilot of a small plane wants to avoid going above the cloud ceiling that is currently at 12,000 feet. If the plane is ascending at a rate of 630 feet per minute, how long can the pilot ascend and still stay at least 1500 feet below the ceiling? (Lesson 7-7)

A practice test for Chapter 7 is available on page 780.

Demonstrate your knowledge by giving a clear, concise, solution to each problem. Be sure to include all relevant drawings and justify your answers. You may show your solutions in more than one way or investigate beyond the requirements of the problem.

1. The table below shows some of the biggest money-losing films.

Film	Loss (millions)
A	48.1
B	47.3
C	47.0
D	44.1
E	38.1

   a. Film A lost $10.3 less than twice what a British film about the Titanic lost. How much did the British film lose?

   b. The sum of what Film F and Film E lost is $5.1 million more than twice what Film F lost. How much did Film F lose?

   c. Film G lost just over $6.3 million more than half what Film B lost. How much did Film G lose? Write your answer as an inequality.

   d. Use the numbers in the table to write a problem that could be solved using a two-step equation or inequality.

2. Terrica is cutting fabric to make a tablecloth for a round table in her living room. The top of the table is 56.25 centimeters across. The table is 1.2 meters tall.

   a. What is the circumference of the top of the table?

   b. What should Terrica make the diameter of the tablecloth be if it goes to the floor on all sides?

   c. If Terrica is going to put lace around the edge of the tablecloth, how much should she buy?

Thinking Critically

Two formulas for the circumference of a circle are $C = 2\pi r$ and $C = \pi d$. If you could only remember one of them, could you always find the circumference of a circle? Explain.

 Portfolio

Select one of the assignments from this chapter that you found especially challenging and place it in your portfolio.

Investigation

STADIUM STAMPEDE

You are the stadium manager for the West Prairie State University, home of the Mighty Mustangs. You oversee the operations of Mustang Stadium and help plan team-spirit activities for the home football games. The university has a large stadium with a seating capacity of 80,000. The stadium has two levels, an upper level and a lower level.

- *In each level, there are 26 sections. Each section has about the same number of seats.*

- *The sections are labeled A through Z, with section A situated at the center of the north end zone of the field.*

- *The rest of the sections are labeled in alphabetical order going around the stadium in a clockwise direction.*

- *In each section of the lower level, there are 90 rows numbered from 1 to 90. They are in sequential order starting with row 1 next to the field.*

- *Each section of the upper level has only 75 rows numbered from 1 to 75. Each seat in the row is also numbered.*

The following ticket stub shows a seat for the big game.

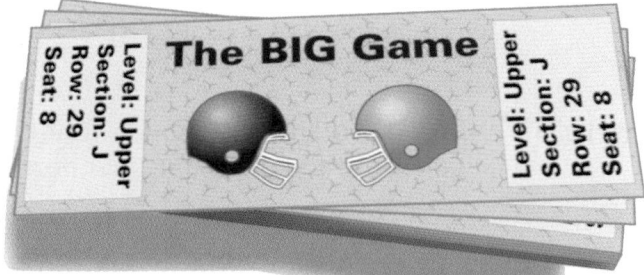

Various groups come to your office for research, advice, and permission for different projects. The spirit squad has been working on keeping the team spirit high this year for the football games. Also, the athletic director has been working on ticket pricing for stadium seating.

Starting Your Investigation ....

To start the research for either the spirit squad or athletic director, you will want to draw a model of the stadium. Since every row in every section of the stadium seats about the same number of people (plus or minus one person), how many seats are in a row? Explain your answer. What is the seating capacity of an upper section in the stadium? What is the seating capacity of a lower section?

On your model, label all the sections and levels in the stadium. Also, draw an enlargement of section A (both upper and lower levels), and label the number of each row in the section. Also, choose one row and label all the seat numbers.

Describe where the best seats are in the stadium. If you could sit wherever you want during the big game, what seat would you pick?

> **Think About . . .**
> The spirit squad wants to do the wave during a time-out. The wave is a type of stadium cheer where the people in each section of the stadium stand, raise their hands, and cheer in a sequence around the stadium. The spirit squad asks you a few questions:
>
> - *How long does it take the crowd to do one rotation around the stadium?*
>
> - *How many times can the wave go around the stadium during a two-minute time-out?*
>
> - *If the wave is done for two minutes during each of six time-outs in the game and for three minutes during each quarter and half-time, how many times would the average fan have to stand up and sit down during a game?*
>
> Reread the above questions throughout this Investigation. As you come up with other questions, write these new questions in the list also.

Research For current information on football stadiums, visit the following web site.

Data Collection and Comparison To share and compare your data with other students in the U.S., visit:

www.glencoe.com/ sec/math/prealg/ mathnet

Investigation Stadium Stampede

Working on the Investigation Lesson 8-2, p. 384

Working on the Investigation Lesson 8-6, p. 404

Working on the Investigation Lesson 9-5, p. 453

Working on the Investigation Lesson 9-9, p. 471

Closing the Investigation End of Chapter 9, p. 476

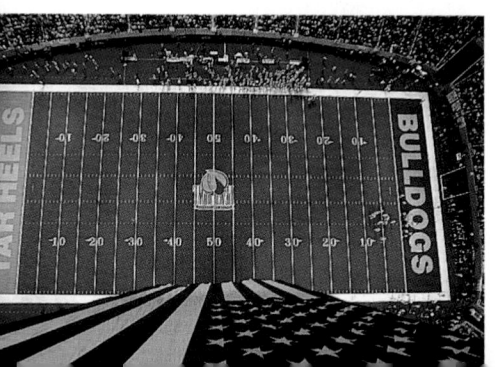

You will continue working on this Investigation throughout Chapters 8 and 9.

Be sure to keep your materials in your Investigation Folder.

Functions and Graphing

In this chapter, you will:

- determine whether relations are functions,

- graph data in scatter plots,

- graph linear equations using slope and intercepts,

- solve systems of equations, and

- graph inequalities.

Why It's Important

Science This scatter plot shows crickets and their chirps. On the horizontal, or *x*-axis, you will see the number of chirps per 15 seconds. On the vertical, or *y*-axis, you will see the corresponding temperature for each number of chirps. This type of graph makes it easy to see that there is some type of relationship between the number of chirps and the temperature. Scientists often use scatter plots to look for patterns and relationships in data. *You will learn how to make scatter plots in Lesson 8-3.*

Crickets and Their Chirps

Temperature (°F)

Chirps in 15 s

PREREQUISITE SKILLS

To be successful in this chapter, you'll need to understand these concepts and be able to apply them. Refer to the lesson in parentheses if you need more review before beginning the chapter.

▶ *Graph points in all quadrants of the coordinate plane.* *(Lesson 2-2)*

On graph paper, draw coordinate axes. Then graph and label each point.

1. $(2, 5)$ **2.** $(-3, 2)$ **3.** $(0, -4)$ **4.** $(4, -1)$

Name the ordered pair for each point graphed on the coordinate plane at the right.

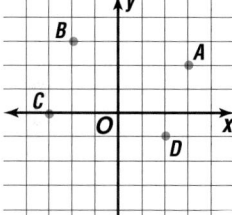

5. A **6.** B

7. C **8.** D

▶ *Add and subtract fractions with unlike denominators.* *(Lesson 5-5)*

Evaluate each expression if $x = \frac{2}{3}$, $y = -\frac{1}{2}$, and $z = 2$.

9. $x + y$ **10.** $x - y$ **11.** $y + z$ **12.** $z - y$

▶ *Multiply fractions.* *(Lesson 6-3)*

Evaluate each expression if $c = \frac{3}{4}$, $d = -\frac{1}{3}$, $a = \frac{1}{2}$, and $x = -4$.

13. cd **14.** cx **15.** ca **16.** dx

READING SKILLS

In this chapter you will learn about **relations, functions,** and **graphing.** A relation shows how two sets of data are related to one another. A function is a special type of relation. You will be graphing linear relations as well. As you may have guessed, a linear relation will look like a line when it is graphed. Functions can also be written as equations with two variables like $y = 2x + 1$.

Application: Ecology

Have you ever thought you couldn't make a difference in landfills by recycling? Think again! The table at the right shows how the United States compares to other nations in terms of the amount of annual household waste. These five countries produce the most waste in the world.

Country	Annual Domestic Waste (thousands of tons)	Equivalent per Person (pounds)
France	15,500	634
Great Britain	15,816	620
Japan	40,225	634
United States	200,000	1925
West Germany	20,780	741

Source: *Encarta*

This data could also be displayed using a set of ordered pairs. Each first coordinate would be the annual domestic waste, and each second coordinate would be the equivalent per person.

{(15,500, 634), (15,816, 620), (40,225, 634), (200,000, 1925), (20,780, 741)}

Learning the Concept

A **relation** is a set of ordered pairs, like the set shown above. The set of first coordinates is called the **domain** of the relation. The set of second coordinates is called the **range** of the relation. So, the domain of the relation above is {15,500, 15,816, 40,225, 200,000, 20,780}, and the range is {634, 620, 1925, 741}. *Notice that 634 is listed only once.*

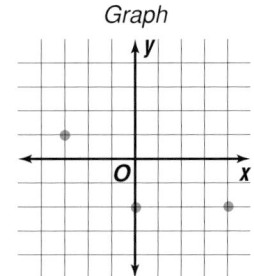

THINK ABOUT IT

What is the domain and range of this relation?

A relation can also be modeled by a table or a graph. For example, the relation {(−3, 1), (0, −2), (4, −2)} can be represented in each of the ways shown at the right.

Table

x	y
−3	1
0	−2
4	−2

Graph

Example **1**   Express the relation shown in the graph below as a set of ordered pairs. Then determine the domain and range.

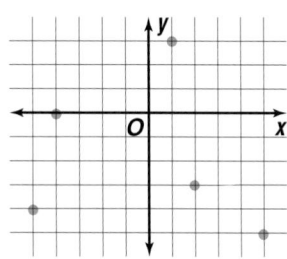

The set of ordered pairs for the relation is $\{(1, 3), (5, -5), (-4, 0), (2, -3), (-5, -4)\}$.

The domain is $\{1, 5, -4, 2, -5\}$.

The range is $\{3, -5, 0, -3, -4\}$.

The relation in Example 1 is a special type of relation called a **function**.

Definition of a Function	A function is a relation in which each element of the domain is paired with exactly one element in the range.

Example **2**   Graph each relation. Then determine whether each relation is a function.

a.
x	6	3	6	-2
y	0	8	-1	3

b.
x	-3	-2	0	2
y	1	-2	1	3

This relation is not a function since the element 6 in the domain is paired with two elements, 0 and −1, in the range.

Since each element of the domain is paired with exactly one element in the range, this relation is a function.

You can use the **vertical line test** to determine if a relation is a function. The relation $\{(-2, 5), (0, 4), (4, 2), (6, 1), (8, 0)\}$ is graphed at the right. Place a straightedge at the left of the graph to represent a vertical line. Slowly move the straightedge to the right across the graph.

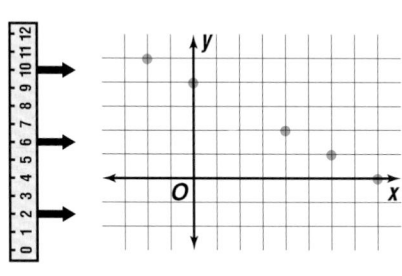

For each value of x, this vertical line passes through no more than one point on the graph. This is true for *every* function.

Example **3**

INTEGRATION

Probability

The graph on the left shows the possible outcomes when two six-sided dice are tossed. The graph on the right shows those outcomes when the faces on the dice show a sum of 7. Is either relation a function? Explain.

Second die

First die

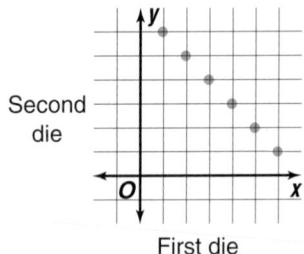

Second die

First die

The relation graphed on the left is *not* a function since it does not pass the vertical line test.

The relation graphed on the right is a function since any vertical line passes through no more than one point of the graph of the relation.

Checking Your Understanding

Communicating Mathematics

Read and study the lesson to answer these questions.

1. **Describe** three different ways to show a relation. Give an example of a relation shown in each way.

2. **Explain** the difference between the domain and the range of a relation.

3. **You Decide**   Marita says that $\{(9, 3), (0, 3), (-5, 0), (-1, 11)\}$ is a function. Terrica disagrees. Who is correct and why?

4. **Graph** the relation described by the annual waste figures in the application at the beginning of the lesson.

Guided Practice

State the domain and range of each relation.

5. $\{(-1.3, 1), (4, -3.9), (-2.4, 3.6)\}$

6. $\left\{\left(-\frac{1}{2}, -\frac{1}{4}\right), \left(1\frac{1}{2}, -\frac{2}{3}\right), \left(3, -\frac{2}{5}\right), \left(5\frac{1}{4}, 6\frac{2}{7}\right)\right\}$

Express the relation shown in each table or graph as a set of ordered pairs. Then state the domain and range of the relation.

7.

x	y
−1	3
0	6
4	−1
7	2

8.

x	y
2	4
0	2
−2	0
−4	2

374 *Chapter 8   Functions and Graphing*

9.

10.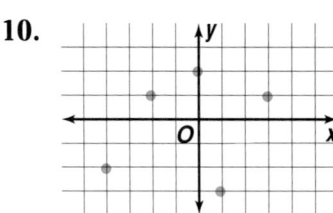

Determine whether each relation is a function.

11. {(5, −2), (3, 2), (4, −1), (−2, 2)}

12.

x	0	2	0
y	4	3	8

13.

14.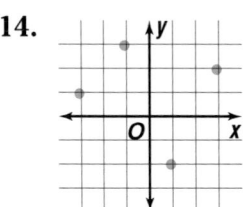

15. Food Lisle buys a dozen bagels for a morning meeting. She knows that the staff prefers blueberry bagels over plain bagels. If she buys at least twice as many blueberry bagels as plain bagels, write a relation to show the different possibilities. (*Hint:* Let the relation consist of ordered pairs of the form (blueberry bagels, plain bagels).)

Exercises: Practicing and Applying the Concept

Independent Practice

Write the domain and range of each relation.

16. {(5, −4), (−2, 3), (5, −1), (2, 3)}   **17.** {(−1, 6), (4, 2), (2, 36), (1, 6)}

18. {(3.1, 2), (7, −4.4), (−3.9, −8.8)}   **19.** {(1.4, 3), (−2, 9.6), (4, 4), (6, −2.7)}

20. $\left\{\left(\frac{2}{5}, \frac{3}{4}\right), \left(98\frac{3}{5}, 37\frac{1}{2}\right), \left(-4, -\frac{7}{12}\right)\right\}$ **21.** $\left\{\left(-\frac{1}{2}, \frac{1}{3}\right), \left(4\frac{2}{3}, -17\right), \left(-12\frac{3}{8}, 66\right)\right\}$

Express the relation shown in each table or graph as a set of ordered pairs. Then state the domain and range of the relation.

22.

x	y
0	5
2	3
1	−4
−3	3
−1	−2

23.

x	y
−4	−2
−2	1
0	2
1	−3
3	1

24.

x	y
−1	5
−2	5
−2	4
−2	1
−6	1

25.

x	y
5	4
2	8
−7	9
2	12
5	14

26.

27.

28.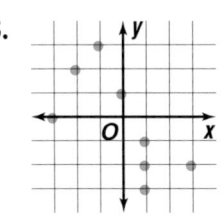

Determine whether each relation is a function.

29.

x	1	2	3
y	3	4	5

30.

x	3	5	7
y	−4	−6	−8

31.

x	1	2	1
y	3	5	−7

32.

x	−2	0	2
y	3	1	3

33. {(4, 0), (6, 0), (8, 0)}

34. {(0, 4), (0, 6), (0, 8)}

35. {(5, 4), (5, −6), (5, 4), (0, 4)}

36. {(−2, 3), (4, 7), (24, −6), (5, 4)}

37.

38.

39.

Critical Thinking

The *inverse* of any relation is obtained by switching the coordinates in each ordered pair of the relation. Determine whether the inverse of each relation is a function.

40. {(4, 0), (5, 1), (6, 2), (6, 3)}

41. {(5, 5), (−3, −1), (−2, 1), (−4, 0)}

Applications and Problem Solving

42. **Traffic Control** The Traffic Engineering Division for Tallahassee studies the flow of traffic at certain intersections to determine if traffic signals are needed. The table at the right shows the actual count of cars and trucks that passed through a particular intersection on one weekday.

Time	Cars	Trucks
3:00P.M.–3:30P.M.	1330	650
3:30P.M.–4:00P.M.	2000	600
4:00P.M.–4:30P.M.	2100	630
4:30P.M.–5:00P.M.	2300	650
5:00P.M.–5:30P.M.	2550	750
5:30P.M.–6:00P.M.	2220	700

a. Use the information in the second and third columns of the table to illustrate a relation that is a function.

b. Use the information to illustrate a relation that is not a function. Explain.

43. **Shipping Rates** When you order something from a mail-order company, you usually pay a shipping fee. The chart at the right relates shipping costs to the total price of merchandise ordered.

Total Price of Merchandise	Shipping Costs
$0.00–$30.00	$4.25
$30.01–$70.00	$5.75
$70.01 and over	$6.95

a. What is the shipping cost for an order of merchandise totaling $75?

b. For what price of merchandise is the shipping cost $5.75?

c. Does the chart represent a function? Explain.

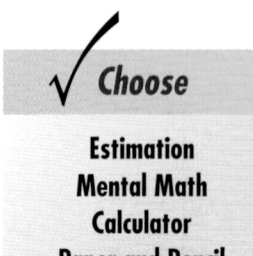

✓ Choose

Estimation
Mental Math
Calculator
Paper and Pencil

44. Water is poured at a constant rate into containers shaped like the ones shown below. Draw a graph of the water level in each of the containers as a function of time.

a.

b.

c.

Mixed Review

45. Consumer Awareness Which has a greater capacity, a 2-liter bottle of cola or a six-pack of cans each containing 354 mL of the same cola? (Lesson 7-8)

46. Solve $-2.2 < \dfrac{b}{-10} - 2.4$. (Lesson 7-6)

47. Physics A ball rebounds $\dfrac{2}{3}$ of its height after every fall. If it is dropped from a height of 48 feet, how high will it bounce at the end of the third bounce? (Lesson 6-8)

48. Solve $z - \dfrac{2}{5} \geq -2$. (Lesson 5-7)

49. Simplify $(-7y^2)(3y^3)$. (Lesson 4-8)

50. Standardized Test Practice Find the perimeter of the rectangle shown. (Lesson 3-5)

6 miles

2 miles

A. 12 miles

B. 16 miles

C. 24 miles

D. 48 miles

For **Extra Practice**, see page 758.

51. Three subtracted from some number is equal to -7. What is the number? (Lesson 2-5)

8-2A Scatter Plots

A Preview of Lesson **8-2**

MATERIALS

- tape measure
- graph paper

Cap size is related to the circumference of your head, and your shoe size is related to the length of your foot. Often it is not easy to establish whether there is a relationship between pairs of numbers by simply looking at them. Graphing ordered pairs on a coordinate system is one way to make it easier to "see" if there is a relationship.

Your Turn   Work with a partner.

▶ Measure, to the nearest inch, the length of your partner's shoe and your partner's armspan. Record your data in a table like the one shown at the right.

Name	Shoe Length (in.)	Armspan (in.)

▶ Extend the table and combine your data with that of your classmates.

▶ On a piece of graph paper, draw a large coordinate axes. Label the axes by writing *Shoe Length* (*in.*) along the horizontal axis and *Armspan* (*in.*) along the vertical axis.

▶ Number the vertical lines, called tick marks, on each axis to make an appropriate scale for the data collected.

▶ Plot a point for each ordered pair (*shoe length, armspan*) using the data your class collected.

The graph you made is called a **scatter plot** of the data you collected.

TALK ABOUT IT

1. Describe the scatter plot.

2. Is there a relationship between shoe length and armspan? If so, write a sentence or two that describes the relationship.

3. Why do you suppose this kind of graph is called a scatter plot?

Extension   4. Using similar methods, determine if there is a relationship between another pair of lengths; for example, the circumferences of your head and wrist, or the circumference of your neck and your armspan.

Integration: Statistics
Scatter Plots

Technology Activity

The sales and advertising costs for ten of the largest U.S. industrial corporations are shown in the chart below. Use a graphing calculator to create a scatter plot of the data. A **scatter plot** is a graph that shows the general relationship between two sets of data.

Company	Sales (millions of dollars)	Advertising Costs (millions of dollars)
A	132,775	1333.6
B	100,786	794.5
C	65,096	185.5
D	62,202	250.7
E	50,157	2024.1
F	36,897	756.6
G	29,890	2165.6
H	22,084	928.6
I	20,577	686.6
J	19,177	186.6

Your Turn   Before you create a scatter plot, clear the statistical memories.

Enter: STAT   4   2nd   L1   ,   2nd   L2   ENTER

Next, enter the sales in L1 and the advertising costs in L2.

Enter: STAT   ENTER   *Accesses the statistical lists.*
    132775   ENTER   100786   ENTER   ...   19177   ENTER
    ▶   1333.6   ENTER   794.5   ENTER   ...   186.6   ENTER

LOOK BACK

You can review setting the viewing window in Lesson 2-2B.

After the data is entered, set the range for the graph. Set a viewing window of [0, 150,000] by [0, 2500] with a scale factor of 10,000 on the x-axis and 250 on the y-axis.

Create the scatter plot by pressing 2nd STAT PLOT   ENTER   and then using the arrow and ENTER keys to highlight "On", the scatter plot, L1 as the Xlist, L2 as the Ylist, and

• as the mark. Press GRAPH to see the scatter plot.

a. Does it appear that there is any relationship between sales and advertising?

b. Are there any clusters of points that are somewhat separated from the other points? If there are, describe the characteristics of each cluster.

Study the scatter plots below.

| If the points appear to suggest a line that slants upward to the right, there is a **positive relationship**. | If the points appear to suggest a line that slants downward to the right, there is a **negative relationship**. | If the points seem to be random, then there is **no relationship**. |

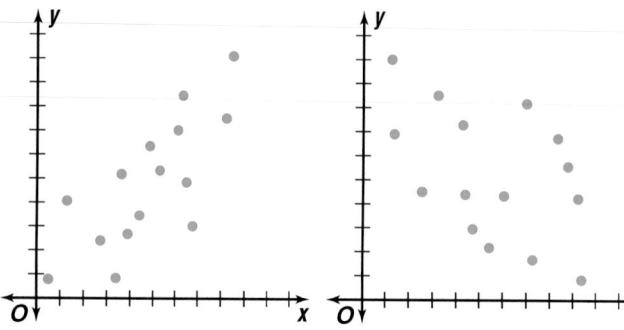

Look at the scatter plot of sales and advertising costs on the graphing calculator. The points in the scatter plot are spread out. There appears to be no relationship between the amount spent on advertising and the amount of sales.

Besides establishing a relationship, scatter plots can be used to study trends or make predictions.

Example

INTEGRATION

Statistics

FYI

Tagalog is an official language of the Philippines along with Spanish and English.

Make a scatter plot of the following data. Then determine if there is a *positive*, *negative*, or *no* relationship. What trends do you see?

Languages Other than English and Spanish Spoken at Home by Americans

Language used at home	Total speakers over 5 years old	
	1980	1990
French	1,572,000	1,703,000
German	1,607,000	1,547,000
Italian	1,633,000	1,309,000
Chinese	632,000	1,249,000
Tagalog	452,000	843,000
Polish	826,000	723,000
Korean	276,000	626,000
Vietnamese	203,000	507,000
Portuguese	361,000	430,000

Source: Bureau of the Census, U.S. Dept. of Commerce

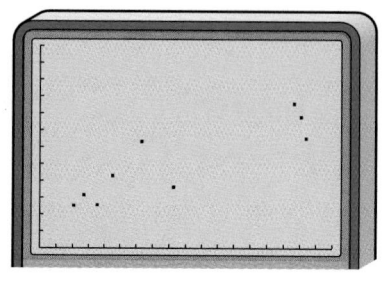

Since the points in the scatter plot are in a pattern that seems to slant upward to the right, the scatter plot shows a *positive* relationship. In general, the greater the number of speakers of a language in 1980, the greater the number of speakers of the same language in 1990.

Communicating Mathematics

Read and study the lesson to answer these questions.

1. **Write** a sentence explaining how a scatter plot can be used.

2. **Describe** how to draw a scatter plot for two sets of data.

3. **Draw** a scatter plot with 15 points that shows a negative relationship.

4. **Write** a short paragraph or two describing what you can determine about the data in the scatter plot shown at the right.

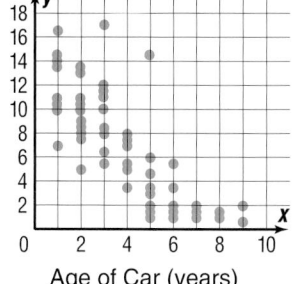

Relationship of Value of Cars and Age of Cars

Guided Practice

What type of relationship, *positive*, *negative*, or *none*, is shown by each scatter plot?

5.

6.

7.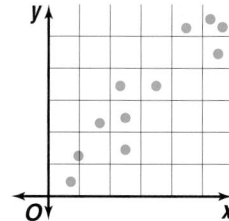

Determine whether a scatter plot of the data for the following might show a *positive*, *negative*, or *no* relationship. Explain your answer.

8. weight, month of birth

9. playing time, points scored

10. **Track and Field**
The scatter plot at the right compares the winning time to the year of the race for the women's Olympic 100-meter dash.

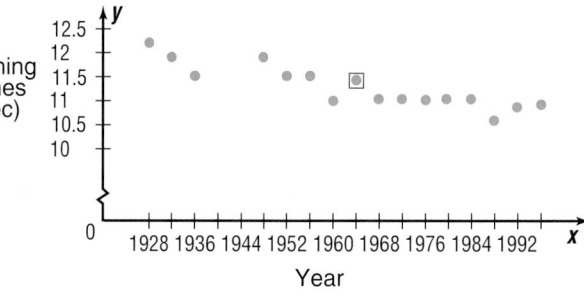

Women's Olympic 100-Meter Dash Times

a. What year and speed does the point with the box around it represent?

b. Find the point(s) that represent(s) the slowest winning time ever run in the Olympic 100-meter dash.

c. If you had trained in the 1950s for the 100-meter race, what time would have been competitive?

d. What conclusion can you make from the scatter plot?

e. Use the scatter plot to predict the winning time in the 2000 Olympics in Sydney, Australia. Check your prediction.

Independent Practice

What type of relationship, *positive*, *negative*, or *none*, is shown by each scatter plot?

11.

12.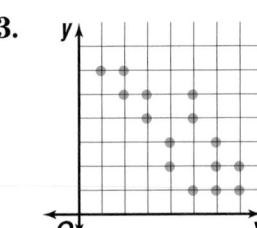

13.

14. A scatter plot of study time and test scores for a class is shown below.

Study Time (Minutes)

a. What study time and test score does the point with the box around it represent?

b. Do the data show a relationship between study time and test scores? If so, is it positive or negative?

c. Where on the plot are the points for people who studied quite a bit and had a fairly high test score?

d. How many students are in the class?

Determine whether a scatter plot of the data for the following might show a *positive*, *negative*, or *no* relationship. Explain your answer.

15. height, weight

16. armspan of student, test scores

17. hair color, height

18. outside temperature, heating bill

19. weight of car, miles per gallon

20. amount of sales, years of experience

21. month of birth, test scores

22. head circumference, height

Graphing Calculator

23. Keyboarding The table below shows the keyboarding speeds in words per minute (wpm) of 12 students. Use a graphing calculator to create a scatter plot of the data.

Experience (weeks)	4	7	8	1	6	3	5	2	9	6	7	10
Speed (wpm)	38	46	48	20	40	30	38	22	52	44	42	55

a. Do the data show a relationship between speed and experience? If so, is it positive or negative?

b. Use this scatter plot to estimate the keyboarding speed of a student with 12 weeks experience.

c. Estimate the experience level of a student whose speed is 33 wpm.

d. What conclusion can you make from the scatter plot?

24. **Sales**   A scatter plot of monthly insulated boot sales and monthly skiing accidents in West Virginia shows a positive relationship.

   a. Why might this be true?

   b. Does a positive relationship necessarily mean that one factor causes the other? Explain.

25. **Education**   The table below shows the average tuition for public colleges in the United States for the years 1989 to 1998.

Year	1989	1990	1991	1992	1993
Cost per Year ($)	1846	2035	2159	2409	2604
Year	1994	1995	1996	1997	1998
Cost per Year ($)	2820	2977	3151	3323	3489

   a. Make a scatter plot of the data.

   b. Do the data show a relationship between year and cost per year? If so, what type of relationship is it?

26. **Geography**   Use an almanac to find the area and the average depth of the ten largest oceans and seas.

   a. Construct a scatter plot to show the relationship between the average depth and area.

   b. What can you determine about the relationship between area and depth of oceans from the scatter plot?

27. **Sports**   The number of points scored and rebounds made by members of the men's professional basketball champion San Antonio Spurs is given in the table below.

   a. Make a scatter plot of the data to show the relationship between the rebounds and points.

   b. Does the scatter plot show any relationship?

   c. Could you predict the number of points a player would have if you were given the number of rebounds for that player? Explain.

   d. Do you think that the position of a player affects the number of rebounds that player will make? Explain.

   e. Do you think the amount of time played affects the number of points scored by a player? Why or why not?

Player	Rebounds	Points
Duncan	571	1084
Robinson	492	775
Elliot	213	561
Johnson	118	487
Elie	137	455
Jackson	99	301
Rose	182	284
Daniels	54	220
Kerr	44	192
Kersey	130	145
Perdue	138	90
King	14	23
Gaze	5	21
Williams	1	2

28. **Geometry**   Make a table with ten possible lengths and widths of rectangles with perimeters of 24 centimeters. Then make a scatter plot of the data. Does there appear to be a positive, negative, or no relationship between length and width?

Mixed
Review

29. Determine the domain and range of the relation {(8, 1), (4, 2), (6, −4), (5, −3), (6, 0)}. (Lesson 8-1)

30. Standardized Test Practice Solve $12x − 24 = −14x + 28$. (Lesson 7-5)

    A. 2   **B.** 3   **C.** 4   **D.** 5

31. Statistics Find the mean, median, and mode for the data set 121, 130, 128, 126, 130, 131. (Lesson 6-6)

32. Patterns Write the next three terms of the sequence 5, 6.5, 8, 9.5. (Lesson 5-9)

33. Biology Cicadas are sometimes called 17-year locusts because they emerge from the ground every 17 years. The number of a certain type of caterpillar peaks every 5 years. If the peak cycles of the caterpillars and the cicadas coincided in 1990, what will be the next year in which they will coincide? (Lesson 4-7)

For **Extra Practice**,
see page 758.

WORKING ON THE

Investigation

Refer to the Investigation on pages 368–369.

Begin working on the Investigation by doing a class experiment. Use a stopwatch to time one student doing the wave. The timer says "go" and starts the stopwatch. The student stands up and then sits down. At this point, the timer stops the stopwatch. Record one person and the time in a table.

Next, time two students doing the wave (each student represents a different section in the stadium). The timer says "go" and starts the stopwatch. The first person stands and then sits. As the first person sits, the second person stands and then sits. The timer stops the stopwatch. Record two people and the time in the table.

Continue the experiment with three people (representing three sections), then eight people, then eleven people, then fourteen people, and so on until you have determined a wave time using every student in the class.

Your table of the class should contain all the data including the number of people (sections) and their corresponding wave times. Use a scatter plot to illustrate the data. The horizontal axis will be the number of people in the wave, and the vertical axis will be the wave time.

Does this data show a relationship between the number of sections and the wave time? Does this scatter plot show a *positive*, *negative*, or *no relationship*?

Write a paper answering the questions raised by the spirit squad on page 369.

Add the results of your work to your Investigation Folder.

Graphing Linear Relations

You'll Learn

- To find solutions for relations with two variables, and

- to graph the solutions.

It's Important

Graphing linear relations can help you determine how far away a thunderstorm is from you.

Application: Science

Did you know that a cricket is nature's thermometer? You can use the equation $t = c + 40$ to find the temperature t if you know the number of chirps c a cricket makes in 15 seconds.

The chart shows the relationship between various numbers of chirps in 15 seconds and the temperature. The domain is {21, 23, 31, 50, 47}, and the relation is {(21, 61), (23, 63), (31, 71), (50, 90), (47, 87)}. The relation can also be shown in a graph or scatter plot like the one shown at the right.

Chirps in 15 seconds	c + 40	Temperature (°F)
21	21 + 40	61
23	23 + 40	63
31	31 + 40	71
50	50 + 40	90
47	47 + 40	87

CULTURAL CONNECTIONS
The first attempts at weather forecasting were made by Greek philosophers in the fourth century B.C.

Learning the Concept

Recall that solving an equation means to replace the variable so that a true sentence results. The solutions of an equation with two variables are ordered pairs.

To find a solution of such an equation, choose any value for x, substitute that value into the equation, and find the corresponding value for y.

It is often convenient to organize the solutions in a table.

Example   a. **Find four solutions for the equation $y = -2x + 1$. Write the solutions as ordered pairs.**
b. **Graph the equation $y = -2x + 1$.**

a. Choose four convenient values for x. Substitute each value for x in the expression $-2x + 1$.

Do the computation to find y.

Four solutions are $(-2, 5)$, $(0, 1)$, $(1, -1)$, and $(3, -5)$.

x	−2x + 1	y
−2	−2(−2) + 1	5
0	−2(0) + 1	1
1	−2(1) + 1	−1
3	−2(3) + 1	−5

b. First, graph the ordered pairs you found in part a.   Then draw the line that contains these points.

The ordered pair for every point on the line in part b of Example 1 is a solution of $y = -2x + 1$. It appears from the graph that $(-1.5, 4)$ and $(2, -3)$ are also solutions. You can check this by substitution.

Check $(-1.5, 4)$.   Check $(2, -3)$.
$y = -2x + 1$ $y = -2x + 1$
$4 \stackrel{?}{=} -2(-1.5) + 1$ $-3 \stackrel{?}{=} -2(2) + 1$
$4 = 4$ ✔ $-3 = -3$ ✔

An equation like $y = -2x + 1$ is called a **linear equation** because its graph is a straight line. An equation with two variables has an infinite number of solutions.

Example ②

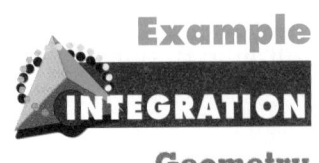

Geometry

In any right triangle, the sum of the measures of the two acute angles is 90°. If x and y represent the measures of the two acute angles, then the equation $x + y = 90$ models this condition.

a. Graph the equation.

b. Use the graph to name another solution of the equation.

c. Is $(-10, 100)$ a solution of the equation?

a. First solve the equation for y in terms of x.

$x + y = 90$
$y = 90 - x$ *Subtract x from each side.*

Now make a table of ordered pairs that satisfy the equation. Since any line can be defined by only two points, only two ordered pairs are needed. Finding a third ordered pair is a good idea to check the accuracy of the first two.

x	$90 - x$	y	(x, y)
45	$90 - 45$	45	(45, 45)
10	$90 - 10$	80	(10, 80)
70	$90 - 70$	20	(70, 20)

Then graph the ordered pairs and connect them with a line.

b. Since (50, 40) is on the graph, it is another solution of the equation.

c. Since $-10 + 100 = 90$, the ordered pair is a solution of the equation. However, since angles in triangles cannot have negative measure, the domain for this particular situation should not include negative numbers. *Should the domain include 0?*

Checking Your Understanding

Communicating Mathematics

Read and study the lesson to answer these questions.

1. **Determine** which ordered pair, (10, 20), (8, 4), or (300, 600), is a solution of $y = 0.5x$.

2. **Explain** why a linear equation has an infinite number of solutions.

3. **Indicate** which equation is graphed at the right.

   a. $x - y = 3$
   b. $x + y = 3$
   c. $y = x + 3$
   d. $y = -x - 3$

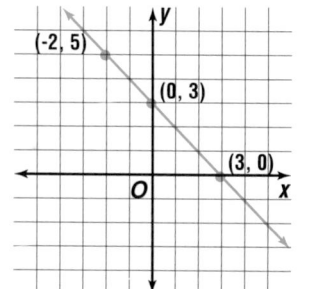

4. **Define** a linear equation.

5. **Describe** how you would graph the equation $x + y = 3$.

MATERIALS

graphing calculator

6. You can graph linear equations on a graphing calculator. First, clear all the STAT PLOTS. Next, press $\boxed{Y=}$ to access the Y= list. Use the $\boxed{CLEAR}$ key to remove any equations that are already in the list. Then enter the equation $y = 3x - 2.1$ in as function Y1.

   Enter: 3 $\boxed{X,T,\theta,n}$ $\boxed{-}$ 2.1

Press $\boxed{ZOOM}$ 6 to select the standard viewing window and complete the graph.

Graph each equation in the standard viewing window.

   a. $y = 13 - 5.5x$
   b. $y = 2x + 3.7$

Guided Practice

7. Which ordered pair(s) is a solution of $x - 3y = -7$?

   a. (2, 4)   **b.** (2, -1)   **c.** (2, 3)   **d.** (-1, 2)

Copy and complete the table for each equation. Then use the results to write four solutions for each equation. Write the solutions as ordered pairs.

8. $y = x + 4$

x	x + 4	y
−1	−1 + 4	
0		
1		
3		

9. $y = 1.5x - 2$

x	1.5x − 2	y
−2	1.5(−2) − 2	
0		
2		
4		

Find four solutions for each equation. Write the solutions as ordered pairs.

10. $y = x - 7$ **11.** $y = 5x + 2.8$ **12.** $y = -2x$

Determine whether each relation is linear.

13.

x	−2	−1	0	1	2
y	4	5	−1	2	0

14.

x	−1	−1	−1	−1	−1
y	−7	−3	0	4	12

Graph each equation.

15. $y = x - 7$ **16.** $y = 2.5x + 2$ **17.** $y = -2$ **18.** $y = \frac{1}{3}x$

19. Forestry Since trees are a renewable resource, foresters replace trees that have been cut down. Each year, more than two billion trees are planted in U.S. forests. About half of the seedlings planted survive until they are full grown. Let x represent the number of seedlings planted, and let y represent the number that become full grown trees.

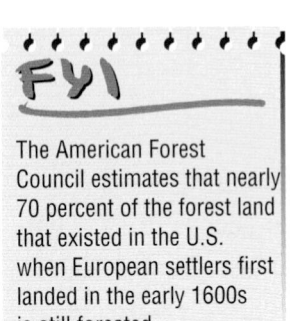

a. Write an equation that shows the relationship described above.

b. Make a table and find four solutions of the equation. Write the solutions as ordered pairs.

c. Why do negative values for x not make sense?

d. Graph the equation.

Exercises: Practicing and Applying the Concept

Independent Practice

Which ordered pair(s) is a solution of the equation?

20. $4x + 2y = 8$ **a.** (2, 0)   **b.** (0, 2)   **c.** (0.5, −3)   **d.** (1, −2)

21. $2a - 5b = 1$ **a.** (−2, −1)   **b.** (2, 1)   **c.** (7, 3)   **d.** (−7, −3)

22. $3x = 8 - y$ **a.** (3, 1)   **b.** (2, 2)   **c.** (4, −4)   **d.** (8, 0)

Find four solutions for each equation. Write the solutions as ordered pairs.

23. $y = 2.5x$

24. $y = 4x$

25. $y = 3x + 7$

26. $y = x$

27. $y = -x - 4$

28. $y = 10x - 1$

29. $x + y = 1$

30. $2x + y = 10$

31. $y - x = 1$

32. $y = \frac{1}{3}x$

33. $y = -\frac{1}{2}x + 3$

34. $y = \frac{2}{3}x - 1$

35. $x + y = 0$

36. $y = 5$

37. $x = -2$

Determine whether each relation is linear.

38. $3x + y = 20$

39. $y = x^2$

40. $y = 5$

Graph each equation.

41. $y = x + 1.5$

42. $y = -\frac{1}{2}x$

43. $y = \frac{2}{3}x + 1$

44. $y = x$

45. $2x + y = 5$

46. $x - y = 4$

47. $x + y = 2$

48. $y = 5$

49. $x = -2$

Translate each sentence into an equation. Then find four solutions for each equation. Write the solutions as ordered pairs. Graph each equation.

50. Some number is four more than a second number.

51. Some number is one-fourth a second number.

52. Four times a number minus a second number is 8.

53. A solution of $y = cx + 1$ is $(2, 5)$. Find the value of c.

54. A solution of $2x + ay = 4$ is $(-1, 2)$. Find the value of a.

55. Find a linear equation that has the following solutions.
$\{(0, 0), (1, -3), (2, -6), (-1, 3)\}$

Critical Thinking

56. Find an ordered pair that is a solution for both $x + y = 15$ and $x - y = -1$.

Applications and Problem Solving

57. Geometry The formula for finding the perimeter of a square with sides s units long is $P = 4s$.

s units

s units

  a. Find five ordered pairs of values that satisfy this condition.

  b. Draw the graph that contains these points.

  c. Why do negative values of s make no sense?

58. Physics As a thunderstorm approaches, you see lightning as it occurs, but you hear the accompanying sound of thunder a short time afterward. The distance y, in miles, that sound travels in x seconds, is given by the equation $y = 0.21x$.

  a. Create a table to find five ordered pairs of values that relate the time it takes to hear thunder and the distance from the lightning.

  b. Graph the equation.

  c. How far away is lightning when the thunder is heard 2.5 seconds after the light is seen?

59. **Weather**   The freezing point for water is 0°C and 32°F, and the boiling point is 100°C and 212°F. The relationship between Fahrenheit and Celsius temperatures is linear.

   a. Write two ordered pairs that relate these equivalent Celsius and Fahrenheit temperatures.

   b. Draw the graph that contains these two points.

   c. How could you use this graph to find other equivalents of other Celsius and Fahrenheit temperatures?

   d. Use the graph to estimate two equivalent Celsius and Fahrenheit temperatures.

Mixed Review

60. **Health**   What relationship (*positive*, *negative*, or *none*) do you think a scatter plot of Calorie intake and weight gained might show? (Lesson 8-2)

61. Solve $2a - 5 > 17$. (Lesson 7-6)

62. Express 0.004976 in scientific notation. (Lesson 6-9)

63. Estimate 2.49×1.9. (Lesson 6-2)

64. **Standardized Test Practice**   The total sales at a restaurant were $392.56 less today than they were yesterday. What were yesterday's sales if today's sales were $1293.59? (Lesson 5-6)

   A. $901.03   **B.** $905.31   **C.** $1685.05   **D.** $1686.15

65. Find $\frac{7^4}{7^2}$. Express the quotient using an exponent. (Lesson 4-8)

66. **Geometry**   The perimeter of any square is 4 times the length of one of its sides. If the perimeter of a square is 56 inches, what is the length of each side of the square? (Lesson 3-5)

67. Graph and label $A(-2.5, 4)$ and $B\left(-1\frac{1}{2}, -6\right)$. (Lesson 2-2)

68. Solve $4b = 36$ mentally. (Lesson 1-6)

For **Extra Practice**, see page 759.

69. Simplify $5n + 9n$. (Lesson 1-5)

70. Find the value of $4[12(22 - 19) - 3 \cdot 6]$. (Lesson 1-2)

From the FUNNY PAPERS

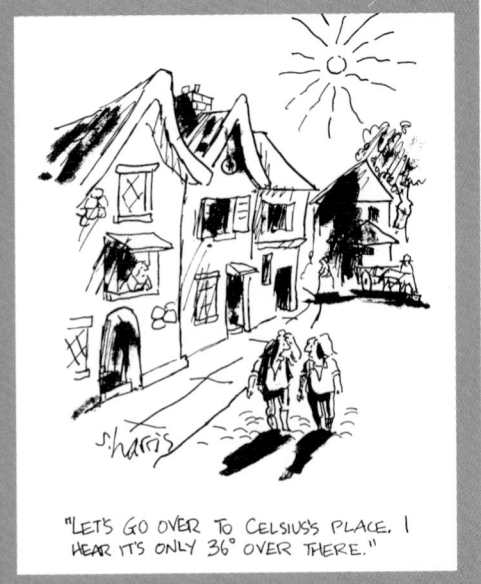

"LET'S GO OVER TO CELSIUS'S PLACE. I HEAR IT'S ONLY 36° OVER THERE."

In the United States, we measure temperature in degrees Fahrenheit, but most other countries use degrees Celsius.

1. Is it really cooler at "Celsius's Place"? Explain.

2. Use the graph you made in Exercise 59 to find the temperature in degrees Fahrenheit if it is 36°C.

8-3B Graphing Parabolas

An Extension of Lesson **8-3**

MATERIALS

graph paper

compass

Not all graphs of equations in two variables are straight lines. In this activity, you will investigate graphs that are nonlinear. In Lesson 8-3, you learned that linear equations have graphs that are straight lines. **Nonlinear equations** have graphs that are not straight lines. For example, equations of the form $y = x^2$ and $y = 2x^2$ are examples of **quadratic equations** and equations of the form $y = x^3$ and $y = 2x^3$ are examples of **cubic equations.**

The formula for the area of a square, $A = s^2$, is a quadratic equation (s is the length of a side). The formula for the volume of a cube, $V = a^3$, is a cubic equation (a is the length of an edge).

Activity

Work with a partner.

▶ On a sheet of graph paper, graph and label points $A(0, 1)$, $B(-2, 2)$, $C(2, 2)$, and $F(0, 2)$ on a coordinate grid.

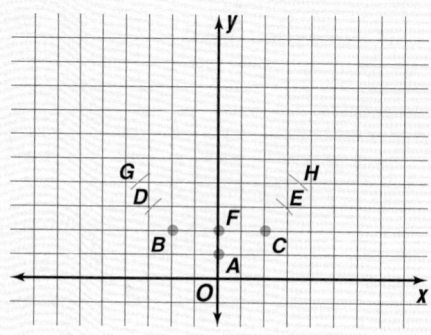

▶ Open the compass to a length of 3 units. With the compass point at $F(0, 2)$, draw an arc that intersects the grid line 3 units above the x-axis on each side of the y-axis. Label the points of intersection D and E.

▶ Open the compass to a length of 4 units. With the compass point at F, draw an arc that intersects the grid line 4 units above the x-axis on each side of the y-axis. Label the points of intersection G and H.

▶ Repeat this process to plot three more pairs of points.

▶ Sketch a smooth curve through the points you plotted.

The curve you sketched is called a **parabola**.

1. Use the compass to compare the distance above the x-axis and the distance from point F for each point you plotted.

2. Based on your answer to Exercise 1, write a definition for a parabola. (*Hint:* The x-axis is a line.)

Extension

3. **a.** Find solutions for the equation $y = x^2 - 2$ if the replacement set for x is $\{-3, -2, -1, 0, 1, 2, 3\}$.

   b. Write the solutions as ordered pairs and then graph the equation by drawing a smooth curve through the points.

   c. How does the graph of $y = x^2 - 2$ compare to the graph above?

4. Graph each nonlinear equation.

   a. the area of a square with varying lengths of the side (Use $A = s^2$.)

   b. the volume of a cube with varying lengths of the edge (Use $V = a^3$.)

8-4 Equations as Functions

- To determine whether an equation represents a function, and

- to find functional values for a given function.

It's Important

Oceanographers often use equations as functions.

MATERIALS: index cards   cardboard

Hands-On Activity

Working with a partner, make a function machine using two blank 3-by-5-inch cards. Near the middle of a card, draw three boxes. Cut out the squares on the left and on the right. Label the left "window" INPUT and the right "window" OUTPUT. Write a function rule such as $\times 2 - 3$ in the center box.

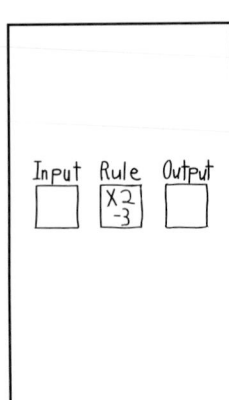

On the other card, list the integers from -5 to 4 in a column about one-half inch from the left edge. Tape this card to a piece of cardboard.

Your Turn   Place the function machine over the number column so that -5 is in the left window. Write the output in the right window. Slide the function machine down so that the input is -4. Write the output in the right window. Continue this process for all 10 inputs.

a. If the input is represented by x and the output is y, write an algebraic equation that summarizes what the function machine does.

b. Suppose you were given the output and had to determine the input. How would you go about finding the input?

Learning the Concept

The equation in the function machine above involved using two variables. Since the solutions of an equation in two variables are ordered pairs, such an equation describes a relation. The set of values of x is the domain of the relation. The set of corresponding values of y is the range.

Example
a. **Solve $y = 3x + 4$ if the domain is $\{-2, -1, 0, 1\}$.**
b. **Graph the equation.**
c. **Is the equation a function?**

a. Make a table of the domain and corresponding range values.

b. Use the ordered pairs to graph the equation.

x	y
-2	-2
-1	1
0	4
1	7

c. Use the vertical line test to determine if the equation is a function.

Since the vertical line passes through no more than one point on the graph for each value of x, the equation is a function.

Letters other than f are also used for names of functions.

Equations that represent functions can be written in **functional notation**, $f(x)$. The symbol $f(x)$ is read "f of x" and represents the value in the range that corresponds to the value of x in the domain. For example, $f(8)$ is the element in the range that corresponds to the element 8 in the domain.

You can determine a functional value by substituting the given value for x into the equation. For example, if $f(x) = 7x - 5$ and $x = 2$, then $f(2) = 7(2) - 5$ or 9.

Example If $g(x) = 3x + 4$, find each of the following.

a. $g(3)$ **b.** $g(-5)$ **c.** $2g(0)$

$$\begin{aligned} g(3) &= 3(3) + 4 \\ &= 9 + 4 \\ &= 13 \end{aligned}$$ $$\begin{aligned} g(-5) &= 3(-5) + 4 \\ &= -15 + 4 \\ &= -11 \end{aligned}$$ $$\begin{aligned} 2g(0) &= 2[3(0) + 4] \\ &= 2[0 + 4] \\ &= 2[4] \\ &= 8 \end{aligned}$$

You can use an equation to represent real-world situations.

Example ③

Real World
APPLICATION

Oceanography

CAREER CHOICES

Oceanographers are needed in all parts of the world. For more information, contact:

National Ocean Industries Assn.
1050 17th Street NW, Suite 700
Washington, D.C. 20036

Oceanographers use sonar to locate fish and other objects beneath the surface of water. Sonar units emit a pulse of sound and calculate the distance to the object using the time it takes for the pulse to reflect back. You can use the function $vt = d$, where v is the velocity, t is the time, and d is the distance to find the time. The speed or velocity of sound in water is 1454 meters per second. How far below the surface is a school of fish if it takes 0.022 seconds for the sound to return to the sonar?

The sound must go to the fish and return, so the time to the fish is $0.022 \div 2$ or 0.011 seconds.

$$vt = d$$
$$(1454)(0.011) = d \quad \text{*Replace } t \text{ with 0.011 and } v \text{ with 1454.*}$$
$$15.994 = d \quad \text{*Use a calculator.*}$$

The fish are about 16 meters from the sonar.

Communicating Mathematics

Read and study the lesson to answer these questions.

1. **Give an example** of a linear equation that is not a function.
2. **Demonstrate** how to find $f(5)$ if $f(n) = 2n - 6$.
3. **Write** a function for which $f(9) = 13$.

Guided Practice

For each equation,
a. **solve for the domain = {−3, 0, 1, 4}, and**
b. **determine if the equation is a function.**

4. $y = 5x + 8$ 5. $g(x) = x^2 + 1$ 6. $x = 7$

Given $h(x) = 4 - \frac{1}{2}x$, find each value.

7. $h(2)$ 8. $h(0)$ 9. $h(-4)$ 10. $h\left(\frac{1}{2}\right)$

11. **Personal Finance**   Carlos is saving money to buy a used car for $2200. He already has $1500 in his savings account. He plans to add $80 each week from the money he earns working at the YMCA. The equation $f(x) = 1500 + 80x$ describes Carlos' total savings $f(x)$ after x weeks. After how many weeks will he have enough to purchase the car?

Exercises: Practicing and Applying the Concept

Independent Practice

For each equation,
a. **solve for the domain = {−2, 0, 3, 5}, and**
b. **determine if the equation is a function.**

12. $x + y = -23$ 13. $y = 4x + 16$ 14. $xy = 360$ 15. $y = 2$

16. $x = -8.8$ 17. $y = 4 - \frac{1}{2}x$ 18. $x = 24 - y$ 19. $x^2 + y = 11$

Given $f(x) = 5x + 3$ and $g(x) = x^2 - 2$, find each value.

20. $g(5)$ 21. $f(-10)$ 22. $f(-4)$ 23. $g\left(\frac{1}{2}\right)$

24. $g(-3.3)$ 25. $5[g(3)]$ 26. $f\left(\frac{1}{3}\right)$ 27. $f(0.25)$

28. $f(2a)$ 29. $g(2b)$ 30. $2[g(b)]$ 31. $f[f(4)]$

Critical Thinking

32. Graph the nonlinear equations $y = x^2 - 4$ and $y = -2x^3$. Are these equations functions? Explain.

Applications and Problem Solving

33. **Science**   A temperature in degrees Fahrenheit has a corresponding Celsius temperature. Let f represent degrees Fahrenheit. Then degrees Celsius can be represented by $C = \dfrac{5(f - 32)}{9}$.
a. If the domain is {0, 32, 72, 98.6}, what is the range?
b. Graph the equation.
c. Is the relation described by $C = \dfrac{5(f - 32)}{9}$ a function? Explain.

34. **Anthropology** When a human skeleton is found, an anthropologist uses the length of certain bones to determine the height of the living person. A *femur* is the bone from the knee to the hip. The height h, in centimeters, of a female with a femur of length x is estimated by $h = 61.412 + 2.317x$.

a. Graph the equation $h = 61.412 + 2.317x$.

b. Is the relation described by $h = 61.412 + 2.317x$ a function? Explain.

c. A woman's femur measuring 49 cm is found in some ruins. What was the height of the person?

Mixed Review

35. **Mining** The deepest mine in the world is located in Carletonville, South Africa. The temperature of the walls of the mine varies with their depth. The temperature in degrees Fahrenheit, y, is estimated by $y = 18x + 66.5$, where x is the depth in kilometers. (Lesson 8-3)

a. Create a table to find five ordered pairs of values that relate the depth of a mine and the temperature of its walls.

b. Graph the equation.

c. What would the temperature of the walls be if the mine is 3.6 kilometers deep?

36. **Geometry** Find the circumference of a circle with a radius of 14 yards. (Lesson 7-4)

37. Find the value of $\left(-\dfrac{3}{4}\right)^2$. (Lesson 6-3)

38. **Standardized Test Practice** Write $\dfrac{15rs^2}{50rs}$ in simplest form. (Lesson 4-6)

  A. $\dfrac{3s}{10}$ **B.** $\dfrac{3s}{10s}$ **C.** $\dfrac{5s}{10}$ **D.** $\dfrac{5rs}{10rs}$

39. Solve $-3a \geq 18$ and graph the solution. (Lesson 3-7)

40. **Geography** The table at the right shows the number of miles of seacoast in selected states. Make a bar graph of the data. (Lesson 1-10)

State	Miles of Seacoast
Alaska	5580
Delaware	28
Florida	1350
Georgia	100
Oregon	296
Maine	228
Mississippi	44
Texas	367

For **Extra Practice**, see page 759.

Problem-Solving Strategy: Draw a Graph

You'll Learn

■ To solve problems by using graphs.

It's Important

Using graphs allow you to see patterns in data.

Imagine you are sitting in your summer clothes and sandals eating lunch and it's 74 degrees below zero! It happens every day on high-flying airplanes that are far from the heat of Earth's surface.

Altitude (thousands of feet)	Temp. (°F)
5	41
10	23
15	4
20	-15
25	-33

The temperature at different altitudes above sea level when the temperature at sea level is 60°F is shown in the table at the left.

The table does not include an altitude for -74°. At what altitude would a jet be flying when the thermometer has this reading? *This problem will be solved in Example 1.*

Learning the Concept

A graph represents the same information as its equation. But, because a graph is visual, it allows you to see patterns that may not be obvious from the equation. A graph is a powerful tool in problem solving.

Example 1

Real World APPLICATION

Air Travel

Refer to the application at the beginning of the lesson. Find the altitude where the temperature is -74°F. Assume that the temperature decreases at a constant rate.

Explore   You need to estimate an altitude that is not in the table. You know the temperature at five different altitudes and that the temperature decreases at a constant rate.

Plan   One way to solve this problem is to graph the given information. Then read the graph to find the altitude that corresponds to a temperature of -74°F.

Solve   Let the horizontal axis of the graph represent altitude. Let the vertical axis represent the temperature. Graph each ordered pair (*altitude, temperature*) using the pairs of values taken from the table. Then draw the line that contains these points.

An estimate of $-74°$ would be about halfway between -70 and -80 on the vertical axis. Estimate the altitude by reading the corresponding value on the horizontal axis. The ordered pair $(36, -74)$ appears to be on the graph, so $-74°$ corresponds to about 36,000 feet.

Examine Is the solution reasonable? The lowest temperature in the table, $-33°$, corresponds to an altitude of 25,000 feet. The temperature drops about 19 degrees for each 5,000 feet increase in altitude. $-74 - (-33°) = -41°$ and -41 is about $2(-19)$. Therefore, the altitude will be about $2 \times 5,000$ or 10,000 feet higher. Since $25,000 + 10,000 = 35,000$, a solution of 36,000 is reasonable.

You can use a graphing calculator to plot points and construct a line without erasing the points. Then you can use the trace function to answer the question.

Example ❷ **The ordered pairs (50.7, 16.47) and (99.92, 25.42) represent data from two rows of a table comparing cubic meters of natural gas used and the cost. What would the bill for 140 cubic meters of natural gas be?**

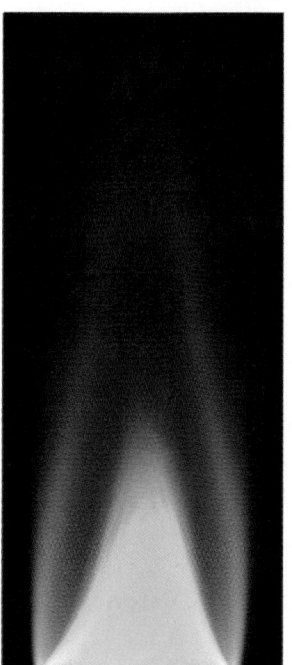

Energy

Use the scatter plot procedure on page 379 to plot the two ordered pairs using ☐ as the mark.

The calculator can draw the graph of the line that passes through the points.

Enter:

Press TRACE and the up arrow key, to trace the line. Then press the right arrow key until the cursor gets to the point on the line where the x-coordinate is about 140. The corresponding value of y is the cost for 140 cubic meters of gas.

The cost will be about $32.80.

X=140.52632   Y=32.803717

Checking Your Understanding

Communicating Mathematics

Read and study the lesson to answer these questions.

1. **Explain** how a graph can provide the same information as a table and at the same time provide more information.

2. **Describe** how you would make a graph using a table of data.

3. Estimate the corresponding value for each item using the graph you created in Example 2.

a. cost: $21.50

b. gas used: 30 cubic meters

4. Forestry An area has been reforested. After five years, one of the trees is 2 meters tall. After six more years, the same tree is 3.2 meters tall. Name two ordered pairs that can be used to graph the information.

5. Oceanography The average atmospheric pressure at sea level is 14.7 pounds per square inch. As divers go deeper into the ocean, the pressure increases as shown in the chart. Use a graph to estimate the pressure at 10,000 feet below sea level. Assume that the rate is constant.

Depth (feet)	Pressure (lb/in^2)
(Sea level) 0	14.7
500	237
1500	683
4500	2019

Exercises: Practicing and Applying the Concept

6. Retail During a storewide sale, a television that usually sells for $600 is on sale for $450. A CD player that usually sells for $200 is on sale for $150. Name two ordered pairs that can be used to graph the information.

Use a graph to solve each problem. Assume that the rate is constant in each problem.

7. Business A gas station sells 87-octane gas for $1.03 a gallon and 93-octane sells for $1.28. What is the price of 89-octane gas?

8. Physical Science A cup is attached to a spring. When a marble is placed into the cup, the spring stretches 0.6 cm. When 6 marbles are in the cup, the spring stretches 4.0 cm. How much will the spring stretch if 10 marbles are in the cup?

9. Energy Natural gas companies use *degree days* to determine the amount of gas to charge for on a calculated bill. In September, there were 300 degree days, and a homeowner used 85 cubic meters of natural gas. In January, there were 700 degree days, and the homeowner used 265 cubic meters of gas. How much gas would be used if there are 450 degree days?

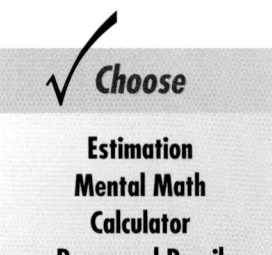

√ Choose

Estimation
Mental Math
Calculator
Paper and Pencil

Solve using the graph. Explain your method.

10. Tamoko knows that her car can travel about 30 miles on one gallon of gasoline. The graph shows this relationship.

a. Tamoko is starting on a 250-mile trip. If she has 6 gallons of gasoline in her car, will she have to buy more gasoline sometime during the trip?

b. If so, approximately when?

Driving Distance

11. **Health** Steve used a chart to see if he was close to the average weight for his height. The chart gave a weight of 130 pounds for a male 60 inches tall. The average weight for a male 66 inches tall was 143 pounds. Steve is 6 feet tall. What should Steve's weight be?

Critical Thinking

12. The graphs of $f(x) = 4x - 6$ and $g(x) = -2x + 6$ are shown at the right. Find an ordered pair that is a solution for both equations. Justify your answer.

Mixed Review

13. If $g(x) = 2x - 1$, find $g\left(\frac{5}{2}\right)$. (Lesson 8-4)

14. **Geometry** The area of a trapezoid can be found using $A = h \cdot \frac{1}{2}(b_1 + b_2)$. A trapezoid has an area 64 square inches, a height of 8 inches, and one base 7 inches long. Find the length of the other base. (Lesson 7-2)

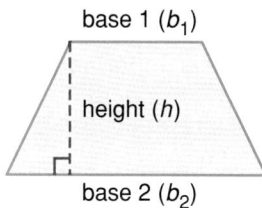

For **Extra Practice**, see page 759.

15. Solve $y = 4\frac{3}{4} - 5\frac{1}{6}$. (Lesson 5-5)

16. **Standardized Test Practice** Evaluate $(a^2 - b)^2$ if $a = 2$ and $b = 4$. (Lesson 4-2)

    A. 0   **B.** 2   **C.** 4   **D.** 6

17. Is the quotient of -42 and 7 positive or negative? (Lesson 2-8)

Self Test

Write the domain and range of the relation. Then determine whether each relation is a function. (Lesson 8-1)

1.
x	4	1	3	6
y	2	3	3	4

2. $\{(4, 2), (-3, 2), (8, 2), (8, 9), (7, 5)\}$

Use the scatter plot to answer each question. (Lesson 8-2)

3. What kind of relationship, if any, is shown by the scatter plot?

4. What does the point with the triangle around it represent?

Weights and Heights of Basketball Players

Find four solutions for each equation. Write the solutions as ordered pairs. Then draw the graph. (Lesson 8-3)

5. $y = 4x - 2$

6. $y = \frac{1}{2}x + 5$

Given $f(x) = 9x - 4$, find each value. (Lesson 8-4)

7. $f(6)$ 8. $f(0)$ 9. $f(-3)$

10. **Physical Science** At 20°C, sound travels 172 meters in 0.5 seconds and 206 meters in 0.6 seconds. Name two ordered pairs that could be used to graph this information. (Lesson 8-5)

You'll Learn

■ To find the slope of a line.

It's Important

Slope is important when designing wheelchair accessible areas.

Real World Application: Construction

In many buildings, the wheelchair ramps are built next to the walls with stairways in the middle. If you were to compare the handrail on the ramp to the handrail on the stairway, you would notice that the handrail on the ramp is not as steep as the one for the stairs.

The steepness of the handrails depends on the vertical change and the horizontal change. It can be expressed as a ratio.

$$\text{steepness} = \frac{\text{vertical change}}{\text{horizontal change}}$$

Learning the Concept

In mathematics, the **slope** m of a line describes its steepness. The vertical change is called the **change in y**, and the horizontal change is called the **change in x**.

$$\text{slope or } m = \frac{\text{change in } y}{\text{change in } x}$$

Example **Find the slope of the line graphed below.**

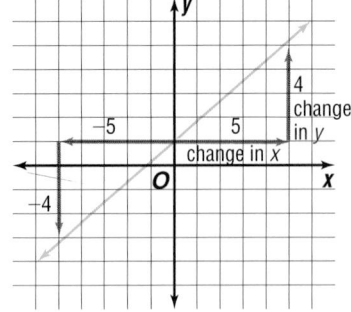

In Quadrant I, the change in y is 4, and the corresponding change in x is 5. Therefore, the slope of the line is $\frac{4}{5}$. Is the slope of the line the same in Quadrant III?

$$\frac{\text{change in } y}{\text{change in } x} \text{ or } \frac{-4}{-5} \text{ or } \frac{4}{5}$$

The slopes are the same.

The slope of a line can be determined by using the coordinates of any two points on the line. The change in y can be found by subtracting the y-coordinates. Likewise, the change in x can be found by subtracting the corresponding x-coordinates.

$$\text{slope} = \frac{\text{change in } y}{\text{change in } x} \text{ or } \frac{\text{difference in } y \text{ coordinates}}{\text{difference in } x \text{ coordinates}}$$

Example **2** **Find the slope of the line that contains $A(-2, 5)$ and $B(4, -5)$. Then graph the line.**

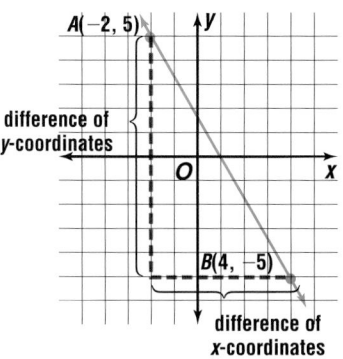

$$\text{slope} = \frac{\text{difference in } y\text{-coordinates}}{\text{difference in } x\text{-coordinates}}$$

$$\text{slope of line } AB = \frac{5 - (-5)}{-2 - 4}$$

$$= \frac{10}{-6} \text{ or } -\frac{5}{3}$$

The slope of the line is $-\frac{5}{3}$.

Graph the two points and draw the line. Use the slope to check your graph by selecting any point on the line and then go down 5 units and right 3 units. This point should also be on the line.

Sometimes the vertical change is referred to as the *rise*, and the horizontal change is referred to as the *run*. You can remember slope as *rise over run*.

$$\text{slope} = \frac{rise}{run}$$

Example **3** **Signs indicating the slope of an upcoming hill are often posted in hilly areas. Find the slope of a road that increases 633.6 feet in 2 miles.**

$$\text{slope} = \frac{rise}{run} = \frac{vertical\ change}{horizontal\ change}$$

Since 1 mile equals 5280 feet, 2 miles equals 10,560 feet.

$$\text{slope} = \frac{633.60}{10,560}$$

$$= \frac{3}{50}$$

The slope of the road is $\frac{3}{50}$. This means that the road rises 3 feet vertically for every 50 feet traveled horizontally.

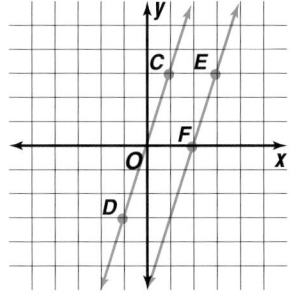

THINK ABOUT IT

Why is the slope of a horizontal line 0?

Why does a vertical line have no slope?

Connection to Geometry

The figure at the right shows the graphs of two lines that will never intersect. These lines are **parallel**. Is there a special relationship between the slopes of parallel lines?

slope of line $CD = \dfrac{3 - (-3)}{1 - (-1)} = \dfrac{6}{2}$ or 3

slope of line $EF = \dfrac{3 - 0}{3 - 2} = \dfrac{3}{1}$ or 3

The slopes are equal.

Lesson 8-6 Slope **401**

Communicating Mathematics

Read and study the lesson to answer these questions.

1. **Explain** what it means when a line has a slope of 3.

2. Match all the positions of the drawing of a ski slope below with its corresponding slope as shown in the table at the right.

Ski Surface	Slope
A. vertical drop	negative
B. downhill	none
C. flat	positive
D. uphill	0

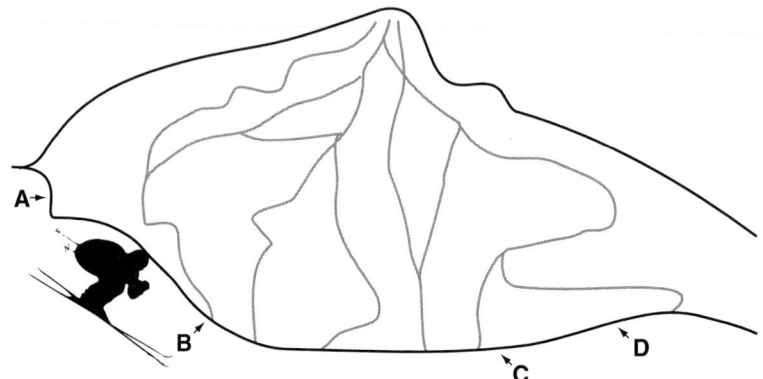

3. **Draw** the graph of a line that has a slope of $\frac{2}{3}$.

4. **Demonstrate** how to find the slope of a line when you know the coordinates of two points on the line.

5. Write a note to someone at home that describes how to tell the difference between the graph of a line with a positive slope and the graph of a line with a negative slope. Include drawings in your note.

Guided Practice

Find the slope of each line.

6.

7.

8.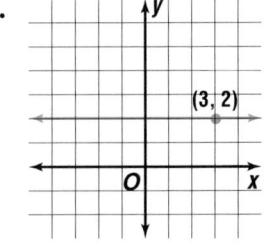

Find the slope of the line that contains each pair of points.

9. $R(9, -2), S(3, -5)$

10. $T(14, 3), U(-11, 3)$

11. $V(-1, -2), X(2, -5)$

12. $B(-6, -4), C(-8, -3)$

13. **Painting** A ladder that reaches a height of 16 feet is placed 4 feet away from the wall. What is the slope of the ladder?

Independent Practice

Determine the slope of each line named below.

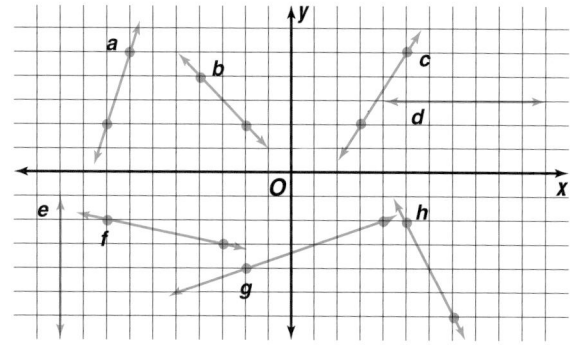

14. *a*

15. *b*

16. *c*

17. *d*

18. *e*

19. *f*

20. *g*

21. *h*

22. Look at lines *b*, *f*, and *h* above. These lines have negative slopes. Make a conjecture about how the absolute value of the slope of the line affects how steeply it slants.

23. **Statistics** What does a scatter plot that shows a positive relationship share with the graph of a line that has a slope of 1?

Find the slope of the line that contains each pair of points.

24. $C(3, 4), D(4, 6)$

25. $E(-3, 6), F(-5, 9)$

26. $G(2, 3), H(-1, 3)$

27. $J(1, -3), K(5, 4)$

28. $M(7, -4), N(9, -1)$

29. $P(5, -2), Q(4, -3)$

30. $Y(0, 0), Z(0.5, 0.75)$

31. $D(5, -1), E(-3, -4)$

32. $F\left(\frac{3}{4}, 1\right), G\left(\frac{3}{4}, -1\right)$

33. $H\left(3\frac{1}{2}, 5\frac{1}{4}\right), J\left(2\frac{1}{2}, 6\right)$

Critical Thinking

34. **Geometry** The figure at the right shows two lines that are perpendicular. Parallel lines have the same slope. Make a conjecture about the slopes of perpendicular lines.

Applications and Problem Solving

35. **Carpentry** In a stairway, the slope of the handrail is the ratio of the riser to the tread. If the tread is 12 inches long and the riser is 8 inches long, what is the slope of the handrail?

36. **Statistics** The graph at the right shows how fruit juice consumption changed over a 5-year period.

   a. Describe the section of the graph that shows when the greatest increase occurred.

   b. Describe what no change in consumption would look like on the graph.

   c. What does a section with a negative slope mean?

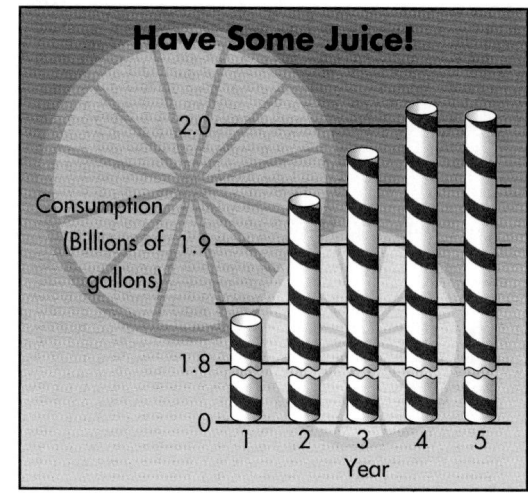

37. Driving The western entrance to the Eisenhower Tunnel in Colorado is at an elevation of 11,160 feet. The roadway has a downward slope of 0.00895 toward the eastern entrance, and its horizontal distance is 8941 feet. What is the elevation of the eastern end of the tunnel?

Mixed Review

38. Travel After 2 hours, Andrea checks the odometer in her car. She has traveled 160 kilometers. After 3 more hours, she has traveled a total distance of 400 kilometers. Use a graph to find the number of hours it will take for Andrea to travel a total distance of 560 kilometers. Assume that the rate is constant. (Lesson 8-5)

39. Food Preparation Suppose you are in charge of baking a ham for your family picnic. The directions tell you to bake the ham at 325° for 12 minutes for each pound. The ham weighs 7.45 pounds. To the nearest minute, how long should you bake the ham? (Lesson 6-5)

For **Extra Practice**, see page 760.

40. Round 24.692 to the nearest tenth. (Lesson 5-2)

41. Is the number 57 *prime* or *composite*? (Lesson 4-4)

42. Solve $a + (-7) = 8$. (Lesson 3-2)

43. Find the value of $6(5 + 9) \div 7$. (Lesson 1-2)

44. Standardized Test Practice After half of the people at a Spanish Club meeting left, one third of those remaining began to plan the club's Cinco de Mayo (5th of May) celebration. The other 18 people were cleaning the room. How many people attended the meeting? (Lesson 1-1)

A. 36   **B.** 48   **C.** 54   **D.** 60

WORKING ON THE

Investigation

Refer to the Investigation on pages 368–369.

- Transcribe the data from your table in Lesson 8-2 onto a coordinate grid using the same type of axes as your scatter plot. Connect the data points.

- If you compared a graph of the data from another class, how could you tell if 11 sections move faster than 20 sections? Explain.
- Below are three graphs of wave cheer data for a stadium with 26 sections.
- Describe each graph separately. Explain what you think happened. Are all these graphs reasonable?

Add the results of your work to your Investigation Folder.

Graph A

Graph B

Graph C

8-6B

Circles and Slope

An Extension of Lesson **8-6**

MATERIALS

- graph paper
- scissors
- tape
- cylindrical objects
- straightedge

Work with a partner.

▶ Copy the coordinate axes on a sheet of graph paper as shown at the right. Cut off the bottom of the paper so that the horizontal axis becomes the edge of the paper.

▶ Affix the graph paper to a wall or chalkboard.

▶ Hold the cylindrical base of an object against the graph paper so its left edge is at the origin and half of the cylindrical object is on either side of the horizontal axis. Mark the right side of the base where it touches the horizontal axis.

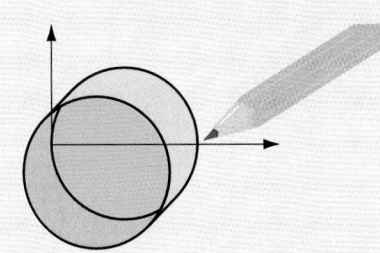

▶ Roll the cylindrical object up one complete rotation. The mark on the cylindrical object will be at the right side. Mark this point on the graph.

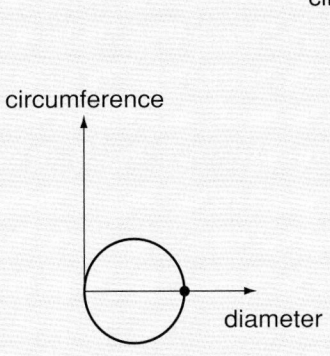

| *Start* | *Rolling* | *All the way around* |

▶ Repeat this procedure with a different cylindrical object.

▶ Write the ordered pair for each point.

▶ Draw a line through the two points marked on the graph paper.

TALK ABOUT IT

1. Estimate the slope of the line using ordered pairs.

2. Determine the slope of the line.

3. Where would you expect a point to be if you repeated the above procedure with a third cylindrical object?

Extension

4. Refer to Lesson 7-4 on circles and circumference. What should the slope of the line be? Explain your reasoning.

5. Graph the length of a semicircle as a function of the radius. Use values of *r* like 1, 2, 3, 4, and so on. What is the slope of this graph?

8-7 Intercepts

You'll Learn

- To graph a linear equation using the *x*- and *y*-intercepts or using the slope and the *y*-intercept.

It's Important

Intercepts can be used to determine the cost of any taxi ride.

Application: Sports

On your mark, get set, go! Sports analysts have found that the world record for the 10,000-meter run has been decreasing steadily since 1940. The trend can be described by the equation $y = 30.18 - 0.07x$, where *x* is the number of years since 1940 and *y* is the record time. The graph of the equation is a line that crosses the *x*-axis at approximately (431, 0) and crosses the *y*-axis at (0, 30.18). The points where a graph crosses the axes are called **intercepts**.

$y = 30.18 - 0.07x$

Learning the Concept

THINK ABOUT IT

What kind of line only crosses the *x*-axis? What kind of line only crosses the *y*-axis?

The graph of a linear function may cross either the *x*-axis, the *y*-axis, or both axes.

The **x-intercept** is the *x*-coordinate of the point where the graph crosses the *x*-axis.

The **y-intercept** is the *y*-coordinate of the point where the graph crosses the *y*-axis.

The line graphed at the right crosses both axes.

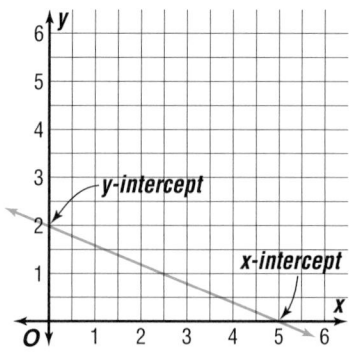

y-intercept

x-intercept

x-intercept   The line graphed above crosses the *x*-axis at (5, 0). Therefore the *x*-intercept is 5. Note that the corresponding *y*-coordinate is 0.

y-intercept   The line above crosses the *y*-axis at (0, 2). So the *y*-intercept is 2. Note that the corresponding *x*-coordinate is 0.

The *y*-intercept and the *x*-intercept can be used to graph a linear equation.

Example **1** **Graph $y = \frac{3}{2}x - 6$ using the x- and y-intercepts.**

Step 1 Find the x- and y-intercepts.

To find the x-intercept, let $y = 0$.

$$y = \frac{3}{2}x - 6$$

$$0 = \frac{3}{2}x - 6$$

$$0 + 6 = \frac{3}{2}x - 6 + 6$$

$$6 = \frac{3}{2}x$$

$$4 = x$$

The x-intercept is 4.
The ordered pair is $(4, 0)$.

To find the y-intercept, let $x = 0$.

$$y = \frac{3}{2}x - 6$$

$$y = \frac{3}{2}(0) - 6$$

$$y = -6$$

The y-intercept is -6.
The ordered pair is $(0, -6)$.

Step 2 Graph the intercepts and draw the line that contains them.

Step 3 Check by choosing some other point on the line and determine whether its ordered pair is a solution of $y = \frac{3}{2}x - 6$. Try $(2, -3)$.

$$y = \frac{3}{2}x - 6$$

$$-3 \stackrel{?}{=} \frac{3}{2}(2) - 6$$

$$-3 \stackrel{?}{=} 3 - 6$$

$$-3 = -3 \quad \checkmark$$

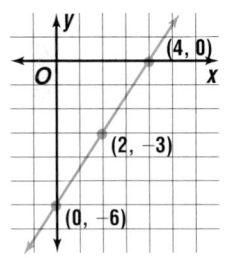

In Example 1, you may have noticed that the y-intercept for the equation was easy to find. Is there a quick way to find the slope? Use the intercepts to find the slope.

$$\text{slope} = \frac{0 - (-6)}{4 - 0} = \frac{6}{4} \text{ or } \frac{3}{2}$$

Notice that $\frac{3}{2}$ is the coefficient of x in the equation $y = \frac{3}{2}x - 6$. When a linear equation is written in the form $y = mx + b$, it is in **slope-intercept form**.

$$y = mx + b$$

slope y-intercept

You can use the slope-intercept form of an equation to graph a line quickly.

Example **2**

Real World
APPLICATION

Child Care

The cost for a child to attend a certain day care center is $19 a day plus a registration fee of $30. Make a graph that can be used to find the cost of any length of attendance.

Explore You know it costs $19 a day and there is a $30 registration fee. Let d represent the number of days. Then $C = 19d + 30$ represents the total cost, C, of attendance.

(continued on the next page)

THINK ABOUT IT

What does the ordered pair (1, 49) represent?

Plan   The *y*-intercept is 30 so the graph contains the point at (0, 30).

The slope is 19 or $\frac{19}{1}$.

Solve $\frac{19}{1} = \frac{\text{change in } y}{\text{change in } x}$

Starting at (0, 30), go to the right 1 unit and up 19 units. This will be the point at (1, 49). Then draw the line that contains (0, 30) and (1, 49).

Examine   Substitute the coordinates of a third point on the line into the equation $C = 19d + 30$ to determine whether it is a solution of the equation.

Checking Your Understanding

Communicating Mathematics

Read and study the lesson to answer these questions.

1. **Record** all the information you can determine from the graph shown at the right.

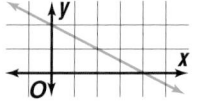

2. **Explain** how to graph a line if the *x*-intercept is -3 and the *y*-intercept is 1.

3. **Explain** how to find the *x*- and *y*-intercept for $y = -\frac{1}{2}x + 8$.

4. **Draw** a graph of a line that has a *y*-intercept, but no *x*-intercept. Describe your line.

5. **Explain** what the intercepts of the graph of the equation involving 10,000-meter run times described at the beginning of the lesson mean.

6. **Illustrate** how to graph a line if you know its *y*-intercept and its slope.

7. **Assess Yourself** Design a personal fitness program for yourself.

   a. Choose an activity from the chart and decide how many hours a week you will do the activity.

   b. Make a graph that shows the number of Calories burned for 0–10 hours.

   c. Use the graph to estimate how many Calories you will burn each week.

Activity	Calories burned per hour
Swimming	288
Walking	300
Jogging	654
Bicycling	174
Aerobics	546
Weight training	756

State the *x*-intercept and the *y*-intercept for each line.

8.

9.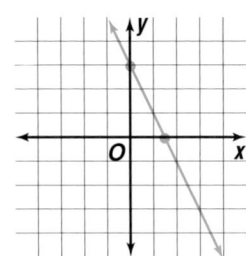

Find the *x*-intercept and the *y*-intercept for the graph of each equation. Then graph the line.

10. $y = x - 8$

11. $y = x + 6$

12. $y = -2x - 1$

13. $y = 6 - 9x$

Graph each equation using the slope and *y*-intercept.

14. $y = 2x + 1$

15. $y = -7x - 4$

16. **Travel** Tom Wishnock is taking a long trip. In the first hour, he drives only 40 miles because of heavy traffic. After that, he averages 60 miles an hour. The equation $y = 60x + 40$, where *x* represents the number of hours that Tom spent driving 60 miles per hour, and *y* represents the total distance traveled.

   a. Name the *x*-intercept and the *y*-intercept of the graph of $y = 60x + 40$.

   b. Graph the equation.

   c. How long will it take Tom to finish his 400-mile trip?

Exercises: Practicing and Applying the Concept

Independent Practice

State the *x*-intercept and the *y*-intercept for each line.

17. *a*

18. *b*

19. *c*

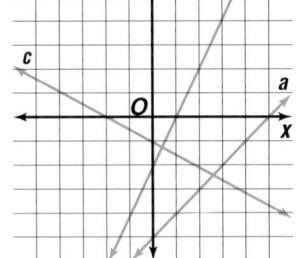

20. *d*

21. *e*

22. *f*

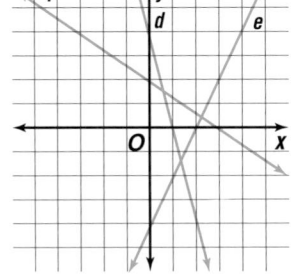

Use the *x*-intercept and the *y*-intercept to graph each equation.

23. $y = x - 2$

24. $y = x + 1$

25. $y = x - 4$

26. $y = 2x - 1$

27. $y = 3x + 4$

28. $y = -5x + 10$

29. $y = \frac{1}{2}x - 5$

30. $y = 3 - 0.5x$

31. $y = \frac{1}{3}x + 2$

32. $y = \frac{2}{3}x + 4$

33. $y = 3x - 4$

34. $y = -5x + 6$

Graph each equation using the slope and y-intercept.

35. $y = \frac{2}{3}x + 3$ **36.** $y = \frac{3}{4}x + 4$ **37.** $y = -\frac{3}{4}x + 4$

38. $-4x + y = 6$ **39.** $-2x + y = 3$ **40.** $3y - 7 = 2x$

Given the slope of a line and its y-intercept, graph the line.

41. slope $= 3$, y-intercept $= 1$ **42.** slope $= -\frac{4}{3}$, y-intercept $= -1$

Critical Thinking

43. Explain why you cannot graph the equation $y = 2x$ by using intercepts only. Then draw the graph.

Applications and Problem Solving

44. Business Suppose you have a lawn-mowing service. You charge a flat fee of $1.50 per mowing for gas, plus a fee of $4.50 per hour for labor. The equation $y = 4.5x + 1.5$ represents the total fee, y, for mowing a lawn that takes x hours to mow.

  a. Graph this equation.

  b. What does the y-intercept represent?

  c. What is your total fee for a lawn that takes 3 hours to mow?

45. Aviation The equation $a = 24,000 - 1500t$, where t is the time in minutes and a is the altitude in feet, represents the steady descent of a jetliner.

  a. Graph this equation.

  b. Name the x-intercept of the graph.

  c. What does the x-intercept represent?

Mixed Review

46. Find the slope of a line that contains the points $S(-7, -3)$ and $T(-4, -5)$. (Lesson 8-6)

47. Statistics Refer to the graph at the right. Let x represent the number of visitors to Harper's Ferry National Park. The expression $2x + 100,000$ is an estimate of the number of visitors at which park? (Lesson 7-5)

Visiting the Past	
Most Visited	**Visitors**
Gettysburg National Military Park, Pa.	1,352,728
Chickamauga and Chattanooga National Military Park, Ga.	851,534
Manassas National Battlefield Park, Va.	767,138
Kennesaw Mountain National Battlefield Park, Ga.	762,422
Harper's Ferry National Historical Park, W. Va.	624,168

For **Extra Practice**, see page 760.

48. Solve $y = 15 - 8.7$. (Lesson 5-3)

49. Find the GCF of 39 and 65. (Lesson 4-5)

50. Solve $-41 > r - (-8)$. (Lesson 3-6)

51. Simplify $-3a + 12a + (-14a)$. (Lesson 2-4)

52. Standardized Test Practice The Mets are losing to the Astros by a score of 5 to 3. Which inequality would tell how many runs would give the Mets the lead? (Lesson 1-9)

  A. $3 + x > 5$ **B.** $3 + x < 5$ **C.** $5 + x > 3$ **D.** $5 + x < 3$

53. What property allows you to say that $(9 + 10) + 7 = 9 + (10 + 7)$? (Lesson 1-4)

Families of Graphs

An Extension of Lesson **8-7**

MATERIALS

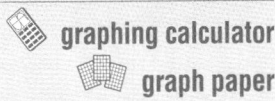
graphing calculator

graph paper

A **family of graphs** is a group of graphs that displays one or more similar characteristics. Many linear graphs are related because they have the same slope or the same *y*-intercept as other functions in the family. You can graph several functions on the same screen and observe if any family traits exist.

Activity ① **Work with a partner.**

▶ Many graphing calculators can graph up to nine functions at one time. Graph $y = 1.5x$, $y = 1.5x + 2$, $y = 1.5x + 4$, and $y = 1.5x - 4$ on the same screen in the standard viewing window.

▶ Begin by clearing the graphics screen. [Y=] Press and clear any equations in the list using the arrow and the [CLEAR] keys.

▶ Enter each equation into the Y= list.

▶ **Enter:** [Y=] 1.5 [X,T,θ,n] [ENTER] 1.5
[X,T,θ,n] [+] 2 [ENTER] 1.5
[X,T,θ,n] [+] 4 [ENTER] 1.5
[X,T,θ,n] [−] 4

▶ The standard viewing window can be selected automatically from the zoom menu. Press [ZOOM] 6 and the graphs will appear automatically.

TALK ABOUT IT

1. Describe the family of graphs.

Activity ② **Work with a partner.**

▶ Clear the graphics screen. Then graph $y = 4x$ and $y = -4x$ on the same screen in the standard viewing window. Sketch the graphs on a piece of graph paper.

▶ Then graph $y = -\frac{1}{4}x$ on the same screen. Sketch the graph on the graph paper.

▶ Finally graph $y = -6x$ and $y = -1.5x$ on the same screen. Add the graphs to the graph paper.

TALK ABOUT IT

2. Explain the difference between the graphs of $y = 4x$ and $y = -4x$.

3. Write the equation of a line whose graph is between the graph of $y = -4x$ and $y = -6x$.

Extension

4. Clear the screen and graph $y = 2x$, $y = 2x + 1$, $y = 2x + 2$, and $y = 2x + 3$. How are the graphs the same? How do they differ?

Systems of Equations

You'll Learn

- To solve systems of linear equations by graphing.

It's Important

Systems of equations can be used to determine the cost of two telephone calling plans.

Technology Activity

Between 1984 and 1994, the number of cellular phone users doubled more than 7 times.

In a questionnaire concerning phone service, potential customers were asked whether they would prefer to pay $10 a month and 10¢ for each local call or $5 a month and 20¢ for each local call. Is there a certain number of calls where the two payment plans result in the same total cost?

The Use and Cost of Cellular Phones

* in constant 1992 dollars

Source: Technology Futures, Inc.

To solve this problem, let n represent the number of local calls made in a month and let c represent the monthly cost. You can write two equations to represent the situation above.

Plan 1 $c = 10 + 0.1n$ **Plan 2** $c = 5 + 0.2n$

Your Turn   You can use a graphing calculator to solve the problem quickly. Enter each equation into the calculator and use the table feature to find the value of n for which the value of n is the same.

Enter: Y= 10 + .1 X,T,θ,n ENTER 5 +
.2 X,T,θ,n ENTER 2nd TABLE

The table on the calculator will show the values of n in the column marked X. Values of c for the plans 1 and 2 will be in columns Y_1 and Y_2 respectively.

a. Describe each output.

b. Locate the row where the outputs are the same. Explain what the identical outputs mean and why it happened.

Learning the Concept

The equations $c = 10 + 0.1n$ and $c = 5 + 0.2n$ together are called a **system of equations**. The **solution** to this system is the ordered pair that is a solution of both equations. Refer to the outputs in the situation above. What is the solution of the system of equations?

Another method for solving a system of equations is to graph the equations on the same coordinate plane. The coordinates of the point where the graphs intersect is the solution of the system of equations.

Example **Solve the system of equations $y = 2x$ and $y = -x + 3$ by graphing.**

Graph $y = 2x$ and $y = -x + 3$ on the same coordinate plane.

The lines intersect at $(1, 2)$. Therefore, the solution to the system of the equations is $(1, 2)$.

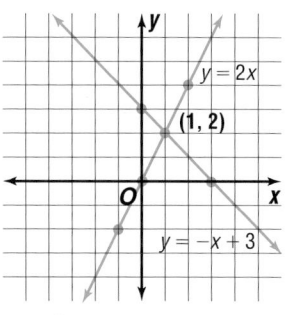

To check, substitute the coordinates into each equation to see whether it is a solution to both equations.

Check: $y = 2x$ $y = -x + 3$

$2 \stackrel{?}{=} 2(1)$ $2 \stackrel{?}{=} -(1) + 3$

$2 = 2$ ✔ $2 = 2$ ✔

Solving systems of equations is also used to solve problems.

Example ② **Refer to the application at the beginning of the lesson. Solve the system of equations $c = 10 + 0.1n$ and $c = 5 + 0.2n$ by graphing.**

FYI

The first test of a cellular phone was in Chicago in 1978.

The graphs intersect at the point $(50, 15)$.

Check:

$c = 10 + 0.1n$ $c = 5 + 0.2n$

$15 \stackrel{?}{=} 10 + 0.1(50)$ $15 \stackrel{?}{=} 5 + 0.2(50)$

$15 = 15$ ✔ $15 = 15$ ✔

The solution is $(50, 15)$. This means that two payment plans result in the same total cost, $15, when 50 local calls are made.

As you can see in the examples above, when the graphs of two linear equations intersect in exactly one point, the system has exactly one ordered pair as its solution. It is also possible for the two graphs to be parallel lines or to be on the same line. When the graphs are parallel, the system of equations does not have a solution. When the graphs are the same line, the system of equations has infinitely many solutions.

Example **Solve each system of equations by graphing.**
a. $y = x + 3$ and $y = x + 4$

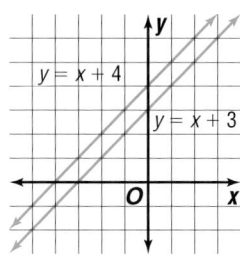

The graphs of the equations are parallel lines. Since they do not intersect, there is no solution to this system of equations.

Notice that the two lines have the same slope but different y-intercepts.

b. $y = -2x + 3$ and $3y = -6x + 9$

Each equation has the same graph. Any ordered pair on the graph will satisfy both equations. Therefore, there are infinitely many solutions to this system of equations.

Notice that the graphs have the same slopes and intercepts.

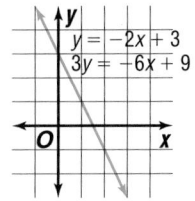

Checking Your Understanding

Communicating Mathematics

Read and study the lesson to answer these questions.

1. **Explain** what is meant by a system of equations and describe its solution.

2. **Compare and contrast** the two local calling plans at the beginning of the lesson. Which plan would you prefer? Explain.

3. **Show** why there is no solution for the system $x + y = 4$ and $y = 8 - x$.

4. **Create and graph** a system of equations that has one solution, another system that has no solution, and a third that has infinitely many solutions.

MATERIALS
- geoboard
- geobands

5. **Geometry** You can use a geoboard to model the first quadrant of the coordinate plane and geobands to model graphs of equations. Let the row of pegs on the bottom represent the x-axis and the column of pegs on the left represent the y-axis. The graph of the equation $y = x + 1$ is shown at the right.

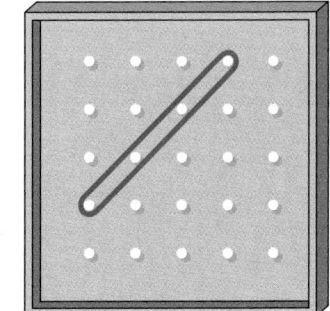

   a. Make the "line" shown at the right on a geoboard.

   b. Use a geoband to show the graph of $y = -x + 3$ on the same geoboard.

   c. What is the coordinate of the peg that is enclosed by both geobands?

   d. What is the solution to the system of equations $y = x + 1$ and $y = -x + 3$?

Guided Practice

State the solution of each system of equations.

6.

7.

8.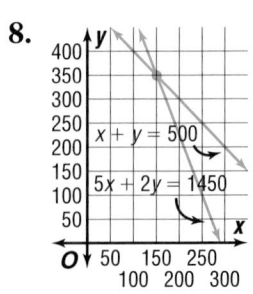

Use a graph to solve each system of equations.

9. $y = x - 8$
$y = -2x - 2$

10. $y = x + 6$
$y = 6 - 9x$

11. $y = -\frac{1}{2}x + 2$
$y = 3x - 5$

12. $y = \frac{3}{8}x - \frac{1}{2}$
$y = \frac{3}{8}x - \frac{21}{8}$

13. **Art** Kirima bought 12 feet of framing material to make a rectangular frame for her oil painting. She uses all the framing material with no waste and the length is twice the width.

a. Write a system of equations that represents this situation.

b. Solve this system of equations by graphing.

c. Explain what the solution means.

Exercises: Practicing and Applying the Concept

Independent Practice

The graphs of several equations are shown at the right. State the solution of each system of equations.

14. a and c

15. b and c

16. c and d

17. b and d

18. a and d

19. a and the x-axis

20. b and the y-axis

21. a, b, and d

22. b, c, and the y-axis

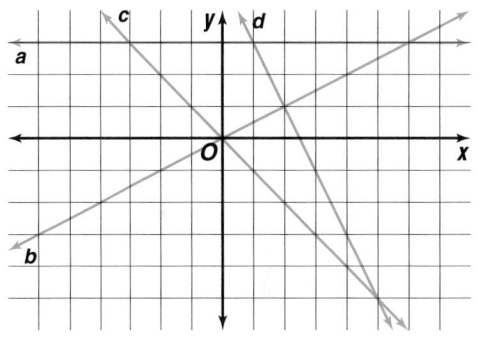

Use a graph to solve each system of equations.

23. $y = x + 2$
$y = -2x + 17$

24. $y = x$
$y = 3$

25. $y = -x + 2$
$y = 2x + 5$

26. $y = -2x - 5$
$y = -2x - 8$

27. $y = x + 6$
$y = 3x$

28. $y = -x$
$y = 4x$

29. $y = -5x - 2$
$y = 2x + 12$

30. $y = -x$
$y = 0.5x + 3$

31. $y = 3x - 8$
$y = 2x - 3$

32. $\frac{1}{2}x + \frac{1}{3}y = 6$
$y = \frac{1}{2}x + 2$

33. $\frac{2}{3}x + \frac{1}{4}y = 4$
$x = -\frac{3}{8}y + 6$

34. $3.4x + 6.3y = 4.4$
$2.1x + 3.7y = 3.1$

Critical Thinking

35. The solution for the system of equations $Ax + y = 6$ and $Bx + y = 7$ is $(1, 3)$. What are the values of A and B?

Applications and Problem Solving

36. **Geometry** The graphs of the equations $y = 2$, $3x + 2y = 1$, and $3x - 4y = -29$ contain the sides of a triangle.

a. Graph the three equations on the same coordinate plane.

b. Find the coordinates of the vertices of the triangle.

37. **Business** The Howell Company has fixed costs of $900 per week. Each item produced by the company costs $2 to manufacture and can be sold for $5. If x is the number of items produced each week, the cost of producing them can be represented by the equation $y = 900 + 2x$. The weekly income from selling the items can be represented by $y = 5x$.

    a. Solve this system of equations by graphing.

    b. The *break-even point* is the point at which the income is the same as the cost of producing the goods or services. Name the break-even point of the system of equations and explain what it means.

38. **Human Growth** Jordan Gossell was born six weeks premature. He weighed 3 pounds at birth and gained an average of 2 pounds per month. Paige Plesich was born on her due date. She weighed 7 pounds at birth and gained an average of 1 pound per month.

    a. Let a represent each baby's age in months, and let w represent each baby's weight in pounds. Write a system of equations that represents this situation.

    b. Solve this system of equations by graphing.

    c. Explain what the solution means.

39. **Law Enforcement** A police K-9 unit spots a prowler 60 meters away. The dog runs toward the prowler at a speed of 15 meters per second, and the prowler runs away at a speed of 5 meters per second.

    a. Write a system of equations that represents this situation. (*Hint*: Let x represent time and y represent distance.)

    b. Solve this system of equations by graphing.

    c. Explain what the solution means.

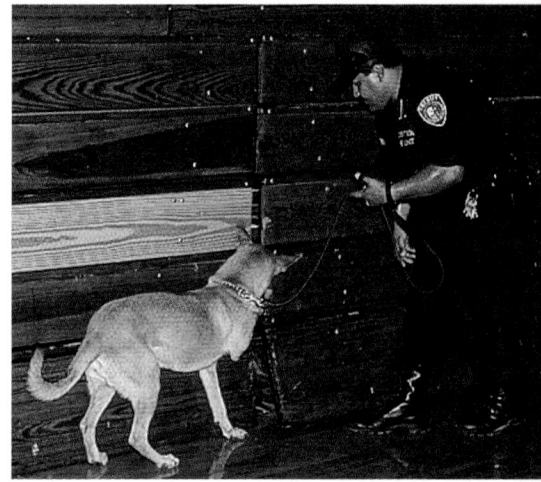

Mixed Review

40. Graph $y = -3x + 2$ by using the x-intercept and the y-intercept. (Lesson 8-7)

41. If 4.05 times an integer is increased by 3.116, the result is between 13 and 25. Use an inequality to find the number. (Lesson 7-6)

42. **Publishing** A page of type is to be divided into three columns. If the page is $6\frac{3}{4}$ inches wide, how many inches wide is each column? (Lesson 6-4)

43. **Standardized Test Practice** Leya has $575.29 in her checking account. She deposits her paycheck for $125.90. The same day, she receives bills for $397.28 and $225.40. If she pays both bills, how much is in her checking account at the end of the day? (Lesson 5-3)

    A. $78.51   **B.** $303.91   **C.** $529.31   **D.** $1072.07

44. Solve $-15 + t \leq 12$. Check your solution. (Lesson 3-6)

45. Evaluate $|-17| - |3|$. (Lesson 2-1)

46. Simplify $4(c + 3)$. (Lesson 1-5)

For **Extra Practice**, see page 760.

8-9A Graphing Inequalities

A Preview of Lesson **8-9**

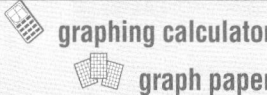
The graph of a linear function separates the coordinate plane into two regions, one above the line and one below it. For the graph of $y = 2x + 5$, shown at the right, points in the region *above* the line are represented by the inequality $y > 2x + 5$. Points in the region *below* the line are represented by the inequality $y < 2x + 5$. You can use a graphing calculator to investigate the graphs of inequalities.

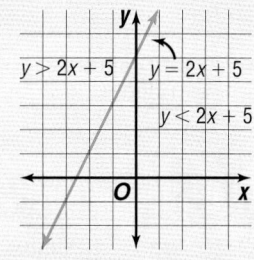

Your Turn

Work with a partner.

▶ Graph $y < 2x + 5$ in the standard viewing window.

The graphing calculator shades between two functions. The first function defines the lower boundary; in this case, Ymin, or −10. The second function defines the upper boundary; in this case, the function $y = 2x + 5$.

Enter: `2nd` `DRAW` 7 `(−)` 10 `,`
2 `X,T,θ,n` `+` 5 `)` `ZOOM`
6 `CLEAR` `ENTER`.

Notice that the graph is a shaded region. This indicates that all ordered pairs in that region satisfy the inequality $y < 2x + 5$.

▶ Clear the drawing that is currently displayed.

Enter: `2nd` `DRAW` `ENTER`.

▶ Now graph $y > 2x + 5$ in the standard viewing window.

The lower boundary is $y = 2x + 5$, and the upper boundary is Ymax or 10.

Enter: `2nd` `DRAW` 7 2 `X,T,θ,n`
`+` 5 `,` 10 `)` `ENTER`.

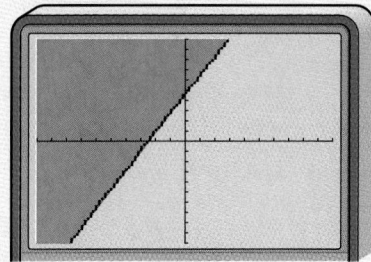

TALK ABOUT IT

1. What is similar about the two graphs?

2. How does the graph of $y < 2x + 5$ differ from the graph of $y > 2x + 5$?

3. How do the graphs differ from the graph of $y = 2x + 5$?

4. Write the keystroke sequence that would graph the following inequalities. Then use a graphing calculator to graph each inequality.

   a. $y < x + 4$ **b.** $y > 2x − 1$

   c. $y > −x + 3$ **d.** $y < −2x + 2$

Graphing Inequalities

You'll Learn

- To graph linear inequalities.

It's Important

Graphing inequalities can help your theater group determine how many tickets must be sold to cover expenses.

MATERIALS: graph paper

Hands-On Activity

The graph of $y = x + 2$ is shown at the right. Copy the axes and the graph on a piece of graph paper.

Your Turn   Graph the following ordered pairs on your coordinate system.

(3, 4)	(3, −2)	(−2, −1)
(−1, 0)	(4, 4)	(0, −3)

a. Where do these points lie in the plane in relation to the graph of $y = x + 2$?

b. Which sentence, $y = x + 2$, $y > x + 2$, or $y < x + 2$, is true for all of the ordered pairs you graphed?

c. Which sentence, $y = x + 2$, $y > x + 2$, or $y < x + 2$, is true for the points located above the graph of $y = x + 2$?

d. How many points belong to the graph of $y < x + 2$?

Learning the Concept

To draw the graph of an inequality such as $y > 2x − 3$, first, draw the graph of $y = 2x − 3$. This graph is a line that separates the coordinate plane into two regions. In the graph at the right, one of the regions is shaded yellow, and the other is shaded blue. The line is the **boundary** of the two regions.

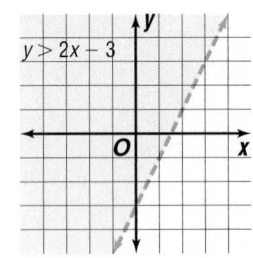

To determine which region is the solution to $y > 2x − 3$, test a point in the blue region. For example, you can test the origin, $(0, 0)$.

$$y > 2x − 3$$
$$0 \overset{?}{>} 2(0) − 3$$
$$0 > −3 \quad ✔$$

Since $0 > −3$ is true, $(0, 0)$ *is* a solution of $y > 2x − 3$.

Points on the boundary are *not* solutions of $y > 2x − 3$, so it is shown as a dashed line. Thus, the graph is all points in the blue region.

Now consider the graph of $y \leq 2x - 3$. The origin, $(0, 0)$, is not part of the graph of $y \leq 2x - 3$ since $0 \leq -3$ is not true.

Since the inequality $y \leq 2x - 3$ means $y < 2x - 3$ or $y = 2x - 3$, points on this boundary *are* part of the solution of $y \leq 2x - 3$. Therefore, the boundary is shown as a solid line on the graph.

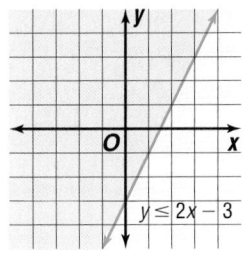

Thus, the graph is all the points in the yellow region and the line $y = 2x - 3$.

Example **Graph $y < -2x + 2$.**

Graph the equation $y = -2x + 2$. Draw a dashed line since the boundary is *not* part of the graph.

The origin, $(0, 0)$, is part of the graph since $0 < -2(0) + 2$ is true. Thus, the graph is all points in the region below the boundary. Shade this region.

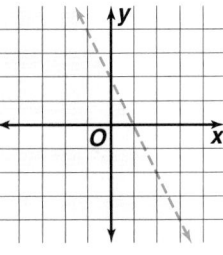

You may have to solve some inequalities for y first to find a solution to the system graphically.

Example

Sales

2 Tickets for *West Side Story* are $66.50 for floor and $47.50 for balcony. In order to cover the expenses for the show, at least $66,500 per show must be made from ticket sales.
  a. Write an inequality to represent the situation.
  b. Use a graph to determine how many of each type of ticket must be sold to cover the expenses.
  c. List three possible solutions.

a. Let x represent the number of $47.50 tickets sold, and let y represent the number of $66.50 tickets sold.

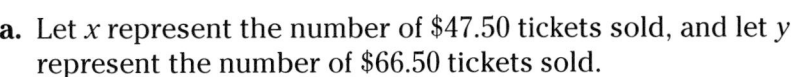

sales of $47.50 tickets	plus	sales of $66.50 tickets	is greater than or equal to	$66,500
$47.5x$	$+$	$66.5y$	$\geq$	$66,500$

b. To graph the inequality, first solve for y.

$$47.5x + 66.5y \geq 66,500$$
$$66.5y \geq 66,500 - 47.5x \quad \text{Subtract } 47.5x \text{ from each side.}$$
$$y \geq 1000 - \frac{5}{7}x \quad \text{Divide each side by } 66.5$$

Graph $y = 1000 - \frac{5}{7}x$ as a solid line since the boundary is part of the graph. The origin is not part of the graph since $0 \geq 1000 - \frac{5}{7}(0)$ is not true. Thus, the graph is all the points in the region that do not contain the origin.

c. Any point in the shaded region represents a solution. However, fractional or negative solutions do not make sense because fractional or negative numbers of tickets cannot be sold. One solution is (500, 700). This corresponds to selling five hundred $47.50 tickets and seven hundred $66.50 tickets for a total cost of 47.5(500) + 66.5(700) or $70,300. (420, 700) and (400, 800) are also solutions.

Checking Your Understanding

Communicating Mathematics

Read and study the lesson to answer these questions.

1. **Write** an inequality that describes the graph shown at the right.

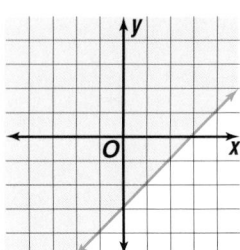

2. **You Decide**   Jeff says that inequalities with $<$ and $\leq$ symbols have dashed lines and those with $>$ and $\geq$ symbols have solid lines. Sanura disagrees. Who is correct and why?

3. **Write** an inequality whose graph is all points below the line $y = 9x + 4$.

4. **Describe** the location of solutions that result in a total cost of exactly $66,500 on the graph in Example 2.

5. **Explain** how to determine which side of the boundary line to shade when graphing an inequality.

Guided Practice

State whether the boundary is included in the graph of each inequality.

6. $y \geq x - 13$ 　　　7. $y \leq 1$ 　　　8. $y > 3x$

Determine which ordered pair(s) is a solution to the inequality.

9. $y < 2x + 1$ 　　**a.** $(-2, 2)$ 　　**b.** $(4, -1)$ 　　**c.** $(3, 1)$
10. $4y > -3x - 2$ 　　**a.** $(-1, -1)$ 　　**b.** $(2, -2)$ 　　**c.** $(-2, 4)$

Determine which region is the graph of each inequality.

11. $y > x - 1$ 　　　12. $y \leq \frac{2}{3}x + 2$ 　　　13. $y < 5x$

　　　　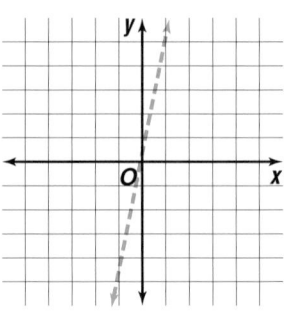

Graph each inequality.

14. $y < x + 2$ 　　15. $y \geq x$ 　　16. $y < -2x$ 　　17. $y \geq 3$

18. **Personal Finance** Silvina wants to earn at least $23 this week. Her father has agreed to pay her $2.25 an hour to weed the garden and $5.00 to mow the lawn.

    a. Graph the inequality $2.25h + 5n \geq 23$ where h is the number of hours she spends weeding the garden and n is the number of times she mows the lawn.

    b. If Silvina mows the lawn once, what is the least number of hours she will need to spend weeding the garden in order to make at least $23?

Exercises: Practicing and Applying the Concept

Independent Practice

Determine which ordered pair(s) is a solution to the inequality.

19. $y < 3 - x$ **a.** $(-2, 2)$ **b.** $(4, -1)$ **c.** $(3, 1)$

20. $3y > -1 - 2x$ **a.** $(2, 1)$ **b.** $(-5, 1)$ **c.** $(1, 1)$

21. $4y - 8 < 0$ **a.** $(0, 2)$ **b.** $(2, 5)$ **c.** $(-2, 0)$

22. $3x + 4y \geq 17$ **a.** $(1, 1)$ **b.** $(4, 2)$ **c.** $(-3, 7)$

23. $2x \geq y - 8$ **a.** $(5, 12)$ **b.** $(4, 8)$ **c.** $(-3, -1)$

24. $5x + 1 \leq 3y$ **a.** $(-2, 3)$ **b.** $(4, 7)$ **c.** $(2, 3)$

Determine which region is the graph of each inequality.

25. $y > 2x - 6$ 26. $y \geq 4 - 3x$ 27. $y > \dfrac{3}{2}x$

Graph each inequality.

28. $y < 2$ 29. $y \geq -2$ 30. $x \leq 2$ 31. $x > -4$

32. $x + y > 2$ 33. $x + y \geq -3$ 34. $y \leq 0.5x + 5$ 35. $x + y < 1$

36. $y > \dfrac{1}{4}x - 8$ 37. $y > -x - 6.5$ 38. $y \leq \dfrac{1}{3}x$ 39. $y \geq 4.5x - 0.5$

40. $-y < -x$ 41. $4y + x \leq 16$ 42. $2x \geq 3y$ 43. $-x > -y$

Critical Thinking

44. The solution to a system of inequalities is the set of all ordered pairs that satisfies *both* inequalities.

    a. Write a system of inequalities for the graph shown at the right.

    b. Check your answers by substituting into the inequalities.

45. Business For a certain business to be successful, its monthly overhead O must be at least \$3000 less than its monthly sales S.

   a. Write an inequality to represent this situation.

   b. Graph the inequality.

   c. Do points above or below the boundary line indicate a successful business? Justify your answer.

   d. Would negative numbers make sense in this problem? Explain.

   e. List two solutions.

46. Pets Biologists have found that for fish to thrive, the total length L of all the fish (in inches) in a tank must be no more than 3.75 times the number of quarts Q of water in the tank.

   a. Write an inequality to represent this situation.

   b. Graph the inequality.

   c. List two solutions where at least five 2-inch fish are in a tank. Explain what each solution means.

47. Consumer Awareness You are shopping for cassettes and CDs. Cassettes cost \$7, CDs cost \$14, and you have \$28 to spend.

   a. Use a graph to determine how many cassettes and CDs you can buy.

   b. List two solutions and explain what they mean.

48. Family Activity Investigate your family's eating habits and the U.S. recommended daily allowances (RDAs) of nutrients. Write inequalities for the recommended amounts of at least three nutrients. Does your diet measure up?

Mixed Review

49. Graph $y - 4x = 3$ and $y = x$. Then find the solution of the system of equations. (Lesson 8-8)

50. Standardized Test Practice Angie's bowling handicap is 7 less than half her average. Her handicap is 53. What is Angie's bowling average? (Lesson 7-3)

   A. 60

   B. 120

   C. 140

   D. 185

51. Order $\frac{3}{5}$, 0.63, and 6.02×10^{-1} from least to greatest. (Lesson 6-1)

52. Weather The people of Waynesburg, PA, celebrate July 26 as Rain Day, because it almost always rains on that day. Is this an example of inductive or deductive reasoning? (Lesson 5-8)

53. Use divisibility rules to determine if 38 is divisible by 2, 3, 5, 6, or 10. (Lesson 4-1)

54. Solve $6b > -31.2$. (Lesson 3-7)

55. Solve $\frac{96}{-8} = m$. (Lesson 2-8)

For **Extra Practice**, see page 761.

56. State whether $3x - 7 > 5$ is *true*, *false*, or *open*. (Lesson 1-9)

Vocabulary

After completing this chapter, you should be able to define each term, property, or phrase and give an example of two of each.

Algebra

boundary (p. 418)
domain (p. 372)
function (p. 373)
functional notation (p. 393)
linear equation (p. 386)
parabola (p. 391)
range (p. 372)
relation (p. 372)
solution of system (p. 412)
system of equations (p. 412)
vertical line test (p. 373)

Geometry

change in x (p. 400)
change in y (p. 400)
family of graphs (p. 411)
intercepts (p. 406)
parallel lines (p. 401)
slope (p. 400)
slope-intercept form (p. 407)
x-intercept (p. 406)
y-intercept (p. 406)

Problem Solving

draw a graph (p. 396)

Statistics

scatter plot (p. 378)
negative relationship (p. 380)
no relationship (p. 380)
positive relationship (p. 380)

_inter_NET
C O N N E C T I O N
Chapter Review For additional lesson-by-lesson review, visit:
www.glencoe.com/ sec/math/prealg/ mathnet

Understanding and Using Vocabulary

Choose the term from the list above that best completes each statement or phrase.

1. A _____ is a relation in which each element of the domain is paired with exactly one element in the range.

2. The set of first coordinates of a relation is called the _____.

3. The _____ is the y-coordinate of the point where a graph crosses the y-axis.

4. The set of second coordinates of a relation is called the _____.

5. A _____ is a set of ordered pairs.

6. _____ lines are always the same distance apart.

7. The _____ is the x-coordinate of the point where the graph crosses the x-axis.

8. In an equation in slope-intercept form, the b represents the _____.

Skills and Concepts

Objectives and Examples	Review Exercises

Upon completing this chapter, you should be able to:

► **use tables and graphs to represent relations and functions.** (Lesson 8-1)

Table

x	y
1	2
6	11
5	11

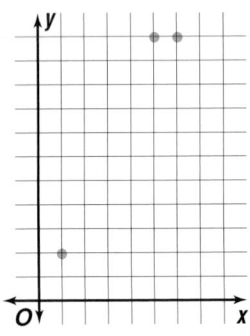

The domain is {1, 6, 5}.

The range is {2, 11}.

Since each element of the domain is paired with exactly one element of the range, this is a function.

Use these exercises to review and prepare for the chapter test.

Write the domain and range of each relation. Then determine whether each relation is a function.

9. {(7, 17), (9, 19), (0, 18), (6, 40)}

10. {(4, 5), (4, 10), (4, 22), (4, 36)}

11. {(6, 2), (8, 2), (15, 2)}

12.
x	1	18	18	35	27
y	2	4	6	4	2

13.

► **construct and interpret scatter plots** (Lesson 8-2)

Relation of years of experience and amount of money earned

The points in the scatter plot are in a pattern that slants upward to the right. So it shows a positive relationship.

14. What type of relationship, *positive, negative,* or *none,* is shown by the scatter plot below?

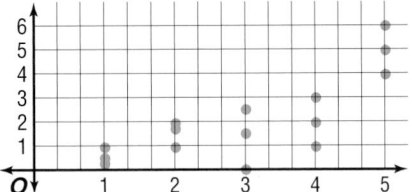

Determine whether a scatter plot of the data might show a *positive, negative,* or *no* relationship. Explain your answer.

15. size of TV set, amount of TV watched

16. number of small children, amount of glass displayed

17. age of a person, volume of the radio

18. amount of income, value of home

19. years of marriage, number of kids

Objectives and Examples

▶ **find solutions for relations with two variables and graph the solution** (Lesson 8-3)

Find four solutions for $y = 2.2x - 1$. Then graph the equation.

x	2.2x − 1	y
0	2.2 (0) − 1	−1
1	2.2 (1) − 1	1.2
2	2.2 (2) − 1	3.4
3	2.2 (3) − 1	5.6

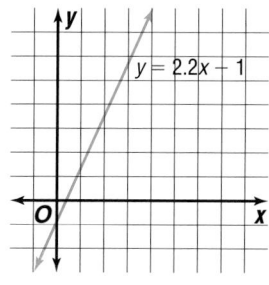

Four solutions are $(0, -1)$, $(1, 1.2)$, $(2, 3.4)$, $(3, 5.6)$.

▶ **find functional values for the given function** (Lesson 8-4)

$$g(x) = 2x^2 - 3$$

$$
\begin{aligned}
g(3) &= 2 \cdot 3^2 - 3 & g(-1) &= 2 \cdot (-1)^2 - 3 \\
&= 2(9) - 3 & &= 2(1) - 3 \\
&= 18 - 3 & &= 2 - 3 \\
&= 15 & &= -1
\end{aligned}
$$

▶ **find the slope of a line** (Lesson 8-6)

Find the slope of the line that contains $A(4, 4)$ and $B(0, 1)$.

$$
\begin{aligned}
\text{slope} &= \frac{4 - 1}{4 - 0} \\
&= \frac{3}{4}
\end{aligned}
$$

▶ **graph a linear equation using the x- and y-intercepts or using the slope and the y-intercept** (Lesson 8-7)

$y = 4x - 1$

slope
$m = 4$

y-intercept
$y = 4(0) - 1$
$y = -1$
$(0, -1)$

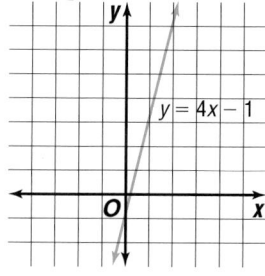

Review Exercises

Find four solutions for each equation. Write the solution as ordered pairs. Then graph each equation.

20. $y = x - 1$
21. $y = 3x + 2$
22. $x + y = 5$
23. $2x + y = 7$
24. $x - 2y = -1$
25. $x = 7$
26. $y = 1$
27. $y = \frac{1}{2}x$

Given $f(x) = x + 5$ and $g(x) = 3x^2 + 1$, find each value.

28. $f(2)$
29. $f(0)$
30. $f(-6)$
31. $g(1)$
32. $g(5)$

Find the slope of the line that contains each pair of points.

33. $A(-3, 5)$, $B(2, 7)$
34. $J(4, 1)$, $K(2, 6)$
35. $W(4, 0)$, $X(7, -3)$
36. $A(6, 18)$, $B(6, 40)$
37. $P(0, 2)$, $Q(3, 2)$
38. $W(3, 8)$, $X(1, 5)$

Use the slope and the y-intercept to graph each equation.

39. $y = 2x + 1$
40. $y = \frac{2}{3}x - 3$
41. $y = -\frac{1}{2}x + 1$
42. $y = 4$
43. $x = 7$
44. $x + 2y = 6$
45. $y = 2 - \frac{1}{2}x$
46. $5x - y = 9$

Objectives and Examples	Review Exercises

▶ **solve systems of linear equations by graphing** (Lesson 8-8)

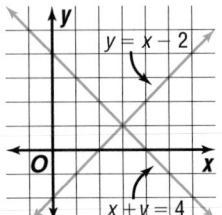

$y = x - 2$
$x + y = 4$

The lines intersect at (3, 1). So the solution to the system of equations is (3, 1).

Use a graph to solve each system of equations.

47. $x + y = 8$
$x - y = -8$

48. $y = 2x - 4$
$2y = 4x - 8$

49. $y = -x + 1$
$y = -x + 2$

50. $y = \frac{1}{2}x$
$y = \frac{3}{2}x - 2$

51. $y = 3 - x$
$y = 5 - x$

52. $x + 2y = 4$
$-2x - 2y = -10$

▶ **graph linear inequalities** (Lesson 8-9)

Graph $y > \frac{1}{3}x + 2$.

The boundary is dashed since the line is not part of the graph. The origin (0, 0) is not part of the graph since $0 < \frac{1}{3}(0) + 2$.

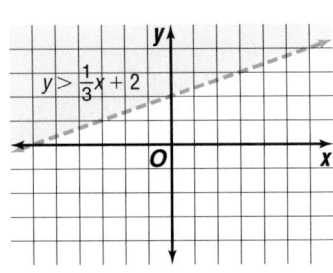

Graph each inequality.

53. $y < x - 2$

54. $y \geq -2x + 1$

55. $y > \frac{1}{2}x$

56. $y \leq -x + 8$

57. $y \leq 3x - 1$

58. $y > \frac{1}{4}x + 2$

Applications and Problem Solving

59. Nutrition This table shows nutritional values of selected foods. (Lesson 8-2)

   a. Draw a scatter plot using Calories from fat as the x values and Calories from carbohydrates as the y values.

   b. Does the graph show a positive, negative, or no relationship?

Food	Percentage of Calories from fat	Percentage of Calories from carbohydrate
Apple (medium, delicious)	9	89
Bagel (plain)	6	76
Banana (medium)	2	93
Bran muffin (large)	40	53
Fruit cookies (four cookies)	18	75
Energy bar	8	76
Candy bar (regular size)	42	51
Diet bar	30	63

60. Living Expenses An investment company charged $495 a month to rent a 2-bedroom townhouse in 1993. In 1995, the rate was $515 per month. Use a graph to determine how much the townhouse rent will be in 2000. Assume that the rate is constant. (Lesson 8-5)

61. Sports In the 1992–1993 season, Dino Ciccarelli played 82 games and scored 41 goals. The next season, he played 66 games and scored 28 goals. Use a graph to determine about how many goals Dino would score in a 76-game season. Assume that the rate is constant. (Lesson 8-5)

A practice test for Chapter 8 is available on page 781.

Demonstrate your knowledge by giving a clear, concise solution to each problem. Be sure to include all relevant drawings and justify your answers. You may show your solutions in more than one way or investigate beyond the requirements of the problem.

The table below lists the top-ranked movies and their box-office revenues.

Rank	Distributor	U.S. Gross Totals (millions)	Foreign Gross Totals (millions)
1	A	$338	$530
2	B	$179	$170
3	C	$118	$185
4	D	$45	$248
5	E	$107	$151

1. Make a graph of the information using the rank as the *x*-coordinate and the world gross totals as the *y*-coordinate. (*Hint*: The world total is the sum of U.S. and foreign revenues.)

2. Give the domain of the relation.

3. Give the range of the relation.

4. Does the chart represent a function?

5. Do you think that the film's distributor affected the amount of money it made?

6. The marketing manager of a film company has determined that a new film must make $700 million in sales between U.S. and foreign sales in order to break even.

   a. Graph an inequality to show this situation.

   b. Would the films listed in the table have broken even if this was their goal?

Thinking Critically

▶ Give two examples of relations that could be graphed on a scatter plot.

▶ Choose one of the relations you wrote. Develop sample data and graph the relation. Then determine whether it shows a positive, negative, or no relation.

 Portfolio

Select one of the assignments from this chapter that you found especially challenging. Revise your work if necessary and place it in your portfolio. Explain why you found it to be a challenge.

Section One: Multiple Choice

There are eleven multiple-choice questions in this section. After working each problem, write the letter of the correct answer on your paper.

1. If $4x - 8 = 28$, what is the value of x?
   - **A.** 5
   - **B.** 9
   - **C.** 20
   - **D.** 36

2. An investor bought 80 shares of stock for $2160 and later sold them for $1920. Which integer represents the profit or loss per share of stock?
   - **A.** -3
   - **B.** -240
   - **C.** 240
   - **D.** 3

3. What is the y-intercept for the graph of $y = x + 2$?
   - **A.** -2
   - **B.** 1
   - **C.** 0
   - **D.** 2

4. A group of divers are at a level of -20 meters. If they descend 15 meters more, at what level will they be?
   - **A.** 5 m
   - **B.** -5 m
   - **C.** 35 m
   - **D.** -35 m

5. How is the product $5 \times 5 \times 5$ expressed in exponential notation?
   - **A.** 5^2
   - **B.** 3^5
   - **C.** 5^3
   - **D.** 5^5

6. A mail-order card company charges 50¢ for each greeting card plus a handling charge of $1.50 per order. Which sentence could be used to find n, the number of cards ordered, if the total charge was $9?
   - **A.** $0.5n + 1.5 = 9$
   - **B.** $0.50n + 1.50n = 9$
   - **C.** $9 = 50n + 1.50$
   - **D.** $9 = (0.5 + 1.50)n$

7. Which number should come next in this pattern?

   $$0.5, 2, 3.5, 5, \ldots$$
   - **A.** 8
   - **B.** 7.5
   - **C.** 7
   - **D.** 6.5

8. Which is the solution of the system of equations $x + y = 4$ and $y = -3x$?
   - **A.** $(6, 2)$
   - **B.** $(1, 3)$
   - **C.** $(-2, 6)$
   - **D.** $(3, 1)$

9. One red blood cell is about 7.5×10^{-4} centimeters long. What is another way to express this measure?
   - **A.** 0.000075 cm
   - **B.** 0.00075 cm
   - **C.** 0.0075 cm
   - **D.** 0.075 cm

10. If $3x + 4(x + 1) = 5x - 8$, what is the value of x?
    - **A.** -6
    - **B.** -4.5
    - **C.** -2
    - **D.** 6

11. Express 65,000 in scientific notation.
    - **A.** 6.5×10^3
    - **B.** 6.5×10^4
    - **C.** 65×10^3
    - **D.** 0.65×10^5

Section Two: Free Response

This section contains twelve questions for which you will provide short answers. Write your answer on your paper.

12. John has $2.50. He spent one third of the money he had this afternoon at lunch. His friend Lakesha borrowed $2, and he spent 50¢ on a pop. How much money did John have when he left this morning?

13. Lee wants to set up a baseball tournament. Each team should have 12 different players. How many people does he need to have 8 teams?

14. Yolanda has $50. She bought jeans for $25. How much money can she spend for each of 4 gifts and have $1 left for parking?

15. The price of Koda's stock this morning was $87\frac{7}{8}$. At the end of today, the price had increased by $\frac{7}{8}$. What was the closing price of Koda's stock?

16. When practicing during the summer on their own, members of the Lincoln High track team must run at least 18 miles a week. If Jaime wants to run the distance in no more than 6 days each week, write and solve an inequality representing how to find the number of miles he needs to run each day.

17. Mei used a chart to see if she was close to the average weight for her height. The chart gave a weight of 118 pounds for a female 60 inches tall. The average weight for a female 64 inches tall was 134 pounds. Mei is 5 feet 8 inches tall. What should Mei's weight be?

18. How many boards, each $3\frac{1}{2}$ feet long, can be cut from a board 21 feet long?

19. Graph the inequality $y \geq 2x + 2$.

20. How can the product $4 \times 4 \times 4 \times 4 \times 4 \times 4$ be written in exponential notation?

21. A contractor is going to paint a room with four walls, each 8 feet tall and 12 feet wide. Each wall will be given two coats of paint. If each can of paint will cover up to 400 square feet, how many cans will the job take?

22. The City Marathon had 3542 participants this year. 1054 runners did not finish the race. How many runners finished the City Marathon this year?

23. A computer catalog offers discount components that customers can add to their computers. Kim chose a printer for $300, a modem for $150, and a CD-ROM drive for $225. If the tax, shipping, and handling was $71.50, how much was the total bill?

Test-Taking Tip Most standardized tests have a time limit, so you must use your time wisely. Some questions will be much easier than others. If you cannot answer a question within a few minutes, go on to the next one. If there is still time left when you get to the end of the test, go back to the questions that you skipped.

Section Three: Open-Ended

This section contains two open-ended problems. Demonstrate your knowledge by giving a clear, concise solution to each problem. Your score on these problems will depend upon how well you do the following.

• Explain your reasoning.
• Show your understanding of the mathematics in an organized manner.
• Use charts, graphs, and diagrams in your explanation.
• Show the solution in more than one way or relate it to other situations.
• Investigate beyond the requirements of the problem.

24. Explain the difference between a relation and a function using an example of each.

25. Make a scatter plot of the areas and populations of the states and territories of Australia as shown in the table below.

State/Territory	Area (thousands of square miles)	Population (thousands)
New South Wales	310	5732
Victoria	88	4244
Queensland	667	2977
Western Australia	975	1586
South Australia	380	1401
Tasmania	26	452
Australia Capital	1	280
Northern Territory	520	175

Source: *World Almanac*

Test Practice For additional test practice questions, visit:
www.glencoe.com/sec/math/prealg/ mathnet

9 Ratio, Proportion, and Percent

What You'll Learn

In this chapter, you will:

- write expressions for ratios, decimals, fractions, and percents,

- find simple probability,

- use ratios and proportions to solve problems,

- use statistics to predict, and

- solve problems by making a table.

Why It's Important

Real World

Geology In the diagram below, notice that three time periods, or eras, are given from 570 million years ago to the present. You can use percents to help you compare the amount of time occupied by each era. For example, the Paleozoic Era lasted from 570 million years ago to 225 million years ago. That is a period of 345 million years. You can find the percent of time occupied by the Paleozoic Era by using a calculator and performing these operations: $345 \div 570 \times 100 \approx 61\%$. *You will learn about using percent equations in Lesson 9-9.*

GEOLOGIC TIME SCALE

ERAS		PERIODS	MAJOR EVENTS
	Present		
Cenozoic		Quaternary	Ice Ages/active faulting in west
Mesozoic		Tertiary	Modern topography
			Oldest fossil humans
Paleozoic	65 million years ago	Cretaceous	Major mountain-building in Rockies
			Giant dinosaurs
		Jurassic	
			Oldest fossil birds
		Triassic	Beginning of opening of the Atlantic Ocean
	225 million years ago	Permian	Last major mountain-building in Appalachians and Quachitas
		Pennsylvanian	Vast coal swamps in North America
Precambrian		Mississippian	
		Devonian	Oldest forest
		Silurian	Oldest fish
		Ordovician	Major mountain-building in New England and Eastern North America
	570 million years ago	Cambrian	Seas covered most of North America
Origin of Earth ±4.5 billion years ago			

(Eras columns labeled vertically: Cenozoic Era, Mesozoic Era, Paleozoic Era)

Focus on...

To be successful in this chapter, you'll need to understand these concepts and be able to apply them. Refer to the lesson in parentheses if you need more review before beginning the chapter.

▶ *Find the slope of a line.* *(Lesson 8-6)*

Find the slope of the line that contains each pair of points.

1. $(1, 1), (-3, 5)$ **2.** $(2.5, 6), (0, 0)$

3. $(22.5, 5.5), (25, 6.5)$ **4.** $(-1, 4), (5, 4)$

▶ *Write fractions as terminating or repeating decimals.* *(Lesson 6-1)*

Write each fraction as a decimal. Use a bar to show a repeating decimal.

5. $\frac{1}{3}$ **6.** $\frac{2}{10}$ **7.** $\frac{225}{375}$ **8.** $\frac{15}{90}$

▶ *Solve equations by using the multiplication and division properties of equality.* *(Lesson 3-3)*

Solve each equation.

9. $6b = -12$ **10.** $\frac{1}{3}x = 15$ **11.** $2 = -\frac{2}{3}z$ **12.** $37 = 2.5c$

▶ *Divide fractions using multiplicative inverses.* *(Lesson 6-4)*

Find each quotient.

13. $\frac{1}{2} \div 3$ **14.** $33\frac{1}{3} \div 100$ **15.** $37\frac{1}{2} \div 100$ **16.** $32\frac{1}{3} \div 100$

Focus on...

READING SKILLS

A **ratio** is a comparison of two numbers. A **proportion** states that two ratios are equal. A proportion can be used to solve many types of problems. Solving problems that involve percent is a very common use of proportion. A **percent** is the ratio of some number to 100. For example, 50% means 50 compared to 100.

Ratios and Rates

What You'll Learn

- To write ratios as fractions in simplest form, and

- to determine unit rates.

Why It's Important

You can use unit rates to compare prices of different brands of the same product at the grocery.

MATERIALS: ☐ tape measure ▨ grid paper ✐ ruler

Hands-On Activity

People come in many different shapes and sizes. But with a closer look, you can find a pattern in some body measurements.

Your Turn Work with a partner. Use a tape measure to find the distance around your wrist and around your head to the nearest $\frac{1}{2}$ inch. Copy the table below and record the data for you and your partner. Graph the data as ordered pairs on a graph similar to the one below.

	Head Measurement	Wrist Measurement
You		
Your Partner		

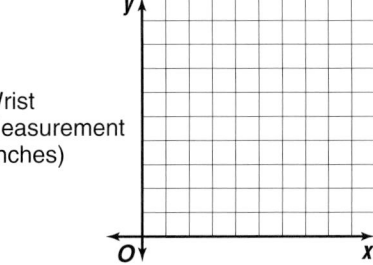

Wrist Measurement (inches)

Head Measurement (inches)

Draw the line between the two points. Then find the slope of the line.

a. Divide each wrist measurement by the corresponding head measurement and record your answers. This is the wrist to head *ratio*. How do the ratios compare to the slope of the line?

b. How do your results compare to the results of other groups?

You can review scatter plots in Lesson 8-2 and slope in Lesson 8-6.

Learning the Concept

A **ratio** is a comparison of two numbers by division. If the measurement around someone's wrist is 5 inches and the measurement of his or her head is 20 inches, the ratio comparing wrist and head measurements could be written as follows.

5 to 20   5:20 $\frac{5}{20}$

Ratios are often expressed as fractions in simplest form or as a decimal.

$$\frac{5}{20} = \frac{1}{4} \quad \textit{The GCF of 5 and 20 is 5.}$$

(÷5)

or $5 \div 20 = 0.25$

Example **Express each ratio as a fraction in simplest form and as a decimal.**
a. 6 free throws made out of 10 attempts

$$\frac{6}{10} = \frac{3}{5}$$

$\div 2$

or $6 \div 10 = 0.6$

$\frac{3}{5}$ or 0.6 of the free-throws were made.

b. $375 in commission to $9375 in sales

$$\frac{375}{9375} = \frac{\overset{1}{\cancel{3}} \cdot \overset{1}{\cancel{5}} \cdot \overset{1}{\cancel{5}} \cdot \overset{1}{\cancel{5}}}{\underset{1}{\cancel{3}} \cdot \underset{1}{\cancel{5}} \cdot \underset{1}{\cancel{5}} \cdot \underset{1}{\cancel{5}} \cdot 5 \cdot 5} \quad \text{or} \quad \frac{1}{25}$$

$375 \div 9375 = 0.04$

$\frac{1}{25}$ or 0.04 of the sales is commission.

A **rate** is a special ratio. It is a comparison of two measurements with different units of measure, like miles and gallons or cents and grams. The ratio $\frac{89 \; cents}{15 \; ounces}$ compares the price and size of a can of soup. To find the price per ounce, simplify the rate so that the denominator is 1 ounce. A rate with a denominator of 1 is called a **unit rate**.

Example

Real World
APPLICATION
Consumerism

Many grocery stores list the unit price of items on the shelf tag so that customers may compare similar items. The cost of a 12-ounce box of Brand A cereal is $3.29. A store brand cereal that is similar to Brand A costs $4.89 for an 18-ounce box. Which cereal has a lower per-ounce cost?

Brand A	**Store brand**
$\div 12$	$\div 18$
$\dfrac{329 \; cents}{12 \; ounces} = \dfrac{\bullet}{1 \; ounce}$	$\dfrac{489 \; cents}{18 \; ounces} = \dfrac{\bullet}{1 \; ounce}$
$\div 12$	$\div 18$

329 ÷ 12 = 27.417   489 ÷ 18 = 27.167

The cost is about 27.4¢ per ounce for Brand A. The store brand is about 27.2¢ per ounce. So the store brand is less expensive per ounce.

The group of Greek mathematicians called the Pythagoreans studied the geometric figures and objects in nature that displayed the **golden ratio**. The great painter, Leonardo da Vinci, and many other artists and architects have used the golden ratio to create beautiful designs for over 4000 years. You can find the value of the golden ratio by comparing measurements of the Pyramid of Khufu.

Example **3**

Real World **APPLICATION**

Architecture

One of the earliest examples of the golden ratio in architecture is the Pyramid of Khufu in Giza, Egypt, built about 2600 B.C. Find the approximate value of the golden ratio by finding the ratio of the slant height of the pyramid to half of the length of one side of the base.

slant height 611.54 ft

base 756.08 ft

$$\frac{slant\ height}{half\ of\ base\ length} = \frac{611.54}{\frac{1}{2}(756.08)}$$

$$= \frac{611.54}{378.04}$$

611.54 $\boxed{\div}$ 378.04 $\boxed{=}$ *1.617659507*

The golden ratio is approximately 1.618.

Checking Your Understanding

Communicating Mathematics

Read and study the lesson to answer these questions.

1. **Compare and contrast** ratios and rates.

2. **Write** a ratio that compares the number of students in your school to the number of teachers. Then express this ratio as a unit rate.

 MATERIALS

☐ tape measure

3. **Describe** and give an example of a unit rate.

4. Use a tape measure to measure the distance around your ankle, knee, and wrist to the nearest one-half inch.

  a. Write a knee/ankle ratio, wrist/knee ratio, and ankle/wrist ratio.

  b. Compare the ratios in part a to those of a classmate.

Guided Practice

Express each ratio or rate as a fraction in simplest form.

5. 100 to 500
6. 325:25
7. 150 miles in 6 gallons
8. 98 cents for 12 ounces
9. Juan Gonzalez' 43 home runs in 1992 to 46 home runs in 1993

Express each ratio as a unit rate.

10. 12 pounds lost in 5 weeks
11. $2.29 for a 6 pack of cola
12. 99¢ for a dozen
13. 6 inches of rain in 24 hours

14. **Consumer Awareness** A magazine reported that the average American commuter burns 190 gallons of gasoline a year driving about 4000 miles to and from work. How many miles per gallon does the average commuter get in his or her car?

Exercises: Practicing and Applying the Concept

Independent Practice

Express each ratio or rate as a fraction in simplest form.

15. 13 out of 169   **16.** 125:50   **17.** 64 to 16

18. 351:117   **19.** 156 to 84   **20.** 35 out of 149

21. 2 girls to 15 boys   **22.** 9 cars to 7 bicycles

23. 5 out of 7 oranges   **24.** 11 pennies out of 99 pennies

25. 24 marbles to 56 marbles   **26.** 15 successes in 35 attempts

√ Choose

**Estimation
Mental Math
Calculator
Paper and Pencil**

Express each ratio as a unit rate.

27. 399.3 miles on 16.5 gallons   **28.** $9.60 for 12 pounds

29. 100 meters in 10.5 seconds   **30.** 27 inches of snow in 9 hours

31. $210 for 30 tickets   **32.** 550 mL for $3.30

Critical Thinking

33. Use ratios to convert the following rates.

    a. 300 cm/min = ? mm/s   **b.** 12 kg/h = ? g/min

    c. 5 oz/min = ? qt/day   **d.** 60 mi/h = ? ft/s

    e. 20 cm/s = ? m/min   **f.** 72 km/h = ? m/s

Applications and Problem Solving

34. Architecture A golden rectangle is a rectangle in which the ratio of the length to the width is the golden ratio. p

    a. Find a golden rectangle in the photo of the Parthenon at the right.

    b. Many everyday objects are golden rectangles. For example, credit cards are approximately golden rectangles. Measure a rectangular object. Is it close to a golden rectangle?

35. Consumer Awareness Best buys in grocery stores are generally found by comparing unit rates such as cents/oz.

    a. Which is the better buy, a 16-oz bag of candy at $2.49 or a 32-oz bag at $3.69? How do you know?

    b. Are there instances in which the better buy might not necessarily be the one you choose to purchase?

36. Family Activity Visit a grocery store. Choose a product such as laundry detergent or pretzels. Find the unit price of a number of different brands. Which one would you choose and why?

37. Patterns Many sequences of numbers follow a pattern in the ratios of a term to the previous term.

    a. The sequence 3, 6, 12, 24, 48, . . . is geometric. Find the ratio between each pair of consecutive terms. What do you observe?

    b. Recall that the sequence 1, 1, 2, 3, 5, 8, 13, 21, . . . is called the Fibonacci sequence. The ratios between consecutive Fibonacci numbers approach a famous number. Find several ratios and make a conjecture about what number the ratios approach.

38. Melissa and Jacob earned some spending money by doing yardwork. The ratio of Melissa's earnings to Jacob's is 3:1. If Melissa gives Jacob $3, their ratio will be 1:1. How much money did Melissa earn?

39. Zoology On your mark, get set, go! The fastest human can run about 26 miles per hour. That may sound fast, but many animals can leave us in the dust.

   a. A pronghorn antelope is the fastest mammal over long distances. It can run 4 miles in about 6.9 minutes. What is a pronghorn's speed in miles per hour? (*Hint:* Remember that 1 hour is 60 minutes.)

   b. The fastest insect, a tropical cockroach, can scurry up to 300 feet in a minute. What is its speed in feet per second?

40. Three people are in a bicycle race to point R and back to the start. Sondra maintains a steady speed of 16 miles per hour for the entire race. Alex averaged 20 miles per hour to point R but only managed 12 miles per hour on the way back. Stephon averaged 12 miles per hour during the first third of the race, but sped up to 16 miles per hour during the second third and finished fast with a speed of 24 miles per hour in the last third. Who won the race and who came in second?

Mixed Review

41. Graph $y \leq 2x - 1$. (Lesson 8-9)

42. Solve $7(a + 2) = 5a - 8$. (Lesson 7-5)

43. Geography The table below shows the area in thousands of square miles of the oceans of the world. Find the mean, median, and mode of the areas. (Lesson 6-6)

Ocean	Area		Ocean	Area
Pacific Ocean	64,186		Hudson Bay	282
Atlantic Ocean	33,420		East China Sea	257
Indian Ocean	28,351		Andaman Sea	218
Arctic Ocean	5106		Black Sea	196
South China Sea	1149		Red Sea	175
Caribbean Sea	971		North Sea	165
Mediterranean Sea	969		Baltic Sea	148
Bering Sea	873		Yellow Sea	114
Gulf of Mexico	582		Persian Gulf	89
Sea of Okhotsk	537		Gulf of California	59
Sea of Japan	391			

Source: *World Almanac*

44. Standardized Test Practice Juice boxes come in packages of four while each package of pretzels contains 10 individually wrapped bags. What is the least number of packages of juice boxes and pretzels you can buy so that each juice box is paired with exactly one pretzel bag? (Lesson 4-7)

   A. Four packages of juice boxes and two packages of pretzels

   B. Four packages of juice boxes and four packages of pretzels

   C. Five packages of juice boxes and four packages of pretzels

   D. Five packages of juice boxes and two packages of pretzels

45. Solve $x - 9 < 19$. Graph the solution on a number line. (Lesson 3-6)

For **Extra Practice**, see page 761.

46. Find $-8 + (-5)$. (Lesson 2-4)

9-2

Problem-Solving Strategy: Make a Table

You'll Learn

■ To solve problems by making a table.

It's Important

You can use tables to organize information in an understandable way.

Application: Business

On average, 5¢ of every dollar an American spends on food is spent in a vending machine. That's a lot of nickels!

A fruit machine accepts dollars and all the fruit costs 65 cents. If the machine gives only nickels, dimes, and quarters, what combinations of coins could it be programmed to give as change for a dollar?

Learning the Concept

You can make a table to show all of the possible combinations of coins. A table allows you to organize information in an understandable way.

Explore   The machine will give back $1.00 − 0.65 or 35 cents in change in a combination that can include nickels, dimes, or quarters.

Plan   Organize data in a table using different combinations of nickels, dimes, quarters. Start with the most quarters.

Solve

quarters	dimes	nickels
1	1	0
1	0	2
0	3	1
0	2	3
0	1	5
0	0	7

Examine   The total for each combination of the coins is 35 cents. Only nickels, dimes, and quarters are used.

Checking Your Understanding

Communicating Mathematics

Read and study the lesson to answer these questions.

1. **Explain** how a table can help you solve problems.

2. What kinds of problems are best solved by making a table or chart?

Guided Practice

Solve by making a table.

3. How many ways can you make change for a 50-cent piece using only nickels, dimes, and quarters?

4. **Number Theory** How many different ways are there to add prime numbers to make a sum of 12? Show all the ways.

5. An octahedron has faces numbered 1 to 8. If two are thrown and the faces down are added, how many ways can you roll a sum less than 8?

Exercises: Practicing and Applying the Concept

Independent Practice

Solve. Use any strategy.

6. Emily had 40 baseball cards. She traded 7 cards for 5 from Dani. She traded 3 more for 4 from Ted and 2 for 1 from Anita. Finally, she traded 11 cards for 8 from Shawna. How many cards does Emily have now?

7. A penny, a nickel, a dime, and a quarter are in a purse. How many amounts of money are possible if you grab two coins at random?

8. Lisa is challenged to find her way through a maze. Each time she walks through a correct opening, she receives $1. In each correct aisle, she is given a reward equal to the total she already has. The correct path passes through 8 openings and 7 aisles. If Lisa takes the correct path, how much money will she have at the end?

9. Ms. Torres sold 100 shares of stock for $5975. When the price per share went down $5, she bought 200 more shares. When the price per share went back up $3, she sold 100 shares. How much did Ms. Torres gain or lose in these transactions?

10. Copy the figure at the right. Place the numbers 1 through 12 at the dots so that the sum of each of the six rows is 26.

11. Patterns Find the next number in the pattern 0, $-1, -3, -6, -10, -15, \ldots$.

12. Tokala has 5 coins with a total value of $1.05. He cannot make change for a dollar, 5 cents, 10 cents, 25 cents, or 50 cents. What five coins does he have?

13. Geometry If the sides of the rectangle are whole numbers of inches, how many different combinations of measures are possible?

Area = 48 in² ω

ℓ

Critical Thinking

14. A programmer is making a program that will create calendars. The program will contain calendar pages with just the days of the week and the dates, no months. The user will insert the name of the month before printing the calendar. If a page will exist for every arrangement of dates in a month, how many pages will need to be designed?

Mixed Review

15. Aviation The first non-stop, around-the-world flight was made by Capt. James Gallagher in March 1949. The 23,452-mile flight was made in 94 hours. What was the average rate in miles per hour? (Lesson 9-1)

16. Standardized Test Practice Solve $3v + 7 < 19$. (Lesson 7-6)
    A. $v < 2$ **B.** $v < 3$ **C.** $v < 4$ **D.** $v < 5$

17. Estimate $90.23 + 89.53$. (Lesson 5-2)

18. Determine whether $6c - d$ is a monomial. Explain why or why not. (Lesson 4-1)

19. Write an inequality using the numbers in the following sentence. *The Montreal Canadiens have won the Stanley Cup 24 times, and the Detroit Red Wings have won it 7 times.* (Lesson 2-3)

For **Extra Practice**, see page 761.

9-3A Fair and Unfair Games

A Preview of Lesson **9-3**

MATERIALS

dice or number cubes

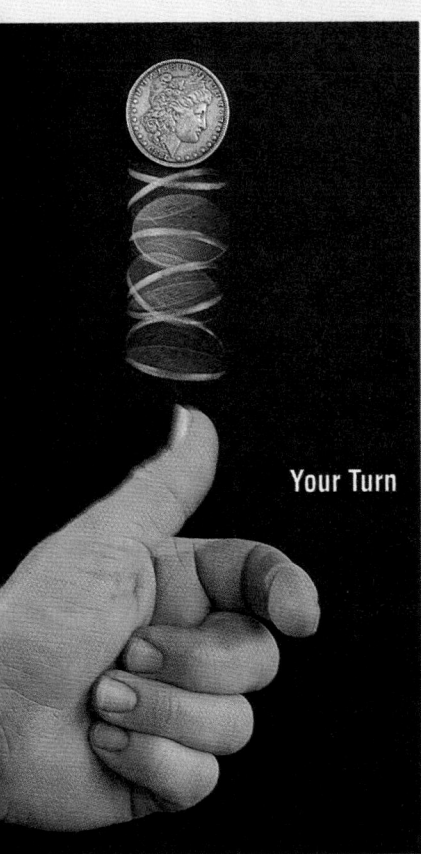

Your Turn

Have you ever heard someone say "Hey, no fair!" while you were playing a game? Most games are designed to be fair; that is, each player has an equal chance of winning. If the game is weighted in some way so that one player has a better chance of winning than the other players, the game is unfair.

Suppose you toss a coin 50 times. Each time the coin shows a head, you win and each time the coin shows tails, you lose.

Wins	Losses
⌿⌿⌿ ⌿⌿⌿ ⌿⌿⌿ ⌿⌿⌿ III	⌿⌿⌿ ⌿⌿⌿ ⌿⌿⌿ ⌿⌿⌿ ⌿⌿⌿ II

Since there are approximately the same number of wins as losses, the game of tossing a coin seems to be fair.

Work with a partner and play games A and B. Copy the table below and record your results in it.

	Win	Lose
Game A		
Game B		

Roll two dice 50 times.

Game A Find the sum of the numbers on the dice. Record an even sum as a win and record an odd sum as a loss.

Game B Find the product of the numbers on the dice. Record an even product as a win and record an odd product as a loss.

TALK ABOUT IT

1. Are the games fair or unfair? Explain.

2. If one of the games is unfair, modify the rules in order to make it fair.

Extension

3. Enriqueta is blindfolded. She picks a marble out of a bag containing 6 blue marbles and 6 red marbles. If the marble is red, she wins. If the marble is blue, she loses. Do you think this is a fair game? Why or why not?

4. Two dice are rolled. If both dice show the same number, you win. If the dice show different numbers, you lose. Do you think this is a fair game? Why or why not?

Integration: Probability
Simple Probability

You'll Learn

■ To find the probability of simple events.

It's Important

Simple probability is used in weather forecasting.

Technology Activity

A radio station in Columbus, Ohio, ran a contest called "Lucky Bucks". If you had a dollar bill that met the criteria announced by the DJ, you could call in to win $100, $1000, or even more. Suppose the DJ said, "If you have a dollar with a serial number ending in 7 or 8, be the first caller and you win!"

Use the graphing calculator program at the right to generate random numbers between 0 and 9 to represent the last digit of your serial number. Generate 50 different digits and record the number of times you qualify to win. Enter 0 as the least integer and 9 as the greatest.

```
PROGRAM: RANDNUM
: ClrHome
: Input "LEAST INTEGER", S
: Input "GREATEST INTEGER", L
: Input "NUMBER OF VALUES", N
: 0 → B
: Lbl 1
: B + 1 → B
: int ((L-S + 1) rand + S) → R
: Disp R
: Pause
: If A ≠ B
: Goto 1
```

a. How many times did you qualify?

b. Compare results with your classmates. How many times did someone in the class qualify?

c. List the digits that qualify to win and the digits that are possible.

d. Compare $\dfrac{number\ of\ times\ qualified\ to\ win}{number\ of\ times\ played}$ and $\dfrac{number\ of\ digits\ that\ qualify}{number\ of\ possible\ digits}$.

e. If you had 10 dollar bills, do you think at least one would qualify? Explain.

f. Predict the outcome for 1 dollar bill and for 100 dollar bills.

Learning the Concept

Probability is the chance some event will happen. It is the ratio of the number of ways an event can occur to the number of possible outcomes.

Definition of Probability	$Probability = \dfrac{number\ of\ ways\ a\ certain\ outcome\ can\ occur}{number\ of\ possible\ outcomes}$

Example **What is the probability that you will roll an even number when you roll a number cube?**

There are 3 possible even outcomes—2, 4, and 6.
There are 6 possible outcomes—1, 2, 3, 4, 5, and 6.

$$P(\text{even}) = \frac{number\ of\ even\ outcomes}{number\ of\ outcomes} = \frac{3}{6} \text{ or } \frac{1}{2}$$

The set of all possible outcomes is called the **sample space**. The sample space in Example 1 was {1, 2, 3, 4, 5, 6}. When you flip a coin, the sample space is heads or tails, or {heads, tails}.

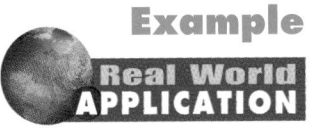

Example 2

Real World
APPLICATION

Games

In the game of backgammon, you get to roll again if you roll doubles. What is the probability of rolling doubles?

Make a table showing all of the combinations that you could roll on a pair of dice as ordered pairs.

	1	2	3	4	5	6
1	(1, 1)	(1, 2)	(1, 3)	(1, 4)	(1, 5)	(1, 6)
2	(2, 1)	(2, 2)	(2, 3)	(2, 4)	(2, 5)	(2, 6)
3	(3, 1)	(3, 2)	(3, 3)	(3, 4)	(3, 5)	(3, 6)
4	(4, 1)	(4, 2)	(4, 3)	(4, 4)	(4, 5)	(4, 6)
5	(5, 1)	(5, 2)	(5, 3)	(5, 4)	(5, 5)	(5, 6)
6	(6, 1)	(6, 2)	(6, 3)	(6, 4)	(6, 5)	(6, 6)

In the sample space, there are 6 outcomes that show doubles and 36 possible outcomes. So the probability of rolling doubles is $\frac{6}{36}$ or $\frac{1}{6}$.

Example 3

In a popular word game, each player chooses seven letter tiles. The 100 tiles are in the distribution shown at the right. Find the probability of each choice.
  a. **an E if no tiles have been chosen**
  b. **an M if 28 tiles have been chosen and 1 of them was an M**
  c. **an X if 42 tiles have been chosen and 1 of them was an X**

Number of tiles	Letters
1	J, K, Q, X, Z
2	B, C, F, H, M, P, V, W, Y, blank
3	G
4	D, L, S, U
6	N, R, T
8	O
9	A, I
12	E

  a. There are 12 E tiles and 100 tiles in all.
    $P(\text{choosing an E}) = \frac{12}{100}$ or $\frac{3}{25}$
  b. There are 2 M tiles. So, if one has been chosen, 1 remains. There are $100 - 28$ or 72 tiles from which to choose.
    $P(\text{choosing an M}) = \frac{1}{72}$
  c. There is only one X tile, so no Xs remain. The probability of choosing an X is 0.

The probability of an event is always between 0 and 1, inclusive. If something cannot happen, its probability is 0. If something is certain, its probability is 1.

0	$\frac{1}{2}$	1
impossible	50–50 chance	certain

Communicating Mathematics

Read and study the lesson to answer these questions.

1. **Define** probability and sample space in your own words.

2. Give an example of an event with a probability of $\frac{1}{6}$.

3. **Compare and contrast** an event with a probability of 0 and an event with a probability of 1.

4. **You Decide** Four coins are tossed. Manuel says it will be fair if he wins if the majority of the coins are heads and Paige wins if the majority of the coins are tails. Paige doesn't think the game is fair. Who is correct and why?

 MATERIALS
⊗ three coins

5. **a.** Use coins to help determine the sample space of the toss of three coins. Make a drawing, chart, or table to show your answer.

   b. What is the probability of two heads?

   c. What is the probability of at least one tail?

Guided Practice

State the probability of each outcome.

6. Today is the 30th of February.

7. Randomly turn to a month in a calendar and it is June.

8. A 3 is rolled on a die.

A spinner like the one at the right is used in a game. Determine the probability of spinning each outcome if the spinner is equally likely to land on each section.

9. a four

10. a one

11. an even number

12. a two or a four

13. not a three

14. less than a six

15. **Government** Recently, 1529 of the 7424 state legislators in the United States were women. If you chose a state legislator at random for an interview, what is the probability that you would choose a man?

Exercises: Practicing and Applying the Concept

Independent Practice

There are 5 red marbles, 7 green marbles, 4 black marbles, and 8 blue marbles in a bag. Suppose you select one marble at random. Find each probability.

16. $P(\text{red})$

17. $P(\text{orange})$

18. $P(\text{blue})$

19. $P(\text{not green})$

20. $P(\text{green or black})$

21. $P(\text{not yellow})$

A die is rolled. Find each probability.

22. $P(\text{odd})$

23. $P(\text{even})$

24. $P(\text{prime})$

25. $P(\text{greater than 6})$

Use the chart of possible outcomes when two dice are rolled in Example 2 to find each probability.

26. P(both odd)

27. P(both even)

28. P(both prime)

29. P(sum greater than 11)

30. P(sum of 7)

31. P(sum of 1)

Graphing Calculator

32. Refer to the application at the beginning of the lesson. What is the probability that you would have qualified to win the Lucky Bucks contest if the DJ asked for dollars that had serial numbers ending in 24? (*Hint:* Use the graphing calculator program on page 440 with numbers between 0 and 99 to simulate the situation.)

Critical Thinking

33. **Complementary events** are two events in which either one or the other must take place, but they cannot both happen at the same time. The sum of their probabilities is 1. Use this definition to solve the following problem. If a batter usually hits 3 out of every 10 pitches, what is the probability that she will *not* hit the next pitch?

Applications and Problem Solving

34. **Demographics** The graph at the right shows the population of the United States by age. Suppose that you are taking a telephone poll to determine how popular a political candidate is with different groups of voters.

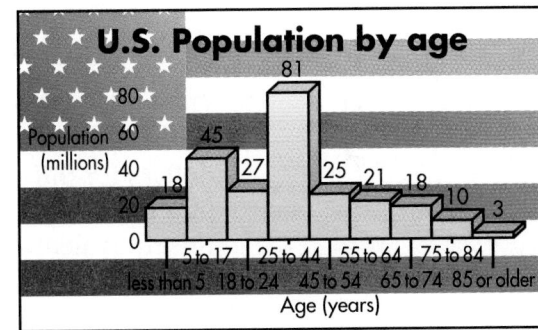

Source: U.S. Census Bureau

a. What is the probability that a randomly chosen person is old enough to vote? (*Hint:* In the United States you must be at least 18 years old to be able to vote.)

b. Find the probability that a randomly-chosen person is 65 years old or older.

35. **Space Exploration** What is the probability that humans will explore the moon on foot? What is the probability that humans will explore the Sun on foot? Explain.

For **Extra Practice**, see page 762.

36. **Family Activity** Devise an experiment using objects in your home such as tossing coins, or choosing objects out of a bag. Write each possible outcome and its probability. Then perform the experiment at least 20 times. How do the probabilities compare with the results?

Mixed Review

37. **Number Theory** How many whole numbers less than 124 are divisible by 2, 3, and 5? (Lesson 9-2)

38. Find the slope of a line that passes through $A(3, 9)$ and $B(1, -3)$. (Lesson 8-6)

39. Solve $6x + 3 = -9$. (Lesson 7-2)

40. **Standardized Test Practice** Solve $n - 9.06 = 2.31$. (Lesson 5-6)
    A. 3.92   B. 6.75   C. 11.37   D. 20.93

41. **Geometry** Find the perimeter and area of a 5.1-by-2.6-meter rectangle. (Lesson 3-5)

9-4 Using Proportions

What You'll Learn

- To use proportions to solve problems.

Why It's Important

Using proportions can help you change the number of servings in a recipe.

Real World

Application: Economics

Imagine paying almost $20 for a 2-liter bottle of soft drink. That's the kind of price that the Russian people have been paying since former President Yeltsin opened their markets to free trade in 1992.

The average monthly wage for a Russian is $110, while the average American takes home about $2197 each month. So if a Russian pays $0.44 for a loaf of bread, that is like an American paying $8.79!

The table at the right shows the cost in U.S. dollars for different items purchased in Russia and the equivalent price in terms of the average American income. What is the U.S. equivalent cost for a bottle of aspirin that costs $6.72 in Russia?

This problem will be solved in Example 3.

Boris Yeltsin

Purchase	Russian cost (U.S. dollars)	Equivalent U.S. cost
eggs (10)	0.69	13.78
blue jeans	48.00	958.56
soft drink (2 liter bottle)	0.95	18.98
gasoline (1 gallon)	1.80	35.96
fast-food hamburger	2.08	41.53
ketchup (17-ounce bottle)	2.31	46.13
American ice cream (2 quarts)	9.24	184.52
oranges (2 pounds)	1.57	31.35

Source: J. Patrick Lewis

Learning the Concept

You can use a proportion to solve problems that relate to ratios. A **proportion** is a statement of equality of two or more ratios. Consider a general proportion $\frac{a}{b} = \frac{c}{d}$.

$$\frac{a}{b} = \frac{c}{d}$$

$$\frac{a}{b} \cdot bd = \frac{c}{d} \cdot bd \quad \textit{Multiply each side by b and d to eliminate the fractions.}$$

$$ad = cb$$

The products ad and cb are called the **cross products** of a proportion. One way to determine if two ratios form a proportion is to check their cross products.

Property of Proportions	**In words:** The cross products of a proportion are equal.
	In symbols: If $\frac{a}{b} = \frac{c}{d}$, then $ad = bc$. If $ad = bc$, then $\frac{a}{b} = \frac{c}{d}$.

Example **Use cross products to determine if each pair of ratios forms a proportion.**

a. $\frac{1}{4}, \frac{4}{16}$

$$\frac{1}{4} \overset{?}{=} \frac{4}{16}$$
$$1 \cdot 16 \overset{?}{=} 4 \cdot 4$$
$$16 = 16$$
So, $\frac{1}{4} = \frac{4}{16}$.

b. $\frac{1.5}{5.0}, \frac{3}{9}$

$$1.5 \times 9 = 13.5$$
$$5 \times 3 = 15$$
$$13.5 \neq 15$$
So, $\frac{1.5}{5.0} \neq \frac{3}{9}$.

You can use cross products to solve proportions.

Example **Solve each proportion.**

a. $\frac{c}{35} = \frac{3}{7}$

$c \cdot 7 = 35 \cdot 3$ *Cross products*

$7c = 105$ *Multiply.*

$\frac{7c}{7} = \frac{105}{7}$ *Divide each side by 7.*

$c = 15$

The solution is 15.

b. $\frac{10}{8.4} = \frac{5}{d}$

$10 \cdot d = 8.4 \cdot 5$ *Cross products*

$\frac{10d}{10} = \frac{40.2}{10}$

$d = 4.2$

Thus, $d = 4.2$.

Proportions can be used to solve many real-life problems.

Example

Real World APPLICATION

Economics

Refer to the application at the beginning of the lesson. Find the U.S. equivalent cost for a bottle of aspirin.

Explore   We know the average monthly income for a Russian and an American and the Russian cost for the aspirin. We need to find the U.S. equivalent cost.

Plan   Write and solve a proportion with the Russian and American incomes and costs. The table shows that the U.S. equivalents are about 20 times the Russian costs. So an estimate of the U.S. equivalent is 20×7 or $140.

Solve   Let a represent the U.S. cost for aspirin.

$$\frac{Russian\ income}{Russian\ cost} = \frac{American\ income}{American\ cost}$$

$$\frac{110}{6.72} = \frac{2197}{a}$$

$$110 \cdot a = 6.72 \cdot 2197 \quad \textit{Find the cross products.}$$

6.72 $\boxed{\times}$ 2197 $\boxed{\div}$ 110 $\boxed{=}$ 134.2167273

The equivalent American cost is about $134.22.

Examine   The cost is close to the estimate, so it is reasonable.

Connection to > Geometry

Proportions are used to relate the size of figures with the same shape.

Example 4

After setting up the proportion, you can estimate to help you decide if your solution is reasonable.

$$\frac{1}{4}x \approx 2 \cdot 20$$

$$\frac{1}{4}x \approx 40$$

$$x \approx 160$$

Blueprints for a house show a room that is $2\frac{5}{12}$ inches long. The key to the blueprints says that $\frac{1}{4}$ inch on the blueprints represents 18 inches in the house. How long is the actual room?

Let x represent the actual room length.

$$\begin{array}{ccc} \text{blueprints} & \rightarrow & \dfrac{\frac{1}{4} \ inch}{2\frac{5}{12} \ inches} = \dfrac{18 \ inches}{x \ inches} & \leftarrow & \text{actual} \\ \text{blueprint length} & \rightarrow & & \leftarrow & \text{actual length} \end{array}$$

$$\frac{1}{4} \cdot x = 2\frac{5}{12} \cdot 18 \quad \text{\textit{Find the cross products.}}$$

$$\frac{1}{4}x = 43\frac{1}{2}$$

$$4 \cdot \frac{1}{4}x = 4 \cdot 43\frac{1}{2} \quad \text{\textit{Multiply each side by 4.}}$$

$$x = 174$$

The room is 174 inches or $14\frac{1}{2}$ feet long.

Checking Your Understanding

Communicating Mathematics

Read and study the lesson to answer these questions.

1. **Identify** the relationship between ratios and proportions.
2. **State** in your own words how to solve a proportion.
3. **Describe** three career areas that might use proportions.
4. **Demonstrate** how cross products can be used to tell whether two fractions could form a proportion.
5. **Assess Yourself** Explain how you use proportions in your life.

MATH JOURNAL

Guided Practice

Replace each ● with = or ≠ to make a true statement.

6. $\dfrac{2}{5}$ ● $\dfrac{4}{10}$

7. $\dfrac{6.25}{5}$ ● $\dfrac{2.5}{2}$

Solve each proportion.

8. $\dfrac{2}{y} = \dfrac{10}{20}$

9. $\dfrac{5}{7} = \dfrac{n}{10.5}$

10. $\dfrac{3}{a} = \dfrac{18}{24}$

11. $\dfrac{7}{16} = \dfrac{x}{4.8}$

Write a proportion that could be used to solve for each variable. Then solve.

12. 3 pounds for $1.50
 x pounds for $4.50

13. 2.7 liters at m dollars
 3 liters at $7.00

14. **Cooking** A recipe calls for $4\frac{1}{2}$ cups of flour for 72 cookies. How many cups of flour would be needed for 48 cookies?

Exercises: Practicing and Applying the Concept

Independent Practice

Replace each ● with = or ≠ to make a true statement.

15. $\dfrac{16}{5}$ ● $\dfrac{4}{2}$

16. $\dfrac{15}{2}$ ● $\dfrac{18}{2.4}$

17. $\dfrac{1}{3}$ ● $\dfrac{8.6}{25.3}$

18. $\dfrac{2.1}{3.5}$ ● $\dfrac{3}{7}$

19. $\dfrac{2}{1}$ ● $\dfrac{15}{7.5}$

20. $\dfrac{3}{8}$ ● $\dfrac{2.4}{0.64}$

Solve each proportion.

21. $\dfrac{5}{m} = \dfrac{25}{35}$ **22.** $\dfrac{r}{3.5} = \dfrac{7.2}{9}$ **23.** $\dfrac{9.6}{t} = \dfrac{1.6}{7}$ **24.** $\dfrac{6.4}{0.8} = \dfrac{8.1}{y}$

25. $\dfrac{5.1}{1.7} = \dfrac{7.5}{t}$ **26.** $\dfrac{2.4}{1.6} = \dfrac{s}{3.4}$ **27.** $\dfrac{18}{12} = \dfrac{24}{g}$ **28.** $\dfrac{7}{45} = \dfrac{x}{9}$

29. $\dfrac{7}{16} = \dfrac{x}{4.8}$ **30.** $\dfrac{3.5}{6.2} = \dfrac{7.35}{b}$ **31.** $\dfrac{3.2}{4.8} = \dfrac{6.8}{y}$ **32.** $\dfrac{t}{0.2} = \dfrac{9.4}{2}$

Write a proportion that could be used to solve for each variable. Then solve the proportion.

33. 5 liters at $6.15
$\quad$ x liters at $8.00

34. 625 bushels for 5 acres
$\quad$ 250 bushels for y acres

35. 8 boxes in 2 crates
$\quad$ z boxes in 5 crates

36. 100 candies in 3 bags
$\quad$ 300 candies in d bags

37. 25 envelopes in 5 boxes
$\quad$ m envelopes in 25 boxes

38. 12 pens in one package
$\quad$ 30 pens in p packages

Solve each proportion.

39. $\dfrac{8}{28} = \dfrac{2x}{7}$ **40.** $\dfrac{1.4}{4} = \dfrac{0.28}{m}$ **41.** $\dfrac{5}{4} = \dfrac{x+5}{16}$ **42.** $\dfrac{15}{m-3} = \dfrac{3}{14}$

Critical Thinking

43. Find two integers such that the ratio of the difference to the sum is 1:7 and the ratio of the sum to the product is 7:24.

Applications and Problem Solving

44. Photography In simple cameras, like the one shown, light from a subject passes through a lens and makes an image on film. The image and subject are always in proportion.

$$\dfrac{\text{image size}}{\text{subject size}} = \dfrac{\text{image distance from lens}}{\text{subject distance from lens}}$$

Find the distance represented by x in the diagram.

45. The amount of paint needed to paint a surface is directly proportional to the area of the surface. If 1 quart of paint is needed to paint a square that is 2 feet on a side, how many quarts must be purchased in order to paint a square that is 3 feet 6 inches on a side?

Mixed Review

46. Games Lin needs to get a jack or higher when she cuts a deck of cards in order to win the deal. If an ace is the highest card possible, what is the probability that Lin will get the deal? (Lesson 9-3)

47. Find the slope of the line that contains $Q(5, 6)$ and $S(2, -2)$. (Lesson 8-6)

48. Solve $-8k - 21 = 75$. (Lesson 7-2)

49. Write 9,412,000 in scientific notation. (Lesson 6-9)

50. Write $0.\overline{7}$ as a fraction in simplest form. (Lesson 5-1)

51. Represent 4^{-3} using positive exponents. (Lesson 4-9)

52. Standardized Test Practice Find the next number in the pattern 1, 2, 4, 5, 7, 8, 10, 11, . . . (Lesson 2-6)

$\quad$ **A.** 12 $\qquad$ **B.** 13 $\qquad$ **C.** 14 $\qquad$ **D.** 15

For **Extra Practice,** see page 762.

9-4B Capture-Recapture

A Extension of Lesson **9-4**

MATERIALS

⬭ small bowls

🫘 dry beans

Did you know that there is a way to estimate how many deer are in a forest? Often naturalists want to know such a population, but it would be impossible or impractical to make an actual count.

One method of estimating a population is the **capture-recapture** technique. In this activity, you will model this technique using dry beans as "deer" and a bowl as the "forest."

Your Turn

▶ CAPTURE   Fill a small bowl with dry beans. Grab a small handful of beans. Mark each bean with an X on both sides. Count the "tagged" beans and record this number in a chart like the one below. This number is the original number captured. Return the "tagged" beans to the bowl and mix well.

Original Number Captured: ___		
Sample	**Recaptured**	**Tagged**
A		
B		
C		

J		
Total		

▶ RECAPTURE   Grab another small handful of beans. Count the total number of beans you grabbed. This number is the number *recaptured*. Count the number of "tagged" beans. Record these numbers. This is sample A. Return all the beans to the bowl and mix.

▶ Repeat RECAPTURE nine more times for samples B through J. Find the total tagged and the total recaptured.

▶ Use the proportion shown below to estimate the number of beans in your bowl.

$$\frac{\text{original number captured}}{\text{number in bowl}} = \frac{\text{total tagged in samples}}{\text{total recaptured}}$$

1. Why is it a good idea to base your prediction on several samples instead of just one sample?

2. What would happen to your estimate if some of your tags fell or wore off?

3. Count the number of beans in your bowl. How does your estimate compare to the actual number?

4. **Research**   Investigate how researchers tag deer and estimate the total number of deer in a forest. Explain how the activity with the bowl and beans simulates their system.

9-5

Using the Percent Proportion

You'll Learn

- To use the percent proportion to write fractions as percents, and

- to solve problems involving percents.

It's Important

The percent proportion is used to make and analyze circle graphs.

Application: Physical Therapy

If you were looking for a bodybuilder, you would probably head for the local gym. But did you think of looking at the medical center? You may not find Arnold Schwarzenegger there, but you will find physical therapists. Physical therapists build bodies by helping people to use exercise to regain their strength after an injury or surgery.

Peggy Harper is a physical therapist for the Findlay City Schools in Findlay, Ohio. She is working with Chris McDaniel, a football player who is recovering from a knee injury. The muscles around Chris's knee are weak from lack of use. Ms. Harper uses a dynamometer to compare the strength of the recovering leg to the strength of the uninjured leg.

Peggy Harper

Ms. Harper determined that Chris can lift 70 foot-pounds with his weak leg and 125 foot-pounds with his strong leg. Using these numbers, she can find the **percent** of strength of the injured leg.

Learning the Concept

A percent is a ratio that compares a number to 100. Percent also means *hundredths*, or *per hundred*. The symbol for percent is %. To find the percent of strength of the weak leg, set up a proportion.

$$\frac{\text{foot-pounds lifted with injured leg}}{\text{foot-pounds lifted with uninjured leg}} = \frac{\text{strength of injured leg}}{100}$$

$$\frac{70}{125} = \frac{x}{100}$$

$70 \cdot 100 = 125x$ *Write the cross products.*

$$\frac{70 \cdot 100}{125} = \frac{125x}{125}$$ *Divide each side by 125.*

$$56 = x$$

So, the injured leg has 56% of the strength of the strong leg.

In the proportion that Ms. Harper used, 70 is the **percentage** (P), 125 is the **base** (B), and the ratio $\frac{70}{125}$ is called the **rate** (r).

$$\frac{70}{125} = \frac{56}{100} \rightarrow \frac{Percentage}{Base} = Rate$$

If r represents the number per hundred, this equation can be rewritten as $\frac{P}{B} = \frac{r}{100}$. This proportion is called the **percent proportion**. You can use the percent proportion to express fractions as percents and to solve percent problems.

Example **Express each fraction as a percent.**

a. $\frac{9}{16}$

$\frac{P}{B} = \frac{r}{100} \quad \rightarrow \quad \frac{9}{16} = \frac{r}{100}$ *Replace P with 9 and B with 16.*

$9 \cdot 100 = 16 \cdot r$ *Find the cross products.*

$\frac{900}{16} = \frac{16r}{16}$ *Divide each side by 16.*

$56\frac{1}{4} = r$

$\frac{9}{16}$ is equivalent to $56\frac{1}{4}\%$.

b. $\frac{15}{4}$

$\frac{P}{B} = \frac{r}{100} \quad \rightarrow \quad \frac{15}{4} = \frac{r}{100}$ *Replace P with 15 and B with 4.*

$15 \cdot 100 = 4 \cdot r$ *Find the cross products.*

$\frac{1500}{4} = \frac{4r}{4}$ *Divide each side by 4.*

$375 = r$

$\frac{15}{4}$ is equivalent to 375%.

TECHNO TIP

You can find rates on a calculator by dividing the percentage by the base and moving the decimal point two places to the right.

There are three basic types of percent problems. Using the proportion $\frac{1}{2} = \frac{50}{100}$, you can see that the types are related as shown below.

$\dfrac{\bullet}{2} = \dfrac{50}{100}$	What number is 50% of 2? $\underbrace{\qquad}$ *Percentage*
$\dfrac{1}{2} = \dfrac{\bullet}{100}$	1 is what percent of 2? $\underbrace{\qquad}$ *Rate*
$\dfrac{1}{\bullet} = \dfrac{50}{100}$	1 is 50% of what number? $\underbrace{\qquad}$ *Base*

Find the percentage.

Find the rate.

Find the base.

Example **2** **Sixty-eight is 20% of what number?**

$\frac{P}{B} = \frac{r}{100} \quad \rightarrow \quad \frac{68}{B} = \frac{20}{100}$ *Replace P with 68 and r with 20.*

$20 \cdot B = 68 \cdot 100$ *Find the cross products.*

$\frac{20 \cdot B}{20} = \frac{68 \cdot 100}{20}$ *Divide each side by 20.*

$B = 340$

68 is 20% of 340.

You can also use the percent proportion to find the percentage when the base and the rate are given.

Example 3

Physical
Therapy

Refer to the application at the beginning of the lesson. Physical therapists usually set a goal of bringing an injured leg up to at least 90% of the strength of the strong leg. How many foot-pounds should Chris expect his injured leg to be able to lift when the physical therapy program is complete?

Use the percent proportion to solve for the number of foot-pounds.

$$\frac{P}{B} = \frac{r}{100} \quad \rightarrow \quad \frac{P}{125} = \frac{90}{100} \qquad \textit{Replace B with 125 and r with 90.}$$

$$P \cdot 100 = 125 \cdot 90 \qquad \textit{Find the cross products.}$$

$$\frac{P \cdot 100}{100} = \frac{125 \cdot 90}{100} \qquad \textit{Divide each side by 100.}$$

$$P = 112.5$$

Chris should expect to be able to lift at least 112.5 foot-pounds with his injured leg when the program is complete.

Checking Your Understanding

Communicating
Mathematics

Read and study the lesson to answer these questions.

1. **Explain** in one or two sentences why the value of r in $\frac{P}{B} = \frac{r}{100}$ represents a percent.

2. **Describe** how to write a fraction as a percent.

3. **Name** an advantage of using a percent instead of a fraction.

4. **Write** a proportion that you could use to find your percentage grade on a test that has 25 questions.

5. **Apply** the percent proportion to write a proportion for each.
   a. 52 is 80% of what number.
   b. What number is 80% of 52?
   c. 52 is what percent of 80?

 MATERIALS
   grid paper
   colored pencils

6. Shade a 5×5 section in one corner of a 10 by 10 grid. Then use a different color to shade each of the other three 5×5 sections.
   a. How many small squares are in the grid?
   b. How many small squares are in each colored section?
   c. What percent is represented by each colored section?
   d. How many 5×5 sections are there?
   e. What fraction of the grid is represented by each colored section?

Guided
Practice

Express each fraction as a percent.

7. $\frac{44}{100}$ 8. $\frac{1}{20}$ 9. $\frac{8}{250}$ 10. $\frac{3}{2}$

Use the percent proportion to solve each problem.

11. Find 15% of 60.

12. Fifty-two is 40% of what number?

13. 37 is what percent of 296?

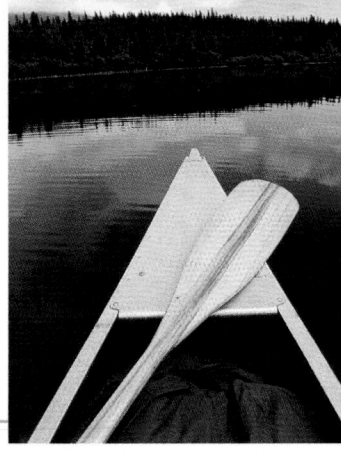

14. **Statistics** The National Park Service issued a report saying that visitors spent 18,735 million hours, or about 27% of the total time spent in national parks, on boating. How many millions of hours did visitors spend in national parks?

Exercises: Practicing and Applying the Concept

Independent Practice

Express each fraction as a percent.

15. $\frac{1}{4}$
16. $\frac{7}{8}$
17. $\frac{3}{5}$
18. $\frac{1}{16}$

19. $\frac{9}{4}$
20. $\frac{24}{25}$
21. $\frac{45.3}{60}$
22. $\frac{24}{3}$

Use the percent proportion to solve each problem.

23. What is 92% of 116?
24. 16 is 40% of what number?
25. 36 is what percent of 80?
26. 21 is 35% of what number?
27. What is 70% of 50?
28. Find 60% of 48.
29. What is 200% of 8?
30. Find 78.5% of 90.
31. 36 is 45% of what number?
32. 28 is 20% of what number?
33. 126 is 10.5% of what number?
34. Find 0.1% of 450.

Critical Thinking

35. Replace the _?_ so that the statement "x is x% of _?_ " is true for all values of x.

36. If $\frac{1}{5}$ is 20% of a number, what percent is $\frac{3}{5}$ of the number?

Applications and Problem Solving

37. **Health** In 1994, the U.S. Food and Drug Administration began requiring food manufacturers to label their products with a nutritional label like the one from a bag of pretzels shown at the right.

a. The label states that the bag contains 1.5 grams of fat, which is 3% of the daily value recommended for a 2000-Calorie diet. How many grams of fat are recommended for a 2000-Calorie diet?

b. The 760 milligrams of sodium (salt) in the pretzels is 32% of the recommended daily value. What is the recommended daily value of sodium?

Nutrition Facts		
Serving Size 1 package (46.8g)		
Servings per container 1		
Amount per serving		
Calories 190 Calories from Fat 15		
		% Daily Value*
Total Fat 1.5g		3%
Saturated Fat 0g		0%
Cholesterol 0mg		0%
Sodium 760mg		32%
Total Carbohydrate 37g		12%
Dietary Fiber less than 1g		2%
Sugars 2g		
Protein 5g		
Vitamin A 0%	•	Vitamin C 0%
Calcium 0%	•	Iron 3%

*Percent Daily Values are based on a 2,000 calorie diet. Your daily values may be higher or lower depending on your calorie needs.

Calories per gram:
Fat 9 • Carbohydrates 4 • Protein 4

38. **Personal Finance** Interest on savings accounts is often figured every quarter of a year. One quarter, Louam's savings account earned $54.84 in interest. This is 2% of her savings.

   a. Find Louam's savings.

   b. The interest in Louam's account is compounded each quarter, so the interest for a quarter is based on the original balance plus the interest earned the previous quarter. Find the interest Louam will earn next quarter if no deposits or withdrawals are made.

39. **Forestry** The table shows the five states with the largest portion of land covered by forests. Find the number of square miles covered by forests in each state.

State	Percent of land covered by forests	Area of state (square miles)
Maine	89.9%	35,387
New Hampshire	88.1%	9351
West Virginia	77.5%	24,231
Vermont	75.7%	9615
Alabama	66.9%	52,423

Source: The U.S. Forest Service

Mixed Review

40. **Jewelry** Jewelers use karats to describe the amount of gold in a piece of jewelry. A piece of jewelry that is 24-karat gold is all gold. How many grams of gold are in a 15-gram piece of jewelry that is labeled as 18-karat gold? (Lesson 9-4)

41. Express the ratio *6 out of 8 free throws* in simplest form. (Lesson 9-1)

42. Solve the system $y = 8x - 1$ and $3x + y = 21$. (Lesson 8-8)

43. **Standardized Test Practice** If $\frac{2}{3}n + 8 = 44$, what is the value of *n*? (Lesson 7-2)

   A. 72   **B.** 54   **C.** 32   **D.** 24

For **Extra Practice**, see page 762.

44. Evaluate the expression $4b$ if $b = 3.2$. (Lesson 6-5)

45. Find the GCF of 65 and 105. (Lesson 4-5)

46. In which quadrant does the graph of $(9, -5)$ lie? (Lesson 2-2)

WORKING ON THE

Investigation

STADIUM STAMPEDE

Refer to the Investigation on pages 368–369.

The ticket prices of the seats are divided into three categories.

Deluxe seats	$33.50
Spectator seats	$22.75
Stadium seats	$15.20

The university has asked you to determine which seats should be designated deluxe, spectator, and stadium. To earn enough gross income, 32% of the seats need to be deluxe, and only 18% should be stadium seats.

They have asked you to prepare a report showing the location of the seats, the price of each seat, the number of seats in each category, and the gross income from a sellout crowd. Prepare this report. Be sure to include any charts or drawings that support your reasoning.

Add the results of your work to your Investigation Folder.

Integration: Statistics
Using Statistics to Predict

Application: Consumerism

How important is the brand name when you choose a product? The International Mass Retail Association surveyed consumers 8 to 17 years old on that subject. The results of the survey are shown in the graph at the right.

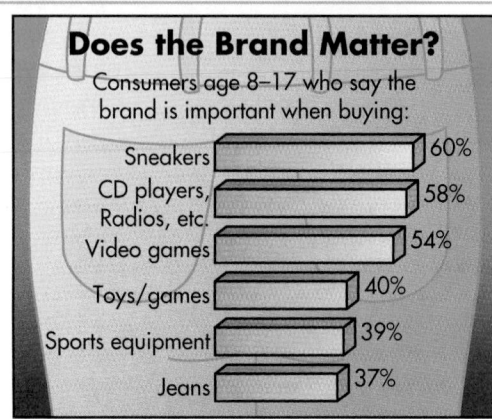

Does the Brand Matter?
Consumers age 8–17 who say the brand is important when buying:

Sneakers	60%
CD players, Radios, etc.	58%
Video games	54%
Toys/games	40%
Sports equipment	39%
Jeans	37%

Source: International Mass Retail Association

Learning the Concept

You can use the preferences or behavior of a group to make predictions about a larger or smaller group that is similar. For example, the survey on brand buying was taken using only a portion of the consumers between 8 and 17 years old. But, the results were used to make statements about all consumers in that age group.

Example **1**

Consumerism

Refer to the application at the beginning of the lesson. How many of the 423 9th-grade students at Kilbourne High School would you expect to say that they consider brand name when buying jeans?

Use the problem-solving plan to find the expected number.

Explore   The graph shows that 37% of the consumers surveyed considered brand name when they purchased jeans. We need to know how many of the freshman class students consider brand name when buying jeans.

Plan   Write and solve a proportion using the percent proportion. Estimate that since 37% is a little over $\frac{1}{3}$, the solution should be a little over $423 \div 3$, or 140.

Solve $\dfrac{\textit{percent considering brand name}}{100} = \dfrac{\textit{number of Kilbourne freshmen who consider brand name}}{\textit{number of Kilbourne freshmen}}$

$$\frac{37}{100} = \frac{x}{423}$$

$$37 \cdot 423 = 100 \cdot x \qquad \textit{37 × 423 ÷ 100 = 156.51}$$

$$x = 156.51$$

Expect 157 students to consider brand when buying jeans.

Examine   The solution is close to the estimate, so it is reasonable.

Example **2**

Real World APPLICATION

Entertainment

inter NET CONNECTION

Data Update For the latest media ratings, visit:

www.glencoe.com/ sec/math/prealg/ mathnet

The television program that began the 1995 television season as the number one show in the media ratings attracted 41% of the media viewers in the first week. If about 91 million viewers watched television that night, about how many were watching the number one show?

Use a proportion.

$$\frac{x}{91} = \frac{41}{100}$$

$x \cdot 100 = 91 \cdot 41$ $91 \cdot 41 \div 100 = 37.31$

$x = 37.31$

About 37.31 million viewers watched the number one show that night.

When taking surveys and using the results to predict the actions of a larger population, be sure that your sample is random and is large enough to represent the population. If the sample is too small, it is likely that the items or people that it includes are not typical of the population. For example, suppose you were told that 4 out of 5 athletes chose a particular type of shoe for basketball. The claim would not be meaningful if only 5 athletes were surveyed.

Checking Your Understanding

Communicating Mathematics

Read and study the lesson to answer these questions.

1. **Explain** how to use the results of a survey to predict the characteristics of a population.

2. **Name** the most important features of a sample.

3. **You Decide**   Refer to the application at the beginning of the lesson. Tamika estimates that 152 of the 253 students in her class consider brand name when choosing sneakers. Gloria estimates only 121 of them do. Who is correct and why?

Guided Practice

Use the survey on favorite drinks to answer the questions.

4. How large is the sample?

5. If this represents a group of 3600, what percent is the sample?

6. What fraction chose fruit drink? What percent is this?

7. Name a location that would not be a good place to conduct this survey.

Favorite Drink	
lemon lime	17
cola	10
root beer	25
fruit drink	12
ginger ale	8

Independent Practice

The owner of a Dairy Dream store surveyed 100 people at her store about their favorite ice cream flavor. The results are in the table below.

8. Give some reasons why this is a valid sample and some reasons why it is not.

9. What percent of the people preferred peanut butter fudge?

10. If there were 800 customers on a Saturday, about how many would prefer chocolate?

Favorite Flavors	
vanilla	20
chocolate	25
strawberry	10
peanut butter fudge	35
mint chocolate chip	10

The results of a survey on the reasons that teens spend time as volunteers are shown in the table below.

11. If you asked 500 teens who volunteer why they do, how many would you expect to answer because they have free time?

Source: Independent Sector

Reason	Percent	Reason	Percent
to help others	47	for a friend	20
enjoy the work	38	religion	19
lots of free time	25	past experience	10
to learn	24	other	7
		don't know	2

12. Out of 250 teen volunteers, how many would you expect volunteer because they enjoy the work?

13. If you surveyed 25 volunteers at a homeless shelter on reasons for volunteering, would you expect the same results as this survey? Explain.

14. **Research** Make a frequency table showing the hair color for students in your class. Use the data to predict the number of students in your school with each hair color. Choose a time and place to observe at least 50 students and record their hair color. Compare that data to your predictions. What are your findings?

Critical Thinking

15. Thirteen percent of the letters used in English words are Es. If you were guessing the letters in a two-letter word of a puzzle, would you guess E? Explain.

Applications and Problem Solving

16. **Music** A survey asked what music people listen to in the car.

   a. How many of the 445 cars in a traffic jam do you predict have rock playing?

   b. If you owned a store that specialized in car stereos, what type of music would you have playing? Explain your choice.

Car Tunes

Country 27.7, Rock 18.9, Pop 15.5

Percent (axis: 0.0 to 30.0)

Music choice

Source: 20/20 Research

17. **Fashion** A magazine surveyed 3600 people about clothing.

   a. 1872 people said they would rather be dressed too casually for an occasion than be dressed too fancy. What percent of the population do you think would feel the same?

   b. What percent of the population would say they were slow to change with the fashions if 2160 of those surveyed said that? **Source:** Cotton Incorporated

18. **Housekeeping** A magazine poll showed that 76% of women and 46% of men surveyed made their bed that morning.

    a. If you used the results of the poll, about how many of the 346 females and 362 males at Centennial High school would you predict made their beds this morning?

    b. The poll questioned adults. Do you think that a poll of teens would have the same results? Explain. **Source:** *TIME* magazine

Mixed Review

19. **Food** The average American eats 10.28 pounds of chocolate in a year. If the total annual candy consumption of the average American is 17.86 pounds, what percent of the candy is chocolate? (Lesson 9-5)

20. **Patterns** State whether the sequence 2, 3.5, 5, 6.5, 8, 9.5, 11, . . . is arithmetic. Then write the next three terms. (Lesson 5-9)

21. **Standardized Test Practice** If 15 more than the product of a number and −2 is greater than 12, which of the following could be the number? (Lesson 3-1)

    A. 6   **B.** 4   **C.** 1.5   **D.** 0

22. **Geometry** Use the coordinate grid at the right to name the point for the ordered pair (−4, 2). (Lesson 2-2)

For **Extra Practice**, see page 762.

23. State whether $14 \geq 2b + 4$ is true or false if $b = 3$. (Lesson 1-9)

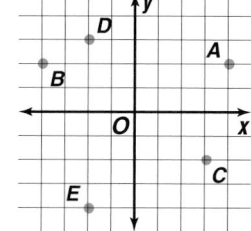

Self Test

Express each ratio as a unit rate. (Lesson 9-1)

  1. 300 feet in 5 minutes

  2. 54 inches of rain in 4 months

  3. How many ways can you make change for a $50 bill using only $5, $10, and $20 bills? (Lesson 9-2)

There are 3 blue pencils, 5 green pencils, 2 black pencils, and 6 red pencils in a drawer. Suppose you grab one pencil at random. Find each probability. (Lesson 9-3)

  4. P(green)

  5. P(blue or red)

Solve each proportion. (Lesson 9-4)

  6. $\dfrac{m}{7} = \dfrac{25}{35}$

  7. $\dfrac{1.5}{2} = \dfrac{s}{2.4}$

Use the percent proportion to solve each problem. (Lesson 9-5)

  8. Find 75% of 400.

  9. 250 is 500% of what number?

The Travel Industry Association polled 1500 adults on their choice of fall vacation destinations. (Lesson 9-6)

  10. How many of those surveyed chose New York as a destination?

  11. If a travel agency books vacations for 300 families this fall, how many should they expect will go to Florida?

Destination	Percent
Florida	35
California	30
Hawaii	20
Nevada	14
New York	11
Colorado	9
Arizona	8
Texas	6
South Carolina	6
Washington	6

Source: Travel Industry Assn.

Fractions, Decimals, and Percents

What You'll Learn

■ To express decimals and fractions as percents and vice versa.

It's Important

You can express your test and homework scores using percents.

TALK ABOUT IT

MATERIALS: grid paper
Hands-On Activity

You can compare quantities using fractions, percents, or decimals. Each of these can be rewritten in the other two forms. The model at the right represents one unit. Since 36 of the 100 cells are shaded, $\frac{36}{100}$ or 0.36 of the model is shaded. Using the definition of percent, we could also say that 36% of the model is shaded.

Your Turn Make models to represent $\frac{1}{2}$, 0.42, and 8%.

a. How could you write $\frac{1}{2}$ as a decimal and a percent?

b. Write 0.42 as a percent. What do you observe about the decimal point?

c. Write a procedure for writing a decimal as a percent and for writing a percent as a fraction.

Learning the Concept

You have written fractions as decimals and percents. Fractions, decimals, and percents are all different names that represent the same number.

To express a decimal as a percent, write the decimal as a fraction with 1 as the denominator. Then write that fraction as an equivalent fraction with 100 as the denominator.

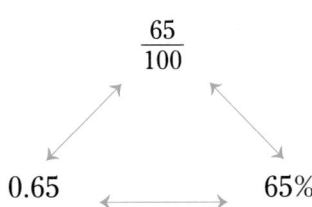

Example **1** **Express each decimal as a percent.**

a. 0.36 → $\overset{\times 100}{\underset{\times 100}{\frac{0.36}{1} = \frac{36}{100}}}$ or 36%   **b.** 0.07 → $\overset{\times 100}{\underset{\times 100}{\frac{0.07}{1} = \frac{7}{100}}}$ or 7%

c. 0.004 → $\overset{\times 100}{\underset{\times 100}{\frac{0.004}{1} = \frac{0.4}{100}}}$ or 0.4%   **d.** 1.8 → $\overset{\times 100}{\underset{\times 100}{\frac{1.8}{1} = \frac{180}{100}}}$ or 180%

Compare each decimal in Example 1 with its equivalent percent. Notice that, in each case, the decimal point moved two places to the right.

To express a fraction as a percent, first write the fraction as a decimal by dividing numerator by denominator. Then write the decimal as a percent.

Example **Express each fraction as a percent.**

a. $\frac{3}{5} = 0.60 \quad \rightarrow \quad 3 \div 5 = 0.6 \quad \rightarrow \quad \frac{60}{100} = 60\%$

b. $\frac{7}{4} = 1.75 \quad \rightarrow \quad 7 \div 4 = 1.75 \quad \rightarrow \quad \frac{175}{100} = 175\%$

c. $\frac{2}{500} = 0.004 \quad \rightarrow \quad 2 \div 500 = 0.004 \quad \rightarrow \quad \frac{0.4}{100} = 0.4\%$

To express a percent as a fraction, write the percent in the form $\frac{r}{100}$ and simplify.

Example **Write each percent as a fraction in simplest form.**

LOOK BACK

You can review dividing fractions in Lesson 6-4.

a. $15\% = \frac{15}{100}$

$= \frac{3}{20}$

b. $87\frac{1}{2}\% = \frac{87\frac{1}{2}}{100}$ *The fraction bar indicates division.*

$= 87\frac{1}{2} \div 100$

$= \frac{\overset{7}{\cancel{175}}}{2} \times \frac{1}{\underset{4}{\cancel{100}}}$

$= \frac{7}{8}$

When you want to express a percent as a decimal, write the percent in the form $\frac{r}{100}$ and then write as a decimal.

Example 4 **Write each percent as a decimal.**

a. $17\% = \frac{17}{100}$
$\qquad\quad = 0.17$

b. $250\% = \frac{250}{100}$
$\qquad\qquad = 2.50$

In Example 4, notice that the decimal point was moved two places to the left to get the equivalent decimal.

Cosmetics

5 **In a magazine survey on favorite perfumes, 26% of those asked used one certain perfume all the time while one-twentieth said that they never use perfume. Which was the larger group?** **Source:** *Self* magazine

Write each number as a decimal to compare.

$26\% = \frac{26}{100}$
$\qquad\quad = 0.26$

$\frac{1}{20} = 0.05$

$\begin{array}{r} 0.05 \\ 20\overline{)1.00} \\ -1\,00 \end{array}$

Since 0.26 is greater than 0.05, the group that said they use the same perfume each day was larger.

Checking Your Understanding

Communicating Mathematics

Read and study the lesson to answer these questions.

1. **Describe** how you would change 24.7% to a decimal.

2. **Write** a fraction, a decimal, and a percent to represent the model at the right.

3. **In your own words, explain** how you can tell if a fraction is greater than 100% or less than 1%.

MATERIALS

🫘 dry beans
📄 paper bag

4. Put 25 beans in a bag. Have your partner grab a handful of beans. Express the portion of the beans that were removed as a fraction, a percent, and a decimal. Repeat. Which handful was a greater portion of the beans?

Guided Practice

Express each decimal as a percent.

5. 0.37   6. 0.475   7. 1.03

Express each fraction as a percent.

8. $\frac{27}{100}$ 9. $\frac{36}{500}$ 10. $\frac{6}{5}$

Express each percent as a fraction.

11. 25%   12. $33\frac{1}{3}\%$ 13. 0.3%

Express each percent as a decimal.

14. 35%   15. $24\frac{1}{2}\%$ 16. 0.4%

17. **Sports** In a survey, 48 of 120 people said they preferred hockey as a spectator sport.
    a. What fraction is this?
    b. What percent is this?

Exercises: Practicing and Applying the Concept

Independent Practice

Express each decimal as a percent.

18. 0.71   19. 0.03   20. 0.543   21. 2.37
22. 0.004   23. 1.32   24. 0.035   25. 0.0004

Express each fraction as a percent.

26. $\frac{3}{16}$ 27. $\frac{7}{12}$ 28. $\frac{5}{4}$ 29. $\frac{7}{3}$
30. $\frac{4}{300}$ 31. $\frac{5}{9}$ 32. $1\frac{1}{4}$ 33. $\frac{3}{40}$

✓ **Choose**

Estimation
Mental Math
Calculator
Paper and Pencil

Express each percent as a fraction.

34. 36%   35. 125%   36. 58%   37. 34.5%
38. 22%   39. $37\frac{1}{2}\%$ 40. $66\frac{2}{3}\%$ 41. $16\frac{1}{3}\%$

Express each percent as a decimal.

42. 35%   **43.** 75%   **44.** 93%   **45.** 39.8%

46. 42.7%   **47.** 0.4%   **48.** 127%   **49.** 235.4%

Choose the greatest number in each set.

50. $\left\{\frac{3}{5}, 0.75, 80\%, 6 \text{ to } 8\right\}$ **51.** $\left\{22\%, 0.022, \frac{1}{11}, 2 \text{ out of } 13\right\}$

Write each list of numbers in order from least to greatest.

52. $\frac{1}{2}$, 31%, 0.05   **53.** 16%, $\frac{1}{4}$, 0.067

Critical Thinking

54. A grocery store carries three different-sized bottles of contact lens solution. Size A is 50% more expensive than size C, but it contains 20% more solution than size B. Size B contains 50% more solution than size C, but it costs 25% more than size A. Which bottle is the most economical choice?

Applications and Problem Solving

55. Retail Sales A pair of basketball shoes is listed at $33\frac{1}{3}\%$ off. What fraction off is this?

56. Health Pediatricians and obstetricians care for newborn babies and their mothers.

     a. Pediatricians estimate that 3 babies out of every 1000 are likely to contract a cold their first month. What percent is this?

     b. Obstetricians have observed that only 4% of babies are born on their due date. What fraction of babies are born on their due dates?

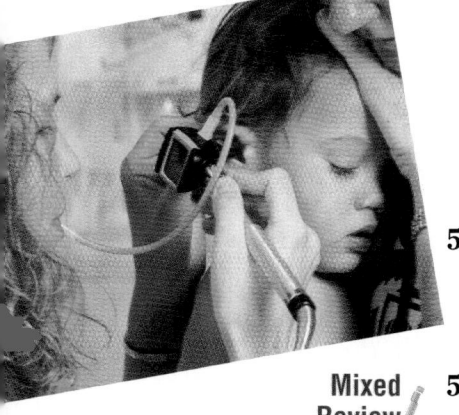

57. Business In an interview with David Letterman, an executive of a marshmallow company said "Marshmallows are 80% air. That's how smart we are, we sell air!" What fraction of a marshmallow is air?

Mixed Review

58. Standardized Test Practice In the favorite perfumes survey, 34% of those asked prefer floral-scented perfume. Of 50 customers at a department store cosmetics counter, how many would you expect to choose a floral perfume? (Lesson 9-6)

     A. 16   **B.** 17   **C.** 34   **D.** 84

59. Find the x- and y-intercept of the graph of $y = 8x - 12$. (Lesson 8-7)

60. Solve $\frac{b}{-2} - 12 \leq 11$. (Lesson 7-6)

61. Find the product of $\frac{4}{9}$ and $\frac{5}{12}$. (Lesson 6-3)

62. Solve $x + 14.7 < 51.2$. (Lesson 5-7)

63. Find the product $n^4 \cdot n^{10}$. (Lesson 4-8)

64. Science In the metric system, the prefix *giga* means 10^9. Write 10^9 as a product of the same factor. (Lesson 4-2)

65. Choose the value of h if $\frac{h}{0.7} = -2.8$. (Lesson 3-1)

     A. -4 **B.** -2.8 **C.** -1.96 **D.** 0.7

66. Solve $r = -9889 \div -319$. (Lesson 2-8)

67. Solve $4 + x = 12$ using an inverse operation. (Lesson 1-8)

For **Extra Practice**, see page 763.

Percent and Estimation

You'll Learn

- To use percents to estimate.

It's Important

You can use percents to estimate the amount of tip to leave at a restaurant.

Application: Health

"You are what you eat." So if you want to be healthy, you need to eat healthful food. Nutritionists suggest that people get no more than 30% of their daily Calorie intake from fat. Do any of the popcorn snacks shown in the table at the right contain about 30% fat Calories or less? *This problem will be solved in Example 2.*

Snack	Calories per serving	Calories from fat per serving
Brand A	140	63
Brand B	110	27
Brand C	142	90
Brand D	108	18
Brand E	100	36
Brand F	130	36
Brand G	160	90

Learning the Concept

Many times when you are working with percents, an exact answer is not needed. In cases like this, you can estimate. We could estimate the percent of the model that is shaded.

Nine of the 20 squares are shaded.

$\frac{9}{20}$ is about $\frac{10}{20}$ or $\frac{1}{2}$

Since $\frac{1}{2} = 50\%$, about 50% of the model is shaded.

There are three methods you can use to estimate a percentage. The table below shows how to estimate 30% of 657 using each method.

Fraction Method	1% Method	Meaning of Percent Method
30% is about 33% or $\frac{1}{3}$.	1% of 660 is 6.6 or about 7.	30% means 30 for every 100 or 3 for every 10. 657 has 6 hundreds and about 6 tens.
$\frac{1}{3}$ of 660 is 220.	30 times 7 is 210.	$(30 \times 6) + (3 \times 6) = 198$
Estimate: 220	Estimate: 210	Estimate: 198

Find the exact amount.

$$\begin{array}{r} 657 \\ \times 0.3 \\ \hline 197.1 \end{array}$$

$657 \times 0.3 = 197.1$

The actual percentage is very close to all of the estimates.

Example **1** **Estimate.**

a. 60% of 996

1% of 996 is 9.96 or
about 10. So 60% of
996 is about 60×10 or 600.

b. 8% of $58

8% is about 10% or $\frac{1}{10}$.
$\frac{1}{10}$ of $58 is $5.80, so
8% of $58 is about $6.00

c. 0.5% of 795

0.5% is half of 1%.
795 is almost 800.
1% means 1 out of 100.
So 1% of 800 is 8, and
$\frac{1}{2}$ of 8 is 4.
0.5% of 795 is about 4.

d. 109% of 62

109% is more than 100%,
so 109% of 62 is more than 62.
109% is almost 110%.
110% = 100% + 10%
62(100% + 10%) = 62 + 6.2 = 68.2
109% of 62 is about 68.

Example **2**

Health

FYI

The average American
eats 71 quarts of popcorn
each year.

Refer to the application at the beginning of the lesson. Do any of the popcorn snacks have less than 30% of their Calories from fat?

Use the meaning of percent method to estimate 30% of the Calories for each snack. 30% means 30 for every 100 Calories and 3 for every 10 Calories.

Snack	Calories	Estimate of 30% of Calories	Fat Calories
Brand A	140	30 + 3(4) = 42	63
Brand B	110	30 + 3(1) = 33	27
Brand C	142	30 + 3(4) = 42	90
Brand D	108	30 + 3(1) = 33	18
Brand E	100	30 + 3(0) = 30	36
Brand F	130	30 + 3(3) = 39	36
Brand G	160	30 + 3(6) = 48	90

Compare the estimate of 30% of the total Calories to the number of Calories from fat in each snack. Brand B, Brand D, and Brand F all have about 30% or less of their Calories from fat.

Estimating percents is a useful skill for everyday situations.

Example **3**

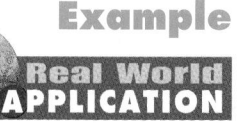

Consumerism

A tip of 15% is standard for good service in a restaurant. If Simone wants to leave a tip of about 15% on a dinner check of $23.85, how much should she leave?

Estimate: $23.85 is about $24.
          10% of $24 is $2.40.
          5% of $24 is $1.20.   *5% is half of 10%.*
          15% is about 2.40 + 1.20 or $3.60.

Simone should leave about $3.60 as a tip.

Communicating Mathematics

Read and study the lesson to answer these questions.

1. **Explain** how you would estimate 27% of 198.

2. **Describe** in your own words one method for estimating using percents.

3. **Estimate** the percent of the model at the right that is shaded.

4. **You Decide** Once a month Angeni and Darlene meet for lunch. The food tax where they live is $5\frac{1}{4}$%. Each of them leaves about a 15% tip. Angeni always finds her tip by taking 10% of the total and then adding that amount to half of that amount. Darlene finds her tip by multiplying the tax by 3. Who is correct and why?

5. **Describe** a situation that would best use each kind of method for estimating using percents.

Guided Practice

Choose the best estimate for the percent shaded.

6.

   a. 10%   **b.** 25%   **c.** 55%

7.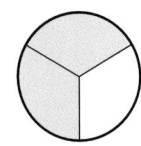

   a. 33%   **b.** 50%   **c.** 66%

Write the fraction, mixed number, or whole number you could use to estimate.

8. 24%   9. 79%   10. 145%

Estimate.

11. 45% of 430   12. 112% of 14.5   13. 0.6% of 325

Estimate each percent.

14. 8 out of 30   15. 9 out of 19   16. $\frac{5}{7}$

17. **Retail Sales** Davon bought a pair of shoes on sale for $29. The regular price was $50.

    a. For about what percent of the selling price did he buy them?

    b. About what percent-off was the sale?

Exercises: Practicing and Applying the Concept

Independent Practice

Choose the best estimate.

18. 39% of 300   **a.** 1.2   **b.** 12   **c.** 120

19. 47% of 605   **a.** 3   **b.** 30   **c.** 300

20. $\frac{1}{3}$% of 240   **a.** 0.8   **b.** 8   **c.** 80

21. 129% of 400   **a.** 50   **b.** 500   **c.** 5000

Write the fraction, mixed number, or whole number you could use to estimate.

22. 67%   **23.** 98%   **24.** 18%   **25.** $2\frac{1}{2}$%

26. 148%   **27.** $8\frac{5}{9}$%   **28.** 0.8%   **29.** 119%

Estimate.

30. 47% of 84   **31.** 28% of 390   **32.** $8\frac{1}{2}$% of 55

33. 98% of 98   **34.** 126% of 198   **35.** 0.9% of 514

36. 15% of $34   **37.** 116% of 18   **38.** 0.05% of 1180

Estimate each percent.

39. 12 out of 15   **40.** 8 out of 35   **41.** 39 out of 79

42. 57 out of 176   **43.** 9 out of 95   **44.** 13 out of 68

45. 1 out of 9   **46.** 3 out of 200   **47.** 7 out of 2445

Critical Thinking

48. In a controversial vote, 40 percent of the Democrats and 92.5 percent of the Republicans voted yes. If all the members of the assembly voted and 68 percent of the voters voted yes, what is the ratio of Democrats to Republicans?

Applications and Problem Solving

49. Nutrition Estimate the percent of fat Calories of each entree in the chart at the right.

50. Research The land area of Georgia is about 60% of the land area of Great Britain.

   a. Find the size (in square miles) of Great Britain.

   b. Estimate the size of Georgia.

Entree	Calories (3.5 oz)	Calories from fat
shrimp	90	7
lobster	90	15
chicken wing, with skin	290	171
ham	220	99
ground chuck	222	81

Source: *The Fat Counter*

51. History The table below shows the populations of the American colonies in 1780. If the total population of the colonies was 2,780,000 in 1780, estimate what percent of the total each colony had.

Colony	Population (thousands)	Colony	Population (thousands)
Connecticut	206.7	North Carolina	270.1
Delaware	45.4	Pennsylvania	327.3
Georgia	56.1	Plymouth and Massachusetts	268.6
Kentucky	45.0		
Maine counties	49.1	Rhode Island	52.9
Maryland	245.5	South Carolina	180.0
New Hampshire	87.8	Tennessee	10.0
New Jersey	139.6	Vermont	47.6
New York	210.5	Virginia	538.0

Source: Bureau of the Census

52. Nutrition A certain hamburger has 560 Calories, and 288 of these are fat Calories.

   a. About what percent of the Calories are fat Calories?

   b. If a person on a 2000-Calorie-per-day diet eats this hamburger, how many Calories and fat Calories remain for the rest of the day?

**Mixed
Review**

53. Express 64% as a fraction. (Lesson 7-8)

54. World Facts On German
highways, called autobahns,
the speed limit is 130
kilometers per hour. What
is the speed limit in meters
per hour? (Lesson 7-8)

55. Patterns State whether
8, 12, 18, 27, 40.5, . . . is a
geometric sequence. If so,
state the common ratio and
list the next three terms.
(Lesson 6-8)

56. Standardized Test Practice Solve $\frac{9}{16} + \frac{11}{16} = y$. (Lesson 5-4)

    A. $1\frac{1}{8}$

    B. $1\frac{1}{6}$

    C. $1\frac{1}{4}$

    D. $1\frac{1}{2}$

57. Solve $5x = 105$ and check your solution. Then graph the solution on a
number line. (Lesson 3-3)

58. Simplify $|-21|$. (Lesson 2-1)

59. Entertainment The table at
the right shows the number of
radios per thousand people in
several countries. Make a bar
graph of the data. (Lesson 1-10)

60. Translate the phrase *twice the
sum of a number and 7* into
an algebraic expression.
(Lesson 1-3)

For **Extra Practice**,
see page 763.

Country	Radios per thousand people
United States	2091
Bermuda	1710
United Kingdom	1240
Australia	1144
Finland	984
New Zealand	902
Virgin Islands (U.S.)	884
France	866
Sweden	842
Canada	828

Source: Duncan's American Radio, Inc.

Using Percent Equations

You'll Learn

■ To solve percent problems using percent equations.

It's Important

You can use percent equations to find the interest earned on your investments.

Technology Activity

Maria Morales has saved some money that she wishes to invest. Her banker has provided the terms, interest rates, and minimum investments on the certificates of deposit, or CDs, that they offer. Maria used the information to prepare the spreadsheet below to compare the options.

Option	Minimum Investment	Monthly Interest Rate (percent)	Term (months)	Interest for Term
1	$2500	0.375	6	
2	$2500	0.45	12	
3	$5000	0.54	36	
4	$7500	0.625	60	

The **principal** is the amount of money in the account. The amount of **interest** earned in a month can be found by using the percent proportion $\frac{interest}{principal} = \frac{interest\ rate}{100}$. Then multiply the monthly interest by the number of months to find the total interest earned.

Your Turn

Use a calculator or spreadsheet to find the interest on each investment.

a. Which option earns the most? What factors contribute to this?

b. Let I represent interest, p the principal, r the interest rate as a decimal, and t the time in months. Write a formula for finding the interest on any account.

Learning the Concept

As you discovered in the activity above, you can find the interest earned on an account using the formula $I = prt$ if you write the rate as a decimal. Writing the rate as a decimal in the percent proportion also allows you to write it as an equation and solve percent problems more quickly.

$$\frac{P}{B} = \frac{r}{100}$$

$$\frac{P}{B} \cdot B = \frac{r}{100} \cdot B \quad \text{\textit{Multiply each side by B.}}$$

$$P = \frac{r}{100} \cdot B$$

Remember $\frac{r}{100}$ is the rate. Let R represent the decimal form of $\frac{r}{100}$.

Therefore, $P = R \cdot B$ or Percentage = Rate · Base.

The form $P = R \cdot B$ is usually easier to use when the rate and base are known.

Example **1**   **a. Find 36% of 65.**   *Estimate: $\frac{1}{3}$ of 60 is about 20.*

What number is 36% of 65?   Write in P = R · B form.

$$P \quad\quad = 0.36 \ \times \ 65$$

$0.36 \times 65 = 23.4$

23.4 is 36% of 65.

b. 15 is what percent of 125?

$15 = R \cdot 125$ *Replace P with 15 and B with 125.*

$\frac{15}{125} = \frac{125R}{125}$ *Divide each side by 125.*

$0.12 = R$

15 is 12% of 125.

Example **2**

Charity

One of the largest charities involved in protecting the environment reported spending $198,722,500 on conservation recently. If this is 72.5% of the money they received, how much money did they receive?

198,722,500 is 72.5% of what number?

$198,722,500 \ = \ 0.725 \ \times \quad\quad B$ $\quad P = R \cdot B$

$\frac{198,722,500}{0.725} = \frac{0.725B}{0.725}$ $\quad\quad$ *Divide each side by 0.725.*

$274,100,000 = B$

The charity raised $274,100,000.

One of the common uses of percents in everyday life is **discounts**.

Example **3**

Retail Sales

The regular price of a pair of jeans is $45.95. If there is a 25% discount and a 6% sales tax (based on the discounted price), how much will the jeans cost?

You can find the sale price in one of two ways.

Method 1	Method 2
First, find the amount of the discount.	First, subtract the percent discount from 100%.
25% of 45.95 = *d*	Since 100% − 25% = 75%, the discount price will be 75% of the original price.
$0.25 \times 45.95 = 11.475$	
The discount is $11.49.	
Then, subtract to find the discount price.	Then, multiply to find the discount price.
$45.95 - 11.49 = 34.46$	$0.75 \times 45.95 = 34.4625$
The discount price is $34.46.	The discount price is $34.46.

Next, find the price including the sales tax.

Method 1	**Method 2**
Find the amount of the tax.	First, add the percent of tax to 100%.
6% of 34.46 = t	
$0.06 \times 34.46 = 2.0676$	Since 100% + 6% = 106%, the total price will be 106% of the discount price.
The tax is $2.07.	
Then, add to find the total price.	Then, multiply to find the total price.
$34.46 + 2.07 = 36.53$	$1.06 \times 34.46 = 36.5276$
The total price is $36.53.	The total price is $36.53.

The jeans will be $36.53 after the discount and the sales tax.

Checking Your Understanding

Communicating Mathematics

Read and study the lesson to answer each question.

1. **Explain** how r and R are different.
2. **Describe** two ways of finding the price of a $60 item after a 25% discount.
3. Under what conditions might you use $P = R \cdot B$ instead of $\frac{P}{B} = \frac{r}{100}$?
4. If $R = 0.32$, what is r?
5. **You Decide**   Miguel and Jalisa are trying to find the monthly interest on a credit card balance of $2500 if the monthly percentage rate is 1.5%. Miguel solves using $P = (1.5)(2500)$. Jalisa uses $P = (0.015)(2500)$. Who is correct? Explain.

Guided Practice

Solve each problem by using the percent equation, $P = R \cdot B$.

6. 75 is 50% of what number?
7. 15% of what number is 30?
8. 18 is what percent of 60?
9. What is 24% of 72?

Find the discount or interest to the nearest cent.

10. $450 TV, 33% off
11. $315 suit, 15% off
12. $3500 at 12% annually for $2\frac{1}{2}$ years
13. **Travel**   About 28% of the Japanese tourists who travel abroad each year visit the United States. If 12 million Japanese people went abroad in 1992, about how many visited the United States?

Exercises: Practicing and Applying the Concept

Independent Practice

Solve each problem by using the percent equation, $P = R \cdot B$.

14. 15 is what percent of 100?
15. What is 20% of 135?
16. What number is 43% of 15?
17. 28 is 40% of what number?
18. 15% of what number is 39?
19. 42% of 75 is what number?
20. 45 is what percent of 150?
21. 42 is what percent of 14?
22. 110% of what number is 880?
23. 5 is what percent of 300?
24. 2.04% of 256 is what number?
25. 18.6 is 15.5% of what number?
26. $33\frac{1}{3}$% of 420 is what number?
27. 500% of what number is 100?

Choose

Estimation
Mental Math
Calculator
Paper and Pencil

Find the discount or interest to the nearest cent.

28. $49.95 sweat suit, 28% off
29. $199.99 tools, 35% off
30. $299.99 VCR, 40% off
31. $1250 refrigerator, 15% off
32. $500 at 0.3% a month for 15 months
33. $250 at $8\frac{1}{2}$% annually for 2 years
34. $1250 at 1.5% monthly for 7 months
35. $945 at 3.5% annually for 15 months

Critical
Thinking

36. Is x% of y always equal to y% of x? Give examples to support your answer.

Applications and
Problem Solving

37. Medicine Doctors divide blood into eight different types. Of the approximately 249 million Americans, 37.4% have type O+. How many Americans have O+ blood?

38. Travel In one year, Americans went on 43.3 million trips. The table at the right shows the number of trips made by different modes of transportation. Find the percent of trips made in each mode of transportation.

39. Advertising A store advertised a 200%-off sale. Explain why you know that their advertisement was incorrect.

Mode of Transportation	Number of Trips (millions)
car, not rented	25.7
airplane	12.0
car, rented	3.3
motor home	1.0
bus	0.7
train	0.6

Source: *American Demographics*

40. Business Sales associates are often paid in **commission**. A commission is usually given as a percent of the amount of the associate's sales. If Computer Generation pays its associates 15% commission, how much commission would be earned from the sale of a $2300 computer?

41. Retail Jackee Johnson receives a 10% discount on all merchandise at the store where she works. She is buying a $439 television that is on sale at 33% off.

a. Find Jackee's price for the television.

b. Does it matter which discount is taken first, the 33% sale price or the 10% employee discount?

c. Could the discount percents be added together first and then taken off or must they be taken off one at a time? Give an example to support your answer.

42. Business Many car dealers offer special interest rates as incentives to attract buyers. How much interest would you pay for the first month of a $5500 car loan with a monthly interest rate of 0.24%?

43. Economics When the prices of products in a country rise over time, the country is experiencing inflation. The rate of inflation in the United States is about 5% per year. A newspaper reported "The rate of inflation in Yugoslavia is 20% a day. In other words, prices double every five days." Are the two statements made in the article the same? Explain. **Source:** *USA TODAY*

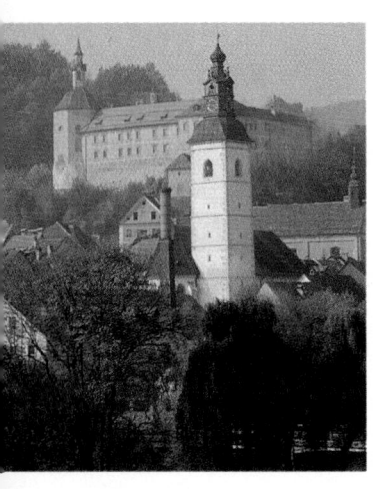

Mixed Review

44. Estimate 68% of 210. (Lesson 9-8)

45. Statistics Does the scatter plot at the right show a positive relationship, a negative relationship, or no relationship? (Lesson 8-2)

46. Determine whether {(6, 1), (8, 4), (−5, 5), (3, 4)} is a function. (Lesson 8-1)

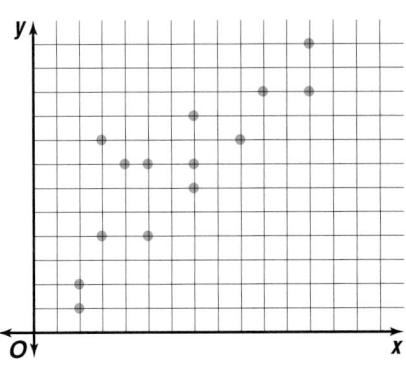

47. Population Vatican City is the least populated country in the world. The population of the Falkland Islands is 298 less than three times the population of Vatican City. If the population of the Falkland Islands is 1916, find the population of Vatican City. (Lesson 7-3)

   a. Write an equation that represents this situation.

   b. Solve the equation to find the population of Vatican City.

48. Write $\frac{5}{16}$ as a decimal. (Lesson 6-1)

49. Find the GCF of 150 and 345. (Lesson 4-5)

For **Extra Practice**, see page 763.

50. Art Pablo Picasso is the artist with the most paintings that have sold for over $1 million, with 148 paintings. This is six more than the number of paintings by Auguste Renoir that have sold for more than $1 million. (Lesson 3-2)

   a. Write an equation to represent the number of Auguste Renoir paintings that have sold for more than $1 million.

   b. Find the number of paintings by Auguste Renoir that have sold for more than $1 million.

51. Standardized Test Practice Find 12 + (−5). (Lesson 2-4)

   A. −17   **B.** −7   **C.** 7   **D.** 17

WORKING ON THE

Refer to the Investigation on pages 368–369.

The Athletic Director of the university called you into her office. She said that the university needs to increase the gross income from the football games. This could be done by either raising the prices of seats or increasing the number of deluxe and/or spectator seats. Conference regulations require that the stadium has at least 18% of its seats priced at under $16.00.

The Athletic Director asked you to prepare three different options to increase the gross income from a game by 15%. Prepare three options and state the advantages and disadvantages of each option.

Add the results of your work to your Investigation Folder.

Percent of Change

You'll Learn

■ To find percent of increase or decrease.

Why
It's Important

Understanding percent of decrease can help you determine discount prices.

Application: Electricity Rates

Rate Hike Announced

Source: New York Times, April 22, 1995

The Consolidated Edison Company of New York announced that they will be raising rates on May 1, 1995. The increase will generate $55.1 million. That represents a 1.1% increase. A typical customer in an apartment that uses 300 kilowatt-hours a month will see an increase in his or her monthly bill from $50.52 to $51.26. A Westchester customer using 450 kilowatt-hours a month will pay $71.05 instead of $70.05.

Find the percent of increase for each customer and compare your answers to the 1.1% amount the company stated. *This problem will be solved in Example 3.*

Learning the Concept

Newspapers often report changes in prices and rates as a **percent of change**. The percent of change is the ratio of the amount of change to the original amount.

Example **Find the percent of change from $120 to $135.**

There are two methods you can use to find the percent of change.

Method 1

Step 1: Subtract to find the amount of change.
$135 - 120 = 15$

Step 2: Solve the percent equation. Compare the amount of increase to the original amount.
$$P = R \cdot B$$
$$15 = R \cdot 120$$
$$\frac{15}{120} = \frac{120R}{120}$$
$$0.125 = R$$

The percent of change is 12.5%.

Method 2

Step 1: Divide the new amount by the original amount.
$$135 \div 120 = 1.125$$

Step 2: Subtract 1 from the result and write the decimal as a percent.
$$1.125 - 1 = 0.125 \text{ or}$$
$$12.5\%$$

The percent of change is 12.5%.

When an amount increases, like in Example 1, the percent of change is a **percent of increase**. When the amount decreases, the percent of change is negative. You can state a negative percent of change as a **percent of decrease**.

Example **2** **Find the percent of decrease from a population of 257 thousand to 243 thousand.**

Method 1	**Method 2**

Method 1

Step 1: Subtract.
$243 - 257 = -14$

Step 2: Solve the percent equation. Compare the amount of change to the original amount.

$$P = R \cdot B$$
$$-14 = R \cdot 257$$
$$\frac{-14}{257} = \frac{257R}{257}$$
$$-0.054 = R$$

The percent of change is -5.4%.
The percent of decrease is 5.4%.

Method 2

Step 1: Divide the new amount by the original amount.

$$243 \div 257 \approx 0.946$$

Step 2: Subtract 1 from the result and write the decimal as a percent.

$0.946 - 1 = -0.054$ or -5.4%

The percent of change is -5.4%.
The percent of decrease is 5.4%.

Example **3** **Refer to the application at the beginning of the lesson. Find the percent increase for each customer.**

Real World APPLICATION

Energy

Apartment bill

Use Method 1.

Step 1: Subtract. $\$51.26 - \$50.52 = \$0.74$ *Estimate:* $\frac{1}{50} = 2\%$

Step 2: Solve the percent equation.

$$P = R \cdot B$$
$$0.74 = R \cdot 50.52$$
$$\frac{0.74}{50.52} = \frac{50.52R}{50.52} \quad \textit{Divide each side by 50.52.}$$
$$0.0146 = R$$

The percent of increase is about 1.5%.

Westchester bill

Use Method 2. *Estimate:* $\frac{71}{70}$ *will be a little over 100%.*

Step 1: Divide the new amount by the original amount.
$$71.05 \div 70.05 \approx 1.014$$

Step 2: Subtract 1 from the result and write the decimal as a percent.
$1.014 - 1 = 0.014$ or 1.4%

The percent of increase is 1.4%.

The apartment bill rose 1.5%, and the Westchester bill rose 1.4%, not the 1.1% estimated by the power company.

THINK ABOUT IT

Could it be that the 1.1% increase in total revenue and a greater increase in customer prices could both be correct? Explain.

Communicating Mathematics

Read and study the lesson to answer each question.

1. **Explain** how to find a percent of change.

2. **Estimate** the percent of change if a temperature goes from 20°C to 25°C.

3. **Name** the amount used as the base when you find a percent of change.

4. The Cardinals football team scored 14 points against the Vikings. The following week, the Cardinals scored 21 points in their game against the Wolves. The coach said the team had improved their score 150%. Is that correct? Explain.

5. Write about the least difficult and most difficult concepts for you in this chapter.

Guided Practice

State whether each percent of change is a percent of increase or a percent of decrease. Then find the percent of increase or decrease. Round to the nearest whole percent.

6. old: $10
   new: $16

7. old: 25 thousand people
   new: 22 thousand people

8. old: 120 pounds
   new: 150 pounds

9. old: 59°F
   new: 55°F

10. **Economics** October 19, 1987 is called Black Thursday because the Dow had its biggest decline, 508.32 points, on that day. If the average was 1738.42 points at closing, what was the percent of decrease of the average that day?

Exercises: Practicing and Applying the Concept

Independent Practice

State whether each percent of change is a percent of increase or a percent of decrease. Then find the percent of increase or decrease. Round to the nearest whole percent.

11. old: 142 pounds
    new: 114 pounds

12. old: $0.59
    new: $0.30

13. old: 425 people
    new: 480 people

14. old: $48
    new: $44

15. old: 96 minutes
    new: 108 minutes

16. old: 14.5 liters
    new: 12.5 liters

17. old: $228
    new: $251

18. old: 29.5 ounces
    new: 26.5 ounces

19. old: $39.99
    new: 42.59

20. old: 124 minutes
    new: 137 minutes

21. old: $106
    new: $250

22. old: 4329.80 points
    new: 4351.50 points

Critical Thinking

23. A tennis outfit is marked down 50% and then reduced at the cash register another 30%.

    a. Is this a total reduction of 80%? Why or why not?

    b. If the outfit is not 80% off, find the actual discount.

24. Kelly bought a CD player, with a pre-sale price of $c, at a 30% discount. Kelly also paid 6% sales tax. After a few months, he sold the CD player to his sister for $d, which was 50% of what he paid originally.

    a. Express d as a function of c.

    b. What did Kelly's sister pay if the original list price was $50?

25. Consumer Awareness Most cars depreciate, or lose value, each year. Kiku paid $7800 for her car. The "blue book" that tells the value of used cars lists the value of a car like Kiku's as $6225. What was the percent of decrease of the value of the car?

26. Merchandising A retail store buys a coat for $85 and marks the price up 65% for sale. Later the price is reduced to $110.
  a. If the coat sells for $110, what percent of the selling price is profit?
  b. What was the percent of discount from the original selling price?

27. Postage When the U.S. Postal Service changed its rates in 1999, the prices of 1-ounce first class letters and two-pound priority mail packages increased. If the price of a priority package was increased at the same percent of increase as a first class letter, what would priority mail have cost?

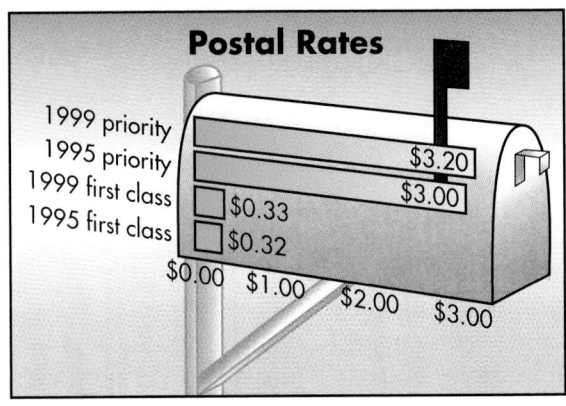

Source: U.S. Postal Service

28. Manufacturing Sometimes manufacturers increase their revenues by selling a product at the same price, but decreasing the quantity sold. For example, a can of fruit juice was changed so that a can which used to hold 6 ounces now contains 5.5 ounces.
  a. What was the percent of decrease?
  b. If the cans were sold in a 6-pack for $1.59, what was the percent of increase in price per ounce?

29. Pets Looking for an unusual pet? How about a hedgehog! They are becoming so popular that some breeders have reported a 250% increase in sales in recent years. If a breeder sold 50 hedgehogs one year before the increase, how many should he or she expect to sell in a year now?

Mixed Review

30. A pair of earrings usually sell for $15.50. Find the sale price of the earrings if they are on sale for 35% off. (Lesson 9-9)

31. Is the equation whose graph is shown at the right a function? Explain. (Lesson 8-4)

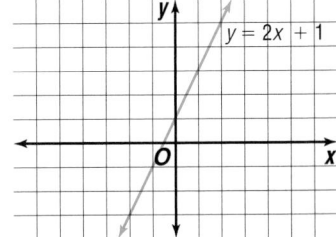

32. Energy The United States consumes about 4,921,000,000 barrels of oil each year. The annual production of crude oil in the United States is about half of what is consumed. About how much oil is produced in the United States each year? (Lesson 6-2)

For **Extra Practice**, see page 764.

33. Standardized Test Practice Which inequality is true? (Lesson 4-7)
  A. $\frac{5}{6} < \frac{7}{8}$ **B.** $\frac{5}{6} > \frac{6}{7}$ **C.** $\frac{6}{7} > \frac{7}{8}$ **D.** $\frac{5}{6} < \frac{4}{5}$

34. Solve $n = 14 - (-3)$. (Lesson 2-5)

Investigation

STADIUM STAMPEDE

Refer to the Investigation on pages 368–369.

The spirit squad has an idea for the homecoming game that will increase fan spirit. They want to pass out pompoms in the university colors. There are several items that they have questions about and they have asked you to prepare a presentation for their next meeting.

- How many pompoms should they pass out? Should they go to every fan or only a certain percent of them? How should they determine that percent and which fans should get them?

- How should the pompoms be passed out? Should they be passed out at all of the gates, certain gates, or placed on random seats prior to the game?

- Can a sponsor be obtained to pay for the pompoms? Would the sponsor pay for all of the pompoms, only a certain percent of them, a specified dollar amount, or a certain number of them? If the sponsor cannot cover all the costs, how will the remaining costs be paid?

- Do you charge for the pompoms, and if so, how much do you charge for each one? What would be a good price? At what point will the cost be too much for the fans to want to buy one?

- Of what type of material should the pompoms be made? Should recycling be considered when determining the material? If so, should containers be at the gates for fans to recycle the pompoms on their way out of the stadium after the game?

Reread the list of questions. Are there other issues that need to be addressed? Prepare a written report for the committee.

PORTFOLIO ASSESSMENT

You may want to keep your work on this Investigation in your portfolio.

Vocabulary

After completing this chapter, you should be able to define each term, property, or phrase and give an example or two of each.

Algebra
base (p. 449)
commission (p. 470)
cross products (p. 444)
discount (p. 468)
interest (p. 467)
percent (p. 449)
percentage (p. 449)
percent of change (p. 472)
percent of decrease (p. 473)
percent of increase (p. 473)
percent proportion (p. 450)
principal (p. 467)
property of proportions (p. 444)
proportion (p. 444)
rate (p. 433, 449)
ratio (p. 432)
unit rate (p. 433)

Geometry
golden ratio (p. 434)

Probability
capture-recapture (p. 448)
probability (p. 440)
sample space (p. 441)

Problem Solving
make a table (p. 437)

Understanding and Using Vocabulary

Determine whether each statement is *true* or *false*.

1. A rate is a special ratio.

2. Percentage is a ratio that compares a number to 1.

3. In the percent proportion $\frac{16}{32} = \frac{50}{100}$, 16 is the base.

4. A sample space is the area of a closed figure.

5. When you use the percent equation $P = R \cdot B$, R represents $\frac{r}{100}$ where r is the rate.

6. A ratio is a comparison of two numbers by multiplication.

7. The property of proportions allows you to compare two proportions.

Skills and Concepts

Objectives and Examples

Upon completing this chapter, you should be able to:

▶ **write ratios as fractions in simplest form and determine unit rates.** (Lesson 9-1)

Express 15 to 21 as a fraction in simplest form.

$$\frac{15}{21} = \frac{5}{7}$$

÷ 3 (top), ÷ 3 (bottom)

Express $1.29 for a dozen as a unit rate.

$$\frac{129 \; cents}{12 \; items} = \frac{\bullet}{1 \; item}$$

÷ 12 (top), ÷ 12 (bottom)

129 [÷] 12 [=] *10.75*

The unit rate is 10.75 cents.

▶ **find the probability of simple events.** (Lesson 9-3)

Find the probability of rolling a number that is a factor of 6 on a die.

There are four ways to roll a factor of 6 −1, 2, 3, and 6, and six possible outcomes −1, 2, 3, 4, 5, and 6.

$P(\text{factor of 6}) = \frac{4}{6} \text{ or } \frac{2}{3}$

▶ **use proportions to solve problems.** (Lesson 9-4)

If 2 pounds of pecans cost $13.90, how much will 5 pounds cost?

$\frac{13.90}{2} = \frac{c}{5}$ *Write a proportion.*

$13.90 \cdot 5 = 2 \cdot c$ *Write cross products.*

$\frac{69.50}{2} = \frac{2c}{2}$ *Divide each side by 2.*

$c = 34.75$

Five pounds will cost $34.75.

Review Exercises

Use these exercises to review and prepare for the chapter test.

Express each ratio as a fraction in simplest form.

8. 10 red to 18 blue
9. 287 yards to 315 yards
10. 15 out of 90
11. 600 to 1000

Express each ratio as a unit rate.

12. 339.2 miles on 10.6 gallons
13. $1.78 for 2 pounds
14. $425 for 17 tickets
15. 142.5 miles in 2.5 hours

There are 6 cans of cola, 4 cans of fruit punch, 9 cans of iced tea, and 2 cans of lemonade in a cooler. Suppose you choose a can randomly. Find each probability.

16. $P(\text{cola})$
17. $P(\text{fruit punch or iced tea})$
18. $P(\text{not carbonated})$
19. $P(\text{cola or lemonade})$
20. $P(\text{iced tea})$
21. $P(\text{neither cola nor fruit punch})$

Solve each proportion.

22. $\frac{8}{6} = \frac{z}{14}$ 23. $\frac{5.1}{1.7} = \frac{7.5}{a}$

Write a proportion that could be used to solve for each variable. Then solve.

24. 20 grams for $5.60
    x grams for $9.80
25. 88.4 miles on 3.4 gallons
    161.2 miles on g gallons

▶ **use the percent proportion to solve problems involving percents.** (Lesson 9-5)

Sixteen is what percent of 20?

$$\frac{P}{B} = \frac{r}{100} \quad \rightarrow \quad \frac{16}{20} = \frac{r}{100}$$
$$16 \cdot 100 = 20 \cdot r$$
$$\frac{1600}{20} = \frac{20r}{20}$$
$$80 = r$$

16 is 80% of 20.

Use the percent proportion to solve each problem.

26. What is 40% of 5?

27. Nineteen is what percent of 25?

28. What is 120% of 50?

29. What is 0.1% of 4000?

▶ **use a sample to predict the actions of a large group.** (Lesson 9-6)

Of the students surveyed, 43% chose pizza as their favorite lunch. How many of the 1294 students in the school would choose pizza?

$$\frac{x}{1294} = \frac{43}{100}$$
$$x \cdot 100 = 1294 \cdot 43$$
$$x = 556.42 \quad \textit{Use a calculator.}$$

About 556 students would choose pizza.

The results of a survey on dessert preferences are shown in the table at the right.

Dessert	Number
cheesecake	7
chocolate cake	48
apple pie	37
ice cream	18

30. How many of the 459 cafeteria customers would choose apple pie?

31. If you asked 500 people which of these desserts they prefer, how many would choose cheesecake?

▶ **express decimals and fractions as percents and vice versa.** (Lesson 9-7)

Express 0.32 as a percent.

$\times\ 100$

$$\frac{0.32}{1} = \frac{32}{100} \text{ or } 32\%$$

$\times\ 100$

Express each decimal as a percent.

32. 0.92   **33.** 0.0056

Express each fraction as a percent.

34. $\frac{63}{100}$ **35.** $\frac{113}{200}$

Express each percent as a fraction.

36. 90%   **37.** 65%

Express each percent as a decimal.

38. 45%   **39.** 235%

▶ **use percents to estimate.** (Lesson 9-8)

Estimate 12% of 54.

12% is about 10%.
10% of 54 is 5.4.
12% of 54 is about 5.4.

Choose the best estimate.

40. 2% of 180   **a.** 36   **b.** 3.6   **c.** 0.36

41. 198% of 5   **a.** 10   **b.** 1.0   **c.** 0.10

42. 8% of 420   **a.** 42   **b.** 5   **c.** 400

43. 73% of 80   **a.** 60   **b.** 40   **c.** 80

44. 352% of 20   **a.** 8   **b.** 40   **c.** 70

45. 0.6% of 620   **a.** 2   **b.** 6   **c.** 12

Objectives and Examples	**Review Exercises**

▶ **solve percent problems using the percent equation.** (Lesson 9-9)

1.8 is 4% of what number?

$$P = R \cdot B$$
$$1.8 = 0.04 \cdot B \quad \textit{Replace P with 1.8 and R}$$
$$\frac{1.8}{0.04} = \frac{0.04B}{0.04} \quad \textit{with 0.04.}$$
$$45 = B$$

1.8 is 4% of 45.

Solve each problem by using the percent equation, $P = R \cdot B$.

46. 54 is 150% of what number?

47. 16% of what number is 3.2?

48. 63 is what percent of 105?

49. 2600 is 65% of what number?

50. 16 is what percent of 2?

51. What percent of 16 is 5?

▶ **find percent of increase or decrease.** (Lesson 9-10)

Find the percent of increase from 159 pounds to 168 pounds.

Subtract: $168 - 159 = 9$
Use the percent equation.

$$9 = R \cdot 159 \quad P = R \cdot B$$
$$\frac{9}{159} = \frac{R \cdot 159}{159}$$
$$0.0566 = R$$

The percent of increase is about 5.7%.

State whether each percent of change is a percent of increase or a percent of decrease. Then find the percent of increase or decrease. Round to the nearest whole percent.

52. old: $0.60
new: $0.75

53. old: 16 ounces
new: 15.8 ounces

54. old: 18.4 grams
new: 17.5 grams

55. old: $15.00
new: $11.75

56. old: $84
new: $118

57. old: 210 pounds
new: 185 pounds

Applications and Problem Solving

58. Records A Micro-Car is listed as the world's smallest car. It is a model of a 1936 sedan. A part of the actual car that is 1 meter long is 0.001 m long on the model. Write the scale of the model as a fraction in simplest form. (Lesson 9-1)

60. Safety Fifty-eight percent of the adults surveyed said they wore their seat belts all the time. If there are 185 million adults in the United States, how many American adults wear their seat belts all the time? (Lesson 9-6)

59. Games A domino has two squares on its face. Each of the two spaces is marked with 1, 2, 3, 4, 5, or 6 dots, or it is blank. A complete set of dominoes includes one domino for each possible combination of dots and blanks. Doubles, like 5 dots in each space, are included. How many dominoes are in a complete set? (Lesson 9-2)

A practice test for Chapter 9 is available on page 782.

Cooperative Learning Project

Work as a group.

1. Develop a multiple-choice questionnaire on any topic. Include at least five questions with three or more choices. Survey 20% of your class for their opinions.

2. Record the data you gather in a table, graph, or scatter plot.

3. What fraction of the people surveyed chose each possible answer? Write a sentence about the probability of randomly choosing a person who chose a certain answer.

4. List the sample space of each question on the questionnaire.

5. Predict the number of people in your class who would choose a certain answer on your questionnaire. Explain how you made your prediction.

6. Write a statement about the percent of your survey group that chose a certain answer.

7. Present the results of your survey in an interesting way. You may wish to make a poster or brochure, including tables or graphs.

Thinking Critically

▶ When you write a decimal as a percent, the first step is to write it as a fraction with a denominator of 1. Then you multiply by 100. Can you write one step that could replace these two?

▶ A weather forecaster said "There is a 50% chance of rain on Saturday and a 50% chance of rain on Sunday. So there is a 100% chance of rain this weekend." Do you think this is correct? Explain.

 Portfolio

Select one of the assignments from this chapter that you found especially challenging and place it in your portfolio.

Investigation

It's Only A Game

What do you do in your spare time? In ancient times people must have had spare time too because archaeologists have found ancient artifacts which they think are game boards and pieces. A board game believed to be about 4500 years old was found at Ur, a city of ancient Sumer (now in Iraq).

The game of chess is believed to have originated in India 1400 years ago. Another game that has been around for some time is dominoes. Historians do not agree on how old dominoes are or who invented them. They believe they were invented in China and spread to Europe in the 1300s.

Just as the playing of games appears to be a natural instinct of the human race, the analysis of games would appear to be a natural instinct of many mathematicians. Games are most commonly classified according to the equipment they use.

The major types of games are:
- *board games*
- *card games*
- *tile games*
- *target games*
- *dice games*
- *table games*
- *paper and pencil games*
- *electronic games*

Starting the Investigation

Work with your group to design a game. Ask friends, teachers, relatives, and neighbors what their favorite game is and why it is their favorite. Describe, in detail, the qualities of each game that led them to their choice. Determine what qualities are common to the favorite games. Then determine which of these qualities you will use in the design of your game.

Think About . . .

- *What will be the age group of the players who will play your game?*
- *For how many players is the game designed?*
- *What is the object of the game?*
- *Will the game be designed for educational purposes? (Will it teach a player something?)*
- *Where will the game be played? (Is it a party game like charades? a street game, such as hopscotch? a lawn game, such as croquet? or a TV game show?)*
- *What type of game do you want to design? Will it be skill-based, pure chance, or a combination of both?*
- *How will players determine who goes first?*
- *How will players take a turn?*
- *What are the basic rules?*
- *Are there any special rules? What are they based on?*
- *When is the game over? What determines the winner?*

Investigation
It's Only a Game

Working on the Investigation
Lesson 10-4, p. 508

Working on the Investigation
Lesson 10-8, p. 528

Working on the Investigation
Lesson 11-7, p. 588

Closing the Investigation
End of Chapter 11, p. 600

You will work on this Investigation throughout Chapters 10 and 11.

Be sure to keep your materials in your Investigation Folder.

inter NET
CONNECTION

Research For current information on game design, visit the following website.

Data Collection and Comparison To share and compare your data with other students in the U.S., visit:

www.glencoe.com/ sec/math/prealg/ mathnet

10 More Statistics and Probability

What You'll Learn

In this chapter, you will:

- display data in stem-and-leaf plots and box-and-whisker plots,

- find measures of variation,

- recognize misleading data,

- count outcomes,

- find permutations and combinations, and

- find probabilities and odds.

Why It's Important

Real World

Meteorology The chart, Normal January Temperatures, shows temperatures for January in 16 cities located in the southern United States. If this data were used in some type of graph, it would be easier to compare the various temperatures. For example, the temperatures could be organized from least to greatest and a bar graph could be drawn. A better type of graph for this data might be a stem-and-leaf plot. *You will learn how to make a stem-and-leaf plot in Lesson 10-1.*

Normal January Temperatures

City	Temp	City	Temp
Asheville, NC	36	Memphis, TN	40
Atlanta, GA	41	Mobile, AL	50
Birmingham, AL	42	New Orleans, LA	51
Charleston, SC	48	Norfolk, VA	39
Jackson, MS	44	Raleigh, NC	39
Jacksonville, FL	52	Richmond, VA	37
Miami, FL	67	Savannah, GA	49
Knoxville, TN	36	Tampa, FL	60

Source: *The World Almanac*

Focus on...

To be successful in this chapter, you'll need to understand these concepts and be able to apply them. Refer to the lesson in parentheses if you need more review before beginning the chapter.

▶ **Find the mean, median, and mode of a set of data.** *(Lesson 6-6)*

Find the mean, median, and mode for each set of data.

1. 12, 18, 11, 7, 23, 8, 21   **2.** 1.3, 0.7, 2.8, 3.1, 1.7, 2.2, 0.9, 2.8

▶ **Find the probability of simple events.** *(Lesson 9-3)*

There are 7 red candies, 10 green candies, 4 orange candies, 5 brown candies and 8 blue candies in a bag. Suppose you select one candy at random. Find each probability.

3. P(green)   **4.** P(white)   **5.** P(brown)   **6.** P(not blue)

▶ **Multiply fractions.** *(Lesson 6-3)*

Find each product.

7. $\frac{1}{2} \cdot \frac{1}{2} \cdot \frac{1}{2}$ **8.** $\frac{5}{12} \cdot \frac{2}{7}$ **9.** $\frac{1}{8} \cdot \frac{2}{3} \cdot \frac{5}{6}$ **10.** $\frac{7}{34} \cdot \frac{10}{33}$

▶ **Add and subtract fractions.** *(Lessons 5-4 and 5-5)*

Find each sum.

11. $\frac{7}{34} + \frac{10}{34} - \frac{8}{34}$ **12.** $\frac{1}{2} + \frac{1}{6}$ **13.** $\frac{45}{100} + \frac{22}{100} - \frac{19}{100}$ **14.** $\frac{1}{10} + \frac{2}{3} - \frac{1}{4}$

Focus on...

READING SKILLS

In Chapter 1, you learned that **statistics** is a branch of mathematics that involves the collection and presentation of data. Statisticians often present data in graphs. The names of graphs can give you clues about their appearance. For example, a bar graph has bars, a circle graph is shaped like a circle, and a box-and-whisker plot looks like a box that has whiskers!

10-1 Stem-and-Leaf Plots

What You'll Learn

■ To display and interpret data in stem-and-leaf plots.

Why It's Important

You can use a stem-and-leaf plot to record a large set of data collected for a science project.

Application: Consumerism

In a taste test reported in two consumer magazines, 35 different chocolate-chip cookies were rated according to flavor, freshness, and texture. For the 27 cookies that were rated good, very good, or excellent, the cost per 30-gram serving is shown below.

10¢	11¢	58¢	10¢	25¢	49¢	24¢	46¢	16¢
13¢	21¢	34¢	16¢	29¢	28¢	13¢	14¢	14¢
20¢	18¢	18¢	21¢	13¢	22¢	11¢	23¢	12¢

If you were asked to organize this set of data and present it in an easy-to-read format, you could construct a **stem-and-leaf plot**.

Learning the Concept

In a stem-and-leaf plot, the greatest place value common to all the data values is usually used for the **stems**. The next greatest place value forms the **leaves**. Follow these steps to construct a stem-and-leaf plot from the data above.

Step 1 Find the least and greatest price.

  The least price is 10¢, and the greatest price is 58¢.

Step 2 Find the stems.

  The least price, 10¢, has a 1 in the tens place. The greatest price, 58¢, has a 5 in the tens place. Draw a vertical line and write the digits in the tens places from 1 to 5 to the left of the line.

```
1|
2|
3|
4|
5|
```

Step 3 Put the leaves on the plot.

  Record each of the prices on the plot by pairing the units digit, or leaf, with the corresponding stem. For example, 10 is plotted by placing the units digit, 0, to the right of the stem 1.

```
1|0 1 0 6 3 6 3 4 4 8 8 3 1 2
2|5 4 1 9 8 0 1 2 3
3|4
4|9 6
5|8
```

Step 4 Rearrange the leaves so they are ordered from least to greatest.

Step 5 Include an explanation or key of the data.

```
1|0 0 1 1 2 3 3 3 4 4 6 6 8 8
2|0 1 1 2 3 4 5 8 9
3|4
4|6 9
5|8        1|3 means 13¢.
```

It is easy to see that nearly all of the prices are between 10¢ and 29¢ because those rows have the most leaves.

Example 1

Real World
APPLICATION

Sports

In the last ten games of the 1995 men's professional basketball playoffs, the points scored by Hakeem Olajuwon were 27, 41, 43, 20, 42, 39, 31, 34, 31, and 35.

a. Make a stem-and-leaf plot of this data.

The stems are 2, 3, and 4.

```
2 | 0 7
3 | 1 1 4 5 9
4 | 1 2 3        2|0 means 20 points.
```

b. In which interval do most of the points scored occur?

From the leaves, you can see that Olajuwon scored in the 30–39 point interval more times than in the other intervals.

A **back-to-back stem-and-leaf plot** is used to compare two sets of data. In this type of stem-and-leaf plot, the leaves for one set of data are on one side of the stem and the leaves for the other set of data are on the other side of the stem. Two keys to the data are needed.

Example 2

Real World
APPLICATION

Nutrition

A nutritional analysis was done for several fast food restaurants. The amount of fat, in grams, of their burgers and chicken, is shown below.

| Burgers | 10, 15, 20, 26, 19, 33, 36, 30 |
| Chicken | 9, 13, 13, 19, 18, 20, 15, 15 |

a. Make a back-to-back stem-and-leaf plot of the data.

The stems are 0, 1, 2, and 3.

```
Burgers    |   | Chicken
           | 0 | 9
     9 5 0 | 1 | 3 3 5 5 8 9
       6 0 | 2 | 0
     6 3 0 | 3 |
   6|3 = 36           2|0 = 20
```

b. If a lower amount of fat is better, which type of fast food is better?

Chicken is better because it has fewer amounts in the 20–39 range than the burgers.

Checking Your Understanding

Communicating
Mathematics

Read and study the lesson to answer these questions.

1. **Explain** why you might use a stem-and-leaf plot.
2. **List** the numbers in the stem for the stem-and-leaf plot at the right.
3. **Explain** how a stem-and-leaf plot is similar to a bar graph. How is it different?
4. **Describe** the information you can; obtain from the stem-and-leaf plot in Example 1.

```
3 | 0
4 | 1 5 7
5 | 3 4 4 7
6 | 0 1 8
   3|0 = 30
```

Work with a partner to complete the following activity.

5. Choose a topic that interests you. Research or gather data about the topic. Topics might include: heights of U.S. presidents, ages of U.S. presidents at inauguration, points scored by basketball team members, time it takes to get to school or do homework.

   a. Construct a stem-and-leaf plot of the data you collected.

   b. Write two or three statements that describe the data.

Guided Practice

State the numbers you would use for each stem in a stem-and-leaf plot of each set of data. Then make the plot.

6. 65, 82, 73, 91, 95, 86, 78, 69, 80, 88

7. 5.7, 5.4, 6.8, 6.3, 7.1, 8.5, 7.5, 6.9, 6.7, 7.7

8. **Basketball** The stem-and-leaf plot at the right shows the scores of the Knox Junior High School's girl's basketball team for the games in the 1996 season.

   a. What were the highest and lowest scores for the Knox team?

   b. What were the highest and lowest scores of their opponents?

   c. What is the median of each set of scores?

Exercises: Practicing and Applying the Concept

Independent Practice

Make a stem-and-leaf plot of each set of data.

9. 64, 60, 72, 61, 73, 80, 68, 70, 65, 67, 70, 80

10. 23, 14, 25, 36, 45, 34, 32, 21, 39, 28, 48, 19, 39, 45, 48, 48, 51

11. 42, 23, 9, 21, 7, 11, 14, 6, 40, 5, 9, 45, 12

12. 83, 94, 54, 92, 85, 54, 96, 89, 75, 117, 68, 99, 116

13. 11.2, 10.4, 12.6, 12.3, 11.1, 14.7, 9.2, 10.8, 12.9, 11.2, 11.7, 13.3, 13.8, 12.8

14. **Architecture** The *World Almanac* lists 15 tall buildings in New Orleans, Louisiana. The number of floors in each of these buildings is listed in the chart at the right.

   a. Make a stem-and-leaf plot of the data.

   b. What is the greatest number of floors?

   c. What is the least number of floors?

   d. How many buildings had 34 or less floors?

   e. How many buildings were in the 30–39 floor range?

51	47	33
53	42	28
45	33	28
39	32	25
36	31	23

15. Retail Sales The table below lists the prices, in dollars, for pairs of running shoes.

Men's	83, 70, 70, 75, 70, 82, 70, 125, 82, 45, 70, 120, 110, 133, 100
Women's	124, 70, 72, 70, 65, 90, 60, 80, 55, 70, 67, 45, 55, 75, 85

    a. Make a back-to-back stem-and-leaf plot of the prices for men's and women's shoes.

    b. What information about the relationship between the prices for men's and women's shoes does it present?

Critical Thinking

16. Describe any restrictions when using a stem-and-leaf plot.

17. Is it possible to estimate the mean from a stem-and-leaf plot? Explain.

Applications and Problem Solving

18. Meteorology The normal January temperatures of 16 southern cities are shown in the table.

    a. Make a stem-and-leaf plot of the temperatures.

    b. If you were touring the south in January, what range of temperatures would you expect?

Normal January Temperatures

Asheville, NC	36	Memphis, TN	40
Atlanta, GA	41	Mobile, AL	50
Birmingham, AL	42	New Orleans, LA	51
Charleston, SC	48	Norfolk, VA	39
Jackson, MS	44	Raleigh, NC	39
Jacksonville, FL	52	Richmond, VA	37
Miami, FL	67	Savannah, GA	49
Knoxville, TN	36	Tampa, FL	60

Source: *The World Almanac*

19. Engineering The table below lists the temperatures at launch time for the first 23 space shuttle launches and how many of those times the O-rings on the solid rocket motors experienced thermal distress or no distress.

Temp (°F)	66	70	69	68	67	72	73	57	63	78	53	75	81	76	79	58
Distress	0	2	0	0	0	0	0	1	1	0	1	1	0	0	0	1
No Distress	1	2	1	1	3	1	1	0	0	1	0	1	1	2	1	0

    a. Make a back-to-back stem-and-leaf plot of the temperatures for distress and no distress.

    b. What information about the relationship between the temperatures of distress and no distress does your back-to-back stem-and-leaf plot make?

For **Extra Practice**, see page 764.

    c. If the temperature on the day of the launch was 31°F, would you expect the O-rings to experience thermal distress? Explain.

Mixed Review

20. Economics The Consumer Price Index (CPI) represents the relative costs of goods and services. If the CPI is 233.2 in April and 236.4 in May, what is the percent of increase from April to May? (Lesson 9-10)

21. Express 45% as a fraction in simplest form. (Lesson 9-7)

22. Solve $x - 3.4 = 9.2$. Check your solution. (Lesson 5-6)

23. Standardized Test Practice If 5 more than the product of a number and -2 is greater than 10, which of the following could be that number? (Lesson 3-1)

    A. 3   **B.** 2   **C.** -1 **D.** -4

24. Find the value of $|-9| + 14$. (Lesson 2-1)

What You'll Learn

■ To use measures of variation to compare data.

Why It's Important

Census bureau employees use measures of variation to describe populations of cities.

10-2 Measures of Variation

Technology Activity

In Chapter 6, you learned that the mean, median, and mode indicate the center of a set of data. Sometimes measures of center don't fully describe a set of data. In the back-to-back stem-and-leaf plot of bicycle helmet prices at the right, both models have the same mean price, the same median price, and the same mode price. But, it is easy to see that all the prices are not the same.

Adult		Youth
	2	5
7 2 1 1	3	4 7 8
3 3 3	4	2 3 3 8
0	5	

1|3 = $31   *2|5 = $25*

You can describe the two lists of helmet prices more accurately by using measures of variation to describe the spread of each set of data. You can use a graphing calculator to find common measures of variation.

Your Turn

▶ First, clear the statistical memory.

Enter: [STAT] 4 [2nd] [L1] [,] [2nd] [L2] [ENTER]

▶ Enter the adult prices in the L1 list and the youth prices in the L2 list.

Enter: [STAT] [ENTER] 31 [ENTER] 31 [ENTER] ...50 [ENTER] [▶]

25 [ENTER] 34 [ENTER] ...48 [ENTER]

▶ Find the median of the lower and upper halves of the adult prices.

Enter: [STAT] [▶] [ENTER] [2nd] [L1] [ENTER]

Press the down arrow to scroll through the list. Record the values for Q_1 and Q_3. Then find the difference between the greatest value, max X, and the least value, min X.

▶ Find the median of the lower and upper halves of the youth prices.

Enter: [STAT] [▶] [ENTER] [2nd] [L2] [ENTER]

Record the values for Q_1 and Q_3. Then find the difference between the greatest and least values.

TALK ABOUT IT

a. How do the differences between the greatest and least values of each list compare?

b. How do the medians of the lower half (Q_1) and upper half (Q_3) of each set of data compare?

Learning the Concept

When you found the difference between the greatest and least values of the helmet prices, you calculated a measure of variation called the **range**.

Definition of Range	The range of a set of numbers is the difference between the least and the greatest number in the set.

Example **Determine the range of each set of data.**

a. 12, 17, 16, 23, 14, 18, 11, 21

First, list the data in order.

11 12 14 16 17 18 21 23

The range is 23 − 11 or 12.

b.

Since the line plot is organized from the least data to the greatest, the range is the difference in the first and last data values of the plot.

The range is 19 − 12 or 7.

TECHNO TIP

The graphing calculator uses Q_1 to represent the lower quartile and Q_3 to represent the upper quartile.

In a large set of data, it is helpful to separate the data into four equal parts called **quartiles**. Recall that the median of a set of data separates the data in half. The median of the lower half of a set of data is called the **lower quartile** and is indicated by LQ. Likewise, the median of the upper half of the data is called the **upper quartile** and is indicated by UQ. Quartiles are used in another measure of variation called the **interquartile range**.

Definition of Interquartile Range	**In words:** The interquartile range is the range of the middle half of a set of numbers.
	In symbols: Interquartile range = UQ − LQ

Example

Olympics

Data Update For the latest on Olympic medals, visit:

www.glencoe.com/ sec/math/prealg/ mathnet

In the thirteen summer Olympic Games from 1948–1996, athletes from the United States earned the following numbers of gold medals.

Year	1948	1952	1956	1960	1964	1968	1972
Gold Medals	38	40	32	34	36	45	33

Year	1976	1980	1984	1988	1992	1996
Gold Medals	34	0	83	36	37	44

a. **Find the range and interquartile range for the number of gold medals.**

First, list the data from least to greatest. Then find the median.

0 32 33 34 34 36 36 37 38 40 44 45 83

                                       ↑

                              Median

(continued on the next page)

FYI:

The Summer Olympics were boycotted by the United States and 63 other countries in 1980. The USSR and 13 other Eastern Bloc nations boycotted the 1984 Summer Olympics.

The upper quartile is the median of the upper half and the lower quartile is the median of the lower half.

0   32   33   34   34   36   36   37   38   40   44   45   83

$$LQ = \frac{33 + 34}{2} \text{ or } 33.5 \qquad \text{Median} \qquad UQ = \frac{40 + 44}{2} \text{ or } 42$$

The interquartile range is $42 - 33.5$ or 8.5.

b. Which measure of variation better summarizes the performance of United States athletes, the range or the interquartile range?

Since the range represents values for years in which the Summer Olympics were boycotted, the interquartile range is more representative of the variability of this set of data.

Checking Your Understanding

Communicating Mathematics

Read and study the lesson to answer these questions.

1. **Compare and contrast** the measures of variation and the measures of central tendency.

2. **Write** a list of at least twelve numbers that has a range of 50 and an interquartile range of 10.

3. What can you say about a group of test scores if it has a small interquartile range?

Guided Practice

Find the range of each set of data.

4. 63, 82, 71, 65, 92, 86, 80, 95, 78, 89

5.
```
                    ×
            × ×      × ×      × ×
    × ×    × × × ×   × ×     × × ×
    +-+--+--+--+--+--+--+--+--+--+--+--+--+
     11  12  13  14  15  16  17  18  19  20  21  22  23  24
```

6. 30.8, 29.9, 30.2, 33.2, 30.1, 30.5, 30.7, 29.8

Use the data in the stem-and-leaf plot shown at the right.

7. What is the range?

8. **a.** Find the median.

   b. What are the upper and lower quartiles?

   c. Find the interquartile range.

4	2 3 9
5	4 5
6	7
7	3 8
8	2
9	1 3 4

4 | 2 = 42

9. **Sales** Discworld's sales of a new CD over a two-week period are shown in the chart at the right.

   a. Organize the data into a line plot or a stem-and-leaf plot.

   b. What is the median?

   c. Find the upper and lower quartiles.

   d. What is the interquartile range?

M	46	80
T	63	94
W	38	42
T	61	39
F	70	63
S	40	84
S	58	62

Independent Practice

Find the range, median, upper and lower quartiles, and the interquartile range for each set of data.

10. 56, 45, 37, 43, 10, 34, 29

11. 30, 90, 40, 70, 50, 100, 80, 60

12. 19°, 21°, 18°, 17°, 18°, 22°, 36°

13. 8, 11, 23, 7, 2, 4, 16, 2, 4

14. 135, 170, 125, 174, 136, 145, 180, 156, 188

15. 211, 225, 205, 207, 208, 213, 180, 200, 210, 229, 199

16.
```
5 | 9
6 | 0 0 1 2 2 3 5
7 | 1 2
```
5|9 = 59¢

17.
```
 7 | 3 5
 8 | 3 8 9
 9 | 0 0 1 2 4 4 7 7 8
10 | 3 6
11 | 4
```
7|3 = 7.3 cm

18.
```
25 | 0 3 7 9
26 | 1 3 4 5 5 6
27 | 1 5 6 6 9
28 | 1 2 3 5 8
29 | 2 5 6 9
```
29|2 = $292

19.

World's Busiest Airports	
City	**Passengers**
Atlanta	52,600,000
Chicago-O'Hare	48,300,000
Dallas-Ft. Worth	38,000,000
Denver	26,000,000
Frankfurt	30,400,000
London Heathrow	41,600,000
Los Angeles	43,000,000
Tokyo	35,900,000
Paris	29,200,000
San Francisco	27,200,000

Source: Airports Council International

20.

Top 10 states with largest Asian-Indian population in 1990	
State	**Population**
California	159,973
Florida	31,457
Illinois	64,200
Maryland	28,330
Michigan	23,845
New Jersey	79,440
New York	140,985
Ohio	20,848
Pennsylvania	28,396
Texas	55,795

Source: U.S. Census Bureau

21. **Home Runs Hit by League Leaders**

```
National League |   | American League
              2 | 2 |
                | 2 |
              1 | 3 | 2 2 2 2 3
  9 9 8 8 8 7 7 7 6 6 6 5 | 3 | 6 6 7 9 9 9
      4 4 4 3 1 0 0 0 0 0 | 4 | 0 0 0 0 1 2 3 3 4 4 4 4
      9 9 8 8 8 7 7 6 6 5 5 | 4 | 5 5 6 6 8 9 9 9 9
                    2 2 | 5 | 0 1
                        | 5 |
                        | 6 | 1
```
2|5 means 52   *6|1 means 61*

Graphing Calculator

Work with a partner to describe the variation in the populations of capital cities of the U.S.

22. Each person should find the population of half of the state capitals. Enter the data in a graphing calculator. Use the sort command to place the populations in order. Find the median, range, and interquartile range of the data. Use this information to describe the population of the capital cities.

23. a. Produce a set of at least ten test scores that has a median of 60 and an interquartile range of 20.

b. Produce a set of at least ten test scores that has a median of 60 and an interquartile range of 50.

c. What conclusions can you draw about the two sets of test scores in parts a and b from comparing the measures of variation?

Applications and Problem Solving

✓ **Choose**

Estimation
Mental Math
Calculator
Paper and Pencil

24. Meteorology The table at the right lists the average winter and summer precipitation for selected cities around the world.

a. Construct a back-to-back stem-and-leaf plot for the precipitation amounts.

b. Find the mean and median for each set of data.

c. How do the means compare?

d. How do the medians compare?

e. How do the interquartile ranges compare?

f. Which month has more consistent precipitation? Explain.

January and July Precipitation (in.)		
City	Jan.	July
Atlanta, GA	4.9	4.7
Berlin, Germany	1.9	3.1
Dallas, TX	1.7	2.0
Dublin, Ireland	2.7	2.8
Indianapolis, IN	2.7	4.3
Kinshasa, Zaire	5.3	0.1
Miami, FL	2.1	6.0
New Orleans, LA	5.0	6.7
Oklahoma City, OK	5.0	6.7
Oslo, Norway	1.7	2.9
Paris, France	1.5	2.1
Rome, Italy	3.3	0.4
San Diego, CA	2.1	0.0
San Francisco, CA	4.5	0.0
Santiago, Chile	0.1	3.0
Singapore	9.9	6.7
Sydney, Australia	3.5	4.6
Washington, DC	2.8	3.9

Source: *Universal Almanac*

25. World Cultures Many North American Indians hold conferences called powwows where they celebrate their culture and heritage through various ceremonies and dances.

▶ The ages of participants in a Menominee Indian powwow were: 20, 18, 12, 13, 14, 72, 65, 23, 25, 43, 67, 35, 68, 13, 56.

▶ The ages of observers of the powwow were: 43, 55, 70, 63, 15, 41, 9, 42, 75, 25, 16, 18, 51, 80, 75, 39, 23, 55, 50, 54, 60, 43.

a. Find the range of each group.

b. Find the interquartile range for each group.

c. Write a paragraph describing the participants.

d. Write a paragraph comparing the participants with the observers.

For **Extra Practice**, see page 764.

26. Statistics The ages of the first twenty people into the museum on Saturday were 17, 9, 12, 25, 8, 39, 27, 14, 29, 40, 36, 8, 15, 41, 28, 29, 30, 31, 29, and 11. Make a stem-and-leaf plot for the data. (Lesson 10-1)

27. Replace the ● in $\frac{2}{3}$ ● $\frac{5}{8}$ with < or > to make a true sentence. (Lesson 6-1)

28. Computers Shantia plans to spend no more than $50 on new software. If she buys a program for $19.95, how much more can she spend? Express your solution as an inequality. (Lesson 5-7)

29. Standardized Test Practice Find the quotient of c^2d^5 and cd. (Lesson 4-8)

    A. cd^2 **B.** cd^3 **C.** cd^4 **D.** cd^5

30. *True* or *false*: The graph of a point with one negative coordinate and one positive coordinate is located in Quadrant III. (Lesson 2-2)

Displaying Data

You'll Learn

- To use box-and-whisker plots, pictographs, and line graphs to display data.

It's Important

Choosing the correct type of graph to display data will help you present information in an understandable way.

Real World

Application: Safety

Students at Horace Mann Middle School were concerned about the speed of traffic on the street in front of their school. The student council asked the city's Traffic Engineering Division to place two radar speed sensors with large digital displays near the street. One display allowed vehicles traveling west to see their speed, and the other was positioned for vehicles traveling east to see their speed.

Sample data on the speeds (in mph) of vehicles traveling west were 25, 35, 27, 22, 34, 40, 20, 19, 23, 25, and 30. Speeds of vehicles traveling east were 26, 22, 31, 36, 22, 27, 15, 50, 32, 29, and 30.

The student council voted to ask the city council to erect a stop sign at either end of the block. When they made their presentation, they wanted to make a display of the data so that the city council members would understand it quickly and easily. They chose a **box-and-whisker plot** because it summarizes data using the median, the upper and lower quartiles, and the highest and lowest, or extreme, values.

Learning the Concept

The speeds of the vehicles traveling west are displayed in the stem-and-leaf plot at the right. The median is marked by a box, and the quartiles are circled. The upper extreme is 40, and the lower extreme is 19.

```
1 | 9
2 | 0 ② 3 5 ⑤ 7
3 | 0 ④ 5
4 | 0        1|9 = 19
```

Here's how to construct a box-and-whisker plot to display the data.

Step 1 Draw a number line for the range of the speeds. Above the number line, mark points for the extreme, median, and quartile values.

Lower Extreme   LQ   Median   UQ   Upper Extreme

```
├┼┼┼┼┼┼┼┼┼┼┼┼┼┼┼┼┼┼┼┼┼┼┼┼┼┼┼┼┼┼┼┼┼┼┼┼┼┼┼┤
10    15    20    25    30    35    40    45    50
```

Step 2 Next, draw a *box* that contains the quartile values. Draw a vertical line through the median value. Finally, extend the *whiskers* from each quartile to the extreme data points.

This display, the box-and-whisker plot, separates the data into fourths, each having an equal amount of data. For example, the whisker from 19 to 22 contains 25% of the data.

Example   **Use the box-and-whisker plot below to answer the questions.**

Ages of Volunteers at the Homeless Shelter

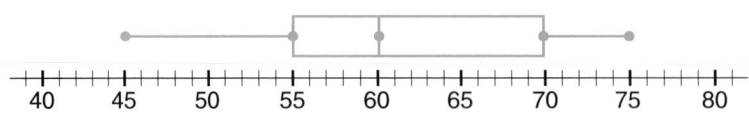

a. What is the age of the youngest volunteer?   45

b. Half of the volunteers are under what age?   60

c. What percent of the volunteers are over 70?   25%

Sometimes the data will have such great variation that one or both of the extreme values will be far beyond the other data. Data that are more than 1.5 times the interquartile range from the quartiles are called **outliers**.

Example

Real World
APPLICATION

Safety

Refer to the application at the beginning of the lesson.
a. Draw a box-and-whisker plot for the speed data of the cars traveling east.

15   22   22   26   27   29   30   31   32   36   50

Step 1 The median is 29, the lower quartile is 22, and the upper quartile is 32. Draw a box to show the median and quartiles.

Step 2 The interquartile range is $32 - 22$ or 10. So, data more than 1.5 times 10 from the quartiles are outliers.

$$1.5(10) = 15$$

Find the limits of the outliers.

Subtract 15 from the lower quartile $22 - 15 = 7$

Add 15 to the upper quartile. $32 + 15 = 47$

The limits for the outliers are 7 and 47. There is one outlier in the data, 50. Plot the outlier with an asterisk. Draw the whiskers to the extreme data that are not outliers.

b. The speed limit in front of the school is 20 mph. Make a double box-and-whisker plot of the east and west speeds to convince the city council that a stop sign is needed.

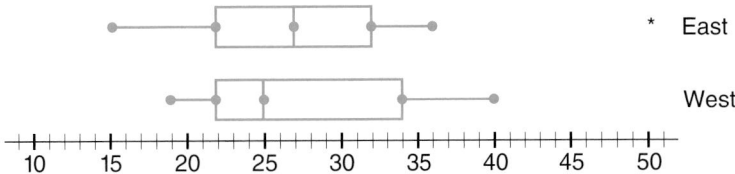

Almost all of the speeds of vehicles traveling west exceed the speed limit, and more than three-fourths of the vehicles traveling east exceed the speed limit. Stop signs at both ends of the block should slow traffic because cars would have to stop twice.

Other statistical graphs that are often used to present data include *line graphs*, *pictographs*, *circle graphs*, and *comparative graphs*.

Source: ABC

Line graphs usually show how values change over a period of time. The line graph at the left shows the number of viewers of football on Monday nights from 1970 to 1994.

Pictographs use pictures or illustrations to show how specific quantities compare. Each symbol represents a convenient number of items to display the data. The pictograph at the right shows how much different age groups in the United States spend on sporting apparel.

Source: Sporting Goods Manufacturing Association

Source: Energy Information Administration

Circle graphs show how parts are related to the whole. The circle graph at the left shows how electricity is generated in the United States.

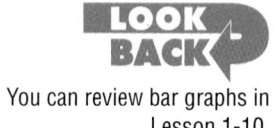

You can review bar graphs in Lesson 1-10.

Comparative graphs like a double-bar graph are used to show trends. They are usually used to compare results of similar groups. The graph at the right shows how the temperatures in the winter of 1995 compared to the 100-year average.

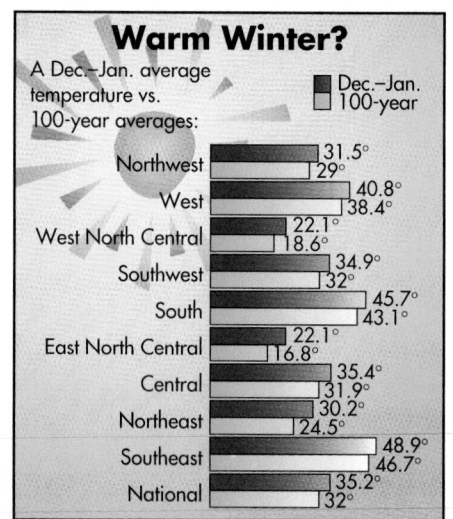

Warm Winter?

A Dec.–Jan. average temperature vs. 100-year averages:

■ Dec.–Jan. □ 100-year

Northwest 31.5° / 29°
West 40.8° / 38.4°
West North Central 22.1° / 18.6°
Southwest 34.9° / 32°
South 45.7° / 43.1°
East North Central 22.1° / 16.8°
Central 35.4° / 31.9°
Northeast 30.2° / 24.5°
Southeast 48.9° / 46.7°
National 35.2° / 32°

Example ③

Meteorology

A newspaper wants to display the high temperatures of the past week. Should they use a box-and-whisker plot, line graph, pictograph, circle graph, or a double-bar graph?

Since the data would show how values change over a period of time, a line graph would give the reader a clear picture of what temperatures were and the changes in temperature.

Checking Your Understanding

Communicating Mathematics

Read and study the lesson to answer these questions.

1. **Explain** how to find the median of the data shown in the box-and-whisker plot at the right.

0 10 20 30 40 50

2. **Describe** the data represented by the box.

3. **Write** the five pieces of information you can learn from the plot.

4. **Describe** how to determine if there are any outliers in a set of data.

5. **Explain** how the information you can learn from a set of data shown in a box-and-whisker plot is different from what you can learn from the same set of data shown in a stem-and-leaf plot.

MATERIALS

🕐 stop watch

🖩 calculator

Work in groups of 4 or 5 to complete the following activity.

6. Make a table like the one shown below.

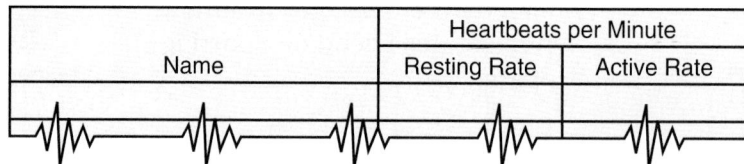

	Heartbeats per Minute	
Name	Resting Rate	Active Rate

a. Find the resting heart rate of each member of your group by taking your pulse for 15 seconds. Then multiply by 4 to find the beats per minute and record the product in your table.

b. In an area designated by your teacher, walk at a brisk rate for 5 minutes without stopping. Then take the pulses again and record them in the table.

c. Extend your table and combine your data of heart rates with that of the other groups.

d. Display the class data in a box-and-whisker plot.

7. Use the box-and-whisker plot below to answer each question.

a. What is the median?　　　　　**b.** What is the upper quartile?

c. What is the lower quartile?　　**d.** What is the interquartile range?

e. What are the extremes?　　　　**f.** What are the outliers, if any?

8. Education　From 1987 to 1996, Richton High School had the following number of graduates.

145, 165, 134, 173, 201, 204, 194, 185, 205, 198

a. Find the range, median, upper and lower quartiles, and the interquartile range for the data.

b. What are the limits for the outliers?

c. Use the data to make a box-and-whisker plot.

Exercises: Practicing and Applying the Concept

9. The box-and-whisker plot below shows the heights in inches of the eleven girls on the Towne High girls' volleyball team.

a. What is the range?　　　　　　**b.** What is the median?

c. What is the upper quartile?　　**d.** What is the lower quartile?

e. What is the interquartile range?　**f.** Are there any outliers?

10. Use the stem-and-leaf plot at the right to answer each question.

a. Make a box-and-whisker plot of the data.

b. What is the median?

c. What is the upper quartile?

d. What is the lower quartile?

e. What is the interquartile range?

f. What are the extremes?

g. What are the limits for outliers?

h. What are the outliers, if any?

4	0 2
5	1 5 9
6	3 5 7 8
7	2 3 7 8
8	0 1

4 | 0 = 40

11. Use the box-and-whisker plot below to answer each question.

Test Scores

a. What percent of the test scores are below the median?

b. What percent of the test scores are represented by the box?

c. What percent of the test scores are below the lower quartile?

d. What percent of the test scores are represented by each whisker?

e. Why isn't the median in the middle of the box?

f. Why is one whisker longer than the other?

For each of the following sets of data, determine whether a bar graph, box-and-whisker plot, comparative graph, line graph, or pictograph is the best way to display the data. Explain your reasoning.

12. The number of people who have different kinds of pets.

13. Points scored by basketball players in 15 games.

14. Number of movies rented from a video store by different age groups.

15. Shoe size of students in your class.

16. Gender of teachers in your school.

17. Students SAT scores by grade level.

18. **The chart at the right shows the number of sunny days for each city.**

   a. Make a bar graph to represent the number of sunny days.

   b. Make a pictograph to represent the number of sunny days.

   c. Describe any advantages the bar graph has over the pictograph.

   d. Describe any advantages the pictograph has over the bar graph.

Sunny Days

City	Number of Days
Albany, NY	60
Burlington, VT	60
Dallas, TX	120
Des Moines, IA	80
Indianapolis, IN	68
Jacksonville, FL	90
Kansas City, MO	100
Lexington, KY	69
Mobile, AL	104
New Orleans, LA	99
Salt Lake City, UT	130

Source: *The World Almanac*

Critical Thinking

19. In making a box-and-whisker plot of the weights of the 23 juniors on the varsity football team, Randy found the median to be 170 pounds, the upper quartile to be 178 pounds, the lower quartile to be 168 pounds, and there were no outliers. Then he discovered four more weights: 200, 175, 169, and 167. When he included these in the data, what happened to the box-and-whisker plot? (*Hint:* You may want to sketch two plots.)

20. Is it possible to have a box-and-whisker plot with only one whisker? Explain your answer.

Applications and Problem Solving

21. **Teaching** Mr. Juarez has two classes of Physical Science. He made the stem-and-leaf plot of the scores for the last test shown at the right.

   a. Make a double box-and-whisker plot for each class. (Use the same number line for both; make one plot above the other.)

   b. What is similar about the data in the two plots?

   c. What is different about the data in the two plots?

   d. Which class had scores that were more evenly spread over the range? Explain.

First Period		Second Period
8 0	2	
8 5	3	
5	4	3 5
0	5	2 4 5 6
0 0	6	0 0 0 8
5 2	7	2 5 8
9 8	8	1 5
8 6 1	9	

1 | 9 = 91   8 | 1 = 81

22. **Consumerism**   The average retail prices for one gallon of unleaded gasoline for ten years are shown in the chart below.

Year	1	2	3	4	5
Price	$1.21	$1.20	$0.92	$0.95	$0.95
Year	6	7	8	9	10
Price	$1.02	$1.16	$1.14	$1.13	$1.11

a. Make a box-and-whisker plot of the prices.

b. What is the median?

c. Are there any outliers? If so, why do you think they occurred?

d. Draw a line graph of the data.

e. Does the line graph present a better picture of the data than the box-and-whisker plot? Explain your answer.

23. **Employment**   The chart below shows the percent of men and women who held two jobs from 1970 to 1995.

Percent Who Hold Down Two Jobs						
Year	1970	1975	1980	1985	1990	1995
Men	7.0	5.8	5.8	5.9	6.4	5.9
Women	2.2	2.9	3.8	4.7	5.9	5.9

a. Make a comparative bar graph to show the percentage of men and women who hold two jobs.

b. What changes have occurred in men's employment?

c. What changes have occurred in women's employment?

d. In general, what conclusion can you make from the data?

Mixed Review

24. **Statistics**   Find the range, median, upper and lower quartiles, and the interquartile range for the set of data in the stem-and-leaf plot shown at the right. (Lesson 10-2)

Words Typed Per Minute

```
4 | 0 2
5 | 1 5 9
6 | 3 5 7 8
7 | 2 3 7 8
8 | 0 1      4 | 0 = 40
```

25. Write the domain and range of the relation $\{(5, -3), (-1, 4), (4.5, 0), (-1, -3)\}$ (Lesson 8-1)

26. **Standardized Test Practice**   How many grams are in 0.046 kilograms? (Lesson 7-8)

   A. 0.46   **B.** 4.6   **C.** 46   **D.** 460

27. Estimate the product of 0.25 and $27.98. (Lesson 6-2)

28. **Nutrition**   One cup of skim milk has 80 Calories, and one cup of whole milk has 150 Calories. If you drink one cup of milk per day, about how many Calories would be saved each year by switching from whole milk to skim? (Lesson 1-1)

a. Which method of computation do you think is most appropriate? Justify your choice.

b. Solve the problem using the four-step plan. Be sure to examine your solution.

For **Extra Practice**, see page 764.

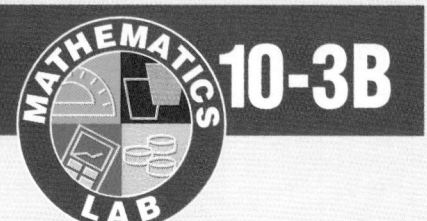

10-3B Making Statistical Graphs

An Extension of Lesson **10-3**

You can use a graphing calculator to construct statistical graphs. A line graph, box-and-whisker plots, and a special kind of bar graph, called a **histogram**, can be created using a graphing calculator.

Activity **Use a graphing calculator to construct a line graph of the data on bicycle sales.**

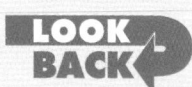

You can review setting the range in Lesson 2-2B and clearing the statistical memory in Lesson 6-6B.

Bicycle Sales in the United States (in millions)										
Year	1983	1984	1985	1986	1987	1988	1989	1990	1991	1992
Sales	9.0	10.1	11.4	12.3	12.6	9.9	10.7	10.8	11.7	11.6

Before you enter any data to graph, you must clear the statistical memory, clear the Y= list, and set the range. Use the range values [1983, 1993] by [0, 15] with a scale factor of 1 on both axes.

Now enter the data into the memory and draw the graph.

Enter: `STAT` `ENTER` *Accesses the statistical lists.*

1983 `ENTER` ... 1992 `ENTER` *Enters the years in L1.*

▶ 9 `ENTER` ... 11.6 *Enters the sales in L2.*

Now choose the type of statistical plot you would like to create.

Enter: `2nd` `STAT PLOT` `ENTER` *Accesses the menu for the first statistical graph.*

Use the arrow and `ENTER` keys to highlight your selections for the graph. Select "On", the line graph icon, "L1" as the Xlist, "L2" as the Ylist, and "•" for the marks.

To create the graph, press `GRAPH`.

Your Turn 1. Use a graphing calculator to create a line graph of the data below. (Enter 1, 2, 3, ..., 13 in list L1 to represent the months.)

Unemployment Rate, 1993–1994												
Month	Nov.	Dec.	Jan.	Feb.	Mar.	Apr.	May	June	July	Aug.	Sep.	Oct.	Nov.
Percent	6.5	6.4	6.7	6.5	6.5	6.4	6.0	6.0	6.1	6.1	5.9	5.8	5.6

2. Based on your graph, do you think the unemployment rate for December, 1994 increased or decreased? Explain.

A *histogram* is a special bar graph that displays the frequency of data that has been organized into equal intervals. The intervals cover all possible values of data. Therefore, there are no spaces between the bars of the histogram.

Activity ② Use a graphing calculator to create a histogram of the number of graduates for each of the states.

Public High School Graduates (in thousands)

39.0	5.5	31.3	25.7	234.2	31.3	27.3	5.2	87.4	60.1
9.0	12.0	103.3	58.6	28.6	24.4	35.8	33.5	13.2	39.0
52.1	88.2	46.5	23.7	46.9	9.0	16.5	9.4	10.1	67.0
15.2	133.6	62.8	7.6	107.5	33.0	24.6	104.8	7.7	33.1
7.1	44.8	174.3	22.2	5.2	58.4	42.5	21.1	49.3	5.7

When you set the range for a histogram, you determine the number of bars (equal intervals) you want to graph. Use the range values [0, 240] by [0, 25] with a scale factor of 30 on the *x*-axis and 5 on the *y*-axis. These settings will create 240 ÷ 30 or 8 bars.

Now enter the data into the memory and draw the graph.

Enter: STAT ENTER *Accesses the statistical lists.*

39.0 ENTER 5.5 ENTER 31.3 ENTER . . . 5.7 ENTER

Enter: 2nd STAT PLOT ENTER *Accesses the menu for the first statistical graph.*

Use the arrow and ENTER keys to highlight "On", the histogram icon, "L1" as the Xlist, and "1" as the frequency. Press GRAPH to view the histogram.

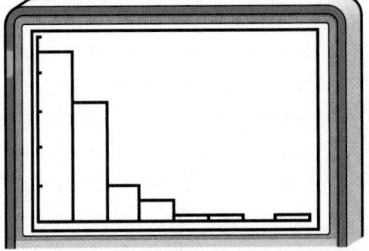

You can make a box-and-whisker plot of the same data.

Enter: 2nd STAT PLOT ENTER

Select the same plot and use the arrow and ENTER keys to highlight the box-and-whisker plot with outliers icon. Press GRAPH to view the box-and-whisker plot.

Your Turn

3. Use a graphing calculator to create a histogram and a box-and-whisker plot for the heights of the United States presidents.

Heights of U.S. Presidents (in inches)

74	66	74.5	64	72	67	73	66	66	72	68	68	69	70
72	76	70	68.5	68.5	70	74	71	66	66	68	74	71	72
69	71	74	70	70	73	75	71.5	72	70	73	74	74.5	

4. If the range of values for a histogram is 35 and you want to create 7 bars, what scale factor should you enter? Explain.

10-4 Misleading Statistics

You'll Learn

- To recognize when statistics are misleading.

It's Important

Recognizing misleading statistics can help you read and interpret advertisements.

Application: Education

Mr. Snyder's sociology class was studying birthrates in the United States. For a project on the history of birthrates from 1960 to 1990, Maria and Jennifer made the two graphs shown below.

▶ Do both graphs show the same information?

▶ Which graph suggests a baby boom from 1975–1990?

▶ Which graph suggests a fairly steady birthrate?

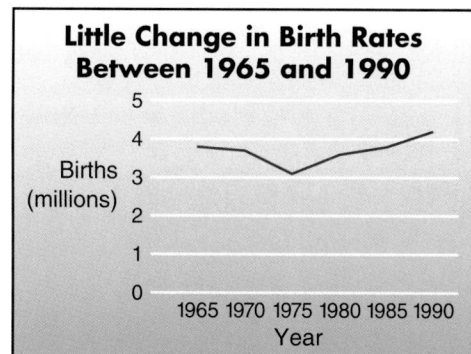

▶ What is the same and what is different about the two graphs?

Learning the Concept

The data used in the graphs above are the same. However, different scales were used to make the data appear to support a particular point of view. The soaring graph at the left is a result of an extended vertical axis and a shortened horizontal axis. A small "break" in the vertical axis is used to show that the axis is not to scale between zero and 2.7. However, it is visually misleading when compared to the complete scale of the graph at the right.

Example 1

The graphs at the right display data about movie attendance.

a. Is the top graph misleading? Explain.

The graph is misleading because there is no title and there are no labels on either scale. Also, the vertical axis does not include zero.

b. Is the bottom graph misleading? Explain.

The graph is not misleading. All the necessary information is present. The scales are labeled, the distance between the units on each axis is uniform, and the vertical axis includes 0.

Theater Admissions Rebound!

You can review measures of central tendency in Lesson 6-6.

Another way statistics can be misleading is by using the inappropriate measure of central tendency.

Example 2

Statistics

After a semester exam in science, Mr. Myers displayed the results to the class. The scores are shown in the stem-and-leaf plot. Mr. Myers told the principal that the average grade on the exam was 77. Ryan told his parents that the class average was 93. Tai told her grandmother the class average was 80. How can all three people think they are correct?

Scores	
9	01333
8	012345
7	023
6	0123579

9|0 = 90

Each person used a different measure of central tendency to describe the average class grade. Mr. Myers used the mean and Tai used the median. Ryan's use of the mode is misleading because 93 is also the highest score on the semester exam.

Checking Your Understanding

Communicating Mathematics

Read and study the lesson to answer these questions.

1. **Describe** two ways graphs can be misleading.

2. **List** some things you might question when reading the results of a survey on employee salaries.

3. **You Decide** Khandi says that the graph at the right is misleading. Carlos disagrees. Who is correct and why?

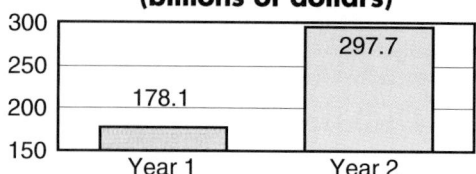
U.S. Holdings of Foreign Stocks (billions of dollars)

4. **Assess Yourself** Do you understand how to determine whether graphs are misleading? Find three graphs in a newspaper or magazine. Determine if the graphs are misleading and explain why or why not.

Guided Practice

5. **Refer to the graph at the right.**

   a. What is the ratio of the number of cars produced in Year 2 to the number of cars produced in Year 1?

   b. How does this ratio compare with the ratio of the sizes of the cars on the graph?

   c. Is this a misleading graph? Explain.

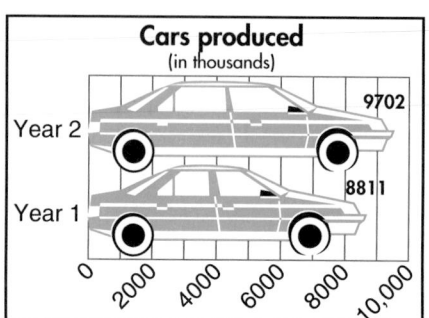
Cars produced (in thousands)

6. **Statistics** Roberta made two graphs of her social studies grades for each grading period.

   a. Explain why these graphs made from the same data look different.

   b. If Roberta wanted to convince her parents that her social studies grades have improved dramatically, which graph would she probably show her parents? Explain.

Exercises: Practicing and Applying the Concept

Independent Practice

7. The houses in the Cambridge section of Anderson were assessed for school taxes. The values are shown in the frequency table at the right.

   a. Which measure of central tendency would you use to find the average cost of a house? Why?

   b. Which average might a real estate agent use with a business executive? Explain.

Value	Number
$150,000	2
$130,000	5
$115,000	4
$95,000	10
$85,000	20
$75,000	2

Use the chart to describe each category. Write the average you used.

	Tashelle	Kim	Dawn	Sita	Bill
Age	40	24	33	28	41
Favorite Entertainment	sports	concerts	theater	movies	sports
Height (cm)	185	163	165	157	193
Salary	$34,700	$18,000	$30,500	$25,500	$47,000

8. average age

9. average height

10. average salary

11. average favorite entertainment

12. How can you describe the "average" person in this group?

13. Find the mean, median, mode, and range of the salaries at each company in the table at the right. Which company would you rather work for? Explain your reasoning.

		Frequency	
Salary	**Company A**	**Company B**
$15,000	18	
$20,000	4	24
$25,000	4	6
$50,000	2	1
$100,000	1	

Critical Thinking

14. Find a graph in a magazine or newspaper. Redraw the graph so that the data will appear to show different results.

Applications and Problem Solving

15. **Conservation** In 1963, there were only 417 pairs of bald eagles living in the contiguous 48 states of the United States. DDT, a pesticide used to kill insects, had poisoned many of the birds. In 1973, the U.S. Congress protected the eagle by banning the use of DDT in areas where eagles live. In 1978, the eagle was named an endangered species. As a result, the eagle population has grown.

a. Do the data in the table represent the growth of the eagle population since the policy changes were made in the United States? Explain.

b. Draw a graph that you might use to accurately represent the data.

Eagle Pairs in Contiguous 48 States of the U.S.

Year	Number of Pairs
1988	2475
1989	2680
1990	3020
1991	3391
1992	3747
1993	4016

Source: U.S. Fish and Wildlife Service

16. **Employment** The salaries of all the employees of Grenwich Bottling Company are shown in the frequency table at the right. They are negotiating a new contract and want a raise because they learned that the average salary for all hourly employees was only $24,000. The president informed them that the average salary for employees was $31,000.

Salary	Number of Employees
$150,000	1
$85,000	2
$60,000	2
$35,000	8
$24,000	25
$15,000	5
$12,000	3

a. Which "average" do you think the union would want to use in its negotiations? Explain.

b. If you were called in to mediate the negotiations, what recommendation would you make and why?

17. The box-and-whisker plot below shows the amount students spent on entertainment for one month. (Lesson 10-3)

    a. What is the range?   **b.** What is the median?

    c. What is the upper quartile?   **d.** What is the lower quartile?

    e. What is the interquartile range?   **f.** What are the extremes?

    g. What are the limits for outliers? **h.** What are the outliers, if any?

18. Which ordered pair(s) is a solution of $3x + 2y = 12$? (Lesson 8-3)

    a. $(4, 0)$ **b.** $(0, 4)$ **c.** $(5, -1.5)$ **d.** $(-5, 1.5)$

19. Geography Multipurpose buildings are common in small towns in Alaska, since the population is small. Between 1980 and 1990, the population of Alaska increased from about 400,000 to 550,000. (Lesson 6-8)

    a. Find the ratio of the 1980 population to the 1990 population.

    b. If the population grows in a geometric pattern over each of the next four decades, what would the population of Alaska be in 2030?

20. Standardized Test Practice A millimeter is $\frac{1}{1000}$ of a meter or 10^{-3} meters. A kilometer is 1000 meters. How many millimeters are in a kilometer? (Lesson 4-8)

    A. 1000   **B.** 10,000   **C.** 100,000   **D.** 1,000,000

For **Extra Practice**, see page 765.

21. Patterns Find the next two integers in the pattern $0, -2, 2, -4, 4, \ldots$ (Lesson 2-6)

WORKING ON THE

Investigation

It's Only A Game

Refer to the Investigation on pages 482–483.

Test your game by having the members of your group play the game several times. Play the game enough times to develop some

different strategies. Have each member of your group use a different strategy. Then have players who are in the intended age group play the game several times.

Use the results of your test to:

- determine the best strategy for your game
- revise any rules
- display the scores or winners using a statistical graph
- determine what makes someone playing the game a "good" player
- make an advertisement for your game

Add the results of your work to your Investigation Folder.

10-5 Counting

What You'll Learn

■ To use tree diagrams or the Fundamental Counting Principle to count outcomes.

Why It's Important

Tree diagrams can be used to analyze games of chance.

Application: Advertising

A bagel bakery makes 9 types of bagels and has 10 flavors of cream cheese. They wanted to advertise the number of different bagel with cream cheese combinations that are possible. How many different possible choices are there? *This problem will be solved in Example 1.*

Learning the Concept

To solve this problem, it may be helpful to solve a simpler problem. Suppose the bagel bakery offers chicken, tuna, and vegetable sandwiches, on plain or onion bagels. How many possible sandwiches are there?

You can draw a diagram to find the number of possible combinations or **outcomes**.

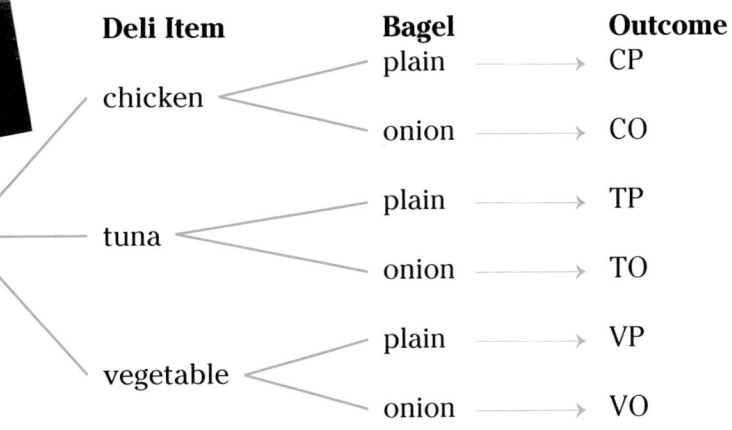

There are 6 possible choices, or outcomes.

The diagram above is called a **tree diagram**. Notice that the product of the number of deli items and the number of bagels, 3 × 2, also equals 6. Thus, you can find the number of possible outcomes by multiplying. This principle is called the **Fundamental Counting Principle**.

Fundamental Counting Principle	If event **M** can occur in *m* ways and is followed by event **N** that can occur in *n* ways, then the event **M** followed by event **N** can occur in *m* · *n* ways.

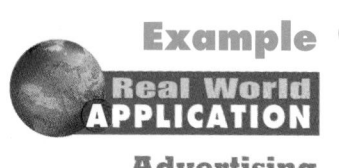
Real World
APPLICATION

Advertising

Refer to the application at the beginning of the lesson. How many different bagel with cream cheese combinations can the bagel bakery advertise are possible?

Explore   You know that there are 9 types of bagels and 10 flavors of cream cheese. You want to know how many different possible choices there are.

Plan   Use the Fundamental Counting Principle.

Solve $\left(\begin{array}{c}\text{number of types}\\\text{of bagels}\end{array}\right) \times \left(\begin{array}{c}\text{number of flavors}\\\text{of cream cheese}\end{array}\right) = \left(\begin{array}{c}\text{number of}\\\text{possible outcomes}\end{array}\right)$

$\qquad\qquad 9 \qquad \times \qquad 10 \qquad = \qquad 90$

There are 90 possible outcomes.

Examine   Make a list or draw a tree diagram to verify the solution.

Connection to ▶ Algebra

If you already know the total number of possible outcomes, you can write and solve an equation to find the number of ways the events can occur.

Example ②

INTEGRATION

Algebra

In literature class, each student must choose one short story and one poem to read for homework. The students must choose from a list of 16 short stories and *h* poems. There are 304 different combinations of short stories and poems possible. Find the number of poems on the list.

From the Fundamental Counting Principle, you know that 16 times *h* is 304.

$16h = 304$

$\dfrac{16h}{16} = \dfrac{304}{16}$ *Divide each side by 16.*

$304 \div 16 = 19 \quad h = 19$

There are 19 poems on the list.

Example ③

INTEGRATION

Probability

A die is rolled twice.

a. How many outcomes are possible?

$\left(\begin{array}{c}\text{number of outcomes}\\\text{for the first roll}\end{array}\right) \times \left(\begin{array}{c}\text{number of outcomes}\\\text{for the second roll}\end{array}\right) = \left(\begin{array}{c}\text{number of}\\\text{possible outcomes}\end{array}\right)$

$\qquad 6 \qquad\qquad \times \qquad\qquad 6 \qquad\qquad = \qquad 36$

There are 36 possible outcomes.

b. What is the probability of rolling a five and then a six?

Only one outcome can be a five and then a six.

$P(\text{five and six}) = \dfrac{\text{number of favorable outcomes}}{\text{number of possible outcomes}}$

$\qquad\qquad\qquad = \dfrac{1}{36}$

The probability of rolling a five and then a six is $\dfrac{1}{36}$.

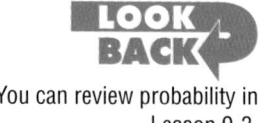

LOOK BACK

You can review probability in Lesson 9-3.

Communicating Mathematics

Read and study the lesson to answer these questions.

1. **a.** **Write** a problem that corresponds to the tree diagram at the right.

   b. **List** the outcomes illustrated in the tree diagram at the right.

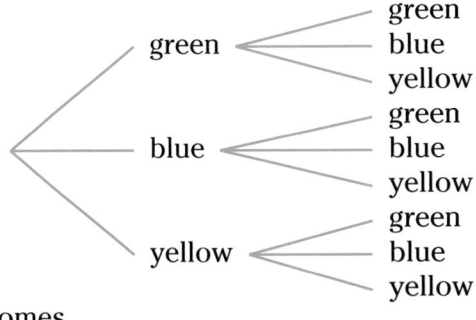

2. **Compare and contrast** using the Fundamental Counting Principle and using a tree diagram to find numbers of outcomes.

MATERIALS

a penny and a nickel

3. **a.** Draw a tree diagram to show the possible outcomes for tossing a nickel and then tossing a penny.

   b. What is the probability of tossing two heads?

   c. Make a frequency table listing the outcomes. Toss a nickel and then a penny and tally the outcome in your frequency table. Repeat for 25 trials.

   d. What fraction of the time did you toss two heads?

   e. Your answer to part b is the *theoretical probability* of tossing two heads. Your answer to part d is the *experimental probability* of tossing two heads. How does the experimental probability compare to the theoretical probability?

   f. Toss the coins for 25 more trials. Compare the probabilities again. Did the experimental probability change? If so, how?

Guided Practice

4. The spinner at the right is spun twice.

   a. Draw a tree diagram that represents the situation.

   b. How many outcomes are possible?

   c. How many outcomes show red and blue?

   d. What is the probability of two oranges?

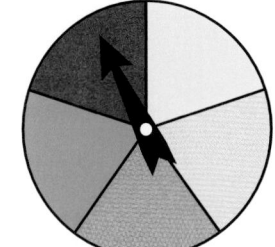

5. A coin is tossed, and a die is rolled.

   a. How many outcomes are possible?

   b. What is the probability of heads and a 4?

6. Three coins are tossed. How many outcomes are possible?

7. **Fashion** Reina has three silver necklaces, three pairs of silver earrings, and two silver bracelets. How many combinations of the three types of jewelry are possible?

Exercises: Practicing and Applying the Concept

Independent Practice

Draw a tree diagram to find the number of outcomes for each situation.

8. Each four-sided die is rolled once.

 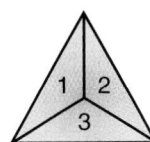

9. Each spinner is spun once.

10. The continental breakfast at Myriad Hotel has toast, muffin, or bagel with coffee, milk or juice. How many combinations of one bread and one beverage are possible?

11. Jarred had five different poses to choose for his senior picture and five different frames. How many different ways could he choose a picture and a frame?

Find the number of possible outcomes for each event.

12. Four coins are tossed.   13. A die is rolled twice.

14. Two coins are tossed and a die is rolled.

15. School sweatshirts come in two colors, gray or white, and in four sizes, small, medium, large, and extra large.

   a. How many outcomes are possible?

   b. What is the probability of selecting a large gray sweatshirt at random?

16. A sedan comes with two or four doors, a four or six cylinder engine, and six exterior colors. How many different autos are possible?

17. A quiz has five true-false questions. How many outcomes for answering the five questions are possible? (*Hint:* One outcome is FFTFF.)

18. An eight-sided die is rolled four times.

   a. How many outcomes are possible?

   b. What is the probability of getting four 7s?

19. A coin is tossed thirty times.

   a. How many outcomes are possible?

   b. What is the probability of getting heads every time?

Critical Thinking

20. A test has all true-false questions. If there are x questions on the test, write an expression for the number of possible answer keys.

Applications and Problem Solving

21. **Telecommunication**   Telephone numbers such as 219-555-1212 are made up of a three-digit area code (219), a three-digit exchange (555), and a four-digit extension (1212). The digits of an exchange cannot be 0 or 1, so it appears that there are $8 \cdot 8 \cdot 8$ or 512 exchanges in an area code. However, there are 26 exchanges that are not available in each area code because of special numbers like 911, 555, and 800. There are actually only 486 exchanges available in each area code.

   a. How many extensions are possible in one exchange?

   b. How many telephone numbers are available in each area code?

   c. Many local telephone companies ran out of numbers during the 1990s. They had to extend the three-digit exchange to include 0 and 1 in the second and third digits. (Zero and one still cannot be used as the first digit.) How many new numbers became available in each area code by allowing 0 and 1 in the second and third digits?

22. **Business**   The Yogurt Oasis advertises that there are 1512 ways to enjoy a one-topping sundae. They offer six flavors of frozen yogurt, six different serving sizes, and several toppings. How many toppings do they offer?

23. Real Estate A builder has house styles that are priced at $80,000, $95,000, $92,000, and $170,000. Which "average" price gives the best description of the houses that the builder offers? Why? (Lesson 10-4)

24. Standardized Test Practice Use the formula $I = prt$ to find the interest to the nearest cent on $500 at 4.25% simple interest for 6 months. (Lesson 9-9)

    A. $9.36   **B.** $10.63   **C.** $11.59   **D.** $12.75

25. Given $f(x) = 2x + 4$, find the value of $f(-5)$. (Lesson 8-4)

26. Solve $2(k - 3) \le 8$ and check your solution. Then graph the solution on a number line. (Lesson 7-6)

27. Sarah earned less than $20 baby-sitting at $3.50 an hour. How many hours did Sarah baby-sit? Define a variable and translate the sentence into an inequality. Then solve. (Lesson 3-8)

For **Extra Practice**, see page 765.

Self Test

The table below shows the normal monthly precipitation, in inches, for selected cities in the United States. Use this data to complete Exercises 1–5.

City	Jan.	Feb.	Mar.	Apr.	May	June	July	Aug.	Sept.	Oct.	Nov.	Dec.
Birmingham, AL	5.1	4.7	6.2	5.0	4.9	3.7	5.3	3.6	3.9	2.8	4.3	5.1
Louisville, KY	2.9	3.3	4.7	4.2	4.6	3.5	4.5	3.5	3.2	2.7	3.7	3.6
Miami, FL	2.0	2.1	2.4	2.9	6.2	9.3	5.7	7.6	7.6	5.6	2.7	1.8
Oklahoma City, OK	1.0	1.3	2.1	2.9	5.5	3.9	3.0	2.4	3.4	2.7	1.5	1.2

1. Make a stem-and-leaf plot of the precipitation for Birmingham. (Lesson 10-1)

2. Find the range and interquartile range of the precipitation for Oklahoma City. (Lesson 10-2)

3. What are the extreme measures of precipitation for Louisville? (Lesson 10-2)

4. Construct a box-and-whisker plot of the precipitation for Miami. (Lesson 10-3)

5. Determine whether a bar graph, box-and-whisker plot, comparative graph, line graph, or pictograph, is the best way to display the data. Explain your reasoning. (Lesson 10-3)

Use the graphs at the right to complete Exercises 6–7. (Lesson 10-4)

6. Explain why these graphs made from the same data look different.

7. Which graph would Brian show to his parents if he wants an increase in his allowance? Explain.

8. Wrapping paper comes in red, pink, green, or blue with white or yellow ribbon. Draw a tree diagram that illustrates how many combinations are possible. (Lesson 10-5)

9. Mark has three pairs of shorts, four shirts, and two pairs of running shoes. How many outcomes are possible? (Lesson 10-5)

10. Three coins are tossed. What is the probability that all 3 coins land heads up? (Lesson 10-5)

10-6A Permutations and Combinations

A Preview of Lesson **10-6**

MATERIALS

 6 index cards

 markers

An arrangement of names, objects, or people in a particular order is called a **permutation**. For example, if you use a triangle, square, and circle, you can arrange them in the following permutations.

△□ ○   △○□   □○△   □△○   ○□△   ○△□

Activity 1

Work with a partner.

Write the letters M, A, T, H on four cards, one letter per card. Shuffle the cards and place them face down. Choose three of the cards and turn them face up. For example, you may have chosen the cards shown at the right. Record the "word" shown; for example, ATM.

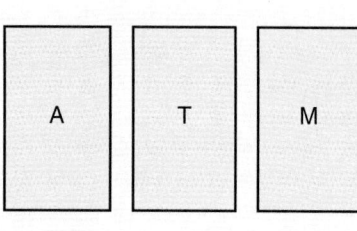

▶ Rearrange the three cards to make another "word." Record that word.

▶ Continue rearranging the cards and recording the words until you have listed all the 3-letter words you can with the cards chosen. Record the total number.

▶ Now use all four cards. Arrange the cards to form all the 4-letter "words" that you can. How many arrangements are there?

Activity 2

The student council at Northern Potter Junior High School is planning to have a sundae sale. They plan to offer six different toppings.

| peanuts | chocolate chips | chocolate syrup |
| raisins | strawberries | butterscotch syrup |

The price depends on the *combinations* of toppings chosen.

Your Turn

Work in groups of two or three.

▶ Write the names of the six toppings on six index cards.

▶ To make a sundae, select any pair of cards. Make a list of all the different combinations of two toppings. Note that the order is not important; that is, *peanuts, raisins,* is the same as *raisins, peanuts.*

▶ Suppose that the student council decides to offer a deluxe sundae with four toppings on it. Make a list of all the different sundaes that are possible if four toppings of the six are used.

1. In Activity One, suppose that you had five cards with five different letters. How many five-letter words do you think can be formed?

2. What is the difference between a combination (like a combination of toppings) and a permutation?

You'll Learn

- To use permutations and combinations.

It's Important

You can use combinations when you choose a small committee from a larger group.

Permutations and Combinations

Application: Swimming

The swim team coach at Clearwater High School is preparing the lineup for the state finals in the 400-meter freestyle relay. She must choose four swimmers from the five that qualified at the district championships. As she looks over the five names, she wonders how many arrangements are possible.

She reasons that any of the 5 swimmers can start the race. Once that swimmer is selected, there are 4 swimmers left who can swim second, 3 who can swim third, and 2 who can swim fourth.

starting swimmers		*2nd swimmer*		*3rd swimmer*		*4th swimmer*		*Possible relay teams*
5	×	4	×	3	×	2	=	120

Using the Fundamental Counting Principle, as shown above, she finds there are 120 possible relay teams.

Learning the Concept

An arrangement or listing in which order is important is called a **permutation**. For the example above, the symbol $P(5, 4)$ represents the number of permutations of 5 things taken 4 at a time.

$$P(5, 4) = 5 \cdot 4 \cdot 3 \cdot 2$$

If one of the five Clearwater swimmers became ill and was unable to swim, the number of relay teams can be expressed as $P(4, 4)$.

$$P(4, 4) = 4 \cdot 3 \cdot 2 \cdot 1$$

The product $4 \cdot 3 \cdot 2 \cdot 1$ can be written using the mathematical notation 4!. The symbol 4! is read *four **factorial***. The expression $n!$ means the product of all counting numbers beginning with n and counting backward to 1. We define 0! as 1.

Lesson 10-6 Permutations and Combinations **515**

Example **Find each value.**

a. $P(8, 3)$

$P(8, 3)$ represents the number of permutations or orders of 8 things taken 3 at a time.

$$P(8, 3) = \underbrace{\text{choices for 1st position}}_{8} \cdot \underbrace{\text{choices for 2nd position}}_{7} \cdot \underbrace{\text{choices for 3rd position}}_{6}$$

$$= 8 \cdot 7 \cdot 6$$
$$= 336$$

b. $6!$

$$6! = 6 \cdot 5 \cdot 4 \cdot 3 \cdot 2 \cdot 1$$
$$= 720$$

Sometimes order is not important. For example, *pepperoni, mushrooms, onions* is the same as *onions, pepperoni, mushrooms* when you buy a pizza.

Arrangements or listings where order is not important are called **combinations**. To find the number of different three-topping pizzas when there are eight toppings to choose from, divide the number of permutations, $P(8, 3)$, by the number of different ways three things can be ordered. There are $3!$ or $3 \cdot 2 \cdot 1$ ways to order three toppings. The symbol $C(8, 3)$ means the number of combinations of 8 things taken 3 at a time.

$$C(8, 3) = \frac{P(8, 3)}{3!} = \frac{8 \cdot 7 \cdot 6}{3 \cdot 2 \cdot 1} \text{ or } 56$$

It is possible to make 56 different pizzas with three toppings when there are eight toppings to choose from.

Example

Geometry

Polygons can be drawn with any number of sides and vertices. Find the number of line segments that can be drawn between two vertices of a hexagon.

Explore   A hexagon has 6 vertices.

Plan   Each line segment connects two vertices, so you must find the combination of 6 vertices taken 2 at a time.

Solve
$$C(6, 2) = \frac{P(6, 2)}{2!}$$
$$= \frac{6 \cdot 5}{2 \cdot 1} \text{ or } 15$$

There are 15 line segments that can be drawn.

Examine   Draw a hexagon and all the line segments and see if there are 15. Be sure to count the sides of the hexagon.

Communicating Mathematics

Read and study the lesson to answer these questions.

1. **Explain** what $P(4, 3)$ means.

2. **Compare** $6 \cdot 5 \cdot 4$ and $6!$.

3. **Write** an expression to represent the number of five-person basketball teams that could be formed from all the students in your math class.

4. Write a short summary explaining the difference between a permutation and a combination of 4 things taken 3 at a time. Draw models to illustrate your written explanation.

Guided Practice

Tell whether each situation represents a *permutation* or *combination*.

5. six people remaining in a game of musical chairs

6. first, second, third place, and honorable mention awards of six students who are finalists in a science fair

How many ways can the letters of each word be arranged?

7. WE   8. STUDY   9. MATH

Find each value.

10. $P(6, 3)$ 11. $5!$

12. $C(5, 4)$ 13. $\frac{7!3!}{5!0!}$

14. How many combinations of 6 different flowers can you choose from one dozen different flowers?

15. **Music**   The Riverside High School Orchestra has been invited to play three pieces at the opening of the Summer Music Festival. The orchestra has eight pieces that they could play. How many different programs can they play?

Independent Practice

Tell whether each situation represents a *permutation* or *combination*.

16. four books in a row

17. six CDs from a group of twenty

18. a team of 6 players from 12

19. ten people in a line to buy tickets

20. position of students at the twelve computers in the computer center

21. ten wooden giraffes placed on display from a large collection

How many ways can the letters of each word be arranged?

22. SHE   23. BOUGHT   24. SANDWICH

Find each value.

25. $P(6, 6)$ **26.** $7!$ **27.** $9!$ **28.** $C(6, 6)$

29. $11!$ **30.** $P(5, 1)$ **31.** $C(7, 3)$ **32.** $P(8, 4)$

33. $0!$ **34.** $C(8, 4)$ **35.** $C(10, 2)$ **36.** $P(12, 12)$

37. $\dfrac{7!3!}{5!1!}$ **38.** $\dfrac{10!0!}{6!2!}$ **39.** $\dfrac{5!2!}{4!3!}$ **40.** $\dfrac{9!5!0!}{11!}$

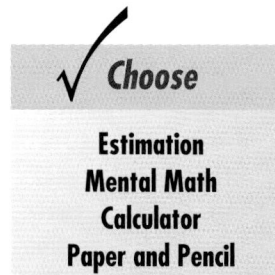

Choose

Estimation
Mental Math
Calculator
Paper and Pencil

41. How many different 6-player starting squads can be formed from a volleyball team of 15 players?

42. How many 3-digit numbers can you write using only the digits 1, 2, and 3 exactly once in each number?

43. How many ways can 4 members of a family be seated in a theater if the mother is seated on the aisle?

44. How many different 5-card hands is it possible to deal from a standard deck of 52 cards?

Critical Thinking

45. There are 720 ways for three students to win first, second, and third place in a debating match. How many students were competing?

46. Ten people are running for chair and vice-chair of the Social Studies Club. After they are selected, three directors will be elected from the remaining candidates. How many different ways can the offices be filled?

Applications and Problem Solving

47. **United Nations** The United States has brought a procedural measure to the United Nations Security Council for a vote. Nine votes are necessary for the measure to pass. There are 5 permanent members (China, France, Russian Federation, United Kingdom, and the United States) and 10 nonpermanent members on the Security Council. How many combinations of votes are possible if the measure passes with 9 votes?

48. **Geometry** Twelve points are marked on a circle. How many different line segments can be drawn between any two of the points?

49. **License Plates** North Carolina issues general license plates with three letters followed by four numerals. (The first numeral cannot be 0.) Numerals can repeat, but letters cannot. How many license plates can North Carolina generate with this format?

50. Draw a tree diagram to find the number of outcomes for rolling two identical number cubes. (Lesson 10-5)

51. A temperature of 32° Fahrenheit corresponds to a temperature of 0° Celsius. A temperature of 100° Celsius corresponds to a temperature of 212° Fahrenheit. About what temperature in degrees Celsius corresponds to a temperature of 0° Fahrenheit? (Lesson 8-5)

52. Geometry The perimeter of a rectangle is 38 meters. Find the dimensions if the length is 5 meters less than twice the width. (Lesson 7-5)

w

$2w-5$

53. Standardized Test Practice Which graph is the solution of the inequality $\frac{7}{9}x > 147$? (Lesson 6-7)

A.
187 189 191 193 195

B.
111 113 115 117 119

C.
187 189 191 193 195

D.
111 113 115 117 119

54. Estimate $1\frac{3}{5} - \frac{5}{12}$. (Lesson 5-2)

55. Evaluate $5x^{-2}$ if $x = 2$. (Lesson 4-9)

56. Travel Juan and his friends are traveling 120 miles to attend a concert. (Lesson 3-4)

    a. If they average 50 mph, how long will the trip take?

    b. If road construction on the trip limits their average speed to 40 mph, how much longer will the trip take?

57. Simplify $3x - 12x$. (Lesson 2-5)

For **Extra Practice**, see page 765.

58. Name the property shown by the statement $(3 + 6) + 8 = 3 + (6 + 8)$. (Lesson 1-4)

10-7 **Odds**

What You'll Learn

- To find the odds of a simple event.

Why It's Important

Understanding odds can help you determine your chances of winning a prize.

 MATERIALS: graph paper / straight pin

Hands-On Activity

On a piece of graph paper, copy the diagram shown at the right. Shade squares gray or black, as shown.

Your Turn

▶ Hold a straight pin about six inches above the center of the diagram. Drop the pin 50 times and record the color of the square where the point of the pin lands. If you miss the diagram, drop the pin again.

 TALK ABOUT IT

a. Would you say that the pin is more likely to land on a white square or a shaded square? Explain?

b. How many ways can the point of the pin land on a black square?

c. How many ways can the point of the pin not land on a gray square?

Learning the Concept

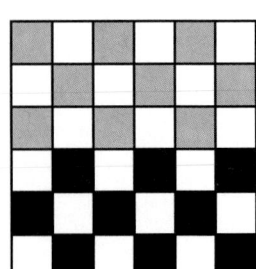

The probability that the point of the pin will land on a gray square is $\frac{9}{36}$ or $\frac{1}{4}$. The probability that the point of the pin will *not* land in a gray square is $\frac{27}{36}$ or $\frac{3}{4}$.

Another way to describe the chance of an event's occurring is with **odds**. The odds in favor of an event is the ratio that compares the number of ways the event can occur to the ways that the event *cannot* occur.

Definition of Odds	The odds in favor of an outcome is the ratio of the number of ways the outcome can occur to the number of ways the outcome cannot occur.
	Odds in favor = number of successes: number of failures
	The odds against an outcome is the ratio of the number of ways the outcome cannot occur to the number of ways the outcome can occur.
	Odds against = number of failures: number of successes

Example **1** **A bag contains 5 green marbles, 2 yellow marbles, and 3 blue marbles. What are the odds of drawing a green marble from the bag?**

There are $10 - 5$ or 5 marbles that are not green.

Odds of drawing a green marble

$$= \underbrace{\textit{number of ways of}}_{\textit{drawing green marble}} : \underbrace{\textit{number of ways of}}_{\textit{drawing other marbles}}$$

$$= \qquad 5 \qquad : \qquad 5$$
$$= 1{:}1 \quad \textit{This is read "1 to 1."}$$

The odds of drawing a green marble are 1:1.

Sometimes when finding odds, you must first find the total number of possible outcomes. This can involve finding permutations or combinations.

Example **2**

Finance

The state of Florida has a *Fantasy 5* drawing three times a week in which 5 numbers out of 39 are drawn at random. The proceeds from the lottery help to finance education in the state. What are the odds of winning the *Fantasy 5* jackpot?

Order is not important. You can find the odds using combinations.

There is only one winning combination, so the number of ways to succeed is 1. There are $C(39, 5)$ ways to draw 5 numbers from a group of 39. So the total number of possible outcomes is $C(39, 5)$ or 575,757. This means that the number of failures (not drawing all 5 numbers) must be $575{,}757 - 1$ or 575,756.

odds of drawing 5 correct numbers
  = number of successes : number of failures
  = $1 : 575{,}756$ *This is read "1 to 575,756."*

The odds of winning the *Fantasy 5* jackpot are 1: 575,756.

Checking Your Understanding

Communicating Mathematics

Read and study the lesson to answer these questions.

1. **Explain** how to find the odds of an event occurring.

2. **You Decide** Ade says that the probability of getting two in one roll of a die is 1 out of 6. Sharla says that the odds of getting two in one roll of a die are 1:5. Who is correct and why?

*M*ATH *J*OURNAL

3. Describe a situation in real life that uses odds.

Guided Practice

Find the odds of each outcome if a die is rolled.

4. a number greater than 5

5. a multiple of 2

6. a number less than 4

7. not a 3

8. **Statistics** The United States Census Bureau reported that the number of unmarried adults nearly doubled from 1970 to 1993. The report stated that 58% of unmarried adults had never been married.

    a. What are the odds that an unmarried adult was never married?

    b. In a group of 200 unmarried adults, how many would you estimate had never been married?

Exercises: Practicing and Applying the Concept

Independent Practice

Find the odds of each outcome if the spinner at the right is spun.

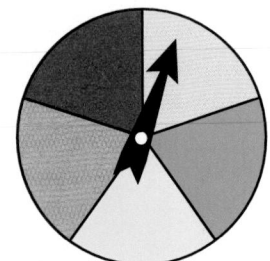

  9. blue

  10. purple

  11. not green

  12. red or orange

Find the odds of each outcome if a pair of fair dice are rolled.

  13. an even sum

  14. an odd sum

  15. a sum that is a multiple of 3

  16. a sum less than 2

  17. a composite number

  18. a prime number greater than 9

  19. *not* a sum of 7

  20. *not* a sum of 11 or 12

  21. a sum of 8 with a 6 on one die

  22. an even sum or a sum less than 6

  23. *neither* an odd sum *nor* a sum greater than 8

Five cards are selected from a standard deck of 52 playing cards.

  24. What are the odds of selecting a black king?

  25. What are the odds of *not* selecting a heart *or* a seven?

Critical Thinking

  26. A red die and a blue die are rolled. What are the odds that the number showing on the red die is less than the number showing on the blue die?

Applications and Problem Solving

  27. **Entertainment** The Durham Bulls baseball team in Durham, North Carolina, invited the community to submit promotion ideas for the opening of their new stadium. One suggestion was that the team call out the last four-digit numbers of a telephone number and award a prize to the first person with those digits in their telephone number to reach the press box. Find the odds that the four digits called out are the last four digits of your telephone number.

  28. **Television** The odds in favor of a person in North America appearing on television sometime in their lifetime is 1:3. If there are 32 students in your class, predict how many will appear on television.

29. Medicine A worker at an aluminum plant in Washington state filed a workers' compensation suit against his employer because he developed cancer. He claimed that 8 of 90 workers in his shop had developed the same type of cancer. The medical odds of a person developing this type of cancer is normally 1:1000.

  a. Based on the medical odds, how many people in his shop should have developed the cancer?

  b. Based on the number of workers and the number of cases of cancer in this shop, estimate the number of people who would develop this cancer if 1000 people worked in this shop.

30. Family Activity During two hours of watching television or listening to the radio with your family, keep track of the commercials. Record the categories, such as food, cars, cosmetics, and so on, of each commercial break. Find the odds that a randomly-selected commercial is from each category.

Mixed Review

31. Find the value of 8!. (Lesson 10-6)

32. Sports The heights (in inches) of members of a volleyball team are 64, 60, 72, 61, 73, 80, 68, 70, 65, 67, 70, and 80. (Lesson 10-1)

  a. Construct a stem-and-leaf plot of the data.

  b. What percent of the heights are between 60 and 70 inches? Why do you think that is true?

33. A penny, a nickel, a dime, and 2 quarters are in a purse. Without looking, Andie picks out three coins. What different amounts of money could she have chosen? (Lesson 9-2)

For **Extra Practice**, see page 766.

34. Geometry Find the circumference of the circle shown at the right. (Lesson 7-4)

5 ft

35. Replace ● with $<$, $>$, or $=$ to make $\frac{1}{6}$ ● 0.16 a true statement. Use a number line if necessary. (Lesson 6-1)

36. Standardized Test Practice State whether the inequality $c + 5 < 16$ is *true*, *false*, or *open*. (Lesson 1-9)

  A. true   **B.** false   **C.** open   **D.** none

From the → FUNNY PAPERS

1. Express a googol in scientific notation.

2. Do you think the odds are expressed correctly here? Why or why not?

3. Research who defined the term *googol*.

Problem-Solving Strategy: Use a Simulation

What You'll Learn

■ To investigate problems using simulation.

Why It's Important

Simulations allow you to gather large amounts of data in a small amount of time.

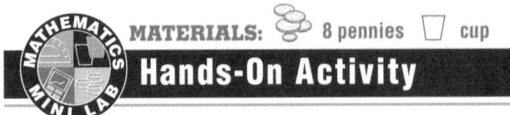

MATERIALS: 8 pennies ☐ cup

Hands-On Activity

The Paiute ('pi- (y)ut) Indians of Pyramid Lake in northwestern Nevada invented the Walnut Shell Game. This game was played with 8 walnut shell halves filled with pitch and powdered charcoal. A player tosses a basket tray containing the walnut shell halves and scores one point if three or five of the walnut shells land with the flat side up. No point is scored for any other combination. The winner is the first player to reach an agreed-on sum. On average, how many times would you have to toss the basket before you get 10 points?

Your Turn Work with a partner to simulate the walnut shell game.

▶ Place eight pennies in a cup. Put your hand over the cup and shake it gently. Then pour the pennies out. Record the number of heads in a frequency table like the one at the right.

Outcome	Tally	Frequency
0 heads		
8 heads		

▶ Repeat the process until the sum of the frequencies for three heads and five heads is 10.

▶ Compare your results to other pairs of students.

TALK ABOUT IT

a. Do you think this process is a reasonable approximation of the walnut shell game? Explain.

b. What is an advantage of finding an answer by acting it out?

Learning the Concept

A simulation is an application of the Acting It Out problem-solving strategy.

When you solve a problem by modeling a situation, like you did in the activity above, you are doing a **simulation**. A simulation acts out the event so that you can see outcomes.

You can conduct a simulation of the outcomes of many problems by using manipulatives such as dice, a coin, or a spinner. The manipulative, such as a die, a coin, a spinner, or a combination of them, should have the same number of outcomes as the number of possible outcomes.

Real-World Event	Simulation
having a baby boy or baby girl	flip a coin
history of making 2 free throws for every 3 attempted	spinning a 3-section spinner with 2 sections the same color
win one of twelve prizes	roll a die and flip a coin

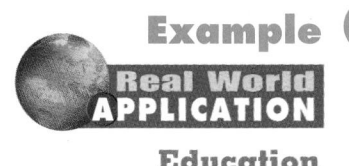

Example ❶ A quiz has 10 true-false questions. The correct answers are T, F, F, T, F, F, T, T, T, F. You need to correctly answer 7 or more questions to pass the quiz. Is tossing a coin to decide the answers a good strategy for taking the quiz?

Explore   The quiz has ten questions. Since two choices are available for each answer, tossing a coin is used to simulate guessing the answers.

Plan   Toss a coin and record the answer for each question. Write T (true) if tails shows and write F (false) if heads shows. Repeat the simulation three times. Shade the cells with the correct answers.

Solve

Answers	T	F	F	T	F	F	T	T	T	F	Number Correct
Simulation 1	F	T	F	F	T	F	F	F	T	F	4
Simulation 2	F	T	F	F	F	T	T	F	T	F	5
Simulation 3	F	F	T	T	T	F	T	T	T	T	6

Since none of the simulations results in a passing grade, this is probably not a good way to take the quiz.

Examine   Try some more simulations to confirm the results.

Sometimes it is necessary to create a manipulative for a simulation.

Example ❷ The Cougars are one point behind in their basketball game. Carmen is fouled as time runs out. If she misses the foul shot, the Cougars lose. If Carmen makes the foul shot, the score is tied and she gets another shot. If she makes the first and second shots, the Cougars win. Carmen has a record of making three out of every four free throws.

a. **Conduct a simulation for 25 trips to the foul line.**

Construct a spinner with four equal sections like the one shown at the right. Spin the spinner and record the results of several trips to the foul line. A sample simulation for 25 trips to the foul line is shown in the chart below.

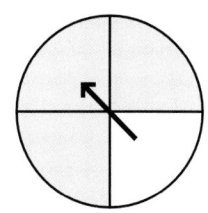

Misses the first shot (0 points)	Makes the first shot, misses the second (1 point)	Makes both shots (2 points)
IIII	IIII	I̶H̶I̶ I̶H̶I̶ I̶H̶I̶ II

b. **What is the probability that Carmen makes both shots and the Cougars win the game?**

The experiment shows that Carmen will make 2 points $\frac{17}{25}$ or 68% of the time. Based on this simulation, the probability that Carmen scores 2 points and the Cougars win the game is 68%.

Communicating Mathematics

Read and study the lesson to answer these questions.

1. **Explain** what a simulation is.

2. **Tell** whether the results of a simulation will be exactly the same as the theoretical probability. Explain your reasoning.

3. **Write** an explanation of how you would modify the spinner in Example 2 to simulate a player who has a record of making one basket out of every three tries.

Guided Practice

4. Using a device other than a coin, conduct three more simulations for the problem in Example 1. Do your simulations verify the results obtained while using a coin?

5. **Football** Faud Reveiz is a placekicker for the Minnesota Vikings. In one year, he made 80% of the field goals he attempted when the distance was between 40 and 49 yards.

   a. How would you simulate his next four field goal attempts between 40 and 49 yards?

   b. Conduct a simulation for 15 field goal attempts between 40- and 49-yards long.

   c. What is the probability that his next two field goal attempts between 40 and 49 yards will be successful?

Exercises: Practicing and Applying the Concept

Independent Practice

Solve. Use a simulation to act out the problem.

6. A fast-food restaurant includes prizes with children's meals. During the summer promotional campaign, six different prizes were available. There is an equally likely chance of getting each prize each time.

   a. Use a die to simulate this problem. Let each number represent one of the prizes. Conduct a simulation until you have one of each number.

   b. According to your simulation, how many children's meals must be purchased in order to get all six different prizes?

7. **Business** Mr. Namura runs a small gourmet restaurant. He has 12 tables. It is a popular restaurant and anyone wanting to eat at 7:00 on Saturday evening must have reservations. Mr. Namura knows that one out of six reservations usually does not show, so he takes reservations for 14 tables for 7:00.

   a. Describe a way to simulate the number of tables that will be filled at 7:00.

   b. Conduct your simulation for part a ten times.

   c. From your simulations, how many times is the restaurant overbooked? Are all 12 tables always filled?

Choose

Estimation
Mental Math
Calculator
Paper and Pencil

Solve. Use any strategy.

8. Unit 2 of a science book starts on page 126 and ends on page 241. How many pages are in the unit?

9. Antonio received a birthday gift of money from his grandmother. He loaned one dollar to his younger brother and spent half of the remaining money. The next day he earned $6. Later, he lost $\frac{1}{6}$ of his money, but he still had $15 left. How much money did his grandmother send him?

10. There are three traffic lights along the route that Yoshie rides from school to her home. The probability that any one of the lights is green is 0.4, and the probability that it is not green is 0.6. If the three lights operate independently, estimate the probability that these lights will all be green on Yoshie's way home from school.

11. **Hockey** There are 16 hockey teams in a single elimination tournament.

   a. How many games will be played during the tournament?

   b. How many ways can the teams win first and second place?

12. The diagram below shows a system of power lines that bring electricity from a power plant to three towns. Recently it has been very windy. For the last 10 days, each line has been knocked down about half the time due to the high winds. What is the probability that Altoona will still be able to get power on any given day even if some of the lines are down?

Graphing Calculator

13. When you have a situation involving numbers that cannot be generated from an object like a coin or a spinner, you can use the random number generator function on a computer or a graphing calculator. A random number generator produces numbers that have no pattern.

The graphing calculator program below will generate 20 random numbers between 1 and the number you enter for T. In order to use the program, you must first enter the program into the calculator's memory. To access the program memory, use the following keystrokes.

Enter: | PRGM | ▶ | ▶ | ENTER |

Run the program. Press | PRGM | and choose the program from the list by pressing the number next to its name.

Press the enter key twenty times. The calculator will display a number from 1 to 6 each time you press the enter key.

PROGRAM: RANDOM
: Input T
: For (N,1,20)
: int ((T*rand) + 1) →D
: Disp D
: Pause
: End

a. Run the program to simulate 20 spins of a spinner that has 7 equal sections.

b. How would you alter the program to generate 50 random numbers?

**Critical
Thinking**

14. In a survey of 240 students, 100 students said they play soccer, and 120 students said they play basketball. If 40 students play both sports, how many students do not play either basketball or soccer?

**Mixed
Review**

15. Probability Find the odds of stopping on 2 if the spinner at the right is spun. (Lesson 10-7)

16. Geometry Find the slope of the line that contains the points (1, 5) and (3, 3). (Lesson 8-6)

For **Extra Practice**,
see page 766.

17. Sales Jeff bought a used 10-speed bike for $20 more than one-half its original price. He paid $110 for the bike. (Lesson 7-3)

    a. Write an equation that represents this situation.

    b. What was the original price of the bike?

18. Write r^4 as the product of the same factor. (Lesson 4-2)

19. Standardized Test Practice The temperature outside was 20°F. The windchill factor made it feel like −15°F. Find the difference between the real temperature and the apparent temperature. (Lesson 2-5)

    A. −35°   **B.** −5°   **C.** 5°   **D.** 35°

WORKING ON THE
Investigation

Refer to the Investigation on pages 482–483.

Look over the list of favorite games that you placed in your Investigation Folder at the beginning of this Investigation.

For this part of the Investigation, your group has been chosen to test a new TV game show called *Don't Lose It All!* This game is for two players. Each player takes a turn spinning two identical wheels at the same time. Each wheel is divided into six equal regions. Five of the regions say "Win $25" and the sixth region says "Lose Your Cash". The object of the game is to accumulate the most money by the end of the show.

A player's turn ends if one or both of the wheels stop on "Lose Your Cash" or the player decides to let his or her opponent spin. On a spin, one of three things can occur.

1. Both wheels stop on "Win $25" and the player is awarded $50.
2. One wheel stops on "Lose Your Cash," and one stops on "Win $25." In this case, a player loses any money made on this turn but keeps money from any previous turns.
3. Both wheels stop on "Lose Your Cash" and the player loses all the money won so far.

• Create two spinners to simulate the wheels in the game. Play the game enough times to develop some different strategies. Have each member of your group use a different strategy and play the game several more times to determine the best strategy.

• Determine a way to simulate the game your group is designing.

Add the results of your work to your Investigation Folder.

10-8B Making Predictions

An Extension of Lesson **10-8**

MATERIALS

📄 paper bag

🔵 **10 marbles of 3 different colors**

In this activity, you will conduct and interpret probability experiments. Probability calculated by making observations or experiments is called **experimental probability**. Experimental probability is an estimate based on the **relative frequency**. Relative frequency is the number of times a certain event actually happened.

Activity **Work with a partner.**

▶ Get a bag of marbles from your teacher. You will try to determine how many marbles there are of each color.

▶ Draw one marble from the bag, record its color, and replace it in the bag. Repeat this process 10 times.

▶ Determine the relative frequency for each color marble.

TALK ABOUT IT

1. Based on the relative frequencies, which color marble do you think is most prevalent in the bag? Predict the colors of the marbles in your bag.

2. Based on your experiment, what is the experimental probability of drawing each color marble?

$$\text{experimental probability} = \frac{\text{relative frequency of color}}{\text{total number of draws}}$$

Activity ② **Repeat both steps above for 20, 30, 40, and 50 draws.**

TALK ABOUT IT

3. Is it possible to have a certain color marble in the bag and never draw that color? Is this situation more likely to happen if you make two draws, ten draws, or fifty draws? Explain.

4. Compute the experimental probability of drawing each color for each number of draws. Write a paragraph that describes how the experimental probability changed as you increased the number of draws.

5. Predict the colors of the marbles in your bag. Did your prediction change from the prediction you made in Exercise 1? Explain why or why not. Open the bag and check your prediction against the marbles.

Extension

6. Make up a fair game for two players to draw marbles from your bag. Assign point values for drawing certain colors of marbles. Make sure each player has an equal chance of winning.

10-9 Probability of Independent and Dependent Events

You'll Learn

- To find the probability of independent and dependent events.

It's Important

When you study genetics in science you will use probability of two events.

Technology Activity

The television meteorologist said that there was a 50% chance of rain on Saturday and a 25% chance of rain on Sunday. Janice was upset because the ninth grade at her high school was planning their class picnic for Saturday and the rain date was Sunday. She thought that the probability for rain on Saturday *and* Sunday was 75%. Kiona said that the probability of rain on both days was just 12.5%.

You can use the graphing calculator program at the right to simulate this situation. Recall that you must first enter the program into the calculator's memory. Access the program memory using the following keystrokes.

Enter: PRGM ▶ ▶

The program generates random numbers for 50 weekends. The first use of the random number function simulates the probability of rain on Saturday. The second use of the random number function simulates the probability of rain on Sunday.

```
:PROGRAM:RAIN
:0→S: 0→R
:Disp "PROBABILITY",
 "OF RAIN ON"
:Input "SATURDAY", A
:Input "SUNDAY", B
:For (N,1,50)
:int ((100/A*rand)
 +1)→T
:If T=1
:S+1→S
:End
:For (M,1,50)
:int ((100/B*rand)
 +1)→G
:If G=1
:R+1→R
:End
:(S/50)*(R/50)→F
:F*100→F
:Disp "PERCENT OF
 TIMES", "IT RAINS
 BOTH", "DAYS IS", F
:Stop
```

Your Turn Run the program several times and record the percent of times that it rains on both days for each 50-weekend simulation.

a. Are your results closer to what Janice thought or to what Kiona said?

b. Run the program several times for a weekend when the probability of rain is 100% for Saturday and 20% for Sunday. What do you observe?

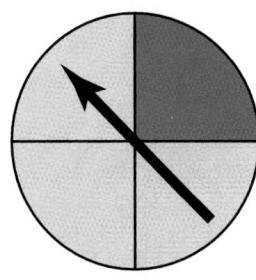

You could also simulate the situation above using manipulatives. This could be done by tossing a coin and spinning the spinner at the right 50 times each. Tossing a head has the same probability as the chance of rain on Saturday and spinning blue on the spinner has the same probability as the chance of rain on Sunday. Does the outcome of tossing the coin affect the outcome of spinning the spinner?

The weather situation in the activity above is an example of **independent events**. Events are independent when the outcome of one event does *not* influence the outcome of a second event.

Using the Fundamental Counting Principle, you know that there are $2 \cdot 4$ or 8 possible outcomes of tossing a coin and spinning the spinner. Of the 8 outcomes, there is one pair (heads, blue) that matches the meteorologist's prediction (50% chance of rain on Saturday, 25% chance of rain on Sunday). Determine the probability.

$$P(\text{heads, blue}) = \frac{\textit{number of ways heads and blue can occur}}{\textit{number of possible outcomes}}$$

$$= \frac{1}{8}$$

Note that $P(\text{heads})$ is $\frac{1}{2}$, $P(\text{blue})$ is $\frac{1}{4}$, and $\frac{1}{2} \cdot \frac{1}{4} = \frac{1}{8}$, which is $P(\text{heads, blue})$.

This suggests that the probability of two independent events is the product of the probability of each event.

Probability of Two Independent Events	**In words:**	The probability of two independent events can be found by multiplying the probability of the first event by the probability of the second event.
	In symbols:	$P(A \text{ and } B) = P(A) \cdot P(B)$

Example ①

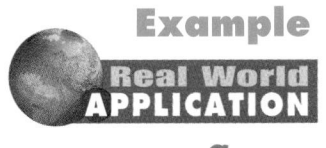

LOOK BACK

You can review the sample space of rolling two dice on page 441.

In a popular board game, you go to jail if you roll three doubles in a row. What is the probability of rolling three doubles in a row?

The events are independent since each roll of the dice does not affect the outcome of the next roll.

There are six ways to roll doubles, (1, 1), (2, 2), and so on, and there are 36 ways to roll two dice. Thus, the probability of rolling doubles on a toss of the dice is $\frac{6}{36}$ or $\frac{1}{6}$.

$$P\left(\begin{array}{c}\textit{doubles on} \\ \textit{three rolls}\end{array}\right) = P\left(\begin{array}{c}\textit{doubles on} \\ \textit{roll one}\end{array}\right) \cdot P\left(\begin{array}{c}\textit{doubles on} \\ \textit{roll two}\end{array}\right) \cdot P\left(\begin{array}{c}\textit{doubles on} \\ \textit{roll three}\end{array}\right)$$

$$= \frac{1}{6} \cdot \frac{1}{6} \cdot \frac{1}{6}$$

$$= \frac{1}{216}$$

The probability of rolling all doubles in three rolls of the dice is $\frac{1}{216}$.

Sometimes the outcome of one event affects the outcome of a second event. When this happens, the events are called **dependent events**. For example, suppose you choose a card from a deck, keep it, and then your friend chooses a card. Since you did not replace your card, your friend has fewer cards from which to choose. These events are dependent.

Probability of Two Dependent Events	**In words:**	If two events, A and B, are dependent, then the probability of both events occurring is the product of the probability of A and the probability of B after A occurs.
	In symbols:	$P(A \text{ and } B) = P(A) \cdot P(B \text{ following } A)$

Example **2**

Bag lunches are prepared for 50 students on a field trip. There are three different desserts in the bags. Twenty bags contain a brownie, 15 contain a chocolate chip cookie, and 15 contain an oatmeal raisin cookie. The bags are arranged in random order. Connie and Tale were the first and second in line. What is the probability that Connie and Tale both select a bag containing a chocolate chip cookie?

For Connie, $P(\text{chocolate chip}) = \dfrac{\textit{number containing chocolate chip}}{\textit{total number of lunches}}$

$$= \frac{15}{50} \text{ or } \frac{3}{10}$$

Assume that Connie chose a bag containing a chocolate chip cookie.

For Tale, $P(\text{chocolate chip}) = \dfrac{\textit{number containing chocolate chip left}}{\textit{number of lunches left}}$

$$= \frac{14}{49} \text{ or } \frac{2}{7}$$

$P(\text{chocolate chip cookies for both}) = \dfrac{3}{10} \cdot \dfrac{2}{7}$

$$= \frac{6}{70} \text{ or } \frac{3}{35}$$

The probability that both students get a chocolate chip cookie is $\dfrac{3}{35}$.

Checking Your Understanding

Communicating Mathematics

Read and study the lesson to answer these questions.

1. **Explain** the difference between independent and dependent events.

2. **Write** an example of two independent and two dependent events. List each sample space.

3. **Describe** how to find the probability of the second of two dependent events.

Determine whether the events are independent or dependent. Explain.

4. rolling a die and then spinning a spinner

5. drawing a marble from a bag and then drawing a second marble without replacing the first one

The chart below lists the number and type of two soft drinks found in a tub filled with ice. Two cans are selected at random. Find the probability of each outcome.

6. a diet orange and a regular lemon-lime

7. a regular orange and a regular lemon-lime

Soda	Regular	Diet
orange	3	4
lemon-lime	7	6

8. a diet lemon-lime and a regular lemon-lime

9. a regular orange and a diet lemon-lime

10. Do the probabilities in Exercises 6–9 represent dependent or independent events? Explain.

11. **Economics** Recently, 44% of full-time, hourly wage earners in the United States were women. If 4% of these women earn the minimum wage, what is the probability that a full-time, hourly wage earner chosen at random for a survey is a woman who is paid the minimum wage?

Exercises: Practicing and Applying the Concept

Determine whether the events are independent or dependent. Explain.

12. choosing a card from a deck of playing cards and then choosing a second card without replacing the first card

13. rolling a die twice

14. selecting a name from the Raleigh telephone book and a name from the Miami telephone book

15. tossing a coin and spinning a spinner

16. as team captain, selecting someone to be on your basketball team and then selecting a second player

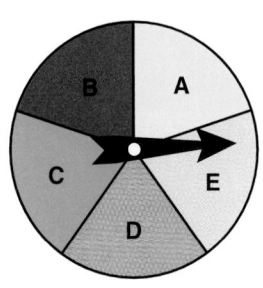

A die is rolled and the spinner is spun. Find each probability.

17. $P(1 \text{ and } D)$

18. $P(\text{an odd number and } C)$

19. $P(\text{a composite number and a vowel})$

20. $P(\text{an even number and a consonant})$

21. $P(\text{a 6 and } F)$

In a bag there are 5 red marbles, 2 green marbles, 4 yellow marbles, and 3 blue marbles. Once a marble is drawn, it is not replaced. Find the probability of each outcome.

22. two green marbles in a row   **23.** a red and then a yellow marble

24. two red marbles in a row   **25.** a green and then a blue marble

26. a red marble, a green marble, and then a blue marble

27. three yellow marbles in a row

Critical Thinking

28. There are 9 marbles in a bag. Some of the marbles are red, some are white, and some are blue. The probability of selecting a red marble, a white marble, and then a blue marble is $\frac{1}{21}$.

   a. How many of each color are in the bag?

   b. Explain why there is more than one correct answer to part a.

Applications and Problem Solving

✓ Choose

Estimation
Mental Math
Calculator
Paper and Pencil

29. TV Game Shows On a popular game show, contestants can win a car if they draw the digits in the price of the car from a bag that contains the five digits in the price of the car and three strikes. Once a number or strike is drawn, it is not replaced.

   a. What is the probability that the first draw is a strike?

   b. A contestant has drawn three digits in the price of the car without drawing a strike. What is the probability that the next two draws will be strikes?

30. Ambulance Service Mercy Hospital has two ambulances that are each available for emergencies 80% of the time.

   a. In an emergency situation, what is the probability that both ambulances will be available?

   b. What is the probability that neither will be available?

   c. Explain why the probabilities for parts a and b do not have a sum of 1.

31. Aerospace Design The electronic telemetry system on a spacecraft has three motion sensors, a signal conditioner, and a transmitter. The probability of failure for each motion sensor and the signal conditioner is 0.0001, and the probability of failure for the transmitter is 0.001. The systems run independently. What is the probability of *success* for the telemetry system?

Mixed Review

32. Sports Colin kicks extra points for the football team. He makes 75% of his attempts for extra points. How could you simulate the results of the next six attempts? (Lesson 10-8)

33. Standardized Test Practice Which is equivalent to 1.9 milliliters? (Lesson 7-8)

   A. 1900 L   **B.** 190 L   **C.** 0.019 L   **D.** 0.0019 L

34. Evaluate $c - a$ if $a = \frac{2}{9}$ and $c = \frac{5}{9}$. Write the result in simplest form. (Lesson 5-4)

35. Solve $14 = b - (-12)$ and check your solution. Then graph the solution on a number line. (Lesson 3-2)

36. State whether the product of $-3(5)(-2)$ is positive or negative. Then find the product. (Lesson 2-7)

For **Extra Practice**, see page 766.

37. Write an expression for *twice the sum of a number and 3.* (Lesson 1-3)

10-10

Probability of Compound Events

MATERIALS: colored pencils

Hands-On Activity

Suppose you are playing a board game with three friends. It's your turn and you need to roll a 6 or 10 to avoid landing on properties owned by the other players and having to pay them rent. What is the probability that you can avoid paying rent?

Your Turn

▶ The chart at the right is a partial list of the possible sums when two dice are rolled. Copy and complete the chart. Circle the combinations with a sum of 6 with a colored pencil. Then circle those with a sum of 10 using a second color.

Second Die

+	1	2	3	4	5	6
1	2	3	4	5	6	7
2	3	4	5			
3	4	5				
4	5					
5	6					
6	7					

First Die

TALK ABOUT IT

a. What is the probability the sum of the dice will be 6?

b. What is the probability the sum of the dice will be 10?

c. What is the probability the sum of the dice will be 6 or 10?

d. What do you notice about the probabilities in parts a, b, and c?

Learning the Concept

When two events cannot happen at the same time, like rolling a 6 or a 10, they are said to be **mutually exclusive**. The probability that one of two mutually exclusive events will occur is the sum of their individual probabilities.

Probability of Mutually Exclusive Events	**In words:**	The probability of one or the other of two mutually exclusive events can be found by adding the probability of the first event to the probability of the second event.
	In symbols:	$P(A \text{ or } B) = P(A) + P(B)$

Example 1

THINK ABOUT IT

What is the probability of two events both occurring when they are mutually exclusive?

The spinner at the right is spun. What is the probability that the spinner will stop on yellow or on 2?

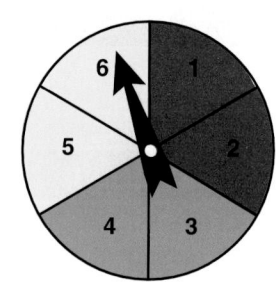

The events are mutually exclusive because the spinner cannot stop on both yellow and 2 at the same time.

$P(\text{yellow or 2}) = P(\text{yellow}) + P(2)$

$$= \frac{1}{3} + \frac{1}{6}$$

$$= \frac{3}{6} \text{ or } \frac{1}{2}$$

The probability that the spinner stops on yellow or 2 is $\frac{1}{2}$.

When two events are **inclusive**, they can happen at the same time. For example, from a standard deck of 52 playing cards, it is possible to draw a card that is *both* a queen and a red card. What is the probability of drawing a queen or a red card?

$P(\text{queen})$	$P(\text{red card})$	$P(\text{red queen})$
$\frac{4}{52}$	$\frac{26}{52}$	$\frac{2}{52}$
1 queen in each suit	*13 hearts and 13 diamonds*	*queen of hearts and queen of diamonds*

The probability of drawing a red queen is counted twice, once for a queen and once for a red card. To find the correct probability, you must subtract $P(\text{red queen})$ from the sum of $P(\text{queen})$ and $P(\text{red card})$.

$$P(\text{queen or red card}) = \frac{4}{52} + \frac{26}{52} - \frac{2}{52} \text{ or } \frac{7}{13}$$

The probability of drawing a queen or a red card is $\frac{7}{13}$.

Probability of Inclusive Events

In words: The probability of one or the other of two inclusive events can be found by adding the probability of the first event to the probability of the second event and subtracting the probability of both events happening.

In symbols: $P(A \text{ or } B) = P(A) + P(B) - P(A \text{ and } B)$

Example 2

Real World APPLICATION

Business

An auto dealer finds that of the cars coming in for service, 70% need a tune up, 50% need a new air filter, and 35% need both. What is the probability that a car brought in for service needs either a tune-up or a new air filter?

Since it is possible to get a tune-up *and* a new air filter, these events are inclusive.

$P(\text{tune-up}) = \frac{70}{100}$ $\qquad$ $P(\text{air filter}) = \frac{50}{100}$

$P(\text{tune-up and air filter}) = \frac{35}{100}$

$P(\text{tune-up or air filter}) = \frac{70}{100} + \frac{50}{100} - \frac{35}{100} \text{ or } \frac{85}{100}$

The probability that a car needs a tune-up or air filter is $\frac{85}{100}$ or 85%.

Communicating Mathematics

Read and study the lesson to answer these questions.

1. **Write** an example of two mutually exclusive events in your own life.

2. **Describe** the difference between mutually exclusive and inclusive events.

3. **Explain** why you must subtract the probability of both events happening when finding the probability of two events that are not mutually exclusive.

Guided Practice

Determine whether each event is *mutually exclusive* or *inclusive*.

4. A student is selected at random from a group of male and female students in the seventh, eighth, or ninth grade.

   a. P(male or ninth grader)   **b.** P(male or female)

   c. P(female or eighth grader)   **d.** P(seventh or eighth grader)

5. The spinner at the right is spun. Find each probability.

   a. P(C or D)

   b. P(E or vowel)

   c. P(B or consonant)

6. **Advertising**   A poll asked 1225 adults across the United States whether they liked a certain television commercial for a long-distance service. Those polled were also asked how effective they thought the commercial was in helping to sell the service. The results of the second question are shown in the graph at the right.

   a. If a person is selected at random, what is the probability that he or she would say the commercial was not effective or somewhat effective?

How effective is the campaign

Not sure 1%   Very effective 11%

Not effective 44%   Somewhat 44%

Source: *USA TODAY*

   b. If a person is selected at random, what is the probability that he or she would say the commercial was somewhat or very effective?

Independent Practice

Determine whether each event is *mutually exclusive* or *inclusive*. Then find the probability.

7. An eight-sided die is tossed.

   a. P(even or less than 5)   **b.** P(6 or odd)

   c. P(4 or prime)   **d.** P(greater than 4 or an 8)

8. A card is drawn from the cards at the right.

   a. P(6 or even)

   b. P(9 or even)

   c. P(odd or even)

   d. P(5 or less than 3)

   e. P(4 or greater than 3)

   | 1 | 3 | 4 | 5 |

   | 6 | 7 | 9 |

Suppose $P(A) = \frac{1}{5}$, $P(B) = \frac{2}{3}$, and A and B are mutually exclusive.

9. What is $P(A \text{ and } B)$?   **10.** What is $P(A \text{ or } B)$?

Suppose $P(A) = \frac{1}{5}$, $P(B) = \frac{2}{3}$, and A and B are independent.

11. What is $P(A \text{ and } B)$?   **12.** What is $P(A \text{ or } B)$?

Two cards are drawn from a standard deck of cards. Find each probability.

13. P(both aces or both jacks)

14. P(both red or both face cards)

15. A bag contains six blue marbles and three red marbles. A marble is drawn, it is replaced and another marble is drawn. What is the probability of drawing a red marble and a blue marble in either order?

An eight-sided die is tossed 40 times. Determine how many times you would expect each outcome.

16. a 5 or a 7   **17.** a 6 or a 3

18. an even number or a 5

19. a prime number or an even number

20. a multiple of 3 or a multiple of 2

21. a multiple of 2 or a multiple of 4

Critical Thinking

22. Design and draw a spinner containing the numbers 1, 2, 3, and 4 so that the probability of spinning a 3 or a 4 is $\frac{2}{3}$.

Applications and Problem Solving

23. **Economics** Thirty-one percent of minimum-wage workers are between 16 and 19 years old. Twenty-two percent of minimum-wage workers are between 20 and 24 years old. If a person who makes minimum wage is selected at random, what is the probability that he or she will be between 16 and 24 years old?

24. **Government** Florida has 23 members in the United States House of Representatives. Eight members are Democrats, and 15 are Republicans. Of the 17 men in the house, 4 are Democrats. The Speaker of the House wants to choose a representative from Florida at random to serve on the agriculture committee. What is the probability that the representative will be a woman or a Democrat?

Mixed Review

25. **Standardized Test Practice** A pair of dice is rolled. What is the probability that both dice will show a 3 or a 4? (Lesson 10-9)

    A. $\frac{1}{9}$ **B.** $\frac{1}{6}$ **C.** $\frac{1}{3}$ **D.** $\frac{2}{3}$

26. **Consumer Awareness** Unit prices are often displayed for products at grocery stores. Suppose a 13-ounce box of breakfast cereal costs $3.69 and a unit price of 0.95¢ per ounce is displayed. Is this the correct unit price? Explain your answer. (Lesson 9-1)

27. Graph $y = x - 3$. (Lesson 8-3)

28. Write an inequality for "the number of students in each class is less than or equal to 28." Then draw a graph of the inequality. (Lesson 3-6)

For **Extra Practice**, see page 767.

29. Find $15 \cdot 20$ mentally using the distributive property. (Lesson 1-5)

Vocabulary

After completing this chapter, you should be able to define each term, property, or phrase and give an example or two of each.

Discrete Mathematics
combination (p. 516)
factorial (p. 515)
Fundamental Counting Principle
   (p. 509)
outcomes (p. 509)
permutation (pp. 514, 515)
tree diagram (p. 509)

Probability
dependent events (p. 532)
experimental probability (p. 529)
inclusive (p. 536)
independent events (p. 531)
mutually exclusive (p. 535)
odds (p. 520)
relative frequency (p. 529)
simulation (p. 524)

Statistics
back-to-back stem-and-leaf plot
   (p. 487)
box-and-whisker plot (p. 495)
circle graph (p. 497)
comparative graph (p. 498)
histogram (p. 502)
interquartile range (p. 491)
leaves (p. 486)
line graph (p. 497)
lower quartile (p. 491)
outliers (p. 496)
pictograph (p. 497)
quartile (p. 491)
range (p. 490)
stem-and-leaf plot (p. 486)
stems (p. 486)
upper quartile (p. 491)

Problem Solving
use a simulation (p. 524)

⬛*inter*NET
C O N N E C T I O N
Chapter Review For additional lesson-by-lesson review, visit:
www.glencoe.com/ sec/math/prealg/ mathnet

Understanding and Using Vocabulary

Choose the letter of the term that best matches each statement or phrase.

1. two events that can happen simultaneously

2. the difference between the least number and the greatest number in a set of numbers

3. an arrangement or listing in which order is important

4. the median of the upper half of a set of data

5. two events that cannot happen at the same time

6. the range of the middle half of a set of data

7. an arrangement or listing where order is not important

A. mutually exclusive
B. relative frequency
C. inclusive
D. upper quartile
E. interquartile range
F. range
G. combination
H. outliers
I. permutation

Skills and Concepts

Objectives and Examples	Review Exercises

Upon completing this chapter, you should be able to:

Use these exercises to review and prepare for the chapter test.

▶ **display and interpret data in stem-and-leaf plots.** (Lesson 10-1)

Display the data 46, 32, 59, 42, 51, 30, 49 in a stem-and-leaf plot.

$$
\begin{array}{c|l}
3 & 2\ 0 \\
4 & 6\ 2\ 9 \\
5 & 9\ 1
\end{array}
\qquad 5|9 = 59
$$

Make a stem-and-leaf plot of each set of data.

8. 3, 4, 6, 8, 17, 19, 7, 21, 40, 43
9. 10.8, 11.0, 9.8, 10.7, 11.5, 10.7, 9.5, 11.9
10. 18, 16, 20, 25, 30, 23, 30, 29, 21, 15, 18
11. 120, 111, 133, 148, 117, 112, 121, 146

▶ **use measures of variation to compare data.** (Lesson 10-2)

$$
\begin{array}{c|l}
4 & 1\ 2\ 0\ 3 \\
5 & 9\ 1\ 8 \\
6 & 6\ 5
\end{array}
\qquad 6|6 = 6.6
$$

Range: 6.6 − 4.0 or 2.6
Median: 4.0 4.1 4.2 4.3 5.1 5.8 5.9 6.5 6.6
                          ↑
                      median
Lower Quartile: 4.15
Upper Quartile: 6.2
Interquartile Range: 6.2 − 4.15 or 2.05

Find the range, median, upper and lower quartiles, and the interquartile range.

12. 23 45 16 51 47
13. 6 10 15 21 28 36
14.
$$
\begin{array}{c|l}
9 & 9\ 7\ 5 \\
10 & 1\ 3\ 6 \\
11 & 2\ 9\ 8\ 3
\end{array}
\qquad 11|3 = 113
$$
15. 205 236 217 214 200 298 240
16.
$$
\begin{array}{c|l}
11 & 0\ 6\ 0 \\
12 & 8\ 5\ 5 \\
13 & 7\ 1\ 3
\end{array}
\qquad 12|8 = 12.8
$$

▶ **use box-and-whisker plots to display data.** (Lesson 10-3)

Range: 49 − 29 or 20
Median: 37.5
Upper Quartile: 43.75
Lower Quartile: 31.25
Interquartile Range: 43.75 − 31.25 or 12.50
Extremes: 29 and 49

17. Use the stem-and-leaf plot below to answer each question.

$$
\begin{array}{c|l}
2 & 4\ 5\ 5\ 7\ 9 \\
3 & 0\ 3\ 4\ 6\ 8 \\
4 & 1\ 1\ 3\ 5 \\
5 & 4\ 4
\end{array}
\qquad 2|4 = 24
$$

   a. Make a box-and-whisker plot of the data.

   b. What is the median?

   c. What are the outliers, if any?

| | Objectives and Examples | Review Exercises |

Objectives and Examples

▶ **recognize when statistics are misleading.**
(Lesson 10-4)

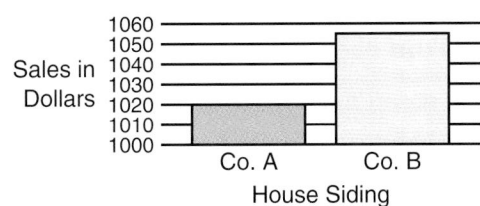

The second graph is misleading because the vertical axis does not include zero.

▶ **use tree diagrams or the Fundamental Counting Principle to count outcomes.**
(Lesson 10-5)

How many ways can you choose a shirt and tie?

shirt choices · tie choices

4 · 3 = 12

There are 12 ways to choose a shirt and tie.

▶ **use permutations and combinations.**
(Lesson 10-6)

Seven people are being arranged for a portrait. Does this represent a permutation or a combination?

This is a permutation because the order that the photographer chooses the people determines how the picture looks.

▶ **find the odds of a simple event.**
(Lesson 10-7)

| Green | Red | White | Blue |

The odds of picking the green box is 1:3.

Review Exercises

Use the following chart to describe each category. Write the type of average you used.

	Tina	Steve	Ryan	Katie
Favorite Sport	football	softball	soccer	soccer
Number of classes	9	10	5	3
Number of family members	3	5	12	6

18. average favorite sport
19. average number of classes.
20. average number of family members
21. How can you describe the "average" person in this group?

Find the number of possible outcomes for each event.

22. An eight-sided die is rolled twice.
23. A multiple-choice quiz has 5 questions each with 4 choices.
24. A 6-sided die and a coin are tossed.
25. A card is drawn and a coin is tossed.

Tell whether each situation represents a *permutation* or a *combination*.

26. five people in a row
27. 32 cassette tapes from a group of 47

Find each value.

28. $P(7,4)$
29. $5!$
30. $C(8, 2)$
31. $\frac{6!3!}{4!2!}$

Find the odds of each outcome if a card is selected from a standard deck of cards.

32. a queen
33. not a face card
34. an ace or six
35. a red card

Objectives and Examples

▶ **find the probability of independent and dependent events.** (Lesson 10-9)

Suppose you are playing a carnival game where each person chooses a plastic duck from a pond. Ten ducks are marked with numbers for prizes. Once a prize is given, the number is removed, and there is one less prize to be won. Is two people winning prizes an example of a dependent or independent event?

It is dependent because the first person winning affects the probability that the second person will win.

▶ **find the probability of mutually exclusive events or inclusive events.** (Lesson 10-10)

Suppose you roll two dice. Are the events *rolling a six* and *rolling a sum of less than seven* mutually exclusive or inclusive?

Since you cannot roll a six and a sum of less than seven in one roll, the events are mutually exclusive.

Review Exercises

Determine whether the events are independent or dependent. Explain.

36. choosing a recipe from a box and then choosing another without replacement

37. on the first turn of a board game, you buy a property and then on the second turn you buy a property

A bag contains 5 red, 3 yellow, and 3 purple marbles. Once a marble is drawn, it is not replaced. Find each probability.

38. P(two green marbles)

39. P(a red and a yellow)

40. P(two yellow)

41. P(a purple and a red)

Determine whether each event is mutually exclusive or inclusive. Then find the probability.

42. P(rolling a six or sum greater than ten)

43. P(rolling an even number or sum is odd)

A die is tossed. Find each probability.

44. P(3 or 4)

45. P(4 or even)

Applications and Problem Solving

46. **Transportation** The numbers of seats in commonly-used passenger airplanes are listed below. Make a box-and-whisker plot of the data. (Lesson 10-3)

398	390	288	281	266	254
248	221	186	185	149	148
144	141	131	124	113	112
100	97	72			

A practice test for Chapter 10 is available on page 783.

47. **Games** Bruno's Restaurant is giving out game cards as a promotional contest. There are eight different cards available. How many different ways could you get cards on two trips to Bruno's? (Lesson 10-5)

48. **Sports** A soccer team plays 16 games during the season. The team manager predicts that the team has a 75% chance of winning each game. Describe a simulation to determine the probability of the team winning three games in a row. (Lesson 10-8)

Cooperative Learning Activity

Work with your group to write an article for your school newspaper entitled "Are you average?"

► Choose the type of information you think students in your school would find interesting. For example, you might investigate students' heights, the numbers of languages they speak, what kind of pets they own, or how long they have lived in the area.

► Next, conduct a survey in the school cafeteria to gather data.

► Present the results of your research in an interesting way. Include some graphs and tables. Be sure to explain what the results mean in the article.

► State the probability that a randomly-chosen student at your school possesses one or more characteristics. For example, how likely is it that a person speaks two languages and owns a hedgehog?

Thinking Critically

Recently, a magazine reported that the average American is a white woman who is 32.7 years old. She is married with children, owns a home, and has two television sets and a VCR.

► Does this describe anyone you know?

► Explain what the magazine meant by "the average American."

 Portfolio

Select an item from your work in this chapter that shows how your understanding of the concepts improved and place it in your portfolio.

Section One: Multiple Choice

There are ten multiple-choice questions in this section. After working each problem, write the letter of the correct answer on your paper.

1. Theo wants to lose 2 pounds per week for 8 weeks. What will the change in his weight be?

    A. −6 pounds   **B.** −8 pounds
    C. +8 pounds   **D.** −16 pounds

2. The high temperatures (°F) for January 15th during the past 10 years are 15°, 18°, 25°, 38°, 6°, 25°, 10°, 30°, 18°, and 25°. What are the lower and upper quartiles?

    A. 18°, 25°   **B.** 15°, 25°
    C. 21.5°, 30°   **D.** 10°, 30°

3. Which fraction is less than $\frac{5}{12}$?

    A. $\frac{1}{2}$ **B.** $\frac{3}{4}$

    C. $\frac{5}{8}$ **D.** $\frac{2}{6}$

4. If $m + 2.1 = 8$, what is the value of m?

    A. 5.9   **B.** 8
    C. 6.9   **D.** 10.1

5. Mr. Garza has 4 black ties, 3 gray ties, 2 blue ties, and 1 brown tie in his closet. If he selects one tie without looking, what is the probability that it will be blue?

    A. $\frac{1}{2}$ **B.** $\frac{1}{5}$

    C. $\frac{1}{10}$ **D.** $\frac{2}{8}$

6. Angie's Place sells made-to-order sandwiches. There are 3 kinds of bread, 3 kinds of cheese, and 4 kinds of meat. How many different combinations of bread, cheese, and meat (one of each) can be ordered?

    A. 10   **B.** 18
    C. 12   **D.** 36

7. If $5y - 4 = 3y + 12$, what is the value of y?

    A. 1   **B.** 4
    C. 2   **D.** 8

8. The number of students in Sam's aerobics class increased from 15 to 24. What was the percent of increase?

    A. 37.5%   **B.** 62.5%
    C. 60%   **D.** 135%

9. A coin is tossed and a die is rolled. What is the probability of tossing tails and rolling a 5?

    A. $\frac{1}{12}$ **B.** $\frac{1}{4}$

    C. $\frac{1}{10}$ **D.** $\frac{1}{3}$

10. Which ordered pairs are solutions for $y = 2x - 3$?

    A. $(-3, -9)$, $(2, 1)$, $(4, 5)$
    B. $(-3, 0)$, $(-1, -5)$, $(3, 3)$
    C. $(-3, 3)$, $(-2, 1)$, $(-1, -1)$
    D. $(-2, -7)$, $(-1, 1)$, $(0, -3)$

Section Two: Free Response

This section contains eight questions for which you will provide short answers. Write your answer on your paper.

11. Barbara needs at least $7\frac{1}{8}$ pounds of fertilizer for her lawn. She has $4\frac{3}{4}$ pounds. How much more does she need?

12. Juanita drove 350 miles and used 14 gallons of gasoline. Express her gas mileage as a unit rate.

13. The circumference of the circle outlining the boundary of where a discus thrower competes is 17.27 feet. What is the diameter of the circle?

14. Find the slope of a line that contains $(3, -4)$ and $(-2, 2)$.

15. As a team, the Eagles made 24 shots in 40 times at the free throw line. What percentage did they make?

16. Matt's parents have agreed to help him buy a car. They will purchase the car and he will make no-interest payments to them over the next three years. Matt can make a down payment of $1500. He cannot afford monthly payments over $200. What is the most he can pay for the car?

17. When rolling a 6-sided die, what are the odds of rolling a number divisible by 3?

18. In how many different ways can four students be arranged in a row?

Test-Taking Tip As part of your preparation for a standardized test, review basic definitions and formulas such as the ones below.

A number is *prime* if it has no factors except itself and 1. For example, 7 is a prime number. The factors for 8 are 1, 2, 4, and 8. Therefore, 8 is not prime since it has factors other than itself and 1.

The area of a circle with radius r units is πr^2.

Section Three: Open-Ended

This section contains two open-ended problems. Demonstrate your knowledge by giving a clear, concise solution to each problem. Your score on these problems will depend on how well you do the following.
- Explain your reasoning.
- Show your understanding of the mathematics in an organized manner.
- Use charts, graphs, and diagrams in your explanation.
- Show the solution in more than one way or relate it to other situations.
- Investigate beyond the requirements of the problem.

19. Describe how you would calculate the sale price of any item. Use a specific example to demonstrate your method.

20. Collect information from your classmates on how long it takes them to get to school. Construct an appropriate statistical graph to display the data. Explain your choice of graph.

interNET
CONNECTION

Test Practice For additional test practice questions, visit:
www.glencoe.com/sec/math/prealg/mathnet

CHAPTER 11 Applying Algebra to Geometry

What You'll Learn

In this chapter, you will:

- use some basic terms of geometry,

- construct a circle graph,

- identify the relationships of intersecting, parallel, and skew lines,

- identify properties of congruent and similar figures,

- classify and draw polygons, and

- identify and draw transformations.

Why It's Important

Architecture Architects need a thorough knowledge of both geometry and algebra as they design and build houses. They use their knowledge of geometry to create attractive and functional designs. They use algebra to find the relationship between the scale in their drawings and the actual measurements of the finished building. If you know the scale in a drawing, such as the one of the room below, you can write and solve a proportion to find the actual length of each wall. *In Lesson 11-7 you will learn how to use algebra and proportions to find lengths in scale drawings.*

Focus on...

To be successful in this chapter, you'll need to understand these concepts and be able to apply them. Refer to the lesson in parentheses if you need more review before beginning the chapter.

▶ **Use the percent proportion to solve problems.** *(Lesson 9-5)*

Use the percent proportion to solve each problem.
1. Find 20% of 150.
2. Find 8% of 230.
3. Find 15% of 360.
4. Find 72% of 360.

▶ **Write fractions as terminating or repeating decimals.** *(Lesson 6-1)*

Write each fraction as a decimal. Use a bar to show a repeating decimal.

5. $\frac{3}{16}$

6. $\frac{2}{3}$

7. $\frac{7}{8}$

8. $\frac{5}{6}$

▶ **Solve equations that involve more than one operation.** *(Lesson 7-2)*

Solve each equation.
9. $4x - 1 = 3$
10. $-8 = 11 - 2z$
11. $-5 + 8a = -29$
12. $38 = -9b + 2$

▶ **Use proportions to solve problems.** *(Lesson 9-4)*

Solve each proportion.

13. $\frac{4}{9} = \frac{8}{3x}$

14. $\frac{5}{a} = \frac{2.5}{7.5}$

15. $\frac{4}{5} = \frac{8}{n}$

16. $\frac{11.7}{3} = \frac{r}{5}$

Focus on...

To be successful in geometry, you need to understand some new terms. Many of the terms are easy if you look carefully at the words. For example, the word **triangle** has the prefix *tri* which means three. The word **quadrilateral** has the prefix *quad* which means four. Triangle and quadrilateral are the names for three and four-sided figures, or **polygons**. You can determine the number of sides of other polygons by looking at the prefixes. In Lesson 11-8, you will learn the names of other polygons.

The Language of Geometry

What You'll Learn

- To identify points, lines, planes, rays, segments, angles, and parallel, perpendicular, and skew lines, and

- to classify angles as acute, right, or obtuse.

Why It's Important

Understanding angles can help air traffic controllers determine airline flight paths.

Application: Construction

Students in Todd Scholl's construction trades classes in Fairplay, Colorado, don't just study. They put their learning into practice by building real houses. In fact, they built the town's Silverheels Middle School, a four-classroom structure.

Before construction of a building can begin, the corners of the building site must be level. A laser interferometer can determine the heights of the corners of the site with great accuracy. A laser is visible because light is being reflected from particles in the air.

We can use ideas from the construction of buildings to model geometric ideas.

Learning the Concept

A point has no dimensions.

Each particle visible in the laser beam has a specific location in space and suggests the idea of a geometric **point**. A point is a specific location in space with no size or shape. A point can be represented by a dot and named with a capital letter.

A •
point A B
•
point B

A line has one dimension, length.

A laser beam travels in a straight path. In geometry, a straight path is called a **line**. A line is a collection of points that extends indefinitely in two directions. Arrowheads are used to show that a line has no endpoints. A line can be named with a single lowercase letter or by two points on the line.

line CD or $\overleftrightarrow{CD}$ or
line DC or $\overleftrightarrow{DC}$ or
line ℓ

THINK ABOUT IT

Can you measure the length of a line? Can you measure a line segment?

Each wooden beam used in the construction of a building site is a model of a **line segment**. A line segment is part of a line containing two endpoints and all points between the endpoints. A line segment is named by its endpoints.

segment EF or $\overline{EF}$
segment FE or $\overline{FE}$

The path of a laser beam as it begins at the generator and extends in one direction is a model of a **ray**. A ray is a portion of a line that extends from one point infinitely in *one* direction. A ray is named using the endpoint first and then any other point on the ray.

ray DF or $\overrightarrow{DF}$

Example **Name two points, a line, two rays, and a line segment in the figure at the right.**

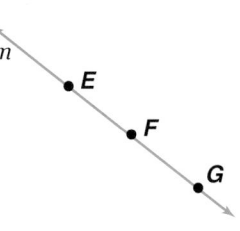

Two of the points are *E* and *F*.

We can use a single lowercase letter or any two of the labeled points to name the line. So we can name it *m*, $\overleftrightarrow{EF}$, $\overleftrightarrow{EG}$, or $\overleftrightarrow{FG}$.

Two of the rays are $\overrightarrow{FG}$ and $\overrightarrow{FE}$.

One of the line segments shown is $\overline{EG}$.

When two rays have the same endpoint, they form an **angle**. The common endpoint is called the **vertex**, and the rays are called **sides** of the angle. An angle is named using three points with the vertex as the middle letter. If no other angle shares the same vertex, the angle is named with just the vertex letter or with a numeral.

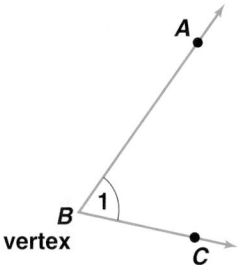

▶ angle *ABC* or ∠*ABC*, angle *CBA* or ∠*CBA*,

▶ angle *B* or ∠*B*,

▶ angle 1 or ∠1

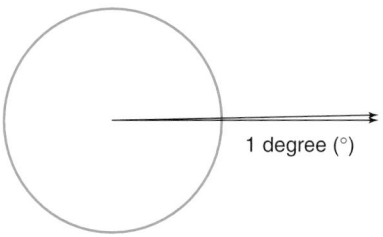

1 degree (°)

The most common unit of measure for angles is the **degree**. A circle can be separated into 360 arcs of the same length. An angle has a measurement of one degree if its vertex is at the center of the circle and its sides contain the endpoints of one of the 360 equal arcs.

You can use a **protractor** to measure angles.

Example **Use a protractor to measure ∠*PQR*.**

Place the protractor so the center mark is at *Q* (the vertex of ∠*PQR*) and the straightedge aligns with $\overrightarrow{QR}$ (one side of the angle). Use the scale that begins with 0 at $\overrightarrow{QR}$. Read where $\overrightarrow{QP}$ (the other side of the angle) crosses this scale. The measure of angle *PQR* is 135 degrees. Using symbols, *m*∠*PQR* = 135°.

You can also use a protractor to draw an angle of a given measure.

Draw an angle *Y* that measures 50°.

Use a straightedge to draw a ray.

Place the protractor with the center mark at *Y* and the 0-mark aligned with the ray.

Use the scale that begins with 0 at the ray, find the mark for 50, and draw a dot.

Use a straightedge to draw the other side of the angle. Angle *Y* measures 50°.

You can classify angles by their degree measure. **Acute angles** have measures greater than 0° and less than 90°. **Right angles** have measures of 90°. **Obtuse angles** have measures greater than 90° but less than 180°. **Straight angles** measure 180°.

Example ④

Real World APPLICATION

Baseball

According to Ted Williams, one of baseball's greatest hitters, the best way to strike a baseball is at a right angle. If you swing too early or too late, the ball will probably go foul. Classify each angle as *acute*, *right*, or *obtuse*.

a.
Incoming Path
Angle > 90°

b.
Angle = 90°

The small square indicates that ∠RST is a right angle.

c.
Angle < 90°

a. *m∠JKL* > 90°. So, ∠*JKL* is obtuse.   *The swing is too early.*
b. *m∠RST* = 90°. So, ∠*RST* is right.
c. *m∠DOT* < 90°. So, ∠*DOT* is acute.   *The swing is too late.*

A plane has two dimensions, length and width, but no thickness.

The level building site described in the application on page 548 is a model of part of a **plane**. A plane is a flat surface with no edges, or boundaries. A plane can be named by a single uppercase script letter or by using any three points of the plane. *The three points must not lie on the same line.*

plane *AMF*
plane *ℰ*

Lines that lie in the *same plane* either intersect or are parallel.

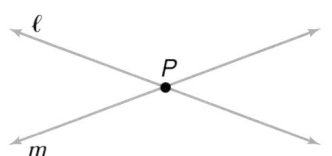

Lines ℓ and m **intersect** at point P.

There is no point of intersection of lines t and n. Lines t and n are **parallel**. Using symbols, $t \parallel n$.

Two lines that intersect to form a right angle are **perpendicular**. Rays and line segments can also be perpendicular. The symbol $\perp$ means *is perpendicular to.*

$m \perp n$

$\overline{RS} \perp \overline{CD}$

$\overrightarrow{AD} \perp \overrightarrow{AQ}$

If two lines do not intersect and are not in the same plane, they are called **skew lines**. The roof line along the front of a rectangular building and the line of the base of the side of the building are an example of skew lines.

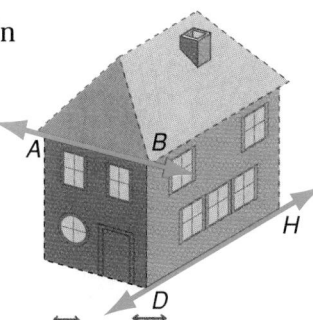

$\overleftrightarrow{AB}$ and $\overleftrightarrow{DH}$ are skew.

Example **5** **Identify each pair of line segments as *intersecting*, *parallel*, or *skew*.**

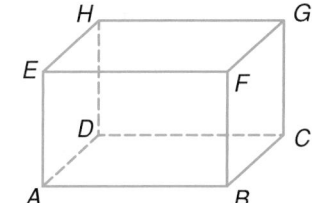

a. $\overline{EH}$ and $\overline{FG}$
These segments are in the same plane (EFG), and they do not intersect. So, $\overline{EH} \parallel \overline{FG}$.

b. $\overline{AE}$ and $\overline{FG}$
These segments do not intersect, and they are not in the same plane. So, $\overline{AE}$ and $\overline{FG}$ are skew.

c. $\overline{BC}$ and $\overline{CD}$
These segments are in the same plane (ABC), and they intersect at point C.

Checking Your Understanding

Communicating Mathematics

Read and study the lesson to answer these questions.

1. **Tell** how many endpoints each of the following has.

   a. a line   **b.** a line segment   **c.** a ray

2. **Compare and contrast** parallel lines and skew lines.

3. **Demonstrate** how a protractor is used to measure angles and how it is used to draw angles.

4. Draw and label a diagram to show a right angle whose sides are $\overrightarrow{QP}$ and $\overrightarrow{QR}$.

5. Explain the difference between an acute angle and an obtuse angle.

6. You can use gumdrops and toothpicks to model geometric concepts. A gumdrop represents a point, and a toothpick represents a portion of a line. The figure at the right shows a model of a line segment. Use gumdrops and toothpicks to model each of the following.

MATERIALS
toothpicks
gumdrops

    a. an angle   **b.** a ray   **c.** a line   **d.** intersecting lines

Guided Practice

Tell whether each model suggests a point, a line, a plane, a ray, or a line segment. Explain your answer.

  7. a wall   **8.** the tip of a needle

Use the figure at the right to name an example of each term.

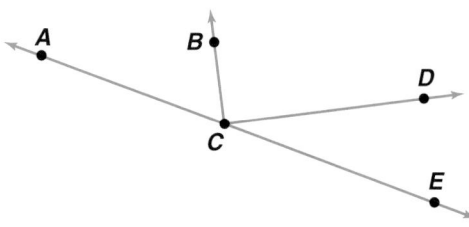

  9. point

10. line segment

11. ray

12. line

In the figure above, use a protractor to find the measure of each angle. Then classify the angle as *acute*, *right*, or *obtuse*.

13. $\angle ACB$ **14.** $\angle BCD$

Use a protractor to draw angles having the following measurements.

15. $78°$ **16.** $145°$

17. Photography Cameras are often mounted on tripods to give stability. Why do tripods give stability?

Exercises: Practicing and Applying the Concept

Independent Practice

Tell whether each model suggests a point, a line, a plane, a ray, or a line segment. Explain your answer.

18. a pencil   **19.** a CD-ROM

20. a telephone wire   **21.** an arrow's path

Draw and label a diagram to represent each of the following.

22. point T **23.** $\angle Q$

24. line ℓ **25.** $\overline{XY}$

26. plane RST **27.** $\overrightarrow{EF}$

28. $\angle BOY$ **29.** $\overleftrightarrow{AB}$

30. lines ℓ and m intersect   **31.** $\overleftrightarrow{AB} \parallel \overleftrightarrow{CD}$

32. $\overleftrightarrow{XY} \perp \overleftrightarrow{MN}$ **33.** lines a and b are skew

In the figure at the right, use a protractor to find the measure of each angle. Then classify the angle as *acute*, *right*, or *obtuse*.

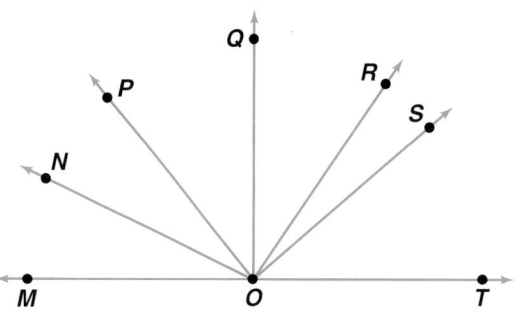

34. $m\angle MOQ$

35. $m\angle NOT$

36. $m\angle TOP$

37. $m\angle QON$

38. $m\angle POS$

39. $m\angle ROS$

Use a protractor to draw angles having the following measurements. Classify each angle as *acute*, *right*, or *obtuse*.

40. $112°$ **41.** $47°$ **42.** $95°$ **43.** $16°$ **44.** $162°$ **45.** $90°$

Critical Thinking

46. *True* or *false*. Explain your answers.

    a. A line is part of a line segment.

    b. The intersection of any two rays is always a point.

Applications and Problem Solving

47. Space Travel   The space shuttle approaches the runway at an angle of $18°$, six times as steep as that of a commercial jet. At what angle does a commercial jet make its approach? Draw a diagram showing both approach angles.

48. Aviation   Airplane flight paths can be described using angles and compass directions. The path of a particular airplane is described as $36°$ west of north. Draw a diagram that represents this path.

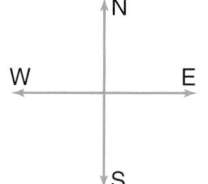

49. Patterns   At certain times of the day, the hands of a clock form a right angle. In twelve hours, how many times do the hands form a right angle?

Mixed Review

50. Probability   A die is tossed. Is P(even or prime) *mutually exclusive* or *inclusive*? Find the probability. (Lesson 10-10)

51. Replace the ● with = or ≠ to make $\frac{1.5}{2}$ ● $\frac{1.8}{2.4}$ true. (Lesson 9-4)

52. Health   What relationship (*positive*, *negative*, or *none*) do you think a scatter plot of number of students with colds and season of year (spring, summer, fall, winter) might show? Explain. (Lesson 8-2)

53. The sum of an even integer and the next greater even integer is more than 98. Find the least pair of such integers. (Lesson 7-7)

54. Solve $n - 6.2 \le 5.8$. Graph the solution on a number line. (Lesson 5-7)

For **Extra Practice**, see page 767.

55. A volleyball team has 6 members. Suppose each member shakes hands with every other member. How many handshakes take place? (Lesson 4-3)

56. Geometry   Find the perimeter and area of a rectangular floor that is 12 feet wide by 15 feet long. (Lesson 3-5)

57. Standardized Test Practice   Evaluate $\frac{d}{8}$ if $d = -96$. (Lesson 2-8)

    A. -12　　　　　**B.** 12　　　　　**C.** -88　　　　　**D.** 88

58. Solve $3d = 24$ mentally. (Lesson 1-6)

11-1B Constructions

An Extension of Lesson **11-1**

MATERIALS

 compass

ruler

protractor

When two line segments have the same length, we say they are **congruent**. Similarly, two angles with the same measure are congruent. You can use a compass and straightedge to **construct** congruent segments and angles. A **straightedge** is any object you can use to draw a straight line, such as an index card or a ruler. You use a **compass** to draw a circle or part of a circle.

Activity 1 **Construct a line segment congruent to a given line segment.**

First, draw a segment and label it $\overline{AB}$. This will be the given segment.

Step 1: Use a straightedge to draw $\overrightarrow{PS}$ so it is longer than $\overline{AB}$.

Step 2: Place the steel tip of the compass at A and the writing tip at B.

Step 3: Keep the same setting on the compass and place the steel tip at P. Draw an arc that intersects $\overrightarrow{PS}$ at Q. $\overline{PQ}$ is congruent to $\overline{AB}$. In symbols, $\overline{PQ} \cong \overline{AB}$.

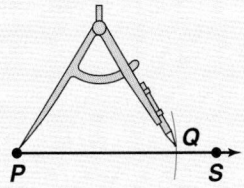

Activity 2 **Construct an angle that is congruent to a given angle.**

First, draw an angle and label it $\angle PQR$. This will be the given angle.

Step 1: Use a straightedge to draw $\overrightarrow{ST}$.

Step 2: Place the steel tip of the compass at Q. Draw an arc that intersects both sides of $\angle PQR$ to locate points X and Y.

Step 3: Keeping the same compass setting, place the steel tip at S and draw an arc that intersects $\overrightarrow{ST}$. Label the intersection point A.

Step 4: Place the compass so one tip is at X and the other is at Y. Keep that setting and place the steel tip at A. Draw an arc that intersects the arc you drew in step 3. Label the intersection point M.

Step 5: With a straightedge, draw $\overrightarrow{SM}$. $\angle MST$ is congruent to $\angle PQR$. In symbols, $\angle MST \cong \angle PQR$.

1. How does the compass work in copying a line segment?

2. Explain how the compass works in copying an angle.

Activity 3 **Construct a line segment that bisects a given line segment.**

Draw a segment and label it $\overline{CD}$. This is the given line segment.

Step 1: Place the steel tip of the compass at C. Extend the compass until the writing tip is more than halfway to D and draw an arc "above" and "below" $\overline{CD}$ as shown.

Step 2: With the same setting on the compass, place the steel tip at D and draw two arcs as shown to locate two new points, Q and R.

Step 3: With a straightedge, draw the line segment determined by these two new points. This segment will intersect $\overline{CD}$ at M.

TALK ABOUT IT

3. Use a ruler to measure $\overline{CM}$ and $\overline{MD}$. This construction *bisects* a segment. What do you think *bisects* means?

4. In this construction, $\overline{QR}$ is perpendicular to $\overline{CD}$. What type of angles are formed by perpendicular segments?

Activity 4 **Construct a ray that bisects a given angle.**

Draw an angle and label it $\angle ABC$. This is the given angle.

Step 1: Place the steel tip of the compass at B and draw an arc that intersects both sides of the angle. Label the points of intersection X and Y.

Step 2: Place the steel tip of the compass at X and open the compass to a length more than halfway to Y. Draw an arc in the interior of $\angle ABC$.

Step 3: Keeping the same compass setting, place the steel tip at Y and draw an arc that intersects the arc you drew in Step 2. Label the point of intersection T.

Step 4: With a straightedge, draw $\overrightarrow{BT}$. $\overrightarrow{BT}$ is the bisector of $\angle ABC$.

TALK ABOUT IT

5. Use a protractor to measure $\angle ABT$ and $\angle TBC$. What is true about their measures?

6. Why do we say BT is the bisector of $\angle ABC$?

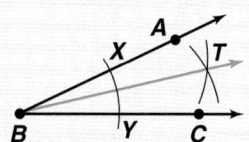

Math Lab 11-1B Constructions **555**

Integration: Statistics
Making Circle Graphs

You'll Learn

■ To make a circle graph to illustrate data.

It's Important

Newspapers often use circle graphs to show results of an opinion survey.

Application: Weather

Have you ever been caught in a thunderstorm and wondered where the safest place to be is? The chart at the right indicates where lightning strikes most often.

A circle graph can be used to display this data. What angles would you use for each section?

Where Lightning Strikes

Fields, Ballparks	28%
Under trees	17%
Bodies of Water	13%
Near Heavy Equipment	6%
Golf Courses	4%
Telephone Poles	1%
Other / Unknown	31%

Source: National Weather Service

Learning the Concept

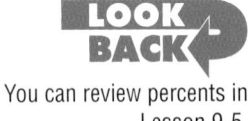

You can review percents in Lesson 9-5.

A **circle graph** is used to compare parts of a whole in a way that helps you visualize the information. As you know, there are 360° in a circle. So, you can multiply to find the number of degrees in each section of the circle graph.

THINK ABOUT IT

Check the sum of the degree measures. What should it be?

Fields/Ballparks   28 % × 360 = 100.8

Under Trees   17 % × 360 = 61.2

Bodies of Water   13 % × 360 = 46.8

Near Heavy Equipment   6 % × 360 = 21.6

Golf Courses   4 % × 360 = 14.4

Telephone Poles   1 % × 360 = 3.6

Other/Unknown   31 % × 360 = 111.6

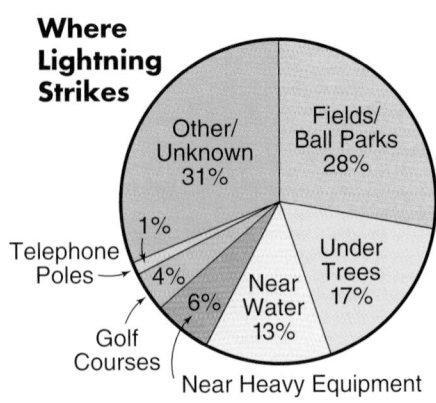

Example 1 shows how to make a circle graph for data that has not been expressed as a percent.

You can use spreadsheets and word processing programs to construct circle graphs from data you enter.

To be president of the United States, an individual must be at least 35 years old. Make a circle graph to display the data on presidents' ages at their first inauguration shown at the right.

President's Age on Inauguration Day	
Age Interval	Number of Presidents
35–43	2
44–52	14
53–61	19
62–70	6

Explore   You know the number of presidents in each age interval.

Plan   Find the total number of presidents. Find the ratio of the number in each interval to the total. Then find the number of degrees for each interval. Finally, draw the graph.

Solve   **Step 1:**   Find the total number of presidents.
$2 + 14 + 19 + 6 = 41$

Step 2:   Find the ratio that compares the number in each interval to the total.

35–43: 2 ÷ 41 = *0.048780*
44–52: 14 ÷ 41 = *0.341463*
53–61: 19 ÷ 41 = *0.463415*
62–70: 6 ÷ 41 = *0.146341*

Step 3:   Multiply to find the number of degrees for each section of the graph. Round to the nearest degree.

35–43: 0.048780 × 360 = *17.5608* → $18°$
44–52: 0.341463 × 360 = *122.92668* → $123°$
53–61: 0.463415 × 360 = *166.8294* → $167°$
62–70: 0.146341 × 360 = *52.68276* → $53°$

Step 4:   Draw a circle graph. Use a compass to draw a circle.

Start with the least number of degrees, in this case 18°. Use a protractor to draw an angle of 18°.

Repeat for the remaining sections. Label each section and give the graph a title.

Presidents' Age at Inauguration

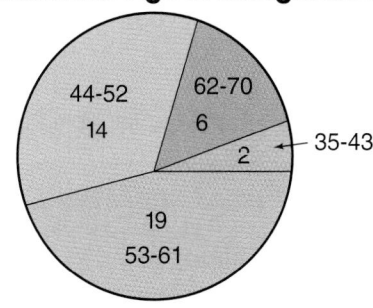

Examine   Add the degree measures of each section. The total is 361°. With rounding, this is reasonable.

7551-6"We wish peace because it is right"—President Roosevelt delivering Inaugural address Washington. Copyright 1905 by Underwood & Underwood

Communicating Mathematics

Read and study the lesson to answer these questions.

1. **Explain** how a circle graph differs from a bar graph and from a line graph. When is it appropriate to display data in a circle graph?

2. **Tell** the steps to make a circle graph if your data is in percent form.

3. **Explain** how to find the circle graph angles for data that is not in percent form.

4. Find a circle graph in a newspaper or magazine and make a copy to put in your journal. Write a short paragraph that explains the data displayed by the graph.

Guided Practice

Use the circle graph at the right to answer each question.

5. Which element makes up approximately $\frac{1}{4}$ of the earth's crust?

6. How much of Earth's crust is composed of oxygen and silicon?

Elements of Earth's Crust

Other 12%
Aluminum
Silicon 28%
8%
Iron 5%
Oxygen 47%

7. The table below shows the Vera family's monthly budget.

Monthly Budget			
Category	**Amount ($)**	**Percent**	**Angle (°)**
Housing	640		
Food	400		
Transportation	450		
Insurance	140		
Savings	90		
Misc.	360		
Totals			

a. What total would you expect in the percent column?

b. What total would you expect in the degree column?

c. Copy and complete the table. Round degrees to the nearest tenth.

d. Make a circle graph to display the data. Use spreadsheet or word processing software if it is available.

8. **Probability** What is the probability of tossing a head when you flip a fair coin? What would a circle graph for all the possible results of this experiment be like?

Independent Practice

9. Count txhe months of the year that are in each of these three categories: 28 or 29 days long, 30 days long, and 31 days long. Make a circle graph to display your findings.

10. In a recent poll, students were asked to name their favorite season of the year. The results of the survey are shown at the right. Make a circle graph to display the data. Use spreadsheet or word processing software if it is available.

Season Preference	
Summer	27%
Autumn	22%
Winter	18%
Spring	33%

11. Probability Three red marbles and 5 green marbles are placed in a bag. In an experiment, you are to select one marble. If the marble you select is red, you will win a silver dollar. If the marble you select is green, you will lose.

a. What is the probability that you will win?

b. Make a circle graph that shows the probability of winning and losing.

Critical Thinking

12. Suppose someone drew the graph at the right to display information on the audience share during a Network A show on a Thursday in October.

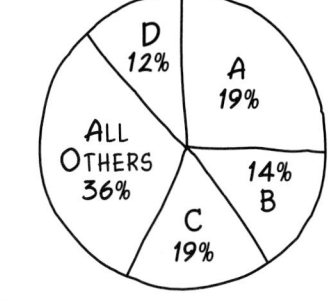

Networks

a. Does the graph accurately represent the data it presents? Why or why not?

b. Why might someone have drawn the graph this way?

Applications and Problem Solving

Make a circle graph to display each set of data.

13. Geography

Areas (in square miles) of the Oceans of the World	
Pacific Ocean	64,186,300
Atlantic Ocean	33,420,000
Indian Ocean	28,350,500
Arctic Ocean	5,105,700

14. Biology

Chemical Composition of the Human Body

Oxygen	65%
Carbon	18%
Hydrogen	10%
Nitrogen	3%
Other	4%

15. Landscaping A seed company prepares a grass seed mixture that is designed specifically for the Black Hills of South Dakota where the growing season is very short. The following mixture is used.

Black Hills Reclamation Mix	
30% VNS Kentucky Bluegrass	25% VNS Timothy
10% Regar Bromegrass	10% Alsike Clover
10% Lincoln Smooth Brome	10% VNS Tall Fescue
5% Medium Red Clover	

a. Make a circle graph to display the data.

b. Find the amount of each seed that would be needed to make 250 pounds of the mixture.

16. Family Activity Choose a set of data to collect, such as colors of cars in a parking lot, or the number of commercials in TV commercial breaks and record the data in a frequency table.

a. Make a circle graph and a bar graph to display the data.

b. Which graph better represents the data and why?

17. Simple interest, which you studied in Lesson 9-9, is paid only on the initial principal of a savings account or a loan. **Compound interest** is paid on the initial principal and on interest earned in the past.

Sample Find the value of a $1000 savings account after one and one-half years if the account pays 5% interest compounded semiannually. 5% interest compounded semiannually means that interest is paid twice a year, or every 6 months. One and one-half years equal three 6-month periods. The interest rate every 6 months is 5% ÷ 2 or 2.5%. Expressed as a decimal, 2.5% = 0.025.

6-month Period	Principal at Beginning of 6-month Period	Interest Earned During 6-month Period	Principal at End of 6-month Period
1st 6 Months	$1000	$1000 · 0.025 = $25	$1000 + $25 = $1025
2nd 6 Months	$1025	$1025 · 0.025 = $25.63	$1025 + $25.63 = $1050.63
3rd 6 Months	$1050.63	$1050.63 · 0.025 = $26.27	$1050.63 + $26.27 = $1076.90

The value of the savings account after one and one-half years is $1076.90.

a. For the Sample problem above, write an expression for the amount of simple interest that would be paid in one and one-half years. Then, find the amount of interest. How does the simple interest compare to the compound interest?

b. Find the value of a $10,000 certificate of deposit (CD) after one year if the CD pays 6% interest compounded quarterly.

c. What is the value of a $1200 savings account after 3 months if the account pays 4% interest compounded monthly? Compare this to the simple interest earned on the same account for the same period of time.

d. Make a conjecture about how the amount of interest earned changes if the compounding occurs more frequently.

e. Mrs. Blaylock borrowed $1000 at 10% interest compounded every six months. How much interest will she owe after 3 years? Compare this to how much she would owe with 10% simple interest.

Mixed Review

18. Standardized Test Practice   The phone company assigns the first three digits of a phone number based on the geographic area of the home or business. The remaining four digits are selected from a database. If any digit from 0 to 9 can be used in any position, how many different combinations of the last four digits are possible? (Lesson 10-6)

A. 100   **B.** 1000   **C.** 10,000   **D.** 100,000

19. Graph $y = \frac{1}{2}x - 4$ by using the slope and y-intercept. (Lesson 8-7)

20. The distance from Kuri's house to school is about 1800 feet. About what part of a mile is this? (Lesson 6-5)

For **Extra Practice**, see page 767.

21. Write a problem that could be solved using the equation $3x = 25$. (Lesson 1-8)

11-3

Angle Relationships and Parallel Lines

What You'll Learn

■ To identify the relationships of vertical, adjacent, complementary, supplementary angles, and angles formed by parallel lines and a transversal.

Why It's Important

When you study Earth's rotation and revolution you will use angles.

Notice that the writing lines on notebook paper are parallel lines. You can use these lines to investigate angles and parallel lines.

Your Turn

▶ Use the lines on notebook paper to draw two parallel lines.

▶ Then draw another line that intersects the parallel lines.

▶ Use a protractor to measure all the angles formed when this line crosses the parallel lines. Make a sketch that includes the measure of each angle.

▶ Repeat this activity using a different line.

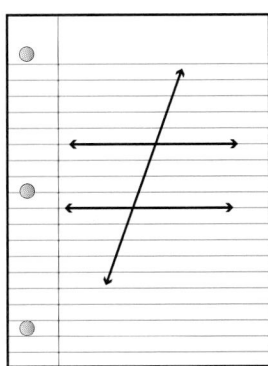

TALK ABOUT IT

a. Not including straight angles, how many angles are formed when a line intersects two parallel lines?

b. What do you notice about the measures of the angles?

c. Describe the pairs of angles that appear to have the same measure.

d. What do you notice about measures of angles that are side by side?

Learning the Concept

In the activity above, you discovered some of the special relationships that *pairs* of angles can have.

When two lines intersect, they form two pairs of "opposite" angles called **vertical angles**. In the figure at the left, angles 1 and 3 are vertical angles as are angles 2 and 4. If you use a protractor to measure the angles, you will discover that the angles in each pair have the same measure. Angles with the same measure are **congruent**. Vertical angles are always congruent.

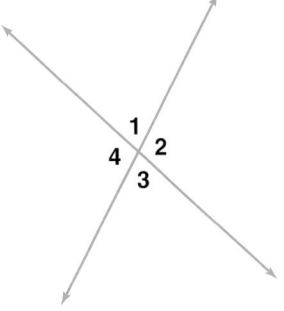

When two angles in a plane have the same vertex, share a common side, and do not overlap, they are called **adjacent angles**. In the figure at the right, ∠1 and ∠2 are adjacent angles.

m∠AOB means the degree measure of angle AOB.

We can find the measure of ∠AOB by adding the measures of ∠1 and ∠2. That is, $m\angle AOB = m\angle 1 + m\angle 2$.

If the sum of the measures of two angles is 90°, the angles are **complementary**.

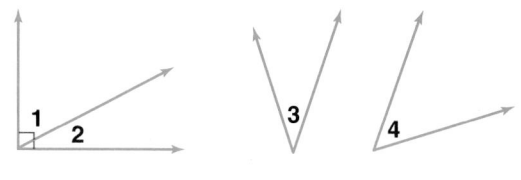

$m\angle 1 + m\angle 2 = 90°$ $m\angle 3 + m\angle 4 = 90°$

If the sum of the measures of two angles is 180°, they are **supplementary**.

$m\angle 1 + m\angle 2 = 180°$ $m\angle 3 + m\angle 4 = 180°$

Example ① **Use angles A, B, and C to solve each problem.**
a. Angles A and B are complementary. If $m\angle A = 70°$, find $m\angle B$.

$$m\angle A + m\angle B = 90 \quad \textit{Definition of complementary angles}$$
$$70 + m\angle B = 90 \quad \textit{Substitute.}$$
$$m\angle B = 20 \quad \textit{Subtract 70 from each side.}$$

The measure of $\angle B$ is 20°.

b. Angles A and C are supplementary. Find $m\angle C$.

$$m\angle A + m\angle C = 180 \quad \textit{Definition of supplementary angles}$$
$$70 + m\angle C = 180 \quad \textit{Replace } m\angle A \textit{ with 70.}$$
$$m\angle C = 110 \quad \textit{Subtract 70 from each side.}$$

The measure of $\angle C$ is 110°.

Connection to ▶ Algebra

Geometric relationships can be expressed using the tools of algebra.

Example ② **Angles PQR and RQT are complementary. If $m\angle PQR = x + 5$ and $m\angle RQT = x - 9$, find the measure of each angle.**

$$m\angle PQR + m\angle RQT = 90 \quad \textit{Definition of complementary angles}$$
$$(x + 5) + (x - 9) = 90 \quad \textit{Substitute.}$$
$$2x - 4 = 90 \quad \textit{Combine like terms.}$$
$$2x = 94 \quad \textit{Add 4 to each side.}$$
$$x = 47 \quad \textit{Divide each side by 2.}$$

Now replace x with 47 in the expression for each angle.

$$m\angle PQR = x + 5 \qquad\qquad m\angle RQT = x - 9$$
$$= 47 + 5 \text{ or } 52 \qquad\qquad = 47 - 9 \text{ or } 38$$

The measure of $\angle PQR$ is 52°, and the measure of $\angle RQT$ is 38°.

Check: $52° + 38° = 90°$ The angles are complementary.

THINK ABOUT IT

Why is 47° not the measure of one angle?

THINK ABOUT IT

Why do you think these angles are referred to as interior and exterior angles?

As you discovered in the modeling activity at the beginning of the lesson, when two parallel lines are intersected by a third line, called a **transversal**, eight angles are formed. Four are *interior* angles and four are *exterior* angles.

Interior angles: $\angle 3, \angle 4, \angle 5, \angle 6$
Exterior angles: $\angle 1, \angle 2, \angle 7, \angle 8$

When we study the relationship between the different angles, we can pair the angles as follows.

▶ **alternate interior angles:** $\angle 4$ and $\angle 6$, $\angle 3$ and $\angle 5$
Alternate interior angles are nonadjacent interior angles found on opposite sides of the transversal.

▶ **alternate exterior angles:** $\angle 1$ and $\angle 7$, $\angle 2$ and $\angle 8$
Alternate exterior angles are nonadjacent exterior angles found on opposite sides of the transversal.

▶ **corresponding angles:** $\angle 1$ and $\angle 5$, $\angle 2$ and $\angle 6$, $\angle 3$ and $\angle 7$, $\angle 4$ and $\angle 8$
Corresponding angles are angles that have the same position on two different parallel lines cut by a transversal.

In the activity at the beginning of the lesson, you may have discovered the following relationships.

Parallel Lines Cut by a Transversal	Corresponding angles are congruent. Alternate interior angles are congruent. Alternate exterior angles are congruent.

Example ③ A transversal crosses two parallel lines so that $m\angle 4 = 75°$.
  a. Which other angles also have a measure of 75°?
  b. Find the measures of the other angles.

a. Since $\angle 2$ and $\angle 4$ are vertical angles, they are congruent. So $m\angle 2 = 75°$.
Since $\angle 4$ and $\angle 6$ are alternate interior angles, they are congruent. So, $m\angle 6 = 75°$.
$\angle 4$ and $\angle 8$ are corresponding angles so they are congruent. Thus, $m\angle 8 = 75°$.
Angles 2, 4, 6, and 8 have a measure of 75°.

b. Both angles 1 and 3 are supplementary to $\angle 4$ so the sums of each of their measures and the measure of angle 4 is 180°.
$180 - 75 = 105$. So, $m\angle 1 = 105°$ and $m\angle 3 = 105°$.
Since $\angle 7$ and $\angle 1$ are alternate exterior angles, they are congruent. So, $m\angle 7 = 105°$.
$\angle 5$ and $\angle 1$ are corresponding angles so they are congruent. Thus, $m\angle 5 = 105°$.

Angles 1, 3, 5, and 7 have a measure of 105°.

Real World
APPLICATION

Carpentry

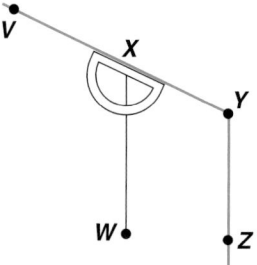

A carpenter uses a protractor and a plumb line (a string with a weight attached) to measure the angle between a slanted ceiling and the wall. If $m\angle WXY = 65°$, find $m\angle XYZ$.

$\angle WXY$ is supplementary to $\angle WXV$.

$m\angle WXY + m\angle WXV = 180$

$65 + m\angle WXV = 180$

$m\angle WXV = 115$

The plumb line and the wall are parallel lines. So, $\angle XYZ$ is congruent to $\angle WXV$. *Corresponding angles are congruent.*

$m\angle XYZ = m\angle WXV$

$m\angle XYZ = 115$

The measure of $\angle XYZ$ is $115°$.

Checking Your Understanding

Communicating Mathematics

Read and study the lesson to answer these questions.

1. **Compare and contrast** vertical angles and adjacent angles.

2. **Draw** $\overline{MN} \parallel \overline{PQ}$ and transversal $\overline{RS}$. Number the angles and name a pair of each.

   a. corresponding angles   **b.** alternate interior angles

   c. alternate exterior angles   **d.** adjacent angles

   e. adjacent, supplementary angles

Draw and label a diagram to show each pair of angles and describe their characteristics.

3. $\angle WXY$ and $\angle YXZ$ are adjacent angles.

4. $\angle GHI$ and $\angle IHJ$ are adjacent, complementary angles.

5. $\angle KLM$ and $\angle MLN$ are adjacent, supplementary angles.

Guided Practice

6. Identify the figure in which angles 1 and 2 are supplementary.

   a.   1   2   **b.**   1 / 2   **c.**   1 / 2   **d.**   1 / 2

Find the value of *x* in each figure.

7.

8.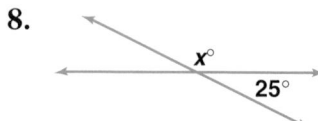

9. Angle C and $\angle D$ are complementary. Find $m\angle C$ if $m\angle D$ is $75°$.

10. Angle P and $\angle Q$ are supplementary. Find $m\angle P$ if $m\angle Q$ is $90°$.

11. Angles X and Y are supplementary. $m\angle X = 2x$ and $m\angle Y = 4x$. Find the measure of each angle.

12. Angles F and G are complementary. $m\angle F = x + 8$ and $m\angle G = x - 10$. Find the measure of each angle.

Refer to the diagram at the right to complete Exercises 13–15.

13. Name the interior angles.

14. Name the exterior angles.

15. **a.** Use a protractor to find the measure of one angle.

   b. Use the measure from **part a** to find all the other angle measures.

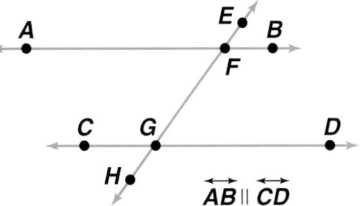

16. **Public Transit** The angle at the corner where two streets intersect is 125°. If a bus cannot make a turn at an angle of less than 70°, can bus service be provided on a route that includes turning that corner in both directions? Explain.

Exercises: Practicing and Applying the Concept

Independent Practice

Find the value of x in each figure.

17.

18.

19.

20.

21.

22.

Find the measure of each angle.

23.

24.

25.

In the figure at the right, $\ell \parallel m$. If the measure of $\angle 1$ is 47°, find the measure of each angle.

26. $\angle 2$ 27. $\angle 3$

28. $\angle 5$ 29. $\angle 4$

30. $\angle 7$ 31. $\angle 8$

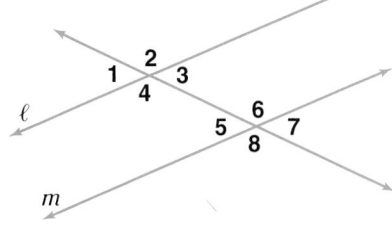

Each pair of angles is complementary. Find each angle measure.

32. $m\angle A = 2x + 15, m\angle B = x - 3$ 33. $m\angle R = x - 9, m\angle S = x + 17$

Each pair of angles is supplementary. Find the measure of each angle.

34. $m\angle T = 2x + 17$ and $m\angle S = 5x - 40$

35. $m\angle F = 3x + 40$ and $m\angle G = 2x + 10$

In the figure at the right, $\ell \parallel m$. Find the value of x for each of the following. *The figure is not drawn to scale.*

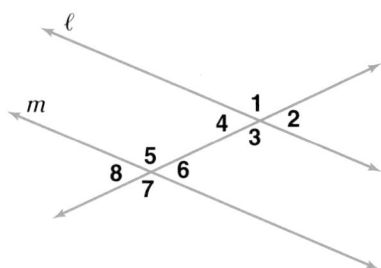

36. $m\angle 1 = 6x - 20$ and $m\angle 2 = x + 40$

37. $m\angle 4 = 3x + 17$ and $m\angle 5 = 2x + 13$

38. $m\angle 6 = 9x - 10$ and $m\angle 7 = 5x + 36$

39. The measure of the supplement of an angle is 15° less than four times the measure of the complement. Find the measure of the angle.

Critical Thinking

Draw a diagram for each situation. Use a protractor to measure the angles formed and answer each question.

40. If two parallel lines are cut by a transversal, how are the interior angles on the same side of the transversal related? Explain.

41. If a transversal is perpendicular to one of two parallel lines, is it perpendicular to the other parallel line also? Explain.

Applications and Problem Solving

42. Physics The drawing at the right shows a beam of light reflected from the surface of a flat mirror. The angle of incidence (the angle between the incident ray and the normal line) and the angle of reflection (the angle between the reflected ray and the normal line) are congruent. The normal line is perpendicular to the surface of the mirror. Suppose a beam of light strikes the mirror at a 52° angle to the mirror. What is the measure of the angle between the incident ray and the reflected ray?

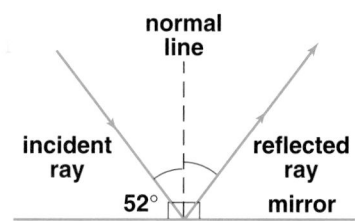

43. Carpentry A standard stair rail is designed to make an angle of 45° with the floor. The first vertical post and the lower rail also form an angle of 45°. If the upper rail is parallel to the lower rail, what angle should it make with the first vertical post?

Mixed Review

44. Retail Sales 17% of Foodtown's sales come from the produce department. Find the measure of the angle to represent produce sales for a circle graph showing Foodtown's sales. (Lesson 11-2)

45. Probability Two dice are thrown. Find $P(\text{sum} > 8)$. (Lesson 9-3)

46. Biology A certain bacteria doubles its population every 12 hours. After 3 full days, there are 1600 bacteria. How many bacteria were there at the beginning of the first day? (Lesson 7-1)

47. Write 56,780 in scientific notation. (Lesson 6-9)

48. Write $\frac{12}{54}$ in simplest form. (Lesson 4-6)

49. Simplify $9b + (-16)b$. (Lesson 2-4)

50. Standardized Test Practice Evaluate $|-5| - |2|$. (Lesson 2-1)
   A. −7   **B.** −3   **C.** 3   **D.** 7

51. Solve $8 + n = 13$ mentally. (Lesson 1-6)

For **Extra Practice**, see page 767.

11-3B Slopes of Parallel Lines

An Extension of Lesson **11-3**

MATERIALS

graph paper

You can determine whether two lines are parallel by finding their slopes. In Lesson 8-6, you learned that parallel lines have the same slope. So, if two lines have the same slope, then they are parallel. The graphing calculator program at the right finds the slope of a line when you enter the coordinates of two points on the line.

```
PROGRAM:SLOPE
: Disp "ENTER"
: Disp "COORDINATES"
: Disp "POINT 1"
: Input A
: Input B
: Disp "POINT 2"
: Input C
: Input D
: If A = C
: Then
: Disp "SLOPE UNDEFINED"
: Stop
: End
: (D-B)/(C-A) → M
: Disp "SLOPE = ", M
```

Your Turn

Work with a partner.

▶ Enter the program in a graphing calculator.

▶ Find the coordinates of two points on line *m* below, Point 1 and Point 2.

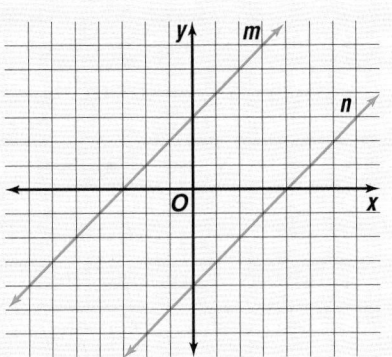

▶ Enter the coordinates in the program. Record the slope of line *m*.

▶ Now find the coordinates of two points on line *n* above.

▶ Enter the coordinates in the program. Record the slope of line *n*.

1. How does the program calculate the slope of the line?

2. How do the slopes of lines *m* and *n* compare?

3. Based on their slopes, are lines *m* and *n* parallel?

Your Turn

▶ Now draw two different lines that you think are parallel on graph paper.

▶ Find the coordinates of two points on each line.

▶ Enter the coordinates in the program to find the slope of each line.

▶ Repeat the steps above for two lines that you think are *not* parallel.

TALK ABOUT IT

4. What were the results for the lines that you drew?

5. Suppose you wanted to draw two lines that were parallel. How could you draw the lines to be sure that they were parallel?

Triangles

You'll Learn

- To find the missing angle measure of a triangle, and

- to classify triangles by angles and sides.

It's Important

Architects and civil engineers use triangles in building designs.

CAREER CHOICES

Civil engineers use geometry to make beautiful and functional designs. For more information on this field, contact:

American Society of Civil Engineers
345 East 47th Street
New York, NY 10017

Application: Engineering

Civil engineers who design and build railroad bridges often choose a *truss* bridge to span canyons and rivers. A truss bridge can span more than 1000 feet and requires less building material than other types of bridges.

Learning the Concept

Trusses get their strength from braces that form triangles. Because triangles are rigid, they are strong. A **triangle** is formed by three line segments that intersect only at their endpoints. A triangle can be named by its vertices. The triangle shown below is triangle *PQR*, or △*PQR*.

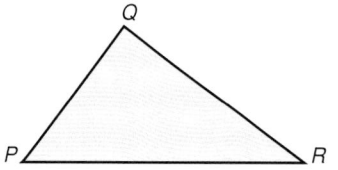

The vertices are P, Q, and R.
The sides are $\overline{PQ}$, $\overline{QR}$, and $\overline{PR}$.
The angles are ∠P, ∠Q, and ∠R.

There is a special relationship among the angles of a triangle.

1. Use a straightedge to draw a triangle on a piece of paper. Cut out the triangle and label the vertices *A*, *B*, and *C*.

2. Fold the triangle so that the point *C* lies on $\overline{AB}$ and the fold is parallel to $\overline{AB}$. Label ∠*C* as ∠2.

3. Fold again so point *A* meets the vertex of ∠2. Label ∠*A* as ∠1.

4. Finally, fold so point *B* also meets the vertex of ∠2. Label ∠*B* as ∠3.

Step 1

Step 2

Step 3

Step 4

What do you notice about the sum of the measures of angles 1, 2, and 3? This activity suggests the following relationship.

Angles of a Triangle	The sum of the measures of the angles of a triangle is 180°.

Example In $\triangle CAT$, the measure of $\angle C$ is 47°, and the measure of $\angle A$ is 59°. Find the measure of $\angle T$.

> *Estimate:* $50 + 60 = 110$ and $180 - 110 = 70$, so $m\angle T$ is about 70°.
>
> $m\angle C + m\angle A + m\angle T = 180$ *The sum of the measures of the angles of a triangle is 180°.*
>
> $47 + 59 + m\angle T = 180$ *Replace $m\angle C$ with 47 and $m\angle A$ with 59.*
>
> $106 + m\angle T = 180$ *Add $47 + 59$.*
>
> $m\angle T = 74$ *Subtract 106 from each side.*

The measure of $\angle T$ is 74°.

Connection to > Algebra

The relationships of the angles in a triangle can be represented using algebra.

INTEGRATION

Algebra

Example The measures of the angles of a certain triangle are in the ratio 2:3:4. Find the measure of each angle.

Let $2x$ represent the measure of one angle, $3x$ the measure of a second angle, and $4x$ the measure of the third angle.

> $2x + 3x + 4x = 180$ *The sum of the measures is 180°.*
>
> $9x = 180$ *Combine like terms.*
>
> $x = 20$ *Divide each side by 9.*
>
> $2x = 40, 3x = 60,$ and $4x = 80.$

The measures of the angles are 40°, 60°, and 80°.   *Check by adding.*

A triangle can be classified by its angles. Every triangle has two acute angles. The third angle can be used to classify the triangle.

Acute	Right	Obtuse
		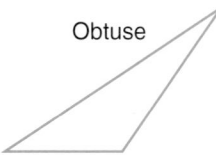
An **acute triangle** has three acute angles.	A **right triangle** has one right angle.	An **obtuse triangle** has one obtuse angle.

Triangles can also be classified by the number of congruent sides.

Equilateral	Isosceles	Scalene
		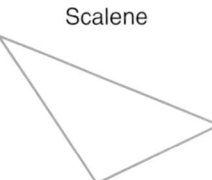
An **equilateral triangle** has three congruent sides.	An **isosceles triangle** has at least two congruent sides.	A **scalene triangle** has no congruent sides.

Communicating Mathematics

Read and study the lesson to answer these questions.

1. If you know the measures of two of the angles in a triangle, how can you find the measure of the third angle?

An A-frame house is shaped like an isosceles triangle. Give a real-world example of the following triangles.

2. obtuse　　　　3. equilateral　　　　4. right

5. Make up memory tools you can use to remember the names of triangles that are classified by sides or angles. Record the tool in your journal.

Guided Practice

Find the value of *x*. Then classify each triangle as *acute*, *right*, or *obtuse*.

6.

7.
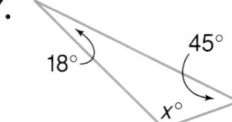

8. **Algebra**　The measures of the angles of a certain triangle are in the ratio 1:3:5. Find the measures of each angle.

Use a ruler to determine the number of congruent sides in each triangle. Then classify the triangle as *scalene*, *isosceles*, or *equilateral*.

9.

10.

11. **Construction**　Find the measure of the missing angle, *x*°, in the roof truss shown at the right.

Independent Practice

Find the value of *x*. Then classify each triangle as *acute*, *right*, or *obtuse*.

12.

13.

14.

Choose

Estimation
Mental Math
Calculator
Paper and Pencil

15. The measures of the angles of a triangle are in the ratio 2:3:5. Find the measure of each angle.

16. The measures of the angles of a triangle are in the ratio 1:4:7. Find the measure of each angle.

Use a ruler to determine the number of congruent sides in each triangle. Then classify the triangle as *scalene*, *isosceles*, or *equilateral*.

17.

18.

19.

Use a protractor to determine the measures of the angles in each triangle. Then classify the triangle as *acute*, *right*, or *obtuse*.

20.

21.

22.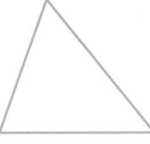

Use the figure at the right to solve each of the following.

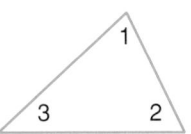

23. Find $m\angle 1$ if $m\angle 2 = 40°$ and $m\angle 3 = 55°$.

24. Find $m\angle 1$ if $m\angle 2 = 60°$ and $m\angle 3 = 60°$.

25. Find $m\angle 1$ if $m\angle 2 = 81°$ and $m\angle 3 = 74°$.

26. Find $m\angle 2$ if $m\angle 1 = 45°$ and $m\angle 3 = 75°$.

27. Find $m\angle 2$ if $m\angle 1 = 47°$ and $m\angle 3 = 48°$.

Find the measures of the angles in each triangle.

28.

29.

30.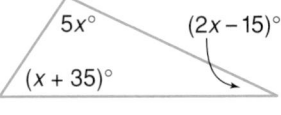

Critical Thinking

31. Explain how you could use a piece of string to convince someone that a given triangle was:

    a. scalene   **b.** isosceles   **c.** equilateral

32. Are the acute angles of a right triangle complementary? Explain.

Applications and Problem Solving

33. Civil Defense The Civil Defense program is a civilian organization created to help people when a disaster occurs. The Civil Defense symbol is shown at the right. If the triangle pictured in the symbol is an equilateral triangle, and all of the angles are congruent, what is the measure of each angle?

34. Number Theory Numbers that can be represented by a triangular arrangement of dots are called *triangular numbers*. The first three triangular numbers are 1, 3, and 6. Find the next four triangular numbers and draw the triangular arrangement of dots for each one.

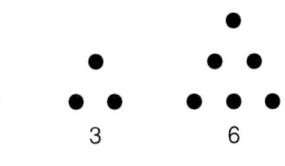

35. Public Utilities   Above-ground electric wires and transformers, are mounted on utility poles. A support cable, called a guy wire, is sometimes attached to give the pole stability. If the guy wire makes an angle of 65° with the ground, what is the measure of the angle formed by the guy wire and the pole? Assume that the ground is level.

Mixed Review

36. Angles X and Y are supplementary. If $m\angle X = 35°$, find $m\angle Y$. (Lesson 11-3)

37. Probability   Find the odds of rolling a sum less than 6 if a pair of dice are rolled. (Lesson 10-7)

38. Graph $y > x + 3$. (Lesson 8-9)

39. Sewing   A curtain pattern says to measure the length of the window and add 4 inches for curtain length. Then it says to add 4 inches for a heading, $2\frac{1}{2}$ inches for a rod casting, and 6 inches for a hem. Mary's window is $42\frac{3}{4}$ inches long. How long should the fabric be? (Lesson 5-5)

For **Extra Practice**, see page 768.

40. Replace each ● with $<$, $>$, or $=$. (Lesson 2-3)

    a. -8 ● -14 **b.** $-|20|$ ● -20 **c.** 0 ● $|-2|$

41. Standardized Test Practice   Simplify $4(x + 3)$. (Lesson 1-5)

    A. $4x + 7$ **B.** $x + 12$ **C.** $4x + 7$ **D.** $4x + 12$

Self Test

Use a protractor to draw an angle with the given measurements. Classify each angle as *acute*, *right*, or *obtuse*. (Lesson 11-1)

  1. 60°   **2.** 115°   **3.** 20°

4. Retail   At Recordtown, 42% of sales are from CDs, 28% are from tapes, 18% are from equipment, and 12% are from other items. Make a circle graph of this data. (Lesson 11-2)

In the figure at the right, ℓ is parallel to m. If the measure of $\angle 3$ is 34°, find the measure of each angle. (Lesson 11-3)

  5. $\angle 1$ **6.** $\angle 4$ **7.** $\angle 5$

Find the measure of each angle using the following information. (Lesson 11-3)

  8. $\angle A$ and $\angle B$ are complementary.   **9.** $\angle F$ and $\angle G$ are supplementary.
     $m\angle A = x + 25$; $m\angle B = 2x - 10$ $m\angle F = 3x - 50$; $m\angle G = 2x - 20$

Find the value of x. Then classify each triangle as *acute*, *right*, or *obtuse*. (Lesson 11-4)

  10.   **11.**   **12.**

11-5 Congruent Triangles

Technology Activity

You can use a graphing calculator to plot points and draw line segments connecting them. By plotting three points and the connecting segments, you can graph triangles.

You will use the line feature from the draw menu. When you enter two points, the calculator graphs the line segment between them.

Your Turn

Graph $\triangle XYZ$ with vertices $X(10, 5)$, $Y(30, 5)$, $Z(5, 25)$ and $\triangle X'Y'Z'$ with vertices $X'(-15, 2)$, $Y'(5, 2)$, and $Z'(-20, 22)$ using the steps below.

▶ Press ZOOM 6 ZOOM 8 ENTER. *Sets the screen to integer values.*

▶ Press 2nd DRAW 2.

▶ Use the arrow keys to move the cursor to $(10, 5)$. The coordinates at the bottom of the screen will help you find the point. Press ENTER.

▶ Then move the cursor to $(30, 5)$. Press ENTER twice.

▶ Move the cursor to $(5, 25)$ and press ENTER twice. Then move the cursor back to $(10, 5)$ to complete the triangle. Press ENTER.

▶ Repeat the third, fourth, and fifth steps above using the coordinates of $\triangle X'Y'Z'$ to draw the second triangle.

TALK ABOUT IT

a. How do the size and shape of $\triangle XYZ$ and $\triangle X'Y'Z'$ compare?

b. If you placed $\triangle X'Y'Z'$ on top of $\triangle XYZ$, which angles would match up?

c. Which sides would match up?

Learning the Concept

Figures that have the same size and shape are **congruent**. The triangles shown below are congruent. Slash marks are used to indicate which *sides* are congruent and arcs are used to indicate which *angles* are congruent. Remember, the symbol $\cong$ means *is congruent to*.

Congruent Angles	Congruent Sides
$\angle P \cong \angle X$	$\overline{PQ} \cong \overline{XY}$
$\angle Q \cong \angle Y$	$\overline{QR} \cong \overline{YZ}$
$\angle R \cong \angle Z$	$\overline{PR} \cong \overline{XZ}$

Parts of congruent triangles that "match" are called **corresponding parts**. For example, in the triangle above, $\angle P$ corresponds to $\angle X$, and $\overline{PQ}$ corresponds to $\overline{XY}$.

Corresponding Parts of Congruent Triangles

If two triangles are congruent, their corresponding sides are congruent and their corresponding angles are congruent.

When you write $\triangle PQR \cong \triangle XYZ$, the corresponding vertices are written in order. So, $\triangle PQR \cong \triangle XYZ$ means that vertex P corresponds to vertex X, vertex Q corresponds to vertex Y, and vertex R corresponds to vertex Z.

Example ❶ **If $\triangle COD \cong \triangle ATM$, name the congruent angles and sides.**

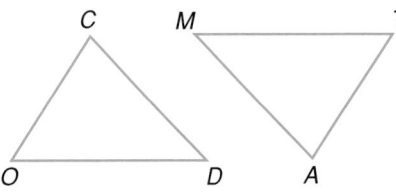

First, name the three pairs of congruent angles by looking at the order of the vertices.
$\angle C \cong \angle A$
$\angle O \cong \angle T$
$\angle D \cong \angle M$

Since C corresponds to A and O corresponds to T, $\overline{CO}$ corresponds to $\overline{AT}$. Similarly $\overline{OD}$ corresponds to $\overline{TM}$, and $\overline{CD}$ corresponds to $\overline{AM}$. The congruent sides are as follows.
$\overline{CO} \cong \overline{AT}$, $\overline{OD} \cong \overline{TM}$, $\overline{CD} \cong \overline{AM}$

You can draw conclusions about congruent triangles based on the congruence statements.

Example ❷ **The corresponding parts of two congruent triangles are marked in the figure. Write a congruence statement for the two triangles.**

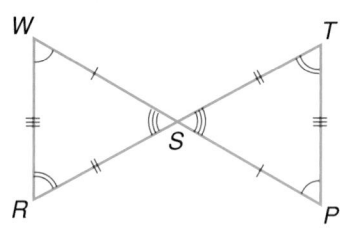

List the congruent angles and sides.

$\angle W \cong \angle P$	$\overline{PS} \cong \overline{WS}$
$\angle R \cong \angle T$	$\overline{TS} \cong \overline{RS}$
$\angle WSR \cong \angle PST$	$\overline{TP} \cong \overline{RW}$

The congruence statement can be written by matching the vertices of the congruent angles. $\triangle TPS \cong \triangle RWS$.

THINK ABOUT IT

How would you know that $\angle WSR \cong \angle PST$ even if they were not marked?

You can use corresponding parts to find measures of angles and sides in a figure that is congruent to a figure with known measures, as shown in Example 3.

Example 3

Construction

Roof trusses allow a roof to withstand the stress of heavy loads. A roof truss for a particular building is made so that the angle on the left measures 35°. The triangular sides of the truss are to be congruent as marked below.

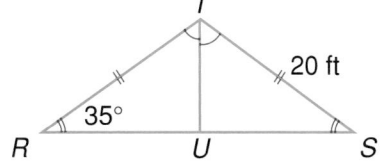

a. What should the measure of angle *TSR* be?

∠R ≅ ∠S
∠R has a measure of 35°,
so ∠S also has a measure of 35°.

b. What should the measure of side *TR* be?

$\overline{TR} \cong \overline{TS}$

$\overline{TS}$ has a given measure of 20 feet. So $\overline{TR}$ also has a measure of 20 feet.

Checking Your Understanding

Communicating Mathematics

Read and study the lesson to answer these questions.

1. **Tell** what it means when one triangle is congruent to another triangle.

2. **Name** the side that corresponds to $\overline{ST}$ and name the angle that corresponds to ∠T if △RST ≅ △WXY.

3. **Tell** which figure appears to be congruent to △ .

   a. **b.** **c.** ▽ **d.** none of these

 MATERIALS

⬚ graph paper
✏ straightedge
✂ scissors

4. On a piece of graph paper, draw two triangles like the ones at the right. Label the vertices as shown and cut out the triangles. Place one triangle over the other and turn until the corresponding parts match. Next, slide the top triangle to the right until the two triangles are side-by-side. Now the corresponding parts are in the same relative positions and you can easily identify them.

 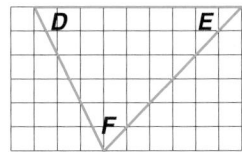

   a. Write a congruence statement for the two triangles.

   b. Name congruence statements for the congruent angles and sides.

Guided Practice

5. Complete the congruence statement for the triangles shown at the right. Then name the corresponding parts.

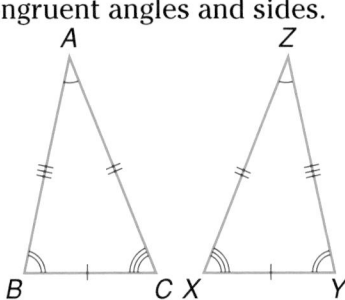

   a. △ABC ≅ ___
   b. $\overline{AB}$ ≅ ___
   c. ∠Y ≅ ___
   d. ∠C ≅ ___
   e. $\overline{BC}$ ≅ ___
   f. $\overline{XZ}$ ≅ ___
   g. ∠Z ≅ ___

Lesson 11-5 Congruent Triangles **575**

6. Write a congruence statement for the triangles shown below.

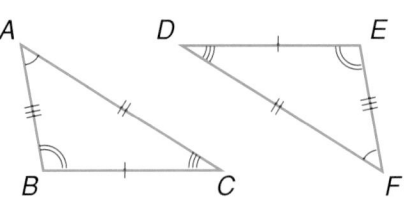

7. If $\triangle DEF \cong \triangle HEG$, what is the measure of $\overline{EH}$?
 a. 6   **b.** 8
 c. 10   **d.** not enough information

8. Landscaping The fence around a triangular garden is 12 meters long. A second garden is congruent to the first. How much fence is needed for the second garden?

Exercises: Practicing and Applying the Concept

Independent Practice

Complete the congruence statement for each pair of congruent triangles. Then name the corresponding parts.

9.

10.

11.

12.

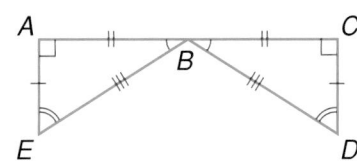

Write a congruence statement for each pair of congruent triangles.

13.

14.

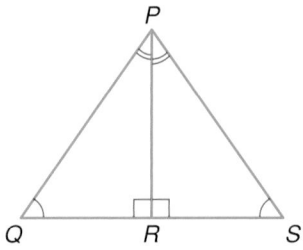

If $\triangle RST \cong \triangle MAP$, name the part congruent to each angle or segment given. (*Hint*: Make a drawing.)

15. $\overline{RS}$ **16.** $\angle P$ **17.** $\overline{MP}$ **18.** $\overline{ST}$ **19.** $\angle R$ **20.** $\angle A$

Find the value of *x* for each pair of congruent triangles.

21.

22.

23.

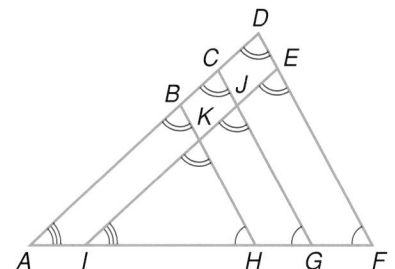

Critical Thinking

24. If two triangles have three pairs of congruent, corresponding angles, are the triangles congruent? Explain your answer by making a drawing.

25. In the figure at the right, two pairs of overlapping triangles appear to be congruent. Write a congruence statement for each pair.

Applications and Problem Solving

26. Construction One pattern for a roof truss is shown at the right. Use the labels in the figure and name the triangle that seems to be congruent to each triangle listed below.

a. △ABC　　　　　**b.** △BDC　　　　　**c.** △ADC

27. Geometry Two triangles are congruent and the perimeter of one triangle is 5 feet. What is the perimeter of the second triangle?

28. Recreation Make a paper airplane of your own design from an $8\frac{1}{2}$- by-11 inch piece of paper. When you have finished, open the paper and observe the pattern made by the folds. Look for pairs of congruent triangles. Use a protractor to measure angles if you wish. Label the vertices and write congruence statements for the pairs of congruent triangles.

Mixed Review

29. In △PQR, $m\angle P = 45°$, $m\angle Q = 40°$. Find the measure of $\angle R$ and classify the triangle as *acute*, *right*, or *obtuse*. (Lesson 11-4)

30. Standardized Test Practice A shirt that normally sells for $45 is on sale for 15% off. What is the amount of savings? (Lesson 9-5)
　A. $6.75　　　**B.** $11.25　　　**C.** $30.00　　　**D.** $38.25

31. Solve $\frac{a}{5} - 6 = 19$. (Lesson 7-2)

32. Name the multiplicative inverse of $1\frac{3}{5}$. (Lesson 6-4)

33. Eliminate Possibilities Liz, Renee, and Pablo each either brought a sack lunch, bought a plate lunch, or bought from the snack bar for lunch. Use the clues to find each person's lunch. (Lesson 3-1)
　▶ Liz did not buy a lunch.
　▶ Renee did not have a plate lunch.

For **Extra Practice**, see page 768.

11-6

Similar Triangles and Indirect Measurement

Hands-On Activity

MATHEMATICS MINI LAB

You'll Learn

■ To identify corresponding parts and find missing measures of similar triangles, and

■ to solve problems involving indirect measurement by using similar triangles.

It's Important

Triangular relationships are found in architecture.

Have you ever used a copy machine to make a reduction or enlargement of something? The original and the copy are the same shape, but different sizes. In mathematics, we call this a **dilation.** You can use graph paper to investigate the relationship of two such figures.

Your Turn

Work with a partner.

▶ Draw and label a triangle on a coordinate plane.

▶ Multiply the coordinates of each vertex by 2. Draw the new triangle.

TALK ABOUT IT

a. Measure the angles in the two triangles. How do they compare?

b. Use a ruler to measure the sides of the two triangles. Write a ratio comparing the measures of corresponding sides. What do you notice?

c. Repeat the activity by multiplying the coordinates of the original triangle by $\frac{1}{2}$. Are the results the same?

Learning the Concept

The triangles you drew in the activity were **similar**. Figures that have the same shape but not necessarily the same size are similar figures.

As with congruent triangles, write the corresponding vertices of similar triangles in the same order.

In the figure below, $\triangle ABC$ is similar to $\triangle PQR$. This is written as $\triangle ABC \sim \triangle PQR$ in symbols. The symbol $\sim$ means *is similar to*.

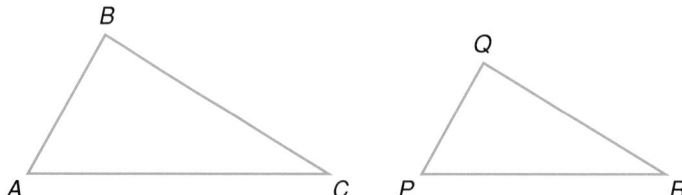

Measure each angle. Compare the corresponding angles. This example and the activity at the beginning of the lesson suggest the following.

It is also true that if the corresponding angles of two triangles are congruent, then the triangles are similar. *You will use this property in Example 2.*

The corresponding sides of congruent triangles also have a special relationship. In the figure below, $\triangle LMN \sim \triangle RST$.

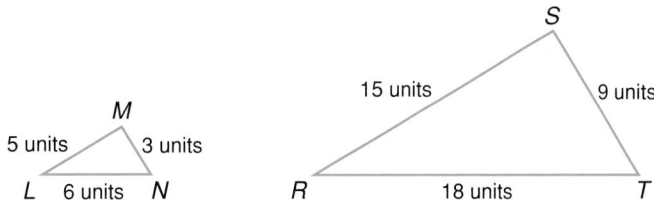

We can compare the measures of corresponding sides by using ratios.

$$\frac{m\overline{LN}}{m\overline{RT}} = \frac{6 \text{ units}}{18 \text{ units}} \text{ or } \frac{1}{3} \qquad \frac{m\overline{MN}}{m\overline{ST}} = \frac{3 \text{ units}}{9 \text{ units}} \text{ or } \frac{1}{3} \qquad \frac{m\overline{LM}}{m\overline{RS}} = \frac{5 \text{ units}}{15 \text{ units}} \text{ or } \frac{1}{3}$$

Notice that the ratios are equivalent. So, the corresponding sides are proportional to each other. This example and the activity at the beginning of the lesson seem to suggest the following.

It is also true that if the corresponding sides of two triangles are proportional, then the triangles are similar.

You can use proportions to find the measures of the sides of similar triangles when some measures are known.

Example **If $\triangle GHI \sim \triangle JKL$, find the value of x.**

Write a proportion using the known measures.

$\dfrac{m\overline{GH}}{m\overline{JK}} = \dfrac{m\overline{HI}}{m\overline{KL}}$ *Corresponding sides are proportional.*

$\dfrac{12}{4} = \dfrac{9}{x}$

$12x = 4 \cdot 9$ *Find the cross products.*

$12x = 36$

$x = 3$ *Divide each side by 12.*

The measure of $\overline{KL}$ is 3 cm.

Example **Use the information in the figure below to find the value of x.**

Since vertical angles are congruent,
∠ACB ≅ ∠DCE.
Angles A and D are marked congruent.
Angles B and E are congruent because
they are both right angles.
So, △ABC ~ △DEC, because their
corresponding angles are congruent.

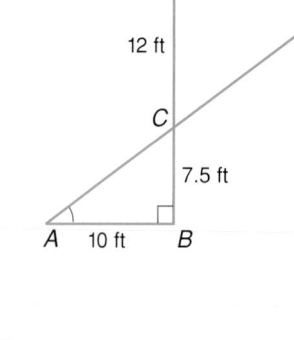

$$\frac{m\overline{AB}}{m\overline{DE}} = \frac{m\overline{BC}}{m\overline{EC}}$$ *Corresponding sides of similar triangles are proportional.*

$$\frac{10}{x} = \frac{7.5}{12}$$

$$10 \cdot 12 = 7.5x$$ *Find the cross products.*

$$120 = 7.5x$$

$$\frac{120}{7.5} = \frac{7.5x}{7.5}$$ *Divide each side by 7.5.*

$$16 = x$$

The measure of $\overline{ED}$ is 16 feet.

You can use what you know about similar triangles to find the measure of objects that are too large to measure directly, as shown in Example 3. This kind of measurement is called **indirect measurement**.

Example **In ancient Egypt, mathematicians used a technique called *shadow reckoning* to determine the heights of tall objects, such as the pyramids. The height of a staff and the length of its shadow are proportional to the height of an object and the length of its shadow. The objects and their shadows form two sides of similar triangles. Use shadow reckoning to find the height of the flagpole in the figure below.**

THINK ABOUT IT

How could you use shadow reckoning to measure the height of your school?

Explore   The meterstick in the drawing is like the staff.

Plan   In the figure, the shadow of the meterstick is 0.7 meters long, and the shadow of the flagpole is 5.6 meters long. Write a proportion comparing corresponding sides of the similar triangles.

Solve $\dfrac{\text{length of meterstick shadow}}{\text{length of flagpole shadow}} = \dfrac{\text{length of meterstick}}{\text{length of flagpole}}$

$$\frac{0.7}{5.6} = \frac{1}{h}$$

$0.7h = 5.6$ *Find the cross products.*

$\dfrac{0.7h}{0.7} = \dfrac{5.6}{0.7}$ *Divide each side by 0.7.*

$h = 8$

The flagpole is 8 meters tall.

Examine You can estimate to check your answer. The meterstick is about $1\frac{1}{2}$ times its shadow, so the flagpole is about $1\frac{1}{2}$ times its shadow.

Checking Your Understanding

Communicating Mathematics

Read and study the lesson to answer these questions.

1. **Explain** what it means for one figure to be similar to another.

2. **You Decide** Shawn thinks the triangles below are similar. Marta thinks they are not. Who is correct and why?

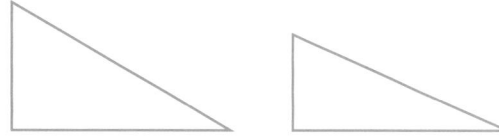

3. **Draw** two similar triangles. Label the vertices and name the corresponding sides and angles.

4. **Explain** how you can find the height of an object that is too large to measure directly.

5. **Make Up a Problem** Write a problem involving the use of similar triangles.

MATERIALS

✎ graphing calculator

6. Use a graphing calculator to graph $\triangle ABC$ with vertices $A(0, 3)$, $B(15, 3)$, and $C(15, 15)$ and $\triangle AXY$ with vertices $A(0, 3)$, $X(30, 3)$, and $Y(30, 27)$. Follow the steps in the procedure on page 573, using the coordinates given in this problem. Do the triangles appear to be similar?

Guided Practice

7. Refer to the similar triangles at the right.

   a. List three proportions that can be written for $\triangle ABE$ and $\triangle ACD$.

   b. What angle of $\triangle CAD$ corresponds to $\angle BAE$?

   c. What side corresponds to $\overline{BE}$?

   d. Find the value of x.

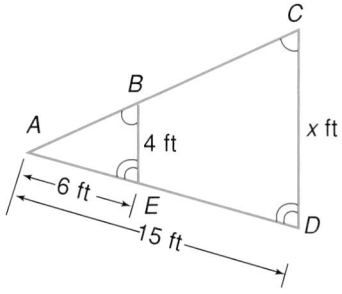

Write a proportion to find each missing measure *x*. Then find the value of *x*.

8.

9.

10. Given: Which of the following is similar to the given triangle?

a. **b.** **c.** **d.** none of these

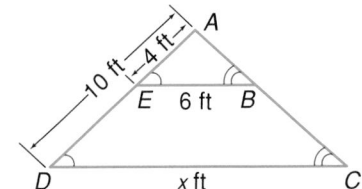

11. Civil Engineering The city of Marion plans to build a bridge across Brandon Lake. Use the information in the diagram at the right to find the distance across Brandon Lake.

Exercises: Practicing and Applying the Concept

Independent Practice

Write a proportion to find each missing measure *x*. Then find the value of *x*.

12.

13.

14.

15.

16.

17.

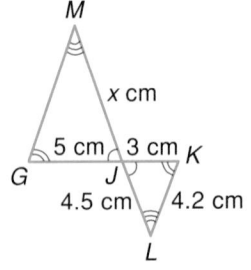

18. On a coordinate grid, graph $A(6, 6)$, $B(2, 2)$, $C(8, 2)$, $D(0, 0)$, and $E(9, 0)$. Draw $\triangle ABC$ and $\triangle ADE$.

  a. Are the triangles similar? How do you know?

  b. Do $\overline{BC}$ and $\overline{DE}$ appear to be parallel?

  c. Will two triangles with a common angle and the sides opposite that angle parallel always be similar? Explain.

Critical Thinking
19. a. Are all congruent triangles also similar? Explain.

  b. Are all similar triangles also congruent? Explain.

20. Photography A photo negative is 52.5 mm wide by 35 mm high. The print is made 15.24 centimeters wide. How many times greater is the area of the print than the area of the negative?

Applications and Problem Solving

21. Construction Find the length of the brace in the roof rafter shown at the right.

22. History Refer to Example 3. The largest known pyramid is Khufu's pyramid. At a certain time of day, a yardstick casts a shadow 1.5 ft long, and the pyramid casts a shadow 241 ft long. Draw a sketch that shows the two similar triangles and use shadow reckoning to find the height of the pyramid. Estimate to check your answer.

23. Surveying A surveyor needs to find the distance across a river and draws the sketch shown at the right. Explain how the surveyor determined the sketch and the known measurements. Then find the distance across the river.

Mixed Review

24. Standardized Test Practice Two triangles are congruent, and the area of one triangle is 15 square feet. What is the area of the second triangle? (Lesson 11-5)

  A. 5 ft^2 **B.** 15 ft^2 **C.** 20 ft^2 **D.** 30 ft^2

25. A quiz has 5 true-false questions. How many outcomes for giving answers to the five questions are possible? (Lesson 10-5)

26. Solve the system of equations $y = 3x$ and $y = x + 4$ by graphing. (Lesson 8-8)

27. Statistics Find the mean, median, and mode for the set of data below. When necessary, round to the nearest tenth. (Lesson 6-6)
26, 32, 54, 34, 36, 48, 39, 40, 54, 89, 78, 45

For **Extra Practice**, see page 768.

28. Find the GCF of $14n^2$, $22p^2$, and $36n^2p$. (Lesson 4-5)

11-7 Quadrilaterals

What You'll Learn

- To find the missing angle measure of a quadrilateral

- to classify quadrilaterals, and

- to explore similar quadrilaterals.

Why It's Important

Identifying quadrilaterals can help you design room layouts.

Application: Quilting

Quilting in America dates back to colonial days. Early quilts were often made using only square or rectangular patches. Later designs used other geometric shapes. The quilt shown at the right was made using a standard one-patch pattern and silk cloth. It is called "Tumbling Blocks" and is on display in the Howell House in Philadelphia, Pennsylvania. Notice how the four-sided patches and the design give the illusion of three dimensions.

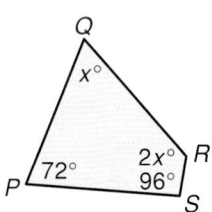

Learning the Concept

Squares and rectangles are simple quadrilaterals. A **quadrilateral** is a closed figure formed by four line segments that intersect only at their endpoints. As with triangles, a quadrilateral can be named by its vertices. The quadrilateral shown below can be named quadrilateral *ABCD*.

When you name a quadrilateral, you can begin at any vertex. But it is important to name consecutive vertices in order.

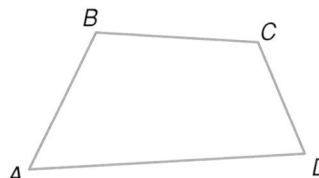

The vertices are A, B, C, and D.
The sides are $\overline{AB}$, $\overline{BC}$, $\overline{CD}$, and $\overline{AD}$.
The angles are $\angle A$, $\angle B$, $\angle C$, and $\angle D$.

If you use a protractor to measure each interior angle of quadrilateral *ABCD*, you will find that $m\angle A = 60°$, $m\angle B = 115°$, $m\angle C = 120°$, and $m\angle D = 65°$. The sum of the measures of the angles of *ABCD* is $360°$.

This example suggests a relationship of the angles of a quadrilateral.

Angles of a Quadrilateral	**The sum of the measures of the angles of a quadrilateral is 360°.**

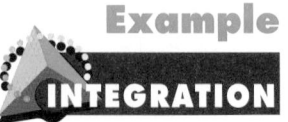

Algebra

Example 1 Find the value of *x*. Then find the missing angle measures.

The sum of the measures of the angles is $360°$.

$72 + 96 + x + 2x = 360$
$168 + 3x = 360$ *Combine like terms.*
$3x = 192$ *Subtract 168 from each side.*
$x = 64$ *Divide each side by 3.*

So $m\angle Q = 64°$, and $m\angle R = 2(64)$ or $128°$.

There are many kinds of quadrilaterals. One way to identify them is to look for pairs of parallel sides.

QUADRILATERALS

Quadrilaterals with no pairs of parallel sides

Parallelogram quadrilateral with 2 pairs of parallel sides

Trapezoid quadrilateral with exactly one pair of parallel sides

Rectangle parallelogram with 4 congruent angles

Rhombus parallelogram with congruent sides

Square parallelogram with congruent sides and congruent angles

In Lesson 11-6, you learned various properties of similar triangles. These same properties also hold for similar quadrilaterals or any pair of similar figures.

Corresponding Angles and Sides of Similar Figures

If two figures are similar, then the angles of one figure are congruent to the corresponding angles of the other figure. If two figures are similar, then their corresponding sides are porportional.

A **scale drawing** is similar to the actual object, but it is generally either smaller or larger. The scale is the ratio of the lengths on the drawing to the actual lengths of the object.

Example ② On a scale drawing of a room, 1 inch represents 2 feet. In the scale drawing, the room is 11 inches long. How long is the actual room?

The scale is the ratio 1 in. to 2 ft or $\frac{1\text{in.}}{2\text{ft}}$. *This is often written 1 in. = 2 ft.*

Write a proportion.

$$\frac{1\text{in.}}{2\text{ft}} = \frac{11\text{ in.}}{x}$$

$1x = 2(11)$ *Find the cross products.*

$x = 22$

The actual room is 22 feet long.

Communicating Mathematics

Read and study the lesson to answer these questions.

1. **Draw** an example of each kind of quadrilateral named in this lesson. Give a real-world example of each.

2. **You Decide** Susan says that all squares are similar. Lynn says they are not. Who is correct? Explain.

3. Which of the following rectangles is similar to the rectangle at the right? Why?

   a.   **b.**   **c.**

MATERIALS

straightedge

4. Work with a partner.

   a. Use a straightedge to draw a quadrilateral.

   b. Draw a line segment that connects any two nonconsecutive vertices. This is a *diagonal*.

   c. How many triangles are formed by the diagonal?

   d. What is the sum of the measures of the angles of a triangle? of two triangles?

   e. How does this activity show that the measures of the angles of a quadrilateral have a sum of 360°?

Guided Practice

5. Find the value of x.

6. Find the value of x. Then find the missing angle measures.

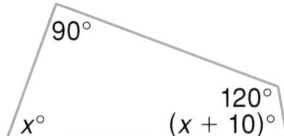

List every name that can be used to describe each quadrilateral. Indicate the name that *best* describes the quadrilateral.

7.

8.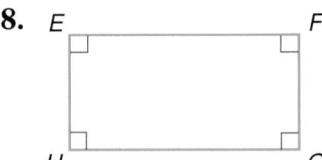

9. **Interior Design** The scale of the drawing at the right is 1 cm = 0.5 m. What is the length of the actual room?

Independent Practice

Find the value of x.

10.   11.

12.

Find the value of x. Then find the missing angle measures.

13.

14.

15.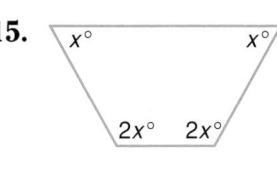

Classify each quadrilateral using the name that *best* describes it.

16. 17. 18.

Make a drawing of each quadrilateral. Then classify each quadrilateral using the name that *best* describes it.

19. In quadrilateral WXYZ, $m\overline{WX}$ = 3 in., $m\overline{XY}$ = 5 in., $m\overline{YZ}$ = 3 in., and $m\overline{WZ}$ = 5 in. Angles W, X, Y, and Z are right angles.

20. In quadrilateral CDEF, $\overline{CD}$ and $\overline{EF}$ are parallel, and $\overline{CF}$ and $\overline{DE}$ are parallel. Angle C is not congruent to angle D.

21. In quadrilateral JKLM, $m\angle J = 90°$, $m\angle K = 50°$, $m\angle L = 90°$, and $m\angle M = 130°$.

Determine whether each statement is *always*, *sometimes*, or *never* true.

22. A rectangle is a parallelogram.   23. A rectangle is a square.

24. A rhombus is a square.   25. A trapezoid is a parallelogram.

26. A square is a parallelogram.   27. A square is a rhombus.

28. Use grid paper to draw each figure. Then enlarge each figure by making each segment twice the length of the original. How does the area of the enlargement compare to the area of the original figure?

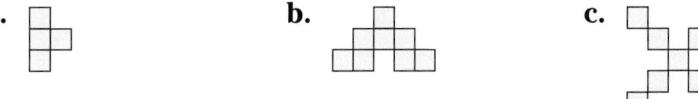

a. b. c.

Critical Thinking

29. In an **equilateral** figure, all sides have the same measure. In an **equiangular** figure, all angles have the same measure.

   a. Is it possible for a quadrilateral to be equi*lateral* without being equi*angular*? Explain.

   b. Is it possible for a quadrilateral to be equiangular without being equilateral? Explain.

30. One angle of a parallelogram measures 40°. What are the measures of the other angles?

Applications and Problem Solving

31. **Sports** Name four sports with playing fields that are quadrilaterals. Find the dimensions of each field and classify the shape.

32. **Construction** In a house blueprint, the actual height of 8 feet is represented by 2 inches. If the actual length of the house is 60 feet, what is the length of the house in the blueprint?

33. **Make a Drawing** A basketball court is 84 feet by 50 feet. Make a scale drawing on $\frac{1}{4}$-inch grid paper. Use the scale $\frac{1}{4}$ inch = 6 feet.

34. **Publishing** Kiona's school newspaper uses 3-inch columns. If a $3\frac{1}{2}$-by-5 inch vertical photograph is reduced to fit in one column, how long will the reduced photograph be?

Mixed Review

35. Find the value of x in △*DEC* at the right. (Lesson 11-6)

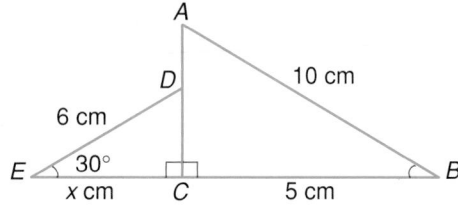

36. Find the measure of ∠*A* in △*ABC* at the right. (Lesson 11-4)

37. **Standardized Test Practice** Express 450 miles on 20 gallons of gas as a unit rate. (Lesson 9-1)

   A. 20 mpg   **B.** 21.5 mpg   **C.** 22.5 mpg   **D.** 23 mpg

38. **Geography** The distance around Earth at the equator is about 25,000 miles. Find the approximate diameter of Earth at the equator. (Lesson 7-4)

39. Viñita noticed the school cafeteria served pizza on the last five Mondays. Viñita decides that the cafeteria always serves pizza on Monday. Is this an example of *inductive* or *deductive* reasoning? (Lesson 5-8)

40. Use mental math, paper and pencil, or a calculator to find at least one number that has four digits and is not divisible by 2, 3, or 5. (Lesson 4-1)

For **Extra Practice,** see page 768.

WORKING ON THE

Investigation

It's Only A Game

Refer to the Investigation on pages 482–483.

Look over the list of favorite games that you placed in your Investigation Folder at the beginning of Chapter 10. Some people may have said that their favorite games are target games, such as darts or horseshoes.

For this part of the Investigation, your group has been hired to design a new target game using sticky balls. The new board should be square and divided into several rectangular regions. Some regions should be marked with an X, and other regions marked with a Y.

- Work with the members of your group to design the board. If a sticky ball is thrown at random at your board, the game should meet the following requirements.
  1. Player A scores points if the ball lands in a region marked X, and player B scores points if the ball lands in a region marked Y.
  2. The game should be fair.
- Write a report explaining how your board meets the requirements. Be sure to use concepts from Chapter 10 in your explanation.

Add the results of your work to your Investigation Folder.

What You'll Learn

- To classify polygons, and

- to determine the sum of the measures of the interior and exterior angles of a polygon.

Why It's Important

Architects use angles when designing buildings.

MATERIALS: computer LOGO software

Technology Activity

If you were to walk a path in the shape of a rectangle, you would make a 90° turn at each corner. In LOGO, the turtle "walks" a path following directions you give. You can use LOGO to make paths of various shapes.

Your Turn

Work with a partner. Type the following commands to make the shapes shown to the right of each set of commands. Sketch each figure and mark the angles. Clear the screen before typing the second set of commands.

FD 100 RT 60	FD 100 RT 60
FD 100 RT 90	FD 80 RT 60
FD 173 RT 120	FD 80 RT 60
FD 173	FD 100 RT 90
What angle	FD 139
must you turn to return the turtle	What angle must
to its starting position?	you turn to return the turtle to
	its starting position?

TALK ABOUT IT

a. In each figure, compare the angle of turn (the exterior angle) to the adjacent angle inside the figure (the interior angle). What pattern do you see?

b. How could you use this pattern to draw figures using LOGO?

c. In each figure, find the sum of the turning (exterior) angles including the final turn to return the turtle to its starting position. What do you notice?

Learning the Concept

The figures you drew above were simple closed figures. A simple, closed figure can be traced in a continuous path without tracing any point other than the starting point more than once. A **polygon** is a simple, closed figure in a plane that is formed by three or more line segments, called **sides**. The segments meet only at their endpoints. These points of intersection are called **vertices** (plural of **vertex**).

THINK ABOUT IT

Can you name a simple, closed figure that is NOT a polygon?

These plane figures are polygons.

These plane figures are not polygons.

 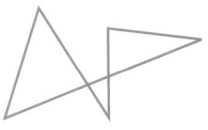

We can classify polygons by the number of sides. Some of the more common polygons are shown below.

Triangle
3 Sides

Quadrilateral
4 Sides

Pentagon
5 Sides

A polygon with n sides is called an **n-gon**.

Hexagon
6 Sides

Octagon
8 Sides

Decagon
10 Sides

A **diagonal** is a line segment that joins two nonconsecutive vertices. You can draw diagonals in any polygon with more than three sides. In the polygons shown below, all possible diagonals from one vertex are shown. The table shows the number of diagonals drawn, and the number of triangles formed in each polygon.

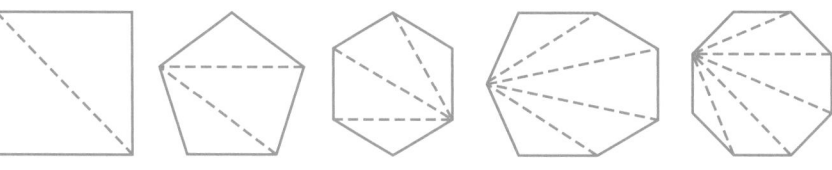

Number of:	Quadrilateral	Pentagon	Hexagon	Heptagon	Octagon
Sides	4	5	6	7	8
Diagonals	1	2	3	4	5
Triangles	2	3	4	5	6

Connection to > Algebra

Compare the number of sides to the number of triangles. What pattern do you see? The number of triangles is always 2 less than the number of sides. You can use diagonals and the property of the sum of the angles of a triangle to find the sum of the measures of the angles of any polygon without using a protractor.

Sum of the Interior Angle Measures in a Polygon	If a polygon has n sides, then $n - 2$ triangles are formed, and the sum of the degree measures of the interior angles of the polygon is $(n - 2)180$.

Example **a. Find the sum of the measures of the interior angles of a pentagon by drawing diagonals.**

Draw all of the diagonals from one vertex.

Three triangles are formed.

The sum of the measures of the interior angles of a triangle is 180°, so the sum of the measures of the interior angles of a pentagon is 3(180) or 540°. *Check by using the formula.*

b. Find the sum of the measures of the interior angles of a hexagon by using the formula.

A hexagon has 6 sides. Therefore, $n = 6$.

$(n − 2)180 = (6 − 2)180$ *Replace n with 6.*
$\quad\quad\quad\quad = 4(180)$ or 720 *You can check this solution by drawing diagonals.*

The sum of the measures of the interior angles of a hexagon is 720°.

A **regular polygon** is a polygon that is **equilateral** (all sides are congruent) and **equiangular** (all angles are congruent). Since the angles of a regular polygon are congruent, their measures are equal.

Example **The Pentagon in Washington, D.C., is named for its unusual shape, a regular pentagon. In a scale model of the Pentagon, what is the measure of each angle?**

Architecture

The scale model and the actual building are similar figures. In similar figures, the angles are congruent. From Example 1a, you know that the sum of the measures of the interior angles of a pentagon is 540°. Since the Pentagon is regular, you can find the degree measure of one angle by dividing 540 by 5.

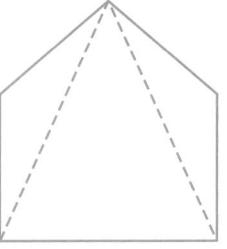

$540 ÷ 5 = 108$

The measure of each angle in a regular pentagon is 108°.

The angles we considered in Examples 1 and 2 are "inside" the polygon. These are the **interior angles**. When a side of a polygon is extended, a special angle is formed, called an **exterior angle**.

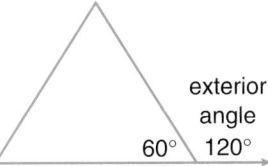

In the activity at the beginning of the lesson, you discovered that the interior and exterior angles of a polygon are supplementary. You also discovered that the sum of the exterior angles of a polygon is 360°. You can use these ideas for another way to find the measure of one angle of a regular polygon.

Example **3** Find the measure of each interior angle in a regular hexagon.

Sketch a regular hexagon.

Draw an exterior angle at each vertex.

There are 6 exterior angles. The sum of the measures of the exterior angles is 360°.

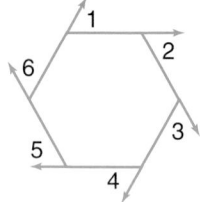

360 ÷ 6 = 60   *Measure of each exterior angle.*

180 − 60 = 120   *Interior and exterior angles are supplementary.*

The measure of each interior angle in a regular hexagon is 120°.

Checking Your Understanding

Communicating Mathematics

Read and study the lesson to answer these questions.

1. **Explain** why a square is a regular polygon.

2. **Explain** the relationship between the number of sides in a polygon and the number of triangles formed by the diagonals.

3. **Tell** the difference between equiangular and equilateral. Give an example of a figure that is equiangular or equilateral but not both.

4. **Draw** △ABC with exterior angle BCD.

 MATH JOURNAL

5. **Assess Yourself** How can knowing the properties of geometric shapes help you in your daily life?

Guided Practice

Classify each polygon below and determine whether it appears to be *regular* or *not regular*.

6.

7.

8. Find the sum of the measures of the interior angles of an octagon.

9. Find the measure of each exterior angle and each interior angle in a regular nonagon (9 sides).

10. **Scale Models** Refer to Example 2. The scale of the model of the Pentagon is to be 1 in. = 50 ft. Each side of the Pentagon is about 1 mile (5280 ft) long. What length should each side of the scale model be?

Exercises: Practicing and Applying the Concept

Independent Practice

Find the sum of the measures of the interior angles of each polygon.

11. heptagon	12. decagon	13. dodecagon
14. nonagon	15. 15-gon	16. 25-gon

Find the measure of each exterior angle and each interior angle of each regular polygon.

17. regular hexagon	18. regular decagon
19. regular dodecagon	20. regular 20-gon

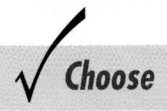
✓ Choose

Estimation
Mental Math
Calculator
Paper and Pencil

21. The measure of one angle of a regular polygon with n sides is $\frac{180(n-2)}{n}$. Find the measure of an interior angle in a regular triangle.

Find the perimeter of each regular polygon.

22. equilateral triangle with sides 27 ft long

23. regular quadrilateral with sides 18 cm long

24. regular pentagon with sides 3 m long

25. regular hexagon with sides 4.5 in. long

26. regular octagon with sides $5\frac{3}{4}$ yd long

Critical Thinking

27. Trace the dot pattern shown at the right. Without lifting your pencil from the paper, draw four line segments that connect all the points.

Applications and Problem Solving

28. Driver's License Exam Part of a driver's license exam includes identifying road signs by color and by shape. The shapes are polygons and some are regular polygons. Identify the shape of each road sign pictured below and tell what it means.

29. Manufacturing Some cafeteria trays are designed so that four people can place their trays around a square table without bumping corners. Each tray is shaped like the one shown at the right. The top and bottom of the tray are parallel.

  a. What shape does the tray have?

  b. Find the measure of each angle of the tray so that the trays will fit side-to-side around the table.

  c. What polygon will be formed at the center of the table if four trays are placed around the table with their sides touching?

Mixed Review

30. Standardized Test Practice A lot for a new house is shaped like a trapezoid, with two right angle corners. If the third corner is an 80° angle, what is the measure of the fourth angle? (Lesson 11-7)
  A. 80°   **B.** 90°   **C.** 100°   **D.** 110°

31. Probability A die is rolled and the spinner is spun. What is P(an even number and a vowel)? (Lesson 10-9)

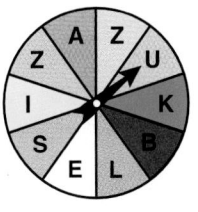

32. Simplify $3.5x + 2.8x + 1.5x$. (Lesson 5-3)

33. Write 1604 in expanded form. (Lesson 4-2)

For **Extra Practice**, see page 768.

34. Solve the equation $-6 - 8 = c$. (Lesson 2-5)

11-8B Tessellations

An Extension of Lesson **11-8**

MATERIALS

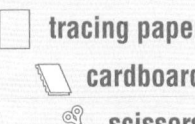

☐ tracing paper

▱ cardboard

✂ scissors

The repetitive pattern of regular polygons shown at the right is an example of a **tessellation**. In a tessellation, the polygons fit together with no holes or gaps.

In this Math Lab, you will investigate tessellations with regular polygons.

Your Turn

Work with a partner.

▶ Trace each regular polygon shown. Cut each shape from the tracing paper and trace it onto a piece of cardboard. Cut each shape from the cardboard.

▶ Use each cardboard piece to try to draw a tessellation for each regular polygon.

▶ For each polygon, calculate the sum of the interior angle measures. Then determine the measure of one angle in each polygon. Record your results in a table.

Equilateral
Triangle

Square

Regular Pentagon

Regular Hexagon

Regular Dodecagon

Regular Octagon

 TALK ABOUT IT

1. How can you tell from the measure of one angle whether or not a single polygon will tessellate?

Your Turn

Try to make tessellations from the combinations below.

a. square and octagon

b. square, triangle, and hexagon

c. square, triangle, and dodecagon

d. another combination you choose

 TALK ABOUT IT

2. How can you tell whether a combination of polygons will tessellate?

You'll Learn

- To identify and draw reflections, translations, and rotations, and

- to identify and draw symmetric figures.

Why
It's Important

When you classify animals in science you will use symmetry.

 TALK ABOUT IT

MATERIALS: 📄 cardboard ✂ scissors
Hands-On Activity

If you look around, you will probably see geometric shapes in repeating patterns. Floors and ceilings are often tiled in patterns. In the activity below, you will investigate one way to create a pattern.

Your Turn

▶ Cut a triangle out of cardboard. Place the triangle on a coordinate grid as shown, so one side lies parallel to the *x*-axis. Trace the triangle on the paper. Label the vertices as shown. Write the coordinates of the vertices.

▶ Slide the cutout triangle 4 units left and 3 units down as shown. Trace the triangle. Label the vertices X', Y', and Z', so they correspond to the first triangle. Write the coordinates of the vertices of $\triangle X'Y'Z'$.

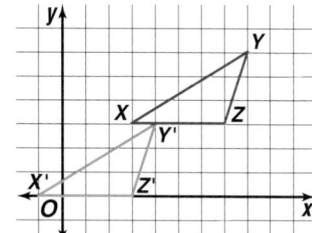

How are the coordinates of $\triangle XYZ$ and $\triangle X'Y'Z'$ related?

Learning the Concept

Transformations are movements of geometric figures. A **translation** is one type of transformation. In a translation, you "slide" a figure horizontally, vertically, or both, as you did in the activity.

Two other types of transformations are rotations and reflections.

In a **rotation**, you turn the figure around a point. Computer drawing packages allow the user to create drawings. After a drawing is created, a command called "rotate right" can be used to rotate the figure as shown at the right.

In a **reflection**, you "flip" a figure over a line. You can use a geomirror to draw the reflection of a figure. Each diagram below shows a figure and its reflection over a line. Place a geomirror on the dashed line in each figure and verify each reflection.

  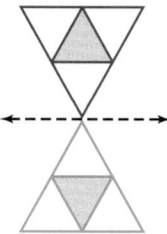

How could you describe the figure and its reflection?

Example 1

Tell whether each transformation is a translation, a rotation, or a reflection.

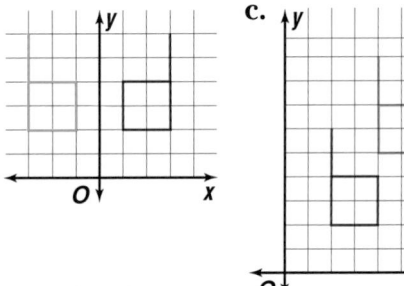

a. The figure is *rotated* about point *A*.

b. The figure is *reflected* about the *y*-axis.

c. The figure is *translated* 2 units horizontally and 3 units vertically.

When you draw a figure and its reflection, you create a figure that is **symmetric**. The line where you placed the geomirror is called a *line of symmetry*. Some common items with line symmetry are shown below.

Some of the items shown above have one line of symmetry. Others have several lines of symmetry. Each line of symmetry separates the figure into *two* congruent parts.

Example 2

Draw all lines of symmetry for each figure.

a.

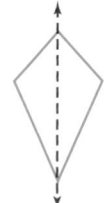

A kite has one line of symmetry.

b.

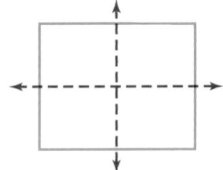

A rectangle has two lines of symmetry.

c.

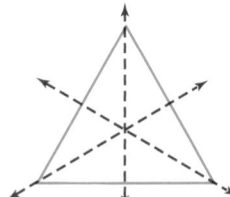

An equilateral triangle has three lines of symmetry.

Artists have used symmetry to create beautiful designs for centuries.

Example 3

Real World APPLICATION

World Culture

In 1895, excavators found Hopi pottery among the ruins of the pueblo known as Sikyatki. The pattern below was found on the exterior of a Sikyatki food bowl. Describe the transformations used to make the pottery design.

The original design is translated to the right and rotated 180° or reflected over a horizontal line.

Communicating Mathematics

Read and study the lesson to answer these questions.

1. **Compare and contrast** a translation, a reflection, and a rotation of the same figure.

2. **Show** how to complete the figure so that $\overleftrightarrow{XY}$ is a line of symmetry.

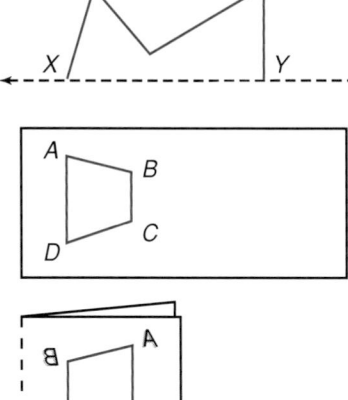

MATERIALS

✐ wax paper
🖍 colored pencils
📏 ruler

3. **a.** On one half of a piece of wax paper, use a straightedge and a colored pencil to draw a quadrilateral. Label the vertices A, B, C, and D.

   b. Fold the wax paper in half and turn it over so $ABCD$ is on the bottom side. Use a different color pencil to trace $ABCD$.

   c. Label the vertices A', B', C', and D' so that A corresponds to A', B to B', and so on.

   d. Unfold the paper. The fold line represents the line of reflection and $A'B'C'D'$ is a reflection of $ABCD$.

   e. Use a ruler to measure $\overline{AA'}$, $\overline{BB'}$, $\overline{CC'}$, and $\overline{DD'}$. At what point does the line of reflection intersect each segment?

Guided Practice

Tell whether each geometric transformation is a translation, a reflection, or a rotation. Explain your answer.

4.

5.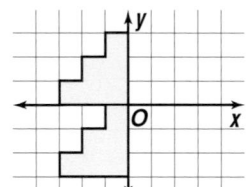

Trace each figure. Draw all lines of symmetry.

6.

7.

8. **Art** A sample of a Navaho weaving is shown below. Describe the transformations and lines of symmetry that exist in the weaving.

Independent Practice

Tell whether each geometric transformation is a *translation*, a *reflection*, or a *rotation*. Explain your answer.

9.

10.

11.

12. A translation moves $\triangle PQR$ 3 units to the right and 3 units down to form $\triangle P'Q'R'$. Write the coordinates of each vertex of $\triangle P'Q'R'$.

13. If the first figure below is rotated about point R, which could be the resulting figure?

 a. **b.**

c.

Trace each figure. Draw all lines of symmetry.

14.

15.

16.

17. Graph $A(-1, 3)$, $B(4, 4)$, and $C(2, 1)$ on a coordinate plane. Draw $\triangle ABC$ and graph its reflection if the *x*-axis is the line of reflection.

18. Graph $P(-5, 1)$, $Q(-3, 4)$, and $R(-1, 3)$ on a coordinate plane. Draw $\triangle PQR$ and graph its reflection if the *y*-axis is the line of reflection.

19. Graph $R(3, -1)$, $S(6, 2)$, and $T(1, 4)$ on a coordinate plane. Draw $\triangle RST$ and translate the triangle 5 units to the left and 4 units down.

Critical Thinking

20. Find the vertices of a triangle after it is translated 2 units to the left and 3 units up and then is reflected across the graph of $y = x + 2$. The original vertices of the triangle are $(2, 0)$, $(3, 2)$, and $(6, 2)$.

Application and Problem Solving

21. **Communication** International code flags can be used by sailors to send messages at sea. Describe the kinds of symmetry that are displayed by the flags and tell which flags have each kind of symmetry.

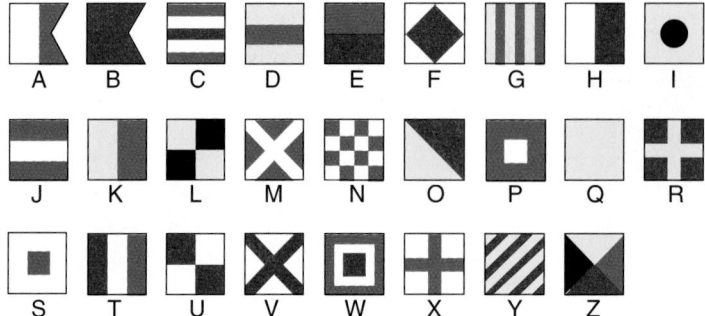

22. **Games** Miniature golf is played with a putter and a golf ball on a carpeted "green" surrounded by side rails. On some holes there is a straight path to the hole. Others are "dog legs", and it is not possible to aim directly at the hole. You can, however, get a "hole in one" by mentally reflecting the hole across the line of the wall and aiming for the reflection.

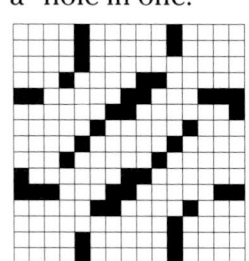

   a. Trace the figure at the top right. Draw a path using a reflection that will allow the ball to reach the hole in one stroke.

   b. A more difficult "dog leg" is shown at the bottom right. Trace the figure and find the path of the ball for a "hole in one."

23. **Games** The crossword puzzle at the right is a 15 by 15 crossword puzzle with half-turn symmetry. That is, when the puzzle is turned upside down it looks exactly the same as when it is right-side up. Draw a 9 by 9 grid. Then fill in at least 18 squares to make a puzzle with half-turn symmetry.

Mixed Review

24. Find the sum of the measures of the angles of a 15-gon. (Lesson 11-8)

25. **Standardized Test Practice** What is the value of x in $5x + 17 = -1 - x$? (Lesson 7-5)

   A. -3 **B.** $\frac{8}{3}$ **C.** 3 **D.** 4

26. Solve $\frac{2}{5}x \leq -0.16$. (Lesson 6-7)

For **Extra Practice**, see page 769.

27. Solve $-84 > 4t$. Then graph the solution on a number line. (Lesson 3-7)

28. Write the coordinates for a point on the x-axis. (Lesson 1-7)

From the → **FUNNY PAPERS**

© 1995 Bill Amend/Dist. by Universal Press Syndicate

TALK ABOUT IT

1. Explain why the comic is funny.

2. What is shown in the second frame of the comic?

3. What calculations did Jason have to make to get the ball to hit Paige?

It's Only A Game

Refer to the Investigation on pages 482–483. Add the results of your work to your Investigation folder.

Plan and give an oral presentation to describe what you know about games and strategy. Your presentation should include the following.

- A report of the survey you conducted at the beginning of the Investigation.

- A report of the results of your work at the end of Lesson 10-4. Include any revisions that were the direct result of the test and an explanation of why those revisions were made.

- A report on how your group determined the best strategy for *Don't Lose It All!* at the end of Lesson 10-8.

- A model of the board your group created at the end of Lesson 11-7 and a brief explanation of the accompanying game.

- A final version of the game your group designed. Include an explanation of how the survey at the beginning of the Investigation affected the type of game your group designed.

Extension

Contact a computer game manufacturer and inquire about how a computer game is created and how long it takes to create it.

PORTFOLIO ASSESSMENT

You may want to keep your work on this Investigation in your portfolio.

Vocabulary

After completing this chapter, you should be able to define each term, property, or phrase and give an example or two of each.

Geometry
acute angle (p. 550)
acute triangle (p. 569)
adjacent angles (p. 561)
alternate interior angles (p. 563)
alternate exterior angles (p. 563)
angle (p. 549)
complementary (p. 562)
congruent (pp. 554, 561, 573)
construct (p. 554)
corresponding angles (p. 563)
corresponding parts (p. 573)
degree (p. 549)
diagonal (p. 590)
equiangular (p. 591)
equilateral (p. 591)
exterior angle (p. 591)
indirect measurement (p. 580)
interior angle (p. 591)
intersect (p. 551)
isosceles triangle (p. 569)
line (p. 548)
line segment (p. 548)
obtuse angle (p. 550)
obtuse triangle (p. 569)
parallel (p. 551)
parallelogram (p. 585)
perpendicular (p. 551)
plane (p. 550)
point (p. 548)

polygon (p. 589)
quadrilateral (p. 584)
ray (p. 548)
rectangle (p. 585)
reflection (p. 595)
regular polygon (p. 591)
rhombus (p. 585)
right angle (p. 550)
right triangle (p. 569)
rotation (p. 595)
scale drawing (p. 585)
scalene triangle (p. 569)
sides (pp. 549, 589)
similar (p. 578)
skew lines (p. 551)
straight angle (p. 550)
square (p. 585)
supplementary (p. 562)
symmetric (p. 596)
tessellation (p. 594)
transformation (p. 595)
translation (p. 595)
transversal (p. 563)
trapezoid (p. 585)
triangle (p. 568)
vertex (pp. 549, 589)
vertical angles (p. 561)

Statistics
circle graph (p. 556)

interNET **CONNECTION**

Chapter Review For additional lesson-by-lesson review, visit:
www.glencoe.com/ sec/math/prealg/ mathnet

Understanding and Using Vocabulary

Choose the letter of the term that best completes each statement.

1. A(n) _?_ angle measures between 0° and 90°.
2. Two angles are _?_ if the sum of their measures is 90°.
3. When a _?_ intersects two parallel lines, the corresponding angles are congruent.
4. When a line is perpendicular to another line, the angles formed are _?_ angles.
5. A(n) _?_ angle measures between 90° and 180°.
6. A parallelogram with four congruent sides is a _?_.
7. Two angles are _?_ if the sum of their measures is 180°.

a. obtuse
b. acute
c. vertical
d. trapezoid
e. rhombus
f. right
g. transversal
h. complementary
i. supplementary

Skills and Concepts

Objectives and Examples	Review Exercises

Upon completing this chapter, you should be able to:

▶ **identify points, lines, planes, rays, segments, angles, and parallel, perpendicular and skew lines.**
(Lesson 11-1)

Draw and label a diagram to represent ∠ABC.

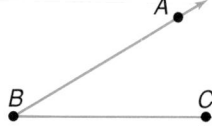

Use these exercises to review and prepare for the chapter test.

Draw and label a diagram to represent each of the following.

8. point *R*   9. ray *TS*

10. plane *XYZ*   11. acute angle *JKL*

12. line *n*   13. $\overline{EF}$

14. $\overrightarrow{HG} \parallel \overleftrightarrow{KL}$ 15. lines *w* and *y* are skew

▶ **make a circle graph to illustrate data.**
(Lesson 11-2)

This circle graph compares the amount of land and water to the whole surface of Earth.

Earth's Surface

30 % × 360 = *108*

70 % × 360 = *252*

16. The table below shows the types of jams and jellies that were entered in state fairs in 1994. Make a circle graph of the data.

Type	Entries
Strawberry	266
Raspberry	266
Blackberry	176
Grape	165
Plum	131
Other	592

17. Make a circle graph of the number of hours each week that you study, talk on the telephone, watch television or listen to music, and sleep.

▶ **identify the relationships of vertical, adjacent, complementary, and supplementary angles, and angles formed by two parallel lines and a transversal.**
(Lesson 11-3)

Find the measure of each labeled angle.

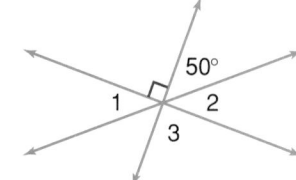

$m\angle 1 = 180 - 90 - 50$ or $40°$
$m\angle 2 = m\angle 1$ or $40°$
$m\angle 3 = 90°$

In the figure below, $a \parallel b$. Find the measure of each angle.

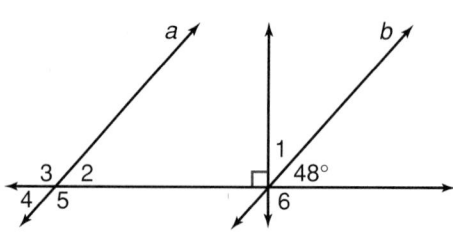

18. ∠1   19. ∠2   20. ∠3

21. ∠4   22. ∠5   23. ∠6

Objectives and Examples	Review Exercises

▶ **find the missing angle measure of a triangle.** (Lesson 11-4)

Find $m\angle 1$ if $m\angle 2 = 50°$ and $m\angle 3 = 80°$. Classify the triangle as *acute*, *right*, or *obtuse*.

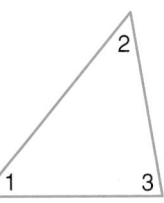

$m\angle 1 = 180 - 50 - 80$ or $50°$
The triangle is acute.

Use the figure below to solve each problem.

24. Find $m\angle 1$ if $m\angle 2 = 35°$ and $m\angle 3 = 90°$. Classify the triangle as *acute*, *right*, or *obtuse*.

25. Find $m\angle 2$ if $m\angle 1 = 42°$ and $m\angle 3 = 46°$. Classify the triangle as *acute*, *right*, or *obtuse*.

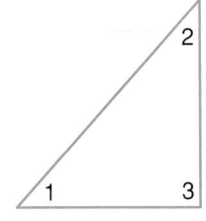

▶ **identify congruent triangles and their corresponding parts.** (Lesson 11-5)

If $\angle ABC \cong \angle FED$, name the parts congruent to angle A and segment AC.

$\angle A$ is congruent to $\angle F$. $\overline{AC}$ is congruent to $\overline{FD}$.

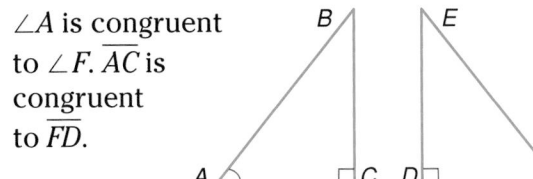

If $\triangle XYZ \cong \triangle ABC$, name the part congruent to each angle or segment.

26. $\angle X$

27. $\angle B$

28. $\overline{YZ}$

29. $\overline{AC}$

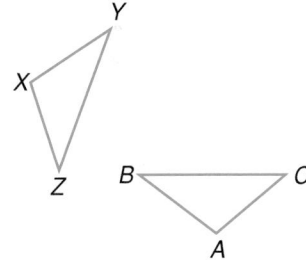

▶ **identify corresponding parts and find missing measures of similar triangles.** (Lesson 11-6)

Find the value of x in the similar triangles.

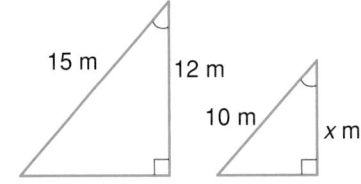

$$\frac{15}{10} = \frac{12}{x}$$
$$120 = 15x$$
$$8 = x$$

Find the value of x in each pair of similar triangles.

30.

31.

▶ **find the missing angle measure of a quadrilateral and classify quadrilaterals.** (Lesson 11-7)

Find the missing angle measures.
$75 + 135 + x + 2x = 360$
$210 + 3x = 360$
$3x = 150$
$x = 50$

The measures are $50°$ and $50 \cdot 2$ or $100°$.

Find the missing angle measures in each figure.

32.

33.

Objectives and Examples	Review Exercises

Objectives and Examples

▶ **classify polygons and determine the sum of the measures of the interior and exterior angles of a polygon.** (Lesson 11-8)

Find the sum of the measures of the angles of an octagon.

$(n - 2)180 = (8 - 2)180$
$= 6(180)$
$= 1080$

Review Exercises

34. Find the sum of the degree measures of the angles of a hexagon.

35. Find the measure of each interior angle in a regular decagon.

36. Find the measure of an exterior angle in a regular pentagon.

▶ **identify and draw reflections, translations, and rotations and identify and draw symmetric figures.** (Lesson 11-9)

Draw all lines of symmetry for the figure below.

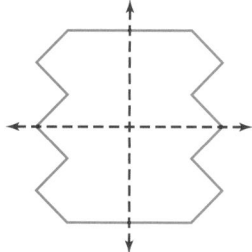

Trace each figure. Draw all lines of symmetry.

37.   **38.**

Tell whether each transformation is a *translation*, a *reflection*, or a *rotation*. Explain your answer.

39.   **40.**

Applications and Problem Solving

41. Statistics Make a circle graph to display the data in the chart. (Lesson 11-2)

Water Use in the U.S.	
Agriculture	36%
Public Water	8%
Utilities	33%
Industry	23%

42. Home Maintenance When a 6-foot ladder is placed against the side of a house at an angle that is reasonable for climbing, it touches the side of the house at a point 5 feet above the ground. If a 15-foot ladder is placed at the same angle to the house, can it be used to reach a point 12 feet above the ground? How high will it reach? (Lesson 11-6)

A practice test for Chapter 11 is available on page 784.

Scale Drawings   Measurements of angles and geometric shapes are used to make scale drawings. For this project, you will need to choose a real-life space and complete the following activities.

1. Choose an indoor or outdoor area of which to make a scale drawing. For example, you might choose a room or the exterior of a house.

2. Make a sketch of the area, marking details such as walls, doors, windows, electrical outlets, and any other details in the space.

3. Make a plan for measuring the details of the space. Consider how you will measure the angles in the space as well as the distances.

4. Make your measurements and record them on the sketch.

5. Select an appropriate scale for your drawing.

6. Prepare a final scale drawing from the information on your sketch.

**Thinking
Critically**

▶ A quadrilateral has two pairs of parallel sides. What kind of figure could it be?

▶ A triangle has angles measuring x, $2x$, and $3x$ degrees. What kind of triangle is it?

▶ The sum of the degree measures of the angles in a regular polygon is $(n - 2)180$. Explain how you can remember this formula.

 Portfolio

Select your favorite activity from this chapter and write a note explaining what you learned from the activity.

Investigation

OH GIVE ME A HOME

MATERIALS

- ruler
- protractor
- graph paper
- cardboard or foam board
- glue
- paint (optional)

Look around the room where you are sitting. What factors do you think had to be considered before any walls were built? If you did not know any mathematics, would you be able to construct the building that you are sitting in now?

Few things require the use of mathematics more than the construction of buildings. Even the most complex buildings are made up of arrangements of geometric shapes.

Suppose you are a member of a group of three architects that has been hired to design a home. You will oversee every detail of the design from selecting the location to planting the last shrub in the yard.

To create an attractive and efficient building, you must bring harmony to the following elements.

- *Appearance*–The outside appearance of a structure is determined not only by its shape but also by the choice of building materials. Many architects have skillfully blended the rough textures of wood and stone with the smoothness of glass and metal.

- *Durability*–A structure should be well built so that it can withstand all kinds of weather and provide protection from temperature extremes without expensive maintenance.

- *Function*–Every structure is designed for a certain purpose. To be functional, a building should fulfill the needs of its users.

Starting the Investigation ....

When someone decides to build a house, the first thing they must do is choose a *lot*, or piece of land. Work with your group to select a location for the dwelling you will design and build. Then check the zoning laws and building codes for that location.

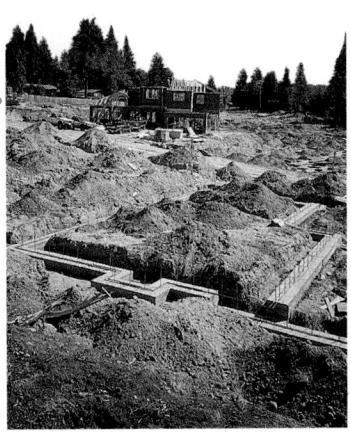

As a group, identify at least four different dwellings in your community. Before you begin, you should familiarize yourself with some styles of houses by referring to an encyclopedia, or consulting with a Realtor. Describe the physical structure of the house and state its purpose. For example, Description: brick 1-story; Purpose: home for a retired couple. Include a sketch of each building.

An *elevation* is a drawing of a building's exterior. An elevation shows the location and size of features such as doors and windows. Typically, four elevation drawings are done, one for each of the four sides of the building. They also show what types of materials will be used on the outside of the building. Choose a style of dwelling that your group would like to build. Then sketch its front elevation.

Research For current information on architecture, visit:

www.glencoe.com/ sec/math/prealg/ mathnet

FRONT ELEVATION

Investigation
Oh Give Me a Home

Working on the Investigation
Lesson 12-5, p. 637

Working on the Investigation
Lesson 12-8, p. 653

Working on the Investigation
Lesson 13-2, p. 671

Closing the Investigation
End of Chapter 13, p. 698

You will work on this Investigation throughout Chapters 12 and 13. Be sure to keep your materials in your Investigation Folder.

12 Measuring Area and Volume

What You'll Learn

In this chapter, you will:

- find the area of polygons and circles,

- investigate the relationship between area and probability,

- find the surface area and volume of prisms, cylinders, pyramids, and cones, and

- solve problems by making a model or drawing.

Why It's Important

Tepees The tepee shown below is the largest in the United States. If you wanted to make a larger tepee, you would need to know how much canvas you would need and probably, if you were on a budget, how much the canvas would cost. A tepee is a cone, so you can use the formula for the surface area of a cone to find the amount of canvas needed. You would need to find the diameter of the base of the tepee and the distance from the tip of the cone to any point on the circumference of the base. *In Lesson 12-6, you will learn how to find the surface area of cones.*

Focus on...

PREREQUISITE SKILLS

To be successful in this chapter, you'll need to understand these concepts and be able to apply them. Refer to the lesson in parentheses if you need more review before beginning the chapter.

▶ **Solve problems by using formulas.** *(Lesson 3-4)*

Solve by replacing the variables with the given values.

1. $A = \frac{1}{2}bh$, if $b = 5$ and $h = 3$

2. $A = \frac{1}{2}h(a + b)$, if $h = 6$, $a = 7$, and $b = 11$

3. $V = \frac{1}{3}s^2h$, if $s = 9$ and $h = 10$

4. $A = 2(\ell w + \ell h + wh)$, if $\ell = 1.5$, $w = 2.7$, and $h = 3.2$

▶ **Find the circumference of a circle.** *(Lesson 7-4)*

Find the circumference of each circle. Round to the nearest tenth.

5. The radius is 3 ft.   **6.** The diameter is 16 cm.

7. The diameter is 5.5 km.   **8.** The radius is 9 in.

▶ **Use proportions to solve problems.** *(Lesson 9-4)*

Solve each proportion. Round to the nearest tenth, if necessary.

9. $\frac{1}{2} = \frac{x}{15}$ **10.** $\frac{11}{14} = \frac{7}{m}$ **11.** $\frac{z}{100} = \frac{13}{64}$ **12.** $\frac{10}{a} = \frac{23}{50}$

▶ **Find the perimeter and area of rectangles and squares.** *(Lesson 3-5)*

Find the perimeter and area of each rectangle.

13. a rectangle that is 6 cm wide and 13 cm long

14. a square that is 20.5 feet on a side

Focus on...

READING SKILLS

In this chapter you will learn about area and volume. When you have three-dimensional objects, like a rectangular prism or box, the area usually refers to the **surface area**. The surface area of a box is the amount of material that is needed to cover the outside of the box. The volume is how much the box will hold. Surface area is always measured in square units while volume is measured in cubic units.

Areas and Geoboards

A Preview of Lesson **12-1**

MATERIALS

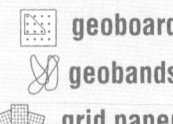

geoboard

geobands

grid paper

In Lesson 3-5, you learned a formula to find the area of a rectangle. But sometimes you need to find the area of shapes that are not rectangular. In this lab, we'll use a geoboard to find the areas of other polygons.

Activity 1 Make these figures on a geoboard. If one square represents one square unit, find the area of each figure.

a.

b.

c.

d.

e.

f.

g. Make a figure on the geoboard and ask your partner to find its area.

TALK ABOUT IT

1. How did you find the areas of the polygons in a–f?
2. Was your partner's polygon made up of whole- and half-squares, or did it include other parts of squares?

Activity 2 Look at the triangle below. Suppose you make another triangle just like it to form a rectangle. You can find the area of the rectangle. Then divide by 2 to find the area of each triangle.

The area of the rectangle is 6 square units. So, the area of each triangle is 3 square units.

Find the area of each triangle.

h.   **i.**   **j.**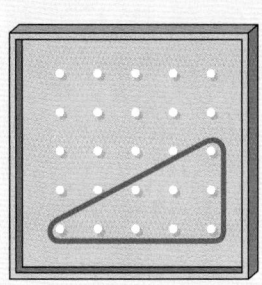

k. Make a right triangle on your geoboard and ask your partner to find the area.

     3. Explain how you found the areas of the triangles.

Extension   Here is another strategy for finding areas. Make the triangle at the right on a geoboard. Now build a rectangle around it as shown in step 2. You can subtract to find the area of the original triangle.

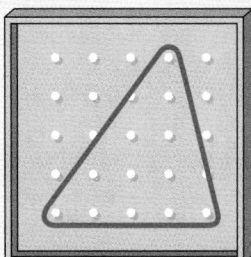

Step 1

The large rectangle has an area of 16 square units. The area of triangle a is half of 4 or 2 square units. The area of triangle b is half of 12 or 6 square units. The area of the original triangle is $16 - 2 - 6$ or 8 square units.

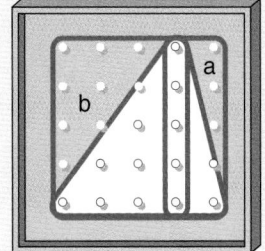

Step 2

Your Turn   **Find the area of each polygon.**

l.   **m.**   **n.**

o.   **p.**   **q.**

r. Make a polygon on your geoboard and ask your partner to find its area.

     4. How did you find the areas of the polygons?

Area: Parallelograms, Triangles, and Trapezoids

MATERIALS: ▦ centimeter grid paper ✂ scissors ⬭ tape

Hands-On Activity

A popular toy starts out as one toy and can be transformed into a different toy by moving its parts. In the activity below, you will transform one geometric shape into another.

Your Turn   Work with a partner.

▶ On centimeter grid paper, draw a rectangle that is 4 cm high and 6 cm long. Cut out the rectangle.

▶ Trace it on a piece of plain paper and label the measures on the sides as shown. Label it rectangle A.

▶ Use scissors to cut a triangle from one side of the original rectangle as shown.

▶ Tape the triangle to the other side of the rectangle to form a parallelogram. Trace the parallelogram on the plain paper next to the rectangle. Label it parallelogram A.

▶ Repeat this activity with two different rectangles (B and C), making two different parallelograms (B and C).

TALK ABOUT IT

a. Compare the area of each rectangle and the parallelogram made from it.

b. How do the measures of the sides of each rectangle relate to the parallelogram formed?

c. How do you find the area of a rectangle?

d. What is the area of a rectangle that is 10 cm long and 2 cm high?

e. Describe a parallelogram with the same area.

Learning the Concept

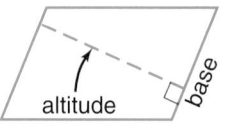

The modeling activity above suggests that you can find the area of a parallelogram by multiplying the measures of the base and the height. The base can be any side of the parallelogram. The height is the length of an **altitude**, a line segment perpendicular to the base with endpoints on the base and the side opposite the base.

<table>
<tr><td rowspan="2">Area of a Parallelogram</td><td>In words:</td><td>If a parallelogram has a base of b units and a height of h units, then the area A is b · h square units.</td></tr>
<tr><td>In symbols:</td><td>$A = bh$</td></tr>
</table>

Example **Find the area of each parallelogram.**

a.

The base is 3 cm. The height is 4 cm.
$A = bh$ *Formula for area of a parallelogram*
$A = 3 \cdot 4$ *Replace b with 3 and h with 4.*
$A = 12$

The area is 12 cm².

b.

$A = bh$
$A = 8.4 \cdot 6.3$
$A = 52.92$

The area is about 53 in².

THINK ABOUT IT

How can you tell that the triangles are congruent?

If you draw a diagonal in a parallelogram, you form two congruent triangles. So the area of each triangle is half the area of the parallelogram. The area of parallelogram *ABCD* at the right is 8 · 4 or 32 square units.

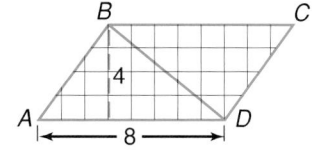

So, the area of $\triangle ABD$ is $\frac{1}{2} \cdot 32$ or 16 square units.

Using the formula for the area of a parallelogram, we can find the formula for the area of a triangle.

<table>
<tr><td rowspan="2">Area of a Triangle</td><td>In words:</td><td>If a triangle has a base of b units and a height of h units, then the area A is $\frac{1}{2}b \cdot h$ square units.</td></tr>
<tr><td>In symbols:</td><td>$A = \frac{1}{2}bh$</td></tr>
</table>

As with a parallelogram, you can use any side of a triangle as a base. The height is the length of the corresponding altitude, a line segment perpendicular to the base from the opposite vertex.

Notice that an altitude can be outside the triangle.

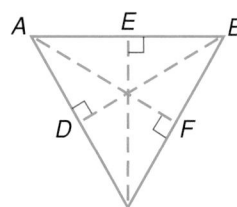

base	altitude
$\overline{AC}$	$\overline{BD}$
$\overline{AB}$	$\overline{CE}$
$\overline{BC}$	$\overline{AF}$

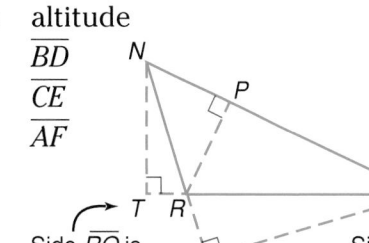

base	altitude
$\overline{NQ}$	$\overline{PR}$
$\overline{NR}$	$\overline{QS}$
$\overline{QR}$	$\overline{NT}$

Side $\overline{RQ}$ is extended to *T*   Side $\overline{NR}$ is extended to *S*

Example **2** **Find the area of each triangle.**

a.

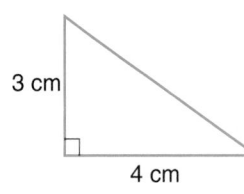

3 cm
4 cm

The base is 4 cm long. The height is 3 cm.

$A = \frac{1}{2}bh$ *Formula for area of a triangle*

$A = \frac{1}{2}(4)(3)$ *Replace b with 4 and h with 3.*

$A = \frac{1}{2}(12)$

$A = 6$

The area of the triangle is 6 cm².

b.

6 cm
5.4 cm

$A = \frac{1}{2}bh$

$A = \frac{1}{2}(6)(5.4)$ *Replace b with 6 and h with 5.4.*

$A = 16.2$ *$0.5 \times 6 \times 5.4 = 16.2$*

The area of the triangle is 16.2 cm².

You can use triangles to find the area of a trapezoid. Remember, a trapezoid is a quadrilateral with exactly two parallel sides called **bases**. The height of a trapezoid is the distance between the bases. The height is the length of an **altitude**, a line segment perpendicular to both bases, with endpoints on the base lines.

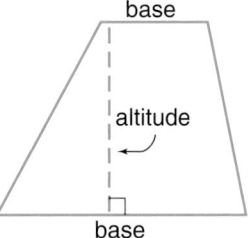

base
altitude
base

Connection to ▷ **Algebra**

On page 613, we drew a diagonal to separate a parallelogram into two congruent triangles. When you draw a diagonal in a trapezoid, the two triangles formed will *not* be congruent. But, the area of the trapezoid is still the sum of the areas of the triangles.

THINK ABOUT IT

Why are the altitudes of the triangles congruent?

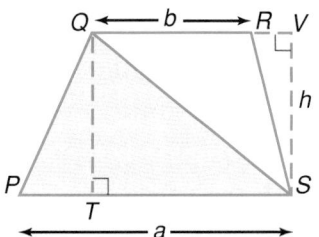

The triangles are △PQS and △QRS.
The altitudes of these triangles, $\overline{QT}$ and $\overline{SV}$, are congruent. Both are h units long.
The base of △PQS, $\overline{PS}$, is a units long.
The base of △QRS, $\overline{QR}$, is b units long.

Area of trapezoid *PQRS* = area of △*PQS* + area of △*QRS*

$$= \quad \frac{1}{2}ah \quad + \quad \frac{1}{2}bh$$

$$= \frac{1}{2}h(a + b) \quad \textit{Distributive property}$$

Area of a Trapezoid

In words: If a trapezoid has bases of *a* units and *b* units and a height of *h* units, then the area *A* of the trapezoid is $\frac{1}{2} \cdot h \cdot (a + b)$ square units.

In symbols: $A = \frac{1}{2}h(a + b)$

Example **Find the area of the trapezoid.**

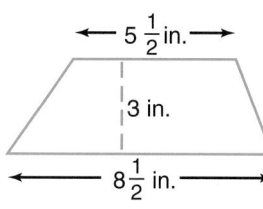

$A = \frac{1}{2}h(a + b)$

$A = \frac{1}{2} \cdot 3\left(8\frac{1}{2} + 5\frac{1}{2}\right)$ *Replace h with 3, a with*

$A = \frac{1}{2} \cdot 3 \cdot 14$ $\qquad$ $8\frac{1}{2}$, *and b with* $5\frac{1}{2}$.

$A = 21$

The area of the trapezoid is 21 in².

Example ④

Landscaping

Marisa works during the summer taking care of several of her neighbors' lawns. In addition to mowing, she also fertilizes the lawns. The label on the bag of the fertilizer she will be using indicates that one bag fertilizes 2000 square feet. How much fertilizer should she buy for the yard shown below?

The yard is in the shape of a trapezoid. The house and driveway are rectangles within the trapezoid that do not need fertilizer. Subtract the area of the rectangles from the area of the trapezoid to find the area that will be fertilized.

Area of the lot

$A = \frac{1}{2}h(a + b)$ $\qquad$ *Formula for area of a trapezoid*

$A = \frac{1}{2} \cdot 80 \cdot (100 + 125)$ $\qquad$ *Replace h with 80, a with 100, and b with 125.*

$A = \frac{1}{2} \cdot 80 \cdot 225$

$A = 9000$

Area of the house

$A = \ell w$ $\qquad$ *Formula for area*
$A = 24 \cdot 30$ $\qquad$ *of a rectangle*
$A = 720$

Area of the driveway

$A = \ell w$
$A = 12 \cdot 50$
$A = 600$

$9000 - 720 - 600 = 7680$

The area to be fertilized is 7680 square feet. Since each bag of fertilizer covers 2000 square feet, Marisa will need 7680 ÷ 2000 or about 4 bags of fertilizer for the yard.

Checking Your Understanding

Communicating Mathematics

Read and study the lesson to answer these questions.

1. What is the difference between an altitude in a triangle and an altitude in a parallelogram or trapezoid?

2. **Draw** and label a diagram to show the three altitudes of a right triangle.

3. **Tell**, in your own words, how to find the area of a trapezoid.

4. **Draw** on centimeter grid paper a parallelogram, a triangle, and a trapezoid that each have an area of 16 square units.

State the measures of the base and height and how to find the area of each figure. Then find each area.

5.

6.

7.

Find the area of each figure described below.

8. triangle: base, 10 cm; height, 4.5 cm

9. parallelogram: base, 12 cm; height, 18 cm

10. trapezoid: bases, $6\frac{1}{2}$ ft and 11 ft; height, 5 ft

11. Geography The state of Indiana is shaped almost like a trapezoid. The distances along the east and west boundaries are about 200 miles and about 280 miles, and the east-west distance is about 140 miles.

   a. Estimate the area of the state.

   b. Look up the actual area in an encyclopedia and compare it to your estimate.

Exercises: Practicing and Applying the Concept

Independent Practice

Find the area of each figure.

12.

13.

14.

15.

16.

17.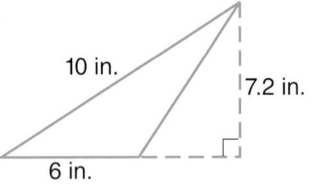

✓ **Choose**

Estimation
Mental Math
Calculator
Paper and Pencil

18.

19.

20.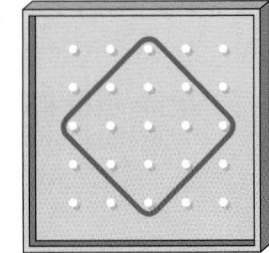

Find the area of each figure described below.

21. trapezoid: height, 2 in.; bases, 3 in. and 6 in.

22. triangle: base, 10 in.; height, 7 in.

23. parallelogram: base $3\frac{1}{4}$ ft; height, 2 ft

24. triangle: base 0.8 km; height, 2 km

25. trapezoid: height, 3.5 m; bases 10 m and 11 m

26. parallelogram: base, 25 yd; height, 11 yd

27. In Exercise 20, you may have found the area of the square by finding the areas of two triangles and adding them. Show that this is the same as $A = \frac{1}{2}d^2$ where d is the length of a diagonal.

Critical Thinking

28. Explain or demonstrate how the formula for the area of a trapezoid can be used to find the areas of parallelograms and triangles.

Applications and Problem Solving

29. Consumerism Carpeting is sold in several standard widths, but since room sizes are not standard, there frequently is wasted carpeting, for which the home owner must pay.

   a. If you are carpeting a room that is 14 feet by 22 feet, what is the area of the room?

   b. If the carpet you have chosen is sold in 12 foot and 15 foot widths, which size will result in less waste?

30. Geography The state of Tennessee is shaped almost like a parallelogram. The distance along the north boundary is 447 miles, and the north-south distance is 116 miles. Estimate the area of the state.

31. Pet Care In many communities, pets must be kept in a fenced area, or on a leash. If you had 48 feet of fencing to fence a rectangular area, describe the shape you would choose in order to give your pet the greatest amount of area. Justify your answer.

32. Advertising Use the formula in Exercise 27 to determine whether the claim about the viewing area in the advertisement below is accurate. Remember that television screens are measured diagonally and that the diagonals of a square are congruent.

For **Extra Practice**, see page 769.

Mixed Review

33. Geometry How many lines of symmetry does a regular pentagon have? (Lesson 11-9)

34. Estimate 26% of 120. (Lesson 9-8)

35. Standardized Test Practice Solve $n = -9.45 \div -4.5$. (Lesson 6-5)

   A. -4.95 **B.** 4.95   **C.** -2.1 **D.** 2.1

36. Write $(-4)(-4)(-4)$ using exponents. (Lesson 4-2)

37. Rewrite $3 + (x + 4)$ using the associative and commutative properties. Then simplify. (Lesson 1-4)

12-1B Fractals

An Extension of Lesson **12-1**

MATERIALS

✐ ruler

⚖ compass

△ protractor

The design shown at the right is called the Koch snowflake. It is made up of three rotated copies of the Koch curve, introduced by Helge von Koch, a mathematician who worked on fractal geometry. Fractals are self-similar. That is, if you look at any part of a fractal, you will see a miniature replica of the larger design. A tree of broccoli is an example of a fractal in nature. Each branch of the tree is a miniature replica of the entire tree.

Your Turn

Work with a partner to construct a Koch curve.

a. Use a ruler to draw a line segment about 6 inches long. This is stage 0 of the curve.

b. Use the ruler to separate the line into three congruent segments.

c. Set the compass to the length of one segment. Then draw arcs from each end of the middle segment as shown at the right. Use the point where the arcs intersect to draw an equilateral triangle. Then erase the base of the triangle. This is stage 1.

d. Let the length of the stage 0 curve be 1 unit. Write an expression to represent the length of the stage 1 curve and record in a table.

e. Repeat step c on the sides of stage 1 to produce a stage 2 figure. Again determine the length and record in the table.

f. Repeat step c once more to produce a stage 3 figure. Determine the length and record in the table.

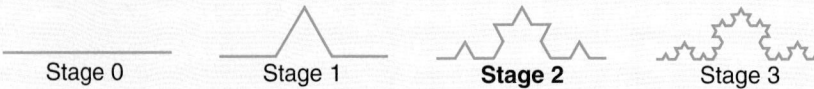

| Stage 0 | Stage 1 | **Stage 2** | Stage 3 |

TALK ABOUT IT

1. Why do you think the Koch curve is considered to be self-similar?

2. What happens to the length in each stage?

3. Suppose the curve is continued indefinitely. Make a conjecture about the length.

4. Research fractal geometry. Then write your own definition of a fractal.

Extension

5. The Koch curve is produced by repeating step c above an infinite number of times. The Koch snowflake is made from three rotated copies of the curve. Stage 0 is an equilateral triangle. Use your constructions to produce a stage 1, stage 2, and stage 3 snowflake. Suppose the area of the stage 0 triangle is 1 square unit. Determine the area of each stage. Then make a conjecture about the area of the Koch snowflake.

12-2

Area: Circles

What You'll Learn

■ To find the areas of circles.

Why It's Important

You can compare the size and prices of pizzas if you can find the area of circles.

MATERIALS: ▦ centimeter grid paper ⟋compass ⟋ruler ◠ protractor

Hands-On Activity

Pizza, anyone? Most pizzas are round and come in different sizes, measured by diameter. Have you ever wondered whether one size would give you more pizza for your money? To find out, you need to find the area of a circle. This activity will help you understand how to find the area of a circle.

Your Turn Work with a partner.

▶ On centimeter grid paper, use a compass to construct a circle with a diameter of 20 centimeters. Do this by putting the point of the compass at the corner of a grid square and adjusting the compass so it is 10 centimeters wide.

▶ Count the grid squares and the parts of grid squares that lie inside the circle. Use your judgment on how to count each partial grid square. Record this estimate of the area of the circle.

▶ With a protractor, straightedge, and pencil, separate the circle into 16 congruent parts as shown in the photograph above. Shade one half of the circle as shown.

▶ Cut out the parts and arrange them in a side-by-side pattern as shown. When you have finished, you should have a shape that looks like a parallelogram.

TALK ABOUT IT

a. What part of the circle makes up the base of the parallelogram?

b. What part of the circle is the height of the parallelogram?

c. Use a ruler to find the approximate measure of the base and height.

d. What is the approximate area of the parallelogram?

e. Compare to the estimate you found by counting grid squares.

Learning the Concept

In the activity, you separated the circle into parts and made a figure that looks like a parallelogram. The circle has the same area as that figure.

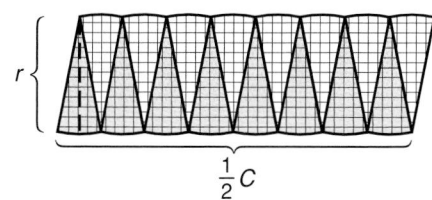

You can use the formula for the area of a parallelogram to find the formula for the area of a circle.

You can review circumference in Lesson 7-4.

$A = b \times h$

$A = \left(\frac{1}{2} \times C\right) \times r$ — *The base of the parallelogram is one-half the circumference.*

$A = \left(\frac{1}{2} \times 2\pi r\right) \times r$ — *Remember, $C = 2\pi r$.*

$A = \pi \times r \times r$

$A = \pi r^2$ — *The formula is $A = \pi r^2$.*

Area of a Circle	**In words:** If a circle has a radius of r units, then the area A is $\pi \cdot r^2$ square units.
	In symbols: $A = \pi r^2$

Example ① Find the area of each circle. Round to the nearest tenth.

a.

4 in.

$A = \pi r^2$ — *Formula for area of a circle*

$A = \pi \cdot 4^2$ — *The radius is 4 in. Estimate: $3 \cdot 16 = 48$*

$\pi \boxed{\times} 4 \boxed{x^2} \boxed{=} \; 50.26548246$

$A \approx 50.3$ — *Compare with the estimate.*

The area is about 50.3 in².

b.

6 cm

$A = \pi r^2$

$A = \pi \cdot 3^2$ — *The radius is one-half the diameter, so $r = \frac{1}{2}(6)$ or 3.*

$A \approx 28.3$ — *Estimate: $3 \times 9 = 27$*

The area is about 28.3 cm².

Example ②

Consumerism

A 10-inch, two-topping pizza sells for $5.99 and a 14-inch, two-topping pizza sells for $11.19. Which is a better buy, one 14-inch or two 10-inch pizzas?

Use $A = \pi r^2$ to find the area of each size of pizza.

10-inch	14-inch
$A = \pi r^2$	$A = \pi r^2$
$A = \pi \cdot 5^2$	$A = \pi \cdot 7^2$
$A \approx 79$	$A \approx 154$

The area of the 14-inch pizza is almost twice as much as the 10-inch. So two 10-inch pizzas are about the same amount of pizza for 2($5.99) or $11.98, as one 14-inch pizza for $11.19. It appears that the 14-inch pizza is the better buy.

Communicating Mathematics

Read and study the lesson to answer these questions.

1. **Explain** how to find the area of a circle if you know the measure of the radius.

2. **Estimate** the area of the circle at the right by counting grid squares. Then find the area using the formula. Compare the difference of the two areas to the actual area to find the percent of error.

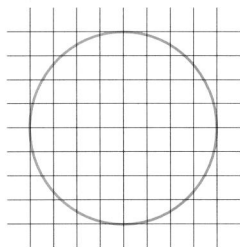

3. **You Decide** Joyce says the area of a circle with a diameter of 10 cm is about 75 cm². Sanjay says the area is about 300 cm². Use estimation to determine who is correct. Explain.

Guided Practice

Find the area of each figure. Round to the nearest tenth.

4.
9 cm

5.
24 in.

6.
21 m
← 42 m →

7. **City Planning** The circular region inside the streets at DuPont Circle in Washington, DC, is 250 feet across. How much area does the grass and sidewalk cover?

Exercises: Practicing and Applying the Concept

Independent Practice

Find the area of each circle. Round to the nearest tenth.

8.
2 in.

9.
8 cm

10.
$35\frac{1}{4}$ ft

11.
6.2 m

12.
$42\frac{1}{2}$ in.

13.
24.6 cm

14. radius, 23 cm

15. diameter, 14.6 in.

16. radius, 24.6 m

Find the area of each figure. Round to the nearest tenth.

17.
5 ft

18.
2.5 m
2.5 m

19.
100 ft

Critical Thinking

20. The area and the circumference of a circle are the same number.
    a. What is the radius of the circle?
    b. What is unique about this radius?

Applications and Problem Solving

21. History Stonehenge, an ancient monument in England, may have been used as a calendar. The stones are arranged in a circle 30 meters in diameter. Find the area of the circle.

22. Landscaping Hiroshi needs to match some pipe used in the sprinkling system where he works. He doesn't want to cut the pipe to find the diameter, but he can measure the circumference. He finds that it is about 6.25 inches. What is the diameter of the pipe?

23. Architecture The largest dome built in ancient times is the Roman Pantheon. The diameter of its dome is 42.7 m. The Houston Astrodome has a diameter of 216.4 m. About how many times more area does the Astrodome cover than the Pantheon?

24. Sports The Athletic Booster Club for a high school is raising funds for artificial grass for the football playing area in the infield of the track. The grass will cost $50 per square yard, including installation. To the nearest dollar, how much money must they raise to pay for the grass?

25. Public Safety A town has installed a tornado warning system. The sound emitted can be heard for a 2-mile radius. Find the area that will benefit from the system.

Mixed Review

26. Real Estate A farmer's land is in the shape of a trapezoid. The north boundary is 220 yards long, the south boundary is 176 yards long, and the east-west distance is 1320 yards. How much land does the farmer have? (Lesson 12-1)

27. Standardized Test Practice Solve $\frac{15}{c} = \frac{8}{9.6}$. (Lesson 9-4)

A. 0.08   **B.** 12.5   **C.** 18   **D.** 1152

For **Extra Practice**, see page 769.

28. Solve $3x - 5 > 3x + 4$. (Lesson 7-6)

29. Express $0.\overline{4}$ as a fraction in simplest form. (Lesson 5-1)

From the → **FUNNY PAPERS**

1. What is the formula the man is trying to write?
2. What does the formula represent?
3. Make up a way of remembering the correct formula.

12-3

Integration: Probability
Geometric Probability

You'll Learn

■ To find probabilities using area models.

It's Important

Finding probabilities using area models can help you determine whether your golf ball will land on the putting green.

In a game at the Johnson Middle School Carnival, you toss a beanbag onto a target like the one shown at the right. You get a prize if your beanbag lands in one of the colored circles. Let's assume that all the beanbags land on the board, and that any beanbag is equally likely to land any place on the board. What is the probability that a beanbag will land in one of the circles? *This problem will be solved in Example 2.*

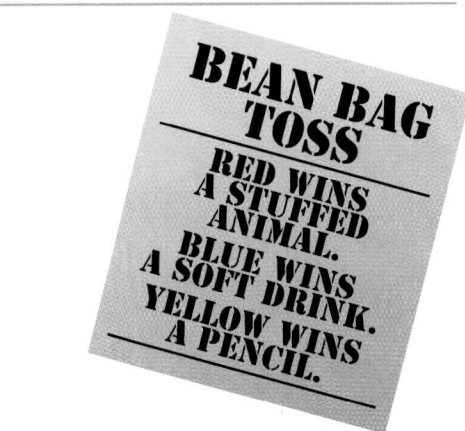

BEAN BAG TOSS
RED WINS A STUFFED ANIMAL.
BLUE WINS A SOFT DRINK.
YELLOW WINS A PENCIL.

36 in.

6 in. 6 in. 6 in.

36 in.

Learning the Concept

Geometric probability uses ideas about area to find the probability of an event. In Lesson 9-3, you calculated probabilities for spinners. A dartboard is similar to a spinner. On the dartboard below, there are ten equally likely outcomes, five of which are shaded. The probability of a dart landing in a shaded section is $\frac{5}{10}$ or $\frac{1}{2}$.

In this lesson, for problems involving dartboards, we will assume that the dart will land on the dartboard and that it is equally likely to land anywhere on the dartboard.

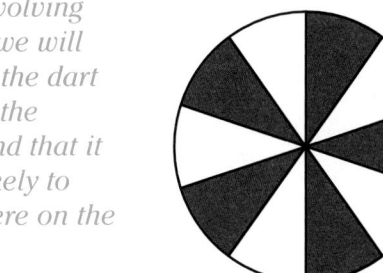

$P(\text{shaded}) = \frac{5}{10}$ or $\frac{1}{2}$

You can also relate probability to the areas of other geometric shapes.

Lesson 12-3 **INTEGRATION** *Probability   Geometric Probability*   **623**

The figure at the right represents a dartboard.

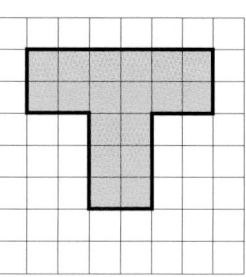

a. Find the probability of landing in the shaded region.

b. Suppose you threw 50 darts. How many would you expect to land in the shaded region?

a. You can count squares to find the area of the shaded region and the area of the whole target.

$$P(\text{shaded}) = \frac{\text{shaded area}}{\text{area of target}}$$

$$= \frac{18}{64} \text{ or } \frac{9}{32}$$

b. Let n represent the darts landing in the shaded region.

$$\frac{n}{50} = \frac{9}{32} \qquad \textit{Write a proportion.}$$

$$50 \cdot 9 = 32n \qquad \textit{Find the cross products.}$$

$$450 = 32n$$

$$14.1 \approx n \qquad \textit{About 14 darts should land in the shaded region.}$$

Refer to the application at the beginning of the lesson. What is the probability of winning a prize on any given toss?

The area within the circles represents the outcomes of winning a prize and the total area of the board (including the circles) represents all the possible outcomes.

First find the area of the entire board and the area within the circles. Then write a fraction to represent the probability of winning.

The board is a 36-inch by 36-inch square.

$$A = s^2 \qquad \textit{Formula for area of a square}$$

$$A = 36^2 \qquad \textit{Replace s with 36.}$$

$$A = 1296$$

The board has an area of 1296 in^2.

The area for winning a prize is a circle with a radius of 15 in. *Why?*

$$A = \pi r^2 \qquad \textit{Formula for area of a circle}$$

$$A = \pi(15)^2 \qquad \textit{Replace r with 15. Estimate: } 3 \times 225 = 675$$

$$A \approx 707$$

The area for winning a prize is about 707 in^2.

So, $P(\text{winning a prize}) = \dfrac{\text{winning area}}{\text{total area}}$

$$\approx \frac{707}{1296}$$

$$\approx 0.546 \qquad \textit{Use a calculator.}$$

The probability of winning is about 55% or $\dfrac{11}{20}$.

Communicating Mathematics

Read and study the lesson to answer these questions.

1. Refer to the application at the beginning of the lesson. Tell how you would find the probability of winning a soft drink or a pencil.

2. **Draw** a dart board for which the probability of landing on a shaded area is $\frac{1}{3}$.

MATERIALS

🗒 grid paper
🫘 dry beans

3. Work with a partner.

▶ Outline two squares as shown at the right.

▶ Hold 25 beans about 3 inches above the paper and drop them onto the paper.

▶ Record the number of beans that land within the small square and the number that land within the large square (including the small square.) If a bean lands on a line, count it in the area where the greater part of the bean lands. Do not count beans that land outside both squares.

▶ Repeat three more times.

▶ Find the ratio $\dfrac{\text{number of beans in small square}}{\text{number of beans in large square}}$.

▶ Find the ratio $\dfrac{\text{area of small square}}{\text{area of large square}}$.

▶ Repeat the activity or combine results with other pairs.

▶ Compare the two ratios. What conclusion can you make about the relationship of area and probability?

Guided Practice

Each figure represents a dartboard. Find the probability of landing in the shaded region.

4.

5.

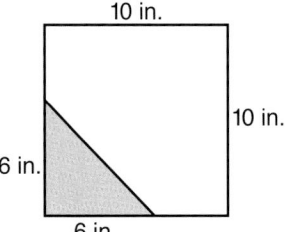

6. Suppose you throw 50 darts at the target in Exercise 4. Predict how many darts will land in the shaded region.

7. Suppose you throw 50 darts at the target in Exercise 5. Predict how many darts will land in the shaded region.

8. **Skydiving** A sky diver parachutes onto a 100-yard by 100-yard square field with a tree in each corner of the field. The diver will not get tangled in tree branches if she stays at least 6 yards away from the tree trunk. Assume the diver lands within the field and that she is equally likely to land anywhere in the field. What is the probability of a clear landing?

Independent Practice

Each figure represents a dartboard. Find the probability of landing in the shaded region.

9.

10.

11.

12.

13.

14.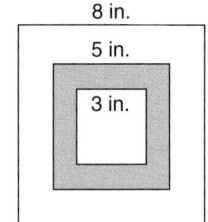

15. Suppose you throw 100 darts at the target in Exercise 9. Predict how many darts will land in the shaded region.

16. Suppose you throw 75 darts at the target in Exercise 12. Predict how many darts will land in the shaded region.

Critical Thinking

17. **Games** At the Johnson Middle School Carnival, you can win a CD by tossing a quarter onto a grid board so that it doesn't touch a line. The sides of the squares of the grid are 40 millimeters long and the radius of a quarter is 12 millimeters. What is the probability of winning? (*Hint:* Find the area where the center of the coin could land so that the edges don't touch a line.)

Applications and Problem Solving

18. **Games** Refer to the application at the beginning of the lesson. What is the probability of winning a stuffed animal?

19. **Mining** A geologist told Sam Antonio that there is oil under 500 acres of his 17,500 acre ranch. If Sam randomly starts drilling, what is the probability that he will strike oil?

20. Computers Computers can be used to simulate problem situations using a random number generator. Suppose a computer program randomly selects two real numbers between 0 and 2. Find the probability that the sum of the two numbers is:

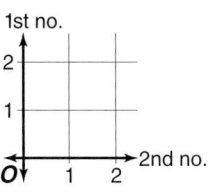

a. less than 2

b. greater than 1

c. between 1 and 2

(*Hint:* Use a coordinate grid.)

Mixed Review

21. Standardized Test Practice The science department chairperson surveyed 350 tenth grade students to determine the number of students who planned to enroll in biology and chemistry next year. The results of the survey are shown in the diagram. How many students plan to take chemistry?

A = set of students surveyed
B = set of students to take biology
C = set of students to take chemistry
(Lesson 12-2)

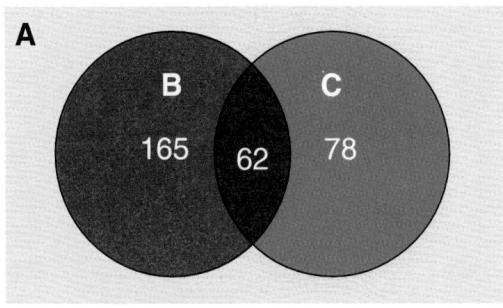

A. 32 **B.** 46 **C.** 78 **D.** 140

22. Geometry The measures of the angles of a triangle are in the ratio 1:3:5. Find the measure of each angle. (Lesson 11-4)

23. Science An inclined plane is a slanted surface, which may be used for raising objects to higher places. Find the slope of the inclined plane pictured below. (Lesson 8-6)

0.3 m

3 m

24. Evaluate xy if $x = \frac{1}{5}$ and $y = \frac{3}{5}$. (Lesson 6-3)

25. Write an inequality using the numbers in the sentences below. Use the symbols $<$ or $>$. (Lesson 2-3)

Water boils at 212°F. It freezes at 32°F.

For **Extra Practice**, see page 769.

12-3B

Geometric Probability

An Extension of Lesson **12-3**

You can use a graphing calculator to experiment with area and probability. The program below generates random ordered pairs.

MATERIALS

graphing calculator

grid paper

straightedge

PROGRAM:RANDPRS
: ClrHome
: Fix 1
: Input "LEAST X", A
: Input "GREATEST X", B
: Input "LEAST Y", C
: Input "GREATEST Y", D
: Input "NUMBER OF PAIRS", P
: 0 →I
: Lbl 1

: I +1 → I
: (B–A)rand+A →X
: Disp X
: (D–C)rand+C →Y
: Disp Y
: Disp " "
: Pause
: If I≠P
: Goto 1
: Stop

Your Turn

Work with a partner.

▶ Use a straightedge to draw a square like the one shown at the right. Shade the figure as shown.

▶ Enter the program above into a graphing calculator.

▶ Use the program to generate 50 ordered pairs. Enter 0 as the least value and 8 as the greatest value for both *x* and *y*.

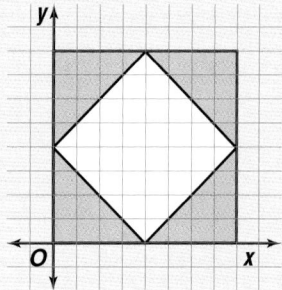

▶ Graph each pair on the grid paper and keep a tally of how many points are in the shaded and unshaded areas.

▶ Write a ratio comparing the points in the shaded area to the total number of points.

▶ Find the areas of the small square and the large squares. Write a ratio comparing the area of the small square to the area of the large square.

TALK ABOUT IT

1. How do the two ratios you wrote compare?

2. Repeat the activity above to generate and graph another 50 ordered pairs. Combine the results with the results from the first run and write the ratio.

3. What effect does more data have on the results?

4. Predict the results of repeating the activity using the figure at the right. Then check your prediction by running the program and graphing the results.

12-4 Problem-Solving Strategy: Make a Model or a Drawing

Real World

Application: Pet Care

June has attached a 20-foot rope to the corner of a 10-foot by 10-foot shed to tie up her horse Harry. How much grazing area does Harry have?

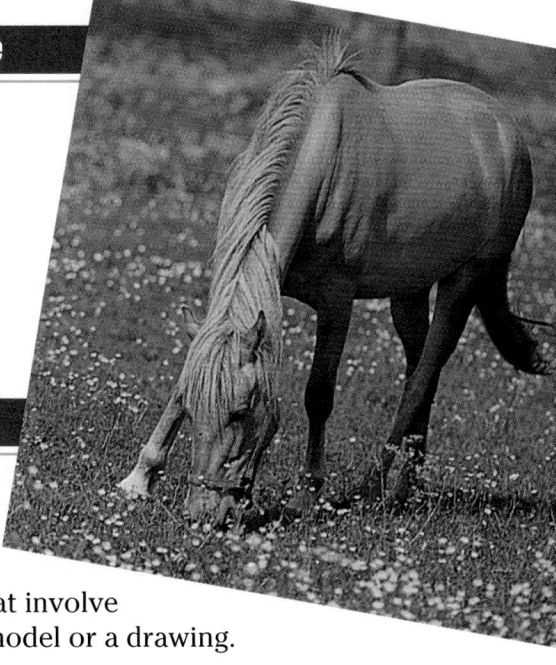

Learning the Concept

Without a drawing, it is difficult to find the area or even visualize the shape of the horse's grazing area. Many problems, especially those that involve geometry, are easier if you make a model or a drawing.

Explore You know the dimensions of the shed and the length of the rope. You need to find the grazing area.

Plan Make a drawing to illustrate the problem. Notice that you have sections of circles. The largest section is $\frac{3}{4}$ of a circle with a radius of 20 feet. The smaller sections are each $\frac{1}{4}$ of a circle with a radius of 10 feet. You can find the areas of each section and add to find the total grazing area.

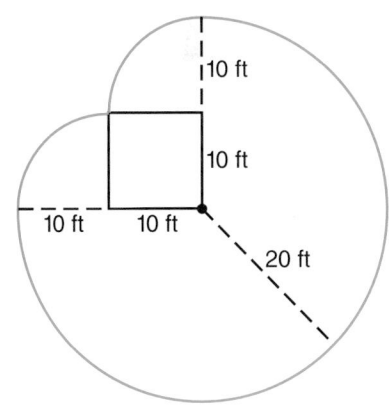

Solve Area of the large section:

Estimate: $3 \times 400 \times \frac{3}{4}$
   $= 900$

$A = \pi r^2 \times \frac{3}{4}$
$A = \pi \times 20^2 \times \frac{3}{4}$
$A \approx 942$

Area of each smaller section:

Estimate: $3 \times 100 \times \frac{1}{4}$
   $= 75$

$A = \pi r^2 \times \frac{1}{4}$
$A = \pi \times 10^2 \times \frac{1}{4}$
$A \approx 79$

Total area = 942 + 2(79) or 1100   *There are 2 smaller sections.*
The horse has a grazing area of about 1100 ft².

Examine An answer of 1100 ft² seems reasonable, compared to your estimate.

Communicating Mathematics

Read and study the lesson to answer these questions.

1. Refer to the application at the beginning of the lesson. Draw a picture of the problem if the rope was 25 feet long. Then explain how the problem is different.

2. How can making a model or drawing help in problem solving?

3. Think of a real-life problem where making a model or drawing can help you solve the problem. Write about the problem and its solution. Include a drawing.

Guided Practice

Solve by making a model or drawing.

4. Refer to the application at the beginning of the lesson. Suppose the rope is only 15 feet long. Find the grazing area.

5. **Framing**   A painting 15 cm by 25 cm is bordered by a mat that is 3 centimeters wide. The frame around the mat is 2 centimeters wide. What is the area of the picture including the frame and mat?

Exercises: Practicing and Applying the Concept

Independent Practice

Solve. Use any strategy.

6. A rectangular swimming pool is 15 feet wide and 30 feet long. One-foot-square tiles are to be installed around the pool to form a walkway 3 feet wide all around the pool. What is the area to be covered by the tiles?

7. There are five ways to place four one-inch squares together so that any two adjoining sides match exactly. These match: These do not match:

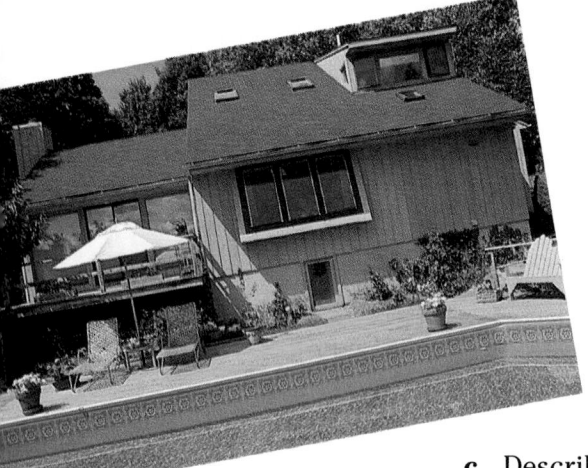

LOOK BACK

You can review perimeter in Lesson 3-5.

   a. Find the five ways.

   b. Find the perimeter of each figure.

   c. Describe the shape or shapes for which the perimeter is greatest.

   d. Describe the shape or shapes for which the perimeter is least.

   e. How many ways can you arrange five squares? Do their perimeters differ?

8. Find the sum of the first 100 *even* positive numbers.

9. The perimeter of a rectangle is 32 meters. Its area is 48 square meters. What are the dimensions of the rectangle?

10. **Sports**   A circular track has a diameter of 400 feet. Les starts out jogging around the track at a rate of 600 feet per minute. One minute later, Antonio starts jogging at the same place. He goes in the same direction and jogs at 700 feet per minute. Will Antonio pass Les on the first, second, third, or fourth quarter of the track?

11. Sandy has 36 identical cubes. Each edge of a cube is 1 inch long. How can Sandy arrange the cubes to have the smallest surface area?

12. Geometry Various views of a solid figure are shown below. The edge of one block represents one unit of length. A dark segment indicates a break in the surface.

a. Use the views to sketch the solid.

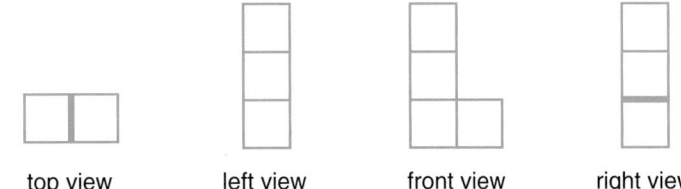

top view   left view   front view   right view

b. Choose a solid object from your home and draw the top, left, front, and right views.

13. Ms. Warren sold 100 shares of stock for $2475. When the price per share went down $4, she bought 200 more shares. When the price per share went back up $3, she sold 100 shares. How much did Ms. Warren gain or lose in her transactions?

14. Keiko's little brother fills an 8-quart pail with sand for a sand castle. He also has empty 3-quart and 5-quart pails. How could he use these to divide the sand into two equal portions?

15. Three different views of the same cube are shown below. What figure is on the side opposite the X?

Critical Thinking

16. The figures you drew in Exercise 7 are called *nets*. Some nets can be folded to form solid figures. Draw all of the nets containing five one-inch squares that could be folded into open boxes. Mark the square that would be the bottom of the box.

Mixed Review

17. Probability The figure at the right represents a dartboard. What is the probability of landing in the shaded region? (Lesson 12-3)

18. Find the value of $P(5, 4)$. (Lesson 10-6)

19. Wages You earn $8 an hour and a $12 bonus. What is the least number of hours you must work to earn at least $100? (Lesson 7-7)

20. Write 2^{-3} using positive exponents. (Lesson 4-9)

21. Standardized Test Practice Which point on the graph has coordinates (3, 1)? (Lesson 1-7)

    A. A

    B. B

    C. C

    D. D

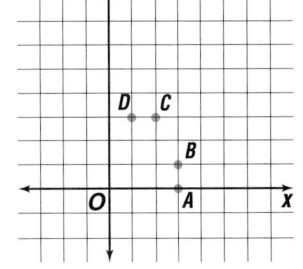

For **Extra Practice**, see page 770.

12-5

Surface Area: Prisms and Cylinders

Application: Postal Service

What You'll Learn

■ To find surface areas of triangular and rectangular prisms and circular cylinders.

Why It's Important

You can use surface area to help you find how much wrapping paper you'll need to wrap a present.

Maria was buying stamps at a U. S. Post Office when she noticed shipping boxes on display. The price list is shown at the right.

Size (length × width × height)	Price
8 in. × 8 in. × 8 in.	$1.25
15 in. × 10 in. × 12 in.	$2.00
20 in. × 14 in. × 10 in.	$2.50

Maria wondered if the price of each box was related to the amount of material needed to make the box. Is the ratio of price to material about the same for each box? *You will solve this problem in Exercise 27.*

Learning the Concept

THINK ABOUT IT

Rectangles *ABCD* and *EFGH* are bases. Could another pair of regions be the bases of this prism? Does a base always need to be on the bottom?

In geometry, solids like the shipping box shown above are called **prisms**. A prism is a solid figure that has two parallel congruent sides, called **bases**. A prism is named by the shape of its bases. The shipping box and the prism shown at the right are **rectangular prisms**.

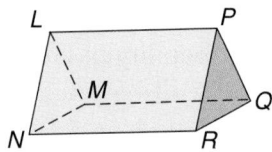

The parallel bases of the prism shown at the left are shaped like triangles, so it is called a **triangular prism**. Can you think of something in everyday life that has the shape of a triangular prism?

The amount of material that it would take to cover a geometric solid is the **surface area** of the solid. If you "open up" or "unfold" a prism, the result is a **net**. Nets help us see the regions or **faces** that make up the surface of the prism. The surface area of a prism is the area of the net.

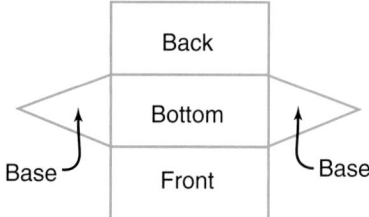

A triangular prism has five faces.

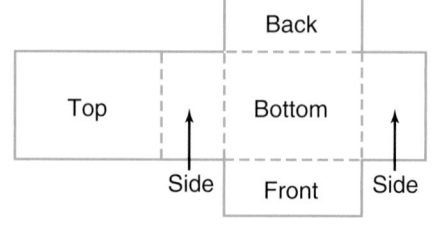

A rectangular prism has six faces.

Example **1** Find the surface area of the 15 in. × 10 in. × 12 in. box in the application at the beginning of the lesson.

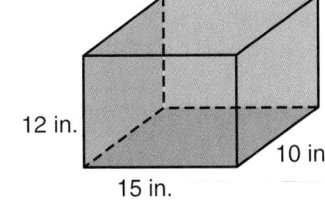

12 in.
10 in
15 in.

Method 1

Sketch the box.
Use the formula $A = \ell w$ to find the areas of the faces.

Front and Back $12 \times 15 \times 2 = 360$

Top and Bottom $10 \times 15 \times 2 = 300$

Two Sides $12 \times 10 \times 2 = 240$

Add to find the total surface area. $360 + 300 + 240 = 900$
The surface area of a 15 in. × 10 in. × 12 in. box is 900 in².

Method 2

Draw a net for the box and label the dimensions of each face. Find the area of each face. Then add to find the total surface area.

10 in.
12 in.
10 in.
15 in.
12 in.
12 in.

Bottom $10 \times 15 = 150$

Side $12 \times 10 = 120$

Top $10 \times 15 = 150$

Side $12 \times 10 = 120$

Back $12 \times 15 = 180$

Front $12 \times 15 = 180$

Total $150 + 120 + 150 + 120$
$\qquad + 180 + 180 = 900$

The surface area of the box is 900 in².

Example **2** Find the surface area of the triangular prism shown below.

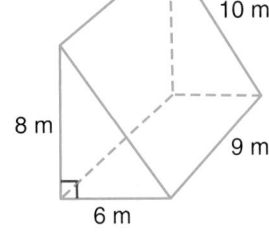

10 m
8 m
9 m
6 m

Use the formula $A = \frac{1}{2}bh$ to find the area of the bases.

$\frac{1}{2} \cdot 6 \cdot 8 = 24$

Use the formula $A = \ell w$ to find the area of the sides.

Front $9 \cdot 10 = 90$
Bottom $6 \cdot 9 = 54$
Back $8 \cdot 9 = 72$

Add to find the total surface area. $2(24) + 90 + 54 + 72 = 264$
The surface area of the triangular prism is 264 m².
Check this answer by drawing a net of the prism.

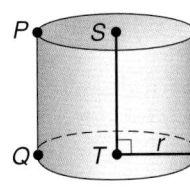

You probably have cans of food in your kitchen at home. These cans are examples of **cylinders**. The bases of a circular cylinder are two parallel, congruent circular regions.

An altitude is any perpendicular line segment joining the two bases. $\overline{ST}$ and $\overline{PQ}$ are altitudes.

If you open the top and bottom of a cylinder and then make a vertical cut in the curved surface, you could lay the cylinder flat in the same way you did with a rectangular prism. You would then have a pattern, or net, for making that cylinder.

From the diagram above, you can see that:

1. The bases of the cylinder are congruent circular regions.
2. The curved surface of the cylinder opens to form a rectangular region.
3. The width of the rectangle is the height of the cylinder, or h.
4. The length of the rectangle is the circumference of a base, or $2\pi r$.

Connection to ⟩ **Algebra**

The area of each base is πr^2. Since the area of a rectangle is ℓw, the area of the curved surface is $2\pi r \cdot h$. So, the surface area of a circular cylinder is $2(\pi r^2) + 2\pi r \cdot h$.

Example ❸

Real World
APPLICATION

Manufacturing

A potato chip can shaped like a cylinder has a diameter of 3 inches and a height of 8.5 inches.
a. How much area must the label cover?
b. What is the total surface area of the can?

Explore   You need to find the area of the label and the total surface area of the can. You know the diameter and the height of the can.

Plan   The label covers the curved surface of the can. Use $A = \ell w$ or $A = 2\pi r \cdot h$ to find the area of the curved surface.

Estimate: $2 \cdot \pi \cdot 1\frac{1}{2} \cdot 8.5 \rightarrow 2 \cdot 3 \cdot \frac{3}{2} \cdot 9 = 81$

The top and bottom bases of the can are circles. Use $A = \pi r^2$ to find the area of each base.

Estimate: $\pi \cdot 1\frac{1}{2} \cdot 1\frac{1}{2} \rightarrow 3 \cdot 1 \cdot 2 = 6$

Then add to find the total surface area.
Estimate: $81 + 6 + 6 = 93$

Solve   **a.** $A = \ell w$ *Area of a rectangle*
          $A = 2\pi r \cdot h$ *Replace ℓ with the expression for*
                                    the circumference of a circle.
          $A = 2 \cdot \pi \cdot 1.5 \cdot 8.5$ *$d = 3$, so $r = 1.5$.*
          $A \approx 80.1$

The label covers 80.1 in².

b. $A = \pi r^2$ *Formula for area of a circle*
       $A = \pi \cdot 1.5^2$
       $A \approx 7.1$ The area of each base is about 7.1 in².

The total surface area is about $80.1 + 2(7.1)$ or about 94.3 in².

Examine   Compared to the estimates, the answers make sense.

Checking Your Understanding

Communicating Mathematics

Read and study the lesson to answer these questions.

1. How is surface area important when you giftwrap a package?

2. **Define** surface area in your own words.

3. **Explain** how to find the surface area of a circular cylinder.

4. **State** an example from your daily life of a rectangular prism, a triangular prism, and a circular cylinder.

5. A cube is a rectangular prism with square faces. Which size of box in the application at the beginning of the lesson is a cube? How do you know?

6. Work with a partner.
   Measure the edges of a small cereal box. Sketch the box and label its dimensions. Then find a way to cut the box along the seams so you can lay it out flat. Sketch the figure formed.

   a. The figure you formed is a net for the box. What shape is each section of the net?

   b. Find the area of the cardboard used to make the box.

Guided Practice

Name each solid. Then find the surface area. Round to the nearest tenth.

7. 5 ft, 10 ft, 4 ft, 6 ft
8. 3.5 m, 12 m, 3.5 m
9. 22 m, 14 m

10. **Building Maintenance**   The walls of a cylindrical tank that holds natural gas need to be painted. If the tank is 40 feet high and has a radius of 50 feet, what is the area that needs to be painted? If a gallon of paint covers 500 ft², how many gallons of paint are needed?

Independent Practice

Find the surface area of each solid. Round to the nearest tenth.

11.
3 in.
7 in.   12 in.

12.
10 ft
20 ft

13.
10 cm
3 cm
5 cm   4 cm

14.
4 m
20 m

15.
5.2 cm
6 cm
6 cm   6 cm
10 cm

16.
2 cm
15 cm

17.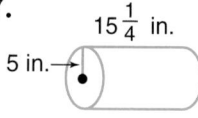
$15\frac{1}{4}$ in.
5 in.

18.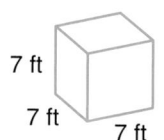
7 ft
7 ft   7 ft

19.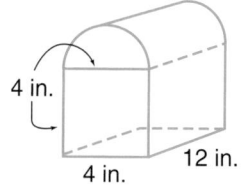
4 in.
4 in.   12 in.

20. Write the equation of the surface area (*S*) of a cube that has edges measuring *a* units. Graph the surface area as a function of *a*. (Use values of *a* like 0, 0.5, 1, 1.5, 2, and so on.)

21. If you add 1 cm to the length of each side of a cube, which of the following is true about the surface area?
   a. It would increase by 1 cm².   **b.** It would increase by 6 cm².
   c. It would remain the same.   **d.** need more information

22. Suppose you double the length of the sides of a cube. How is the surface area affected?

Critical Thinking

23. A cube with 12-inch sides is placed on a cube with 15-inch sides. Then a cube with 9-inch sides is placed on the 12-inch cube.
   a. What is the surface area of the three-cube tower?
   b. Does the way each cube is placed on the one below it affect the surface area? Why or why not?

Applications and Problem Solving

24. Business Claire works after school at her family's tent company. One of their best-selling tents is an A-frame tent that is 4 feet high and has a rectangular bottom 4 feet wide by 6 feet long. The sides of the tent are 4.5 feet long. How much canvas is needed to make the tent?

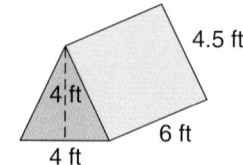
4.5 ft
4 ft
6 ft
4 ft

25. Manufacturing A carton of canned fruit cocktail holds 24 cans. Each can has a diameter of 7.6 cm and a height of 10.8 cm. How much paper is needed to make the labels for the 24 cans?

26. Home Improvement Juan's uncle wants to put drywall on his garage walls and ceiling. The garage is 24 feet by 20 feet, and the walls are $12\frac{1}{2}$ feet high. The drywall he will be using is sold in sheets that are 4 feet wide by 12 feet high by $\frac{1}{2}$ inch thick. How many sheets of drywall will he need to purchase?

27. Postal Service Refer to the application at the beginning of the lesson. How much cardboard does each box take if there are pieces that overlap completely at the top and bottom? Is the ratio of price to material about the same for each box?

28. Architecture A hallway in a Spanish-style house has a semi-circular arch above the walls. The width of the hall and the diameter of the arch are 6 feet. The walls are eight feet from the floor to the beginning of the arch. The hall is 20 feet long. If a gallon of paint covers about 400 square feet, how much paint will be needed to paint the hallway with two coats?

8 ft

20 ft

6 ft

29. Family Activity Choose a room in your home. Sketch the room and measure the walls, windows, and doors.

For **Extra Practice**, see page 770.

   a. Find the area of the walls that would be painted or wallpapered.

   b. Visit a paint store and select a paint that would be appropriate for the room. Find the cost for two coats of paint for the room.

Mixed Review

30. Farming A farmer attaches a 25-foot rope to the corner of a 10-foot by 15-foot shed to tie up the family's pet goat. How much grazing area does the goat have? (Lesson 12-4)

31. Geometry Draw an angle that measures 75°. (Lesson 11-1)

32. Find the value C(12, 6). (Lesson 10-6)

33. Write 3.605×10^3 in standard form. (Lesson 6-9)

34. Standardized Test Practice Which sequence is arithmetic? (Lesson 5-9)

   A. 1, 2, 4, 8,...   **B.** 1, 3, 6, 10,...   **C.** 10, 7, 4, 1,...   **D.** $\frac{1}{2}, \frac{1}{3}, \frac{1}{4}, \frac{1}{5}$,...

WORKING ON THE

Investigation

Refer to the Investigation on pages 606–607.

Rooms in a home can be separated into *private*, *service*, and *social* zones. Bedrooms and bathrooms are part of the private zone.

The service zone includes the kitchen, laundry, and garage. The social zone is an area where members of the household gather to spend time together.

The three zones don't have to be in different rooms. Many kitchen tables are used to do homework, thus making it a private zone. A kitchen becomes a social room when people gather to eat a meal.

Begin working on the Investigation by doing a group activity to design a model of a 2-bedroom house. The house you design must include all three zones.

Add the results of your work to your Investigation Folder.

Surface Area: Pyramids and Cones

Application: Architecture

The pyramids of Egypt were built thousands of years ago as burial tombs for the pharaohs and their relatives. Pyramids have also inspired modern architects. The noted Chinese-American architect I. M. Pei completed renovation of the Louvre Museum in Paris in 1990. For the new entrance, he designed a huge square glass pyramid. It has a square base that is 116 feet on each side. How much glass did it take to cover the pyramid? *This problem will be solved in Example 2.*

Learning the Concept

In geometry, a **pyramid** is a solid figure that has a polygon for a base and triangles for the sides. We name pyramids by the shape of their bases. Notice that a pyramid has just one base. All the other faces of a pyramid, the *lateral* faces, intersect at a point called the **vertex**. The pyramid shown at the right is a **square pyramid**.

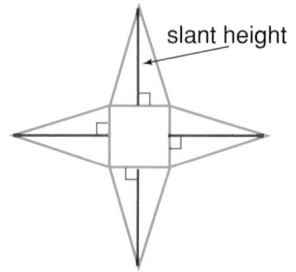

The pyramid shown at the left is a **triangular pyramid**. Its base is an equilateral triangle, so its lateral faces are congruent triangles.

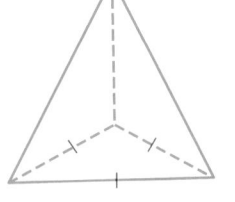

Let's look at a net of a square pyramid. Notice that the lateral faces are triangles. Because the base is a square, the triangles are all congruent, and their altitudes all have the same length. The length of any one of these altitudes is called the **slant height** of the pyramid.

Square Pyramid   Net of Square Pyramid

slant height

The surface area of a pyramid is the sum of the areas of its base and its lateral faces.

Example **Find the surface area of the pyramid.**

6 in.

$5\frac{1}{2}$ in.

The base is a square. Use the formula $A = s^2$ to find the area. *Estimate: $5 \times 5 = 25$ and $6 \times 6 = 36$. The area is about 30 in².*

Use the formula $A = \frac{1}{2}bh$ to find the area of each triangular side.

Estimate: $\frac{1}{2} \cdot 5 \cdot 6 = 15$. The area of the four sides is about 60 in².

Base	**Each Triangular Side**
$A = s^2$	$A = \frac{1}{2}bh$
$A = \left(5\frac{1}{2}\right)^2$	$A = \frac{1}{2} \cdot 5\frac{1}{2} \cdot 6$
$A = 30\frac{1}{4}$ or 30.25	$A = 16\frac{1}{2}$ or 16.5

Add to find the total surface area.

area of base + *area of 4 triangular surfaces* = *surface area*

30.25 + 4×16.5 = 96.25

The surface area is 96.25 in². *Compare with the estimate.*

In the application at the beginning of the lesson, we need to find the **lateral surface** of the pyramid, the surface not including the base.

Example **Real World APPLICATION**

Architecture

2

The Louvre pyramid in the application at the beginning of the lesson has a slant height of about 92 feet. Its square base is 116 feet on each side. Find the amount of glass it took to cover the pyramid.

You are covering the lateral surface of the pyramid. Use the formula $A = \frac{1}{2}bh$ to find the area of each triangular face.

$A = \frac{1}{2}bh$

$A = \frac{1}{2} \cdot 116 \cdot 92$ *Replace b with 116 and h with 92.*

$A = 5336$

There are four triangular faces. Multiply to find the lateral surface area.

$4 \times 5336 = 21,344$

It took 21,344 ft² of glass to cover the pyramid.

In this textbook, when we say cone, assume we are talking about a circular cone.

A cone is another three-dimensional shape. You see cones in everyday life such as the cone for some ice cream cones, a holder for cotton candy, or some paper drinking cups. Most of the cones that you see are called **circular cones** because the base is a circle.

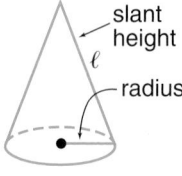

slant height ℓ

radius

The net of a cone shows the regions that make up the cone. You can see that the base of the cone is a circle, with area πr^2. If you cut the lateral surface into sections, you can form a parallelogram. The base of the parallelogram is half the circumference of the cone, or $\frac{1}{2} \cdot 2\pi r$. Its height is the slant height of the cone, ℓ. So the area of the lateral surface is $A = \frac{1}{2} \cdot 2\pi r \cdot \ell$, or $\pi r\ell$. You can find the total surface area SA of a cone using the formula $SA = \pi r^2 + \pi r\ell$.

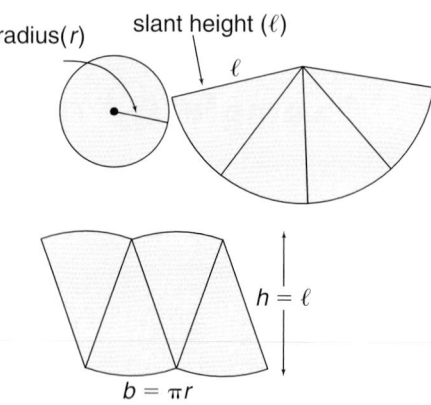

radius(r)   slant height (ℓ)

$h = \ell$

$b = \pi r$

Example ③ Find the surface area of a circular cone with a slant height of 10 cm and a base with a radius of 5 cm. Round to the nearest tenth.

Base	**Lateral Surface**
$A = \pi r^2$	$A = \pi r\ell$
$A = \pi \cdot 5^2$	$A = \pi \cdot 5 \cdot 10$
$A \approx 78.5$	$A \approx 157.1$

10 cm

5 cm

Add to find the total surface area.

$78.5 + 157.1 = 235.6$

The surface area is about 235.6 cm^2.

Checking Your Understanding

Communicating Mathematics

Read and study the lesson to answer these questions.

1. **Describe** the slant height of a pyramid and of a cone.

2. **Explain** why the expression $2\pi r$ can be used to find the perimeter of the base of a circular cone.

3. **State** how to find the area of the lateral surfaces of a pyramid.

MATH JOURNAL

4. **Assess Yourself** Write about the meaning of the term *surface area*. Then list some situations in everyday life when you might use surface area.

Guided Practice

Name each shape. Then find its surface area. Round to the nearest tenth.

5.

8.2 m

6 m

6 m

6.

5 cm

12 cm

7.

6 in.

5 in.

6 in.

6 in.

4 in.

8. **Construction** A church spire is being built in the shape of a regular octagonal pyramid. Each edge of the base is 8 feet long. The slant height of the spire is about 75.3 feet. How much roofing will be needed to cover the spire?

Independent Practice

Find the surface area of each solid. Round to the nearest tenth.

9.
6.3 ft
4 ft   4 ft

10.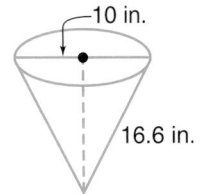
10 in.
16.6 in.

11.
11.4 m
7 m

12.
10.3 ft
8 ft
8 ft
8 ft

13.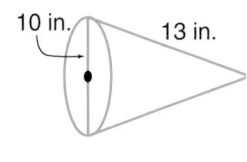
10 in.   13 in.

14.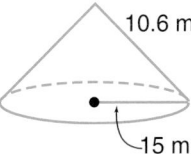
10.6 m
15 m

15.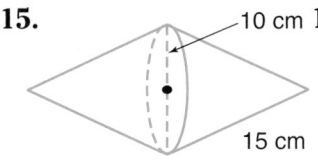
10 cm
15 cm

16.
10.6 m
A = 48 m²
8.3 m

17.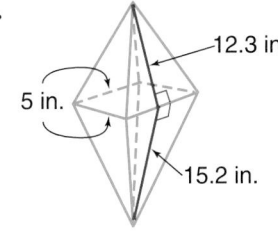
12.3 in.
5 in.
15.2 in.

Critical Thinking

18. Manufacturing A bar of lead 13 inches by 2 inches by 1 inch is melted and recast into 100 conical fishing sinkers that have a diameter of 1 inch and a height of 1 inch. The slant height of each sinker is about 1.1 inches. How does the total surface area of all the sinkers compare to the surface area of the original lead bar?

Applications and Problem Solving

19. History The monument of Cestius in Rome is a square pyramid. It is 121.5 feet high and has a base 98.4 feet long on each side. The slant height is about 131 feet. What is its lateral surface area?

20. Architecture I. M. Pei also designed smaller square pyramids that stand near the entrance of the Louvre. These pyramids have bases 26 feet on a side and a slant height of about 20.6 feet. How much glass did it take to cover one of these pyramids?

21. Tepees The largest tepee in the United States has a diameter of 42 feet and a slant height of about 47.9 feet. The tepee belongs to Dr. Michael Doss of Washington, D.C. Dr. Doss is a member of Montana's Crow Tribe. How much canvas was used to make the cover of the tepee?

22. Food An ice cream shop uses sugar cones that are 5 inches long and have a diameter of 2 inches. The manager would like to order special napkins that would fit around the cone twice. Sketch the shape of the napkin and label its dimensions.

23. Geometry Find the surface area of
the rectangular prism at the right.
(Lesson 12-5)

12 cm

15 cm

10 cm

24. Probability If two coins are tossed,
what is the probability that one will
land heads up and the other tails up?
(Lesson 9-3)

For **Extra Practice**,
see page 770.

25. Define a variable and write and solve an equation for the following
situation. Four times a number plus eight is 20. What is the number?
(Lesson 7-3)

26. Standardized Test Practice Solve $w + \frac{1}{6} = 3\frac{2}{3}$. (Lesson 5-6)

   A. $3\frac{1}{6}$ **B.** $3\frac{1}{4}$ **C.** $3\frac{1}{3}$ **D.** $3\frac{1}{2}$

27. Simplify $(5x)(-3y)$. (Lesson 2-7)

Self Test

Find the area of each figure. (Lessons 12-1 and 12-2)

1.

4 m

3 m

6 m

2.

8 ft

6 ft

10 ft

3.

12 in.

Name each shape. Then find its surface area. Round to the nearest tenth. (Lessons 12-5
and 12-6)

4.

5 mm

4 mm

7 mm

5.

8 in.

12 in.

6.

9 m

8 m

8 m

Solve.

1

2

5 6

8 7

4

3

7. Probability In a dart game, you win points based on the region in
which the dart lands. The diameter of the whole target is 24 inches.
The diameter of the inner circle is 8 inches. What is the probability
that the dart lands in the 8-point region? (Assume the dart lands on
the target.) (Lesson 12-3)

8. Make a Drawing A circular fountain has a diameter of 8 feet. A flower garden is planted
around the fountain. The garden extends 12 feet all around the fountain. If a pound of
organic fertilizer covers about 300 square feet, how much fertilizer is needed for the
garden? (Lesson 12-4)

9. How much paper would be used to make an open paper cone with a diameter of 5 inches
and a slant height of 6 inches? (Lesson 12-6)

12-7A Volume

A Preview of Lesson **12-7**

MATERIALS

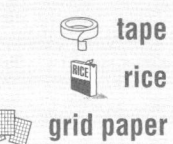 5 × 8 inch index cards

tape

rice

grid paper

In this activity, you will investigate volume by making containers of different shapes and comparing how much each container holds.

Your Turn   Work with a partner.

▶ Use three 5 × 8 cards to make three containers with a height of 5 inches as shown. Make one with a square base (with 2 in. sides.), one with a triangular base (with sides of 2 in., 3 in., and 3 in.) and one with a circular base.

▶ Then tape one end of each container to another card as a bottom, but leave the top open.

▶ Estimate which container would hold the most (have the greatest volume) and which would hold the least (have the least volume). Do you think each container would hold the same amount?

▶ Use rice to fill the container that you believe holds the least amount. Then put the rice into another container. Does the rice fill this container? Continue the process until you find out which, if any, container has the least volume and which has the greatest.

 TALK ABOUT IT

1. Which container held the most? Which held the least?

2. How do the heights of the three containers compare? What is each height?

3. Compare the perimeters of the bases of each container. What is each base perimeter?

4. Trace the base of each container onto grid paper. Estimate the area of each base.

5. Which container has the greatest base area?

6. Does there appear to be a relationship between the area of the base and the volume? Explain.

Volume: Prisms and Cylinders

You'll Learn

■ To find volumes of prisms and circular cylinders.

It's Important

Finding the surface area of cylinders can help product designers determine the amount of sheet metal needed to make a can.

Technology Activity

The program below computes the volume of a rectangular prism when you enter the dimensions. Use this program to discover how to calculate the volume of a container.

Your Turn Enter the program in a programmable calculator.

```
PROGRAM: VOLUME
:  Input "LENGTH =", L
:  Input "WIDTH =", W
:  Input "HEIGHT =", H
:  L*W→A
:  L*W*H→V
:  Disp "AREA OF BASE ="
:  Disp A
:  Disp "VOLUME ="
:  Disp V
:  Stop
```

▶ Sketch a rectangular prism like the one shown at the right. Choose measurements for each side and label the sketch.

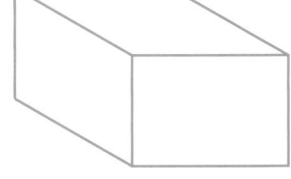

▶ To run the program, press ⬛PRGM⬛. Use the down arrow to highlight the program. Press ⬛ENTER⬛ twice. Enter the data asked for in the program. Record the results on your sketch.

▶ Repeat for at least three different rectangular prisms. Sketch each prism and record the dimensions and the results of the program. (To run the program again, press ⬛ENTER⬛.)

TALK ABOUT IT

a. What dimensions did the program ask for in a rectangular prism?

b. What is the relationship between the dimensions and the volume?

c. How does the area of the base of the prism relate to its volume?

Learning the Concept

The amount a container will hold is called its capacity, or **volume**. Volume is usually measured in cubic units. Two common units of measure for volume are the cubic centimeter (cm^3) and the cubic inch (in^3). The volume of a prism can be found by multiplying the area of the base times the height.

<table>
<tr><td>Volume of a Prism</td><td>In words:</td><td>If a prism has a base area of B square units and a height of h units, then the volume V is $B \cdot h$ cubic units.</td></tr>
<tr><td></td><td>In symbols:</td><td>$V = Bh$</td></tr>
</table>

Example **Find the volume of each prism.**

a.

5 in.
2 in.
3.5 in.

$V = Bh$ — *Formula for volume of a prism*

$V = \ell wh$ — *Since the base of the prism is a rectangle, $B = \ell w$.*

$V = 2 \cdot 3.5 \cdot 5$ — *Replace ℓ with 2, w with 3.5, and h with 5.*

$V = 35$

The volume is 35 in³.

b. A prism has bases that are right triangles as shown below. The height of the prism is 10 cm.

8 cm
10 cm
6 cm

$V = Bh$

$V = \left(\dfrac{1}{2} \cdot 6 \cdot 8\right) \cdot 10$ — *The base of the prism is a triangle. The area of the triangle is $\dfrac{1}{2} \cdot 6 \cdot 8$.*

$V = 24 \cdot 10$

$V = 240$

The volume is 240 cm³.

Since the base of a circular cylinder is a circle and its area is equal to πr^2, we can substitute πr^2 for B to find the volume of a cylinder.

<table>
<tr><td>Volume of a Circular Cylinder</td><td>In words:</td><td>If a circular cylinder has a base with a radius of r units and a height of h units, then the volume V is $\pi r^2 h$ cubic units.</td></tr>
<tr><td></td><td>In symbols:</td><td>$V = \pi r^2 h$</td></tr>
</table>

Example **A can of tomato juice has a diameter of 14 cm and a height of 20 cm. What is its volume?**

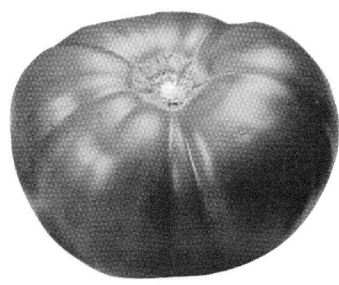

14 cm
20 cm

$V = \pi r^2 h$ — *Formula for volume*

$V = \pi \cdot 7^2 \cdot 20$ — *The diameter of the can is 14 cm; so the radius is 7 cm.*

$V \approx 3079$

The volume is about 3079 cm³.

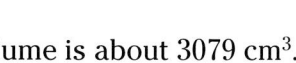

If you know the volume of a cylinder, you can solve the formula for *h* or *r* to find the height or radius of the cylinder.

Example ③

Real World APPLICATION

Engineering

A cylindrical natural gas storage tank is being manufactured to hold at least 1,000,000 cubic feet of natural gas and have a diameter of no more than 80 feet. What height should the tank be?

Explore   The tank is a cylinder. You know the volume and the diameter. You must find the height.

Plan   The formula for the volume of a cylinder is $V = \pi r^2 h$. Solve the formula for *h* to find the height.

Solve

$V = \pi r^2 h$
$1{,}000{,}000 = \pi \cdot (40)^2 \cdot h$ *V = 1,000,000, r = 40*
$1{,}000{,}000 = \pi \cdot 1600 \cdot h$
$1{,}000{,}000 \boxed{÷} \boxed{\pi} \boxed{÷} 1600 \boxed{=}$ *198.9436789*
$198.9 \approx h$

The tank should have a height of at least 199 feet.

Examine   If the tank is 199 feet high with a diameter of 80 feet, will it have a volume of at least 1,000,000 cubic feet?
Is $\pi \cdot (40)^2 \cdot 199 \geq 1{,}000{,}000$?

$\pi \boxed{×} 40 \boxed{x^2} \boxed{×} 199 \boxed{=}$ *1000283.101*

The answer checks.

THINK ABOUT IT

Why should the result be rounded *up* in this case?

Checking Your Understanding

Communicating Mathematics

Read and study the lesson to answer each question.

1. **Explain** why you can use the same formula to find the volume of a prism or a cylinder.

2. **Tell** what each variable represents in the formula $V = Bh$.

3. **Compare and contrast** surface area and volume.

4. **You Decide**   Jake says that doubling the length of each side of a cube doubles the volume. Carlos says the volume is eight times greater. Who is correct? Explain your answer.

 MATH JOURNAL

5. Write a problem from an everyday situation in which you will need to find the volume of a cylinder or a rectangular prism. Explain how to solve the problem.

Guided Practice

Find the volume of each prism or cylinder. Round to the nearest tenth.

6.

8 cm
4 cm   5 cm

7.
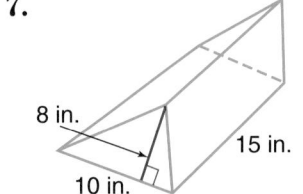
8 in.
15 in.
10 in.

8.

5 cm
15 cm

Find the volume of each prism or cylinder. Round to the nearest tenth.

9. rectangular prism: length 3 in.; width 5 in.; height 15 in.

10. octagonal prism; base area 25 m²; height of 1.5 m

11. circular cylinder: radius 2 ft; height $2\frac{1}{4}$ ft

12. **Pet Care** Tina has an old fish tank that is a circular cylinder. The tank is 2 feet in diameter and 6 feet high. How many cubic feet of water does it hold?

Exercises: Practicing and Applying the Concept

Independent Practice

Find the volume of each solid. Round to the nearest tenth.

Choose

Estimation
Mental Math
Calculator
Paper and Pencil

13.
5 ft
5 ft   5 ft

14.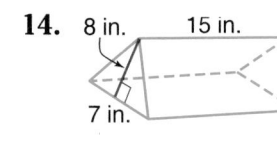
8 in.   15 in.
7 in.

15.
4 cm
8 cm
16 cm

16.
10 m
A = 4.5 m²

17.
8 ft
8 ft

18.
14 in.
6 in.

19.
6 cm
20 cm

20.
2.7 m
30 m

21.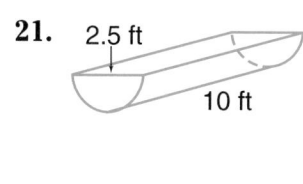
2.5 ft
10 ft

22. rectangular prism: length 6 in.; width 6 in.; height 6 in.

23. rectangular prism: length 4.2 cm; width 3.2 cm; height 6.2 cm

24. pentagonal prism: base area 45.2 cm²; height 47.8 cm

25. circular cylinder: radius 10 yd; height 15 yd

26. triangular prism; base of triangle 8 in.; altitude of triangle 15 in.; height of prism 6.5 in.

27. circular cylinder: diameter 2.6 m; height 3.5 m

Graphing Calculator

28. Refer to the program in the activity at the beginning of the lesson.
   a. Change the program to find the volume of a circular cylinder.
   b. Sketch three circular cylinders with the same height and different diameters. Run the program to find the volume of each cylinder.
   c. Graph the diameters and volumes on a coordinate plane. What pattern do you observe?

29. Suppose you roll an $8\frac{1}{2}$- by 11-inch piece of paper to form a cylinder. Will the volume be greater if you roll it so the height is $8\frac{1}{2}$ inches or 11 inches, or will the volumes be the same?

30. Write the nonlinear function for each of the following. Then graph each nonlinear function.

  a. the volume, V, of a rectangular prism as a function of a fixed height of 2 and a square base of varying lengths, s

  b. the volume, V, of a triangular prism as a function of a fixed height of 0.5 and an equilateral triangle base of varying lengths, b

Applications and Problem Solving

31. **Chemistry** A quartz crystal is a hexagonal prism. It has a base area of 1.41 cm^2 and a volume of 4.64 cm^3. What is its height?

32. **Baking** A rectangular cake pan is 30 cm by 21 cm by 5 cm. A round cake pan has a diameter of 21 cm and a height of 4 cm. Which will hold more batter, the rectangular pan or two round pans?

33. **Recreation** The dimensions of three swimming pools are given below. If each pool is shaped like a rectangular prism, which one holds about 45,000 gallons of water when full? (1 ft^3 ≅ 7.481 gal)

  a. 50 ft by 20 ft by 6 ft **b.** 12 ft by 15 ft by 12 ft **c.** 45 ft by 12 ft by 80 ft

34. **Make a Drawing** A twelve-pack of ice tea cans is sold in a rectangular box. The cans are 6 inch high cylinders with a diameter of 3 inches.

  a. What are the dimensions of the box if the cans are arranged in a 4 by 3 array?

  b. How much wasted space is there in the box?

35. **Consumer Awareness** Firewood is usually sold by a measure known as a cord. A full cord may be a stack $8 \times 4 \times 4$ feet or $8 \times 8 \times 2$ feet.

  a. What is the volume of a full cord?

  b. A "short cord" is $8 \times 4 \times$ the length of the logs. What part of a full cord is a short cord of $2\frac{1}{2}$ foot logs?

Mixed Review

36. Find the surface area of a square pyramid with a base 14 yards long and a height of 3 yards. (Lesson 12-6)

37. **Geometry** Find the value of x in $\triangle ABC$ at the right. (Lesson 11-6)

38. **Geometry** Angles X and Y are supplementary. If $m\angle X = 58°$, find $m\angle Y$. (Lesson 11-3)

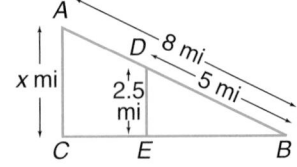

For **Extra Practice**, see page 770.

39. If $f(x) = 5x - 3$, find $f(2)$. (Lesson 8-4)

40. **Work Backward** Drew has $2.25. Half of the money he had when he left home this morning was spent on lunch. He loaned Cassie two dollars after lunch. How much money did Drew start with? (Lesson 7-1)

41. **Standardized Test Practice** If you toss three 4-sided dice, how many different combinations of tosses are possible? (Lesson 4-3)

  A. 12 **B.** 24 **C.** 48 **D.** 64

42. Write a problem that could be solved using the equation $3x = 25$. (Lesson 1-8)

12-8

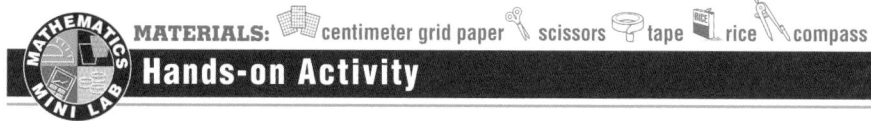

Volume: Pyramids and Cones

Hands-on Activity

You'll Learn

- To find volumes of pyramids and cones.

It's Important

Knowing how to find the volume of cones can help highway maintenance workers determine how much roadway can be salted from a pile of salt.

Look at the nets shown at the right. They will fold into models of a square prism (a cube) with one open side and a square pyramid. The base of both solids is the same size. If you fill each model with rice, you can compare their volumes.

Your Turn   Work with a partner.

▶ Copy the two nets on centimeter grid paper.

▶ Cut them out and fold on the dashed lines.

▶ Tape the edges together to form models of the solids.

▶ Estimate how much larger the volume of the cube is than the volume of the pyramid.

▶ Make an opening in the base of the pyramid so you can put rice into it.

▶ Fill the pyramid with rice. Then pour this rice into the cube. Repeat until the cube is filled.

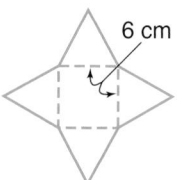

TALK ABOUT IT

a. How many pyramids of rice did it take to fill the cube?

b. What do you know about the areas of the bases of each solid?

c. Compare the heights of the cube and the pyramid.

d. Compare the volume of the cube and the pyramid.

Learning the Concept

In the modeling activity, you found that the volume of the pyramid was $\frac{1}{3}$ the volume of the prism. Since the volume of a prism is Bh, the volume of the pyramid is $\frac{1}{3}Bh$. The height, h, of a pyramid is the length of a segment from the vertex, perpendicular to the base.

Volume of a Pyramid	**In words:** If a pyramid has a base of B square units and a height of h units, then the volume V is $\frac{1}{3} \cdot B \cdot h$ cubic units. **In symbols:** $V = \frac{1}{3}Bh$

Example **Find the volume of the pyramid.**

$V = \frac{1}{3}Bh$ *Formula for volume of a pyramid*

$V = \frac{1}{3}\ell wh$ *Replace B with ℓw.*

$V = \frac{1}{3} \cdot 4 \cdot 5 \cdot 15$

$V = 100$

The volume is 100 m³.

height
15 m

5 m   4 m

The relationship of the volumes of a cone and a cylinder is similar to the relationship of the volumes of a pyramid and a prism. The volume of a cone is $\frac{1}{3}$ the volume of a cylinder. Since the volume of a cylinder is $\pi r^2 h$, the volume of a cone is $\frac{1}{3}\pi r^2 h$. *You'll verify this formula in Exercise 4.*

Volume of a Cone	**In words:** If a cone has a radius of *r* units and a height of *h* units, then the volume *V* is $\frac{1}{3} \cdot \pi \cdot r^2 \cdot h$.
	In symbols: $V = \frac{1}{3}\pi r^2 h$

Example **2** **Find the volume of the cone to the nearest whole number.**

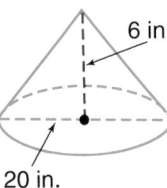
6 in.

20 in.

$V = \frac{1}{3}\pi r^2 h$ *Formula for volume of a cone*

$V = \frac{1}{3} \cdot \pi \cdot 10^2 \cdot 6$ *Since d = 20, r = 10; h = 6*

$V \approx 628$

The volume is about 628 in³.

Example **3**

Many communities use a salt and sand mixture on snowy or icy roads. When the mixture is dumped from a truck, it forms a mound with a conical shape.

a. If the mound has a diameter of 15 feet and a height of 10 feet, how much salt-sand mixture is there? *Estimate:* $\frac{1}{3} \cdot 3 \cdot 8^2 \cdot 10 = 640$

b. If the mixture is spread at a rate of 1 cubic foot per 500 square feet of roadway, how many square feet of roadway can be salted by the mixture in one mound?

a. $V = \frac{1}{3}\pi r^2 h$ *Formula for volume of a cone*

$V = \frac{1}{3} \cdot \pi \cdot 7.5^2 \cdot 10$ *Since d = 15, r = 7.5; h = 10*

$V \approx 589$

The volume is 589 ft³.

b. Each cubic foot of mixture will salt 500 square feet of roadway.
589 × 500 = 294,000

294,000 square feet of roadway can be salted by the mixture in the mound.

Checking Your Understanding

Communicating Mathematics

Read and study the lesson to answer these questions.

1. **Explain** why you can use πr^2 to find the area of the base of a cone.
2. **State** the formula for finding the volume of a pyramid. Then tell what each variable represents.
3. **State** the formula for finding the volume of a cone. Then tell what each variable represents.

 MATERIALS

compass
scissors
tape
rice

4. Work with a partner.

   ▶ Draw a circle with a radius of at least 2 inches. Cut out the circle.

   ▶ Remove about $\frac{1}{3}$ of the circle as shown at the right. Tape the edges of the remaining part of the circle to make an open cone.

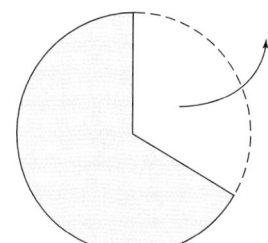

   ▶ Trace the base of the cone on another piece of paper and cut out the circle.

   ▶ Cut a strip of paper as wide as the cone is high. Use the strip and the circle to make a model of a cylinder that has the same size base and the same height as the cone.

   ▶ Estimate how many times larger the cylinder is than the cone. Use rice to check your estimate.

Guided Practice

Explain how to find the volume of each pyramid or cone. Then find the volume. Round to the nearest tenth.

5.
6.
7.

8. circular cone: diameter 12 m; height 15 m
9. hexagonal pyramid: base area 125 in²;
   height $6\frac{1}{2}$ in.

10. **Historical Monuments** The top of the Washington Monument is a square pyramid 54 feet high and 34 feet long on each side. What is the volume of the room under the top?

Washington Monument

Independent Practice

Find the volume of each solid. Round to the nearest tenth.

11.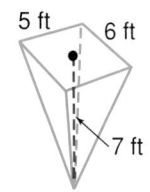
5 ft 6 ft 7 ft

12.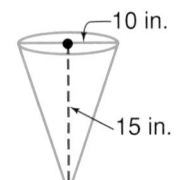
10 in. 15 in.

13.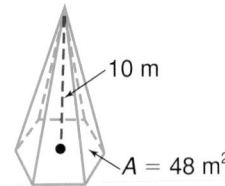
10 m $A = 48$ m^2

14.
9 mm 14 mm

15.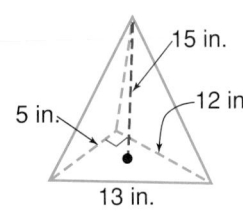
15 in. 12 in. 5 in. 13 in.

16.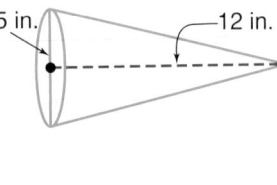
5 in. 12 in.

17.
39 m 15 m

18.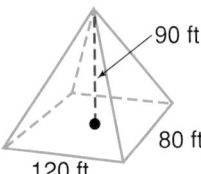
90 ft 80 ft 120 ft

19.
8 cm 10 cm 20 cm

20. rectangular pyramid; length, 9 in.; width, 7 in.; height, 18 in.

21. square pyramid; length, 5 cm; height, 6 cm

22. circular cone; radius, 3 ft; height, 14 ft

23. circular cone; radius 10 m; height, 18 m

24.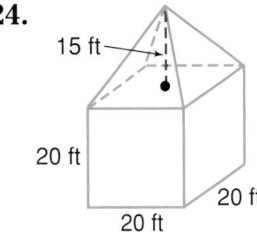
15 ft 20 ft 20 ft 20 ft

25.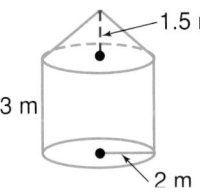
1.5 m 3 m 2 m

26.
42 cm 57 cm 42 cm 43 cm

Critical Thinking

27. **Consumer Awareness** Popcorn at the Bijou Theatre used to be sold in rectangular prism containers. Recently the theater changed to a rectangular pyramid container. Both containers have the same height and the same size base. Popcorn in the new box sells for half the former price. Is this a good deal for the customer? Why or why not?

28. **a.** Suppose you double the height of a cone. How does the volume change?

b. Suppose you double the radius of the base of a cone. How does the volume change?

Applications and Problem Solving

29. **History** The Great Pyramid of Khufu in Egypt was originally 481 feet high and had a square base 756 feet on a side. What was its volume? Use an estimate to check your answer.

30. **Geology** A stalactite in the Endless Caverns in Virginia is shaped like a cone. It is 4 feet long and has a diameter at the roof of 1.5 feet.

    a. Find the volume of the stalactite to the nearest tenth.

    b. The stalactite is made of calcium carbonate, which weighs 131 pounds a cubic foot. What is the weight of the stalactite?

Mixed Review

31. **Construction** Marty's Masonry makes concrete steps for new homes. The steps for one home are to be 5 feet long, with three steps. Each step will be 0.5 foot high and 1 foot wide. How much concrete will be needed? (*Hint:* Make a drawing.) (Lesson 12-7)

32. **Probability** A die is tossed 30 times. How many times would you expect an outcome of a 1 or a 6? (Lesson 10-10)

33. **Standardized Test Practice** Eighteen is what percent of 90? (Lesson 9-5)

    A. 5%   **B.** 10%   **C.** 15%   **D.** 20%

34. Solve $x = 3\frac{3}{5} - 1\frac{3}{10}$. Write the solution in simplest form. (Lesson 5-5)

35. **Geometry** Find the perimeter and area of the rectangle at the right. (Lesson 3-5)

For **Extra Practice**, see page 771.

36. Solve the equation $\frac{-144}{6} = k$. (Lesson 2-8)

WORKING ON THE

Investigation

Refer to the Investigation on pages 606–607.

Once your group has completed the basic layout of the rooms in your house, you are ready to prepare the *working drawings*. Working drawings are used to show specific information that is necessary to build the structure.

The working drawing that shows the location and size of the rooms is called the *floor plan*. A floor plan shows how the rooms would look if the ceiling were removed and you looked down from above.

Using your basic layout, design your house so that it is no larger than 1200 square feet. Each

room should have at least 1 window. Sketch each floor plan on graph paper to a scale of $\frac{1}{4}$ inch = 1 foot.

Add the results of your work to your Investigation Folder.

HANDS-ON ACTIVITY

12-8B

Similar Solid Figures

An Extension of Lesson **12-8**

MATERIALS

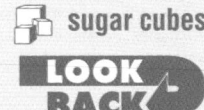 sugar cubes

LOOK BACK

You can review similar figures in Lesson 11-6.

Have you or someone you know ever built a model car or boat? A scale model is an exact replica of the real object, but is larger or smaller than the original. The dimensions of the model and the original are proportional. The number of times that you increase or decrease the linear dimensions is called the **scale factor**. In this activity, you will find out how changing the dimensions of a solid affects the surface area and volume of the solid.

Activity ❶ Work with a partner.

Your Turn

▶ If each edge of a sugar cube is 1 unit long, then each face is 1 square unit and the volume of the cube is 1 cubic unit.

▶ Use cubes to make a larger cube that has sides twice as long as the original cube.

▶ How many small cubes did you use?

TALK ABOUT IT

1. What is the area of one face of the original cube?

2. What is the area of one face of the cube that you built?

3. What is the volume of the original cube?

4. What is the volume of the cube that you built?

Activity ❷ Work with a partner.

Your Turn

▶ Now use cubes to build a cube that has sides three times as long as the original cube.

▶ How many small cubes did you use?

TALK ABOUT IT

5. What is the area of one face of the cube?

6. What is the volume of the cube?

7. Complete the table below.

Scale Factor	Length of a Side	Area of a Face	Volume
1			
2			
3			

8. Study the table. What happens to the area of a face when the length of a side is doubled? tripled?

9. If the scale factor is x, what is the area of the face?

10. What happens to the volume of a cube when the length of a side is doubled? tripled?

11. Let the scale factor be x. Write an expression for the cube's volume.

12. What would be the area of a face and the volume of a cube with sides 4 times as long as the original cube?

Vocabulary

After completing this chapter, you should be able to define each term, property, or phrase and give an example or two of each.

Geometry

altitude (pp. 612, 614, 633)
area of a circle (p. 620)
area of a parallelogram (p. 613)
area of a trapezoid (p. 614)
area of a triangle (p. 613)
base (pp. 614, 632)
circular cone (p. 639)
cylinder (p. 633)
face (p. 632)
lateral surface (p. 639)
net (p. 632)
prism (p. 632)
pyramid (p. 638)
rectangular prism (p. 632)
scale factor (p. 654)
slant height (p. 638)

square pyramid (p. 638)
surface area (p. 632)
triangular prism (p. 632)
triangular pyramid (p. 638)
vertex (p. 638)
volume (p. 644)
volume of a circular cylinder (p. 645)
volume of a cone (p. 650)
volume of a prism (p. 645)
volume of a pyramid (p. 649)

Probability

geometric probability (p. 623)

Problem Solving

make a model or a drawing (p. 629)

interNET
C O N N E C T I O N

Chapter Review For additional lesson-by-lesson review, visit:
www.glencoe.com/ sec/math/prealg/ mathnet

Understanding and Using Vocabulary

Choose the letter of the term or terms that best completes each statement.

1. The _?_ is a common unit of measure for area.

2. The _?_ is a common unit of measure for volume.

3. The _?_ of a solid is the sum of the areas of its faces.

4. _?_ is the amount of space that a solid contains.

5. The _?_ of a parallelogram is a segment perpendicular to the bases with endpoints on the bases.

6. The _?_ of a pyramid is the length of an altitude of one of its lateral faces.

7. The height of a pyramid is the length of a segment from the vertex _?_ to the base of the pyramid.

8. A _?_ is named by the shape of its base.

A. altitude
B. cubic centimeter
C. perpendicular
D. prism
E. pyramid
F. slant height
G. square centimeter
H. surface area
I. volume

Objectives and Examples	Review Exercises

Upon completing this chapter, you should be able to:

Use these exercises to review and prepare for the chapter test.

▶ **find areas of parallelograms, triangles, and trapezoids** (Lesson 12-1)

Find the area of each figure.

Find the area of the trapezoid.

$A = \frac{1}{2}h(a + b)$

$A = \frac{1}{2} \cdot 13(10 + 24)$

$A = 221 \text{ m}^2$

9.

10.

11.

▶ **find the area of circles** (Lesson 12-2)

Find the area of each circle described below. Round answers to the nearest tenth.

Find the area of the circle.

$A = \pi r^2$

$A = \pi \cdot (3)^2$

$A \approx 28.3 \text{ in}^2$

12.

13. diameter, 4.2 m

▶ **find probabilities using area models** (Lesson 12-3)

The probability of landing in the shaded region is $\frac{1}{4}$.

14. The figure below represents a dartboard. Find the probability of landing in the shaded region.

▶ **solve problems using a model or a drawing** (Lesson 12-4)

Making a model or drawing can help you visualize a problem so you can solve it.

Solve by making a drawing.

15. The radius of a circular flower garden and its border is 2.8 meters. Without the border, the radius is 2.3 meters. Find the area of the border.

▶ **find the surface area of rectangular prisms and circular cylinders** (Lesson 12-5)

Find the surface area of the prism.

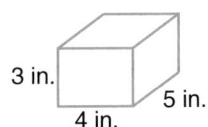

3 in.

5 in.

4 in.

Top and bottom: $2 \times 4 \times 5 = 40$
Front and back: $2 \times 3 \times 4 = 24$
Sides: $2 \times 3 \times 5 = 30$
Surface area $= 40 + 24 + 30$ or 94 in^2

Find the surface area of each figure.

16.

1 cm

1 cm

1 cm

17.

$\frac{5}{8}$ ft

$\frac{1}{2}$ ft

$1\frac{1}{2}$ ft

$\frac{3}{4}$ ft

18. circular cylinder: height, 2.3 m; radius, 1 m

19. circular cylinder: height, 41 ft; diameter, 40 ft

▶ **find the surface areas of pyramids and cones** (Lesson 12-6)

Find the surface area of a cone with radius 5 inches and height 8 inches.

Base: Curved Surface:
$A = \pi r^2$ $A = \pi r \ell$
$A = \pi \cdot (5)^2$ $A = \pi \cdot 5 \cdot 8$
$A \approx 78.5$ $A \approx 125.7$
Total surface area: $78.5 + 125.7$ or 204.2 in^2

Find the surface area of each figure. Round decimal answers to the nearest tenth.

20.

$8\frac{1}{4}$ in.

10 in.

10 in.

21.

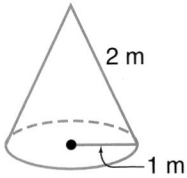

2 m

1 m

▶ **find the volumes of prisms and circular cylinders** (Lesson 12-7)

Find the volume of the cylinder.

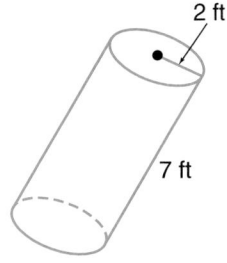

2 ft

7 ft

$V = \pi r^2 h$
$V = \pi \cdot (2)^2 \cdot 7$
$V = 88.0 \text{ ft}^3$

Find the volume of each figure described below. Round decimal answers to the nearest tenth.

22. triangular prism: base of triangle, 5 cm; height, 3.4 cm; prism height, 12 cm

23. hexagonal prism: area of base, 166.25 in^2; height, 20 in.

24. rectangular prism: length, 4 m; width, 4 m; height, 16 m

25. cylinder: radius, 7 ft; height, 18 ft

26. cylinder: diameter, 10 m; height, 22 m

Objectives and Examples	Review Exercises

▶ **find the volumes of pyramids and cones**
(Lesson 12-8)

Find the volume of the
pyramid.

$V = \frac{1}{3}Bh$

$V = \frac{1}{3} \cdot (3 \cdot 3) \cdot 4$

$V = 12m^3$

4 m
3 m
3 m

Find the volume of each figure. Round decimal answers to the nearest tenth.

27.

21 m
8 m
9 m

28.

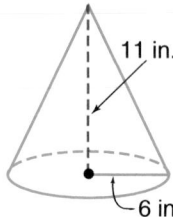
11 in.
6 in.

Applications and Problem Solving

29. **Entertainment** The Eastside High School Drill Team is having new flags made. Some of the flags will be shaped like triangular pennants. The base of the pennants will be 36 inches and the height will be 48 inches. What is the area of one of these flags?
(Lesson 12-1)

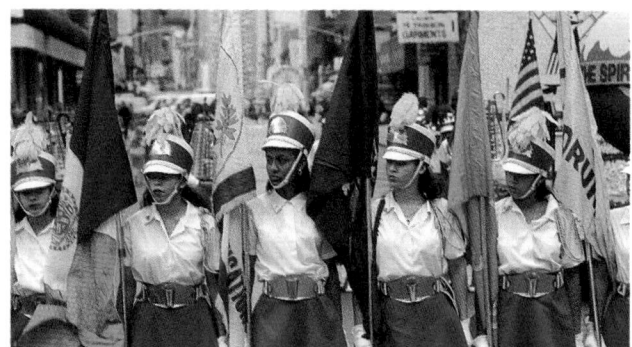

31. **Manufacturing** A leading manufacturer of sugar cubes packs a rectangular box so that there are six cubes along one edge, eleven cubes along a second edge, and three cubes along the third edge. How many sugar cubes are in the box?
(Lesson 12-7)

33. **Agriculture** A water trough is half of a circular cylinder with a radius of 6 feet and a height of 3 feet. If one gallon of water has a volume of about 230 in^3, about how many gallons of water will the trough hold?
(Lesson 12-7)

30. **Agriculture** A silo has a cylindrical shape. The radius of the base is about 20 feet, and the silo is 60 feet high. A company has been hired to paint the silo.
(Lesson 12-5)

a. What is the area of the curved surface?

b. How many gallons of paint would be needed to apply one coat of paint to the silo if one gallon of paint covers about 450 square feet?

c. How much would it cost to paint the silo if one gallon costs $10?

d. How much would it cost to apply two coats of paint to the silo?

32. **Science** Jenny knows that 1 mL of water has a volume of 1 cm^3. She needs 225 mL of water for an experiment. She has an unmarked cylindrical container with a diameter of 12 cm. To what height should she fill the container to measure the water for the experiment? (Lesson 12-7)

34. **Mining** An open pit mine in the Elk mountain range is shaped like a cone. The mine is 420 feet across and 250 feet deep. What volume of material was removed?
(Lesson 12-8)

A practice test for Chapter 12 is available on page 785.

Cooperative Learning Project

Work with a group.

1. Research how Egyptian pyramids were constructed and what tools were used.

2. Choose one pyramid. Find the dimensions of that pyramid.

3. Plan a way to make a scale model of the pyramid. Include the following information in your plan.

   a. Decide how big to make the model.

   b. Determine the scale factor for the model.

   c. Select materials for making the model.

4. Construct the model.

5. Prepare a report on your findings, your procedures, and any problems you encountered. Include information on the surface area and volume of the original pyramid and of the model. You may wish to make a poster or brochure to accompany your model.

Thinking Critically

▶ The edge of a cube and the diameter and height of a cylinder all have the same measure. Which has the least surface area? Which has the least volume?

▶ Suppose you are designing a solid figure with two congruent parallel bases and the distance around the base and the height must be the same for whatever solid you design. What shape base would give the greatest surface area? the greatest volume?

▶ Design three different nets that could be used to make a rectangular prism 4 cm long, 3 cm wide, and 2 cm high. Which net will allow you to cut the greatest number of patterns from a 20-cm × 30-cm grid?

 Portfolio

Select one of the assignments from this chapter that you found especially challenging and place it in your portfolio.

Section One: Multiple Choice

There are eleven multiple-choice questions in this section. After working each problem, write the letter of the correct answer on your paper.

1. Geometry What is the area of the trapezoid?

A. 45 ft²
B. 64 ft²
C. 96 ft²
D. 112 ft²

2. Sports Sandy's handicap score is found using the formula: $s = g + c$, where s is the handicap score, g is the game score, and c is the handicap. What was Sandy's game score if her handicap score is 186 and her handicap is 15?

A. 161
B. 171
C. 191
D. 201

3. Geometry A cylindrical can is 12 cm tall, and the ends each have a diameter of 10 cm. Which expression can be used to find the area the can label covers?

A. $\pi \times 10 \times 12$
B. $\pi \times 5 \times 5 \times 12$
C. $2(\pi \times 5^2) + \pi \times 10 \times 12$
D. $\pi \times 5^2 + 2(\pi \times 10 \times 12)$

4. Which is equivalent to $a^5 \cdot a^2$?

A. $2a^{10}$
B. $2a^7$
C. a^{10}
D. a^7

5. Geometry What is the volume of the triangular prism?

A. 3600 cm³
B. 4800 cm³
C. 30,000 cm³
D. 36,000 cm³

6. Which fractions and decimals are in order from least to greatest?

A. $0.5, \frac{1}{3}, \frac{1}{4}$
B. $\frac{3}{5}, 0.7, \frac{9}{12}$
C. $0.75, \frac{5}{8}, \frac{9}{10}$
D. $\frac{5}{6}, \frac{5}{8}, 0.5$

7. Lee is practicing for a 2500-meter run. He runs the distance twice a day for 5 days. How many kilometers has he run?

A. 25 km
B. 25,000 km
C. 250 km
D. none of these

8. Geometry Which shows a step in constructing an angle congruent to a given angle?

A.

B.

C.

D.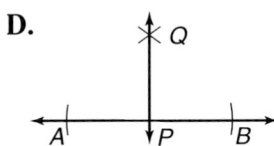

9. In how many different ways can all four students be arranged in a row?

Suki   Janet   Marcus   Bill

A. 4 ways
B. 12 ways
C. 24 ways
D. 36 ways

10. Geometry $\triangle QRS$ is similar to $\triangle XYZ$. What is the length of $\overline{RQ}$?

A. 16 units
B. 18 units
C. 20 units
D. 22 units

11. Find the circumference of a circle that has a diameter 1.2 meters long.

A. 4.5 meters
B. 7.5 meters
C. 3.8 meters
D. 1.1 meters

Section Two: Free Response

This section contains eight questions for which you will provide short answers. Write your answer on your paper.

12. **Geometry**   Find the value of *x*. Then classify the triangle as acute, obtuse, or right.

13. A sweatshirt is on sale for $16. This is 80% of the regular price. What is the regular price?

14. A rectangular swimming pool is 6 meters wide and 12 meters long. A concrete walkway is poured around the pool. The walkway is 1 meter wide and 0.2 meters deep. What is the volume of the concrete?

15. On a trip, the Changs drive 104 miles in 2 hours. If they continue at the same rate, how long will it take them to complete a 550 mile trip?

16. From a 52-card deck, Susanna draws 2 cards. Find the probability they will both be kings.

17. **Geometry**   In the figure below, ℓ is parallel to *m*. If the measure of $\angle 1$ is 108°, find the measure of the remaining angles.

18. **Geometry**   Find the value of *x*.

Test-Taking Tip   Although the basic formulas that you need to answer the questions on a standardized test are often given to you in the test booklet, it is a good idea to review the formulas beforehand.

Area of a circle $= \pi r^2$
Circumference of a circle $= 2\pi r$
Area of a square $= s^2$
Perimeter of a rectangle $= 2(\ell + w)$
Area of a rectangle $= \ell w$

19. **Sports**   The Lakers scored 52 points in the first half of a recent game against the Suns. The Suns average 106 points scored per game. Write and solve an inequality that shows how many points the Lakers need to score in the second half to beat the Suns' average points scored per game.

Section Three: Open-Ended

This section contains two open-ended problems. Demonstrate your knowledge by giving a clear, concise solution to each problem. Your score on these problems will depend upon how well you do the following.

- Explain your reasoning.
- Show your understanding of the mathematics in an organized manner.
- Use charts, graphs, and diagrams in your explanation.
- Show the solution in more than one way or relate it to other situations.
- Investigate beyond the requirements of the problem.

20. **Probability**   Give two examples where geometric probability can be used.

21. **Geometry**   Explain what the difference is between a line, a line segment, and a ray.

Test Practice   For additional test practice questions, visit:
www.glencoe.com/sec/math/prealg/ mathnet

Applying Algebra to Right Triangles

What You'll Learn

In this chapter, you will:

- find squares and square roots,
- solve equations by finding square roots,
- use the Pythagorean theorem,
- use trigonometric ratios to solve problems, and
- use Venn diagrams to solve problems.

Why It's Important

Real World

Fire Fighting The diagram shows a fire tower, a fire, and two right triangles that have been drawn showing known and unknown measures. The angle marked 2° is called the angle of depression. To find the distance to the fire, marked d, you can use an algebraic equation and *trigonometric* ratios. Trigonometry is a branch of mathematics often used to find distances and angle measures that cannot be measured directly. *In Lesson 13-7, you will learn how to use algebra and trigonometry to solve problems like finding the distance from the tower to the fire.*

125 ft   125 ft

Focus on...

To be successful in this chapter, you'll need to understand these concepts and be able to apply them. Refer to the lesson in parentheses if you need more review before beginning the chapter.

▶ **Use the perimeter and area of rectangles and squares to solve problems.** *(Lesson 3-5)*

Find the missing dimensions of each rectangle.

	Length	Width	Perimeter	Area
1.	25 ft			875 sq ft
2.		10 m	40 m	
3.	6 yd		42 yd	
4.		45 cm		2025 sq cm

▶ **Rename decimals as fractions.** *(Lesson 5-1)*

Express each decimal as a fraction or mixed number in simplest form.

5. 0.75   **6.** $0.\overline{25}$ **7.** $4.\overline{3}$ **8.** 8.08

▶ **Find the missing angle measure of a triangle.** *(Lesson 11-4)*

Use the figure at the right to solve each of the following.

9. Find $m\angle 1$ if $m\angle 2 = 30°$ and $m\angle 3 = 60°$

10. Find $m\angle 1$ if $m\angle 2 = 50°$ and $m\angle 3 = 50°$

11. Find $m\angle 2$ if $m\angle 1 = 75°$ and $m\angle 3 = 42°$

▶ **Write ratios as fractions in simplest form.** *(Lesson 9-1)*

Express each ratio as a fraction in simplest form.

12. 12 to 15   **13.** 24:48   **14.** 14 to 7   **15.** 17:51

Focus on...

Trigonometry is the study of triangle properties and relationships. If you examine the word trigonometry, you can determine its meaning. The beginning of the word, *tri*, is the same as the beginning of triangle. The ending of the word, *metry,* means "the process or science of measuring something." Trigonometry, then, means "the process of measuring triangles."

13-1 Squares and Square Roots

Application: Balloning

Have you ever wondered how far you can see out from an airplane or from the top of a hill? How far you can see depends on the curvature of Earth and your height above it. There is a formula that will help you to estimate how far you can see. The formula states that the view in miles V equals 1.22 times the square root of the altitude in feet A.

$$V = 1.22 \times \sqrt{A}$$

Lucia is riding in a hot air balloon about 158 feet above the ground. She looks off to the horizon and can just see her school in the distance. About how far away is she from the school?
You will solve this problem in Example 4.

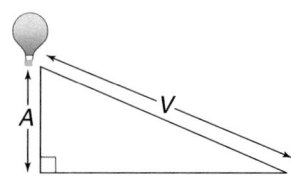

Learning the Concept

A **square root** is one of two equal factors of a number. For example, the square root of 49 is 7 since $7 \cdot 7$ or 7^2 is 49. It is also true that $-7 \cdot (-7) = 49$. So -7 is another square root of 49.

Definition of Square Root	**In words:** The square root of a number is one of its two equal factors.
	In symbols: If $x^2 = y$, then x is a square root of y.

The symbol $\sqrt{}$, called the **radical sign**, is used to indicate a nonnegative square root.

$\sqrt{49}$ indicates the nonnegative square root of 49.

$$\sqrt{49} = 7$$

$-\sqrt{49}$ indicates the negative square root of 49.

$$-\sqrt{49} = -7$$

Example ① **Find each square root.**
   a. $\sqrt{25}$

The symbol $\sqrt{25}$ represents the nonnegative square root of 25. Since $5 \cdot 5 = 25$, $\sqrt{25} = 5$.

b. $-\sqrt{64}$

The symbol $-\sqrt{64}$ represents the negative square root of 64.
Since $8 \cdot 8 = 64$, $-\sqrt{64} = -8$.

Squaring a number and finding the square root of a number are closely related to the square.

 Connection to Geometry

The square shown at the right has sides that are 6 meters long. The area of the square is found by squaring 6. If you know that the area of the square is 36 square meters, then the length of a side is found by finding the square root of 36.

6 m

 Example 2

INTEGRATION

Geometry

The area of a square is 100 square inches. Find its perimeter.

First, find the length of each side.

$A = s^2$
$100 = s^2$ *Replace A with 100.*

What number s when multiplied by itself is 100?

Both 10 and -10 when multiplied by themselves are 100. However, since length cannot be negative, s must be 10.

The length of each side is 10 inches.

Now find the perimeter.

$P = 4 \cdot s$
$P = 4 \cdot 10$ *Replace s with 10.*
$P = 40$ The perimeter is 40 inches.

Area = 100 in²

 LOOK BACK

You can review area and perimeter in Lesson 3-5.

Numbers like 25, 49, and 64 are called **perfect squares** because when you take the square root you get an answer that is a whole number. What if the number is not a perfect square? How do you find the square root of a number like 125?

List some squares of numbers that are close to 125.

$10^2 = 100$ $11^2 = 121$ $12^2 = 144$

The number 125 is not a perfect square; that is, 125 has no square root that is a whole number. However, we know that 125 is greater than 121 or 11^2 and less than 144 or 12^2. So the square root of 125 should be greater than 11 and less than 12.

$$\sqrt{121} < \sqrt{125} < \sqrt{144}$$
$$\sqrt{11^2} < \sqrt{125} < \sqrt{12^2}$$
$$11 < \sqrt{125} < 12$$

Since 125 is closer to 121 than to 144, the best whole number estimate for $\sqrt{125}$ would be 11.

Example ③

TECHNO TIP

Some calculators have a *square root key* labeled $\sqrt{}$ or $\sqrt{x}$. When you press this key, the number in the display is replaced by its nonnegative square root.

Find the best integer estimate for each square root.

a. $\sqrt{59}$

49 and 64 are the closest perfect squares.

$$\sqrt{49} < \sqrt{59} < \sqrt{64}$$
$$\sqrt{7^2} < \sqrt{59} < \sqrt{8^2}$$
$$7 < \sqrt{59} < 8$$

Since 59 is closer to 64 than 49, the best integer estimate is 8.

b. $-\sqrt{38}$

36 and 49 are the perfect squares closest to 38.

$$-\sqrt{49} < -\sqrt{38} < -\sqrt{36}$$
$$-\sqrt{7^2} < -\sqrt{38} < -\sqrt{6^2}$$
$$-7 < -\sqrt{38} < -6$$

Since 36 is closer to 38 than 49, the best integer estimate is -6.

Many applications in science and other fields use square roots in important formulas.

Example ④

Refer to the application at the beginning of the lesson. How far away is Lucia from the school?

Explore   You know that Lucia is about 158 feet above the ground. You also know that she can just see her school in the distance.

Plan   Find the distance she is from the school. Use the formula $V = 1.22 \times \sqrt{A}$

Solve   Substitute values into the formula.
$$V = 1.22 \times \sqrt{A}$$
$$V = 1.22 \times \sqrt{158}$$
Use your calculator.

Enter: 1.22 $\boxed{\times}$ 158 $\boxed{\sqrt{x}}$ $\boxed{=}$ *15.335162*

Lucia is about 15.3 miles from the school.

Examine   The closest perfect square to 158 is 169 or 13^2. $13 \times 1.22 = 15.86$. The answer is reasonable.

Checking Your Understanding

Communicating Mathematics

Read and study the lesson to answer each question.

1. **Explain** why a positive number has two different square roots.

2. **Draw** a square that has an area of 64 square centimeters.

3. **Explain** how to find the measure of the side of a square that has an area of 49.

4. **Explain** why finding the square root and squaring are inverse operations.

MATERIALS

square tiles

5. Arrange 40 tiles into the largest square possible.
   a. How many tiles did you use? How many are left over?
   b. Add tiles until you have the next larger square. How many tiles did you add?
   c. Between what two whole numbers is the square root of 40?

Guided Practice

Find each square root.

6. $-\sqrt{9}$ 7. $\sqrt{16}$ 8. $-\sqrt{4}$ 9. $\sqrt{81}$

Find the best integer estimate for each square root. Then check your estimate using a calculator.

10. $\sqrt{79}$ 11. $-\sqrt{53}$

12. $\sqrt{29}$ 13. $\sqrt{120}$

14. **Aviation** An airplane flies from New York to London at an altitude of 60,000 feet. How far can the pilot see when he or she looks out the window?

Exercises: Practicing and Applying the Concept

Independent Practice

Find each square root.

15. $\sqrt{4}$ 16. $\sqrt{64}$ 17. $\sqrt{9}$

18. $-\sqrt{25}$ 19. $-\sqrt{16}$ 20. $-\sqrt{81}$

21. $\sqrt{100}$ 22. $\sqrt{1}$ 23. $-\sqrt{144}$

24. $\sqrt{169}$ 25. $\sqrt{400}$ 26. $\sqrt{2.25}$

✓ *Choose*

Estimation
Mental Math
Calculator
Paper and Pencil

Find the best integer estimate for each square root. Then check your estimate using a calculator.

27. $\sqrt{89}$ 28. $-\sqrt{44}$ 29. $\sqrt{200}$

30. $-\sqrt{170}$ 31. $-\sqrt{97}$ 32. $-\sqrt{6.76}$

33. $\sqrt{2600}$ 34. $\sqrt{625}$ 35. $\sqrt{118}$

36. $\sqrt{13.69}$ 37. $-\sqrt{26.79}$ 38. $-\sqrt{156.25}$

Simplify.

39. $\sqrt{a^2}$ 40. $\sqrt{169} - (-\sqrt{121})$ 41. $\sqrt{\sqrt{81}}$

Critical Thinking

42. Without a calculator, determine which is greater: $\sqrt{363}$ or 19. Explain your reasoning.

43. In $x^2 = y$, could x be negative? Why or why not?

Applications and Problem Solving

44. Sports The ability of a hang glider to fly depends on the area of its wing. The relationship between wingspan and wing area is called the aspect ratio. The formula for the aspect ratio, R, is $R = \frac{s^2}{A}$, where s represents the wingspan and A represents the wing area.

    a. Solve $R = \frac{s^2}{A}$ for s to write a formula for wingspan when the aspect ratio and wing area are known.

    b. An advertisement for a hang glider says the aspect ratio is 2.7 and the wing area is 30 square feet. What is the wingspan of the glider?

45. Construction Bloomington city code requires that a party house must allow at least 4 square feet for each person on the dance floor. Reston's Hotel Restaurant is going to add a dance floor that is square and is large enough for 100 people. How long should it be on each side?

46. Plumbing Every water heater must have a pipe connecting it to the water meter supplied by the utility company. An engineer who is designing a new water heater has determined that the proper amount of water will be supplied if the opening of the pipe has an area of 0.442 square inches. Pipe is usually described in terms of its interior diameter. Use the formula for the area of a circle, $A = \pi r^2$, to find the proper size of pipe for the water heater.

Mixed Review

47. Geometry Find the volume of a rectangular pyramid with length 8 inches, width 8 inches, and height 9 inches. (Lesson 12-8)

48. How many ways can the letters of the word GYMNAST be arranged? (Lesson 10-6)

49. Express 28% as a fraction in simplest form. (Lesson 9-7)

50. Standardized Test Practice How many boards, each $2\frac{1}{2}$ feet long, can be cut from an 18-foot board? (Lesson 6-4)

    A. 9

    B. 8

    C. 7

    D. 6

51. Estimate 9.2×3.85. (Lesson 6-2)

52. Find the prime factorization of 2160. (Lesson 4-4)

53. Solve $a = -15 + (-5)$. (Lesson 2-4)

54. Agriculture Make a bar graph of the data about the fruit production in the table below. (Lesson 1-10)

_inter_NET CONNECTION

Data Update For the latest U.S. fruit production, visit:

www.glencoe.com/ sec/math/prealg/ mathnet

Annual Fruit Production	
Fruit	**Millions of Metric Tons**
Grapes	60.7
Bananas	49.6
Apples	43.1
Coconuts	41.0
Plantains	26.8

Source: *Top Ten of Everything*

For **Extra Practice**, see page 771.

Problem-Solving Strategy: Use Venn Diagrams

Application: Surveys

The guidance counselor surveyed 80 eighth-graders to find out their interest in learning Japanese or Spanish. Twenty-two students expressed an interest in both languages. Eighteen expressed interest in Japanese only, and five were not interested in learning any foreign language. How many of the students who were surveyed were interested in Spanish only?

Learning the Concept

You can use a **Venn diagram** to illustrate data. A rectangle is used to represent all the data. A circle inside the rectangle is used to represent one group of data. Data that is common to more than one group is represented by the region where intersecting circles overlap.

Draw two intersecting circles in a rectangle to represent Japanese and Spanish. Write the number of students who are interested in both languages and the number of students who are interested in only Japanese in the appropriate regions. Write the number of students who are not interested in learning either language outside the circles but inside the rectangle. There are 22 + 18 + 5 or 45 students represented in the Venn diagram. There are 80 − 45 or 35 students that are interested in learning only Spanish.

Foreign Language Preference

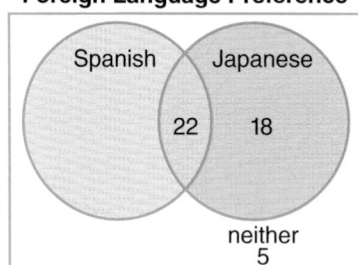

Spanish Japanese

22   18

neither
5

Example

The Venn diagram at the right shows students' music preferences.

a. How many students like all three types of music?

The area where all three circles overlap contains a 5. There are 5 students who like all three types.

Music Preference

Rock 63   Oldies 14

7

5

21   9

Rap 19

b. How many students like oldies?

There are 14 students who only like oldies, 9 who like oldies and rap, 7 who like oldies and rock, and 5 who like all three. There are 14 + 9 + 7 + 5 or 35 students who like oldies.

c. How many students like rap and rock?

The area where the circles for rap and rock overlap contains 5 and 21. There are 5 + 21 or 26 students who like rap and rock.

Checking Your Understanding

Communicating Mathematics

Read and study the lesson to answer each question. The Venn diagram at the right represents the states that produce more than 100 million bushels of corn, wheat, or soybeans per year.

Grain Production

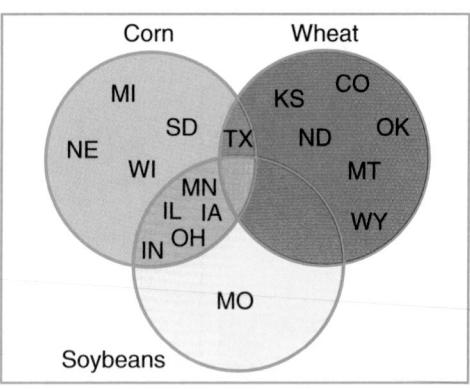

1. **Write** what the section where all three circles overlap represents.

2. **Determine** how many states produce more than 100 million bushels of corn or wheat.

Guided Practice

Solve using a Venn diagram.

3. At a birthday party, 9 people chose just cake for dessert, and 4 people chose just ice cream. Five people chose both cake and ice cream. If each person chose at least one dessert, how many people were at the birthday party?

4. **Recycling**   The Venn diagram at the right shows the number of communities serviced by EarthCare Recycling that choose to have curbside recycling of glass, newspaper, and aluminum. How many communities does EarthCare service?

EarthCare Communities

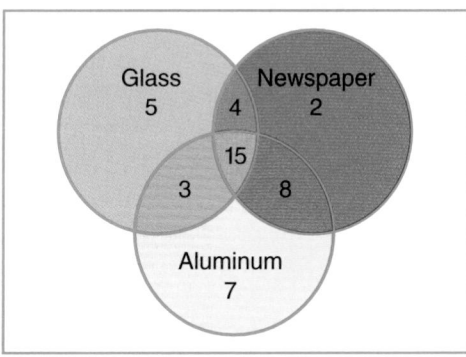

Exercises: Practicing and Applying the Concept

Independent Practice

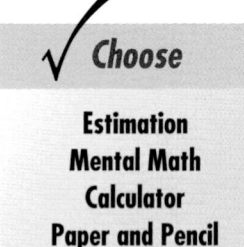

✓ **Choose**

Estimation
Mental Math
Calculator
Paper and Pencil

Solve. Use any strategy.

5. **Geometry**   There are 2 tiles. One is in the shape of a square, and the other is in the shape of an equilateral triangle. The sides of the square are congruent to the sides of the triangle. The tiles are put together to form a pentagon. If one side of the triangle is 20 cm long, what is the perimeter of the pentagon?

6. Of 150 TV viewers one evening, 64 watched Channel A, 36 watched Channel B, and 10 watched both. How many did not watch either?

7. Chef Martino made a huge pan of lasagna for a banquet. He makes 6 cuts along the length of the rectangular pan and 10 cuts along the width. How many pieces does he have?

8. Jaria must read a book for a book report. She read half of her book on Tuesday. On Wednesday she read another 30 pages. On Thursday she read 6 more, and on Friday she read half of what was left. If she still has 20 pages yet to read, how many pages does the book have?

9. A school survey of 250 girls showed that 85 read Magazine A and 65 read Magazine B. 110 girls read neither. How many read both?

For **Extra Practice**, see page 771.

10. There are 26 students in a math class. The class takes a survey on pets and finds that 14 students have dogs, 10 students have cats, and 5 students have birds. Four students have dogs and cats, 3 students have dogs and birds, and 1 student has a cat and a bird. If no one has all three of these animals, how many students have none of these animals?

11. In a geography class of 30 students, 16 said they wanted to visit France, 16 wanted to visit Germany, and 11 wanted to visit Italy. Five said they wanted to visit both France and Italy, and of these, 3 wanted to visit Germany as well. Five wanted only Italy, and 8 wanted only Germany. How many students wanted France only?

Critical Thinking

12. **Geometry** A geometry teacher drew some quadrilaterals on the chalkboard. There were 5 trapezoids, 12 rectangles, 5 squares, and 8 rhombuses. What is the least number of figures the teacher could have drawn?

13. **Geometry** Draw a Venn diagram to show the relationship between squares and parallelograms. Let the rectangle represent polygons.

Mixed Review

14. Find $-\sqrt{121}$. (Lesson 13-1)

15. **Geometry** Draw an example of a rhombus and describe its characteristics. (Lesson 11-7)

16. **Finance** If $800 is deposited in a savings account for six months at 5.75% annual interest, how much interest will it earn? (Lesson 9-9)

17. **Standardized Test Practice** Solve $2x - 5 = 48$. (Lesson 7-2)

    A. 21.5   **B.** 26.5   **C.** 45   **D.** 86

18. Solve $n - (-20) \geq 14$ and graph the solution on a number line. (Lesson 3-6)

WORKING ON THE
Investigation

Refer to the Investigation on pages 606–607.

Look at the floor plan(s) of the house that you placed in your Investigation Folder at the end of Lesson 12-8. Use the floor plan(s) to make elevation drawings from four different directions. Be sure to use the same scale on the elevation drawing that you used on the floor plan. An example of an elevation drawing is shown at the beginning of Chapter 12.

Using your working drawings, construct a cardboard or foam board model of your house. It will probably be easier to work with a scale of $\frac{1}{2}$ inch = 1 foot instead of $\frac{1}{4}$ inch = 1 foot. (*Hint:* Cut doors and windows out before you erect the walls. You can also trace window openings on wax paper. Then draw in the frames and tape the wax paper on the inside wall.)

Make a roof for your model. Don't glue the roof on so that you can remove it to show the arrangement of the rooms.

Place your model on a base that is larger than the house. The excess can be used to show the yard, sidewalks, and driveway. You can also show trees and shrubs.

If possible, take a photograph of your model.

Add the results of your work to your Investigation Folder.

What You'll Learn

- To identify numbers in the real number system, and

- to solve equations by finding square roots.

Why It's Important

Using real numbers can help you estimate distance and time.

Application: Law Enforcement

When investigating an accident, police officers often need to determine how fast a car was traveling before it skidded to a stop. They use the formula $s = \sqrt{30df}$, where s is the speed in miles per hour, d is the length of the skid marks in feet, and f is the coefficient of friction. What was the approximate speed of a car if the skid marks are 80 feet long and the coefficient of friction is estimated to be 1.2?

$s = \sqrt{30df}$

$s = \sqrt{30 \cdot 80 \cdot 1.2}$ *Replace d with 80 and f with 12.*

$s = \sqrt{2880}$

$s = 53.66563146\ldots$

The car was traveling at about 54 miles per hour.

Learning the Concept

You can review rational numbers in Lesson 5-1.

Recall that numbers that can be expressed as $\frac{a}{b}$, where a and b are integers and b is not 0, are rational numbers. Rational numbers can always be expressed as decimals that terminate or repeat. For example, 7, $\frac{1}{3}$, and -4.9 are rationals. The number $\sqrt{2880}$, which is $53.66563146\ldots$, is not a repeating or terminating decimal. This kind of number is called an **irrational number**.

Definition of Irrational Number	An irrational number is a number that cannot be expressed as $\frac{a}{b}$, where a and b are integers and b does not equal 0.

Other examples of irrational numbers are shown below.

$\sqrt{3} = 1.7320508\ldots$ $\qquad\qquad$ $\pi = 3.14159\ldots$

Example **Determine whether each number is rational or irrational.**
a. 0.22222 . . .

The three dots means that the 2s keep repeating. This is a repeating decimal, so it can be expressed as a fraction.

$0.22222\ldots = \frac{2}{9}$

Thus, it is a rational number.

b. 0.35

This is a terminating decimal. It can be expressed as $\frac{35}{100}$ or $\frac{7}{20}$. It is also a rational number.

c. 0.52522522252222 . . .

This decimal does not repeat or terminate and is therefore an irrational number. It does have a pattern to it, but since the 5s are separated by an increasing number of 2s, there is no exact repetition.

The set of rational numbers and the set of irrational numbers together make up the set of **real numbers**. The Venn diagram at the right shows the relationships among whole numbers, integers, rational numbers, irrational numbers, and real numbers.

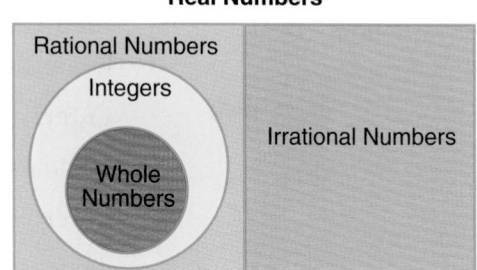

The following chart shows the sets of numbers to which several numbers belong.

Number	Whole Number	Integer	Rational	Irrational	Real
-9		✔	✔		✔
$\sqrt{35}$				✔	✔
$\sqrt{49}$	✔	✔	✔		✔
0.58585858 . . .			✔		✔
0.121231234 . . .				✔	✔
9	✔	✔	✔		✔

The equations you have encountered so far have had rational number solutions. Some equations have irrational number solutions.

You can solve some equations that involve squares by taking the square root of each side.

Example **2** **Solve each equation. Round decimal answers to the nearest tenth.**

a. $x^2 = 81$

$x^2 = 81$

$x = \sqrt{81}$ or $x = -\sqrt{81}$ *Take the square root of each side.*

$x = 9$ or $x = -9$

Check: $9^2 = 81$ and $(-9)^2 = 81$, so the solution checks.

b. $y^2 = 75$

$y^2 = 75$

$y = \sqrt{75}$ or $y = -\sqrt{75}$ *Take the square root of each side.*

75 8.660254038

$y \approx 8.7$ or $y \approx -8.7$

The solutions to many real-world problems are irrational numbers.

Example ③
Real World
APPLICATION
Electricity

The voltage, *V*, in a circuit is given by the formula $V = \sqrt{PR}$. In this formula, *V* is in volts, *P* is the power in watts, and *R* is the resistance in ohms. An electrician has a circuit that produces 1500 watts of power. She wants the voltage in the circuit to be no more than 110 volts. Should she design the circuit with a resistance of a 7.5 ohms or 7.8 ohms?

For this circuit, $P = 1500$ and $V = 110$. To determine which resistance to use, you can use a calculator to evaluate $\sqrt{1500R}$ for $R = 7.5$ and for $R = 7.8$ to find out which one will produce no more than 110 volts.

Evaluate $\sqrt{1500R}$ for $R = 7.5$.

Enter: 1500 $\boxed{\times}$ 7.5 $\boxed{=}$ $\boxed{\sqrt{x}}$ *106.0660172*

Evaluate $\sqrt{1500R}$ for $R = 7.8$.

Enter: 1500 $\boxed{\times}$ 7.8 $\boxed{=}$ $\boxed{\sqrt{x}}$ *108.1665383*

She can design the circuit with *either* resistance. If she wants the voltage in the circuit to be as close to 110 volts as possible, then she should design the circuit with a resistance of 7.8 ohms.

Checking Your Understanding

Communicating Mathematics

Read and study the lesson to answer each question.

1. **Compare and contrast** rational and irrational numbers.
2. **You Decide** Jaya says that $\sqrt{16}$ can be both an integer and a rational number. Do you agree? Explain.
3. **Give an example** of an irrational number that is less than -10.

Guided Practice

Name the sets of numbers to which each number belongs: the whole numbers, the integers, the rational numbers, the irrational numbers, and/or the real numbers.

4. $0.3333\ldots$　　5. 0　　6. $\sqrt{11}$　　7. $-\dfrac{1}{2}$

Solve each equation. Round decimal answers to the nearest tenth.

8. $m^2 = 49$　　9. $r^2 = 361$　　10. $t^2 = 1$　　11. $n^2 = 17$

12. **Hang Gliding** Using $R = \dfrac{s^2}{A}$, find the wingspan *s* of Kesse's hang glider if the aspect ratio *R* is 2.5 and the area of the wing *A* is 120 square feet.

Exercises: Practicing and Applying the Concept

Independent Practice

Name the sets of numbers to which each number belongs: the whole numbers, the integers, the rational numbers, the irrational numbers, and/or the real numbers.

13. 8　　14. -5　　15. $\dfrac{2}{3}$

16. $\sqrt{5}$　　17. -3.5　　18. $-\sqrt{25}$

19. 0.121121112…   **20.** $\frac{5}{4}$ **21.** $\sqrt{18}$

22. $\frac{-9}{3}$ **23.** $2.\overline{5}$ **24.** 3.14359265…

Solve each equation. Round decimal answers to the nearest tenth.

25. $r^2 = 36$ **26.** $x^2 = 64$ **27.** $y^2 = 12$ **28.** $169 = m^2$

29. $n^2 = 120$ **30.** $f^2 = 200$ **31.** $180 = j^2$ **32.** $p^2 = 1.44$

33. $0.0004 = s^2$ **34.** $h^2 = 240$ **35.** $400 = q^2$ **36.** $a^2 = 90{,}000$

37. $(-b)^2 = 81$ **38.** $c^2 - 3^2 = \sqrt{16^2}$ **39.** $\sqrt{81} = d^2$

40. Geometry The length of a rectangle is three times its width. What are the dimensions of the rectangle if its area is 192 ft²?

Critical Thinking

41. Can the product of two irrational numbers be rational? Support your answer with examples.

Applications and Problem Solving

42. Meteorology You can use the formula $t^2 = \frac{d^3}{216}$ to estimate the amount of time that a thunderstorm will last. In this formula, t is the time in hours, and d is the diameter of the storm in miles. If a thunderstorm is 6 miles wide, how long will the storm last?

43. Geometry You can use Heron's formula to find the area of a triangle if you know the measures of its sides. If the measures of the sides are a, b, and c, the area A equals $\sqrt{s(s - a)(s - b)(s - c)}$, where s is one-half the perimeter. Suppose you have a triangle with sides 80 feet, 60 feet, and 100 feet long. Find the area of the triangle.

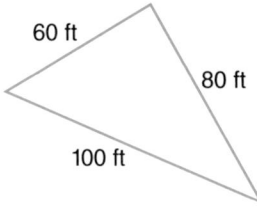

44. Physics If an object is dropped, the time t in seconds it takes to reach a given distance can be estimated by using the formula $d = 0.5gt^2$. In this formula, d is the free-fall distance, and g is the acceleration due to gravity, 32 ft/s². If Rachelle drops a ball from the top of a 55-foot building, how long does it take for the ball to hit the ground?

45. Family Activity Drop a ball from a second or third story window several times and time its fall to the ground. Find the average time. See how that time compares to the answer using the formula given in Exercise 44 by measuring the actual falling distance.

Mixed Review

46. Standardized Test Practice A survey of 120 people showed that from 7:00 to 8:00 one evening, 46 people watched news programs, 34 watched game shows, and 15 watched both. How many did not watch either? (Lesson 13-2)

A. 55 people   **B.** 40 people   **C.** 35 people   **D.** 25 people

47. Geometry Find the volume of a circular cylinder with diameter 25 cm and height 10 cm. Round to the nearest tenth. (Lesson 12-7)

48. Graph $y = 2x + 3$. (Lesson 8-3)

49. Solve $v = \left(\frac{3}{7}\right)\left(-\frac{14}{15}\right)$. (Lesson 6-3)

50. Find the GCF of 36 and $-30ab$. (Lesson 4-5)

51. Evaluate $4cd$ if $c = -2$ and $d = -9$. (Lesson 2-7)

52. Simplify $7(a + b) - 2(3a + 4b)$. (Lesson 1-5)

For **Extra Practice**, see page 771.

FYI

In A.D. 60, Heron, an Egyptian mathematician and scientist, invented the *dioptra* for surveying and making astronomical observations. A dioptra could measure horizontal and vertical angles.

The Pythagorean Theorem

What
You'll Learn

- To use the Pythagorean theorem to find the length of the side of a right triangle and to solve problems.

Why
It's Important

You can use the Pythagorean theorem to find out how far a catcher needs to throw a baseball to reach second base.

MATERIALS: dot paper

Hands-On Activity

Your Turn

Work with a partner.

▶ Draw squares with sides of 3, 4, and 5. Use the three squares to form a triangle as shown.

a. Classify the triangle that is formed.

b. What is the area of each of the three squares?

c. What relationship exists among the three areas?

▶ Repeat the procedure above for squares with sides of 5, 12, and 13. Describe what you discover.

▶ Find other sets of three numbers that have the same properties as 3, 4, 5 and 5, 12, 13.

Learning the Concept

FYI

In the movie *The Wizard of Oz*, with Judy Garland, when the scarecrow gets his brain he recites the Pythagorean theorem . . . but he says it incorrectly!

The sides of a right triangle that are adjacent to the right angle are called the **legs**. The side opposite the right angle is called the **hypotenuse**.

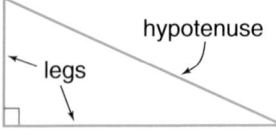

The relationship between the lengths of the legs and the hypotenuse that you noticed in the activity at the beginning of the lesson was proven in the fifth century B.C. by a Greek mathematician, Pythagoras, and his followers. This relationship is called the **Pythagorean theorem**. It is true for *any* right triangle.

Example ❶

Geometry

A rectangle has sides of 6 and 8 inches. How long is a diagonal?

A diagonal and two adjacent sides of the rectangle form a right triangle. The legs of the right triangle are sides of the rectangle, and the hypotenuse of the right triangle is a diagonal of the rectangle.

THINK ABOUT IT

Why can you ignore the negative value of c in this problem?

$$c^2 = a^2 + b^2 \quad \textit{Pythagorean theorem}$$
$$c^2 = 6^2 + 8^2 \quad \textit{Replace a with 6 and b with 8.}$$
$$c^2 = 36 + 64$$
$$c^2 = 100$$
$$c = \sqrt{100} \quad \textit{Take the square root of each side.}$$
$$c = 10$$

The length of the diagonal is 10 inches.

The Pythagorean theorem can also be used to find the length of any side of a right triangle if the lengths of the other two sides are known.

Example ❷

Find the length of the third side of the right triangle.

$$c^2 = a^2 + b^2$$
$$27^2 = a^2 + 25^2$$
$$729 = a^2 + 625$$
$$729 - 625 = a^2 + 625 - 625$$
$$104 = a^2$$
$$\sqrt{104} = a$$
$$104 \;\boxed{\sqrt{x}}\; \textit{10.19803903}$$

The length of the leg is about 10.2 cm.

You can use the Pythagorean theorem to see whether a triangle is right.

Example ❸

The measurements of three sides of a triangle are given. Determine whether each triangle is a right triangle. *Remember, the hypotenuse is the longest side.*

a. 5 ft, 7 ft, 8 ft

$$c^2 = a^2 + b^2$$
$$8^2 \stackrel{?}{=} 5^2 + 7^2$$
$$64 \stackrel{?}{=} 25 + 49$$
$$64 \neq 74$$

The triangle is *not* right.

b. 10 cm, 24 cm, 26 cm

$$c^2 = a^2 + b^2$$
$$26^2 \stackrel{?}{=} 10^2 + 24^2$$
$$676 \stackrel{?}{=} 100 + 576$$
$$676 = 676$$

The triangle is right.

The Pythagorean theorem can be applied to solve problems that occur in real life.

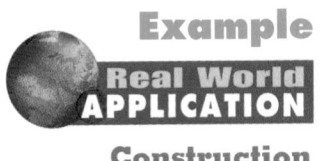

Example 4

Real World
APPLICATION

Construction

A Forest Service warehouse facility is used as a supply depot and a fire-fighting center. A security fence is needed to protect a helicopter, some tanker trucks, and other fire-fighting equipment that must be stored outdoors. A diagram of the region to be fenced is shown at the right. How much fencing must be purchased to surround the region?

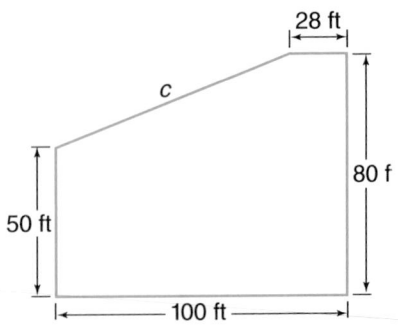

Explore   You know what the boundary of the region to be fenced looks like. You must find the perimeter of the region.

Plan   To find the perimeter, you must find the unknown length, c. You can separate the region into more familiar shapes as shown below.

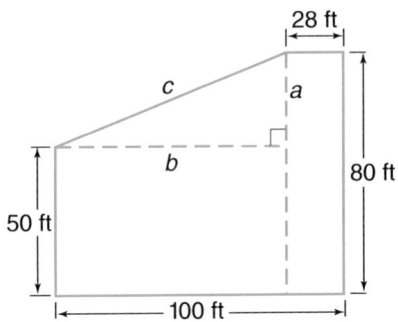

The hypotenuse of the right triangle is the side with length c. Find the measures of the legs of the right triangle and then use the Pythagorean theorem to find the length of the hypotenuse.

Solve   First find the measures of the legs.

$a = 80 - 50$ $b = 100 - 28$
$a = 30$ $b = 72$

Next, apply the Pythagorean theorem.

$c^2 = a^2 + b^2$
$c^2 = 30^2 + 72^2$ *Replace a with 30 and b with 72.*
$c^2 = 900 + 5184$
$c^2 = 6084$
$c = \sqrt{6084}$ *Take the square root of each side.*

6084 $\boxed{\sqrt{x}}$ *78*

$c = 78$

Finally, find the perimeter.

$28 + 80 + 100 + 50 + 78 = 336$

So, 336 feet of fencing is needed.

Examine   If the region included the corner area, it would be a rectangular shape. Its perimeter would be $2 \cdot 100 + 2 \cdot 80$ or 360 feet. The answer should be less than this. Thus, the answer is reasonable.

Communicating Mathematics

Read and study the lesson to answer each question.

1. **Explain** how to find the length of a leg of a right triangle if you know the length of the hypotenuse and the length of the other leg.

2. **Draw** and label a triangle with sides that measure 6 meters, 8 meters, and 10 meters. Is it a right triangle? Explain.

3. The roof of a house is 24 feet above the ground. You have a 26-foot ladder. Use the Pythagorean theorem to explain why the base of the ladder must be placed no more than 10 feet from the house in order to reach the roof.

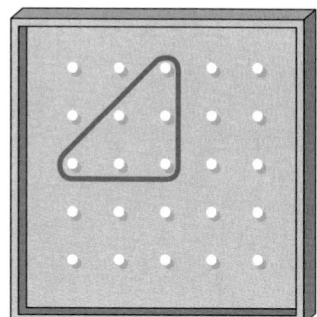

MODELING MATHEMATICS

MATERIALS

dot paper

4. Use dot paper to draw squares on each side of a triangle like the one at the right. Write an equation you could use to find the length of the hypotenuse.

Guided Practice

Write an equation you could use to solve for x. Then solve. Round decimal answers to the nearest tenth.

5.

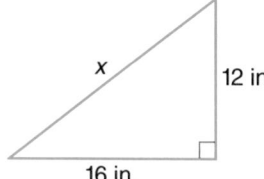

x 12 in.

16 in.

6.

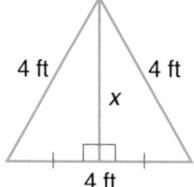

4 ft   4 ft

x

4 ft

Write an equation that can be used to answer each question. Then solve. Round decimal answers to the nearest tenth.

7. How long is the lake?

21 km   30 km

8. How high is the TV screen?

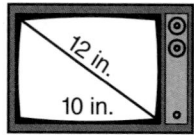

12 in.

10 in.

In a right triangle, if a and b are the measures of the legs and c is the measure of the hypotenuse, find each missing measure. Round decimal answers to the nearest tenth.

9. $a = 3, b = 4$

10. $a = 3, c = 7$

11. $b = 12, c = 35$

12. $a = 15, b = 16.7$

The measurements of three sides of a triangle are given. Determine whether each triangle is a right triangle.

13. 9 cm, 12 cm, 15 cm

14. 6 ft, 7 ft, 12 ft

15. **Baseball**   A baseball diamond is actually a square. The distance between bases is 90 feet. When a runner on first base tries to steal second, a catcher has to throw from home to second base. How far is it between home and second base?

90 ft   90 ft

90 ft   90 ft

Independent
Practice

Write an equation you could use to solve for *x*. Then solve. Round decimal answers to the nearest tenth.

16.

17.

18.

Write an equation that can be used to answer each question. Then solve. Round decimal answers to the nearest tenth.

19. How high is the kite?

20. How long is a suspension line?

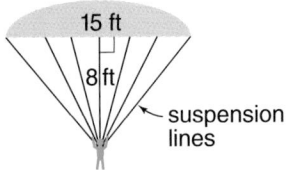

21. How high does the ladder reach?

22. How far apart are the planes?

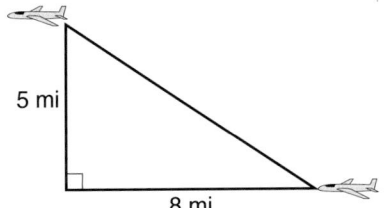

In a right triangle, if *a* and *b* are the measures of the legs and *c* is the measure of the hypotenuse, find each missing measure. Round decimal answers to the nearest tenth.

23. $a = 12, b = 16$

24. $b = 21, c = 29$

25. $a = 2, b = 5$

26. $a = 5, c = 10$

27. $a = 7, c = 9$

28. $b = 3, c = 7$

29. $a = 7, b = 7$

30. $b = 36, c = 85$

31. $a = 14, b = 15$

32. $a = 180, c = 181$

33. $a = \sqrt{11}, c = 6$

34. $b = 13, c = \sqrt{233}$

The measurements of three sides of a triangle are given. Determine whether each triangle is a right triangle.

35. 8 ft, 9 ft, 10 ft

36. 10 m, 24 m, 26 m

37. 15 in., 12 in., 9 in.

38. 6 mi, 7 mi, 8 mi

39. 18 cm, $\sqrt{24}$ cm, 30 cm

40. 15 ft, 16 ft, $\sqrt{31}$ ft

41. Geometry Find the length of a diagonal of a square if its area is 72 m².

42. Geometry Find the length of the diagonal of a cube if each side of the cube is 3 feet long.

43. The points $C(5, 1)$ and $D(8, 5)$ are graphed at the right. Find the distance between C and D. (*Hint:* Notice the right triangle. First find the measures of $\overline{BC}$ and $\overline{BD}$.)

Applications and
Problem Solving

✓ **Choose**

Estimation
Mental Math
Calculator
Paper and Pencil

44. Safety The National Safety Council recommends placing the base of a ladder one foot from the wall for every three feet of the ladder's length.

　a. How far away from a wall should you place a 15-foot ladder?

　b. How high can the ladder safely reach?

45. Hiking Gracia hikes 10 kilometers south, 15 kilometers east, and then 10 kilometers south again. How far is Gracia from the starting point of her hike?

46. Sailing A rope from the top of a mast on a sailboat is attached to a point 6 feet from the base of the mast. If the rope is 24 feet long, how high is the mast?

Mixed
Review

47. Solve $x^2 = 2.5$ to the nearest tenth. (Lesson 13-3)

48. Geometry Draw and label a diagram to represent perpendicular lines k and m. (Lesson 11-1)

49. If $g(x) = -x + 4$, find $g(-1)$. (Lesson 8-4)

50. Solve $5 - 2n = 4n + 41$. (Lesson 7-5)

51. Solve $\left(\frac{2}{3}\right)^2 = x$. (Lesson 6-3)

52. Standardized Test Practice A football team averaged -1 yards a play for 3 plays. What was their total yardage for the three plays? (Lesson 2-7)

　A. -3 yd　　　**B.** -2 yd　　　**C.** -1 yd　　　**D.** 3 yd

For **Extra Practice**, see page 771.

53. Solve $6b = 120$ mentally. (Lesson 1-6)

54. Find the value of the expression $3[4(6 - 2) - 5]$. (Lesson 1-2)

Self Test

Find each square root. (Lesson 13-1)

1. $\sqrt{36}$　　　　　　　　**2.** $\sqrt{121}$　　　　　　　　**3.** $-\sqrt{100}$

4. In a recent survey of 120 students, 60 students said they play tennis and 50 students said they play softball. If 20 students play both sports, how many students do not play either tennis or softball? (Lesson 13-2)

Solve each equation. Round decimal answers to the nearest tenth. (Lesson 13-3)

5. $x^2 = 144$　　　　　　　**6.** $y^2 = 50$　　　　　　　**7.** $g^2 = 40$

In a right triangle, if a and b are the measures of the legs and c is the measure of the hypotenuse, find each missing measure. Round answers to the nearest tenth. (Lesson 13-4)

8. $a = 12, b = 16$　　　　**9.** $b = 63, c = 65$　　　　**10.** $a = 15, c = 39$

13-4B Graphing Irrational Numbers

An Extension of Lesson **13-4**

MATERIALS

compass

straightedge

You already know how to graph integers and rational numbers on a number line. How would you graph an irrational number like $\sqrt{10}$?

Activity

▶ Draw a number line. At 3, construct a perpendicular line segment 1 unit in length. Draw the line segment shown in color. Label it c.

▶ Using the Pythagorean theorem, you can show that the hypotenuse is $\sqrt{10}$ units long.

$$c^2 = a^2 + b^2$$
$$c^2 = 1^2 + 3^2$$
$$c^2 = 10$$
$$c = \sqrt{10}$$

▶ Open the compass to the length of the segment you drew in color. With the tip of the compass at 0, draw an arc that intersects the number line at B. The distance from 0 to B is $\sqrt{10}$ units.

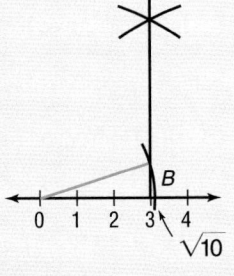

Your Turn Graph $\sqrt{8}$ on a number line. Think of $\sqrt{8}$ as $\sqrt{2^2 + 2^2}$.

TALK ABOUT IT

1. Explain how to graph $-\sqrt{10}$.
2. Explain how to graph $\sqrt{2}$.
3. Describe two different ways to graph $\sqrt{5}$.
4. Explain how the graph of $\sqrt{2}$ can be used to locate the point that represents $\sqrt{3}$.

Extension 5. Graph $\sqrt{12}$ on a number line. (*Hint:* Think of $\sqrt{12}$ as $\sqrt{4^2 - 2^2}$.)

Special Right Triangles

MATERIALS: \ compass protractor scissors ruler

Hands-On Activity

Most people know that Alexander Graham Bell invented the telephone. But the drawings that Bell needed to get a patent were done by Lewis Howard Latimer (1848–1928), an African-American drafter and engineer.

Drafters and engineers often use a compass and straightedge to copy measurements accurately from one location to another.

Your Turn

Work with a partner.

▶ Construct and cut out a square.

▶ Fold the square along the diagonal to form an isosceles right triangle.

▶ Measure each leg and each angle of the triangle.

▶ Use the Pythagorean theorem to determine the length of the hypotenuse.

▶ Use a calculator to divide the length of the hypotenuse by $\sqrt{2}$.

▶ Repeat the above steps for several other squares.

a. Describe a method for finding the length of the hypotenuse of a right isosceles triangle if you know the length of its legs.

Your Turn

▶ Construct an equilateral triangle. Cut out the triangle. Fold the triangle to form a right triangle.

▶ Measure each side and each angle of the right triangle.

▶ Repeat the above steps for several other equilateral triangles.

b. What is the relationship between the shortest side and the angle having the smallest measure?

c. What is the relationship between the length of the hypotenuse and the length of the shortest side?

Learning the Concept

The triangle formed in the first activity above is called a **45°–45° right triangle**. The triangle in the second activity is called a **30°–60° right triangle**. The relationships you discovered above are true for any 45°–45° or 30°–60° right triangle.

In a 45°–45° right triangle, you can find the length of the hypotenuse by multiplying the length of a leg by $\sqrt{2}$.

Example **Find the length of $\overline{AC}$ in $\triangle ABC$.**

$c = a\sqrt{2}$

$c = 5\sqrt{2}$ *Replace a with 5.*

5 ☒ 2 ☒√x☒ ☒=☒ 7.071067812

The length of $\overline{AC}$ is about 7.1 cm.

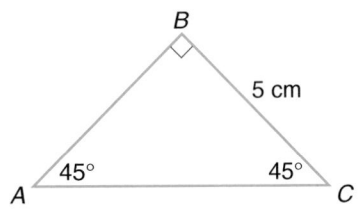

In a 30°–60° right triangle, the length of the side opposite the 30° angle is one-half the length of the hypotenuse.

Example **Find the length of $\overline{PQ}$ in $\triangle PQR$.**

$a = \frac{1}{2}c$

$a = \frac{1}{2}(12)$ *Replace c with 12.*

$a = 6$

The length of $\overline{PQ}$ is 6 inches.

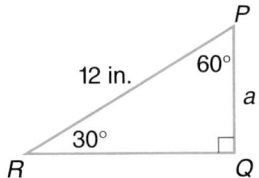

Also, in a 30°–60° right triangle, you can find the length of the side opposite the 60° angle by multiplying the length of the other leg by $\sqrt{3}$.

Example **Large doors, like those on barns and airplane hangars, are often reinforced with a diagonal brace to prevent warping. A barn door is designed so that the brace forms a 30° angle as shown. If the width of the door is 7 feet, how high is the door?**

The brace forms two congruent right triangles. Draw a model of one triangle to find the measures.

Let a be the measure of the side opposite the 30° angle and b be the measure of the side opposite the 60° angle.

$b = a\sqrt{3}$

$b = 7\sqrt{3}$

$b \approx 12.1$

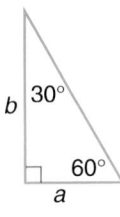

The door is about 12 feet high.

Checking Your Understanding

Communicating Mathematics

Read and study the lesson to answer each question.

1. **Explain** how to find the length of the hypotenuse of an isosceles right triangle without using the Pythagorean theorem.

2. **Draw** and label all the sides and angles of a 30°–60° right triangle with the shortest side having a measurement of 5 meters.

The length of a leg of a 45°–45° right triangle is given. Find the length of the hypotenuse. Round decimal answers to the nearest tenth.

3. 1 yd

4. 9.5 m

The length of a hypotenuse of a 30°–60° right triangle is given. Find the length of the side opposite the 30° angle.

5. 18 m

6. $6\frac{1}{2}$ in.

Find the lengths of the missing sides in each triangle. Round decimal answers to the nearest tenth.

7.

8.

9.

10. Camping A rope tied to the top of a tent pole makes a 60° angle with the ground and is anchored 4 feet from the base of the pole.

   a. How long is the rope?

   b. How tall is the tent pole?

Exercises: Practicing and Applying the Concept

11. One leg of a 45°–45° right triangle is 15 centimeters long. What is the length of the other leg?

12. The shorter leg of a 30°–60° right triangle is 3 inches long.

   a. What is the length of the hypotenuse?

   b. What is the length of the other leg?

The length of a leg of a 45°–45° right triangle is given. Find the length of the hypotenuse. Round decimal answers to the nearest tenth.

13. 15 cm

14. 7 yd

15. 5.2 m

16. $2\frac{1}{2}$ ft

17. 6.9 in.

18. 4.1 mm

The length of a hypotenuse of a 30°–60° right triangle is given. Find the length of the side opposite the 30° angle.

19. 48 in.

20. 11 yd

21. $\frac{1}{4}$ mi

22. 3000 m

23. 4.63 cm

24. 13 mm

Find the lengths of the missing sides in each triangle. Round decimal answers to the nearest tenth.

25.

26.

27.

28.

29.

30.

31.

32.

33.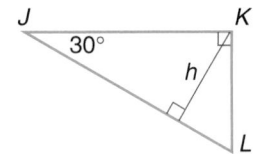

34. The hypotenuse of a 45°–45° right triangle is 12 cm long. Find the length of each of the other sides.

35. In triangle *JKL* shown at the right, the measure of ∠*LJK* is 30°, and the altitude *h* is 5 yards long. Find the length of $\overline{KL}$.

Critical Thinking
36. Draw a diagram for each problem. Then solve each problem.

   a. Find the area and perimeter of a regular hexagon inscribed in a circle that has a radius of 20 centimeters.

   b. Find the volume and surface area of a square pyramid whose lateral faces are equilateral triangles with 8-inch sides.

Applications and Problem Solving

37. Recreation   Two poles at the end of a swing set form a triangle with the ground. Each pole is 10 feet long and forms a 45° angle with the ground. What is the height of the swing set?

38. Construction   An A-frame house is one that looks like the letter A when viewed from the front. An architect is designing an A-frame that has sides that are as long as the width of the front of the house at its base. If the building is to be 26 feet wide, what is the height of the house?

Mixed Review
39. Construction   Harold and Marge are staking an area for their patio. The patio is to be a rectangle 20 feet long by 12 feet wide. To make sure they have the corners square, they measure the diagonals. What should the measure of both diagonals be? (Lesson 13-4)

40. Geometry   Find the area of a circle with a diameter of $8\frac{1}{2}$ inches. Round to the nearest tenth. (Lesson 12-2)

41. Architecture   An architect is designing a gazebo in the shape of a regular pentagon. What will be the measure of each interior angle of the gazebo? (Lesson 11-8)

42. Express $\frac{5}{6}$ as a percent. (Lesson 9-7)

43. Standardized Test Practice   Evaluate $3c^2$ if $c = -5$. (Lesson 4-2)
   A. −75   **B.** 28   **C.** 75   **D.** 225

44. Solve $f + 19 = 4$ and graph the solution on a number line. (Lesson 3-2)

For **Extra Practice**, see page 772.
45. Translate *a score not less than 50 is a win* into an inequality. (Lesson 1-9)

13-6A Ratios in Right Triangles

A Preview of Lesson **13-6**

MATERIALS

protractor

metric ruler

Right triangles and the relationships among their sides have been studied for thousands of years.

When working with right triangles, the side **opposite** an angle is the side that is not part of the angle. In the triangle shown at the right, side r is opposite $\angle R$. The side that is not opposite an angle and not the hypotenuse is called the **adjacent** side. In the triangle at the right, side s is adjacent to $\angle R$.

Your Turn

Work in groups of three.

▶ Each person should copy the table at the right.

▶ Each person in your group should draw a right triangle XYZ in which $m\angle X = 40°$, $m\angle Y = 50°$, and $m\angle Z = 90°$.

▶ Measure the leg opposite the 40° angle. Record the measurement to the nearest millimeter.

	40°angle
Length of leg opposite	
Length of leg adjacent	
Length of hypotenuse	
Ratio 1	
Ratio 2	
Ratio 3	

▶ Measure the leg adjacent to the 40° angle. Record the measurement to the nearest millimeter.

▶ Measure the hypotenuse and record the measurement to the nearest millimeter.

▶ Use your measurements and a calculator to find each ratio for the 40° angle. Record each ratio to the nearest hundredth.

$$\text{ratio } 1 = \frac{\text{opposite leg}}{\text{adjacent leg}} \qquad \text{ratio } 2 = \frac{\text{opposite leg}}{\text{hypotenuse}} \qquad \text{ratio } 3 = \frac{\text{adjacent leg}}{\text{hypotenuse}}$$

▶ Add a column to your table and repeat the above procedure for the 50° angle.

1. Compare your ratios with the other members of your group. How do they compare?

2. Make a conjecture about the ratio of the sides of any 40°–50° right triangle.

3. Repeat the activity for 25°–75° right triangles. What do you find?

13-6

The Sine, Cosine, and Tangent Ratios

You'll Learn

■ To find missing sides and angles of triangles using the sine, cosine, and tangent ratios.

It's Important

Trigonometric ratios are often used to find distances or lengths that cannot be measured directly.

Technology Activity

If you know certain measures of a right triangle, ratios can be used to find the measures of the remaining parts. These ratios are used in the study of **trigonometry**.

Your Turn

Work with a partner.

▶ In triangle ABC, the measure of side a is 1, the measure of side b is 1.732, and the measure of side c is 2.

▶ Find and record the ratio of a to c. Using a calculator, enter 30 [SIN].

▶ Find and record the ratio of b to c. Using a calculator, enter 30 [COS].

▶ Find and record the ratio of a to b. Using a calculator, enter 30 [TAN].

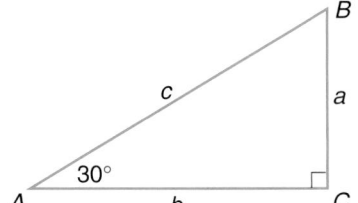

TALK ABOUT IT

a. The first ratio compares the leg opposite $\angle A$ to the hypotenuse. Describe the relationship between the measures of the second ratio and $\angle A$.

b. What is the relationship between the measures of the third ratio and $\angle A$?

c. How did each ratio compare to each calculator display that followed?

Learning the Concept

The word *trigonometry* means triangle measurement. The ratios of the measures of the sides of a right triangle are called **trigonometric ratios**.

Three common trigonometric ratios are defined below.

Definition of Trigonometric Ratios	**In words:** If $\triangle ABC$ is a right triangle and A is an acute angle, $$\text{sine of } \angle A = \frac{\text{measure of the leg opposite } \angle A}{\text{measure of the hypotenuse}}$$ $$\text{cosine of } \angle A = \frac{\text{measure of the leg adjacent to } \angle A}{\text{measure of the hypotenuse}}$$ $$\text{tangent of } \angle A = \frac{\text{measure of the leg opposite } \angle A}{\text{measure of the leg adjacent to } \angle A}$$

In symbols: $\sin A = \frac{a}{c}$, $\cos A = \frac{b}{c}$, $\tan A = \frac{a}{b}$

Sine, cosine, and tangent are abbreviated as sin, cos, and tan, respectively.

The value of the trigonometric ratios depends only on the measure of the angle. The size of the triangle does not affect the value of the ratio.

Example **Find sin S, cos S, and tan S to the nearest thousandth.**

$$\sin S = \frac{\text{measure of leg opposite } \angle S}{\text{measure of hypotenuse}}$$

$$= \frac{12}{13} \text{ or about } 0.923$$

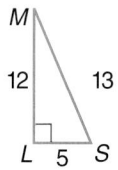

$$\cos S = \frac{\text{measure of leg adjacent to } \angle S}{\text{measure of hypotenuse}}$$

$$= \frac{5}{13} \text{ or about } 0.385$$

$$\tan S = \frac{\text{measure of leg opposite } \angle S}{\text{measure of leg adjacent to } \angle S}$$

$$= \frac{12}{5} \text{ or } 2.4$$

A calculator can be used to find the sine, cosine or tangent ratio for an angle with a given degree measure.

Example **Find each value to the nearest ten thousandth.**

a. sin 37°

37 SIN 0.601815023

Rounded to the nearest ten thousandth, sin 37° ≈ 0.6018.

b. cos 56°

56 COS 0.559192903

Rounded to the nearest ten thousandth, cos 56° ≈ 0.5592.

c. tan 72°

72 TAN 3.077683537

Rounded to the nearest ten thousandth, tan 72° ≈ 3.0777.

You can also use a calculator to find the degree measure of an angle if the sine, cosine, or tangent ratio is known.

Example **a. Find the measure of ∠S given that sin S = 0.3569.**

.3569 SIN⁻¹ 20.909936

The measure of ∠S is about 21°.

b. Find the measure of ∠C given that cos C = 0.2831.

.2831 COS⁻¹ 73.55469014

The measure of ∠C is about 74°.

c. Find the measure of ∠T given that tan T = 4.703.

4.703 TAN⁻¹ 77.99596131

The measure of ∠T is about 78°.

TECHNO
TIP

To use SIN⁻¹, COS⁻¹, and TAN⁻¹, you have to press the 2nd or INV key and then SIN, COS, or TAN.

You can use the trigonometric ratios to find missing measures.

Example **4**
Real World APPLICATION
Transportation

The end of an exit ramp from an interstate highway is 22 feet higher than the highway. If the ramp is 630 feet long, what angle does it make with the highway?

Let $m\angle A = x°$

 leg opposite $\angle A$ = 22 feet 22 ft

 hypotenuse = 630 feet

Now substitute these values into the definition of sine.

$$\sin x° = \frac{22}{630} \quad \begin{array}{l} \leftarrow side\ opposite \\ \leftarrow hypotenuse \end{array}$$

Use a calculator to find the value of x.

22 ÷ 630 = SIN⁻¹ 2.0012119

The ramp makes an angle of about 2° with the road.

Checking Your Understanding

Communicating Mathematics

Read and study the lesson to answer each question.

1. **Write** a definition of the sine ratio.

2. **Compare and contrast** the sine ratio with the cosine ratio.

3. **Describe** the procedure for using a calculator to find the degree measure of the angle that corresponds to a given tangent ratio.

4. **Draw** a right triangle such that $\sin D = \frac{4}{5}$, $\cos D = \frac{3}{5}$, and $\tan D = \frac{4}{3}$.

5. **You Decide** Kelli says that the value of $\sin A$ is greater than the value of $\sin D$. Derice says that $\sin A$ and $\sin D$ are equal. Who is correct and why?

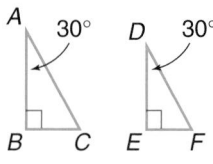

M_{ATH} J_{OURNAL}

6. *Some Old Horse - Caught A Horse - Taking Oats Away* is a helpful mnemonic device for remembering the trigonometric ratios. S, C, and T represent sin, cos, and tan respectively, while O, H, and A represent opposite, hypotenuse, and adjacent respectively. Develop your own mnemonic device for remembering the ratios.

Guided Practice

Refer to △XYZ. Express each ratio as a fraction in simplest form.

7. $\sin X$ 8. $\sin Y$

9. $\cos X$ 10. $\cos Y$

11. $\tan X$ 12. $\tan Y$

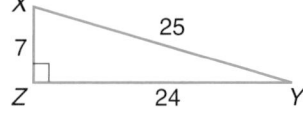

Use a calculator to find each ratio to the nearest ten thousandth.

13. $\cos 71°$ 14. $\tan 2°$ 15. $\sin 25°$

Use a calculator to find the angle that corresponds to each ratio. Round to the nearest degree.

16. $\sin M = 0.4$ 17. $\cos N = 0.18$ 18. $\tan F = 0.64$

19. Public Access The angle that a wheelchair ramp makes with the ground cannot exceed 6°. If a ramp needs to rise 2 feet, how long does the ramp need to be?

Exercises: Practicing and Applying the Concept

Independent Practice

For each triangle, find sin *B*, cos *B*, and tan *B* to the nearest thousandth.

20.

21.

22.

23.

24.

25.
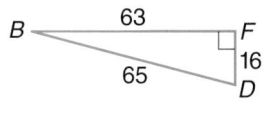

Use a calculator to find each ratio to the nearest ten thousandth.

26. tan 45°   **27.** sin 30°   **28.** cos 60°

29. cos 25°   **30.** tan 31°   **31.** sin 71°

Use a calculator to find the angle that corresponds to each ratio. Round answers to the nearest degree.

32. tan *J* = 0.6   **33.** sin *R* = 0.8   **34.** cos *F* = 0.866

35. sin *E* = 0.6897   **36.** cos *B* = 0.4706   **37.** tan *K* = 1.8

For each triangle, find the measure of the marked acute angle to the nearest degree.

38.

39.

40.

41.

42.

43.
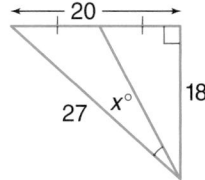

Critical Thinking

44. Write a conjecture about the relationship between the sine and cosine of complementary angles.

45. Scale Drawing   The Leaning Tower of Pisa is 55 meters tall and tilts 5 meters off the perpendicular. Maria wants to make a scale drawing for a project in history class. What should be the measure of the angle she draws to represent the tilt of the tower?

55 m

5 m

46. Transportation   The steepest streets in the United States are Filbert Street and 22nd Street in San Francisco. They rise 1 foot for every 3.17 feet of horizontal distance. Find the angle these streets form with the horizontal.

Mixed Review

47. The hypotenuse of a 30°–60° right triangle is 20 yards long. What is the length of the side opposite the 30° angle? (Lesson 13-5)

48. Construction   A rectangular patio is designed to have a concrete border around a tiled area. The tiled area is to be rectangular, and the border is to extend 2 feet around the tile on all sides. The entire patio will be 24 feet by 14 feet. What is the area to be covered by tile? (Lesson 12-4)

49. There are 3 routes between Jean's house and the school. How many ways can Jean go from her house to the school and back home again? (Lesson 10-5)

50. How many milligrams are in 2.5 grams? (Lesson 7-8)

51. Geometry   If two angles of a triangle have the same measure, the triangle is an isosceles triangle. The triangle shown at the right is an isosceles triangle. Is this an example of inductive or deductive reasoning? (Lesson 5-8)

52. Find the product of $(2b^3)$ and $(-6b^4)$. (Lesson 4-8)

53. Solve $22 = \dfrac{n}{-11}$. Then graph the solution on a number line. (Lesson 3-3)

54. Standardized Test Practice   How many dots are in the next figure in this pattern? (Lesson 2-6)

For **Extra Practice**, see page 772.

A. 15 **B.** 18 **C.** 20 **D.** 25

13-6B Slope and Tangent

An Extension of Lesson **13-6**

Be sure to turn off any functions in the Y= list and any statistical plots before starting to draw.

You can use a graphing calculator to draw line segments. The right triangle shown was drawn in the viewing window [0, 14] by [0, 10] with a scale factor of 1 on both axes. The vertices are (0, 3), (6, 3), and (6, 6).

Enter: [2nd] [DRAW] 2 0 [,] 3 [,] 6 [,] 3 [)] [ENTER] [CLEAR]

[2nd] [DRAW] 2 6 [,] 3 [,] 6 [,] 6 [)] [ENTER] [CLEAR]

[2nd] [DRAW] 2 6 [,] 6 [,] 0 [,] 3 [)] [ENTER]

To find the measure of the angle formed by the horizontal line and the hypotenuse, you can use the tangent ratio. The leg opposite the angle is 3 units long, and the leg adjacent to the angle is 6 units long.

$$\tan a = \frac{\text{leg opposite}}{\text{leg adjacent}}$$

$$= \frac{3}{6} \text{ or } 0.5$$

Use a calculator to find a.

Enter: [2nd] [TAN⁻¹] .5 [ENTER] *26.56505118*

$a \approx 26.6°$

Your Turn ▶ Find the slope of the hypotenuse of the triangle shown at the beginning of the lab.

▶ Draw the triangle shown above on a graphing calculator. Then graph the line that contains the hypotenuse, $y = 0.5x + 3$.

Enter: [Y=] .5 [X,T,θ,n] [+] 3 [GRAPH]

 TALK ABOUT IT

1. What is the slope of the line?

2. What is the slope of the hypotenuse?

3. How does the slope of the hypotenuse compare to the tangent of the angle formed by the hypotenuse and the horizontal leg?

4. Graph $y = 3x + 4$ on your graphing calculator. What is the slope of the line?

5. Sketch the right triangle formed by the graph of $y = 3x + 4$, the *x*-axis, and the *y*-axis. Label it $\triangle ABC$, so that $\angle A$ is between the *x*-axis and the graph of $y = 3x + 4$. What is tan A?

6. What is the measure of the angle that the graph of $y = 3x + 4$ forms with the *x*-axis?

13-7

Using Trigonometric Ratios

What You'll Learn

■ To solve problems by using the trigonometric ratios.

Why It's Important

Using trigonometric ratios can help a pilot determine how far his or her plane is from the airport.

MATERIALS: ⬭ protractor 🗂 index card ⫽ straw 📎 paper clip (large) 〰 string 🎯 tape

Hands-On Activity

Many problems that can be solved using trigonometric ratios deal with angles of elevation. An **angle of elevation** is formed by a horizontal line and a line of sight above it.

A hypsometer is an instrument that measures angles of elevation. You can make a model of a hypsometer.

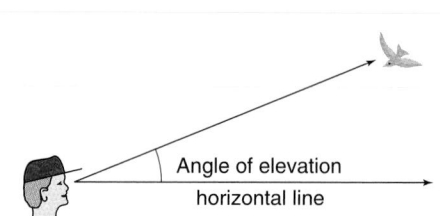

Angle of elevation
horizontal line

Your Turn

Work with a partner.

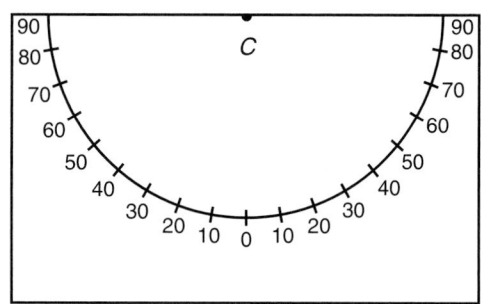

► Tape a protractor on an index card so that both zero points align with the edge of the card. Mark the center of the protractor on the card and label it *C*. Then mark and label every 10° on the card like the one shown at the right.

► Tie a piece of string to a large paper clip. Attach the other end of the string to the index card at *C*.

► Tape a straw to the edge of the index card that contains *C*.

► To use the hypsometer, look through the straw at the object. Have your partner read the angle from the scale.

► If possible, use your hypsometer to find the angle of elevation to a tree on your school property. Also measure the distance from the hypsometer to the ground and from the hypsometer to the base of the object.

Straw

String

Paper Clip

a. Draw a diagram of the situation with the hypsometer and the tree. If you do not have your own measurements, use 35° as the angle of elevation, 4 feet as the distance from the hypsometer to the ground, and 46 feet as the distance from the hypsometer to the tree base.

b. How could you find the height of a tree? Solve.

694 *Chapter 13* *Applying Algebra to Right Triangles*

The trigonometric ratios and the Pythagorean theorem can be used to find the measure of any side or angle of a right triangle if the measure of one side and any other side or acute angle are known.

Example **Find the measure of $\angle J$ in $\triangle JKL$.**

The measure of the leg opposite $\angle J$ and the measure of the hypotenuse are known. Use the sine ratio.

$$\sin J = \frac{\text{measure of leg opposite } J}{\text{measure of hypotenuse}}$$

$$\sin J = \frac{5}{8} \qquad \textit{Replace the measure of the leg opposite } J \textit{ with 5 and the measure of the hypotenuse with 8.}$$

$$\sin J = 0.625$$

$$J \approx 38.7 \qquad \textit{Use a calculator.}$$

The measure of $\angle J$ is about $38.7°$.

A person on a tower or a cliff must look down to see an object below. An **angle of depression** is formed by a horizontal line and another line of sight below it.

Example ② **A fire is sighted from a fire tower in the Black Hills National Forest. Using a hypsometer, the ranger found the angle of depression to be $2°$. If the ranger's eyes are 125 feet above the ground, how far is the fire from the base of the tower?**

Fire Fighting

Let d represent the distance from the fire to the base of the fire tower.

In the diagram, a line from the fire perpendicular to the horizontal line of sight forms a rectangle. Therefore, both lengths can be labeled d. Notice that 125 feet

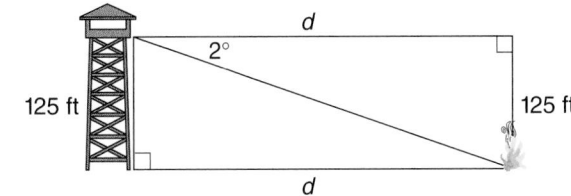

is the length of the leg opposite the $2°$ angle and d is the length of the leg adjacent to the $2°$ angle. The tangent ratio should be used to solve this problem.

$$\tan 2° = \frac{\text{measure of leg opposite}}{\text{measure of leg adjacent}}$$

$$\tan 2° = \frac{125}{d} \qquad \textit{Replace the measure of the leg opposite with 125 and the measure of the leg adjacent with } d.$$

$$d \tan 2° = 125 \qquad \textit{Multiply each side by } d.$$

$$d = \frac{125}{\tan 2°} \qquad \textit{Divide each side by } \tan 2°.$$

$$d \approx 3580 \qquad \textit{Use a calculator.}$$

The fire is about 3580 feet from the base of the tower.

Communicating Mathematics

Read and study the lesson to answer each question.

1. **Draw** a diagram and write a few sentences to explain what is meant by an angle of elevation.

2. **Explain** how to decide whether to use sine, cosine, or tangent when you are finding the measure of an acute angle in a right triangle.

3. **Assess Yourself** Describe a measure in your home or school that you could find using an angle of elevation. If possible, use the hypsometer you made at the beginning of the lesson to find the measure. Then write a few sentences explaining how you found the height.

Guided Practice

Write an equation that you could use to solve for *x*. Then solve. Round decimal answers to the nearest tenth.

4.

5.

6.

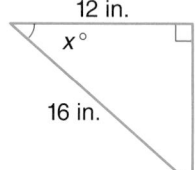

7. **Home Maintenance** A painter props a 20-foot ladder against a house. The angle it forms with the ground is 65°. How far up the side of the house does the ladder reach?

Exercises: Practicing and Applying the Concept

Independent Practice

Write an equation that you could use to solve for *x*. Then solve. Round decimal answers to the nearest tenth.

8.

9.

10.

11.

12.

13.

14.

15.

16.

Use trigonometric ratios to solve each problem.

17. Martina is flying a kite on a 50-yard string. The string is making a 50° angle with the ground. How high above the ground is the kite?

18. A flagpole casts a shadow 25 meters long when the angle of elevation of the sun is 40°. How tall is the flagpole?

19. A guy wire is fastened to a TV tower 50 feet above the ground and forms an angle of 65° with the tower. How long is the wire?

20. Find the area of a right triangle in which one acute angle measures 35° and the leg opposite that angle is 80 cm long.

21. A surveyor is 85 meters from the base of a building. The angle of elevation to the top of the building is 26.5°. If her eye level is 1.6 meters above the ground, find the height of the building.

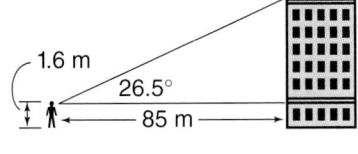

22. Danica is in the observation area of the Sears Tower in Chicago overlooking Lake Michigan. She sights two sailboats going due east from the tower. The angles of depression to the boats are 25° and 18°. If the observation deck is 1353 feet high, how far apart are the boats?

Critical Thinking

23. Find the missing measures in the triangle. (*Hint:* Draw an altitude from *B*.)

Applications and Problem Solving

24. **Aviation**   A plane is 3 miles above the ground. The pilot sights the airport at an angle of depression of 15°. What is the distance between the plane and the airport?

25. **Mountain Climbing**   A mountain climber is planning to scale a vertical rock wall. From a point 10 feet from the base, she estimates the angle of elevation to the top of the wall to be 75°. How high is the vertical wall to the nearest foot?

26. **Meteorology**   The cloud ceiling is the lowest altitude at which solid cloud is present. To find the cloud ceiling at night, a searchlight is shone vertically onto the clouds. If a searchlight is 100 meters from a weather office and the angle of elevation to the spot on the clouds is 70°, how high is the cloud ceiling?

Mixed Review

27. Use a calculator to find the measure of ∠*B* given that sin *B* = 0.8829. (Lesson 13-6)

28. **Collect Data**   Find out how many students are in each grade in your school. Make a circle graph to display your findings. (Lesson 11-2)

29. Find the range, median, upper and lower quartiles, and the interquartile range for the set of data {45, 62, 72, 51, 47, 68, 69, 50, 75}. (Lesson 10-2)

30. **Standardized Test Practice**   Thirty is 15% of what number? (Lesson 9-5)

   A. $\frac{1}{2}$ **B.** $4\frac{1}{2}$ **C.** 50   **D.** 200

For **Extra Practice**, see page 772.

31. Determine whether {(2, −1), (3, −3), (4, −5)} is a function. (Lesson 8-1)

OH GIVE ME A HOME

Architects often give presentations to potential buyers to show and explain their designs. Plan and give an oral presentation to describe what you know about architectural styles and designing homes. Your presentation should include the following.

- A brief architectural history of your community. You might include old and new photographs, drawings, or maps.

- Examples from your Investigation Folder to help you describe and define building terms you used most often in this Investigation.

- Floor plan(s) and elevation drawings of the house your group designed.

- The model of the house your group designed.

- The reasoning behind your design; that is, an explanation that tells why you designed the house the way you did.

- An explanation of what you would do differently if you had to do this Investigation over again.

Extension

Schedule an interview with an architect. Make a list of questions to ask him or her.

- Be sure to ask about the importance of mathematics to architecture.

- Include a question about what courses high school students interested in becoming architects should take.

PORTFOLIO ASSESSMENT

You may want to keep your work on this Investigation in your portfolio.

CHAPTER 13 Highlights

Vocabulary

After completing this chapter, you should be able to define each term, property, or phrase and give an example or two of each.

Algebra

irrational number (p. 672)
perfect squares (p. 665)
Pythagorean theorem (p. 676)
radical sign (p. 664)
real numbers (p. 673)
square root (p. 664)

Geometry

adjacent leg (p. 687)
angle of depression (p. 695)
angle of elevation (p. 694)
cosine (p. 688)
45°–45° right triangle (p. 683)
hypotenuse (p. 676)
leg (p. 676)
opposite side (p. 687)
sine (p. 688)
tangent (p. 688)
30°–60° right triangle (p. 683)
trigonometric ratios (p. 688)
trigonometry (p. 688)

Problem Solving

Venn diagram (p. 669)

inter NET CONNECTION

Chapter Review For additional lesson-by-lesson review, visit:
www.glencoe.com/ sec/math/prealg/ mathnet

Understanding and Using Vocabulary

Choose the correct term to complete each sentence.

1. The (square, square root) of 6 equals 36.
2. In the expression $\sqrt{25}$, the symbol $\sqrt{\ }$ is called a (radical sign, square).
3. The number $\sqrt{121}$ belongs to the set of (irrational, rational) numbers.
4. The longest side of a right triangle is the (hypotenuse, leg).
5. A right triangle with a leg that is half the length of the hypotenuse is a (30°–60°, 45°–45°) right triangle.
6. The angle formed by a horizontal line and another line of sight above it is an angle of (depression, elevation).

Skills and Concepts

Objectives and Examples	**Review Exercises**

Upon completing this chapter, you should be able to:

Use these exercises to review and prepare for the chapter test.

▶ **find squares and square roots** (Lesson 13-1)

Find $-\sqrt{196}$.

The symbol $-\sqrt{196}$ represents the negative square root of 196. Since $14 \cdot 14 = 196$, $-\sqrt{196} = -14$.

Find each square root.

7. $\sqrt{9}$
8. $\sqrt{121}$
9. $-\sqrt{25}$
10. $-\sqrt{81}$
11. $\sqrt{225}$
12. $-\sqrt{900}$

▶ **find and use squares and square roots** (Lesson 13-1)

Find the best integer estimate for $-\sqrt{29}$. 25 and 36 are the closest perfect squares.

$$-\sqrt{36} < -\sqrt{29} < -\sqrt{25}$$
$$-\sqrt{6^2} < -\sqrt{29} < -\sqrt{5^2}$$
$$-6 < -\sqrt{29} < -5$$

Since -29 is closer to -25 than -36, the best integer estimate is -5.

Find the best integer estimate for each square root. Then check your estimate using a calculator.

13. $\sqrt{83}$
14. $\sqrt{54}$
15. $\sqrt{220}$
16. $-\sqrt{900}$
17. $-\sqrt{39}$
18. $\sqrt{9.61}$

▶ **identify numbers in the real number system** (Lesson 13-3)

0.25 can be expressed as $\frac{1}{4}$ so it is a rational number.

The number $\sqrt{6} = 2.449489743\ldots$, which is not a repeating or terminating decimal, is an irrational number.

Name the sets of numbers to which each number belongs: the whole numbers, the integers, the rational numbers, the irrational numbers, and/or the real numbers.

19. 7
20. $\frac{3}{4}$
21. -2.6
22. $\sqrt{11}$

▶ **solve equations by finding square roots** (Lesson 13-3)

Solve $b^2 = 49$.

$$b^2 = 49$$
$$b = \sqrt{49} \quad \text{or} \quad b = -\sqrt{49}$$
$$b = 7 \quad \text{or} \quad b = -7$$

Solve each equation. Round decimal answers to the nearest tenth.

23. $x^2 = 196$
24. $n^2 = 160$
25. $t^2 = 15$
26. $a^2 = 0.04$

▶ **use the Pythagorean theorem to find the length of the side of a right triangle and to solve problems** (Lesson 13-4)

Find the length of the third side of the right triangle.

$$c^2 = a^2 + b^2$$
$$8^2 = 5^2 + b^2$$
$$64 = 25 + b^2$$
$$39 = b^2$$
$$\sqrt{39} = b$$
$$6.2 \approx b$$

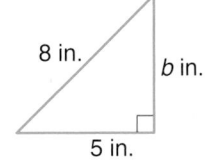

8 in.   b in.   5 in.

The length of the leg is about 6.2 inches.

In a right triangle, if a and b are the measures of the legs and c is the measure of the hypotenuse, find each missing measure. Round answers to the nearest tenth.

27. $a = 12$, $b = 16$

28. $a = 14$, $b = 40$

29. $a = 30$, $b = 16$

30. $a = 8$, $c = 15$

31. $b = 63$, $c = 65$

32. $a = 15$, $c = 39$

▶ **find missing measures in 30°–60° and 45°–45° right triangles** (Lesson 13-5)

In a 30°–60° right triangle, the hypotenuse is twice as long as the length of the side opposite the 30° angle.

In a 45°–45° right triangle, the hypotenuse is $\sqrt{2}$ times the length of a leg.

Find the lengths of the missing sides in each triangle. Round decimal answers to the nearest tenth.

33.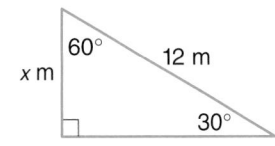
60°   x m   12 m   30°

34.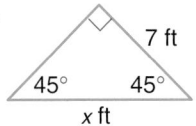
7 ft   45°   45°   x ft

35.
30°   x cm   60°   1.5 cm

▶ **find missing sides and angles of triangles using the sine, cosine, and tangent ratios** (Lesson 13-6)

$$\sin \angle A = \frac{\text{measure of the leg opposite } \angle A}{\text{measure of the hypotenuse}}$$

$$\cos \angle A = \frac{\text{measure of the leg adjacent to } \angle A}{\text{measure of the hypotenuse}}$$

$$\tan \angle A = \frac{\text{measure of the leg opposite } \angle A}{\text{measure of the leg adjacent to } \angle A}$$

For each triangle, find sin X, cos X, and tan X to the nearest thousandth.

36.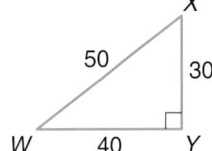
X   50   30   W   40   Y

37.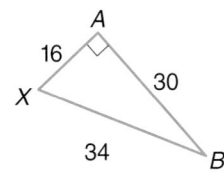
A   16   X   30   34   B

Use a calculator to find the angle that corresponds to each ratio. Round to the nearest degree.

38. $\tan C = 2.145$

39. $\sin H = 0.9945$

40. $\cos M = 0.2588$

Objectives and Examples	Review Exercises

▶ **solve problems using the trigonometric ratios** (Lesson 13-7)

In △*ABC*, find the length of $\overline{BC}$.

$$\sin A = \frac{\text{measure of leg opposite } \angle A}{\text{measure of the hypotenuse}}$$

$$\sin 28° = \frac{x}{12}$$

$$(12)(\sin 28°) = x$$

$$5.6 \approx x$$

The measure of $\overline{BC}$ is about 5.6 feet.

Write an equation that you could solve for *x*. Round decimal answers to the nearest tenth.

41.
15 ft
35°
x ft

42.
6 m
x°
11 m

43.
x in.
15 in.
25°

Applications and Problem Solving

44. Recreation A class survey of 32 students about favorite weekend activities showed that 20 students chose a movie and 15 students chose going to a shopping mall. There were 8 students who chose neither activity. How many students chose both activities? (Lesson 13-2)

45. Communication A telephone pole is 28 feet tall. A wire is stretched from the top of the pole to a point on the ground that is 5 feet from the bottom of the pole. How long is the wire? Round to the nearest tenth. (Lesson 13-3)

46. Geometry Write a formula for the length of the diagonal of a square if the sides are *s* units long. (Lesson 13-5)

47. Navigation From the top of a lighthouse, the angle of depression to a buoy is 25°. If the top of the lighthouse is 150 feet above sea level, find the distance from the buoy to the foot of the lighthouse. (Lesson 13-7)

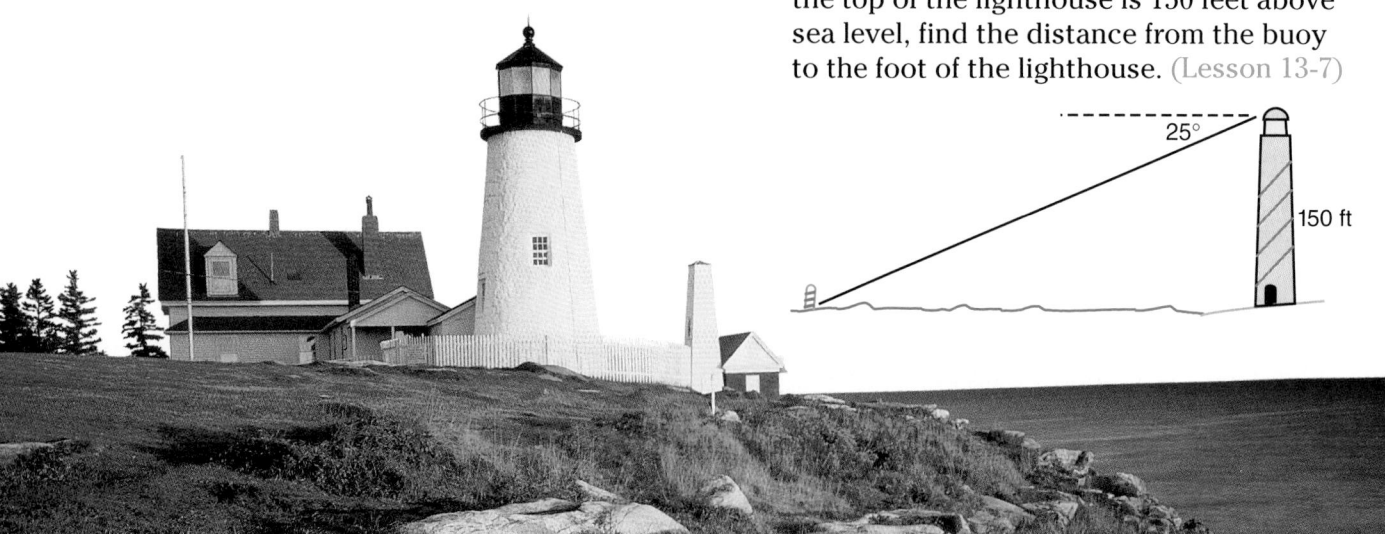

A practice test for Chapter 13 is available on page 786.

Performance Task

Demonstrate your knowledge by giving a clear, concise solution to each problem. Be sure to include all relevant drawings and justify your answers. You may show your solutions in more than one way or investigate beyond the requirements of the problem.

1. You and your group are hiking in the woods. You leave camp and walk 15 kilometers due north. Then you walk 6 kilometers due east. Your job in the group is to chart the group's movement and to find how far you are from the starting point. Vectors are line segments with both magnitude (length) and direction. Draw vectors to represent the group's movement. Then use what you have learned in this chapter to find the direction and distance the hikers are from their starting point.

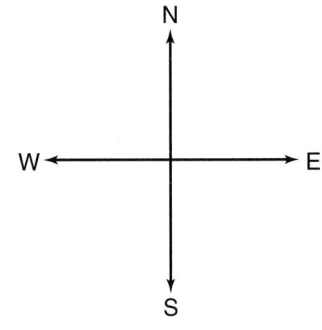

2. A soccer player kicks the ball 12 feet to the west. Another player kicks the ball 18 feet to the north. What is the direction and distance from the starting point of the ball? Round your answer to the nearest whole number.

Thinking Critically

Use a calculator to find the sine, cosine, and tangent of several acute angles.

▶ When is $\cos A > \sin A$?

▶ When is $\cos A = \sin A$?

▶ Describe the value of $\tan A$ when $m\angle A > 45°$.

▶ Describe the range of values for $\cos A$ and $\sin A$.

 Portfolio

Place your favorite word problem from this chapter in your portfolio along with your solution and attach a note explaining why it is your favorite.

What You'll Learn

In this chapter, you will:

- identify and classify polynomials, and

- add, subtract, and multiply polynomials.

Why It's Important

Real World Sometimes you can use a model to help you understand a mathematical idea. The diagram shows a rectangle made of algebra tiles. The dimensions of this rectangle are $2x$ units and $x + 3$ units. You can find the area of the rectangle by counting the number of x^2 tiles and the number of x tiles in the model. You can also find the area of this rectangle by multiplying the length and width. *In Lesson 14-5, you will learn how this algebra tile model relates to multiplying a polynomial by a monomial.*

Focus on...

PREREQUISITE SKILLS

To be successful in this chapter, you'll need to understand these concepts and be able to apply them. Refer to the lesson in parentheses if you need more review before beginning the chapter.

▶ *Determine whether an expression is a monomial.* *(Lesson 4-1)*

Determine whether each expression is a monomial. Explain why or why not.

1. $45abc$　　　　**2.** $x + 3y$　　　　**3.** $-2z^2$　　　　**4.** $7 - abc$

▶ *Simplify algebraic expressions using the distributive property.* *(Lesson 1-5)*

Restate each expression using the distributive property. Do not simplify.

5. $12(x + 3)$　　**6.** $5a + 5b$　　**7.** $(2 + c)7$　　**8.** $6d + 7d$

▶ *Multiply powers that have the same base.* *(Lesson 4-8)*

Find each product. Express your answer in exponential form.

9. $2^2 \cdot 2^5$　　　　**10.** $z^3 \cdot z^4$　　　　**11.** $x^3(xy^5)$　　　　**12.** $9b^4(a^3b^2c)$

▶ *Use negative exponents in expressions.* *(Lesson 4-9)*

Write each expression using positive exponents.

13. 2^{-3}　　　　**14.** 10^{-5}　　　　**15.** x^{-2}　　　　**16.** $a^{-3}b^{-4}$

Focus on...

READING SKILLS

In this chapter, you will learn about polynomials. You have already learned about monomials in Chapter 4. The prefix *mono* means one, so a monomial has one term. A **polynomial** is an algebraic expression that is the sum of one, two, three, or many monomials. The prefix *poly* means many. The prefix *bi* means two, so a **binomial** has two terms and *tri* means three, so a **trinomial** has three terms.

14-1 Polynomials

You'll Learn

- To identify and classify polynomials and find their degrees.

It's Important

Understanding polynomials can help you determine how much money you have in your savings account.

Application: Construction

On construction sites, workers and visitors are required to wear hard hats for protection. If an object is dropped from above, it may injure an unprotected person on the ground. The velocity v of a dropped object can be determined by using the formula $v^2 = 2gs$, where g represents the acceleration due to gravity and s represents the distance the object falls.

Learning the Concept

LOOK BACK

You can review monomials in Lesson 4-1.

As you know, expressions like v^2 and $2gs$ are monomials. Monomials can be numbers, variables, or products of numbers and variables. Expressions that involve variables in the denominator of a fraction or variables with radical signs are not monomials. For example, $3x + 2$, $\frac{5}{m}$, and $\sqrt{ab}$ are not monomials.

An algebraic expression that contains one or more monomials is called a **polynomial**. A polynomial is a sum or difference of monomials. A polynomial with two terms is called a **binomial**, and a polynomial with three terms is called a **trinomial**.

Example **Determine whether the given expression is a polynomial. If it is, classify it as a *monomial*, a *binomial*, or a *trinomial*.**

a. $3x^2 + 7x - 5$

The expression is a polynomial because it is the sum of three monomials. Since there are three terms, it is a trinomial.

b. $3y - \frac{1}{y^2}$

The expression is *not* a polynomial because $\frac{1}{y^2}$ is not a monomial.

The degree of a monomial is the sum of the exponents of its variables. The following chart shows how to find the degree of a monomial.

Monomial	Variables	Exponents	Degree
z	z	1	1
$-3c^2$	c	2	2
$\frac{2}{5}m^3n^4$	m, n	3, 4	3 + 4 or 7
$0.62p^3q^3$	p, q	3, 3	3 + 3 or 6

Remember that $z = z^1$.

A monomial like 7 that does not have a variable associated with it is called a **constant**. The degree of a nonzero constant is 0. The constant 0 has *no* degree.

A polynomial also has a degree. The degree of a polynomial is the same as that of the term with the greatest degree.

Example **Find the degree of $x^2 + xy^2 + y^4$.**

First, find the degree of each term.

x^2 has degree 2.
xy^2 has degree 1 + 2 or 3.
y^4 has degree 4.   *Greatest degree*

So, the degree of $x^2 + xy^2 + y^4$ is 4.

Example 3

APPLICATION

Medicine

Doctors can study the heart of a potential heart attack patient by injecting dye in a vein near the heart. In a normal heart, the amount of dye in the bloodstream after t seconds is given by the polynomial $-0.006t^4 + 0.140t^3 - 0.53t^2 + 1.79t$. Find the degree of the polynomial.

$-0.006t^4$ has degree 4.
$0.140t^3$ has degree 3.
$0.53t^2$ has degree 2
$1.79t$ has degree 1.

$-0.006t^4 + 0.140t^3 - 0.53t^2 + 1.79t$ has degree 4.

You can evaluate polynomials when you know the values of the variables involved.

Example **Evaluate $6x^2 - 5xy$ if $x = -2$ and $y = -3$.**

$$6x^2 - 5xy = 6 \cdot (-2)^2 - 5(-2)(-3) \quad \textit{Replace x with } -2 \textit{ and y with } -3.$$
$$= 6 \cdot 4 - 5 \cdot 6 \qquad\qquad\quad \textit{Use the order of operations.}$$
$$= 24 - 30$$
$$= -6$$

Checking Your Understanding

Communicating Mathematics

Read and study the lesson to answer these questions.

1. **Determine** if $\dfrac{3w^2}{5y}$ is a monomial. Explain your answer.

2. **Write** three binomial expressions. Explain why they are binomials.

3. **Explain** how to find the degree of a monomial and a polynomial.

M_{ATH} J_{OURNAL}

4. Write about a way that you can remember the definitions of monomial, binomial, trinomial, and polynomial.

State whether each expression is a polynomial. If it is, classify it as a *monomial*, *binomial*, or *trinomial*.

5. $\frac{x}{2}$

6. -5

7. $\frac{ab}{c} - c$

8. $3x^2 + 4x - 2$

Find the degree of each polynomial.

9. $-2xy^2$

10. $12x^3y^2$

11. $x^3 + 7x$

12. $x^2 + xy^2 - y^4$

Evaluate each polynomial if $x = 3$, $y = -5$, and $z = -1$.

13. $3x - 5y + z^2$

14. $x^3 + 2y^2 - z$

15. $x^3 - 2xy$

16. $2xy + 6yz^2$

17. Business The cost of an order of 12 burgers, 11 fries, and 5 shakes can be represented by the polynomial $12b + 11f + 5s$. Determine the cost of the order if burgers are \$1.79 each, fries are \$0.99 each, and shakes are \$1.29 each.

Exercises: Practicing and Applying the Concept

State whether each expression is a polynomial. If it is, classify it as a *monomial*, *binomial*, or *trinomial*.

18. $-\frac{1}{9}r^2$

19. $c^2 + 3$

20. $x^2 - 4$

21. $1 + 3x + 5x^2$

22. $\frac{6}{a} + b$

23. $5xy$

24. $3x^2y$

25. $ab^2 + 3a - b^2$

26. $x + y$

27. $-\sqrt{49}$

28. $\frac{e}{6}$

29. $x^2 - \frac{1}{2}x + \frac{1}{3}$

Find the degree of each polynomial.

30. $11c^2 + 4$

31. $3x + 5$

32. 121

33. $4x^3 + xy - y^2$

34. $x^6 + y^6$

35. $d^4 + c^4d^2$

36. $16y^2 + mnp$

37. $x^3 - x^2y^3 + 8$

38. $-7x^3y^4$

39. $-x^2yz^4$

40. $14c^5 - 16c^6d$

41. $2x^5 + 9x + 1$

Evaluate each polynomial if $a = 2$, $b = -3$, $c = 4$, and $d = -5$.

42. $a^3 - 2bc$

43. $ab + cd$

44. $2abc + 3a^2b$

45. $b^3 - 2ac + d^2$

46. $b^2 + a^2b$

47. $a^5 + bd^2$

48. $a^2 + b^2 - c^2 + d^2$

49. $abcd - 25 + a$

50. $5ad^2 - 2a + (bc)^2$

51. $-(-ac)^2 - cd$

52. $\sqrt{c} - b^2d$

53. $d - abc^2 - \sqrt{-27b}$

54. Find the degree of the polynomial $a^{x+3} + x^{x-2}b^3 + b^{x+2}$.

55. Landscaping Lee's Garden Shop wants to place hosta plants along the perimeter of a garden.

    a. Write a polynomial that represents the perimeter of the garden in feet.

    b. What is the degree of the polynomial?

56. Ecology In the early 1900s, the deer population of the Kaibab Plateau in Arizona increased rapidly because hunters had reduced the number of predators. However, the food supply was not great enough to support the increased population. So, the population eventually began to decline. The deer population for the years 1905 to 1930 can be approximated by the polynomial $-0.125x^5 + 3.125x^4 + 4000$, where x is the number of years from 1900.

   a. Find the degree of the polynomial.

   b. Evaluate the polynomial to find the deer population in 1920.

57. Family Activity The amount of money in a savings account can be written as a polynomial. Suppose that your grandparents placed $100 in your savings account each year on your birthday. On your fifth birthday, there would have been $100x^4 + 100x^3 + 100x^2 + 100x + 100$ dollars, where x is the annual interest rate plus 1. Research the interest rate at your family's bank. How much money would you have on your next birthday at their passbook savings rate?

Mixed Review

58. Geometry Find the length of $\overline{XY}$ in $\triangle WXY$. Round to the nearest tenth. (Lesson 13-7)

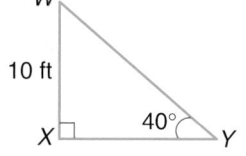

59. Catering A caterer is ordering a tent for outdoor parties. The area enclosed by the tent is 8 meters by 8 meters. The top is a square pyramid with a slant height of 4.2 meters. The hanging sides are 2.8 meters tall on each side. If there is no floor, what is the area of the canvas required to make the tent? (Lesson 12-6)

60. The stem-and-leaf plot shows the number of canned goods collected by each class for the food drive. (Lesson 10-2)

1	98
2	10 56 78 80 85
3	08 12 60 62 75 80
4	15
5	25

             1 | 98 = 198 cans

   a. What is the range?

   b. What is the median?

   c. What is the interquartile range?

   d. Are there any outliers?

61. Retail Sales The retail price of a camera is $98. The sales tax rate is 6.5% of the price. What is the total cost of the camera? (Lesson 9-5)

62. Solve $\frac{2x}{5} - 3 > -5$. (Lesson 7-6)

63. State whether 1, 3, 9, 27, . . . is a geometric sequence. If so, state the common ratio and list the next three terms. (Lesson 6-8)

64. Geometry A rectangle has a length of 3.5 meters and a width of 1.75 meters. What is the perimeter of the rectangle? (Lesson 5-3)

65. Standardized Test Practice Natalie is going to purchase carpet for a square room 16 feet wide. How many square feet is this? (Lesson 3-5)

   A. 8   **B.** 32   **C.** 166   **D.** 256

For **Extra Practice**, see page 772.

66. Find the quotient of 48 and -3. (Lesson 2-8)

67. Simplify $|-8| + |5|$. (Lesson 2-1)

Representing Polynomials with Algebra Tiles

An Extension of Lesson **14-1**

algebra tiles

You can use the concept of area to model monomials. A model of a monomial can be made using tiles or drawings like the ones shown below.

Let ▢ ^{1} be a 1×1 square. The square has an area of 1.

Let ▭ ^{1} be a $1 \times x$ rectangle. The rectangle has an area of x.

Let ▢ be an $x \times x$ square. The square has an area of x^2.

Red tiles are used to represent -1, $-x$, and $-x^2$.

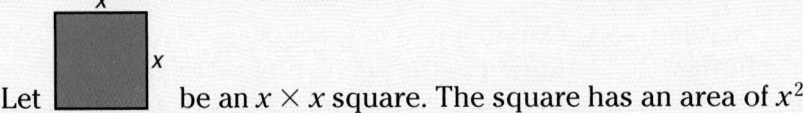

You can use these tiles to make a model of a polynomial.

$$2x^2 \qquad - \qquad 3x \qquad + \qquad 4$$

Your Turn ▶ Model $-3x^2$, $5x + 3$, $4x^2 - x$, and $5x^2 + 2x - 3$ using algebra tiles.

1. Explain how you can tell if an expression is a monomial, binomial, or trinomial by looking at the area tiles.

2. Name the polynomial represented by the model below.

3. Explain how you would find the degree of a polynomial using algebra tiles.

710 *Chapter 14 Polynomials*

14-2

You'll Learn

■ To add polynomials.

It's Important

Adding polynomials is useful in carpentry.

Adding Polynomials

MATERIALS: 📖 algebra tiles

Hands-On Activity

Algebra tiles can be used to model addition of polynomials.

▶ The numerical part of a monomial is called the **coefficient**. For example, the coefficient of $5x^2$ is 5.

▶ When monomials are the same or differ only by their coefficients they are called **like terms**. For example, $2x$ and $7x$, and $-5x^2y$ and $9x^2y$ are like terms; $6a$ and $-5b$, and $7a^3b^2$ and $7a^2b$ are unlike terms. Like terms are represented by tiles that are the same shape and size.

▶ A zero pair is formed by pairing one tile with its opposite.

▶ You can remove or add zero pairs without changing the value of the polynomial.

The model below represents $(2x^2 + 2x - 4) + (x^2 + 3x + 6)$.

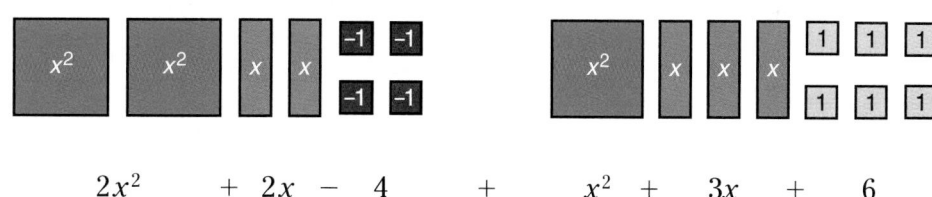

$$2x^2 \quad + \quad 2x \quad - \quad 4 \quad + \quad x^2 \quad + \quad 3x \quad + \quad 6$$

▶ To add the polynomials, combine like terms and remove all zero pairs.

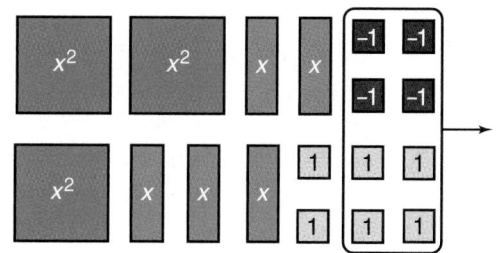

▶ Write the polynomial for the tiles that remain.
Therefore, $(2x^2 + 2x - 4) + (x^2 + 3x + 6) = 3x^2 + 5x + 2$.

Your Turn Model $(x^2 + 4x + 2) + (7x^2 - 2x + 3)$ using algebra tiles.

a. What is the sum of $(x^2 + 4x + 2)$ and $(7x^2 - 2x + 3)$?

b. Compare and contrast finding sums of polynomials using algebra tiles to finding the sum of integers using counters.

Learning the Concept

You can use the principles and properties you learned about integers and rational numbers to add polynomials.

Example 1

Find each sum.

a. $(5x + 7) + (3x + 3)$

Method 1: Add vertically

$$
\begin{array}{r}
5x +\ 7 \quad \text{\textit{Notice that the terms are aligned and added.}} \\
(+)\ 3x +\ 3 \\
\hline
8x + 10
\end{array}
$$

Method 2: Add horizontally

$$
\begin{aligned}
(5x + 7) + (3x + 3) & \\
= (5x + 3x) + (7 + 3) \quad & \text{\textit{Associative and commutative}} \\
& \text{\textit{properties of addition}} \\
= (5 + 3)x + (7 + 3) \quad & \text{\textit{Distributive property}} \\
= 8x + 10 &
\end{aligned}
$$

The sum is $8x + 10$.

b. $(3x^2 + x + 2) + (x^2 + 2x + 4)$

$$
\begin{aligned}
(3x^2 + x + 2) + (x^2 + 2x + 4) &= (3x^2 + x^2) + (x + 2x) + (2 + 4) \\
&= 4x^2 + 3x + 6
\end{aligned}
$$

The sum is $4x^2 + 3x + 6$.

c. $(2x^2 + 5xy + 7y^2) + (4x^2 - 3y^2)$

$$
\begin{array}{l}
2x^2 + 5xy + 7y^2 \\
(+)\ 4x^2 \qquad\ \ - 3y^2 \\
\hline
6x^2 + 5xy + 4y^2 \qquad \text{The sum is } 6x^2 + 5xy + 4y^2.
\end{array}
$$

Polynomials are often used to represent measures of geometric figures.

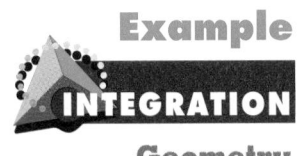

Example 2

The lengths of the sides of golden rectangles are in a special ratio. The rectangle at the right shows the lengths of the sides of a general golden rectangle.

x

1.618x

a. Find the perimeter of the rectangle.
b. Find the length, the width, and the perimeter of the rectangle if x is 5.2.

a. $x + 1.618x + x + 1.618x = (1 + 1.618 + 1 + 1.618)x$
$$= 5.236x$$

b. If x is 5.2, then the width is 5.2 units, the length is 1.618(5.2) units, and the perimeter is 5.236(5.2) units.

Estimate the length and perimeter first and then use a calculator.

Length: (1.618)(5.2) is about $2 \cdot 5$ or 10 units.

$$1.618 \times 5.2 = 8.4135$$

Perimeter: (5.236)(5.2) is about $5 \cdot 5$ or 25 units.

$$5.236 \times 5.2 = 27.2272$$

Communicating Mathematics

Read and study the lesson to answer these questions.

1. **Name** the like terms in $(x^2 + 7x + 1) + (2x^2 - 3x + 5)$.

2. **Demonstrate** how to find the sum of $5x^2 + 6x + 4$ and $2x^2 + 3x + 1$.

3. **You Decide** Caroline says that $4xyz$ and $3zyx$ are like terms. Nikita says that they are not. Who is correct? Explain.

 MATERIALS

algebra tiles

Use algebra tiles to find each sum.

4. $(3x^2 - 2x + 1) + (x^2 + 5x - 3)$ 5. $(-2x^2 + x - 5) + (x^2 - 3x + 2)$

6. $(x^2 - 3x + 6) + (2x^2 + 5x - 4) + (x^2 + x + 1)$

Guided Practice

Find each sum.

7. $(x + 3) + (2x + 5)$

8. $(4x + 3) + (x - 1)$

9. $(3x - 5) + (x + 9)$

10. $(2x - 3) + (x - 1)$

11. $\begin{array}{r} 2x^2 + 4x + 5 \\ (+)\ x^2 -\ x - 3 \\ \hline \end{array}$

12. $\begin{array}{r} 2x^2 - 5x + 4 \\ (+)\ 3x^2 + 8x - 1 \\ \hline \end{array}$

13. **Geometry** Find the perimeter of the rectangle shown at the right.

$x + 5$

$2x + 7$

Independent Practice

Find each sum.

14. $(6y - 5r) + (2y + 7r)$

15. $(3x + 9) + (x + 5)$

16. $(11x + 2y) + (x - 5y)$

17. $(8m - 2n) + (3m + n)$

18. $(13x - 7y) + 3y$

19. $(2x^2 + 5x) + (9 - 7x)$

20. $(3r + 6s) + (5r - 9s)$

21. $(5m + 3n) + 12m$

22. $(3x^2 - 9x + 5) + (5x^2 + 5x - 11)$ 23. $(5x^2 - 7x + 9) + (3x^2 + 4x - 6)$

24. $(6x^2 + 15x - 9) + (5 - 8x - 8x^2)$

25. $(a^3 - b^3) + (3a^3 + 2a^2b - b^2 + 2b^3)$

26. $\begin{array}{r} -6y^2 + 7b - 5 \\ (+)\ 2y^2 - 9b + 8 \\ \hline \end{array}$ 27. $\begin{array}{r} 3a + 5b - 4c \\ 2a - 3b + 7c \\ (+) -a + 4b - 2c \\ \hline \end{array}$ 28. $\begin{array}{r} 3x^2 - 7x + 9 \\ -2x^2 +\ x - 4 \\ (+)\ x^2 + 3x - 1 \\ \hline \end{array}$

29. $(3a + 5ab - 3b^2) + (7b^2 - 8ab) + (2 - 5a)$

30. $(x^2 + x + 5) + (3x^2 - 4x - 2) + (2x^2 + 2x - 1)$

31. $(3x^2 + 7) + (4x - 2) + (x^2 - 3x - 6)$

Find each sum. Then, evaluate if $a = -3$ and $b = 4$.

32. $(3a + 5b) + (2a - 9b)$

33. $(a^2 - 3ab + b^2) + (3a^2 - 2b - 5b^2)$

34. $(a^2 + 7b^2) + (5 - 3b^2) + (2a^2 - 7)$

Critical Thinking

35. If $(4r + 3s) + (6r - 5s) = (10r - 2s)$, find $(10r - 2s) - (6r - 5s)$.

Applications and Problem Solving

36. Geometry The measures of the angles of a triangle are shown in the figure at the right.

a. Find the sum of the measures of the angles.

b. The sum of the measures of the angles in any triangle is $180°$. Find the value of x.

c. Find the measure of each angle.

37. Construction A standard measurement for a window is the *united inch*. You can find the united inches of a window by adding the length of the window to the width. If the length of a window is $3x - 5$ inches and the width is $x + 7$, what is the size of the window in united inches?

Mixed Review

38. State whether $a^2 - \frac{1}{4}$ is a polynomial. If it is, classify it as a *monomial, binomial,* or *trinomial.* (Lesson 14-1)

39. Standardized Test Practice Find the area of the triangle at the right. (Lesson 12-1)

A. 3.75 cm^2 B. 7.5 cm^2

C. 8 cm^2 D. 15 cm^2

40. Consumer Awareness At a 20%-off sale, Antjuan bought a book that regularly sells for $15.98. How much is the discount? (Lesson 9-9)

41. Weather When the sum of the temperature (in °F) and humidity exceeds 130, people generally find the weather to be uncomfortable. (Lesson 8-9)

a. Write an inequality to represent this situation. Then graph the inequality.

b. Which area of the graph indicates uncomfortable weather?

c. Humidity cannot exceed 100%. How does this affect the graph?

42. Consider the sequence 2, 6, 11, 17, 24, . . . (Lesson 5-9)

a. Is the sequence arithmetic?

b. Write the next three terms of the sequence.

43. Geometry A rectangle has a perimeter of 50 feet and an area of 150 square feet. Its width is 10 feet. What is its length? (Lesson 3-5)

44. Statistics The table at the right shows the percent of American homes that have the most popular pets. Make a bar graph of the data. (Lesson 1-10)

45. Sports Two hundred thirty-six women were members of the United States Olympic team for 1992. That number grew to 298 in 1996. How many more women were on the 1996 team than were on the 1992 team? (Lesson 1-6)

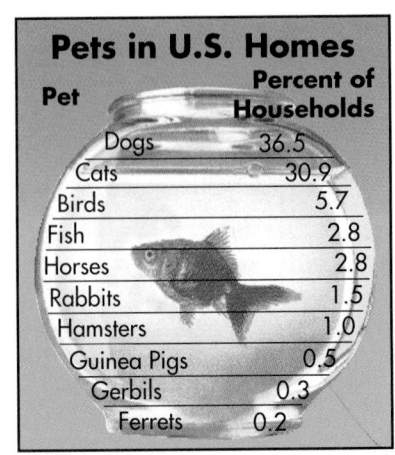

Pets in U.S. Homes

Pet	Percent of Households
Dogs	36.5
Cats	30.9
Birds	5.7
Fish	2.8
Horses	2.8
Rabbits	1.5
Hamsters	1.0
Guinea Pigs	0.5
Gerbils	0.3
Ferrets	0.2

Source: American Veterinary Medical Association

For **Extra Practice**, see page 773.

Subtracting Polynomials

Application: Running

Tegla Loroupe of Kenya was the first woman to cross the finish line of the 1994 New York Marathon. She completed the 26.2-mile course in 2 hours 27 minutes 37 seconds. In 1995, she won again with a time of 2 hours 28 minutes 6 seconds. How much faster was her 1994 time?

To find the difference of the times you must subtract.

$$
\begin{array}{c}
2 \text{ hours } 28 \text{ minutes } 6 \text{ seconds} \\
- \ 2 \text{ hours } 27 \text{ minutes } 37 \text{ seconds}
\end{array}
\quad \rightarrow \quad
\begin{array}{c}
2 \text{ hours } 27 \text{ minutes } 66 \text{ seconds} \\
- \ 2 \text{ hours } 27 \text{ minutes } 37 \text{ seconds} \\
\hline
\phantom{2 \text{ hours } 27 \text{ minutes }} 29 \text{ seconds}
\end{array}
$$

Miss Loroupe's 1994 time was 29 seconds faster.

Learning the Concept

In the application above, like units were subtracted from each other. Similarly, in algebra, when one polynomial is subtracted from another, only like terms can be subtracted.

Algebra tiles were used to help you understand how to add polynomials. They can also be used to help you understand how to subtract polynomials.

Example 1 Find $(4x^2 + 7x + 4) - (x^2 + 2x + 1)$ by using algebra tiles.

Make a model of $4x^2 + 7x + 4$ as shown in the diagram below. Then remove 1 x^2-tile, 2 x-tiles, and 1 1-tile.

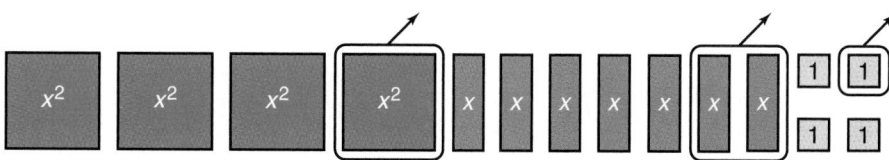

To find the difference, count the number of x^2-tiles, x-tiles, and 1-tiles that remain. There are 3 x^2-tiles, 5 x-tiles, and 3 1-tiles remaining.

$(4x^2 + 7x + 4) - (x^2 + 2x + 1) = 3x^2 + 5x + 3.$

To subtract polynomials, you can use the same methods you used to add them.

Example **2** Find $(5x + 9) - (3x + 6)$.

Method 1: Subtract Vertically	**Method 2: Subtract Horizontally**

Method 1: Subtract Vertically

$$\begin{array}{r} 5x + 9 \\ (-)\ 3x + 6 \\ \hline 2x + 3 \end{array}$$

The difference is $2x + 3$.

Method 2: Subtract Horizontally

$(5x + 9) - (3x + 6)$
$= 5x + 9 - 3x - 6$
$= (5x - 3x) + (9 - 6)$
$= (5 - 3)x + (9 - 6)$
$= 2x + 3$

Recall that you can subtract a rational number by adding its additive inverse. You can also subtract a polynomial by adding its additive inverse.

To find the additive inverse of a polynomial, it is necessary to multiply the entire polynomial by -1. Study the examples in the table below.

Polynomial	Multiply by -1	Additive Inverse
$-y$	$-1(-y)$	y
$4x - 3$	$-1(4x - 3)$	$-4x + 3$
$2xy + 5y$	$-1(2xy + 5y)$	$-2xy - 5y$
$2x^2 - 3x - 5$	$-1(2x^2 - 3x - 5)$	$-2x^2 + 3x + 5$

Example **3** Find each difference.

a. $(7x + 5) - (3x + 2)$
  $= (7x + 5) + (-1)(3x + 2)$ *Add the additive inverse of $3x + 2$.*
  $= 7x + 5 + (-3x - 2)$ *$(-1)(3x + 2) = -3x - 2$*
  $= 7x + 5 - 3x - 2$
  $= 7x - 3x + 5 - 2$ *Commutative property of addition*
  $= 4x + 3$

b. $(3x^2 - 5xy + 7y^2) - (x^2 - 3xy + 4y^2)$
  Align like terms and add the additive inverse of $x^2 - 3xy + 4y^2$.

$$\begin{array}{r} 3x^2 - 5xy + 7y^2 \\ (-)\ x^2 - 3xy + 4y^2 \\ \hline \end{array} \rightarrow \begin{array}{r} 3x^2 - 5xy + 7y^2 \\ (+)\ -x^2 + 3xy - 4y^2 \\ \hline 2x^2 - 2xy + 3y^2 \end{array}$$

Example **4**

Geometry

Lashonda plans to mat and frame a picture. The area inside the frame is $3x^2 - 5$ square inches. The area of the picture is $x^2 + 5$ square inches. How much matting will Lashonda need?

Explore You know the area inside the frame and the area of the picture.

Plan The area of the picture is smaller than the area inside the frame. Subtract the area of the picture from the area inside the frame to find the area needed for the matting.

Solve $(3x^2 - 5) - (x^2 + 5) = 3x^2 - 5 + (-1)(x^2 + 5)$
$= 3x^2 - 5 + (-x^2 - 5)$
$= 3x^2 - 5 - x^2 - 5$
$= 3x^2 - x^2 - 5 - 5$
$= 2x^2 - 10$

Lashonda will need $2x^2 - 10$ square inches of matting.

Examine Check by aligning like terms and subtracting.

Communicating Mathematics

Read and study the lesson to answer each question.

1. **Describe** how subtraction and addition of polynomials are related.

2. **Explain** how to find the additive inverse of a given polynomial. What is the additive inverse of $3x^2 - 5x + 7$?

3. **Write** the sum of a polynomial and its additive inverse.

4. **Write** a polynomial subtraction problem with a difference of $3m^2 - 5m + 4$. Check your work by adding.

 MATERIALS

algebra tiles

Use algebra tiles to find each difference.

5. $(5x^2 + 6x + 4) - (3x^2 + 2x + 1)$ 6. $(2x^2 + 5x + 3) - (x^2 - 2x + 3)$

7. $(3x^2 + 2x - 5) - (2x^2 + x - 4)$ 8. $(8x^2 + 2x - 1) - (2x^2 + x + 5)$

Guided Practice

State the additive inverse of each polynomial.

9. $2abc$ 10. $3x + 2y$ 11. $x^2 + 5x + 1$

12. $3x^2 - 2x + 5$ 13. $-8m + 7n$ 14. $-4h^2 - 5hk - k^2$

Find each difference.

15. $(3x + 4) - (x + 2)$ 16. $(4x + 5) - (2x + 3)$

17. $(5x - 7) - (3x - 4)$ 18. $(2x - 5) - (3x + 1)$

19. $5x^2 + 4x - 1$ 20. $3x^2 + 5x + 4$
     $(-)\, 4x^2 + \ x + 2$ $(-)\, x^2 \qquad - 1$

21. **Geometry** Find how much longer the hypotenuse is than the shorter leg of the triangle shown at the right.

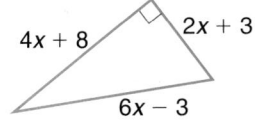

Independent Practice

Find each difference.

22. $(9x + 5) - (4x + 3)$ 23. $(2x + 5) - (x + 8)$

24. $(3x - 2) - (5x - 4)$ 25. $(6x - 5) - (4x + 3)$

26. $(2x + 3y) - (x - y)$ 27. $(9x - 4y) - (12x - 9y)$

28. $(5x^2 - 3) - (2x^2 - 7)$ 29. $(x^2 + 6x) - (3x^2 + 7)$

30. $5a^2 + 7a + 9$ 31. $6m^2 - 5m + 3$
     $(-)\, 3a^2 + 4a + 1$ $(-)\, 5m^2 + 2m - 7$

32. $5x^2 - 4xy$ 33. $9m^2 \qquad + 7$
     $(-) \qquad - 3xy + 2y^2$ $(-)\, -6m^2 + 2m - 3$

34. $15x^2y^2 + 11xy - 9$ 35. $14a + 10b - 18c$
     $(-)\ \ 9x^2y^2 - 13xy + 6$ $(-)\, 5a + \ 7b - 11c$

36. $(10x^2 + 8x - 6) - (3x^2 + 2x - 9)$

37. $(5y^2 + 9y - 12) - (-3y^2 + 5y - 7)$

38. $(6a^2 + 7ab - 3b^2) - (2a^2 + 3ab - b^2)$

39. $(x^3 - 3x^2y + 4xy^2 + y^3) - (7x^3 - 9xy^2 + x^2y + y^3)$

Critical Thinking

40. Suppose that A and B represent polynomials. If $A + B = 3x^2 + 2x - 2$ and $A - B = -x^2 + 4x - 8$, find A and B.

Applications and Problem Solving

Data Update For the latest basketball scores, visit:
www.glencoe.com/ sec/math/prealg/ mathnet

41. Basketball On December 13, 1983, the Denver Nuggets and the Detroit Pistons broke the record for the highest total score in a basketball game. The total points scored was 2 points more than twice what the Nuggets scored. The Pistons scored 186 points in the game.
  a. What was the Nuggets final score?
  b. How many points did the two teams score?

42. Geometry The perimeter of the isosceles trapezoid shown at the right is $16x + 1$ units. Find the length of the missing base of the trapezoid.

Mixed Review

43. Find the sum $(2a^2 + 3a - 4) + (6a^2 - a + 5)$. (Lesson 14-2)

44. Solve $n^2 = 18$ to the nearest tenth. (Lesson 13-3)

45. Geometry In $\triangle CDE$, $m\angle C = 35°$ and $m\angle D = 55°$. Find the measure of $\angle E$ and classify the triangle as *acute*, *right*, or *obtuse*. (Lesson 11-4)

46. Graph $y = -x + 5$ by using the x-intercept and the y-intercept. (Lesson 8-7)

For **Extra Practice**, see page 773.

47. Patterns Two groups share the same meeting room. If the debate team meets every 3rd school day and the dance committee meets every 4th school day, how often will both groups need the room on the same day? (Lesson 4-7)

48. Solve $x = 32 + 56 + (-18)$. (Lesson 2-4)

49. Standardized Test Practice In which quadrant does the point at $(-5, 3)$ lie? (Lesson 2-2)
  A. I   **B.** II   **C.** III   **D.** IV

50. Solve $8c = 72$ mentally. (Lesson 1-6)

Self Test

State whether each expression is a polynomial. If it is, classify it as a *monomial*, a *binomial*, or a *trinomial*. Then find the degree of each polynomial. (Lesson 14-1)

  1. $ax^2 + 6x$ **2.** $4b^2 + c^3d^4 + x$ **3.** $a^3b^4c^5$

Find each sum. (Lesson 14-2)

  4. $(4x + 5y) + (7x - 3y)$ **5.** $(2x^2 + 5) + (-3x^2 + 7)$ **6.** $(9m - 3n) + (10m + 4n)$

7. Geometry The perimeter of the triangle shown at the right is $7x + 2y$ units. Find the length of the third side of the triangle. (Lessons 14-1 and 14-2)

Find each difference. (Lesson 14-3)

  8. $(11p + 5r) - (2p + r)$ **9.** $(7a + 6d) - (6a - 7d)$ **10.** $(4t + 11r) - (t + 2r)$

Technology Activity

Recall that when you multiply powers that have the same base, you add the exponents. For example, $3^2 \cdot 3^3 = 3^{2+3}$ or 3^5. But how do you find the power of a power like $(3^2)^3$? You can use a calculator to discover a pattern.

Your Turn

Copy the table below. Use a calculator to find each value and complete the table.

Power	Value	Power	Value
3^1	3	$(3^2)^1$	9
3^2		$(3^2)^2$	
3^3		$(3^2)^3$	
3^4		$(3^2)^4$	
3^5		$(3^2)^5$	
3^6		$(3^2)^6$	

The ∧ or y^x key allows you to evaluate expressions with exponents.

a. Compare the values of 3^6 and $(3^2)^3$. What do you observe?

b. Compare the exponents of 3^6 and $(3^2)^3$. What do you observe?

c. Guess the value of $(3^4)^2$. Use a calculator to check your answer.

d. Write a rule for finding the value of a power of a power.

Learning the Concept

You can review multiplying powers in Lesson 4-8.

Consider the following powers.

$$(5^4)^2 = 5^4 \cdot 5^4 \quad \textit{Definition of exponent} \quad (x^2)^3 = (x^2) \cdot (x^2) \cdot (x^2)$$
$$= 5^{4+4} \quad \textit{Product of powers} \quad\quad = x^{2+2+2}$$
$$= 5^8 \quad\quad \textit{Substitution} \quad\quad\quad = x^6$$

These and other similar examples suggest that you can find a *power of a power* by multiplying the exponents.

Power of a Power	For any number a and positive integers m and n, $(a^m)^n = a^{mn}$.

Simplify $(p^6)^2$.

$$(p^6)^2 = p^{6 \cdot 2} \quad \textit{Power of a power}$$
$$= p^{12}$$

Check: $(p^6)^2 = p^6 \cdot p^6$
$$= p^{6+6} \text{ or } p^{12} \checkmark$$

Connection to ⟩ Geometry

The volume of a cube can be found by multiplying its length, width, and height. The volume of the small cube shown at the right is $x \cdot x \cdot x$ or x^3. If the length of each edge of the cube is doubled, the volume of the new cube would be $(2x)^3$.

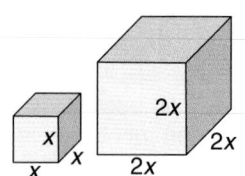

$$(2x)^3 = (2x)(2x)(2x) \quad \textit{Definition of exponent}$$
$$= (2 \cdot 2 \cdot 2)(x \cdot x \cdot x) \quad \textit{Commutative and}$$
$$\textit{associative properties}$$
$$= 2^3 x^3 \quad \textit{Exponential notation}$$
$$= 8x^3 \quad 2^3 = 8$$

Notice that when $2x$ was raised to the third power, *both* the 2 and the x were raised to the third power. This example suggests that the power of a product can be found by multiplying the individual powers.

Power of a Product	For all numbers a and b and positive integer m, $(ab)^m = a^m b^m$.

The formula for the volume of a pyramid is $V = \frac{1}{3}\ell wh$.
a. **Find the measure of the volume of the pyramid.**

$$V = \frac{1}{3}\ell wh$$

$$= \frac{1}{3}(4x)(4x)(4x)$$

$$= \frac{1}{3}(4 \cdot 4 \cdot 4)(x \cdot x \cdot x)$$

$$= \frac{1}{3}(4^3)(x^3)$$

$$= \frac{1}{3}(4x)^3 \quad \textit{Power of a product}$$

The measure of the volume of the pyramid is $\frac{1}{3}(4x)^3$.

b. **If x is 5 cm, what is the volume of the pyramid?**

$$\frac{1}{3}(4x)^3 = \frac{1}{3}(4 \cdot 5)^3 \quad \textit{Replace x with 5.}$$

$$= \frac{1}{3}(20)^3 \quad \textit{Estimate: } \frac{1}{3} \cdot 20 \cdot 20 \cdot 20 = 7 \cdot 20 \cdot 20 \text{ or } 2800$$

20 $\boxed{y^x}$ 3 $\boxed{\div}$ 3 $\boxed{=}$ *2666.6666*

The volume is about 2667 cm³. *Compare to the estimate.*

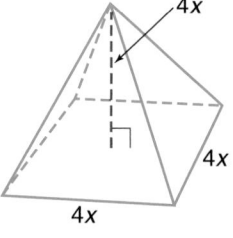

Sometimes the rules for the *power of a power* and the *power of a product* are combined into one rule.

Power of a Monomial	For all numbers a and b and positive integer m, n, and p, $(a^m b^n)^p = a^{mp} b^{mp}$.

Example **Simplify $(x^2 y^3)^4$.**

$$(x^2 y^3)^4 = (x^2)^4 (y^3)^4 \quad \textit{Power of a monomial}$$
$$= x^{2 \cdot 4} y^{3 \cdot 4}$$
$$= x^8 y^{12}$$

Check: $(x^2 y^3)^4 = (x^2 y^3)(x^2 y^3)(x^2 y^3)(x^2 y^3)$
$$= (x^2 \cdot x^2 \cdot x^2 \cdot x^2)(y^3 \cdot y^3 \cdot y^3 \cdot y^3)$$
$$= (x^{2+2+2+2})(y^{3+3+3+3})$$
$$= (x^8)(y^{12}) \quad \text{or} \quad x^8 y^{12} \quad ✔$$

These rules can also be used with negative exponents.

Example **Simplify $(x^3 y^2 z)^{-2}$.**

$$(x^3 y^2 z)^{-2} = x^{3(-2)} y^{2(-2)} z^{1(-2)} \quad \textit{Power of a monomial}$$
$$= x^{-6} y^{-4} z^{-2}$$

You can review negative exponents in Lesson 4-9.

Check: $(x^3 y^2 z)^{-2} = \dfrac{1}{(x^3 y^2 z)^2}$

$$= \dfrac{1}{(x^3 y^2 z)(x^3 y^2 z)}$$
$$= \dfrac{1}{(x^3 \cdot x^3)(y^2 \cdot y^2)(z \cdot z)}$$
$$= \dfrac{1}{x^{3+3} y^{2+2} z^{1+1}}$$
$$= \dfrac{1}{x^6 y^4 z^2} \quad \text{or} \quad x^{-6} y^{-4} z^{-2} \quad ✔$$

Checking Your Understanding

Communicating Mathematics

Read and study the lesson to answer each question.

1. **You Decide**   Nelia says that $5x^3$ and $(5x)^3$ are equivalent expressions. Julie says they are not. Who is correct? Explain.

2. **Analyze** the square shown at the right. What is its area? What would its area be if the length of each side were doubled or tripled?

3. **Assess Yourself**   How well do you understand the rules given in this lesson? For each of the following, write the rules in your own words and then give an example of it.

   a. Power of a power

   b. Power of a product

   c. Power of a monomial

Simplify.

4. $(y^5)^3$ **5.** $(2^4)^3$ **6.** $(2m)^4$

7. $(3xy)^3$ **8.** $(a^2b^3)^2$ **9.** $(-xy)^2$

Evaluate each expression if $a = -1$ and $b = 3$.

10. a^2b **11.** $-ab^2$ **12.** $(-4a^3b)^2$

13. Geometry Find the volume of the rectangular prism below.

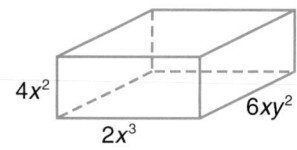

$4x^2$ $2x^3$ $6xy^2$

Exercises: Practicing and Applying the Concept

Simplify.

14. $(7^3)^2$ **15.** $(yz)^4$ **16.** $(-3w)^3$

17. $(-2rs)^4$ **18.** $[(-4)^2]^2$ **19.** $(m^2)^5$

20. $(-y^3)^6$ **21.** $(-2x^2)^5$ **22.** $(x^2y)^3$

23. $(-xy^3)^4$ **24.** $(2a^3b)^5$ **25.** $3x(2x)^2$

26. $-5x(x^3)^2$ **27.** $(4x^2y^3)^2$ **28.** $[(-3)^2]^3$

29. $(t^3)^{-4}$ **30.** $(-w^2)^{-5}$ **31.** $-2(x^3y)^{-2}$

Evaluate each expression if $a = -2$ and $b = 3$.

32. $2ab^2$ **33.** $-3a^2b$ **34.** $-(ab^2)^2$

35. $(-2ab^3)$ **36.** $(a^3b)^2$ **37.** $-(3a^2b)^2$

38. $-3b(2a)^{-2}$ **39.** $2b^2(-3ab)^{-3}$ **40.** $2(3a^{-2})^2$

41. You can use a calculator to confirm the power of a product rule. Copy and complete the following table using a calculator.

Monomial	Value	Monomial	Value
$(2 \cdot 3)^2$		$2^2 \cdot 3^2$	
$(4 \cdot 5)^3$		$4^3 \cdot 5^3$	
$(6 \cdot 7)^4$		$6^4 \cdot 7^4$	
$(8 \cdot 3)^5$		$8^5 \cdot 3^5$	

    a. Compare the values in each row. What do you observe?

    b. Compare the monomials in each row. What do you observe?

    c. Write a rule for finding the power of a product in your own words.

42. Are $(2^6)^4$ and $(4^6)^2$ equal? Explain why or why not. (*Hint:* Write both expressions as a power of 2.)

Applications and Problem Solving

43. Finance Nuna opens a savings account with $1000. It earns 1% interest every 3 months. So, in one year, the balance is $1000(1.01)(1.01)(1.01)(1.01)$ or $1000(1.01)^4$. If she keeps the money in the account for 3 years, the deposit is worth $1000(1.01^4)^3$.

    a. Write an expression that represents what the deposit is worth in 6 years.

    b. What is the deposit worth in 6 years?

44. Biology It takes 2 hours for a culture with 1 bacterium to grow to two bacteria. To grow to 32, or 2^5, bacteria, the culture must double five times. If the culture doubles another five times, there will be $(2^5)^2$ bacteria.

    a. Write $(2^5)^2$ as a power of 2.

    b. $2^5 = 32$. Use this information to estimate $(2^5)^2$.

45. Chemistry Chemicals like salt and magnesium can be extracted from sea water. To determine if such a venture would be profitable, a company would like to know how much salt or magnesium could be extracted from a cubic mile of sea water. How many cubic feet of sea water are in a cubic mile of sea water? (*Hint:* There are 5280 feet in a mile.)

Mixed Review

46. Geometry The perimeter of the triangle at the right is $5x + 2$ units. Find the length of the base of the triangle. (Lesson 14-3)

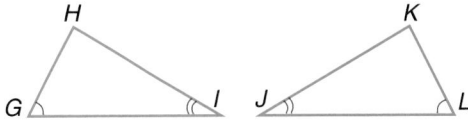

47. Geometry In a right triangle, if the measures of the legs are 5 inches and 12 inches, find the measure of the hypotenuse. (Lesson 13-4)

48. Geometry Write a congruence statement for the pair of congruent triangles at the right. Then name the corresponding parts. (Lesson 11-5)

49. Probability Betsy plays on a basketball team. Her free throw average is 60%. How could you simulate the probable results of her next five free throw attempts? (Lesson 10-8)

For **Extra Practice,** see page 773.

50. Solve $\frac{2x}{-5} = -24$. (Lesson 7-2)

51. On graph paper, draw coordinate axes. Then graph and label each point. Name the quadrant in which each point is located. (Lesson 2-2)

    a. $X(-4, 6)$ **b.** $Y(2, 5)$ **c.** $Z(-3, 0)$

52. Standardized Test Practice Evaluate $|2x| + yz$ if $x = -2$, $y = 4$, and $z = -1$. (Lesson 2-1)

    A. 0

    B. 1

    C. 2

    D. 3

53. Write an expression that means the same as $\frac{xy}{5}$. (Lesson 1-3)

14-5A Multiplying Polynomials

A Preview of Lesson **14-5**

MATERIALS

⊐ algebra tiles

▢ product mat

Algebra tiles are named based on the area of the rectangle. The area of the rectangle is the product of the width and length.

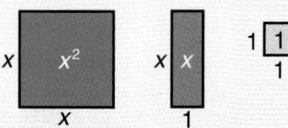

You can use algebra tiles to model more complex rectangles. These rectangles will help you understand how to find the product of simple polynomials. The width and length each represent a polynomial being multiplied. The area of the rectangle represents their product.

Your Turn   **Work with a partner to find $x(x + 2)$.**

▶ Make a rectangle with a width of x and a length of $x + 2$. Use algebra tiles to mark off the dimensions on a product mat.

▶ Using the marks as a guide, fill in the rectangle with algebra tiles.

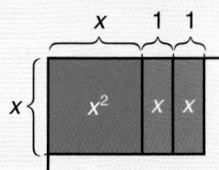

▶ The area of the rectangle is $x^2 + x + x$. In simplest form, the area is $x^2 + 2x$. Therefore, $x(x + 2) = x^2 + 2x$.

TALK ABOUT IT   **Tell whether each statement is *true* or *false*. Justify your answer with algebra tiles.**

1. $x(2x + 3) = 2x^2 + 3x$
2. $2x(3x + 4) = 6x^2 + 4x$

Find each product using algebra tiles.

3. $x(x + 5)$
4. $2x(x + 2)$
5. $3x(2x + 1)$

6. Suppose you have a square garden plot that measures x feet on a side.

   a. If you double the length of the plot and increase the width by 3 feet, how large will the new plot be? Write two expressions for the area of the new plot.

   b. If the original plot was 10 feet on a side, what is the area of the new plot?

Multiplying a Polynomial by a Monomial

14-5

You'll Learn

- To multiply a polynomial by a monomial.

It's Important

Multiplying polynomials can help you solve surface area problems.

Application: Construction

A chain-link fence is going to be installed around the school tennis courts. To determine how much fencing is needed, a representative from the fencing company measures the length and the width of the region that has to be enclosed. The total length of fencing can be determined by finding the perimeter P of the rectangular region.

$P = \ell + w + \ell + w$
$\quad = 2\ell + 2w$ *Combine like terms.*
$\quad = 2(\ell + w)$ *Distributive property*

In the expression $2(\ell + w)$, a polynomial is being multiplied by a monomial. In general, the distributive property can be used to multiply a monomial and a polynomial.

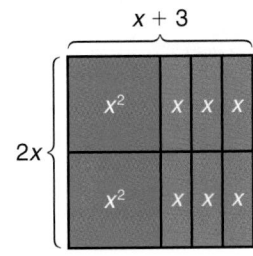

Learning the Concept

As you saw in Lesson 14-5A, you can use algebra tiles to multiply a polynomial and a monomial.

The figure at the right is a rectangle whose length is $x + 3$ and whose width is $2x$.

The area of any rectangle is the product of its length and its width. The area of this rectangle can also be found by adding the areas of the tiles.

Area: Formula	**Area: Algebra tiles**
$A = \ell w$	$A = x^2 + x^2 + x + x + x + x + x + x$
$A = (x + 3)2x$	$A = 2x^2 + 6x$

Therefore, $(x + 3)2x = 2x^2 + 6x$.

Example ❶   **Find each product.**
 a. **$7(2x + 5)$**
     $7(2x + 5) = 7(2x) + 7(5)$ *Distributive property*
            $= 14x + 35$
 b. **$(3x - 7)4x$**
     $(3x - 7)4x = (3x)4x - (7)4x$ *Distributive property*
            $= 12x^2 - 28x$
 c. **$2ab(-9a^2 + 5ab - 4b^2)$**
     $2ab(-9a^2 + 5ab - 4b^2)$
          $= 2ab(-9a^2) + 2ab(5ab) - 2ab(4b^2)$ *Distributive property*
          $= -18a^3b + 10a^2b^2 - 8ab^3$ *Multiply monomials.*

Sometimes problems can be solved by simplifying polynomial expressions.

Example 2

INTEGRATION

Algebra

The world's largest swimming pool is the Orthlieb Pool in Casablanca, Morocco. It is 30 meters longer than 6 times its width. If the perimeter of the pool is 1110 meters, what are the dimensions of the pool?

Explore   You know the perimeter of the pool. You want to find the dimensions of the pool.

Plan   Let w represent the width of the pool. Then $6w + 30$ represents the length. Then write an equation.

$$\underbrace{Perimeter}\ \underbrace{equals}\ \underbrace{twice}\ \underbrace{the\ sum\ of\ the\ length\ and\ width.}$$
$$\quad P \qquad = \qquad 2 \qquad\qquad (\ell + w)$$

Solve
$$P = 2(\ell + w)$$
$$1110 = 2((6w + 30) + w) \qquad \textit{Replace P with 1110 and } \ell \textit{ with } 6w + 30.$$
$$1110 = 2(7w + 30) \qquad\qquad \textit{Combine like terms.}$$
$$1110 = 14w + 60 \qquad\qquad \textit{Distributive property}$$
$$1050 = 14w \qquad\qquad\qquad \textit{Subtract 60 from each side.}$$
$$\frac{1050}{14} = w$$
$$75 = w$$

The width is 75 meters and the length is $6w + 30$ or 480 meters.

Examine   If the width is 75 meters, the length should be 30 meters longer than 6 times 75, or 480 meters. The perimeter is $2(75 + 480)$ or 1110 meters. The solution checks.

Checking Your Understanding

Communicating Mathematics

Read and study the lesson to answer these questions.

1. **Explain** how you would find the product of x and $2x - 1$.

2. **State** the product of x and $2x + 3$ using the rectangle at the right.

3. **Explain** why $x(2x + 3)$ and $(2x + 3)x$ are equivalent expressions.

MATERIALS

☐ algebra tiles
☐ product mat

Use algebra tiles to find each product.

4. $5(x + 2)$ 5. $x(x + 4)$ 6. $2x(x - 1)$

Guided Practice

Find each product.

7. $3(x + 4)$ 8. $x(x + 5)$ 9. $3(x - 2)$

10. $2x(x - 8)$ 11. $4x(3x + 7)$ 12. $x(5x - 12)$

13. **Sports** The perimeter of a football field is 1040 feet. The length of the field is 120 feet less than 3 times the width. What are the dimensions of the football field?

Exercises: Practicing and Applying the Concept

Independent Practice

Find each product.

14. $7(3x + 5)$

15. $-2(x + 8)$

16. $y(y - 9)$

17. $3x(2x - 1)$

18. $-3x(x - 5)$

19. $c(a^2 + b)$

20. $4m(m^2 - m)$

21. $pq(pq + 8)$

22. $-3x(x^2 - 7x)$

23. $-2a(9 - a^2)$

24. $7(-2a^2 + 5a - 11)$

25. $-5(3x^2 - 7x + 9)$

26. $-3y(6 - 9y + 4y^2)$

27. $4c(c^3 + 7c^2 - 10)$

28. $-5x^2(3x^3 - 8x - 12)$

29. $6x^2(-2x^3 + 8x^2 - 7)$

30. $2x(5x^3 - 4x^2 + 6x - 9)$

31. $-x^2(x^3 - x^2 + 3x - 5)$

Solve each equation.

32. $-3(2a - 12) + 48 = 3a + 3$

33. $2(5w - 12) = 6(-2w + 3) + 2$

Critical Thinking

34. Use algebra tiles to make a model of $2x + 6$. Then, form a tile rectangle and find the factors of $2x + 6$. Finally, outline a procedure for finding the factors of any binomial.

Applications and Problem Solving

35. **Manufacturing** The figure at the right shows a pattern for a cardboard box before it has been cut and folded.

    a. Find the area of each rectangular region and add to find a formula for the number of square inches of cardboard needed.

    b. Find the surface area if x is 2.5 inches.

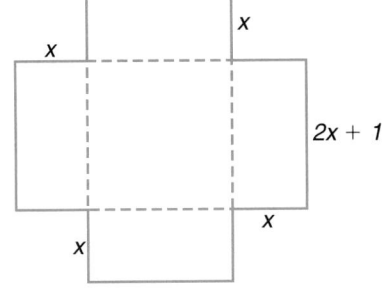

36. **Geometry** Find the measure of the area of the shaded region in simplest terms.

Mixed Review

37. Simplify $-4b(2b)^3$. (Lesson 14-4)

38. Find the best integer estimate for $-\sqrt{150}$. (Lesson 13-1)

39. **Statistics** Describe two ways a graph of sales of several brands of cereal could be misleading. (Lesson 10-4)

40. Solve the system of equations $y = x - 3$ and $y = x - 4$ by graphing. (Lesson 8-8)

41. **Draw a Diagram** The Avarillo's backyard is 130 by 90 feet. If their sprinkler can water an area 30 by 30 feet, how many times will it need to be moved to water the entire yard? (Lesson 4-3)

42. Is 528 divisible by 2, 3, 5, 6, or 10? (Lesson 4-1)

43. **Standardized Test Practice** Find the distance traveled if you drive at 55 mph for $3\frac{1}{2}$ hours. (Lesson 3-4)

    A. 15.71 miles **B.** 165 miles **C.** 176 miles **D.** 192.5 miles

For **Extra Practice**, see page 773.

14-6 Multiplying Binomials

You'll Learn

■ To multiply binomials.

It's Important

Multiplying binomials is useful when designing buildings.

Application: Genetics

Punnett squares are used to show possible ways that genes can combine at fertilization. In a Punnett square, *dominant* genes are shown with capital letters. *Recessive* genes are shown with the lowercase of the same letter.

The Punnett square below represents a cross between two hybrid tall pea plants. A hybrid trait is the result of a combination of a dominant and a recessive gene. The parent plants are tall because the dominant trait masks the recessive trait.

T hybrid tall t

	T	t
T hybrid tall	TT	Tt
t	Tt	tt

Letters representing the parents' genes are placed on the outer sides of the Punnett square.

Letters inside the boxes of the square show the possible gene combinations for their offspring.

Let *T* represent the dominant gene for tallness and *t* represent the recessive gene for shortness.

When the parents' genes are combined, you see all of the possible combinations of the genes of the offspring.

$$(T + t)(T + t) = TT + Tt + Tt + tt$$
$$= TT + 2Tt + tt$$

Learning the Concept

$(T + t)(T + t)$ is an example of the product of two binomials. Algebra tiles can also be used to help you understand how to multiply two binomials.

Consider the binomials $x + 3$ and $x + 2$. To multiply these binomials, mark off a rectangle on a product mat that has dimensions $x + 3$ and $x + 2$. Then fill in the rectangle with algebra tiles.

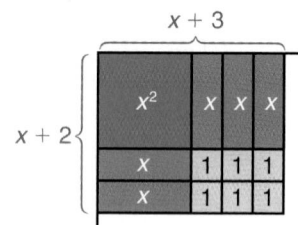

Area: Formula	Area: Algebra Tiles
$A = \ell w$	$A = x^2 + x + x + x + x + x + 1 + 1 + 1 + 1 + 1 + 1$
$\quad = (x + 3)(x + 2)$	$\quad = x^2 + 5x + 6$

The product $(x + 3)(x + 2)$ represents the area of the rectangle. The area is also the sum of the areas of the algebra tiles, $x^2 + 5x + 6$. Therefore, $(x + 3)(x + 2) = x^2 + 5x + 6$.

You can also use the distributive property to multiply binomials.

Example ① **Find each product.**
  a. $(2x + 3)(3x + 5)$

$$\begin{aligned}(2x + 3)(3x + 5) &= (2x + 3)3x + (2x + 3)5 &&\textit{Distributive property}\\ &= (2x)3x + (3)3x + (2x)5 + (3)5\\ &= 6x^2 + 9x + 10x + 15 &&\textit{Multiply polynomials.}\\ &= 6x^2 + 19x + 15 &&\textit{Combine like terms.}\end{aligned}$$

  b. $(2x + 1)(5x - 3)$

$$\begin{aligned}(2x + 1)(5x - 3) &= (2x + 1)5x + (2x + 1)(-3) &&\textit{Distributive}\\ &&&\textit{property}\\ &= (2x)5x + (1)5x + (2x)(-3) + (1)(-3)\\ &= 10x^2 + 5x + (-6x) + (-3) &&\textit{Multiply}\\ &&&\textit{polynomials.}\\ &= 10x^2 - x - 3 &&\textit{Combine like terms.}\end{aligned}$$

Example ②

Real World
APPLICATION

Gardening

A rectangular garden is 5 feet longer than twice its width. It has a sidewalk 3 feet wide on two of its sides. The area of the sidewalk is 213 square feet. Find the dimensions of the garden.

Explore Use algebra tiles to model the situation. Let x represent the width of the garden. Then, $2x + 5$ represents the length of the garden, $x + 3$ the width of the garden and sidewalk, and $2x + 8$ the length of the garden and sidewalk.

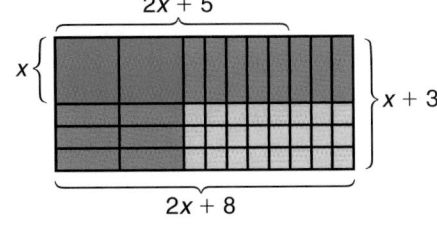

Then, $x(2x + 5) =$ the area of the garden and $(x + 3)(2x + 8) =$ the area of the garden and the sidewalk.

Plan

$$\underbrace{\text{area of garden}}_{\text{and sidewalk}} - \underbrace{\text{area of}}_{\text{garden}} = \underbrace{\text{area of}}_{\text{sidewalk}}$$

$$(x + 3)(2x + 8) - x(2x + 5) = 213$$

Solve
$$\begin{aligned}2x^2 + 6x + 8x + 24 - 2x^2 - 5x &= 213 &&\textit{Count the tiles.}\\ (2x^2 - 2x^2) + (6x + 8x - 5x) + 24 &= 213\\ 9x + 24 &= 213 &&\textit{Simplify.}\\ 9x &= 189\\ x &= 21\end{aligned}$$

The width is 21 feet, and the length is $2x + 5 = 2(21) + 5$ or 47 feet.

Examine   Total Area − area of garden $\stackrel{?}{=}$ area of sidewalk

$$\begin{aligned}(24)(50) - (21)(47) &\stackrel{?}{=} 213\\ 1200 - 987 &\stackrel{?}{=} 213\\ 213 &= 213 \quad ✔\end{aligned}$$

Lesson 14-6 Multiplying Binomials **729**

Communicating Mathematics

Read and study the lesson to answer these questions.

1. **Draw** a rectangular region that represents the product of $x + 3$ and $2x + 1$. Then find the product.

2. **Explain** how the procedure used for multiplying two binomials is similar to the procedure for multiplying a binomial and a monomial. Then explain how it is different.

MATERIALS

algebra tiles

product mat

Use algebra tiles to find each product.

3. $(x + 1)(x + 2)$ 4. $(x + 3)(x + 4)$ 5. $(2x + 3)(x + 2)$

Guided Practice

For each model, name the two binomials being multiplied and then give their product.

6.   7.

Find each product.

8. $(x + 3)(2x + 5)$ 9. $(2x + 3)(x + 1)$

10. $(2x + 3)(3x + 2)$ 11. $(5x - 3)(2x + 1)$

12. **Sports** A shuffleboard court is 10 feet longer than it is wide. A brick path 3 feet in width surrounds the court. Express the total area of the court and path algebraically if the court is w feet wide.

Exercises: Practicing and Applying the Concept

Independent Practice

For each model, name the two binomials being multiplied and then give their product.

13.   14.

15.   16.

Find each product.

17. $(x + 4)(x + 3)$ 18. $(x + 5)(x + 2)$ 19. $(x - 6)(x + 2)$
20. $(x + 7)(x - 5)$ 21. $(x - 9)(x + 4)$ 22. $(x + 2)(x + 2)$
23. $(2x + 3)(x - 4)$ 24. $(3x - 1)(x + 8)$ 25. $(x + 3)(x + 3)$
26. $(2x + 5)(3x + 1)$ 27. $(5x + 2)(2x - 3)$ 28. $(x - 5)(x + 5)$
29. $\left(3x - \frac{1}{4}\right)\left(6x - \frac{1}{2}\right)$ 30. $(x - 2)(x^2 + 2x + 4)$

Critical Thinking

31. Write the multiplication problem modeled by the figure at the right. Name the product.

Applications and Problem Solving

32. **Volunteer Work**   As a project, a senior citizens group makes baby quilts and full-size quilts for victims of natural disasters. The baby quilts are $x + 2$ feet long by x feet wide. The full-size quilts are 4 feet longer and 3 feet wider than the baby quilts.

   a. Draw a rectangle using algebra tiles to represent a baby quilt.

   b. Draw additional tiles to increase the rectangular region to represent a full-size quilt.

   c. Write a product for the full-size quilt and another for the baby quilt.

   d. What is the difference between the area of the full-size quilt and the area of the baby quilt?

33. **Geometry**   The length of the smaller rectangle at the right is 1 inch less than twice its width. Both the dimensions of the larger rectangle are 2 inches longer than the smaller rectangle. The area of the shaded region is 86 inches.

   a. What are the dimensions of the smaller rectangle?

   b. What is the area of the smaller rectangle?

   c. What is the area of the larger rectangle?

Mixed Review

34. Find the product $-2(4c^2 - 3c + 5)$. (Lesson 14-5)

35. **Measurement**   On a sunny day, a tree casts a shadow 60 feet long. At the same time, a yardstick casts a shadow 2 feet long. How tall is the tree? (Lesson 11-6)

36. **Standardized Test Practice**   In which figure are $\angle 1$ and $\angle 2$ supplementary? (Lesson 11-3)

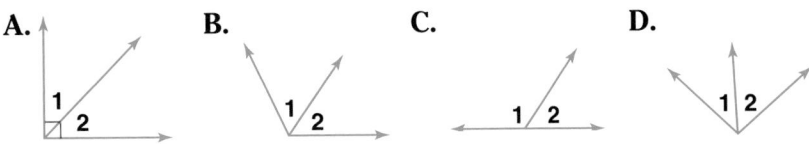

37. **Geometry**   The radius of a circle is 0.5 m. Find its circumference. (Lesson 7-4)

38. What is the GCF of 60 and 25? (Lesson 4-5)

39. Solve $6x < -84$. (Lesson 3-7)

For **Extra Practice**, see page 773.

14-6B Factoring

An Extension of Lesson **14-6**

MATERIALS

⬚ algebra tiles

☐ product mat

Consider a rectangle with a length of $x + 1$ and a width of $x + 3$. From the previous lesson, you know that $(x + 1)(x + 3) = x^2 + 4x + 3$. The binomials $(x + 1)$ and $(x + 3)$ are **factors** of the trinomial $x^2 + 4x + 3$.

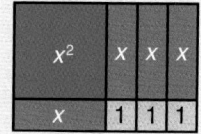

You can use algebra tiles as a model for factoring simple trinomials. If a rectangle with a length and width greater than 1 cannot be formed to represent the trinomial, then the trinomial is not factorable.

Your Turn   **Work with a partner to factor $x^2 + 3x + 2$.**

▶ Model the polynomial.

▶ Place the x^2-tile and the 1-tiles as shown.

▶ Complete the rectangle with the x-tiles.

▶ Since a rectangle can be formed, $x^2 + 3x + 2$ is factorable. The rectangle has a width of $x + 1$ and a length of $x + 2$. So the factors of $x^2 + 3x + 2$ are $x + 1$ and $x + 2$.

TALK ABOUT IT   **Tell whether each polynomial is factorable. Justify your answers with algebra tiles.**

1. $x^2 + 6x + 8$

2. $x^2 + 5x + 6$

3. $x^2 + 7x + 3$

4. $3x^2 + 8x + 5$

5. $5x^2 - x + 16$

6. $8x^2 - 31x - 4$

7. Write a paragraph that explains how you can determine whether a trinomial can be factored. Include an example of one trinomial that can be factored and one that cannot.

Vocabulary

After completing this chapter, you should be able to define each term, property, or phrase and give an example or two of each.

Algebra

binomial (p. 706)

coefficient (p. 711)

constant (p. 707)

factors (p. 732)

like terms (p. 711)

polynomial (p. 706)

power of a monomial (p. 721)

power of a power (p. 719)

power of a product (p. 720)

trinomial (p. 706)

inter NET CONNECTION

Chapter Review For additional lesson-by-lesson review, visit:

www.glencoe.com/ sec/math/prealg/ mathnet

Understanding and Using Vocabulary

Choose the correct term to complete each sentence.

1. The degree of a monomial is the (sum, product) of the exponents of its variables.

2. A (binomial, trinomial) is the sum or difference of three monomials.

3. The degree of a polynomial is the same as that of the term that has the (least, greatest) degree.

4. The numerical part of a monomial is called the (coefficient, variable).

5. To subtract two polynomials, you add the (additive inverse, multiplicative inverse) of the second polynomial.

6. The (commutative property, distributive property) is used in multiplying a polynomial by a monomial.

Determine whether each statement is *true* or *false*.

7. Every polynomial is a monomial.

8. The coefficient of $6ab^2$ is 2.

9. The additive inverse of $(3x + 6y)$ is $(-3x - 6y)$.

10. The power of a product property says that $(8m)^3 = 8^3m^3$.

Skills and Concepts

Objectives and Examples

Upon completing this chapter, you should be able to:

▶ **identify and classify polynomials**
(Lesson 14-1)

The expression $5a^2 - 3a + 4$ is a polynomial. It is a trinomial because it is the sum of three monomials.

The expression $\dfrac{2}{a}$ is not a polynomial because it has a variable in the denominator.

Review Exercises

Use these exercises to review and prepare for the chapter test.

State whether each expression is a polynomial. If it is, identify it as a *monomial*, *binomial*, or *trinomial*.

11. $3x^4 - x$ **12.** $\dfrac{4}{ax}$

13. ax^2 **14.** $9b^2 + b - 1$

15. $8x - \dfrac{5}{3}$ **16.** $k^3 - \dfrac{2}{k}$

▶ **find the degree of polynomials**
(Lesson 14-1)

Find the degree of $x^3y + 3xy - 6y - 1$.

Find the degree of each term.

x^3y has degree $3 + 1$ or 4.
$3xy$ has degree $1 + 1$ or 2.
$6y$ has degree 1.
1 has degree 0.

So, the degree of $x^3y + 3xy - 6y - 1$ is 4.

Find the degree of each polynomial.

17. $4x$

18. $5a^2b$

19. $3x + y^2$

20. $19m^2n^3 - 14mn^4$

21. $x^2 - 6xy + xy^2$

22. $12rs^2 + 3r^2s + 5r^4s$

▶ **add polynomials** (Lesson 14-2)

Find the sum $(b^2 + 3b + 8) + (2b^2 + 5b - 5)$.

$$
\begin{array}{r}
b^2 + 3b + 8 \\
(+)\ 2b^2 + 5b - 5 \\
\hline
3b^2 + 8b + 3
\end{array}
$$

Find each sum.

23. $(2x^2 - 5x) + (3x^2 + x)$

24. $(a^2 - 6ab) + (3a^2 + ab)$

25. $(x^2 - 5x + 3) + (4x - 3)$

26. $(-3y^2 + 2) + (4y^2 - 5y - 2)$

27.
$$
\begin{array}{r}
4x^2 + 3x + 2 \\
(+)\ x^2 \qquad - 1 \\
\hline
\end{array}
$$

28.
$$
\begin{array}{r}
16x^2y - 2xy + xy^2 \\
(+)\ 4x^2y + 6xy - 8xy^2 \\
\hline
\end{array}
$$

Objectives and Examples

▶ **subtract polynomials** (Lesson 14-3)

Find the difference
$(4y^2 - 2y + 3) - (y^2 + 2y - 4)$.

Align like terms and add the additive inverse of $y^2 + 2y - 4$.

$$\begin{array}{r} 4y^2 - 2y + 3 \\ (-)\ y^2 + 2y - 4 \\ \hline \end{array} \longrightarrow \begin{array}{r} 4y^2 - 2y + 3 \\ (+)\ -y^2 - 2y + 4 \\ \hline 3y^2 - 4y + 7 \end{array}$$

▶ **find powers of monomials** (Lesson 14-4)

$$\begin{aligned} (x^5)^3 &= x^{5 \cdot 3} \\ &= x^{15} \end{aligned}$$

$$\begin{aligned} (2d)^4 &= 2^4 \cdot d^4 \\ &= 16d^4 \end{aligned}$$

$$\begin{aligned} (a^3 b^2)^3 &= a^{3 \cdot 3} \cdot b^{2 \cdot 3} \\ &= a^9 b^6 \end{aligned}$$

▶ **multiply a polynomial by a monomial**
(Lesson 14-5)

$$\begin{aligned} -5y(2y^2 - 4) &= -5y(2y^2) - (-5y)(4) \\ &= -10y^3 - (-20y) \\ &= -10y^3 + 20y \end{aligned}$$

Review Exercises

Find each difference.

29. $(7a - 11b) - (3a + 4b)$

30. $(6y - 8z) - (6y + 4z)$

31. $(3a^2 - b^2 + c^2) - (a^2 + 2b^2)$

32. $(14a^2 - 3a) - (6a^2 + 5a + 17)$

33. $\begin{array}{r} 18x^2 + 3x - 1 \\ (-)\ \ 2x^2 + 4x + 6 \\ \hline \end{array}$

34. $\begin{array}{r} 12m^2 -\ \ mn + 9n^2 \\ (-)\ \ 7m^2 + 2mn - 4n^2 \\ \hline \end{array}$

Simplify.

35. $(a^2)^3$

36. $(-2x)^3$

37. $(p^2 q)^3$

38. $-5c(2cd)^3$

39. $4y(y^2 z)^3$

40. $(12a^5)^2 b^3$

41. $6a(-ab)^7$

42. $(-2d^2)^6(-3)^3$

Find each product.

43. $4d(2d - 5)$

44. $x(-5x + 3)$

45. $a^2(2a^3 + a - 5)$

46. $3y(-y^2 - 8y + 4)$

47. $-2g(g^3 + 6g + 3)$

48. $-3az(2z^2 + 4az + a^2)$

► **multiply binomials** (Lesson 14-6)

$(x - 4)(2x + 3)$
$= (x - 4)2x + (x - 4)(3)$
$= x(2x) + (-4)(2x) + x(3) + (-4)(3)$
$= 2x^2 - 8x + 3x - 12$
$= 2x^2 - 5x - 12$

49. For the model below, name the two binomials being multiplied and then give their product.

x^2	x^2	x
x	x	1
x	x	1
x	x	1

Find each product.

50. $(x + 3)(x + 1)$

51. $(2x + 1)(x + 1)$

52. $(3x + 2)(2x + 2)$

Applications and Problem Solving

53. Architecture Norma Merrick Sklarek was the first African-American woman registered as an architect in the United States. Suppose Ms. Sklarek drew the floor plan for the first floor of a house shown below. Using the measurements given, write a polynomial that represents the total area of the first floor. (Lesson 14-1)

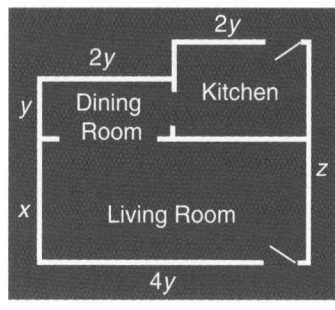

54. Geometry The perimeter of the triangle shown below is $4x - 2y$ units. Find the length of the third side of the triangle. (Lessons 14-1 and 14-2)

$x + y$ $x - y$

55. Geometry Find the volume of the triangular prism below. (Lesson 14-4)

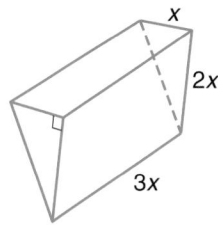

56. Sports The perimeter of a soccer field is 1040 feet. The length of the field is 40 feet more than 2 times the width. What are the dimensions of the field? (Lesson 14-5)

A practice test for Chapter 14 is available on page 787.

Cooperative Learning Project

The triangular arrangement of numbers below is called Pascal's Triangle. Many patterns exist in Pascal's Triangle. One of the patterns is that the rows in the triangle can be used to write powers of binomials.

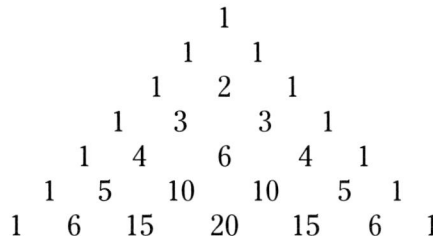

```
              1
           1     1
        1     2     1
     1     3     3     1
  1     4     6     4     1
1     5    10    10     5     1
1   6    15    20    15    6    1
```

Work with a partner to find the powers of the binomial $(x + y)$ and complete the table below.

Power	Product	Coefficients of the terms
$(x + y)^2$		1, 2, 1
$(x + y)^3$		
$(x + y)^4$		
$(x + y)^5$	$x^5 + 5x^4y + 10x^3y^2 + 10x^2y^3 + 5xy^4 + y^5$	

Compare the last column of the table to the rows of Pascal's Triangle. Describe the relationship you find. Predict the product of $(x + y)^6$. Then multiply to verify your product.

Thinking Critically

▶ For all numbers a and b such that $b \neq 0$, and any integer m, is $\left(\dfrac{a}{b}\right)^m = \dfrac{a^m}{b^m}$ a true sentence?

▶ A trapezoid has an area of 425 square inches and a height of 10 inches. The lower base is 5 inches less than twice the upper base. Find the length of each base.

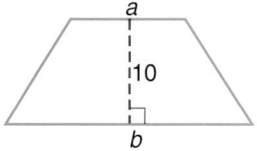

Portfolio

Review the items in your portfolio. Make a list of the items, noting why each item was chosen. Replace any items that are no longer appropriate.

Section One: Multiple Choice

There are ten multiple-choice questions in this section. After working each problem, write the letter of the correct answer on your paper.

1. Which is equivalent to $\frac{3^6}{3^2}$?

    A. 3^3 **B.** 3^4

    C. 3^8 **D.** 3^{12}

2. In a right triangle, the tangent ratio is the measure of the leg opposite an angle to the measure of the leg adjacent to the angle. What is the tangent ratio of angle D?

    A. $\frac{3}{5}$

    B. $\frac{4}{5}$

    C. $\frac{4}{3}$

    D. $\frac{5}{3}$

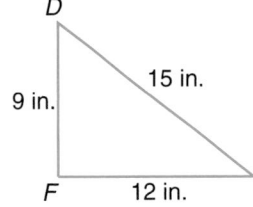

3. Paul Olsen is designing a cereal box. The height of the box is 12 inches, the width is 2 inches, and the length is 6 inches. What is the surface area of the cardboard he needs for one box?

    A. 40 in^2

    B. 144 in^2

    C. 216 in^2

    D. 288 in^2

4. Which is equivalent to $(7x^2 + 3y) - (3x^2 + 5y)$?

    A. $4x^2 - 2y$ **B.** $4x^4 - 2y^2$

    C. $4x^2 + 8y$ **D.** $4x^4 + 8y^2$

5. A sweater that normally sells for \$35 is on sale at 25% off. Which is the best estimate of the sale price?

    A. \$9.   **B.** \$32.

    C. \$26.   **D.** \$43.

6. A rectangular window is 40 inches high and 30 inches wide. What is the length of a diagonal?

    A. 60 in.   **B.** 35 in.

    C. 50 in.   **D.** 25 in.

7. Find the surface area of a cone with a diameter of 12 meters and a slant height of 18 meters. Round your answer to the nearest tenth.

    A. 75.4 m^2 **B.** 452.2 m^2

    C. 94.2 m^2 **D.** 678.2 m^2

8. Which is equivalent to $(-2x)^3$?

    A. $-8x^3$ **B.** $-2x^3$

    C. $8x^3$ **D.** $2x^3$

9. $\triangle EFG$ is similar to $\triangle KML$. What is the measure of $\angle M$?

    A. $40°$

    B. $60°$

    C. $70°$

    D. $80°$

10. $\triangle ABC$ is a right triangle. What is the length of the hypotenuse?

    A. 21 in.

    B. 26 in.

    C. 30 in.

    D. 52 in.

Section Two: Free Response

This section contains twelve questions for which you will provide short answers. Write your answer on your paper.

11. Find the perimeter of a regular hexagon with sides 12.6 meters long.

12. A nickel, a dime, a quarter, and a half dollar are in a purse. Without looking, Alex picks up two coins. What are the different amounts he could choose?

13. Joan and Miguel are flying kites in the park. When Joan is 70 feet from Miguel, her kite forms an angle of 56° with the ground. If the kite is directly over Miguel, how high is the kite to the nearest tenth of a foot?

14. Evaluate $a^4 - 2a^3 + 4b^2 - c$ if $a = 2$, $b = -6$, and $c = -2$.

15. Find the area of the circle shown below.

12 cm

16. There are 25 students in a math class. The class takes a survey and finds that 15 people like pepperoni, 12 people like sausage, and 7 people like both equally. How many people like neither pepperoni nor sausage?

17. In a survey of favorite music types, 15 out of 50 people preferred country. In a group of 1000 people, how many would you expect to like country?

18. Find the sum of $(4x^3 + 2x^2) + (-2x^3 - 7x^2)$.

19. Approximate the value of $\sqrt{45}$.

20. The largest country in the world is Russia, with an area of 6,590,876 square miles. The second largest country is Canada. The area of Russia is 110,416 square miles less than twice the area of Canada. What is the area of Canada?

21. Write 2.25 as a mixed number.

22. Find the sum of the measures of the interior angles of a trapezoid.

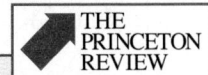
Test-Taking Tip On standardized tests, guessing can improve your score if you make educated guesses. First, find out if there is a penalty for incorrect answers. If there is no penalty, a guess can only increase your score, or at worst, leave your score the same. If there is a penalty, try to eliminate enough choices to make the probability of a correct guess greater than the penalty for an incorrect answer. That is, if there is a quarter point penalty for an incorrect response and you have a choice of three responses, the probability of making a correct guess is 1 out of 3, which is better than the penalty.

Section Three: Open-Ended

This section contains three open-ended problems. Demonstrate your knowledge by giving a clear, concise solution to each problem. Your score on these problems will depend on how well you do the following.

- Explain your reasoning.
- Show your understanding of the mathematics in an organized manner.
- Use charts, graphs, and diagrams in your explanation.
- Show the solution in more than one way or relate it to other situations.
- Investigate beyond the requirements of the problem.

23. Show that $4n(-2n^2 + n - 12) = -8n^3 + 4n^2 - 48n$. Use values for n to justify your answer.

24. Explain the difference between integers and whole numbers.

25. Write a problem that can be simulated by rolling a die.

inter**NET**
CONNECTION

Test Practice For additional test practice questions, visit:
www.glencoe.com/sec/math/prealg/mathnet

Extra Practice

Lesson 1-1

1. **Postal Service** The U.S. Postal Service offers airmail service to other countries. The rates for International Air Mail letters and packages are shown in the table at the right. Determine the air mail rate for a package that weighs 5.5 ounces.
   a. Write the *Explore* step. What do you know and what do you need to find?
   b. Write the *Plan* step. What strategy will you use? What do you estimate the answer to be?
   c. *Solve* the problem using your plan. What is your answer?
   d. *Examine* your solution. Is it reasonable? Does it answer the question?

Weight not over (ounces)	Rate
0.5	$0.50
1.0	$0.95
1.5	$1.34
2.0	$1.73
2.5	$2.12
3.0	$2.51
3.5	$2.90
4.0	$3.29

2. **Postal Service** In 1995, the state of Florida celebrated the 150th anniversary of its statehood. The U.S. Postal Service issued a stamp, the first to bear the 32-cent price, to honor the occasion. Ninety million of the commemorative stamps were issued. About how much postage did the stamps represent?
   a. Which method of computation do you think is most appropriate for this problem? Justify your choice.
   b. Solve the problem using the four-step plan. Be sure to examine your solution.

Lesson 1-2

Find the value of each expression.

1. $8 + 7 + 12 \div 4$
2. $20 \div 4 - 5 + 12$
3. $(25 \cdot 3) + (10 \cdot 3)$
4. $36 \div 6 + 7 - 6$
5. $30 \cdot (6 - 4)$
6. $(40 \cdot 2) - (6 \cdot 11)$
7. $\dfrac{86 - 11}{11 + 4}$
8. $\dfrac{12 + 84}{11 + 13}$
9. $\dfrac{5 \cdot 5 + 5}{5 \cdot 5 - 15}$
10. $(19 - 8)4$
11. $75 - 5(2 \cdot 6)$
12. $81 \div 27 \times 6 - 2$

Lesson 1-3

Evaluate each expression if $a = 2$, $b = 4$, and $c = 3$.

1. $ba - ac$
2. $4b + a \cdot a$
3. $11 \cdot c - ab$
4. $4b - (a + c)$
5. $7(a + b) - c$
6. $8a + 8b$
7. $\dfrac{8(a + b)}{4c}$
8. $36 - 12c$
9. $\dfrac{9(b + a)}{c - 1}$
10. $abc - bc$
11. $28 - bc + a$
12. $a(b - c)$

Translate each phrase into an algebraic expression.

13. nine more than a
14. eleven less than k
15. three times p
16. the product of some number and five
17. twice Shelly's score decreased by 18
18. the quotient of 16 and n

Lesson 1-4

Name the property shown by each statement.

1. $1 \cdot 4 = 4$

2. $6 + (b + 2) = (6 + b) + 2$

3. $9(6n) = (9 \cdot 6)n$

4. $8t \cdot 0 = 0 \cdot 8t$

5. $0(13n) = 0$

6. $7 + t = t + 7$

Find each sum or product mentally.

7. $6 + 8 + 14$

8. $5 \cdot 18 \cdot 2$

9. $0(13 \cdot 6)$

10. $8 + 4 + 12 + 16$

11. $8 \cdot 20 \cdot 10$

12. $4 \cdot 14 \cdot 5$

13. History Many flags have represented the United States since the first colonists arrived. The current flag, called the Stars and Stripes, was originated by the Marine Committee of the Second Continental Congress in a resolution made on June 14, 1777. The flag now has fifty stars in the field to represent the fifty states of the Union. Each row of the flag has either 5 or 6 stars in each row. Rows of 5 stars alternate with rows of 6 stars. How many rows are there with 6 stars?

Lesson 1-5

Simplify each expression.

1. $8k + 2k + 7$

2. $3 + 2b + b$

3. $t + 2t$

4. $9(3 + 2x)$

5. $4(xy + 2) - 2$

6. $(6 + 3e)4$

7. $4 + 9c + 3(c + 2)$

8. $5(7 + 2s) + 3(s + 4)$

9. $9(f + 2) + 14f$

10. Business The Columbus Dispatch newspaper can be ordered for delivery on weekdays or Sundays. A weekday paper is 35 cents and the Sunday Edition is $1.50. The Stadlers ordered delivery of the weekday papers. The month of March had 23 weekdays and April had 20. How much should the carrier charge the Stadlers for those two months?

Lesson 1-6

Identify the solution to each equation from the given list.

1. $16 - f = 7$; 5, 7, 9

2. $9 = \dfrac{72}{m}$; 8, 9, 11

3. $4b + 1 = 17$; 3, 4, 5

4. $17 + r = 25$; 6, 7, 8

5. $9 = 7n - 12$; 3, 5, 7

6. $67 = 98 - q$; 21, 26, 31

Solve each equation mentally.

7. $131 - u = 120$

8. $88 = 11d$

9. $\dfrac{84}{h} = 12$

10. $5t = 0$

11. $\dfrac{x}{2} = 8$

12. $88 + y = 96$

13. $13g = 39$

14. $23 = w + 6$

15. $9z = 45$

Lesson 1-7

Use the grid at the right to name the point for each ordered pair.

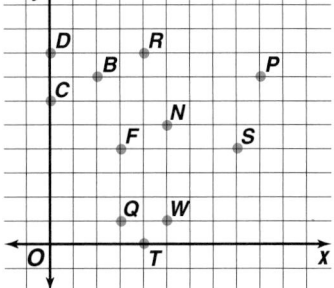

1. $(9, 7)$ **2.** $(5, 5)$

3. $(3, 1)$ **4.** $(2, 7)$

5. $(8, 4)$ **6.** $(4, 0)$

Use the grid at the right to name the ordered pair for each point.

7. R **8.** P

9. W **10.** C

11. D **12.** F

Lesson 1-8

Solve each equation using the inverse operation. Use a calculator when needed.

1. $37 = b + 22$ **2.** $6 + u = 14$ **3.** $p - 19 = 3$

4. $6x = 30$ **5.** $3v = 99$ **6.** $48c = 192$

7. $r \div 7 = 4$ **8.** $12 = \frac{z}{3}$ **9.** $6.2 = 5.8 + g$

Translate each sentence into an equation.

10. The sum of a number and 8 is 14.

11. Twelve less than a number is 50.

12. The product of a number and ten is seventy.

13. A number divided by three is nine.

Lesson 1-9

State whether each inequality is *true*, *false*, or *open*.

1. $8 > 0$ **2.** $18 \leq 18$ **3.** $10 < x$

4. $24 > 2w$ **5.** $9 < 7$ **6.** $13 \leq 55$

State whether each inequality is *true* or *false* for the given value.

7. $5 \geq 2t - 12; t = 11$ **8.** $7 + n < 25; n = 4$ **9.** $6r - 18 > 0; r = 3$

10. $3n + 2 < 26, n = 3$ **11.** $h - 19 < 13; h = 28$ **12.** $20m \geq 10; m = 0$

13. Literature William Shakespeare, who lived from 1564 to 1616, was England's best known poet, actor, and playwright. Each year several productions of Shakespeare's plays are performed around the world. Some of his plays have been made into films, including a 1993 version of *Much Ado About Nothing* and a 1990 version of *Hamlet*. There are currently more than 50 versions of *Hamlet* on film. Write an inequality for the number of versions of Hamlet.

Lesson 1-10

The table at the right shows the albums that have stayed on the U.S. music charts the longest as of December 31, 1993.

Artist/title	Weeks on chart
Carole King, *Tapestry*	302
Johnny Mathis, *Heavenly*	295
Johnny Mathis, *Johnny's Greatest Hits*	490
Pink Floyd, *Dark Side of the Moon*	741
Tennessee Ernie Ford, *Hymns*	277
Various, Original cast recording of *Camelot*	265
Various, Original cast recording of *My Fair Lady*	480
Various, Original cast recording of *The Sound of Music*	276
Various, Soundtrack of *The King and I*	277
Various, Soundtrack of *Oklahoma!*	305

1. Which album has been on the charts the greatest number of weeks?

2. How many weeks was *Tapestry* by Carole King on the charts?

3. How much longer was Johnny Mathis' *Greatest Hits* on the charts than his *Heavenly* album?

4. Make a bar graph of the data on albums.

Lesson 2-1

Simplify.

1. $|-3| + |9|$
2. $|-18| - |5|$
3. $|12 + 7|$
4. $-|6|$
5. $|-8| + |4|$
6. $-|-20|$
7. $|15 - 12|$
8. $|8 + 9|$
9. $-|4| \cdot |-5|$
10. $|-6| \cdot |8|$
11. $-|12| \cdot |9|$
12. $-||-16| + |-22||$

Lesson 2-2

Use the coordinate grid at the right to name the point for each ordered pair.

1. $(-6, 8)$
2. $(1, -2)$
3. $(9, 2)$
4. $(1, 4)$
5. $(-3, -4)$
6. $(2, 5)$

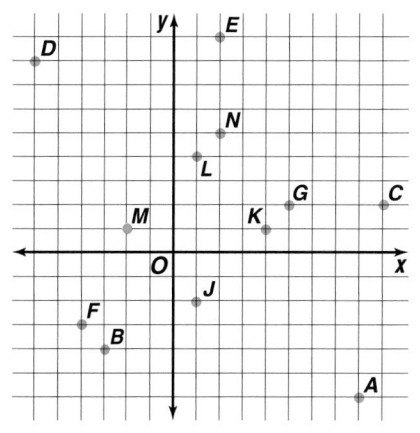

On graph paper, draw coordinate axes. Then graph and label each point. Name the quadrant in which each point is located.

7. $H(-2, -5)$
8. $P(1, 5)$
9. $R(-3, 1)$
10. $M(4, -2)$
11. $K(-4, 5)$
12. $G(3, -5)$

Lesson 2-3

Order the integers in each set from least to greatest.

1. $\{-1, 2, -5\}$
2. $\{0, -2, 8, 5, -9\}$
3. $\{100, -34, -86, 21, 0\}$
4. $\{-1, 16, -43, 8, 27, -40\}$
5. $\{0, -23, 75, -15, 24\}$
6. $\{-6, 6, -5, 18\}$

Write an inequality using the numbers in each sentence. Use the symbols < or >.

7. **World** Finland covers 130,119 square miles and is 3 times greater than Ohio which covers 40,953 square miles.

8. **Sports** Atlanta, Georgia hosted the 1996 Summer Olympics. The point with the highest altitude in the city is 1050 feet above sea level. The lowest point is 940 feet above sea level.

9. **Statistics** The oldest dog lived to be 24 years old. The oldest cat lived to be 26 years old.

10. **Statistics** The slowest animal in the world is the garden snail that moves 0.03 mph. The fastest animal is the cheetah that can run 70 mph.

11. **Astronomy** Mars is 227.9 million kilometers from the Sun. Earth is 149.6 million kilometers from the Sun.

Lesson 2-4

Solve each equation.

1. $n = 5 + (-6)$
2. $-17 + 24 = y$
3. $k = 15 + (-29)$
4. $m = -6 + 13$
5. $50 + (-14) = x$
6. $w = -21 + (-4)$
7. $30 + (-7) = t$
8. $z = (-3) + (-10)$
9. $-15 + 26 = q$
10. $b = -17 + 4 + (-2)$
11. $d = 50 + (-16) + (-11)$
12. $-17 + 8 + (-14) = f$

Lesson 2-5

Solve each equation.

1. $8 - 17 = f$
2. $-15 - 3 = r$
3. $10 - 21 = a$
4. $20 - (-5) = m$
5. $5 - (-9) = g$
6. $-12 - (-7) = v$
7. $-19 - (-6) = k$
8. $-16 - (-23) = b$
9. $-56 - 32 = z$
10. $-49 - (-52) = h$
11. $-6 - 9 - (-7) = d$
12. $6 - (-10) - 7 = n$

Lesson 2-6

Solve. Look for a pattern.

1. Find the next two integers in the pattern 2, 3, 5, 9, 17, ? , ? .

2. If FOG = $6 - 15 - 7$ and LOG = $12 - 15 - 7$; what does JOG equal by the same logic?

3. The digits 0 to 9 have been used to make the multiplication problem below. Find the missing numbers that make the multiplication correct.

$$\begin{array}{r} ?\,0\,2 \\ \times\quad ?\,9 \\ \hline 1\,?\,?\,?\,8 \end{array}$$

Lesson 2-7

Solve each equation.

1. $n = -4(2)$
2. $-8(-5) = h$
3. $r = 13(-4)$
4. $-5 \cdot 6 \cdot 10 = c$
5. $-6(-2)(-14) = w$
6. $u = 18(-3)(6)$

Evaluate each expression.

7. $-6t$, if $t = 15$
8. $7p$, if $p = -9$
9. $-4k$, if $k = -16$
10. aw, if $a = 0$ and $w = -72$
11. dk, if $d = -12$ and $k = 11$
12. st, if $s = -8$ and $t = -10$
13. $3hp$, if $h = 9$ and $p = -3$
14. $-5bc$, if $b = -6$ and $c = 2$
15. $-4wx$, if $w = -1$ and $x = -8$

Lesson 2-8

Divide.

1. $-36 \div 9$
2. $112 \div (-8)$
3. $-72 \div 2$
4. $-26 \div (-13)$
5. $-144 \div 6$
6. $-180 \div (-10)$
7. $304 \div (-8)$
8. $-216 \div (-9)$
9. $80 \div (-5)$
10. $-105 \div 15$
11. $120 \div (-30)$
12. $-200 \div (-8)$

Lesson 3-1

Solve by eliminating possibilities.

1. Lindsay, Lee, Ann, and Marcos formed a study group. Each one has a favorite subject that is different from the others. The subjects are art, math, music, and physics. Use the following information to match each person with his or her favorite subject.

   A. Lindsay likes subjects where she can use her calculator.
   B. Lee does not like music or physics.
   C. Ann and Marcos prefer classes in the cultural arts.
   D. Marcos plans to be a professional cartoonist.

2. Five friends decided to run in a 5-kilometer run to benefit an education association. Nancy beat Pete. Jeff was not last. Gina was beaten by Tom and Jeff. Jeff crossed the finish line just after Tom did. Tom lost to Pete. Choose the correct order in which the friends finished the race.

   a. Nancy, Pete, Gina, Tom, Jeff
   b. Tom, Jeff, Nancy, Pete, Gina
   c. Nancy, Pete, Tom, Jeff, Gina
   d. Nancy, Gina, Jeff, Tom, Pete

Lesson 3-2

Solve each equation and check your solution. Then graph the solution on a number line.

1. $y + 49 = 26$
2. $d - (-31) = -24$
3. $q - 8 = 16$
4. $x - 16 = 32$
5. $40 = a + 12$
6. $b + 12 = -1$
7. $21 = u - (-6)$
8. $-52 = p + 5$
9. $-14 = 5 - g$
10. $121 = k + (-12)$
11. $-234 = m - 94$
12. $110 = x + 25$
13. $f - 7 = 84$
14. $y - 864 = -652$
15. $475 + z = -18$

Lesson 3-3

Solve each equation and check your solution. Then graph the solution on a number line.

1. $-y = -32$

2. $7r = -56$

3. $\dfrac{t}{-3} = 12$

4. $4 = \dfrac{s}{-14}$

5. $\dfrac{b}{47} = -2$

6. $64 = -4n$

7. $-144 = 12q$

8. $\dfrac{r}{11} = -132$

9. $-5g = -385$

10. $-16x = -176$

11. $-21 = \dfrac{y}{-4}$

12. $-372 = 31k$

13. $84 = \dfrac{k}{5}$

14. $-b = 19$

15. $\dfrac{y}{112} = -9$

Lesson 3-4

Solve by replacing the variables with the given values.

1. $A = \pi \cdot r^2$, if $\pi = 3.14$ and $r = 5$

2. $\dfrac{5}{9}(F - 32) = C$, if $F = 86$

3. $d = r \cdot t$, if $d = 366$ and $t = 3$

4. $S = (n - 2) \cdot 180$, if $n = 8$

5. $A = \dfrac{1}{2}bh$, if $A = 36$ and $h = 12$

6. $P = 4s$, if $P = 108$

7. $V = \dfrac{1}{3}(B \cdot H)$, if $B = 27$ and $H = 5$

8. $h = 69 + 2.2F$, if $F = 42$

Lesson 3-5

Find the perimeter and area of each rectangle.

1. a rectangle 23 centimeters long and 9 centimeters wide
2. a 16-foot by 14-foot rectangle
3. a rectangle with a length of 31 meters and a width of 3 meters
4. a square with sides 7.5 meters long

Find the missing dimension of each rectangle.

	Length	Width	Area	Perimeter
5.	9 ft		126 ft²	46 ft
6.		18 in.	108 in²	48 in.
7.	13 yd		273 yd²	68 yd
8.		12 cm	168 cm²	52 cm
9.		3 m	162 m²	114 m
10.	18 mi		720 mi²	116 mi

Lesson 3-6

Solve each inequality and check your solution. Then graph the solution on a number line.

1. $m + 9 < 14$

2. $k + (-5) < -12$

3. $-15 < v - 1$

4. $-7 + f \geq 47$

5. $r > -15 - 8$

6. $18 \geq s - (-4)$

7. $38 < r - (-6)$

8. $z - 9 \leq -11$

9. $-16 + c \geq 1$

10. $24 < a + -3$

11. $-52 \geq d + (-6)$

12. $24 < -2 + b$

13. $n - (-17) \geq 12$

14. $31 < x - 24$

15. $-20 \leq n + (-3)$

16. $p + (-11) < -37$

17. $-40 \leq -72 + w$

18. $72 > a + 88$

Lesson 3-7

Solve each inequality and check your solution. Then graph the solution on a number line.

1. $6p < 78$

2. $\frac{m}{-3} > 24$

3. $-18 < 3b$

4. $-5k \geq 125$

5. $-75 > \frac{a}{5}$

6. $\frac{w}{6} < -5$

7. $\frac{y}{-13} > -20$

8. $14t < 266$

9. $\frac{g}{-25} \geq 8$

10. $-216 \leq 9h$

11. $\frac{g}{-9} < -8$

12. $-18d > 108$

13. $2268 < -63a$

14. $52 \leq \frac{p}{4}$

15. $42n \leq -210$

16. $-60 > -4d$

17. $\frac{k}{-3} \geq -21$

18. $\frac{v}{-8} < 0$

Lesson 3-8

Define a variable and translate each sentence into an equation or inequality. Then solve.

1. Three times a number is equal to thirty-six.

2. The sum of a number and 5 is less than 12.

3. **Sports**   In the 1993–1994 professional basketball season, David Robinson was the leading scorer. Shaquille O'Neal had 6 points less than David Robinson. If O'Neal had 2377 points for the season, how many points did David Robinson get?

4. **College**   The enrollment at the Columbus campus of The Ohio State University is more than six times as large as the enrollment at Stanford University. If there are 38,958 students enrolled at Ohio State, how many students are currently enrolled at Stanford?

Lesson 4-1

Using divisibility rules, state whether each number is divisible by 2, 3, 5, 6, or 10.

1. 98   **2.** 243   **3.** 800

4. 252   **5.** 105   **6.** 210

7. 1001   **8.** 2016   **9.** 2475

Determine whether each expression is a monomial. Explain why or why not.

10. $-6h$ **11.** $9 - v$ **12.** $4g^3$

Lesson 4-2

Write each multiplication expression using exponents.

1. $(-6)(-6)(-6)(-6)(-6)$

2. $(y \cdot y \cdot y) \cdot (y \cdot y \cdot y \cdot y)$

3. 9

4. $3q \cdot 3q \cdot 3q \cdot 3q \cdot 3q \cdot 3q$

5. $\underbrace{n \cdot n \cdot n \cdot \ldots \cdot n}_{17 \ factors}$

6. $8 \cdot 8 \cdot 8 \cdot 8$

Write each power as a multiplication expression of the same factor.

7. 13^3 **8.** $(-4)^5$

9. k^9 **10.** $(2 - w)^2$

11. $(3m)^{15}$ **12.** $(-x)^4$

Lesson 4-3

Solve. Use a diagram.

1. Four people are eligible to be officers in the National Honor Society. The positions available are President, Vice President, Treasurer, and Secretary. How many different ways can the offices be filled?

2. Copy the magic square at the right. Use the numbers 0, 4, 8, and 12 to make all the rows, columns, and diagonals add up to 24.

3. Beverly and Pam are painting a fence. For every 3 sections Pam paints, Beverly paints 5. If Beverly paints 20 sections in 2 hours how long will it take Pam to paint 18 sections?

4. Determine how many games have to be played to determine a champion from 32 teams in a single elimination softball tournament.

0		12	0
4			4
			12

Lesson 4-4

Determine whether each number is prime or composite.

1. 59   2. 369   3. 116

Factor each number or monomial completely.

4. 40   5. $630a$ 6. 187
7. 310   8. 510   9. 1589
10. $-18ab^2$ 11. $-117x^3$ 12. $435j^2k^5$

Lesson 4-5

Find the GCF of each set of numbers or monomials.

1. 112, 216   2. 120, 245   3. $84k, 108k^2$
4. $135ab, -171b$ 5. $185fg, 74f^2g$ 6. $44m, 60n$
7. $90gh, 225k$ 8. $8, -28h$ 9. $-16w, -28w^3$
10. $24a, 30ab, 66a^2$ 11. $-13z, 39yz, 52y$ 12. $60a^3b, 150a^2b^2, 36a^2b$

Lesson 4-6

Write each fraction in simplest form. If the fraction is already in simplest form, write *simplified*.

1. $\frac{3}{54}$ 2. $\frac{3}{16}$ 3. $\frac{6}{58}$
4. $\frac{15}{55}$ 5. $\frac{10}{90}$ 6. $\frac{20}{49}$
7. $\frac{8}{20}$ 8. $\frac{99}{9}$ 9. $\frac{18}{54}$
10. $\frac{21}{64}$ 11. $\frac{40}{76}$ 12. $\frac{49}{56}$

Lesson 4-7

Find the least common multiple (LCM) of each set of numbers or algebraic expressions.

1. 30, 18   2. $3m, 12$ 3. $6a, 17a^5$ 4. 2, 5, 7

Find the least common denominator (LCD) for each pair of fractions.

5. $\frac{2}{5}, \frac{6}{25}$ 6. $\frac{3}{12}, \frac{4}{5}$ 7. $\frac{4}{6}, \frac{7}{9}$ 8. $\frac{1}{4}, \frac{5}{6}, \frac{2}{9}$

Replace each ● with < or > to make a true statement.

9. $\frac{5}{11}$ ● $\frac{6}{13}$ 10. $\frac{15}{34}$ ● $\frac{4}{8}$ 11. $\frac{8}{18}$ ● $\frac{5}{16}$ 12. $\frac{3}{14}$ ● $\frac{5}{20}$

Lesson 4-8

Find each product or quotient. Express your answer in exponential form.

1. $r^4 \cdot r^2$

2. $\dfrac{2^9}{2^3}$

3. $\dfrac{b^{18}}{b^5}$

4. $12^3 \cdot 12^8$

5. $x \cdot x^9$

6. $(2s^6)(4s^2)$

7. $w^3 \cdot w^4 \cdot w^2$

8. $(-2)^2(-2)^5(-2)$

9. $\dfrac{4^7}{4^6}$

10. $3(f^{17})(f^2)$

11. $(5k)^2 \cdot k^7$

12. $\dfrac{6m^8}{3m^2}$

Find each missing exponent.

13. $(2^\bullet)(2^2) = 2^6$

14. $y(y^3)(y^4) = y^\bullet$

15. $\dfrac{5^7}{5^3} = 5^\bullet$

16. $12^9 \cdot 12^\bullet = 12^{14}$

17. $\dfrac{b^{10}}{b^\bullet} = b$

18. $\dfrac{2^7}{8^2} = 2^\bullet$

Lesson 4-9

Write each expression using positive exponents.

1. y^{-9}

2. $3m^{-4}$

3. $5^{-3} \cdot (-a)$

4. $\left(\dfrac{2}{x}\right)^{-7}$

Write each fraction as an expression using negative exponents.

5. $\dfrac{1}{p^4}$

6. $\dfrac{6a}{b^9}$

7. $\dfrac{2}{27}$

8. $\dfrac{17}{a}$

Evaluate each expression.

9. 5^n if $n = -2$

10. $(2a^{-2}b)^2$ if $a = 3$ and $b = 6$

11. $15n^{-3}$ if $n = 5$

12. $9x^{-6}$ if $x = -3$

Find each product or quotient. Express using positive exponents.

13. $(x^8)(x^{-2})$

14. $(g^{-12})(g^9)$

15. $\dfrac{r^{14}}{r^{-2}}$

16. $\dfrac{y^7}{y^{10}}$

Lesson 5-1

Name the set(s) of numbers to which each number belongs.

1. $\dfrac{7}{10}$

2. 41

3. -19

Replace each ● with $<$, $>$, or $=$ to make each sentence true. Use a number line if necessary.

4. $\dfrac{3}{4} \bullet \dfrac{2}{5}$

5. $\dfrac{-13}{25} \bullet \dfrac{-3}{5}$

6. $\dfrac{9}{10} \bullet \dfrac{7}{8}$

7. $5\dfrac{2}{9} \bullet \dfrac{47}{9}$

8. $\dfrac{5}{26} \bullet \dfrac{4}{13}$

9. $\dfrac{-11}{4} \bullet -2\dfrac{1}{2}$

Express each decimal as a fraction or mixed number in simplest form.

10. 0.38

11. 2.346

12. $-0.\overline{4}$

Lesson 5-2

Round to the nearest whole number.

1. 6.5   **2.** 5.193   **3.** 42.09

Round each fraction to 0, $\frac{1}{2}$, or 1.

4. $\frac{9}{10}$ **5.** $\frac{1}{14}$ **6.** $\frac{89}{101}$

Estimate each sum or difference.

7. $9.9 - 3.2$ **8.** $16.2 + 9.31$ **9.** $1.3 + 2.8 - 3.4$

10. $8\frac{3}{5} - 2\frac{1}{6}$ **11.** $19\frac{4}{10} + 13\frac{8}{11}$ **12.** $863\frac{1}{12} - 241\frac{189}{221}$

Lesson 5-3

Solve each equation.

1. $b = 5.8 + 9.3$ **2.** $s = 12.4 - 4.52$

3. $-4.9 + 8.4 = k$ **4.** $-5.2 - 7.8 = q$

5. $p = 14.8 - 29.46$ **6.** $-4.25 + 11.2 = t$

7. $21.4 - 9.2 = z$ **8.** $45.26 - (-6.1) = y$

9. $x = -9.27 - 8.1$ **10.** $-28.94 + 3.48 = u$

Lesson 5-4

Solve each equation. Write the solution in simplest form.

1. $\frac{2}{7} + \frac{3}{7} = g$ **2.** $\frac{8}{15} - \frac{4}{15} = y$ **3.** $\frac{3}{7} + \frac{4}{7} = w$

4. $\frac{5}{6} - \frac{1}{6} = t$ **5.** $\frac{7}{12} - \frac{5}{12} = r$ **6.** $\frac{5}{12} + \frac{11}{12} = w$

Simplify each expression.

7. $12\frac{7}{8}s - 7\frac{3}{8}s + 2\frac{5}{8}s$ **8.** $-6\frac{4}{9}t - \left(-4\frac{5}{9}t\right) + 3\frac{2}{9}t$

9. $6\frac{1}{4}g + \left(-6\frac{3}{4}g\right)$ **10.** $7\frac{2}{5}n - \left(-4\frac{2}{5}n\right)$

Lesson 5-5

Solve each equation. Write each solution in simplest form.

1. $\frac{1}{5} + \frac{2}{7} = d$ **2.** $a = \frac{4}{5} + \frac{7}{9}$ **3.** $\frac{1}{9} - \frac{7}{12} = n$

4. $z = \frac{8}{11} - \frac{4}{5}$ **5.** $\frac{7}{12} - \left(\frac{-4}{11}\right) = y$ **6.** $\ell = -\frac{9}{14} + \frac{15}{16}$

7. $3\frac{2}{5} + 2\frac{4}{7} = k$ **8.** $r = -4\frac{1}{8} + 2\frac{5}{9}$ **9.** $-3\frac{3}{7} - 5\frac{1}{14} = g$

Lesson 5-6

Solve each equation. Check your solution.

1. $a - 4.86 = 7.2$
2. $n + 6.98 = 10.3$
3. $87.64 = f - (-8.5)$
4. $x - \dfrac{2}{5} = -\dfrac{8}{15}$
5. $3\dfrac{3}{4} + m = 6\dfrac{5}{8}$
6. $4\dfrac{1}{6} = r + 6\dfrac{1}{4}$
7. $7\dfrac{1}{3} = c - \dfrac{4}{5}$
8. $-4.62 = h + (-9.4)$
9. $w - 1\dfrac{1}{5} = \dfrac{2}{9}$

Lesson 5-7

Solve each inequality and check your solution. Graph the solution on a number line.

1. $h + 5.7 > 21.3$
2. $78.26 \le v - (-65.854)$
3. $\dfrac{2}{3} \le a - \dfrac{5}{6}$
4. $-13.2 > w - 4.87$
5. $a + \dfrac{5}{12} \ge \dfrac{7}{18}$
6. $7\dfrac{1}{2} < n - \left(-\dfrac{7}{8}\right)$
7. $t - 8.5 > -4.2$
8. $-7.42 \le d - 5.9$
9. $m - (-18.4) < -17.6$
10. $s - \dfrac{2}{3} \ge 9\dfrac{4}{5}$

Lesson 5-8

State whether each is an example of *inductive* or *deductive* reasoning. Explain your answer.

1. Every day that Felicia takes off from work it rains. She is going to take her birthday off, so she thinks it will rain on her birthday.
2. The local radio station calls monthly with a survey. There is less than a week left in the month. They will be calling this week.
3. If you have a C or better in all of your classes, you are eligible to play sports. Hank has an A or a B in all of his classes. Hank is eligible to play sports.
4. The credit card company charges finance charges if the customer doesn't pay the balance in full. Carla could only afford to pay half of her bill this month. Carla will be charged a finance charge.

Lesson 5-9

State whether each sequence is arithmetic. Then write the next three terms of each sequence.

1. $3.5, 4.3, 5.1, \ldots$
2. $5, 10, 20, \ldots$
3. $\dfrac{1}{2}, \dfrac{5}{6}, 1\dfrac{1}{6}, \ldots$
4. $\dfrac{1}{4}, \dfrac{1}{2}, 1, 2, \ldots$
5. $23, 18, 13, \ldots$
6. $45, 43, 39, 33, \ldots$

Lesson 6-1

Write each fraction as a decimal. Use a bar to show a repeating decimal.

1. $\frac{6}{10}$ 2. $-4\frac{7}{12}$ 3. $\frac{8}{11}$ 4. $3\frac{4}{18}$

5. $-\frac{3}{16}$ 6. $8\frac{36}{44}$ 7. $\frac{9}{37}$ 8. $\frac{6}{15}$

Replace each ● with <, >, or = to make a true sentence.

9. $\frac{7}{8}$ ● $\frac{5}{6}$ 10. 0.04 ● $\frac{5}{9}$ 11. $\frac{1}{3}$ ● $\frac{2}{7}$ 12. $\frac{3}{5}$ ● $\frac{12}{20}$

13. $\frac{1}{2}$ ● 0.75 14. 0.3 ● $\frac{1}{3}$ 15. $\frac{2}{3}$ ● 0.64 16. $\frac{2}{20}$ ● 0.10

Lesson 6-2

Estimate each product or quotient.

1. 16.38×1.5 2. $35.54 \div 4.1$ 3. $6\frac{4}{9} \cdot 7.09$ 4. $18.24 \cdot 3.25$

5. $\frac{6}{13} \times 150$ 6. $\left(\frac{1}{4}\right)(15)$ 7. $78 \div 1\frac{11}{12}$ 8. $1\frac{7}{8} \cdot 40$

9. $\frac{1}{3} \cdot 37$ 10. $75 \div 1\frac{7}{16}$ 11. $88 \div \frac{3}{8}$ 12. $71.99 \div 5.7$

13. Estimate $\frac{4}{9}$ times 20.

14. Estimate the quotient of 65.46 and 5.6.

15. Estimate 32 times $5.49.

16. Estimate the quotient of 45 and $\frac{6}{13}$.

Lesson 6-3

Solve each equation. Write each solution in simplest form.

1. $d = \frac{2}{5} \cdot \frac{3}{16}$ 2. $u = 3\frac{1}{4} \cdot \frac{2}{11}$ 3. $\left(\frac{3}{5}\right)\left(-\frac{5}{12}\right) = g$ 4. $t = \left(\frac{4}{5}\right)^3$

5. $s = 2\frac{2}{6} \cdot 6\frac{2}{7}$ 6. $2\left(-\frac{7}{12}\right) = r$ 7. $1\frac{3}{7} \cdot \left(-9\frac{4}{5}\right) = c$ 8. $h = \left(-\frac{6}{7}\right)^2$

Evaluate each expression if $r = -\frac{1}{5}$, $s = \frac{2}{3}$, $x = 1\frac{1}{4}$, and $y = -2\frac{1}{8}$.

9. rx 10. $5r^2$ 11. $s(x + y)$

12. $8y + 12x$ 13. $x^2(s + 2)$ 14. $-x(x - 2s)$

Lesson 6-4

Name the multiplicative inverse for each rational number.

1. $-\dfrac{5}{9}$

2. $5\dfrac{3}{8}$

3. 0.7

4. 2.35

5. -18

6. $\dfrac{a}{b}$

Estimate the solution to each equation. Then solve. Write the solution in simplest form.

7. $w = \dfrac{3}{4} \div \dfrac{15}{16}$

8. $16 \div 1\dfrac{7}{8} = m$

9. $q = 2\dfrac{1}{6} \div 1\dfrac{1}{5}$

10. $y = -11 \div 3\dfrac{1}{7}$

11. $a = \dfrac{8}{45} \div \dfrac{10}{27}$

12. $220 \div \left(-5\dfrac{1}{2}\right) = p$

Lesson 6-5

Solve each equation.

1. $7.3017 \div 0.57 = a$

2. $13.42 \div 67.1 = d$

3. $x = 80 \div (-3.2)$

4. $m = -2.016 \div (-0.13)$

5. $3.8 \cdot 2.9 = k$

6. $85 \cdot 0.07 = w$

7. $r = 15.32(0.0015)$

8. $(16.2)(0.013) = b$

9. $c = 5.4 \cdot 9.7$

Evaluate each expression.

10. $4y^3$, if $y = 0.6$

11. xy, if $x = 0.348$ and $y = -6.4$

12. $\dfrac{3a}{w}$, if $a = 0.4$ and $w = 2$

Lesson 6-6

Find the mean, median, and mode for each set of data. When necessary, round to the nearest tenth.

1. 82, 79, 93, 91, 95

2. 88, 85, 76, 94, 85, 97

3. 0.57, 12.81, 12.6, 0.96, 6.1, 14.3, 4.1, 12.81, 0.96

Use the table to answer Exercises 4–6.

4. What is the mean percent of students who plan to attend college?

5. What is the median percent of students who plan to attend tech school?

6. What is the mode percent of students who plan to start a career after high school?

Percent of Independence High School Seniors with After-High School Plans			
Year	College	Tech School	Career
1990	67.7	23.9	11.5
1991	68.2	20.3	11.5
1992	71.1	22.7	6.2
1993	70.4	24.1	5.5
1994	63.9	20.5	10.2
1995	67.4	22.2	10.4

Lesson 6-7

Solve each equation or inequality. Check your solution.

1. $3.5a = 7$

2. $0.8 = -0.8b$

3. $8 < \frac{2}{3}c$

4. $\frac{m}{13} \geq 0.5$

5. $-9 = \frac{3}{4}g$

6. $0.4y > -2$

7. $-\frac{1}{2}d \leq -5\frac{1}{2}$

8. $-3.5 = 0.07z$

9. $-\frac{1}{6}s = 15$

Solve each equation or inequality and graph the solution on a number line.

10. $\frac{2}{7}t < 4$

11. $8.37 = 2.7d$

12. $\frac{1}{5}m \geq 4\frac{3}{5}$

Lesson 6-8

State whether each sequence is a geometric sequence. If so, state the common ratio and list the next three terms.

1. 2, 4, 8, 16, . . .

2. 125, 75, 45, . . .

3. 100, 75, 50, . . .

4. $\frac{1}{5}$, 1, 5, 25, . . .

5. 2401, 49, 7, . . .

6. $-\frac{4}{5}$, 2, -5, $12\frac{1}{2}$, . . .

7. Write the first five terms in a geometric sequence with a common ratio of 3. The first term is -4.

8. Write the first five terms of a geometric sequence if $a = -12$ and $r = \frac{1}{2}$.

9. In a certain geometric sequence, $a = 1.8$ and $r = -3$. Write the first five terms of the sequence.

10. Use the expression $ar^{(n-1)}$ to find the eighth term in the geometric sequence -8, 12, -18, ...

Lesson 6-9

Write each number in scientific notation.

1. 6,184,000

2. 27,210,000

3. 0.00004637

4. 0.00546

5. 500,300,100

6. 0.00000321

Write each number in standard form.

7. 9.562×10^{-3}

8. 8.2453×10^{-7}

9. 8.2×10^4

10. 9.102040×10^2

11. 2.41023×10^6

12. 4.21×10^{-5}

Lesson 7-1

Solve by working backward.

1. Doug brought some milk bottle caps to Antonio's house to trade. Doug traded Antonio a third of his caps in exchange for 2 designer caps. Then Doug gave Antonio's little sister Maria 6 caps as a gift. If Doug left Antonio's house with 42 milk caps, how many caps did he bring to Antonio's house?

2. Kwan had some cookies. She gave a fourth of her cookies to Justin. Justin then gave half of his cookies to Simon. Simon gave a third of his cookies to Sandy. If Sandy has 3 cookies, how many cookies did Kwan have in the beginning?

3. A certain bacteria doubles its population every 12 hours. After 3 full days, there are 1600 bacteria in a culture. How many bacteria were there at the beginning of the first day?

4. Sam and Akita are playing a game. On five turns, Sam's piece advanced 5 spaces, moved back 4 spaces, advanced 2 spaces, moved back 8 spaces, and advanced two spaces. How far did Sam's piece move from its original place on the board in these turns?

Lesson 7-2

Solve each equation. Check your solution.

1. $3t - 13 = 2$
2. $-8j - 7 = 57$
3. $9d - 5 = 4$
4. $6 - 3w = -27$
5. $\frac{k}{6} + 8 = 12$
6. $-4 = \frac{q}{8} - 19$
7. $15 - \frac{n}{7} = 13$
8. $7.25 = 3r - 6.25$
9. $21.63 - h = -32.7$
10. $-19 = 11b - (-3)$
11. $6 = 20 + \frac{x}{3}$
12. $8.12 + 3a = -3.25$

Lesson 7-3

Define a variable and write an equation for each situation. Then solve.

1. The sum of 29 and 3 times a number is 44. What is the number?

2. The opposite of twice a number less 3 is 17. What is the number?

3. Find five consecutive numbers whose sum is 95.

4. The lengths of the sides of a quadrilateral are consecutive even integers. The perimeter of the quadrilateral is 108 yards. What are the lengths of the sides of the quadrilateral?

Lesson 7-4

Copy and complete the table below by finding the radius, diameter, or circumference of the circle using the information provided.

	Radius	Diameter	Circumference
1.	5 cm		
3.		14 in.	
5.			≈ 56.52 ft
7.		17 m	
9.	6 yds		
11.	20 mm		

	Radius	Diameter	Circumference
2.		25 mi	
4.			≈ 65.94 mi
6.	8 in.		
8.			≈ 100.48 ft
10.		22 cm	
12.			≈ 81.64 in.

Lesson 7-5

Solve each equation. Check your solution.

1. $-7h - 5 = 4 - 4h$

2. $5t - 8 = 3t + 12$

3. $w + 6 = 2(w - 6)$

4. $m + 2m + 1 = 7$

5. $3.21 - 7y = 10y - 1.89$

6. $3(b + 1) = 4b - 1$

7. $\frac{5}{9}g + 8 = \frac{1}{6}g + 1$

8. $\frac{s - 3}{7} = \frac{s + 5}{9}$

Lesson 7-6

Solve each inequality and check your solution. Graph the solution on a number line.

1. $2m + 1 < 9$

2. $-3k - 4 \leq -22$

3. $-2 > 10 - 2x$

4. $-6a + 2 \geq 14$

5. $3y + 2 < -7$

6. $\frac{d}{4} + 3 \geq -11$

7. $\frac{x}{3} - 5 < 6$

8. $-5g + 6 < 3g + 20$

9. $-3(m - 2) > 12$

10. $\frac{r}{5} - 6 \leq 3$

11. $\frac{3(n + 1)}{7} \geq \frac{n + 4}{5}$

12. $\frac{n + 10}{-3} \leq 6$

Lesson 7-7

Define a variable and write an inequality for each situation. Then solve.

1. If 8 times a number is decreased by 2, the result is less than 15. What is the number?

2. George plans to spend at most $40 for shirts and ties. He bought 2 shirts for $13.95 each. How much can he spend for ties?

3. Mia is buying a boat. She can make a down payment of $2,200. She wants to pay off the boat in 3 years. She cannot afford monthly payments over $350. What is the most she can pay for the boat?

Lesson 7-8

Complete each sentence.

1. 8.2 mm = __?__ cm
2. 6.7 km = __?__ m
3. 8.4 kg = __?__ mg
4. 18 cm = __?__ m
5. 250 ml = __?__ L
6. 4 g = __?__ mg

Write which metric unit you would probably use to measure each item.

7. length of a car
8. weight of a boy
9. capacity of a can of motor oil
10. weight of a container of deodorant

Lesson 8-1

Write the domain and range of each relation. Then determine whether each relation is a function.

1. {(3, 6),(35, 64),(1, 1),(21, 7)}
2. {(32, 24),(27, 24),(36, 24),(45, 24)}
3. {(2, 9),(3, 18),(4, 27),(2, 36)}
4. $\left\{\left(\frac{1}{2}, 3\right),\left(\frac{1}{4}, 5\right),\left(\frac{1}{6}, 7\right),\left(\frac{1}{8}, 9\right),\left(\frac{1}{10}, 11\right)\right\}$
5. {(1, 0),(1, 9),(1, 18)}
6. {(5, 5),(6, 6),(7, 7),(8, 7)}

7.

x	8	15	22	29
y	8	8	51	22

8.

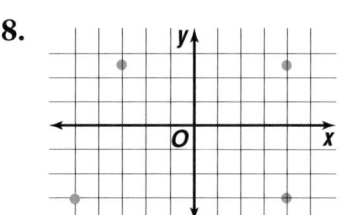

Lesson 8-2

Determine whether a scatter plot of the data for the following might show a *positive*, a *negative*, or *no* relationship. Explain your answer.

1. Size of the bag, number of potato chips
2. Number of team members, amount of playing time
3. Number of flowers in a garden, number of bees
4. Hair color, number of dates
5. Candy consumed, number of cavities
6. Number of meals, weight
7. Number of classes, hours studied

Lesson 8-3

Find four solutions for each equation. Write your solutions as ordered pairs.

1. $x = 4$
2. $y = 0$
3. $x + y = 2$
4. $y = 2x - 6$
5. $x - y = 5$
6. $3x - y = 8$
7. $y = \frac{1}{2}x - 3$
8. $y = \frac{1}{3}x + 1$
9. $2x + y = -2$
10. $2x + 3y = 12$
11. $x + 2y = -4$
12. $2x - 4y = 8$

Lesson 8-4

Determine whether each equation is a function.

1. $y = 2x - 3$
2. $x = 17$
3. $y = |x| - 3$
4. $x + |y| = 2$

Given $f(x) = 3x + 2$ and $g(x) = 2x^2 - x$, determine each value.

5. $f(3)$
6. $g(-5)$
7. $g\left(\frac{1}{2}\right)$
8. $f\left(\frac{2}{3}\right)$
9. $g(-1)$
10. $g(2.5)$
11. $f(-2)$
12. $2[f(3)]$

Lesson 8-5

Use a graph to solve each problem. Assume that the rate is constant in each problem.

1. **Climate** The average high temperature in Vancouver, Canada in January is 41°. In July, the average high temperature is 74°. What would the average high temperature be in April?

2. **Economy** A 23 minute phone call from Columbus, Ohio to Dayton, Ohio costs $1.65. A similar phone call lasting 18 minutes costs $1.30. How much would a 20 minute phone call cost?

3. **Measurement** A distance of 5 miles corresponds to the distance of 8 kilometers. A distance of 30 miles corresponds to the distance of 48.2 kilometers. If you traveled 50 miles, how many kilometers would you have traveled.

Lesson 8-6

Find the slope of the line that contains each pair of points.

1. $P(3, 8), Q(4, -3)$
2. $D(4, 5), E(-3, -9)$
3. $L(-1, 2), M(0, 5)$
4. $J(6, 2), K(6, -4)$
5. $B(8, -3), C(-4, 1)$
6. $D(1, 5), E(3, 10)$
7. $H(7, 2), I(-2, -2)$
8. $K(2, -4), L(5, -19)$
9. $G(5, 6), H(7, 6)$
10. $A(-6, -3), B(-9, 4)$
11. $P(-1, -6), Q(-5, -10)$
12. $B(5, 9), C(-4, -5)$

Lesson 8-7

Use the *x*-intercept and the *y*-intercept to graph each equation.

1. $2x + y = 6$
2. $-4x + y = 8$
3. $3x - 3y = -12$
4. $x + 2y = -4$
5. $y = -\frac{1}{2}x - 6$
6. $y = \frac{5}{2}x - 1$

Graph each equation using the slope and *y*-intercept.

7. $y = 3x - 2$
8. $x - 3y = 9$
9. $y = \frac{1}{2}x + 4$
10. $y = -\frac{2}{3}x - 1$
11. $x - y = -4$
12. $2x + 4y = -4$

Lesson 8-8

Use a graph to solve each system of equations.

1. $x + 2y = 6$
   $y = -0.5x + 3$
2. $y = -2$
   $4x + 3y = 2$
3. $2x = 3y$
   $y = 4 + \frac{2}{3}x$
4. $y = x - 2$
   $y = -\frac{1}{3}x + 2$
5. $x + y = 4$
   $y = 2$
6. $2x + y = 8$
   $2x - y = 0$
7. $x = 3$
   $y = 4$
8. $y = x + 2$
   $2y + x = 1$

Lesson 8-9

Graph each inequality.

1. $y > 2x - 2$
2. $y \geq x$
3. $y < 1$
4. $x + y \leq -1$
5. $y + 3x \leq 0$
6. $x < -3$
7. $2x + 3y \geq 12$
8. $-2x + y > -1$

Lesson 9-1

Express each ratio or rate as a fraction in simplest form.

1. 15 out of 240
2. 140:12
3. $\frac{98}{14}$
4. 6 pens to 14 pens
5. 30 cats to 6 cats
6. 18 out of 45
7. 321 to 96
8. 3 cups to 3 quarts

Express each ratio as a unit rate.

9. 343.8 miles on 9 gallons
10. $7.95 for 5 pounds.
11. $52 for 8 tickets
12. $43.92 for 4 CDs

Lesson 9-2

Solve by making a table.

1. Susan Lee has $1.15 made up of six United States coins. However, she cannot make change for a dollar, a half-dollar, a dime, or a nickel. What six coins does Susan have?

2. How many ways can you add eight prime numbers to get a sum of 20? You may use a number more than once.

3. Anne Zody buys a car for $100, sells it for $110, buys it back for $120, and sells it again for $130. How much does Ms. Zody make or lose?

4. Dwight bought the books he needed for the fall semester for $230. He dropped a class and sold the book back for $18. He added a new class and bought a book for $35. At the end of the semester, he sold his books back for $62. How much did Dwight make or lose?

5. A new computer company begins with 48 employees. After the first year, 8 employees leave for more stable jobs, 2 are transferred to other offices, and 17 new employees are hired. Over the next five years the company's profits increase, resulting in the hiring of 21 more people with the loss of only 3 employees. How many employees are employed by the company after 6 years?

Lesson 9-3

There are 4 blue marbles, 5 red marbles, and 3 green marbles in a bag. Suppose you select one marble at random. Find each probability.

1. $P(\text{green})$ **2.** $P(\text{blue})$ **3.** $P(\text{red})$

4. $P(\text{not green})$ **5.** $P(\text{white})$ **6.** $P(\text{blue or red})$

7. $P(\text{neither red nor green})$ **8.** $P(\text{not orange})$

Lesson 9-4

Solve each proportion.

1. $\dfrac{7}{k} = \dfrac{49}{63}$

2. $\dfrac{s}{4.8} = \dfrac{30.6}{28.8}$

3. $\dfrac{6}{11} = \dfrac{19.2}{g}$

4. $\dfrac{8}{13} = \dfrac{b}{65}$

Write a proportion that could be used to solve for each variable. Then solve the proportion.

5. 6 plums at $1
10 plums at d

6. 8 gallons at $9.36
f gallons at $17.55

7. 3 packages at $53.67
7 packages at m

8. 10 cards at $7.50
p cards at $18

Lesson 9-5

Use the percent proportion to solve each problem.

1. What is 81% of 134?

2. 52.08 is 21% of what number?

3. 11.18 is what percent of 86?

4. What is 120% of 312?

5. 140 is what percent of 400?

6. 430.2 is 60% of what number?

Lesson 9-6

William Wonker wants to find out favorite candy bars. He surveyed 100 people at a candy store.

1. Give some reasons why this is a valid sample.

2. Give some reasons why this is not a valid sample.

3. What percent of the people preferred Giggles Bar?

4. If 600 customers were surveyed, how many would prefer a Venus Bar?

5. How many would prefer a Galaxy Bar?

6. What is the mode?

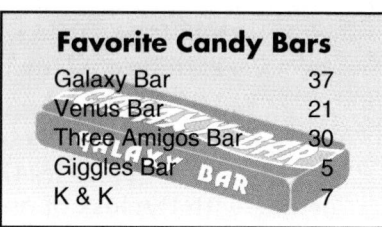

Favorite Candy Bars	
Galaxy Bar	37
Venus Bar	21
Three Amigos Bar	30
Giggles Bar	5
K & K	7

Lesson 9-7

Express each decimal as a percent.

1. 0.06   **2.** 0.374   **3.** 0.0095   **4.** 56.71

Express each fraction as a percent.

5. $\frac{3}{4}$ **6.** $3\frac{1}{4}$ **7.** $\frac{45}{50}$ **8.** $\frac{3}{1000}$

Express each percent as a fraction.

9. 17%   **10.** 0.8%   **11.** 5268%   **12.** $\frac{15}{4}$%

Lesson 9-8

Choose the best estimate.

1. 28% of 500   **a.** 1.5   **b.** 15   **c.** 150

2. 96% of 900   **a.** 90   **b.** 900   **c.** 9000

3. 148% of 350   **a.** 5.25   **b.** 52.5   **c.** 525

4. $\frac{1}{3}$% of 360   **a.** 1.08   **b.** 10.8   **c.** 108

Estimate.

5. 72% of 250   **6.** 47% of 198

7. 0.8% of 380   **8.** 98% of 32

9. $12\frac{1}{2}$% of 130   **10.** 122% of 84

Estimate each percent.

11. 12 out of 20   **12.** 14 out of 40   **13.** 3 out of 75

14. 75 out of 179   **15.** 19 out of 96   **16.** 1.6 out of 88

17. 5 out of 9   **18.** 6 out of 210   **19.** 12 out of 2250

Lesson 9-9

Solve each problem by using the percent equation, $P = R \cdot B$.

1. 9.28 is what percent of 58?   **2.** What number is 43% of 110?

3. 88% of what number is 396?   **4.** What number is 61% of 524?

5. 126 is what percent of 90?   **6.** 52% of what number is 109.2?

Find the discount or interest to the nearest cent.

7. $64.98 dress at 32% off.   **8.** $1000 at 4% annually for 2 years.

9. $589 sofa at 20% off.   **10.** $500 at 2.5% a month for 18 months.

11. $800 at $7\frac{1}{2}$% annually for 15 months.   **12.** $2148 bedroom set at 40% off.

Lesson 9-10

State whether each percent of change is a percent of increase or a percent of decrease. Then find the percent of increase or decrease. Round to the nearest whole percent.

1. old: $56
   new: $42

2. old: $26
   new: $29.64

3. old: $22
   new: $37.18

4. old: $137.50
   new: $85.25

5. old: $455
   new: $955.50

6. old: $3
   new: $15

7. old: 750.75
   new: $765.51

8. old: $953
   new: $476.5

9. old: $101.25
   new: $379.69

10. old: $836
    new: $842.27

Lesson 10-1

Make a stem-and-leaf plot of each set of data.

1. 37, 44, 32, 53, 61, 59, 49, 69
2. 3, 26, 35, 8, 21, 24, 30, 39, 35, 5, 38
3. 15.7, 7.4, 0.6, 0.5, 15.3, 7.9, 7.3
4. 172, 198, 181, 182, 193, 171, 179, 186, 181
5. 101.6, 101.8, 100.5, 102.1, 101.0, 100.1, 100.5, 100.5

Lesson 10-2

Find the range, median, upper and lower quartiles, and the interquartile range for each set of data.

1. 44, 37, 23, 35, 61, 95, 49, 96
2. 30, 62, 35, 80, 12, 24, 30, 39, 53, 38
3. 7.15, 4.7, 6, 5.3, 30.1, 9.19, 3.2
4. 271, 891, 181, 193, 711, 791, 861, 818

Lesson 10-3

Use the stem-and-leaf plot at the right to answer each question.

1. Make a box-and-whisker plot of the data.
2. What is the range?
3. What is the median?
4. What is the upper quartile?
5. What is the lower quartile?
6. What is the interquartile range?
7. What are the extremes?
8. Are there any outliers? If so, what are they?
9. What are the limits for outliers?

Stem	Leaf
1	2
2	
3	
4	3
5	0
6	2 3 7 9
7	0 3
8	0 2 2 4 5 7
9	0 7

4 | 3 = 43

Lesson 10-4

Use the graphs to answer the following questions.

1. Explain why these graphs made from the same data look different.
2. Which graph shows that the family income is consistent? Why?
3. Which graph shows why a son can't have a raise in his allowance? Explain.

Lesson 10-5

Find the number of possible outcomes for each event.

1. Engagement rings come in silver, gold, and white gold. The diamond can weigh $\frac{1}{2}$ karat, $\frac{1}{3}$ karat, or $\frac{1}{4}$ karat. The diamond can have 4 possible shapes.
2. A dress can be long, tea-length, knee-length, or mini. It comes in 2 colors and the dress can be worn on or off the shoulders.
3. The first digit of a 7 digit phone number is a 2. The last digit is a 3.
4. A chair can be a rocker, recliner, swivel, straight back, or a combination of any of the first three features.

Lesson 10-6

1. Seven people are running for student council. There will be four people on the board and one alternate. How many ways can the students be elected?
2. How many ways can the letters of the word ISLAND be arranged?
3. How many ways can five candles be arranged in three candlesticks?

Find each value.

4. $7!$
5. $P(3, 2)$
6. $C(9, 4)$
7. $P(10, 5)$
8. $10!$
9. $\frac{6!2!}{5!}$

Lesson 10-7

Find the odds of each outcome if the spinner below is spun.

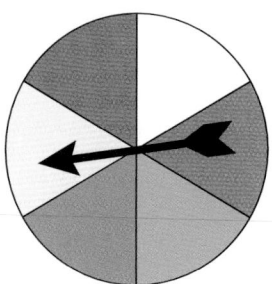

1. blue
2. a color with less than 5 letters
3. a color that begins with a consonant
4. red, yellow, or blue

Find the odds of each outcome if a 10-sided die is rolled.

5. number less than 7
6. odd number
7. composite number
8. number divisible by 3

Lesson 10-8

Solve. Use a simulation.

1. Laneeda is a forward on the freshman basketball team. She usually makes $\frac{2}{3}$ of her shots from the field and $\frac{5}{6}$ of her free throws. If she averages 10 shots a game and 5 free throws, describe a simulation that could give her probable number of points in the next game.

2. A certain restaurant gives out game cards. One out of six cards wins a small soft drink. Describe a simulation to predict how many soft drinks Ted will win with 14 game cards.

Lesson 10-9

Determine whether the events are independent or dependent.

1. go to the movies on Monday and then going back to the same theater on Tuesday
2. rolling a die and spinning a spinner
3. watching a 1 p.m. football game and watching a 4 p.m. football game
4. choosing a person to be quarterback from a roster of 45 people and then choosing a person to be wide receiver
5. selecting a name for your daughter and then selecting a name for your son

A deck of Euchre cards consists of 4 nines, 4 tens, 4 jacks, 4 queens, 4 kings, and 4 aces. Once a card is selected, it is not replaced. Find the probability of each outcome.

6. 3 nines in a row
7. a black jack and then a red queen
8. a nine of clubs, a black king, and a red ace
9. 4 face cards in a row
10. 2 cards lower than a jack

Lesson 10-10

Determine whether each event is *mutually exclusive* or *inclusive*. Then find the probability. A number from 6 to 19 is drawn.

1. *P*(12 or even)
2. *P*(13 or less than 7)
3. *P*(even or odd)
4. *P*(14 or greater than 20)
5. *P*(even or less than 10)
6. *P*(odd or greater than 10)
7. *P*(divisible by 3 or even)
8. *P*(prime or even)

Lesson 11-1

In the figure at the right, use a protractor to find the measure of each angle. Then classify the angle as *acute*, *right*, or *obtuse*.

1. *m∠PQW*
2. *m∠VQW*
3. *m∠TQW*
4. *m∠SQW*
5. *m∠SQR*
6. *m∠VQR*

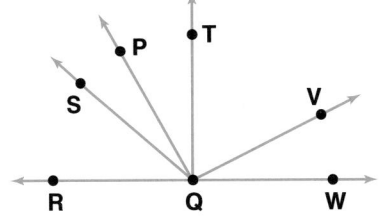

Lesson 11-2

Make a circle graph to display each set of data.

1.

Items Donated for Food Drive	
Canned Goods	169
Pasta	86
Cereal	70
Peanut Butter	42
Condiments	18

2.

Davis Family Monthly Expenditures	
Housing	$800
Taxes	$600
Food	$350
Clothing	$200
Insurance	$150
Savings	$100

Lesson 11-3

In the figure at the right, ℓ is parallel to *m*. If the measure of ∠2 is 38°, find the measure of each angle.

1. ∠1
2. ∠3
3. ∠4
4. ∠5
5. ∠6
6. ∠8

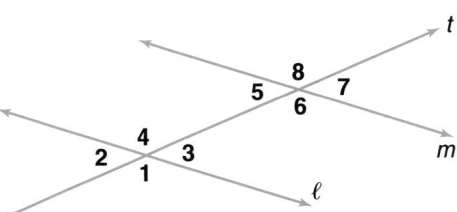

Lesson 11-4

1. Find the measure of each angle of a triangle if they are in the ratio 1:2:3.
2. Find the measure of each angle of a triangle if they are in the ratio 1:1:2.
3. Find the measure of each angle of a triangle if they are in the ratio 1:9:26.

Lesson 11-5

If $\triangle JKL \cong \triangle DGW$, name the part congruent to each angle or segment given.

1. $\angle K$ 2. $\overline{WG}$ 3. $\angle D$ 4. $\overline{KL}$ 5. $\overline{DG}$

Lesson 11-6

Write a proportion to find each missing measure x. Then find the value of x.

1.

2.
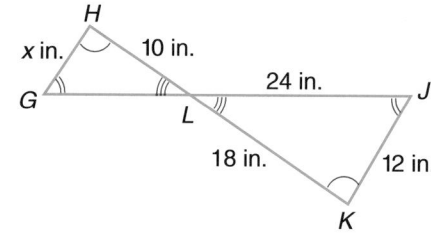

Lesson 11-7

Find the value of x. Then find the missing measures.

1.

2.
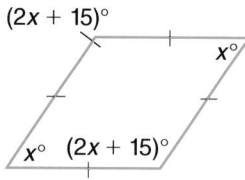

3.

Lesson 11-8

Find the measure of each exterior angle and each interior angle of each regular polygon.

1. equilateral triangle   2. regular pentagon
3. regular octagon   4. regular nonagon

Lesson 11-9

Trace each figure. Draw all lines of symmetry.

1.

2.

3.

4.

Lesson 12-1

Find the area of each figure.

1.

2.

3.

Lesson 12-2

Find the area of each circle. Round to the nearest tenth.

1. diameter, 10 in.

2. radius, 8 mm

3. diameter, 15.4 m

Lesson 12-3

Each figure represents a dart board. Find the probability of landing in the shaded region.

1.

2.

3.

Lesson 12-4

Solve by making a model or drawing.

1. A theater section is arranged so that each row has the same number of seats. Andrew is seated in the fifth row from the front and the third row from the back. His seat is sixth from the left and second from the right. How many seats are in this section?

2. Sixteen players are competing in a bowling tournament. Each player in the competition bowls against another opponent and is eliminated after one loss. How many games does the winner of the tournament play?

3. An artist is creating a pyramid piece of art using differently-colored marble cubes. There is one cube on the top, two cubes in the next layer, three cubes in the next layer, and so on until 10 layers are completed. How many cubes did the artist use?

Lesson 12-5

Find the surface area of each solid. Round decimal answers to the nearest tenth.

1.

2.

3.

Lesson 12-6

Find the surface area of each pyramid or cone. Round decimal answers to the nearest tenth.

1.

2.

3.

Lesson 12-7

Find the volume of each prism or cylinder. Round decimal answers to the nearest tenth.

1.

2.

3.

Lesson 12-8

Find the volume of each pyramid or cone. Round decimal answers to the nearest tenth.

1.
3 ft
6.6 ft

2.
3 m
7 m

3.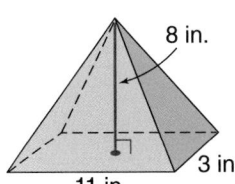
8 in.
3 in.
11 in.

Lesson 13-1

Find the best integer estimate for each square root. Then check your estimate with a calculator.

1. $\sqrt{21}$ **2.** $-\sqrt{85}$ **3.** $\sqrt{7.3}$ **4.** $\sqrt{1.99}$

5. $-\sqrt{62}$ **6.** $\sqrt{74.1}$ **7.** $\sqrt{810}$ **8.** $-\sqrt{88.8}$

Lesson 13-2

Solve using a Venn diagram.

1. A supermarket survey showed that 83 customers chose wheat cereal, 83 chose rice, and 20 chose corn. Six customers bought a box of corn cereal and a box of wheat cereal and 10 others bought a box of rice cereal and a box of corn cereal. Four customers bought all three. How many customers bought a box of rice cereal and a box of wheat cereal if 61 bought just a box of wheat cereal.

2. Of the thirty members in a cooking club, 20 like to mix salads, 17 prefer baking desserts, and 8 like to do both.

   a. How many like to do salads, but not bake desserts?

   b. How many do not like either baking desserts or mixings salads?

Lesson 13-3

Solve each equation. Round decimal answers to the nearest tenth.

1. $x^2 = 14$ **2.** $y^2 = 25$ **3.** $34 = p^2$ **4.** $55 = h^2$

5. $225 = k^2$ **6.** $324 = m^2$ **7.** $d^2 = 441$ **8.** $r^2 = 25,000$

Lesson 13-4

In a right triangle, If a and b are the measures of the legs and c is the measurement of the hypotenuse, find each missing measure.

1. a, 7 m; b, 24 m   **2.** a, 18 in.; c, 30 in.   **3.** b, 10 ft; c, 20 ft

4. a, 3 cm; c, 9 cm   **5.** b, 8 m; c, 32 m   **6.** a, 32 yd; c, 65 yd

Lesson 13-5

Find the lengths of the missing sides in each right triangle. Round decimal answers to the nearest tenth.

1.

2.

3.

4.

Lesson 13-6

For each triangle, find the measure of the marked acute angle to the nearest degree.

1.

2.

3.

4.

Lesson 13-7

Use trigonometric ratios to solve each problem.

1. In a sightseeing boat near the base of Horseshoe Falls at Niagara Falls, a passenger estimates that angle of elevation to the top of the falls to be $30°$. If Horseshoe Falls is 173 feet high, what is the distance from the boat to the base of the falls?

2. While picnicking in San Jacinto Battlefield Park, a student 800 feet from the base of the monument estimates the angle of elevation to the top of the monument to be $25°$. From this information, estimate the height of the monument.

3. From a boat in the ocean, a cliff is sighted through the fog. The angle of elevation is $42°$ and the height of the cliff is 135 meters. How far is the boat from the cliff?

Lesson 14-1

Find the degree of each polynomial.

1. $4x$

2. $a^2 - 6$

3. $11r + 5s$

4. $3y^2\, 4y - 2$

5. $9cd^3 - 5$

6. $-5p^3 + 8q^2$

7. $w^2 + 2x - 3y^3 - 7z$

8. $\dfrac{x^3}{6} - x$

9. $-17n^2p - 11np^3$

Lesson 14-2

Find each sum.

1. $(3a + 4) + (a + 2)$

2. $(8m - 3) + (4m + 1)$

3. $(5x - 3y) + (2x - y)$

4. $(8p^2 - 2p + 3) + (-3p^2 - 2)$

5. $(-11r^2 + 3s) + (5r^2 - s)$

6. $(3a^2 + 5a + 1) + (2a^2 - 3a - 6)$

Lesson 14-3

Find each difference.

1. $(3n + 2) - (n + 1)$

2. $(-3c + 2d) - (7c - 6d)$

3. $(4x^2 + 1) - (3x^2 - 4)$

4. $(5a - 4b) - (-a + b)$

5. $\quad 6x^2 - 4x + 11$
$\underline{(-)\ 5x^2 + 5x - 4}$

6. $\quad 8n^2 + 3mn$
$\underline{(-)\ 4n^2 + 2mn - 9}$

Lesson 14-4

Simplify.

1. $(a^2)^3$

2. $(-2x)^3$

3. $(p^2q)^3$

4. $-5c(2cd)^3$

5. $4y(y^2z)^3$

6. $6a(-ab)^7$

Lesson 14-5

Find each product.

1. $4n(5n - 3)$

2. $-3x(4 - x)$

3. $6m(-m^2 + 3)$

4. $-5x(2x^2 - 3x + 1)$

5. $7r(r^2 - 3r + 7)$

6. $-3az(2z^2 + 4az + a^2)$

Lesson 14-6

Find each product.

1. $(2x + 2)(3x + 1)$

2. $(x + 4)(3x + 1)$

3. $(7x + 4)(3x - 11)$

4. $(x + 3)(x + 1)$

5. $(2x + 1)(x + 1)$

6. $(3x + 2)(2x + 2)$

Chapter 1 Practice Test

Solve using the four-step plan.

1. The distance between Jackie's house and Roberta's house is 90 feet. If it takes Jackie 3 seconds to walk 10 feet, how long will it take her to walk to Roberta's house?

Find the value of each expression.

2. $7 \cdot 4 + 6 \cdot 5$

3. $8 + 3(16 - 12)$

4. $6[4 \cdot (72 - 63) \div 3]$

5. $6[5 \times (41 - 36) - (8 + 14)]$

Evaluate each expression if $a = 8$, $b = 4$, and $c = 3$.

6. $2c + 6bc - 7$

7. $15 \div c + 7b$

8. $6b \div a + 5c$

Name the property shown by each statement.

9. $4 \cdot (6 \cdot 0) = 4 \cdot (0 \cdot 6)$

10. $5 + (9 + 3) = (5 + 9) + 3$

11. $0 + h = h$

12. $8(a + 4) = 8a + 8(4)$

13. $8(6 \cdot 9) = (8 \cdot 6)9$

14. $r \cdot 1 = r$

Simplify each expression.

15. $(b + 12) + 15$

16. $6(v \cdot 2)$

17. $18yz + 13yz$

18. $5(2w + 3) + 7(w + 13)$

Solve each equation mentally.

19. $5x = 25$

20. $8 + g = 8$

21. $\frac{s}{14} = 3$

22. $22 - k = 17$

Name the ordered pair for each point graphed on the coordinate plane at the right.

23. B

24. E

25. F

26. H

27. N

28. K

Solve each equation using the inverse operation. Use a calculator when needed.

29. $x + 48 = 55$

30. $v - 57 = 72$

31. $\frac{b}{2} = 18$

32. $672 = 21t$

State whether each inequality is *true* or *false* for the given value.

33. $16 - t > 7, t = 14$

34. $4p - 6 \geq 9, p = 4$

35. $2v + v > 21, v = 7$

36. $8m - 2m \leq 7m, m = 0$

The frequency table at the right contains data about the number of dogs of different breeds registered in the American Kennel Club in 1992.

37. Which breed has the most dogs registered?

38. Write an equation to represent the difference between the number of Maltese and the number of springer spaniels registered. Then find the difference.

Breed	Dogs (thousands)	Breed	Dogs (thousands)
Beagle	61	Maltese	18
Chow Chow	43	Newfoundland	3
Collie	17	Shar-Pei	90
Dalmatian	39	Springer Spaniel	22
Labrador Retriever	121	Weimaraner	5

Source: American Kennel Club

39. How many dogs of these ten breeds are registered?

40. Make a bar graph of the data.

Write an integer for each situation.

1. a gain of 15 pounds

2. a rent increase of $10

3. a bank overdraft of $25

4. a loss of 6 yards

Simplify.

5. $|-8|$

6. $|-3| + |2|$

7. $|15| - |-3|$

8. $|-20| + |-19|$

Name the ordered pair for each point graphed on the coordinate plane at the right.

9. A

10. C

11. E

12. R

13. S

14. M

15. T

16. V

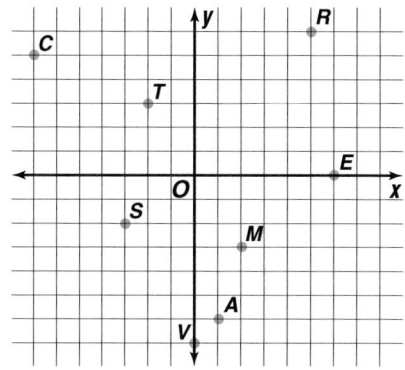

On graph paper, draw coordinate axes. Then graph and label each point. Name the quadrant in which each point is located.

17. $A(8, -5)$

18. $J(-4, -18)$

19. $W(-1, 6)$

20. $F(6, 17)$

21. $M(-8, 5)$

22. $S(-8, -5)$

Replace each ● with <, >, or =.

23. $-6 ● -12$

24. $18 ● 18$

25. $-5 ● -4$

26. $|-7| ● 7$

Simplify each expression.

27. $3x + 5x$

28. $6p - (-4p)$

29. $-9m - (-7m)$

30. $-8k + (-12k)$

31. $-15a - 6a$

32. $10d + (-3d)$

Solve. Look for a pattern.

33. Use the pattern at the right to find $123,456 \times 9 + 7$

34. The following is referred to as an alphametric problem. Each letter stands for one digit 0–9.

$$\begin{array}{r} \text{TRIED} \\ + \text{DRIVE} \\ \hline \text{RIVET} \end{array}$$

$$\begin{array}{rcl} 1 \times 9 + 2 &=& 11 \\ 12 \times 9 + 3 &=& 111 \\ 123 \times 9 + 4 &=& 1111 \end{array}$$

Solve each equation.

35. $q = -13(10)$

36. $-40 \cdot (-6) = h$

37. $-5(-8)(3) = r$

38. $b = \dfrac{-15}{3}$

39. $\dfrac{-12}{-6} = a$

40. $115 \div (-5) = k$

PRACTICE TEST

Solve by eliminating possibilities.

1. Dave is younger than Angie, but older than Bob and Dwight. Angie is older than Dwight. Dwight is younger than Andy. Donna is older than Angie, but younger than Andy. Andy is older than Dave. Who is the oldest?

2. Three people are standing in a straight line as follows: The teenager is to the right of the baby and the adult. The blonde is to the left of the baby. The brunette is to the right of the redhead. State the order of the people and their hair color.

Solve each equation and check your solution. Then graph the solution on a number line.

3. $19 = f + 5$

4. $-15 + z = 3$

5. $x - (-27) = -40$

6. $j - 7 = -13$

7. $-8y = 72$

8. $-9k = -162$

9. $-42 = \dfrac{p}{3}$

10. $\dfrac{n}{-30} = -6$

Solve by replacing the variables with the given values.

11. $P = 2(s + b)$, if $s = 4$ and $b = 16$

12. $S = 2\ell w + 2wh + 2\ell h$, if $\ell = 3$, $w = 5$, and $h = 8$

13. $E = mc^2$, if $E = 325$ and $c = 5$

14. $C = \dfrac{5}{9}(F - 32)$, if $F = 212$

Find the perimeter and area of each rectangle.

15.
48 m
20 m

16.
100 yd
75 yd

17.
20 cm
20 cm

Solve each inequality and check your solution. Then graph the solution on a number line.

18. $g - 8 > -5$

19. $-61 \le w + 50$

20. $17 \ge a - (-19)$

21. $3t \ge -42$

22. $\dfrac{p}{-12} > 6$

23. $-5c > 30$

Define a variable and write an equation or inequality. Then solve.

24. **Animals** According to the International Union for Conservation of Nature and Natural Resources, there are just under 6000 types of endangered species. If 2574 are invertebrates, what is the maximum number of vertebrates?

25. **Statistics** In 1990, there were 308,000 Cherokee Indians and 44,000 Creek Indians. How many times more Cherokee Indians are there than Creek Indians?

Chapter 4 Practice Test

Using divisibility rules, state whether each number is divisible by 2, 3, 5, 6, or 10.

1. 315

2. 400

3. 216

Write each multiplication expression using exponents.

4. $(-7)(-7)(-7)(-7)(-7)$

5. 40

6. $x \cdot x \cdot x \cdot x \cdot x \cdot x \cdot x \cdot x$

Solve. Use a diagram.

7. Six teams are to play each other in a round-robin volleyball tournament, that is, each team will play all five of the other teams. How many games will be played?

Factor each number or monomial completely.

8. 280

9. $882x^3$

10. $-696f^2g$

Find the GCF of each set of numbers or monomials.

11. 63, 231

12. $9xy, 27x^3y^4$

13. $12z^4, 72z^3, 120z^2$

Write each fraction in simplest form. If the fraction is already in simplest form, write *simplified*.

14. $\frac{95}{155}$

15. $\frac{120}{210}$

16. $\frac{65}{104}$

Replace each ⬤ with $<$, $>$, or $=$ to make a true statement.

17. $\frac{3}{7}$ ⬤ $\frac{6}{10}$

18. $\frac{5}{11}$ ⬤ $\frac{3}{5}$

19. $\frac{7}{15}$ ⬤ $\frac{1}{3}$

Find each product or quotient. Express your answer in exponential form.

20. $(2x^4)(-15x^5)$

21. $\frac{q^6}{q}$

22. $18 \cdot 18^7$

Write each expression using positive exponents.

23. $3^{-2}xy^{-3}$

24. $(st)^{-1}$

25. $\frac{3^{-4}}{2^{-2}}$

Express each decimal as a fraction or mixed number in simplest form.

1. -9.8

2. 0.125

3. $1.\overline{6}$

Name the set(s) of numbers to which each number belongs.

4. $\dfrac{2}{3}$

5. -7

6. 51

Estimate each sum or difference.

7. $\$10.03 + \$5.84$

8. $44.03 - 32.9$

9. $20.3 + 59.7 + 62.8$

Simplify each expression.

10. $4.75m - 6.2m + 9$

11. $6.16q - 19 - (-7.32q)$

Solve each equation. Write the solution in simplest form.

12. $a = \dfrac{7}{9} + \dfrac{5}{9}$

13. $\dfrac{27}{18} - \dfrac{9}{18} = h$

14. $2\dfrac{3}{8} - 1\dfrac{5}{8} = p$

15. $\dfrac{1}{6} + \dfrac{5}{24} = p$

16. $5\dfrac{1}{2} - 2\dfrac{2}{3} = r$

17. $\dfrac{-6}{7} - \left(-2\dfrac{10}{21}\right) = y$

Solve each equation or inequality. Check your solution.

18. $u - 7.28 = 14.9$

19. $a + \dfrac{2}{3} = 2$

20. $-9.3 > y - (-4.8)$

21. $j + \dfrac{7}{12} \le 7\dfrac{1}{2}$

State whether each is an example of *inductive* or *deductive* reasoning. Explain your answer.

22. For the past 3 years of school when Joe wears his black shoes to play football, his team wins. On the nights Joe wears his white shoes, his team loses. Sherry noticed that Joe was wearing his white shoes, so she figured they were going to lose.

23. At 32°F water freezes. It is 15° outside. Kelly knows that it is going to snow instead of rain.

State whether each sequence is arithmetic. Then write the next three terms of each sequence.

24. $\dfrac{1}{2}, \dfrac{3}{4}, 1 \ldots$

25. $5040, 720, 120, 24, \ldots$

Write each fraction as a decimal. Use a bar to show a repeating decimal.

1. $\frac{3}{8}$

2. $\frac{-6}{11}$

3. $4\frac{8}{9}$

Estimate each product or quotient.

4. $14.34 \cdot 5.8$

5. $81.2 \div 15.57$

6. $\frac{3}{7} \cdot 98$

Solve each equation. Write each solution in simplest form.

7. $k = -5\left(\frac{-4}{3}\right)$

8. $1\frac{4}{5} \cdot -1\frac{5}{6} = n$

9. $t = \left(\frac{2}{5}\right)^3$

Evaluate each expression.

10. $a \div b$, if $a = \frac{-8}{9}$ and $b = \frac{-4}{3}$

11. $r \div s$, if $r = -1\frac{1}{2}$ and $s = \frac{21}{30}$

Solve each equation.

12. $(2.01)(0.04) = d$

13. $2.13 \div (-0.3) = a$

14. $a = (81)(0.02)(1.5)$

15. $x = 27.9 \div 0.31$

16. $y = 2.3(-0.004)$

17. $-51.408 \div (-5.4) = g$

Find the mean, median, and mode for each set of data. When necessary, round to the nearest tenth.

18. 36, 37, 41, 43, 43

19. 0.2, 0.4, 0.1, 0.6, 1.2, 1.1

20. 2, 8, 16, 21, 3, 8, 9, 6, 7

21. 44, 48, 55, 56, 55, 68, 70

Solve each equation or inequality. Check your solution.

22. $\frac{7}{8}p \geq -4\frac{1}{8}$

23. $0.25w = 6\frac{4}{5}$

24. $-0.5h < 12.5$

State whether each sequence is a geometric sequence. If so, state the common ratio and list the next three terms.

25. 384, 96, 24, . . .

26. $-7, 0, 7, 14, . . .$

27. 20, 16, 12.8, . . .

Write each number in scientific notation.

28. 0.0021

29. 87,500,000

30. 0.00000743

Write each number in standard form.

31. 3.9×10^3

32. 5.32×10^{-4}

33. 7.02×10^0

Solve.

1. Diana makes a salary of $250 a week. In addition to her weekly salary, she receives a bonus of $50 for every 25 boxes of books she sells. If she earned a total of $450 in salary and bonuses this week, how many boxes did she sell?

2. Frank walks 10 blocks to work every day. It takes him between 15 and 20 minutes to get to work. At this rate, what is the least amount of time it would take for Frank to walk 50 blocks?

Solve each equation. Check your solution.

3. $25 = 2d - 9$

4. $4w + (-18) = -34$

5. $\frac{p}{6} - (-21) = 8$

6. $-7 = \frac{d}{-5} + 1$

Define a variable and write an equation for each situation. Then solve.

7. The quotient of a number and 8, decreased by 17, is -15.

8. Find the two consecutive odd integers whose sum is 76.

Find the circumference of each circle.

9.

4.3 cm

10.

$21\frac{1}{2}$ inches

Solve each equation or inequality. Check your solution.

11. $7y + 6 = 3y - 14$

12. $3(x + 2) - 6 = 3x$

13. $\frac{7}{9}d - 7 = \frac{5}{9}d - 3$

14. $-3j - 4 < -22$

15. $-3.2 + 14s > 15s$

16. $3(y - 2) \geq 5(y - 7)$

Define a variable and write an inequality for each situation. Then solve.

17. Linda plans to spend at most $85 on jeans and shirts. She bought 2 shirts for $15.30 each. How much can she spend on jeans?

Complete each sentence.

18. $6 \text{ m} = \underline{\ ?\ } \text{ cm}$

19. $8.3 \text{ L} = \underline{\ ?\ } \text{ ml}$

20. $0.7 \text{ kg} = \underline{\ ?\ } \text{ g}$

Write the domain and range of each relation. Then determine whether each relation is a function.

1. $\{(8, 9), (7, 6.5), (10, 6)\}$

2. $\{(3, 4), (5, 6), (5, 7), (8, 3)\}$

3. $\{(4, 6), (18, 40), (1.6, 4), (6, 6)\}$

4. $\{(10, 9), (4, 8.5), (17, 3), (4, 7)\}$

5.

x	5	8	10	2
y	1	3	3	1

6. $\{(-4, 30), (-4, -5), (6, 10)\}$

Determine whether a scatter plot of the data for the following might show a *positive*, *negative*, or *no* relationship. Explain your answer.

7. income, amount of clothes owned

8. hair color, weight

9. month of birth, years of college

10. outside temperature, frostbite

Find four solutions for each equation. Write your solution as ordered pairs.

11. $x + y = 8$

12. $x = -2$

13. $y = \frac{1}{3}x - 2$

For each equation,
a. solve for the domain $\{-4, -2, 0, 2, 4\}$, and
b. determine if the equation is a function.

14. $y = 4x + 5$

15. $x = 9$

16. $x^2 - y = 6$

Solve by using a graph. Assume the rate is constant.

17. Of the top 25 collegiate football teams on Nov. 26, 1995, Duke University was rated the number 1 academic football college having a 95% graduation rate. Penn State University was rated 8th having a 77% rate. The Ohio State University was ranked 15th. What was OSU's graduation rate?

Find the slope of the line that contains each pair of points.

18. $K(3, 5), L(7, 4)$

19. $G(9, -3), H(-1, 6)$

20. $L(-3, -4), M(-5, 2)$

21. $J(-6, 7), K(25, 4)$

Use the *x*-intercept and *y*-intercept to graph each equation.

22. $y = x - 6$

23. $y = -3 + x$

24. $y = -2x - 9$

Graph each equation using the slope and *y*-intercept.

25. $y = \frac{1}{4}x - 3$

26. $2x - y = -1$

27. $x + 3y = 6$

Use a graph to solve each system of equations.

28. $y = 0.4x$
    $y = 0.4x - 3$

29. $x + y = 2$
    $2x - y = -2$

30. $y = 3x - 6$
    $-6x + 2y = -12$

Graph each inequality.

31. $y \geq 6$

32. $x + y < 4$

33. $y > \frac{1}{3}x + 3$

Express each ratio or rate as a fraction in simplest form.

1. 16 pencils to 72 pencils   **2.** 2 pints out of 1 gallon   **3.** 88 out of 564 students

Solve by making a table.

4. How many ways can you make change for a $50-bill using only $5-, $10-, and $20-bills?

A standard deck of playing cards is placed on the table. Suppose you select one card at random. Find each probability.

5. P(a red card)   **6.** P(a six)   **7.** P(a face card)

Write a proportion that could be used to solve for each variable. Then solve the proportion.

8. 13 cans for $2.99
    x cans for $4.83

9. 3 teaspoons for 24 cookies
    7 teaspoons for y cookies

Use the percent proportion to solve each problem.

10. What is 40% of 60?   **11.** Find 37.5% of 80.

12. Twenty-one is 35% of what number?   **13.** Seventy-five is what percent of 250?

14. Fifty-two is what percent of 80?   **15.** Thirty-six is 45% of what number?

Solve.

16. Eight of the 72 students surveyed at Perry Middle School said they would like to take a music appreciation class. The school's policy says there must be at least 30 interested students in the class. If there are 567 students in the school, predict how many students would be interested. Is this enough for the class?

Express each decimal or fraction as a percent.

17. $\frac{85}{120}$ **18.** 0.42   **19.** $\frac{9}{5}$ **20.** 0.086

Express each percent as a decimal and as a fraction.

21. 67.5%   **22.** $33\frac{1}{3}\%$ **23.** 10.17%

Choose the best estimate.

24. 26% of 49   **a.** 13   **b.** 1.3   **c.** 130

25. 47% of 550   **a.** 26   **b.** 2.6   **c.** 260

Find the discount or interest to the nearest cent.

26. $1550 at 4% for 2 years   **27.** $138 dress, 15% off

28. $143 suit, 10% off   **29.** $1250 at 10.5% for 3 months

State whether each percent of change is a percent of increase or a percent of decrease. Then find the percent of increase or decrease. Round to the nearest whole percent.

30. old: $32
    new: $19

31. old: $245
    new: $315

32. old: $178
    new: $201

33. old: $1010
    new: $1075

The ages of Miss Rebar's seventh grade class were 13.3, 12.5, 12.8, 12.5, 12.8, 13.1, 13.2, 12.2, 12.9, 13.3, and 12.6.

1. Make a stem-and-leaf plot of the data.

2. How young was the youngest student?

Use the data in the stem-and-leaf plot shown at the right.

3. What is the range?

4. Find the median.

5. What is the upper quartile?

6. What is the lower quartile?

7. Find the interquartile range.

8. What are the outliers, if any?

9. Make a box-and-whisker plot of the data.

```
2 | 5 8 8
3 | 2 3 3 6 9
4 | 2 3 5 7
5 | 1 1 3
```
$5 | 1 = 51$

The graphs below show how many sports cards Doug buys each month.

 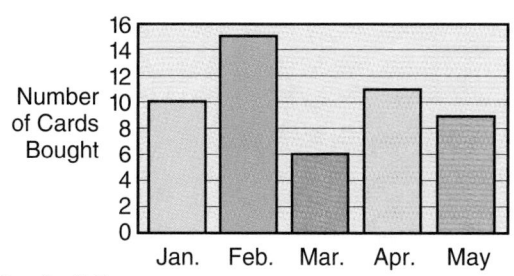

10. Explain why these graphs made from the same data look different.

11. If Doug wanted to convince his wife that he isn't buying very many cards which graph would he use? Explain.

Find the number of possible outcomes for each event.

12. A die is rolled and a coin is tossed.

13. Four dice are rolled.

Tell whether each situation represents a *permutation* or a *combination*.

14. eight notes in a song

15. eight new CDs from a group of thirty

16. five outfits from 15 to be used for a window display

Find the odds of each outcome if a fair eight-sided die is rolled.

17. A number less than five.

18. A composite number.

Heather makes three out of five 10-foot putts. To simulate her chances of making a putt, she puts 40 marbles in a box. A red marble represents a putt made, and a blue marble represents a putt missed. After a marble is drawn, it is replaced in the box.

19. How many blue marbles does Heather need?

20. In 120 drawings, how many red marbles can she expect?

An eight-sided die is rolled and the spinner is spun. Find each probability.

21. P(green and 6)

22. P(orange and an odd number)

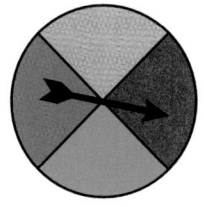

Determine whether each event is *mutually exclusive* or *inclusive*. Then find the probability. A letter of the alphabet is chosen.

23. P(a vowel or a letter following R)

24. P(a consonant or the letter E)

25. P(a vowel or a consonant)

Draw and label a diagram to represent each of the following.

1. point C **2.** $\overline{XY}$ **3.** plane GHI **4.** $\overleftrightarrow{RS}$

5. Make a circle graph to display the data in the chart at the right.

Water Usage in the United States	
Agriculture	36%
Public water	8%
Utilities	33%
Industry	23%

In the figure at the right, $\overline{KG} \parallel \overline{AE}$. If the measure of $\angle HFG$ is 45°, find the measure of each angle.

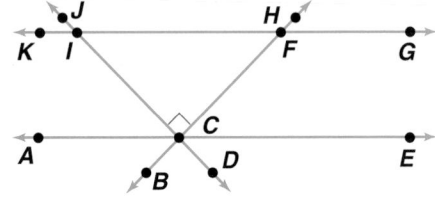

6. $\angle FCE$ **7.** $\angle BCD$

8. $\angle JIF$ **9.** $\angle KIC$

Find the value of x. Then classify the triangle as *acute*, *right*, or *obtuse*.

10.

11.

12.

If $\triangle XYZ \cong \triangle ABC$, name the part congruent to each angle or segment.

13. $\angle X$ **14.** $\angle B$ **15.** $\overline{YZ}$ **16.** $\overline{AC}$

Write a proportion to find the missing measure x. Then find the value of x.

17.

18.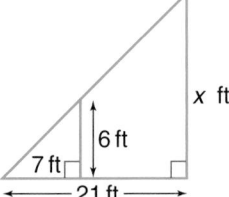

Find the value of x.

19.

20.

21.

Find the sum of the measures of the interior angles of each polygon.

22. hexagon   **23.** 20-gon

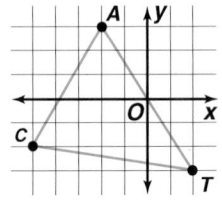

24. A translation moves $\triangle CAT$ 6 units to the right and 3 units up to form $\triangle C'A'T'$. Write the coordinates of each vertex of $\triangle C'A'T'$.

25. Draw $\triangle C'A'T'$ on a coordinate plane and graph its reflection if the x-axis is the line of reflection.

Find the area of each figure.

1.
22 mm 15 mm
37 mm

2.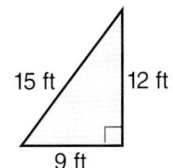
15 ft 12 ft
9 ft

3.
32 m
24 m 17 m

Find the area of each circle. Round to the nearest tenth.

4. radius, 16 cm

5. diameter, 42 in.

6. radius, $3\frac{1}{2}$ in.

7. The figure at the right represents a dartboard. Find the probability of a dart landing in the shaded region.

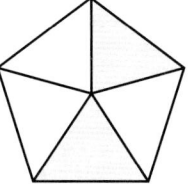

Solve by making a model or drawing.

8. Half of a garden is planted in corn. Half of the remaining garden is planted in strawberries. A third of the part not planted in corn or strawberries is planted in tomatoes. If 6 m² is planted in tomatoes, what is the area planted in corn?

Find the surface area of each solid.

9.
8 in.
10 in.
14 in.

10.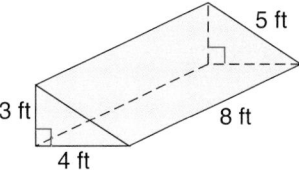
5 ft
3 ft 8 ft
4 ft

11.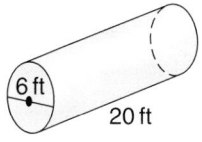
6 ft 20 ft

12.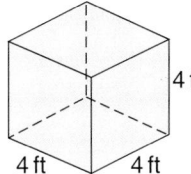
4 ft
4 ft 4 ft

13.
45 cm
73 cm
73 cm

14.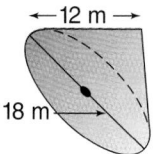
←12 m→
18 m

Find the volume of each solid. Round to the nearest tenth.

15. rectangular prism: length, 4 m; width, 7 m; height, 9 m

16. triangular prism: base of triangle, $2\frac{1}{2}$ ft; altitude, 1 ft; prism height, 2 ft

17. cylinder: diameter, 12 m; height, 9 m

18. rectangular pyramid: length, 18 cm; width, 15 cm; height, 20 cm

19. square pyramid: length, 8 in.; height, 9 in.

20. circular cone: radius, 9 cm: height, 16 cm

Find the best integer estimate for each square root. Then check your estimate using a calculator.

1. $\sqrt{90}$ **2.** $\sqrt{20.25}$ **3.** $-\sqrt{62}$ **4.** $-\sqrt{71}$

Solve using a Venn diagram.

5. Exquisite Interiors has 40 sample floor tiles in the shapes of various polygons. Twenty tiles are regular polygons, 14 tiles are quadrilaterals, and 5 tiles are squares. How many tiles are not regular polygons?

Solve each equation. Round decimal answers to the nearest tenth.

6. $s^2 = 121$ **7.** $g^2 = 40$ **8.** $f^2 = 11$ **9.** $y^2 = 60$

In a right triangle, if a and b are the measures of the legs and c is the measure of the hypotenuse, find each missing measure. Round decimal answers to the nearest tenth.

10. b, 77 m; c, 85 m   **11.** a, 15 ft; c, 17 ft   **12.** b, 91 yd; c, 109 yd

Find the lengths of the missing sides in each triangle. Round decimal answers to the nearest tenth.

13.

14.

15.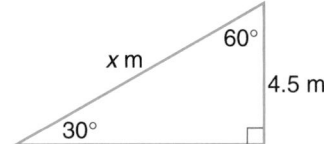

For each triangle, find the measure of the marked acute angle to the nearest degree.

16.

17.

18.

Use trigonometric ratios to solve each problem.

19. What is the angle of elevation of the sun when a 100-foot water tower casts a shadow 165 feet long?

20. **Building Design** A ramp is designed to help people in wheelchairs move more easily from one level to another. If a ramp 16 feet long forms an angle of 6° with the level ground, what is the vertical rise?

Chapter 14 Practice Test

State whether each expression is a polynomial. If it is, identify it as a *monomial*, *binomial*, or *trinomial*.

1. $4a^2$

2. $-2y + 3$

3. $\dfrac{a}{5}$

4. $r^3 - s^2$

5. $-\dfrac{3}{(x + y)}$

6. $7mt^2 + 3m - 2t$

Find the degree of each polynomial.

7. $5a^2b$

8. $12rs^2 + 3r^2s + 5r^4s$

9. $x^2 + 6xy + xy^2$

Find each sum or difference.

10. $(2x^2 + 5x) + (3x^2 + x)$

11. $(7a - 11b) - (3a + 4b)$

12. $(x^2 - 5x + 3) + (4x - 3)$

13. $(3a^2 - b^2 + c^2) - (a^2 + 2b^2)$

14. $\quad -3y^2 \qquad\ + 2$
$\underline{(+)\ 4y^2 - 5y - 2}$

15. $\quad 14a^2 - 3a$
$\underline{(-)\ 6a^2 + 5a + 17}$

Simplify.

16. $(5c)^3$

17. $(4x^2y^3)^2$

18. $-3x(4xy)^2$

Find each product.

19. $x(-5x + 3)$

20. $a^2(2a^3 + a - 5)$

21. $-2g(g^3 + 6g + 3)$

22. $(2x + 1)(3x + 2)$

23. $(3x + 3)(x - 1)$

24. $(-x + 3)(3x - 2)$

Solve.

25. Geometry For the rectangular prism at the right, compute the area of the top, bottom, front, back, left, and right sides. Then add to find the total surface area.

$n + 3$

$2n$

$5n$

A

absolute value (67)   The number of units a number is from zero on the number line.

acute angle (550)   An angle with a measure greater than 0° and less than 90°.

acute triangle (569)   A triangle that has three acute angles.

addition property of equality (125)   If you add the same number to each side of an equation, the two sides remain equal. For any numbers a, b, and c, if $a = b$, then $a + c = b + c$.

addition property of inequality (147)   Adding the same number to each side of an inequality does not change the truth of the inequality. For all numbers a, b, and c
   1. If $a > b$, then $a + c > b + c$.
   2. If $a < b$, then $a + c < b + c$.

additive inverses (89)   An integer and its opposite. The sum of an integer and its additive inverse is zero.

adjacent angles (561)   Two angles that have a common side and the same vertex, but they do *not* overlap.

adjacent leg (687)   The leg of a right triangle that is not opposite that angle or the hypotenuse.

algebraic expression (16)   A combination of variables, numbers, and at least one operation.

algebraic fraction (197)   A fraction with variables in the numerator or denominator.

alternate exterior angles (563)   Nonadjacent exterior angles found on opposite sides of the transversal.
   $\angle 1$ and $\angle 7$, and $\angle 2$ and $\angle 8$ are alternate exterior angles.

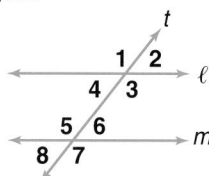

alternate interior angles (563)   Nonadjacent interior angles found on opposite sides of the transversal.
   $\angle 4$ and $\angle 6$, and $\angle 3$ and $\angle 5$ are alternate interior angles.

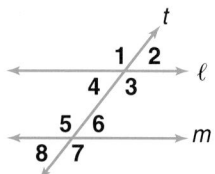

altitude (612, 614, 633)   A line segment that is perpendicular to the base of a figure with endpoints on the base and the side opposite the base.

angle (549)   Two rays with a common endpoint form an angle. The rays and vertex are used to name an angle.

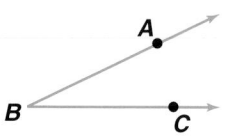
angle *ABC* or $\angle ABC$

angle of depression (695)   An angle of depression is formed by a line of sight along the horizontal and another line of sight below it.

angle of elevation (694)   An angle of elevation is formed by a line of sight along the horizontal and another line of sight above it.

area (138)   The number of square units needed to cover a surface enclosed by a geometric figure.

arithmetic sequence (258)   A sequence in which the difference between any two consecutive terms is the same.

associative property of addition (22, 234)   For any rational numbers a, b, and c,
   $$(a + b) + c = a + (b + c).$$

associative property of multiplication (22, 296)   For any numbers a, b, and c,
   $$(ab)c = a(bc).$$

axes (73)   Two intersecting number lines that form a coordinate system.

 B

back-to-back stem-and-leaf plot (487)   Used to compare two sets of data. The leaves for one set of data are on one side of the stem and the leaves for the other set of data are on the other side.

bar graph (52)   A graphic form using bars to make comparisons of statistics.

bar notation (225)   In repeating decimals the line or bar placed over the digits that repeat. For example, $2.\overline{63}$ indicates the digits 63 repeat.

base (175)   In 5^3, the base is 5. The base is used as a factor as many times as given by the exponent (3). That is,
   $$5^3 = 5 \times 5 \times 5.$$

base (449)   In a percent proportion, the number to which the percentage is compared.
   $$\frac{\text{percentage}}{\text{base}} = \text{rate}$$

base (614) The base of a rectangle, a parallelogram, or a triangle is any side of the figure. The bases of a trapezoid are the parallel sides.

base (632) The bases of a prism are the two parallel congruent sides.

binomial (706) A polynomial with exactly two terms.

boundary (418) A line that separates a graph into half-planes.

box-and-whisker plot (495) A diagram that summarizes data using the median, the upper and lower quartiles, and the extreme values. A box is drawn around the quartile value and whiskers extend from each quartile to the extreme data points.

center (341) The given point from which all points on the circle are the same distance.

circle (341) The set of all points in a plane that are the same distance from a given point called the center.

circle graph (497, 556) A type of statistical graph used to compare parts of a whole.

circular cone (639) A shape in space that has a circular base and one vertex.

circumference (341) The distance around a circle.

coefficient (711) The numerical part of a monomial.

combination (516) An arrangement or listing in which order is not important.

commission (470) A fee paid to a sales associate based on a percent of the total sales.

common difference (258) The difference between any two consecutive terms in an arithmetic sequence.

common multiples (200) Multiples that are shared by two or more numbers. For example, some common multiples of 2 and 3 are 6, 12, and 18.

common ratio (312) The ratio between any two successive terms in a geometric sequence.

commutative property of addition (22, 234) For any rational numbers a and b,
$$a + b = b + a.$$

commutative property of multiplication (22, 296) For any rational numbers a and b, $ab = ba$.

comparative graph (498) A graph used to compare results of similar groups to show trends.

compass (554) An instrument used for drawing circles or parts of circles.

compatible numbers (280) Numbers that have been rounded so, when the numbers are divided by each other, the remainder is zero.

complementary (562) Two angles are complementary if the sum of their measures is $90°$.

composite number (184) Any whole number greater than 1 that has more than two factors.

congruent (554, 561, 573) Line segments that have the same length, or angles that have the same measure, or figures that have the same size and shape.

constant (707) A monomial that does not contain a variable.

coordinate (67, 126) A number associated with a point on the number line.

coordinate system (36, 73) A coordinate plane separated into four regions by two perpendicular number lines called axes.

corresponding angles (563) Angles that have the same position on two different parallel lines cut by a transversal. $\angle 1$ and $\angle 5$, $\angle 2$ and $\angle 6$, $\angle 3$ and $\angle 7$, and $\angle 4$ and $\angle 8$ are corresponding angles.

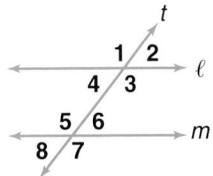

corresponding parts (573) Parts on congruent or similar figures that match.

cosine (688) If $\triangle ABC$ is a right triangle and A is an acute angle,
$$\text{cosine } \angle A = \frac{\text{measure of the side adjacent to } \angle A}{\text{measure of the hypotenuse}}$$
$$\cos A = \frac{b}{c}$$

cylinder (633) A 3-dimensional shape that has two parallel, congruent bases (usually circular) and a curved side connecting the bases.

data (50) Numerical information gathered for statistical purposes.

defining a variable (42) Choosing a variable and a quantity for the variable to represent in an equation.

degree (549) The most common unit of measure for angles.

dependent events (532) Two or more events in which the outcome of one event does affect the outcome of the other event(s).

diagonal (590) A line segment that joins two nonconsecutive vertices of a polygon.

diameter (341)   The distance across a circle through its center.

discount (468)   The amount deducted from the original price.

discrete mathematics (258)   A branch of mathematics that deals with discontinuous or finite sets of numbers. Some fields that deal with discrete mathematics are logic and statistics.

distributive property (26)   For any numbers a, b, and c, $a(b + c) = ab + ac$ and $(b + c)a = ba + ca$.

divisible (170)   A number is divisible by another if, upon division, the remainder is zero.

division property of equality (130)   For any numbers a, b, and c, with $c \neq 0$, if $a = b$, then $\frac{a}{c} = \frac{b}{c}$.

division properties of inequality (152)   For all integers a, b, and c, where $c > 0$,

if $a > b$, then $\frac{a}{c} > \frac{b}{c}$.

For all integers a, b, and c, where $c < 0$,

if $a > b$, then $\frac{a}{c} < \frac{b}{c}$.

domain (372)   The set of all first coordinates from each ordered pair.

empty set (347)   A set with no elements shown by the symbol { } or $\varnothing$.

equation (32)   A mathematical sentence that contains the equals sign, $=$.

equiangular (591)   All angles are congruent.

equilateral (591)   All sides are congruent.

equivalent (275)   The same value represented in different forms. For example, $\frac{3}{5}$ and 0.6 are equivalent.

equivalent equations (124)   Two or more equations with the same solution. For example, $x + 3 = 5$ and $x = 2$ are equivalent equations.

evaluating an expression (11)   Replacing the variables with numbers and finding the numerical value of the expression.

expanded form (176)   A number expressed using place value to write the value of each digit in the number.

experimental probability (529)   An estimated probability based on the relative frequency of positive outcomes occurring during an experiment.

exponent (175)   In 5^3, the exponent is 3. The exponent tells how many times the base, 5, is used as a factor.

$$5^3 = 5 \times 5 \times 5$$

exterior angle (591)   The angle formed when a side of a polygon is extended.

face (632)   A face of a prism is any surface that forms a side or a base of the prism.

factorial (515)   The expression $n!$ is the product of all counting numbers beginning with n and counting backward to 1.

factors (170)   Numbers that divide into a whole number with a remainder of zero.

factor trees (185)   A diagram showing the prime factorization of a composite number. The factors branch out from the previous factors until all the factors are prime numbers.

Fibonacci sequence (263)   A list of numbers in which the first two numbers are both 1 and each number that follows is the sum of the previous two numbers, 1, 1, 2, 3, 5, 8, . . .

formula (134)   An equation that states a rule for the relationship between certain quantities.

45°–45° right triangle (683)   A right triangle that has two angles that each measure 45°.

frequency table (51)   A chart that indicates the number of values in each interval.

function (373)   A function is a relation in which each element of the domain is paired with exactly one element in the range.

functional notation (393)   Equations that represent functions can be written in *functional notation*, $f(x)$. The symbol $f(x)$ is read "f" of "x" and represents the value in the range that corresponds to the value of x in the domain.

Fundamental counting principle (509)   If event M can occur in m ways and is followed by event N that can occur in n ways, then the event M followed by event N can occur in $m \cdot n$ ways.

geometric probability (623)   Probability that uses area to find the probability of an event.

geometric sequence (312)   A sequence in which the ratio between any two successive terms is the same.

gram (359)   A unit of mass in the metric system.

graph (73, 126)   A dot marking a point that represents a number on a number line or an ordered pair on a coordinate plane.

greatest common factor (GCF) (190)   The GCF of two or more integers is the greatest factor that is common to each of the integers.

histogram (502)   A bar graph whose columns are next to each other to show how the data are distributed.

hypotenuse (676)   The side opposite the right angle in a right triangle.

identity property of addition (23, 234)   For any number a, $a + 0 = a$ and $0 + a = a$.

identity property of multiplication (23, 296)   For any number a, $a \times 1 = a$.

inclusive (536)   When two events can happen at the same time.

independent events (531)   Two or more events in which the outcome of one event does *not* affect the outcome of the other event(s).

indirect measurement (580)   Finding a measurement by using similar triangles and writing a proportion.

inequality (46)   A mathematical sentence that contains $<$, $>$, $\neq$, $\leq$, or $\geq$.

integers (66, 224)   The whole numbers and their opposites.
$$\ldots, -3, -2, -1, 0, 1, 2, 3, \ldots$$

intercepts (406)   The points where a graph crosses the axes.

interior angle (591)   An angle inside a polygon.

interquartile range (491)   The range of the middle half of a set of numbers.
$$\text{Interquartile range} = UQ - LQ.$$

intersect (552)   Lines that cross or touch each other at a point.

inverse operations (41)   Operations that undo each other, such as addition and subtraction.

inverse property of addition (234)   For every rational number a, $a + (-a) = 0$.

inverse property of multiplication (289)   For every nonzero number $\frac{a}{b}$, where $a, b \neq 0$, there is exactly one number $\frac{b}{a}$ such that $\frac{a}{b} \times \frac{b}{a} = 1$.

irrational number (672)   A number that cannot be expressed as $\frac{a}{b}$, where a and b are integers and b does not equal 0. For example, $\sqrt{3} = 1.7320508\ldots$ is an irrational number.

isosceles triangle (569)   A triangle that has at least two congruent sides.

lateral surface (639)   The lateral surface of a prism, cylinder, pyramid, or cone is all the surface of the figure except the base or bases.

least common denominator (LCD) (201)   The least common multiple of the denominators of two or more fractions.

least common multiple (LCM) (200)   The least of the nonzero common multiples of two or more numbers. The LCM of 2 and 3 is 6.

leaves (486)   The next greatest place value forms the leaves in a stem-and-leaf plot.

leg (676)   Either of the two sides that form the right angle of a right triangle.

like terms (27, 711)   Expressions that contain the same variables, such as $3ab$ and $7ab$.

line (548)   A never-ending straight path. A representation of a line CD ($\overleftrightarrow{CD}$) is shown below.

linear equation (386)   An equation for which the graph is a straight line.

line graph (497)   A type of statistical graph used to show how values change over a period of time.

line segment (548)   Two endpoints and the straight path between them. A line segment is named by its endpoints. A representation of line segment ST ($\overline{ST}$) is shown below.

liter (359)   The basic unit of capacity in the metric system.

lower quartile (491)   The median of the lower half of a set of numbers indicated by LQ.

mean (300)   The sum of the numbers in a set of data divided by the number of pieces of data.

measures of central tendency (300)   Numbers or pieces of data that can represent the whole set of data.

median (300)   In a set of data, the median is the number in the middle when the data are organized from least to greatest.

meter (359)   The basic unit of length in the metric system.

metric system (360)   A system of measurement using the basic units: meter for length, gram for mass, and liter for capacity.

mixed number (239)   The sum of a whole number and a fraction. For example, $3\frac{1}{2}$.

mode (300)   The number or item that appears most often in a set of data.

monomial (172)   An expression that is either a real number, a variable, or a product of real numbers and variables.

multiple (200)   The product of the number and any whole number.

multiplication properties of inequality (152)
  1. For all integers a, b, and c, where $c > 0$, if $a > b$, then $ac > bc$.
  2. For all integers a, b, and c, where $c < 0$, if $a > b$, then $ac < bc$.

multiplication property of equality (130)   For any numbers a, b, and c, if $a = b$, then $ac = bc$.

multiplicative inverses (289)   Two numbers whose product is 1. The multiplicative inverse of $\frac{3}{4}$ is $\frac{4}{3}$.

multiplicative property of zero (23)   For any number a, $a \times 0 = 0$.

mutually exclusive (535)   Two or more events such that no two events can happen at the same time.

negative (66)   A value less than zero.

negative relationship (380)   If points in a scatter plot appear to suggest a line that slants downward to the right, there is a *negative relationship*.

net (632)   The shape that is formed by "unfolding" a prism. The net shows all the faces that make up the surface area of a prism.

no relationship (380)   Points that seem to be random in a scatter plot are said to have no relationship.

null set (347)   A set with no elements shown by the symbol { } or $\varnothing$.

obtuse angle (550)   Any angle that measures greater than 90° but less than 180°.

obtuse triangle (569)   A triangle with one obtuse angle.

odds (520)   A way to describe the chance of an event occurring.
odds in favor = number of successes : number of failures
odds against = number of failures : number of successes

open sentence (32)   An equation that contains variables.

opposite side (687)   The side that is not part of the angle in a right triangle.

ordered pair (37)   A pair of numbers used to locate a point in the coordinate system.

order of operations (12, 176)
1. Simplify the expressions inside grouping symbols.
2. Evaluate all powers.
3. Then do all multiplications and divisions from left to right.
4. Then do all additions and subtractions from left to right.

origin (36)   The point of intersection of the x-axis and y-axis in a coordinate system.

outcomes (509)   Possible results of a probability event. For example, 4 is an outcome when a die is rolled.

outliers (496)   Data that are more than 1.5 times the interquartile range from the quartiles.

parallel (551)   Lines in the same plane that do not intersect. The symbol $\parallel$ means parallel.

parallelogram (585)   A quadrilateral with two pairs of parallel sides.

percent (449)   A ratio with a denominator of 100. For example, 77% and $\frac{77}{100}$ name the same number.

percentage (449)   In a percent proportion, a number that is compared to another number called the base.
$$\frac{percentage}{base} = rate$$

percent of change (472)   The ratio of the amount of change to the original amount.

percent of decrease (473)   The ratio of an amount of decrease to the previous amount, expressed as a percent.

percent of increase (472)   The ratio of an amount of increase to the previous amount, expressed as a percent.

percent proportion (450)
$$\frac{percentage}{base} = rate \text{ or } \frac{P}{B} = \frac{r}{100}$$

perfect squares (665)   Rational numbers whose square roots are whole numbers. 25 is a perfect square because $\sqrt{25} = 5$.

perimeter (138)   The distance around a geometric figure.

permutation (515)   An arrangement or listing in which order is important.

perpendicular (551)   Two lines that intersect to form a right angle.

pi, π (341)   The ratio of the circumference of a circle to the diameter of a circle. Approximations for π are 3.14 and $\frac{22}{7}$.

pictograph (497)   A type of statistical graph that uses pictures or illustrations to show how specific quantities compare. Each symbol represents a specific number.

plane (550)   A flat surface that has no edges or boundaries. A plane can be named by a single uppercase script letter or by using any three noncollinear points of the plane.

point (548)   A specific location in space.

polygon (589)   A simple closed figure in a plane formed by three or more line segments.

polynomial (706)   A monomial or the sum or difference of two or more monomials.

positive relationship (380)   In a scatter plot, if the points appear to suggest a line that slants upward to the right, there is a *positive relationship*.

power (175)   A number that can be written using an exponent.

power of a monomial (721)   For all numbers a and b, and positive integer m, n, and p, $(a^m b^n)^p = a^{mp} b^{np}$.

power of a power (719)   For any number a and positive integers m and n, $(a^m)^n = a^{mn}$.

power of a product (720)   For all numbers a and b, and positive integer m, $(ab)^m = a^m b^m$.

prime factorization (185)   Expressing a composite number as the product of its prime factors.

prime number (184)   A whole number greater than 1 that has exactly two factors , 1 and itself.

principal (467)   The amount of money in an account.

prism (632)   A figure in space that has two parallel and congruent bases in the shape of polygons.

probability (440)   The chance that some event will happen. It is the ratio of the number of ways a certain event can occur to the number of possible outcomes.
$$\text{probability} = \frac{\text{number of favorable outcomes}}{\text{number of possible outcomes}}$$

probability of inclusive events (536)   The probability of one or the other of two inclusive events can be found by adding the probability of the first event to the probability of the second event and subtracting the probability of both events happening.
$P(A \text{ or } B) = P(A) + P(B) - P(A \text{ and } B)$

probability of mutually exclusive events (535)   The probability of one or the other of two mutually exclusive events can be found by adding the probability of the first event to the probability of the second event.
$P(A \text{ or } B) = P(A) + P(B)$

probability of two dependent events (532)   If two events A and B are dependent, then the probability of both events occurring is the product of the probability of A and the probability of B after A occurs.
$P(A \text{ and } B) = P(A) \times P(B \text{ following } A)$

probability of two independent events (531)   The probability of two independent events can be found by multiplying the probability of the first event by the probability of the second event.
$P(A \text{ and } B) = P(A) \times P(B)$

product of powers (205)   For any number a and positive integers m and n, $a^m \times a^n = a^{m+n}$.

property of proportions (444)
If $\frac{a}{b} = \frac{c}{d}$, then $ad = bc$.
If $ad = bc$, then $\frac{a}{b} = \frac{c}{d}$.

proportion (444)   A statement of equality of two or more ratios.

protractor (549)   An instrument used to measure angles.

pyramid (638)   A solid figure that has a polygon for a base and triangles for sides.

Pythagorean theorem (676)   In a right triangle, the square of the length of the hypotenuse is equal to the sum of the squares of the lengths of the legs. $c^2 = a^2 + b^2$

quadrants (73)   The four regions into which two perpendicular number lines separate the plane.

quadrilateral (584, 585)   A polygon having four sides.

quartile (491)   One of four equal parts of data from a large set of numbers.

quotient of powers (206)   For any nonzero number a and whole numbers m and n,
$\frac{a^m}{a^n} = a^{m-n}$.

radical sign (664)   The symbol used to indicate a nonnegative square root. $\sqrt{}$

radius (341)   The distance from the center of a circle to any point on the circle.

range (372)   The range of a relation is the set of all second coordinates from each ordered pair.

range (490)   The difference between the least and greatest numbers in the set.

rate (433)   A ratio of two measurements having different units.

rate (449)   In a percent proportion the ratio of a number to 100.

ratio (196, 432)   A comparison of two numbers by division. The ratio of 2 to 3 can be stated as 2 out of 3, 2 to 3, 2:3, or $\frac{2}{3}$.

rational numbers (224)   Numbers of the form $\frac{a}{b}$ where a and b are integers and $b \neq 0$.

ray (548)   A part of a line that extends indefinitely in one direction.

real numbers (673)   The set of rational numbers together with the set of irrational numbers.

reciprocal (289)   Another name for a multiplicative inverse.

rectangle (585)   A quadrilateral with four congruent angles.

rectangular prism (632)   A prism with rectangular bases.

reflection (595)   A type of transformation where a figure is flipped over a line.

regular polygon (591)   A polygon having all sides congruent and all angles congruent.

relation (372)   A set of ordered pairs.

repeating decimal (225)   A decimal whose digits repeat in groups of one or more. Examples are 0.181818 . . . and 0.8333. . . .

rhombus (585)   A parallelogram with four congruent sides.

right angle (550)   An angle that measures $90°$.

right triangle (569)   A triangle with one right angle.

rotation (595)   When a figure is turned around a point.

sample (50, 51)   A group that is used to represent a whole population.

sample space (441)   The set of all possible outcomes.

scale drawing (585)   A drawing that is similar but either larger or smaller than the actual object.

scalene triangle (569)   A triangle with no congruent sides.

scatter plot (378)   In a scatter plot, two sets of data are plotted as ordered pairs in the coordinate plane.

scientific notation (317)   A way of expressing a number as the product of a number that is at least 1 but less than 10 and a power of 10. For example, $687{,}000 = 6.87 \times 10^5$.

sequence (258)   A list of numbers in a certain order, such as, 0, 1, 2, 3, or 2, 4, 6, 8.

sides (549)   The two rays that form an angle are called the sides.

sides (589)   The line segments that form a polygon.

similar (578)   Figures that have the same shape but not necessarily the same size.

simplest form (28)   An expression in simplest form has no like terms and no parentheses.

simplest form (196)   A fraction is in simplest form when the GCF of the numerator and the denominator is 1.

simulation (524)   The process of acting out a problem.

sine (688)   If $\triangle ABC$ is a right triangle and A is an acute angle,

$$\text{sine } \angle A = \frac{\text{measure of the side opposite } \angle A}{\text{measure of the hypotenuse}}$$

$$\sin A = \frac{a}{c}$$

skew lines (551)   Two lines that do not intersect and are not in the same plane.

slant height (638)   The length of the altitude of a lateral face of a regular pyramid.

slope (400)   The slope of a line is the ratio of the change in y to the corresponding change in x.

$$\text{slope} = \frac{\text{change in } y}{\text{change in } x}$$

slope-intercept form (407)   The slope-intercept form of a linear equation of a line is $y = mx + b$. The slope of the line is m, and the y-intercept is b.

solution (32)   A value for the variable that makes an equation true. The solution for $12 = x + 7$ is 5.

solution of system (412)   The ordered pair that is the solution of both equations.

solving an equation (32)   The process of finding a solution to an equation.

square (585)   A parallelogram with all sides congruent and all angles congruent.

square pyramid (638)   A pyramid with a square base.

square root (664)   One of the two equal factors of a number. A square root of 144 is 12 since $12^2 = 144$.

standard form (176)   The standard form for seven hundred thirty-nine is 739.

statistics (51)   Information that has been collected, analyzed, and presented in an organized fashion.

stem-and-leaf plot (486)   A system used to condense a set of data where the greatest place value of the data forms the stem and the next greatest place value forms the leaves.

stems (486)   The greatest place value common to all the data values is used for the stem of a stem-and-leaf plot.

straightedge (554)   Any object that can be used to draw a straight line.

substitution property of equality (17)   For all numbers a and b, if $a = b$, then a may be replaced with b.

subtraction property of equality (124)   If you subtract the same number from each side of an equation, the two sides remain equal. For any numbers a, b, and c, if $a = b$, then $a - c = b - c$.

subtraction property of inequality (147)   Subtracting the same number from each side of an inequality does not change the truth of the inequality.
For all numbers a, b, and c:
1. If $a > b$, then $a - c > b - c$.
2. If $a < b$, then $a - c < b - c$.

supplementary (562)   Two angles are supplementary if the sum of their measures is 180°.

surface area (632)   The sum of the areas of all the surfaces (faces) of a 3-dimensional figure.

symmetric (596)   A figure created by a shape and its reflection.

system of equations (412)   A set of equations with the same variables. For example, $c = 10 + 0.1n$ and $c = 5 + 0.2n$ are called a *system of equations*.

tangent (688)   If $\triangle ABC$ is a right triangle and A is an acute angle,

$$\text{tangent } \angle A = \frac{\text{measure of the side opposite } \angle A}{\text{measure of the side adjacent to } \angle A}$$

$$\tan A = \frac{a}{b}$$

term (27)   A number, a variable, or a product of numbers and variables.

term (258)   Each number within a sequence is called a term.

terminating decimal (225, 275)   A decimal whose digits end. Every terminating decimal can be written as a fraction with a denominator of 10, 100, 1000, and so on.

tessellation (594)   A repetitive pattern of polygons that fit together with no holes or gaps.

30°–60° right triangle (683) A right triangle that has one angle measuring 30° and another angle measuring 60°.

transformation (595) A movement of a geometric figure.

translation (595) One type of transformation where a figure is slid horizontally, vertically, or both.

transversal (563) A line that intersects two parallel lines to form eight angles.

trapezoid (585) A quadrilateral with exactly one pair of parallel sides.

tree diagram (509) A diagram used to show the total number of possible outcomes in a probability experiment.

triangle (568) A polygon having three sides.

triangular prism (632) A prism with triangular bases.

triangular pyramid (638) A pyramid with an equilateral triangle as a base.

trigonometric ratios (688) Ratios that involve the measures of the sides of right triangles. The tangent, sine, and cosine ratios are three trigonometric ratios.

trigonometry (688) The study of triangle measurement.

trinomial (706) A polynomial with three terms.

truncate (275) An answer being cut off by a calculator at a certain place-value position, ignoring the digits that follow.

unit rate (433) A rate with a denominator of 1.

upper quartile (491) The median of the upper half of a set of numbers.

variable (16) A placeholder in a mathematical expression or sentence.

Venn diagram (669) A diagram consisting of circles inside a rectangle which is used to show the relationships of sets.

vertex (549) A vertex of an angle is the common endpoint of the rays forming the angle.

vertex (75, 589) A vertex of a polygon is a point where two sides of the polygon intersect.

vertex (638) The vertex of a pyramid is the point where all the faces except the base intersect.

vertical angles (561) Congruent angles formed by the intersection of two lines. In the figure, the vertical angles are ∠1 and ∠3, and ∠2 and ∠4.

vertical line test (373) If any vertical line drawn on the graph of a relation passes through no more than one point of its graph, then the relation is a function.

volume (644) The number of cubic units needed to fill a container.

whole numbers (224) The set of numbers {0, 1, 2, 3, . . .}. It also includes any number that can be written as a whole number, such as $\frac{5}{5}$ or $\frac{9}{1}$.

x-axis (36) The horizontal number line which helps to form the coordinate system.

x-coordinate (37) The first number of an ordered pair.

x-intercept (406) The x-coordinate of the point where the graph crosses the x-axis.

y-axis (36) The vertical number line which helps to form the coordinate system.

y-coordinate (37) The second number of an ordered pair.

y-intercept (406) The y-coordinate of the point where the graph crosses the y-axis.

zero pair (82) The result of pairing one positive counter with one negative counter.

absolute value/valor absoluto (67)   El número de unidades que un número dista de cero en la recta numérica.

acute angle/ángulo agudo (550)   Un ángulo cuya medida es mayor que $0°$ y menor que $90°$.

acute triangle/triángulo agudo (569)   Un triángulo que tiene tres ángulos agudos.

addition property of equality/propiedad de adición de la igualdad (125)   Si sumas el mismo número a ambos lados de una ecuación, los dos lados permanecen igual. Para cualquiera de los números a, b y c, si $a = b$, entonces $a + c = b + c$.

addition property of inequality/propiedad de adición de la desigualdad (147)   El sumar el mismo número a cada lado de una desigualdad, no altera la validez de la desigualdad. Para cualquiera de los números a, b y c:
1. Si $a > b$, entonces $a + c > b + c$.
2. Si $a < b$, entonces $a + c < b + c$.

additive inverses/inversos aditivos (89)   Un número entero y su opuesto. La suma de un número entero y su inverso aditivo es cero.

adjacent angles/ángulos adyacentes (561)   Dos ángulos que tienen un lado común y el mismo vértice, pero que *no* se traslapan.

adjacent leg/cateto adyacente (687)   El cateto de un triángulo rectángulo que no es el opuesto a un ángulo y que no es la hipotenusa.

algebraic expression/expresión algebraica (16)   Una combinación de variables, números y por lo menos una operación.

algebraic fraction/fracción algebraica (197)   Una fracción con variables en el numerador o en el denominador.

alternate exterior angles/ángulos externos alternos (563)   Ángulos exteriores no adyacentes que se encuentran en lados opuestos de la transversal. $\angle 1$ y $\angle 7$, y $\angle 2$ y $\angle 8$ son ángulos externos alternos.

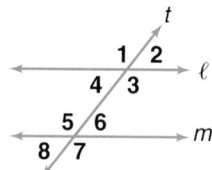

alternate interior angles/ángulos internos alternos (563)   Ángulos internos no adyacentes que se encuentran en lados opuestos de la transversal. $\angle 4$ y $\angle 6$, y $\angle 3$ y $\angle 5$ son ángulos internos alternos.

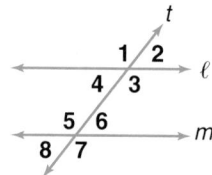

altitude/altitud (612, 614, 633)   Un segmento de recta perpendicular a la base de una figura con extremos en la base y el lado opuesto a la base.

angle/ángulo (549)   Dos rayos con un punto en común forman un ángulo. Los rayos y los vértices se usan para nombrar el ángulo.

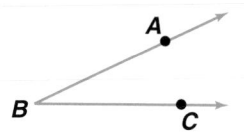

ángulo ABC o $\angle ABC$.

angle of depression/ángulo de depresión (695)   El ángulo de depresión se forma por una línea visual a lo largo de la horizontal y otra línea visual por debajo de la misma.

angle of elevation/ángulo de elevación (694)   Un ángulo de elevación se forma por una línea visual a lo largo de la horizontal y otra línea visual por encima de la misma.

area/área (138)   El número de unidades cuadradas necesarias para cubrir una superficie no encerrada por una figura geométrica.

arithmetic sequence/secuencia aritmética (258)   Una secuencia en que la diferencia entre cualquier par de términos consecutivos es la misma.

associative property of addition/propiedad asociativa de la adición (22, 234)   Para cualquiera de los números racionales a, b y c,
$$(a + b) + c = a + (b + c).$$

associative property of multiplication/propiedad asociativa de la multiplicación (22, 296)   Para cualquiera de los números a, b y c,
$$(ab)c = a(bc).$$

axes/ejes (73)   Dos rectas numéricas que se intersecan forman un sistema de coordenadas.

back-to-back stem-and-leaf plot/diagrama de tallo y hojas consecutivo (487)   Se usa para comparar dos conjuntos de datos. Las hojas de uno de los conjuntos de datos están a un lado del tallo y las hojas para el otro conjunto de datos están en el otro lado.

bar graph/gráfica de barra (52)   Una forma gráfica que usa barras para comparar estadísticas.

bar notation/notación de barra (225)   En decimales que se repiten, la línea o barra que se coloca sobre los dígitos que se repiten. Por ejemplo, $2.\overline{63}$ indica que los dígitos 63, se repiten.

base/base (175)   En 5^3, la base es 5. La base se usa como un factor tantas veces como lo indique el exponente (3). Es decir,
$$5^3 = 5 \times 5 \times 5.$$

base/base (449)   En una proporción de porcentaje, el número con el cual se compara el por ciento.
$$\frac{\text{porcentaje}}{\text{base}} = \text{proporción}$$

base/base (614)   La base de un rectángulo, un paralelogramo o un triángulo es cualquier lado de la figura. Las bases de un trapecio son los lados paralelos.

base/base (632)   Las bases de un prisma son los dos lados congruentes paralelos.

binomial/binomio (706)   Un polinomio con exactamente dos términos.

boundary/frontera (418)   Una línea que separa una gráfica en mitades de planos.

box-and-whisker plot/diagrama de caja y patillas (495) Una representación gráfica que resume datos usando la mediana, los cuartillos superior e inferior y los valores extremos. Se dibuja una caja alrededor del valor cuartílico y las patillas se extienden desde cada cuartillo hasta los puntos extremos de los datos.

center/centro (341)   El punto dado desde el cual todos los puntos en un círculo están equidistantes.

circle/círculo (341)   El conjunto de todos los puntos en un plano que están equidistantes de un punto dado llamado centro.

circle graph/gráfica de círculo (497, 556)   Un tipo de gráfica estadística que se usa para comparar partes de un todo.

circular cone/cono circular (639)   Una figura en el espacio que tiene una base circular y un vértice.

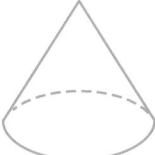

circumference/circunferencia (341)   La distancia alrededor de un círculo.

coefficient/coeficiente (711)   La parte numérica de un monomio.

combination/combinación (516)   Un arreglo o lista en el cual el orden no es importante.

commission/comisión (470)   Un estipendio que se le paga a un vendedor o a una vendedora basado en un porcentaje del total de las ventas.

common difference/diferencia común (258)   La diferencia entre cualquiera par de términos consecutivos en una secuencia aritmética.

common multiples/múltiplos comunes (200)   Múltiplos compartidos por dos o más números. Por ejemplo, algunos múltiplos comunes de 2 y 3 son 6, 12 y 18.

common ratio/proporción común (312)   La proporción entre cualquier par de términos consecutivos en una secuencia geométrica.

commutative property of addition/propiedad conmutativa de la adición (22, 234)   Para cualquiera de los números racionales a y b,
$$a + b = b + a.$$

commutative property of multiplication/propiedad conmutativa de la multiplicación (22, 296)   Para cualquiera de los números racionales a y b,
$$ab = ba.$$

comparative graph/gráfica comparativa (498)   Una gráfica que se usa para comparar resultados de grupos similares y mostrar tendencias.

compass/compás (554)   Un instrumento que se usa para dibujar círculos o partes de círculos.

compatible numbers/números compatibles (280)   Números que han sido redondeados de forma que cuando se dividen uno entre otro, el residuo es cero.

complementary/complementario (562)   Dos ángulos son complementarios si la suma de sus medidas es 90°.

composite number/número compuesto (184)   Cualquier número entero mayor de uno que posee más de dos factores.

congruent/congruente (554, 561, 573)   Segmentos de recta que tienen la misma longitud, o ángulos que tienen la misma medida, o figuras que tienen la misma forma y tamaño.

constant/constante (707)   Un monomio que no contiene una variable.

coordinate/coordenada (67, 126)   Un número asociado con un punto en la recta numérica.

coordinate system/sistema de coordenadas (36, 73)   Un plano de coordenadas separado en cuatro regiones iguales por dos rectas numéricas que se intersecan, llamadas ejes.

corresponding angles/ángulos correspondientes (563)   Ángulos que tienen la misma posición en dos rectas paralelas diferentes cortadas por una transversal. $\angle 1$ y $\angle 5$, $\angle 2$ y $\angle 6$, y $\angle 3$ y $\angle 7$, y $\angle 4$ y $\angle 8$ son ángulos correspondientes.

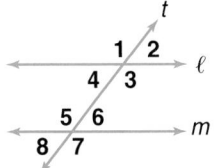

corresponding parts/partes correspondientes (573)   Partes en figuras congruentes o similares que encajan.

cosine/coseno (688)   Si $\triangle ABC$ es un triángulo rectángulo y A es un ángulo agudo,
$$\text{coseno } \angle A = \frac{\text{medida del lado adyacente a } \angle A}{\text{medida de la hipotenusa}}$$
$$\cos A = \frac{b}{c}$$

cylinder/cilindro (633)   Una figura tridimensional que posee dos bases congruentes paralelas (generalmente circulares) y un lado curvo que conecta las bases.

data/datos (50)   Información numérica que se reúne con propósitos estadísticos.

defining a variable/definir una variable (42)   El escoger una variable y una cantidad para la variable para representarla en una ecuación.

degree/grado (549)   La unidad de medida más común para ángulos.

dependent events/eventos dependiente (532)   Dos o más eventos en los cuales el resultado de un evento afecta el resultado de otro(s) evento(s).

diagonal/diagonal (590)   Un segmento de recta que une dos vértices no consecutivos de un polígono.

diameter/diámetro (341)   La distancia a través del centro de un círculo.

discount/descuento (468)   La cantidad que se rebaja del precio original.

discrete mathematics/matemáticas discretas (258)   Una rama de las matemáticas que tiene que ver con los conjuntos de números finitos o interrumpidos. Algunas disciplinas que tratan con las matemáticas discretas son la lógica y la estadística.

distributive property/propiedad distributiva (26)   Para todos los números reales a, b y c, $a(b + c) = ab + ac$ y $(b + c)a = ba + ca$.

divisible/divisible (170)   Un número es divisible entre otro si el residuo de la división es cero.

division property of equality/propiedad de división de la igualdad (130)   Para cualquiera de los números a, b y c, con $c \neq 0$, si $a = b$, entonces $\frac{a}{c} = \frac{b}{c}$.

division properties of inequality/propiedades de división de la desigualdad (152)   Para todos los números enteros a, b y c, en que $c > 0$, si $a > b$, entonces $\frac{a}{b} > \frac{b}{c}$. Para todos los números enteros a, b y c, en que $c < 0$, si $a > b$, entonces $\frac{a}{c} < \frac{b}{c}$.

domain/dominio (372)   El conjunto de todas las coordenadas de cada par ordenado.

empty set/conjunto vacío (347)   Un conjunto sin ningún elemento, el cual se representa con el símbolo { } o ∅.

equation/ecuación (32)   Un enunciado matemático abierto que contiene el signo de igualdad, =.

equiangular/equiangular (591)   Todos los ángulos son congruentes.

equilateral/equilátero (591)   Todos los lados son congruentes.

equivalent/equivalente (275)   El mismo valor representado de diferentes formas. Por ejemplo, $\frac{3}{5}$ y 0.6 son equivalentes.

equivalent equations/ecuaciones equivalentes (124)   Dos o más ecuaciones con la misma solución. Por ejemplo, $x + 3 = 5$ y $x = 2$ son ecuaciones equivalentes.

evaluating the expression/evaluando una expresión (11)   Reemplazar las variables con números y hallar el valor numérico de la expresión.

expanded form/forma desarrollada (176)   Un número que se expresa usando el valor de posición para escribir el valor de cada dígito en el número.

experimental probability/probabilidad experimental (529)   Una posibilidad estimada con base en la frecuencia relativa de los resultados positivos que ocurren durante un experimento.

exponent/exponente (175)   En 5^3, el exponente es 3. El exponente indica cuántas veces la base, 5, se usa como factor.
$$5^3 = 5 \times 5 \times 5$$

exterior angle/ángulo externo (591)   El ángulo que se forma cuando se extiende un lado de un polígono.

face/cara (632)   La cara de un prisma es cualquier superficie que forma un lado o la base del prisma.

factorial/factorial (515)   La expresión $n!$ es el producto de todos los números contables comenzando con n y contando al revés hasta 1.

factors/factores (170)   Números que se dividen entre un número entero y cuyo residuo es cero.

factor trees/árboles de factores (185)   Un diagrama que muestra la factorización prima de un número compuesto. Los factores se ramifican a partir de los factores previos hasta que todos los factores son números primos.

Fibonacci sequence/secuencia de Fibonacci (263)   Un listado de números en el que los dos primeros números son ambos 1 y cada número que sigue es la suma de los dos números previos, 1, 1, 2, 3, 5, 8, . . .

formula/fórmula (134)   Una ecuación que expresa una regla para la relación entre ciertas cantidades.

45°–45° right triangle/triángulo rectángulo de 45°–45° (683)   Un triángulo rectángulo que posee dos ángulos que miden 45° cada uno.

frequency table/tabla de frecuencia (51)   Una tabla que indica el número de valores en cada intervalo.

function/función (373)   Una función es una relación en que cada elemento del dominio se aparea exactamente con un elemento de la amplitud.

functional notation/notación funcional (393)   Las ecuaciones que representan funciones se pueden escribir en *forma de notación*, $f(x)$. El símbolo $f(x)$ se lee "f" de "x" y representa el valor en la amplitud que corresponde al valor de x en el dominio.

fundamental counting principle/principio fundamental de contar (509)   Si el evento M puede ocurrir de m maneras y es seguido por el evento N que puede ocurrir de n maneras, entonces el evento M seguido del evento N puede ocurrir en $m \cdot n$ maneras.

geometric probability/probabilidad geométrica (623)   Probabilidad que usa el área para hallar la probabilidad de un evento.

geometric sequence/secuencia geométrica (312)   Una secuencia en que la proporción entre cualquier par de números consecutivos es la misma.

gram/gramo (359)   Una unidad de masa del sistema métrico.

graph/gráfica (73, 126)   Un puntito que representa un número en una recta numérica o un par ordenado en un plano de coordenadas.

greatest common factor (GCF)/máximo factor común (MFC) (190)   El MFC de dos o más números enteros es el máximo factor que es común a cada uno de los números.

histogram/histograma (502)   Una gráfica de barra cuyas columnas están una al lado de la otra para mostrar cómo están distribuidos los datos.

hypotenuse/hipotenusa (676)   El lado opuesto al ángulo recto en un triángulo rectángulo.

identity property of addition/propiedad de identidad de la adición (23, 234)   Para cualquier número *a*, $a + 0 = a$ y $0 + a = a$.

identity property of multiplication/propiedad de identidad de la multiplicación (23, 296)   Para cualquier número *a*, $a \times 1 = a$.

inclusive/inclusivo (536)   Cuando dos eventos pueden suceder al mismo tiempo.

independent events/eventos independientes (531)   Dos o más eventos en los cuales el resultado de uno de ellos *no* afecta el resultado de ninguno de los otros.

indirect measurement/medida indirecta (580)   Hallar una medida mediante el uso de triángulos similares y escribiendo una proporción.

inequality/desigualdad (46)   Un enunciado matemático que contiene $<, >, \neq, \leq$ o $\geq$.

integers/números enteros (66, 224)   Los números enteros y sus opuestos.
$$\ldots, -3, -2, -1, 0, 1, 2, 3, \ldots$$

intercepts/intersecciones (406)   Los puntos en donde una gráfica cruza los ejes.

interior angles/ángulos internos (591)   Un ángulo dentro de un polígono.

interquartile range/amplitud intercuartílica (491)   La amplitud de la mitad central de un conjunto de datos.
$$\text{Amplitud intercuartílica} = UQ - LQ,$$

intersect/intersecar (552)   Líneas que se cruzan o se tocan unas a otras en un punto.

inverse operations/operaciones inversas (41)   Operaciones que se anulan entre sí, tales como la adición y la sustracción.

inverse property of addition/propiedad del inverso de la adición (234)   Para cualquier número racional *a*, $a + (-a) = 0$.

inverse property of multiplication/propiedad del inverso de la multiplicación (289)   Para todo número no cero $\frac{a}{b}$, donde *a*, $b \neq 0$, hay exactamente un número $\frac{b}{a}$, tal que $\frac{a}{b} \times \frac{b}{a} = 1$.

irrational number/número irracional (672)   Un número que no se puede expresar como $\frac{a}{b}$, donde *a* y *b* son números enteros y *b* no es igual a 0. Por ejemplo, $\sqrt{3} = 1.7320508\ldots$ es un número irracional.

isosceles triangle/triángulo isósceles (569)   Un triángulo que tiene por lo menos dos lados congruentes.

lateral surface/superficie lateral (639)   La superficie lateral de un prisma, cilindro, pirámide o cono es toda la superficie de la figura excepto la base o las bases.

least common denominator (LCD)/mínimo común denominador (MCD) (201)   El mínimo común múltiplo de los denominadores de dos o más fracciones.

least common multiple (LCM)/mínimo común múltiplo (MCM) (200)   El menor número diferente de cero de los múltiplos comunes de dos o más números. El MCM de 2 y 3 es 6.

leaves/hojas (486)   El valor de posición que le sigue al valor mayor forma las hojas en una gráfica de tallo y hojas.

leg/cateto (676)   Cualquiera de los dos lados que forman el ángulo recto de un triángulo rectángulo.

like terms/términos semejantes (27, 711)   Expresiones que contienen las mismas variables, tales como $3ab$ y $7ab$.

line/recta (548)   Una trayectoria recta sin fin. Una representación de la recta CD $(\overleftrightarrow{CD})$ se muestra a continuación.

linear equation/ecuación lineal (386)   Una ecuación cuya gráfica es una recta.

line graph/gráfica lineal (497)   Un tipo de gráfica estadística que se usa para mostrar cómo los valores cambian a través de un período de tiempo.

line segment/segmento de recta (548)   Dos extremos y la trayectoria recta entre ellos. Un segmento de recta se denomina por sus extremos.

liter/litro (359)   La unidad básica de capacidad en el sistema métrico.

lower quartile/cuartillo inferior (491)   La mediana de la mitad inferior de un conjunto de números indicados por *LQ*.

mean/media (300)   La suma de los números en un conjunto de datos dividida entre el número de datos.

measures of central tendency/medidas de tendencia central (300)   Datos o grupos de datos que pueden representar el conjunto total de datos.

median/mediana (300)   En un conjunto de datos, la mediana es el número en el centro cuando los datos se organizan en orden de menor a mayor.

meter/metro (359)   La unidad básica de longitud en el sistema métrico.

metric system/sistema métrico (360)   Un sistema de medida que usa las unidades básicas: metro para longitud, gramo para masa y litro para capacidad.

mixed numbers/números mixtos (239)   La suma de un número entero y una fracción. Por ejemplo, $3\frac{1}{2}$.

mode/modal (300)   El número o artículo que aparece con más frecuencia en un conjunto de datos.

monomial/monomio (172)   Una expresión algebraica que es un número real, una variable o el producto de números reales y variables.

multiple/múltiplo (200)   El producto del número y cualquier número entero.

multiplication properties of inequality/propiedades multiplicativas de la desigualdad (152) **1.** Para todos los números enteros *a*, *b* y *c*, en que *c* > 0, si *a* > *b*, entonces *ac* > *bc*. **2.** Para todos los números enteros *a*, *b* y *c*, en que *c* < 0, si *a* < *b*, entonces *ac* > *bc*.

multiplication property of equality/propiedad multiplicativa de la igualdad (130) Para cualquiera de los números reales *a*, *b* y *c*, si *a* = *b*, entonces *ac* = *bc*.

multiplicative inverses/inversos multiplicativos (289) Dos números cuyo producto es 1. El inverso multiplicativo de $\frac{3}{4}$ es $\frac{4}{3}$.

multiplicative property of zero/propiedad multiplicativa de cero (23) Para cualquier número *a*, $a \times 0 = 0$.

mutually exclusive/exclusivo mutuamente (535) Dos o más eventos tales que ningún par de eventos pueden suceder al mismo tiempo.

negative/negativo (66) Un valor menor que cero.

negative relationship/relación negativa (380) Si los puntos en un diagrama de dispersión parecen sugerir una recta que se inclina hacia abajo y hacia la derecha, entonces existe una *relación negativa*.

net/red (632) La figura que se forma al "desdoblar" un prisma. La red muestra todas las caras que componen el área de un prisma.

no relationship/no relacionado (380) Se dice que los puntos que parecen ser aleatorios en un diagrama de dispersión son no relacionados.

null set/conjunto vacío (347) Un conjunto sin elementos y el cual se representa con el símbolo { } o ∅.

obtuse angle/ángulo obtuso (550) Cualquier ángulo que mide más de 90° pero menos de 180°.

obtuse triangle/triángulo obtuso (569) Un triángulo con un ángulo obtuso.

odds/posibilidades (520) Una manera de describir la posibilidad de que un evento ocurra.
posibilidades en favor = número de éxitos: número de fracasos
posibilidades en contra = número de fracasos: número de éxitos

open sentence/enunciado abierto (32) Una ecuación que contiene variables.

opposite side/lado opuesto (687) El lado que no forma parte del ángulo en un triángulo rectángulo.

ordered pair/par ordenado (37) Un par de números utilizados para ubicar un punto en el sistema de coordenadas.

order of operations/orden de operaciones (12, 176)
  1. Simplifica las expresiones dentro de los símbolos de agrupación.
  2. Evalúa todas las potencias.
  3. Luego, realiza todas las multiplicaciones y las divisiones de izquierda a derecha.
  4. Finalmente, realiza todas las sumas y las restas de izquierda a derecha.

origin/origen (36) El punto de intersección del eje *x* y del eje *y* en un sistema de coordenadas.

outcomes/resultados (509) Posibles resultados de un evento de probabilidad. Por ejemplo, 4 es un resultado cuando se lanza un dado.

outliers/valores atípicos (496) Datos que están a más de 1.5 veces que la amplitud intercuartílica de los cuartillos.

parallel/paralelo (551) Rectas en el mismo plano que no se intersecan. El símbolo ∥ quiere decir paralelo.

parallelogram/paralelogramo (585) Un cuadrilátero con dos pares de lados paralelos.

percent/por ciento (449) Una proporción con un denominador de 100. Por ejemplo, 77% y $\frac{77}{100}$ son el mismo número.

percentage/porcentaje (449) En una proporción de por ciento, un número que se compara con otro llamado base.
$$\frac{\text{porcentaje}}{\text{base}} = \text{tasa}$$

percent of change/porcentaje de cambio (472) La proporción de la cantidad de cambio comparada con la cantidad inicial.

percent of decrease/porcentaje de disminución (473) La proporción de una cantidad de disminución comparada con la cantidad previa, expresada en forma de porcentaje.

percent of increase/porcentaje de aumento (472) La proporción de una cantidad de aumento comparada con la cantidad previa, expresada en forma de porcentaje.

percent proportion/proporción de porcentaje (450)
$$\frac{\text{porcentaje}}{\text{base}} = \text{tasa o } \frac{P}{B} = \frac{r}{100}$$

perfect squares/cuadrados perfectos (665) Números racionales cuyas raíces cuadradas son números enteros. 25 es un cuadrado perfecto porque $\sqrt{25} = 5$.

perimeter/perímetro (138) La distancia alrededor de una figura geométrica.

permutation/permutación (515) Un arreglo o listado en que el orden es importante.

perpendicular/perpendicular (551) Dos rectas que se intersecan para formar un ángulo recto.

pi, π/pi, π (341) La proporción de la circunferencia de un círculo con el diámetro del círculo.
Aproximaciones para π son 3.14 y $\frac{22}{7}$.

pictograph/pictografía (497) Un tipo de gráfica estadística que usa láminas o ilustraciones para mostrar comparación entre cantidades específicas. Cada símbolo representa un número específico de artículos.

plane/plano (550) Una superficie plana sin aristas o fronteras. Un plano se puede nombrar con una sola letra mayúscula o mediante el uso de tres puntos no colineales del plano.

point/punto (548) Una ubicación específica en el espacio.

polygon/polígono (589) Una figura cerrada simple en un plano formado por tres o más segmentos de recta.

polynomial/polinomio (706) Un monomio o la suma o diferencia de dos o más monomios.

positive relationship/relación positiva (380) En una gráfica de dispersión, si los puntos parecen sugerir una recta que se inclina hacia arriba y a la derecha, entonces existe una *relación positiva*.

power/potencia (175)   Un número que se puede escribir usando un exponente.

power of a monomial/potencia de un monomio (721)   Para todos los números *a* y *b*, y los números enteros positivos *m*, *n* y *p*, $(a^m b^n)^p = a^{mp} b^{np}$.

power of a power/potencia de una potencia (719)   Para cualquier número *a* y cualquiera de los números enteros *m* y *n*, $(a^m)^n = a^{mn}$.

power of a product/potencia de un producto (720)   Para todos los números *a* y cualquier número entero positivo *m*, $(ab)^m = a^m b^m$.

prime factorization/factorización prima (185)   El expresar un número compuesto como un producto de sus factores primos.

prime number/número primo (184)   Un número entero mayor que 1 cuyos únicos factores son 1 y el número mismo.

principle/capital (467)   La cantidad de dinero en una cuenta.

prism/prisma (632)   Figura en el espacio que tiene dos bases paralelas y congruentes en forma de polígonos.

probability/probabilidad (440)   La posibilidad de que algún evento pueda suceder. Es la proporción del número de veces que cierto evento puede ocurrir comparada con el número de resultados posibles.

Probabilidad = $\dfrac{\text{número de resultados favorables}}{\text{número de resultados posibles}}$.

probability of inclusive events/probabilidad de eventos inclusivos (536)   La probabilidad de que uno u otro de dos eventos inclusivos ocurra se puede hallar sumando la probabilidad del primero a la probabilidad del segundo evento y restando la probabilidad de que ambos eventos ocurran.
$$P(A \text{ o } B) = P(A) + P(B) - P(A \text{ y } B)$$

probability of mutually exclusive events/probabilidad de eventos exclusivos mutuamente (535)   La probabilidad de que el primero o el segundo de dos eventos mutuamente exclusivos ocurra se puede hallar sumando la probabilidad del primero a la probabilidad del segundo evento.
$$P(A \text{ o } B) = P(A) + P(B)$$

probability of two dependent events/probabilidad de dos eventos dependientes (532)   Si dos eventos *A* y *B* son dependientes, entonces la probabilidad de que ambos eventos ocurran es el producto de la probabilidad de *A* y de la probabilidad de *B* después de que *A* ocurra.
$$P(A \text{ y } B) = P(A) \times P(B \text{ seguida de } A)$$

probability of two independent events/probabilidad de dos eventos independientes (531)   La probabilidad de dos eventos independientes se puede hallar multiplicando la probabilidad del primero por la probabilidad del segundo evento.
$$P(A \text{ y } B) = P(A) \times P(B)$$

product of powers/producto de potencias (205)   Para cualquiera de los números *a* y los números enteros positivos *m* y *n*, $a^m \times a^n = a^{m+n}$.

property of proportions/propiedad de las proporciones (444)

Si $\dfrac{a}{b} = \dfrac{c}{d}$, entonces $ad = bc$.

Si $ad = bc$, entonces $\dfrac{a}{b} = \dfrac{c}{d}$.

proportion/proporción (444)   Un enunciado de la igualdad de dos o más razones.

protractor/transportador (549)   Un instrumento que se usa para medir ángulos.

pyramid/pirámide (638)   Una figura sólida que tiene un polígono por base y triángulos por lados.

Pythagorean theorem/teorema de Pitágoras (676)   En un triángulo rectángulo, el cuadrado del largo de la hipotenusa es igual a la suma de los cuadrados de los largos de los catetos.
$$c^2 = a^2 + b^2$$

quadrants/cuadrantes (73)   Las cuatro regiones en que dos rectas numéricas perpendiculares separan el plano.

quadrilateral/cuadrilátero (584, 585)   Un polígono con cuatro lados.

quartile/cuartillo (491)   Una de las cuatro partes iguales de datos de un conjunto grande de números.

quotient of powers/cociente de potencias (206)   Para todo número *a* diferente de cero y los números enteros *m* y *n*, $\dfrac{a^m}{a^n} = a^{m-n}$.

radical sign/signo radical (664)   El símbolo que se usa para indicar una raíz cuadrada no negativa. $\sqrt{}$

radius/radio (341)   La distancia desde el centro de un círculo hasta cualquier punto en el círculo.

range/amplitud (372)   La amplitud de una relación es el conjunto de todas las segundas coordenadas de cada par ordenado.

range/amplitud (490)   La diferencia entre los valores mayor y menor en un conjunto de datos.

rate/razón (433)   La proporción de dos medidas que tienen diferentes unidades.

rate/tasa (449)   En una proporción de por ciento, la razón de un número a 100.

ratio/razón (196, 432)   Una comparación de dos números mediante división. La razón de 2 a 3 se puede expresar como 2 de 3, 2 a 3 ó $\frac{2}{3}$.

rational numbers/números racionales (224)   Números que se pueden expresar como $\frac{a}{b}$, en que *a* y *b* son números enteros y $b \neq 0$.

ray/rayo (548)   Una parte de una recta que se extiende infinitamente en una dirección.

real numbers/números reales (673)   El conjunto de números racionales unido con el conjunto de números irracionales.

reciprocal/recíproco (289)   Otro nombre para el inverso multiplicativo.

rectangle/rectángulo (585)   Un cuadrilátero con cuatro ángulos congruentes.

rectangular prism/prisma rectangular (632)   Un prisma con bases rectangulares.

reflection/reflexión (595)   Un tipo de transformación en que una figura se voltea alrededor de una línea.

regular polygon/polígono regular (591)   Un polígono que tiene todos los lados y todos los ángulos congruentes.

relation/relación (372)   Un conjunto de pares ordenados.

repeating decimal/decimal periódico (225)   Un decimal cuyos dígitos se repiten en grupos de uno o más. Por ejemplo: 0.181818 . . . y 0.8333. . . .

rhombus/rombo (585)   Un paralelogramo con cuatro lados congruentes.

right angle/ángulo recto (550)   Un ángulo que mide 90°.

right triangle/triángulo rectángulo (569)   Un triángulo con un ángulo recto.

rotation/rotación (595)   Cuando se hace girar una figura alrededor de un punto.

sample/muestra (50, 51)   Un grupo que se usa para representar una población entera.

sample space/espacio muestral (441)   El conjunto de todos los resultados posibles.

scale drawing/dibujo a escala (585)   Un dibujo que es similar, pero que es o más grande o más pequeño que el objeto en sí.

scalene triangle/triángulo escaleno (569)   Un triángulo sin lados congruentes.

scatter plot/diagrama de dispersión (378)   En un diagrama de dispersión, se grafican dos conjuntos de datos como pares ordenados en el plano de coordenadas.

scientific notation/notación científica (317)   Una manera de expresar un número como el producto de un número que es por lo menos 1, pero menos de 10, y una potencia de 10. Por ejemplo, $687{,}000 = 6.87 \times 10^5$.

sequence/secuencia (258)   Una lista de números en cierto orden, como por ejemplo: 0, 1 2 3 ó 2, 4, 6, 8.

sides/lados (549)   Los dos rayos que forman un ángulo se llaman lados.

sides/lados (589)   Los segmentos de recta que forman un polígono.

similar/similar (578)   Figuras que tienen la misma forma, pero no necesariamente el mismo tamaño.

simplest form/forma reducida (28)   Una expresión en forma reducida no tiene términos semejantes ni paréntesis.

simplest form/forma reducida (196)   Una fracción está en forma reducida cuando el máximo factor común (MFC) del numerador y del denominador es 1.

simulation/simulación (524)   El proceso de representar un problema.

sine/seno (688)   Si $\triangle ABC$ es un triángulo rectángulo y A es un ángulo agudo,

$$\text{seno } \angle A = \frac{\text{medida del lado opuesto a } \angle A}{\text{medida de la hipotenusa}}$$

$$\text{sen } A = \frac{a}{c}.$$

skew lines/rectas alabeadas (551)   Dos rectas que no se intersecan y no están en el mismo plano.

slant height/altura inclinada (638)   El largo de la altura de una cara lateral de una pirámide regular.

slope/pendiente (400)   La pendiente de la recta es la razón del cambio en y al cambio correspondiente en x.

$$\text{Pendiente} = \frac{\text{cambio en } y}{\text{cambio en } x}$$

slope-intercept form/forma de pendiente-intersección (407)   La forma de pendiente-intersección de una ecuación lineal de una recta es $y = mx + b$. La pendiente de la recta es m y la intersección con el eje y es b.

solution/solución (32)   El valor de una variable que hace que una ecuación sea válida. La solución de $12 = x + 7$ es 5.

solution of system/solución de un sistema (412)   El par ordenado que resuelve ambas ecuaciones en el sistema.

solving an equation/resolviendo una ecuación (32)   El proceso de hallar una solución a una ecuación.

square/cuadrado (585)   Un paralelogramo con todos los lados congruentes y todos los ángulos congruentes.

square pyramid/pirámide cuadrada (638)   Una pirámide con una base cuadrada.

square root/raíz cuadrada (664)   Uno de los dos factores iguales de un número. La raíz cuadrada de 144 es 12 puesto que $12^2 = 144$.

standard form/forma estándar (176)   La forma estándar de setecientos treinta y nueve es 739.

statistics/estadística (51)   Información que ha sido reunida, analizada y presentada de una manera organizada.

stem-and-leaf plot/gráfica de tallo y hojas (486)   Un sistema que se usa para condensar un conjunto de datos en el cual el mayor valor de posición de los datos forma el tallo y el siguiente valor de posición forma las hojas.

stems/tallos (486)   El mayor valor de posición común a todos los valores de los datos y que forma el tallo en una gráfica de tallo y hojas.

straightedge/regla (554)   Cualquier objeto que se puede usar para dibujar una recta.

substitution property of equality/propiedad de sustitución de la igualdad (17)   Para todos los números a y b, si $a = b$, entonces, a se puede reemplazar por b.

subtraction property of equality/propiedad de sustracción de la igualdad (124)   Si restas el mismo número de cada lado de una ecuación, los dos lados permanecen iguales. Para cualquiera de los números a, b y c, si $a = b$, entonces $a - c = b - c$.

subtraction property of inequality/propiedad de sustracción de la desigualdad (147)   Si restas el mismo número de cada lado de una desigualdad, la validez de la desigualdad no cambia.
Para todos los números a, b y c:
1. si $a > b$, entonces $a - c > b - c$;
2. si $a < b$, entonces $a - c < b - c$.

supplementary/suplementario (562)   Dos ángulos son suplementarios si la suma de sus medidas es 180°.

surface area/área de superficie (632)   La suma de las áreas de todas las superficies (caras) de una figura tridimensional.

symmetric/simétrico (596)   Una figura creada por la figura y su reflexión.

system of equations/sistema de ecuaciones (412) Un conjunto de ecuaciones con las mismas variables. Por ejemplo: $c = 10 + 0.1n$ y $c = 5 + 0.2n$ se llaman un *sistema de ecuaciones*.

tangent/tangente (688) Si $\triangle ABC$ es un triángulo rectángulo y A es un ángulo agudo,

tangente $\angle A = \dfrac{\text{medida del lado opuesto a } \angle A}{\text{medida del lado adyacente al } \angle A}$

$\tan A = \dfrac{a}{b}$.

term/término (27) Un número, una variable o un producto de números y variables.

term/término (258) Cada número en una secuencia se llama un término.

terminating decimal/decimal terminal (225, 275) Un decimal cuyos dígitos terminan. Cada decimal terminal se puede escribir en forma de fracción con un denominador de 10, 100, 1000 y así sucesivamente.

tessellation/teselación (594) Un patrón repetitivo de polígonos que encajan juntos sin dejar espacios vacíos.

30°–60° right triangle/triángulo rectángulo de 30°–60° (683) Un triángulo rectángulo que tiene un ángulo que mide 30° y otro que mide 60°.

transformation/transformaciones (595) Movimientos de figuras geométricas.

translation/traslación (595) Un tipo de transformación en la cual una figura se desliza horizontal, verticalmente o en ambas direcciones.

transversal/transversal (563) Una recta que interseca dos líneas paralelas para formar ocho ángulos.

trapezoid/trapezoide (585) Un cuadrilátero con exactamente un par de lados paralelos.

tree diagram/diagrama de árbol (509) Un diagrama que se usa para mostrar el número total de posibles resultados en un experimento de probabilidad.

triangle/triángulo (568) Un polígono con tres lados.

triangular prism/prisma triangular (632) Un prisma con bases triangulares.

triangular pyramid/pirámide triangular (638) Una pirámide cuya base es un triángulo equilátero.

trigonometric ratios/razones trigonométricas (688) Razones que involucran las medidas de los lados de triángulos rectángulos. La tangente, el seno y el coseno son tres razones trigonométricas.

trigonometry/trigonometría (688) El estudio de la medición de triángulos.

trinomial/trinomio (706) Un polinomio con tres términos.

truncate/truncado (275) Una respuesta que la calculadora interrumpe en una posición determinada, ignorando los dígitos que siguen.

unit rate/razón unitaria (433) Una razón cuyo denominador es 1.

upper quartile/cuartillo superior (491) La mediana de la mitad superior de un conjunto de datos.

variable/variable (16) Símbolo que ocupa un lugar en expresiones matemáticas.

Venn diagram/diagrama de Venn (669) Un diagrama que consiste en círculos dentro de un rectángulo y el cual se usa para mostrar las relaciones de conjuntos.

vertex/vértice (549) El vértice de un ángulo es el extremo común de los rayos que forman el ángulo.

vertex/vértice (75, 589) El vértice de un polígono es un punto en donde se intersecan dos lados del polígono.

vertex/vértice (638) El vértice de una pirámide es el punto en donde se intersecan todas las caras, excepto la base.

vertical angles/ángulos verticales (561) Ángulos congruentes formados por la intersección de dos rectas. En la figura, los ángulos verticales son $\angle 1$ y $\angle 3$, y $\angle 2$ y $\angle 4$.

vertical line test/prueba de recta vertical (373) Si cualquier recta vertical dibujada en la gráfica de una relación pasa por un solo punto de esa gráfica, entonces la relación es una función.

volume/volumen (644) El número de unidades cúbicas que se necesitan para llenar un recipiente.

whole numbers/números enteros (224) El conjunto de números {0,1,2,3, . . .}. También incluye cualquier número que se pueda escribir como un número entero, por ejemplo, $\dfrac{5}{5}$ ó $\dfrac{9}{1}$.

x-axis/eje x (36) La recta numérica horizontal que ayuda a formar el plano de coordenadas.

x-coordinate/coordenada x (37) El primer número en un par ordenado.

x-intercept/intersección con el eje x (406) La coordenada x del punto sobre el cual una gráfica cruza el eje x.

y-axis/eje y (36) La recta numérica vertical que ayuda a formar el plano de coordenadas.

y-coordinate/coordenada y (37) El segundo número en un par ordenado.

y-intercept/intersección con el eje y (406) La coordenada y del punto sobre el cual una gráfica cruza el eje y.

zero pair/par cero (82) El resultado de aparear un contador positivo con un contador negativo.

SELECTED ANSWERS

Chapter 1 Tools for Algebra and Geometry

Pages 9–10 Lesson 1-1

5a. We know the bonus for 2, 4, 6, and 8 packages over plan. We need to know the bonus for 16 packages over plan. **5b.** Extend the pattern. The bonus should be about $250.

5c.

10	$200
12	$225
14	$250
16	$275

5d. The answer is reasonable and answers the question. **7.** yes **9.** $1.87

11a.

11b. 2, 4, 6, 8, 10, 12 **11c.** 1, 2, 3, 4, 5, 6 **11d.** The number of cuts is half the number of pieces.

Pages 14–15 Lesson 1-2

7. ÷, 61 **9.** ×, 5 **11.** −, 3 **13.** 8 **15.** 9 **17.** 12
19. 4 **21.** 4 **23.** 1 **25.** 145 **27.** 60 **29.** 128
31. $71 - (17 + 4) = 50$ **33.** $18 \div (3 + 6) + 12 = 14$
35a. 3($330) + 2($247) + 4($229) **35b.** $2400
37. $252 **39.** 2024, 3024, 7224

Pages 19–20 Lesson 1-3

5. 19 **7.** 12 **9.** 10 **11.** $3 + v$ **13.** $6n$ **15a.** $6x$
15b. 204 pounds **17.** 2 **19.** 14 **21.** 4 **23.** 12
25. 3 **27.** 10 **29.** $92 \div c$ **31.** $p - 5$ **33.** $u - 10$
35. $88 \div b$ **37.** $3k$ **39.** $2s + 2$ **41.** 16 less some number **43.** the quotient of some number and 5
45. twice some number plus 1 **47a.** $c \div 4 + 37$
47b. 68° **49a.** 4($6.25) + 2($3.75) **49b.** $32.50
51. 29

Page 21 Lesson 1-3B

1. 14 **3.** 7 **5.** 55 **7.** Use REPLAY then change the expression.

Pages 24–25 Lesson 1-4

5. associative addition **7.** commutative addition
9. 29 **11.** 48 **13.** $(7 \cdot 6)z; 42z$ **15.** $8.35
17. commutative multiplication **19.** multiplicative zero **21.** commutative addition **23.** commutative addition **25.** commutative addition
27. associative addition **29.** 360 **31.** 0 **33.** 130
35. 660 **37.** $9 + 5$ **39.** $18w + 9$ **41.** $b(6 \cdot 5); 30b$
43. $(13 + 11) + m; 24 + m$ **45.** $p(7 \cdot 4); 28p$
47. $15 + (4w + w); 15 + 5w$ **49b.** Division is not commutative. **53.** 6 **55.** C

Pages 28–30 Lesson 1-5

5. 5(7) + 5(8) **7.** 4(x) + 4(3) **9.** $13a + 7$
11. $7c + 7$ **13.** $20ab$ **15.** 3(11) + 3(12)
17. $t(6 + 11)$ **19.** $2(r + 6s)$ **21.** $9(v) + 8(v)$
23. $2(4x) + 2(9y)$ **25.** $38a + 45$ **27.** $25c + 16$
29. $33y + 35$ **31.** $32b + 60$ **33.** $72 + 25f$
35. $11x + 15y$ **37.** 12,000 cubic meters
39. $8 + (19 + 17)$ **41.** $n \div 9$ **43.** B **45.** yes

Page 30 Self Test

1. 1 5 10 10 5 1
   1 6 15 20 15 6 1
3. 59 **5.** 2 **7.** commutative addition **9.** $62.50

Page 31 Lesson 1-5B

1. F **3.** T **5.** F

Pages 34–35 Lesson 1-6

7. 13 **9.** 5 **11.** 50 **13a.** $220 + d = 360$
13b. 140 people **15.** 39 **17.** 32 **19.** 1 **21.** 6
23. 120 **25.** 8 **27.** 9 **29.** 11 **31.** 7 **33.** 7
35a. $3973.05 + c = 4003.33$ **35b.** 30.28 **37.** $7x + 1$
39. B **41.** 5

Pages 38–40 Lesson 1-7

5. (3, 2) and (3, 3) **7.** State Capitol **9.** no **11.** M
13. L **15.** T **17.** J **19.** (6, 4) **21.** (5, 0) **23.** (9, 2)
25. (10, 8) **27a.** (a, 0) **27b.** (0, b) **27c.** origin
29a.

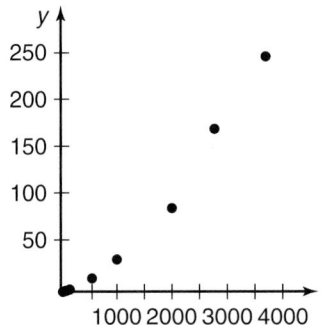

29b. about 78 years **31a.** 50 cars are waiting at the light at 7:00 A.M. and 10 cars are waiting at 9:00 A.M. **31b.** The workday started at most businesses in the area. **31c.** Sample answer: Add point (5:15, 50). About the same number of cars will take people out from work that took them there in the morning. **33.** $10m + 1$ **35.** 12 **37.** 1

Pages 43–45 Lesson 1-8

5. Sample answer: $12 - 9 = 3$ **7.** Sample answer: $4 + 12 = 16$ **9.** 14 **11.** 11.88 **13.** $n + 27 = 31$; $n = 4$ **15.** Sample answer: Evan missed 7 questions on the quiz. If his score was 28, how many points were possible? **17.** $14 = 11 + 3$
19. $14 = 28 \div 2$ **21.** $h = 48 \div 6$ **23.** 22 **25.** 88
27. 120 **29.** 0.83 **31.** 3.77 **33.** 0.6 **35.** $3x = 18$; 6
37. $n - 7 = 22$; 29 **39.** $n \div 4 = 14$; 56 **41.** Sample answer: Jack scored 16 points in this week's game. That is two more than he scored last week.

How many points did Jack score last week?
43. Sample answer: The product of any number and 0 is 0 according to the multiplicative property of zero. Thus, we could rewrite $5 = a \cdot 0$ as $5 = 0$. $5 \neq 0$, so division by zero does not make sense.
45. 60 times **47.** 19 **49.** associative property of addition **51.** $5 \cdot t$ **53a.** Sample answer: calculator **53b.** about 317 miles

Pages 48–49 Lesson 1-9
5. false **7.** true **9.** true **11.** $d \geq 27$ **13.** true
15. open **17.** true **19.** true **21.** open **23.** true
25. true **27.** true **29.** $t \geq \$100$ **31.** $\ell \geq 3$
33. $p < 15$ **35.** $f > 80{,}000$ **37a.** $t \leq 10$ years
37b. $t \leq 1$ year **39.** 11 **41.** $(8 \cdot 7)c$; $56c$ **43.** 9

Page 50 Lesson 1-10A
1. 9–11 hours

Pages 54–55 Lesson 1-10
7. Miami Metrozoo and Toledo **9.** 13.6 million
11. 15 **13.** yes, the total of the states is 51
15. Most states will issue a driver's license to a 16-year-old, all to an 18-year-old.
17.

Height	Buildings
0–100	0
101–200	0
201–300	0
301–400	4
401–500	8
501–600	4
601–700	1
701–800	3
801–900	0
901–1000	1

19.

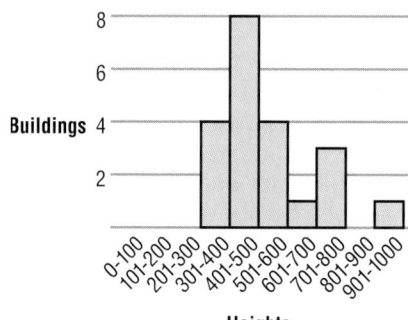

Seattle, WA Buildings

21a. 41 **21b.** 72–74 in. **21c.** You cannot tell from the frequency table how many presidents were exactly six feet tall. The table shows how many presidents fell in different height ranges, not exact heights. **21d.** Leadership is associated with tallness. **23.** $s \geq \$210{,}070$

25.

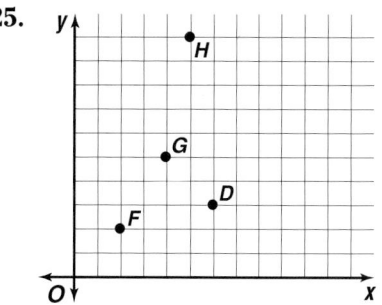

27. B

Page 57 Chapter 1 Highlights
1. algebraic expression **3.** in simplest form
5. addition **7.** like terms

Pages 58-61 Study Guide and Assessment
9. 9 **11.** 16 **13.** 6 **15.** 60 **17.** 4 **19.** 10 **21.** 1
23. 24 **25.** $10n$ **27.** $b + 10$ **29.** $r - 12$
31. commutative, $+$ **33.** associative, $\times$
35. multiplicative identity **37.** $8n$ **39.** $18x + 8$
41. $10x + 26y$ **43.** 6 **45.** 9 **47.** 3 **49.** 5 **51.** N
53. (5, 6) **55.** (0, 4) **57.** 12 **59.** 28.35 **61.** 108
63. false **65.** false **67.** $p < 60$
69.

Representatives	States
1–10	38
11–20	7
21–30	3
31–40	1
41–50	0
51–60	1

71a. Sample answer: Estimation because an exact answer is not needed. **71b.** about 28,000 miles

Chapter 2 Exploring Integers

Pages 69–70 Lesson 2-1
7. -5 **9.** $+13$ **11.** 5 **13.** 10 **15.** 12
17a. $-15, -24$

17b.

19.

21.

23. $+400$ **25.** -3 **27.** $+5$ **29.** 11 **31.** 0 **33.** 8
35. 9 **37.** 12 **39.** -36 **41.** 2 **43.** 6 **45.** 2
47. 8, 0, -8 **49.** It is sometimes true. If A and B are both positive, both negative, or one is 0, it is always true. If one number is negative and the other is positive, it is never true.

51.

53. 17–18

55.

Physical Science Test Scores

57. 180 square feet   **59.** $3x - 8$

Page 71   Lesson 2-1B

1. −1   **3.** Sample answer: There seems to be a cluster between −2 and 1.   **5.** Sample answer: Northern states are losing House members while southern states are gaining House members.

Pages 74–76   Lesson 2-2

5. B   **7.** D   **9.** IV   **11.** none   **13a.** I   **13b.** III **13c.** II   **15.** $(4, -2)$ **17.** $(1, -4)$ **19.** $(4, 3)$ **21.** $(2, 2)$ **23.** $(-1, 5)$ **25.** II   **27.** III   **29.** IV **31.** I   **33.** III   **35.** none

37.

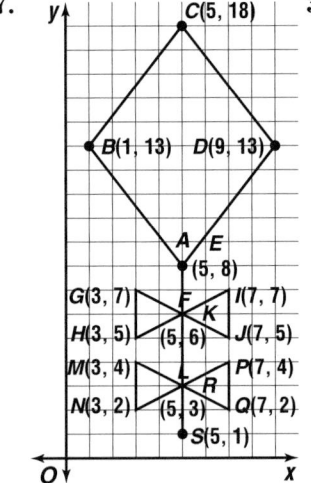

39.

41a. $P(2, 2)$, $Q(2, 5)$, $R(7, 5)$, $S(7, 2)$ **41b.** Rectangle *PQRS* shifted 4 units down   **41c.** A rectangle shaped like *PQRS* in quadrant III   **43.** 12 **45.** $s < 250,000$ **47.** 22   **49.** eleven 2-point; eight 1-point

Pages 77–79   Lesson 2-2B

1.

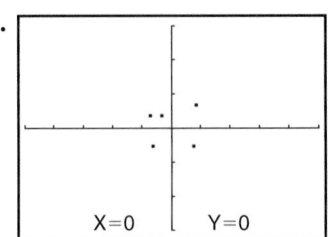

3.

5a. $[-10, 10]$ by $[-10, 10]$ **5b.** $[-47, 47]$ by $[-31, 31]$

Pages 79–81   Lesson 2-3

5. $2 > -5$ **7.** $<$ **9.** $=$ **11.** $\{-88, -9, -4, 0, 3, 43, 234\}$ **13.** $-2 < 3, 3 > -2$ **15.** $>$ **17.** $=$ **19.** $>$ **21.** $<$ **23.** $<$ **25.** $<$ **27.** $3 > 2$ **29.** $60 < 75$ **31.** $565 > 344$ **33.** Sample answer: $66 < 265$ **35.** $-8 < -3; -3 > -8$ **37.** $0 > -8; -8 < 0$ **39.** $\{-11, -3, 8\}$ **41.** $\{-65, -6, 1, 29, 56\}$ **43.** $\{-65, -53, 48, 87, 199\}$ **45b.** Lead   **45c.** Helium

47–48.

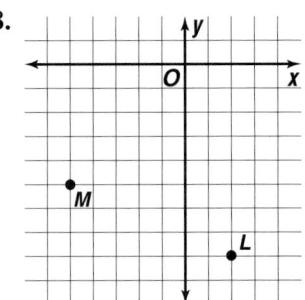

49. −6; 120   **51.** Sample answer: $(6 \times 20) + (6 \times 6)$ **53.** 4 adults

Page 82   Lesson 2-4A

1. 5   **3.** −1   **5.** −3   **7.** −8   **9.** Sample answer: First, place the appropriate number and kind of counters on the mat. Then match all positive counters with all the possible negative counters. These pairs can then be removed because zero does not affect the value of the set on the mat. The remaining counters give you the sum.

Pages 85–87   Lesson 2-4

7. $8 + (-5) = 3$ **9.** $-4 + 9 = 5$ **11.** positive, 9 **13.** negative, −4   **15.** −111   **17.** $-6a$ **19.** 13 **21.** 7   **23.** −4   **25.** 12   **27.** −66   **29.** 26   **31.** −9 **33.** −7   **35.** $-7 + 12; 5$ **39.** 21   **41.** −27   **43.** 63 **45.** $3y$ **47.** $-15m$ **49.** $-20d$ **51.** 7666   **53.** false

55a.

Breed	Change
Akita	−191
Beagle	−390
Chow Chow	+8846
Dachshund	+1473
Labrador Retriever	−4020
Pug	+286

55b. +6004   **57.** false   **59.** 3 m/s   **61.** $r − 9 = 15$; $24   **63.** $n + (8 + 9)$; $n + 17$ **65.** false

Page 88   Lesson 2-5A
1. 5   **3.** 1   **5.** −15   **7.** 4   **9.** Sample answer: The answers to each set of exercises are the same. The exercises differ in that the second term of the subtraction exercises is the additive inverse of the second term of the addition exercises.

Pages 91–93   Lesson 2-5
7. 8   **9.** −9x   **11.** $−7 + (−11) = x$; −18   **13.** $8 + 3 = b$; 11   **15.** $−14 + 19 = y$; 5   **17.** −27x   **19.** −12   **21.** −8   **23.** −14   **25.** −3   **27.** 32   **29.** 9   **31.** 37   **33.** −29   **35.** −63   **37.** −46   **39.** −20   **41.** −27   **43.** −5   **45.** 26   **47.** −10x   **49.** 15p   **51.** 24cd   **53.** −8a   **55.** Sample answer: Every integer has an additive inverse because every integer has an opposite. The sum of an integer and its opposite is zero. Zero is its own additive inverse because $0 + 0 = 0$.   **57.** −$2133   **59.** 36m   **61.** $4 < 6$ **63.** C   **65.** $3y + 23$ **67.** Sandy Koufax   **69.** $k + 3$

Page 93   Self Test
1. 8   **3.** −3, 0, 2, 5

4–6.

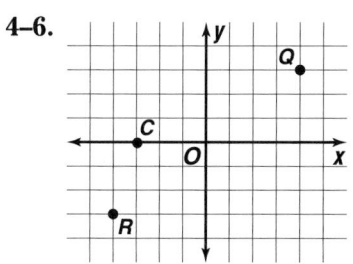

7. >   **9.** =   **11.** 10   **13.** −15

Pages 95–97   Lesson 2-6
5. 120, 720   **7.** 204   **9.** 123,454,321   **11.** 4 touchdowns   **13.** 1:33 P.M.

15a.

15b. 2, 5, 9, 14; 27   **17.** 36 segments   **19.** 55   **21.** 1001   **23.** 6   **25.** $11y + 21$

Page 98   Lesson 2-7A
1. −6   **3.** 6   **5.** −6   **7.** −16   **9.** Sample answer: $−2 \times 4$ means to remove 2 sets of 4 positive 1-tiles after putting in 8 zero pairs.

Pages 102–103   Lesson 2-7
5. −, −56   **7.** +, 518   **9.** +, 54   **11.** −27   **13.** −320   **15.** −132   **17.** −40x   **19.** −56 feet   **21.** 28   **23.** −26   **25.** 75   **27.** −585   **29.** 168   **31.** −308   **33.** 56   **35.** −48   **37.** −36   **39.** −12b   **41.** 48b   **43.** −50rs   **45.** −3ab   **47.** If there is an even number of negative numbers being multiplied, the product will be positive. If there is an odd number of negative numbers being multiplied, the product will be negative.   **49.** −36,000 feet or about −6.8 miles   **51.** 49°F   **53.** Sample answer: $65 > 34$ **55.** 39,177   **57.** commutative, +

Pages 107–108   Lesson 2-8
5. −; −7   **7.** −; −6   **9.** −; −15   **11.** 40   **13.** 7   **15.** 13   **17.** −3   **19.** $−60 ÷ 4 = y$; −15   **21.** −4   **23.** 6   **25.** 13   **27.** −111   **29.** −9   **31.** −12   **33.** −9   **35.** 4   **37.** −19   **39.** −5   **41.** 7   **43.** −9   **45.** −17   **47.** 12   **49.** −7   **51.** −3   **53.** Sample answer: $x = −144$; $y = 12$; $z = −12$ **55.** about $415   **57.** 10:31, 10:35, 11:21   **59.** 41   **61.** 1014   **63.** 6   **65.** 7   **67.** 27

Page 109   Chapter 2 Highlights
1. F   **3.** C   **5.** E   **7.** A

Pages 110–112   Study Guide and Assessment

9.

11.

13. 24   **15.** −40   **17.** 29   **19.** 12   **21.** III   **23.** II   **25.** IV   **27.** =   **29.** >   **31.** <   **33.** $17 < 35$ or $35 > 17$ **35.** −16, −4, 2, 3   **37.** −300, −33, 9, 124, 210   **39.** −3   **41.** −6   **43.** −15   **45.** −10   **47.** −30   **49.** 21   **51.** −7b   **53.** 6r   **55.** −8   **57.** 65   **59.** 36cd   **61.** −300yz   **63.** 7   **65.** 15   **67.** −16   **69.** 19   **71.** −18°F   **73.** 7623

Chapter 3   Solving One–Step Equations and Inequalities

Pages 120–122   Lesson 3–1
5. 6087   **7.** 24   **9.** 5 people   **11.** 501   **13.** You would buy 4 pairs of $70 shoes at a cost of $280 before the $200 pair wore out. It is less expensive to buy the $200 pair of shoes that lasts longer. But, if the shoes are uncomfortable or go out of style, it may make more sense to buy the less expensive shoes.   **15.** blue   **17.** 21   **19.** $24k + 3$

Page 123   Lesson 3-2A
1. 4   **3.** 3   **5.** 7   **7.** 12   **9.** 11

Pages 127–128   Lesson 3-2

7. −25
```
    ●────┼────┼────┼────┼────┼────┼
  −25  −20  −15  −10  −5   0    5
```

9. 37
```
  ┼────┼────┼────┼────┼────●────┼────┼
  32   33   34   35   36   37   38   39
```

11. −1
```
  ┼────┼────●────┼────┼────┼
  −3   −2   −1   0    1    2
```

13. 322 = 264 + e, $58 million

15. 38
```
  ┼────┼────┼────●────┼────┼────┼
  35   36   37   38   39   40   41
```

19. −32
```
  ┼────┼────┼────●────┼────┼────┼────┼
−56 −48 −40 −32 −24 −16 −8   0
```

23. −4
```
  ┼────●────┼────┼────┼────┼────┼
  −5   −4   −3   −2   −1   0    1
```

27. 50
```
  ┼────┼────┼────┼────┼────┼────●────┼
−10   0   10   20   30   40   50   60
```

31. −1384
```
  ┼───●────┼────┼────┼────┼
−1385      −1380       −1375
```

33. a **35.** 5 **37.** −8 **39.** 12 **41.** 25 years old
43. −162 **45.** −3, −1, 3, 15 **47a.** v = Calories
burned in an hour of volleyball **47b.** $v + 120 = 384$
47c. $v = 264$

Pages 131–133   Lesson 3-3

5. 9
```
  ┼────┼────┼────┼────┼────┼────┼────┼────┼────●
  0    1    2    3    4    5    6    7    8    9
```

7. −12
```
  ┼────●────┼────┼────┼────┼────┼────┼
−14  −12  −10  −8   −6   −4   −2    0
```

9. 17
```
  ┼────┼────┼────┼────●────┼────┼────┼────┼
  10        15             17        20        25
```

11. 24 dollars

13. −19
```
  ┼────┼────●────┼────┼────┼────┼
    −20              −15
```

17. 168
```
  ┼────┼────┼────●────┼────┼────┼
    165              170
```

21. 8
```
  ┼────┼────┼────┼────┼────┼────┼────┼────●
  0    1    2    3    4    5    6    7    8
```

25. 936
```
  ┼────●────┼────┼────┼────┼
    935              940
```

29. 128
```
  ┼────┼────┼────●────┼────┼
    125              130
```

33. −13,432
```
  ┼───●────┼────┼────┼────┼
 −13,432 −13,430
```

39a. $6x = 300{,}000{,}000$ **39b.** $15x = 300{,}000{,}000$
39c. Most species lay between 20 and 50 million
eggs. **41.** 18 cm **43.** D **45.** 17 **47.** 10
49a. Estimation; an exact answer is not needed.
49b. Yes, $7 is enough.

Pages 135–137   Lesson 3-4
5. 45 **7.** 165 miles **9.** 360 **11.** 13 **13.** 3.4
15. 14 mph **17.** 10.9 hrs **19.** $s = \ell - d$
21a. 30 sq in. **21b.** 6 cm **23a.** 0.8 hours or
48 minutes **23b.** 8.9 mph **25a.** $a = \dfrac{f - s}{t}$
25b. 2 m/s^2 **25c.** −2 m/s^2 **25d.** negative
27. −27 **29.** $x + 11 = 25$; 14

Page 137   Self Test
1. Stiers–Ratcliffe, Gibson–Smith,
Means–Powhatan, Hunt–Willow,
Bedard–Pocahontas

3. 6
```
  ┼────┼────┼────┼────┼────┼────●────┼────┼────┼
  −1   0    1    2    3    4    5    6    7    8
```

5. −7
```
  ┼────┼────●────┼────┼────┼────┼────┼────┼
−10 −8  −6  −4  −2   0    2    4    6    8
```

7. 100
```
  ┼────┼────┼────┼────┼────●────┼────┼────┼
  0   20   40   60   80  100 120 140 160
```

9. $\dfrac{13}{2}$
```
  ┼────┼────┼────┼────┼────┼────●────┼────┼
  0    1    2    3    4    5    6    7    8
```

Page 138   Lesson 3-5A
1. 18 units

3. 32 units

5. The two figures have the same perimeter
because the squares still share the same number
of sides.

Pages 142–144   Lesson 3-5
5. 14 cm, 12 sq cm **7.** 42 ft; 90 sq ft **9.** 12 m
11. 70 ft; 264 sq ft **13.** 34 km; 30 sq km
15. 15 m; 9.86 sq m **17.** 3.6 m; 0.81 sq m
19. 12.4 cm; 9.61 sq cm **21.** 19.2 mm, 23.04 sq mm
23. 5 m **25.** 11 yd **27.** 39 ft **29.** 225 sq in.
31a.
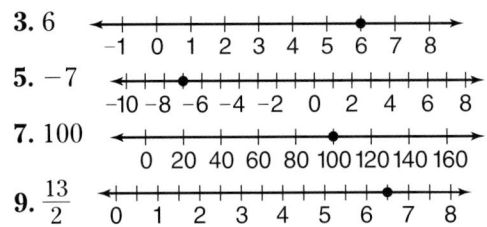

The perimeter is
four times the
length of a side.

31b.

The area is the square
of the length of a side.

31c. perimeter: $8x$ units; area:
$4x^2$ square units **33a.** 150 ft **33b.** 900 sq ft

37. 20

39. $375 > 210$　**41.** $20d + 28$

Page 145　Lesson 3-5B
1. 9 sq units　**3.** 32 sq units　**5.** Both give the same result.

Pages 148–150　Lesson 3-6
7. $y \le -1$　**9.** $d < 75$

11. $m < -13$

13. $y < -4$

15. $a \le -31$

17. $x < 6$　**19.** $y > 1$

21. $t \le 14$

25. $m < 19$

29. $x > 27$

33. $m > -9$

37. $k < -26$

41. $c \ge -25$

45. $x + 285 \ge 375$; $x \ge \$90$　**47.** $x \le 137$ minutes or 2 hours 17 minutes　**49.** 100　**51.** $-2a$
53a. 49 million

53b.

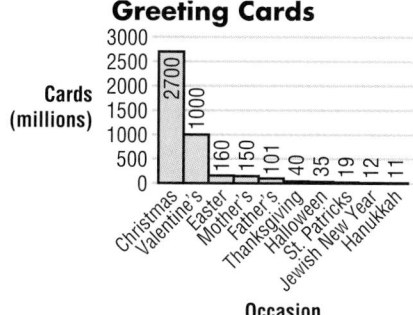

Greeting Cards

55. Identity, $\times$

Pages 154–155　Lesson 3-7
5. 5, no　**7.** -3, yes

9. $x \le 2$

11. $p < 105$

15. $y > 9$

19. $p \ge -14$

23. $m \ge 14$

27. $r > -21$

31. $x < -12$

33. 5　**35.** 800　**37.** $t > 6$ years　**39.** $m < 28$　**41.** D
43. $-56y$
45a. $[(3 \times 4) + (2 \times 3) + (1 \times 2) + (1 \times 1)] \div 7$
45b. 3.0

Pages 158–159　Lesson 3-8
5. n = number; $4n < 96$; $n < 24$　**7.** Sample answer: The total cost of a car repair was $187. If the labor cost $169, how much did the parts cost?
9. n = number; $\frac{n}{-3} > 5$; $n < -15$　**11.** a = account balance; $a + 50 > 400$; $a > \$350$　**13.** p = original purchase price; $4p \le 85{,}000$; $p \le 21{,}250$
15. Sample answer: The low temperature was 16° F on Tuesday. This is at least 9° more than the Wednesday low. What was the low temperature on Wednesday?
17. $|x| < 4$; $-3, -2, -1, 0, 1, 2, 3$
19. a = average; $35a < 600$; $a < 17.1$ points per game　**21.** c = changes in price; $8c = 32$; $c = 4$ times the 1971 price/oz.　**23.** h = height; $5(8)h = 440$; $h = 11$ centimeters　**25.** x = sample scores; $80 - 6 \le x \le 80 + 6$; $74 \le x \le 86$ strokes
27. $h > 13$;

29. $-7y$　**31.** 24

Page 161　Chapter 3 Highlights
1. true　**3.** false　**5.** true　**7.** false

Pages 162–164　Study Guide and Assessment
9. 51

13. 19

17. 126

21. 330

25. 24 **27.** 56.52 **29.** 36 in., 72 sq in. **31.** 40 ft;
75 sq ft

33. $x < -12$
[number line with points at -16, -12, -8, -4, 0; open circle at -12]

37. $f > -4$
[number line with points at -5, 0; open circle at -4]

41. $t < 132$
[number line with points at 130, 135; open circle]

45. $f \geq 15$
[number line with points at 0, 5, 10, 15, 20, 25; closed circle at 15]

47. $3x < 36$; less than \$12 **49.** more than
224 boxes **51.** 103

Chapter 4 Exploring Factors and Fractions

Pages 173–174 Lesson 4-1
7. 2 **9.** 2, 3, 5, 6, 10 **11.** yes **13.** yes **15.** 3
17. 2, 5, 10 **19.** 5 **21.** 3 **23.** none **25.** 2
27. yes **29.** yes **31.** no **33.** no **35.** 0, 4, or 8
37. 1035 **39.** 5555 **41b.** 1996, 2008, 2020, 2032,
2044 **43a.** 6 **43b.** 28 **45.** -267 **47.** -8 **49.** 33

Pages 177–179 Lesson 4-2
7. m^3 **9.** $2 \cdot 2 \cdot 2$ **11.** $10 \cdot 10 \cdot 10 \cdot 10 \cdot 10 \cdot 10$
13. 54 **15a.** $6(5^2)$ or 150 in^2 **15b.** 5^3 or 125 in^3
15c. The surface area is multiplied by 4 and the
volume is multiplied by 8. **17.** n^4 **19.** 14^1 **21.** 3^{15}
23. b^4 **25.** $12 \cdot 12$ **27.** $(-7)(-7)(-7)$
29. $\underbrace{6 \cdot 6 \cdot \ldots \cdot 6 \cdot 6}_{20 \text{ factors}}$ **31.** $\underbrace{1 \cdot 1 \cdot \ldots \cdot 1 \cdot 1}_{55 \text{ factors}}$
33. $(-1)(-1)(-1)(-1)$ **35.** 31 **37.** 162
39. -78 **41.** $(1 \times 10^2) + (4 \times 10^1) + (9 \times 10^0)$
43. 23,405 **45a.** -1 **45b.** $\frac{1}{2}$ **45d.** $\frac{1}{3}$ **47.** The
volume of a cube with sides s units long is s^3.
49a. $2^0, 2^1, 2^2$ **49b.** 2^3 **49c.** 2^{29} cents
49d. 536,870,912 **49e.** $2^0 + 2^1 + 2^2 + \ldots + 2^{29}$
49f. No, the reward will cost \$10,737,418.23 over
the thirty-day period.

51. $x > -12$
[number line from -14 to 4 with open circle at -12]

53. 18°F **55.** $-8p$ **57.** $b + 4$

Page 180 Lesson 4-2B
1. 62 **3.** 2 **5.** 1024 **7.** The values are not the
same because the order of operations is different.
The value of the first expression is -24 and the
value of the second expression is 0.

Pages 182–183 Lesson 4-3
5.

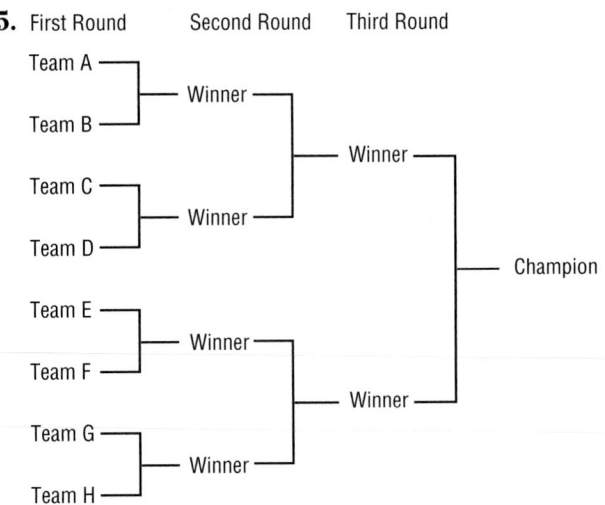

7. $(4 + 3) \times 6 + 3 = 45$ **9.** 1:23 P.M.
11.

4	15	1	14
9	6	12	7
16	3	13	2
5	10	8	11

13. about \$243 **15.** 30 feet **17.** $x \leq -9$ **19.** 4

Pages 187–188 Lesson 4-4
5. composite **7.** $2 \cdot 19$ **9.** $2 \cdot 2 \cdot 2 \cdot 7$
11. $-1 \cdot 2 \cdot 3 \cdot 7 \cdot a \cdot b \cdot c$ **13.** Damon is correct
that 3067 is prime, but it has two factors, 1 and
3067. **15.** prime **17.** prime **19.** $3 \cdot 17$ **21.** $1 \cdot 41$
23. $2 \cdot 5 \cdot 11$ **25.** $3 \cdot 3 \cdot 3 \cdot 3$ **27.** $1 \cdot 13$
29. $2 \cdot 3 \cdot 7 \cdot x \cdot y \cdot y$ **31.** $3 \cdot 7 \cdot x \cdot y \cdot y \cdot y$
33. $2 \cdot 2 \cdot 7 \cdot f \cdot f \cdot g$ **35.** $2 \cdot 3 \cdot 5 \cdot 7 \cdot m \cdot n \cdot n \cdot n$
37. $3 \cdot 5 \cdot 5 \cdot m \cdot m \cdot k$ **39.** 211 **41.** 31 and 53
43a. 23 **43b.** yes **45.** 16 **47.** 52 inches; 153
square inches **49.** $>$ **51.** $x + 15$

Page 189 Lesson 4-4B
1. Sample answer: 67, 121, 55, 81, 63, 66

Pages 192–194 Lesson 4-5
5. 2 **7.** 1 **9.** b **11.** 12 inches **13.** 8 **15.** 12
17. 36 **19.** $8x$ **21.** $14b$ **23.** 9 **25.** 6 **27.** 6
29. $6a$ **31.** yes **33.** no **35.** no **37.** 15 **39.** 18
shelves **41.** $2 \cdot 2 \cdot 2 \cdot 5 \cdot 7 \cdot 11$ **43.** 10 **45.** 135
47. 1728

Page 194 Self Test
1. 3 **3.** 3 **5.** 2^3 **7.** 121 **9.** $-1 \cdot 2 \cdot 13$ **11.** 14
13. 9

Page 195 Lesson 4-6A
1. Shade 2 of 6 sections. **3.** They are equal in size.
5. Yes; sample answer $\frac{1}{4}$ **7.** yes

Pages 198–199 Lesson 4-6

7. $\frac{1}{7}$ **9.** $\frac{11}{15}$ **11.** $\frac{5}{7}$ **13.** $\frac{9}{22}$ **15.** $\frac{1}{11t}$

17. simplified **19.** $\frac{1}{7}$ **21.** $\frac{17}{19}$ **23.** $\frac{1}{3}$ **25.** $\frac{5}{8}$

27. simplified **29.** simplified **31.** $\frac{62}{111}$ **33.** $\frac{4x}{5y}$

35. $\frac{41}{7}$ **37.** $\frac{z^3}{x}$ **39.** $\frac{10p}{13q}$ **41.** simplified

43. $\frac{45}{119}$ **45.** $\frac{22}{60 \cdot 24}$ or $\frac{11}{720}$ **47a.** 2:6

47b. $\frac{1}{3}$ **47c.** Sample answer: 1 oz. alcohol, 3 oz. water **49.** 108 **51.** 68d **53.** 12

Pages 202–204 Lesson 4-7

5. 70 **7.** 60a **9.** 630 **11.** 10 **13.** 100 **15.** <
17a. East Central **17b.** West Central **17c.** New England **19.** 18 **21.** 288 **23.** 84 **25.** 84y **27.** 40
29. 180 **31.** 105t^2 **33.** 16 **35.** 35 **37.** 220
39. 25b **41.** > **43.** < **45.** < **47.** <
49. $\frac{3}{8} < \frac{5}{12}$ **51.** dog, $\frac{7815}{12} > \frac{4723}{15}$ **53a.** video game
53b. VCR **53c.** on-line services **55.** $x > 576$
57. −98 **59.** Utah

61.

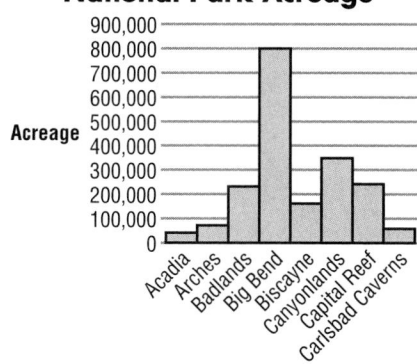

National Park Acreage

Pages 207–209 Lesson 4-8

5. 10^7 **7.** 3^{10} **9.** 8^6 **11.** y^2 **13.** about 10 times more intense **15.** w^6 **17.** b^7 **19.** 10^7 **21.** a^4
23. $20x^7$ **25.** f^{12} **27.** $12a^5$ **29.** a^3b^4 **31.** t
33. k^2m **35.** w^4y^2 **37.** $3n^8$ **39.** 9 **41.** 3 **43.** 1
45a. 100 times **45b.** base **47.** < **49.** 2

51. −7

$$\longleftarrow\!+\!+\!+\!\bullet\!+\!+\!+\!+\!+\!+\!+\!\longrightarrow$$
$$-10 \;\; -8 \;\; -6 \;\; -4 \;\; -2 \;\;\;\; 0 \;\;\;\; 2$$

53. II **55.** $2x - 4$

Pages 212–214 Lesson 4-9

5. $\frac{1}{15^5}$ **7.** 10^{-7} **9.** 3^{-2} **11.** $\frac{3}{16}$ **13.** $\frac{1}{4}$ **15.** $\frac{1}{(-2)^5}$
17. $\frac{1}{s^2t}$ **19.** $\frac{2}{(xy)^2}$ **21.** 5^{-3} **23.** 2^{-4} or 16^{-1}
25. $2 \cdot 3^{-2}$ **27.** fg^{-3} **29.** $\frac{1}{27}$ **31.** $\frac{2}{3}$ **33.** a^3 **35.** c^7
37. a^{2b-a} **39a.** $\frac{6.67}{10^{11}}$Nm^2/kg^2
39b. 0.0000000000667 Nm^2/kg^2 **41.** $-50x^4$ **43.** 7^5
45. $2230 + x = 4750$; \$2520 **47.** $n - 8 = 3$; 11

Page 215 Chapter 4 Highlights

1. greatest common factor **3.** factors **5.** multiple
7. least common denominator **9.** ratio

Pages 216–218 Study Guide and Assessment

11. 5 **13.** 3 **15.** 2, 3, 6 **17.** 2, 3, 6 **19.** $(c + 2)^3$
21. 1^3 **23.** $k \cdot k \cdot k \cdot k \cdot k \cdot k \cdot k \cdot k \cdot k$
25. $32 \cdot 32 \cdot 32 \cdot 32 \cdot 32 \cdot 32 \cdot 32 \cdot 32 \cdot 32 \cdot 32 \cdot 32$
27. $-1 \cdot 2 \cdot 3 \cdot 19$ **29.** $-1 \cdot 3 \cdot 5 \cdot 5 \cdot 11 \cdot x \cdot x \cdot y$
31. $-1 \cdot 3 \cdot 3 \cdot 7 \cdot 29 \cdot j \cdot k \cdot k \cdot k$
33. $2 \cdot 5 \cdot 5 \cdot 11 \cdot m \cdot n \cdot q \cdot q$ **35.** 13 **37.** $35a^2b^2$
39. 5 **41.** 1 **43.** $\frac{19}{20}$ **45.** $\frac{8}{11}$ **47.** $\frac{1}{2}$ **49.** simplified
51. 60 **53.** 30jk **55.** 180 **57.** 36$abcd$ **59.** >
61. > **63.** d^5 **65.** $(-4)^4$ **67.** $2x^{22}$ **69.** $110a^{10}b^3$
71. $\frac{4d}{c^2}$ **73.** $\frac{4x}{(yz)^2}$ **75.** $3f^9$ **77.** 21 games
79. home improvement

Chapter 5 Rationals: Patterns in Addition and Subtraction

Pages 227–228 Lesson 5-1

7. $\frac{1}{20}$ **9.** $\frac{2}{9}$ **11.** rationals **13.** none **15.** =

17a. $\frac{1}{16}$ **17b.** Rational; it is a fraction. **19.** $-\frac{7}{10}$

21. $\frac{57}{100}$ **23.** $\frac{3}{100}$ **25.** $-\frac{1}{3}$ **27.** $\frac{25}{99}$ **29.** $2\frac{34}{99}$
31. rationals **33.** whole numbers, integers, rationals **35.** none **37.** rationals **39.** rationals

41. none **43.** > **45.** = **47.** > **49.** < **51.** $\frac{2}{7}$

53. Yes; $\frac{6}{2.4} = \frac{5}{2}$ or $\frac{60}{24}$. **55.** $\frac{1}{40}$ **57a.** $\frac{1}{30}$ **57b.** Yes; it can be written as a fraction. **59.** $60x^2$
61. $b < -18$ **63.** −19 **65.** B

Pages 231–233 Lesson 5-2

5. 6 **7.** 803 **9.** 11 **11.** 1 **13.** $\frac{1}{2}$ **15.** a **17.** 40

19. 9 **21.** 29 **23.** 60 **25.** 21 **27.** 1 **29.** $7\frac{1}{2}$

31. 1 **33.** 5 **35.** 110 **37.** $28\frac{1}{2}$ **39.** 50 **41.** \$5

43. \$3 **45.** No, the change should be about \$7.

47. Sample answer: $11\frac{1}{10} + 1\frac{7}{15}$ **49.** \$59

51. more than half **53.** $\frac{13}{50}$ **55.** −16 **57.** false
59a. $2(6.25 + 2.25)$ or $2(6.25) + 2(2.25)$ **59b.** \$17

Pages 236–238 Lesson 5-3

5. 3.68 **7.** −4.91 **9.** 77.33 **11.** −6.4 **13.** 43.9
15. $-4.1m$ **17.** 14.1 **19.** 1.5 **21.** 63.431 **23.** 2.3
25. −26.4 **27.** −51.73 **29.** $12.3m$ **31.** $6.9y$
33. $11.5x - 5$ **35.** 29.4 **37.** 12.56 **39.** 12.67
41. 24.45 left; no; third side must be less than 20.55 feet **43.** 378.5 **45a.** 24.3 bags per 10,000 passengers **45b.** Sample answer: Bad weather causes changes in flight schedules. **47.** yes; $2 + 4 + 1 + 2 = \$9$ **49.** $2 < a$ **51.** −11 **53.** 12

Pages 241–243 Lesson 5-4

5. $1\frac{2}{7}$ **7.** $\frac{1}{3}$ **9.** 1 **11a.** $\frac{9}{25}$ **11b.** $\frac{1}{5}$ of each dollar

13. $\frac{1}{5}$ **15.** $\frac{4}{9}$ **17.** $1\frac{1}{3}$ **19.** $1\frac{1}{3}$ **21.** $2\frac{1}{3}$ **23.** 1

25. $\frac{1}{2}$ **27.** $\frac{2}{3}$ **29.** $-\frac{1}{6}$ **31.** $-2r$ **33.** $\frac{2}{5}a$ **35.** $\frac{1}{3}b$

37. Sample answer: $\frac{2x}{5} - \frac{1}{5}$ **39.** $5\frac{3}{8}$ inches
41. $3\frac{1}{2}$ yards **43a.** $\frac{25}{170}$ or $\frac{5}{34}$ **43b.** $\frac{1}{28}$ **45.** $c \geq -13$
47. $-22a$

Pages 246–247   Lesson 5-5
7. $\frac{1}{8}$ **9.** $-\frac{1}{8}$ **11.** $\frac{11}{14}$ **13.** $\frac{5}{9}$ **15.** $-\frac{1}{12}$ **17.** $\frac{15}{26}$
19. $1\frac{5}{6}$ **21.** $3\frac{1}{6}$ **23.** $4\frac{1}{4}$ **25.** $16\frac{10}{21}$ **27.** $10\frac{1}{15}$
29. $-\frac{17}{90}$ **31.** $3\frac{5}{24}$ **33.** $-1\frac{3}{8}$ **35.** $1\frac{5}{24}$ **37.** Sample
answer: $\frac{1}{18} + \frac{1}{6}$ **39.** $\frac{11}{102}$ **41.** $\frac{t^3}{s^2}$ **43.** -5 **45.** 13
47. $18x$

Page 247   Self Test
1. $-\frac{1}{6}$ **3.** $1\frac{4}{9}$ **5.** 1 **7.** 14.52 **9.** 30.6 **11.** $5\frac{1}{7}$
13. $8\frac{3}{4}$ inches

Pages 249–250   Lesson 5-6
5. $-\frac{1}{15}$ **7.** -5.28 **9.** 3.25 **11.** -8.7 **13.** 21.9 miles
15. 6.4 **17.** $15\frac{3}{4}$ **19.** 13.4 **21.** $-\frac{31}{36}$ **23.** $1\frac{39}{40}$
25. -2.8 **27.** 8.01 **29.** $12\frac{17}{18}$ **31.** $\frac{19}{312}$
33. 688.2 million barrels **35.** 0.92 million or
920,000 metric tons **37.** $4\frac{11}{12}$ **39.** $d \geq -84$ **41.** 18
43. 15

Pages 253–254   Lesson 5-7
5. $x \leq 2\frac{23}{30}$
9. $y < 1.92$
13. $y < -\frac{1}{6}$
17. $d < 2\frac{5}{6}$
19. $r \leq 1\frac{7}{8}$
23. $h < 15.96$
27. $t > \frac{5}{108}$

29. x is between 0 and 1 or x is less than -1.
31. $x > 9.7$ **33a.** Sample answer: The average
salary of a high school graduate is slightly less
than half of the average salary of a college
graduate. **33b.** Sample answer: No, if it was that
low the inequality would be written differently. The
inequality means that it is less than but close to
one-half. **35.** $3\frac{3}{4}$ **37.** 2 **39.** $-30abc$ **41a.** Los
Angeles **41b.** 20 cents per mile

Pages 256–257   Lesson 5-8
5. inductive **7.** inductive **9.** 18, 21
11. $1111^2 = 1{,}234{,}321$ **13a.** ground you; deductive
13b. No; there could be another reason.
15a. 1010 **15b.** 110 **15c.** Inductive reasoning
does not always work, so it would not be wise to
use it to prove something. **17.** pink **19.** $z \leq 0.91$
21. 1 **23.** 10

Pages 260–262   Lesson 5-9
5. yes; 17, 20, 23 **7.** no; 2, -4, -11 **9.** no; 23, 30,
38 **11a.** 9.95, 12.90, 15.85, 18.80, 21.75, 24.70
11b. \$36.50 **13.** yes; 2.5, 2.0, 1.5 **15.** no; 25, 36, 49
17. no; 32, 64, 128 **19.** yes; 55, 46, 37 **21.** yes; 3,
-1, -5 **23.** yes; 9.55, 10.57, 11.59 **25.** no; 46, 59,
74 **27.** no; $\frac{7}{8}, \frac{8}{9}, \frac{9}{10}$ **29.** 7; 11 **31.** 100, 125, 150,
175, 200, 225 **33a.** 50 s **33b.** $d = \frac{1}{5}t$
35. 320 feet/second **37.** deductive **39.** 6

41.

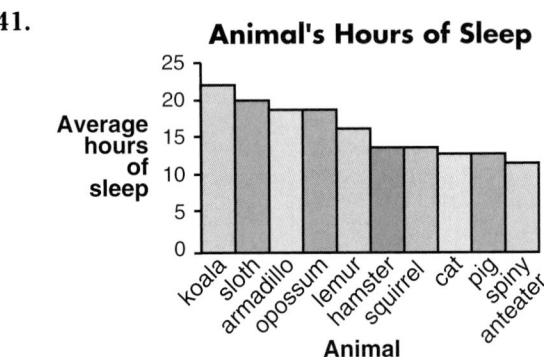

Animal's Hours of Sleep

Page 263   Lesson 5-9B
1. 13 **3.** 2, 5, 13, 34, ... The sequence is the 3rd,
5th, 7th, . . . terms of the Fibonacci sequence.

Page 265   Chapter 5 Highlights
1. B **3.** I **5.** C **7.** A

Pages 266–268   Study Guide and Assessment
9. integers, rationals **11.** none **13.** $-9\frac{9}{20}$
15. $-\frac{13}{100}$ **17.** $4\frac{8}{33}$ **19.** $8\frac{89}{100}$ **21.** 8 **23.** 10 **25.** 1
27. $2\frac{1}{2}$ **29.** 11.23 **31.** -7.68 **33.** -17.95 **35.** $13\frac{7}{9}$
37. $\frac{1}{4}$ **39.** $-\frac{3}{5}$ **41.** $-\frac{1}{15}$ **43.** $6\frac{11}{12}$ **45.** $5\frac{1}{36}$
47. -8.8 **49.** $1\frac{13}{16}$ **51.** $1\frac{5}{6}$

53. $f \leq 4\frac{1}{8}$
55. $q > 5.3$
57. $m > -6.151$

59. $a \leq \dfrac{9}{10}$

61. yes; 13, 5, -3 **63.** no; $\dfrac{4}{5}, \dfrac{2}{3}, \dfrac{0}{1}$ **65a.** $\dfrac{35}{1200}$ or $\dfrac{7}{240}$

65b. Rational, it is a fraction.   **67.** $6\dfrac{5}{8}$ inches

69a. 25 units   **69b.** The lengths are the squares of 1, 2, 3, ... It is not an arithmetic sequence because there is no common difference.

Chapter 6   Rationals: Patterns in Multiplication and Division

Pages 277–279   Lesson 6-1

7. $2.\overline{6}$ **9.** 0.125   **11.** $-\dfrac{3}{4}$ **13.** $-\dfrac{3}{4}$ **15.** 0.7   **17.** 0.5
19. 0.7   **21.** 0.3125   **23.** $-3.\overline{36}$ **25.** -4.3125
27. $>$ **29.** $<$ **31.** $<$ **33.** $<$ **35.** $>$ **37.** $<$
39. .7777777   **41.** 0.7777778   **43.** $0.\overline{3}$, 0.25, 0.7, 0.8,
$0.41\overline{6}$, $0.\overline{571428}$, 0.12   **43a.** terminating: $\dfrac{1}{4}, \dfrac{7}{10}, \dfrac{4}{5}, \dfrac{3}{25}$;
repeating $\dfrac{1}{3}, \dfrac{5}{12}, \dfrac{4}{7}$ **43b.** $3 = 1 \times 3$; $4 = 2^2$;
$10 = 2 \times 5$; $5 = 1 \times 5$; $12 = 2^2 \times 3$; $7 = 1 \times 7$; $25 = 5^2$
43c. 2 and 5   **45.** 0.33   **47.** \$82.38; \$46.88; \$36.75
49. 64, 58, 52   **51.** b^{11} **53.** C   **55.** $r > 7$ or
$r \geq 8$ **57.** 27

Pages 282–283   Lesson 6-2

5. c   **7.** d   **9.** $18 \div 3 = 6$ **11.** $8 \div 1 = 8$
13. $20 \times \dfrac{1}{4} = 5$ **15.** 160   **17.** 4   **19.** 5   **21.** 2
23. 6   **25.** 20   **27.** 150   **29.** 5   **31.** 10   **33.** If the numbers were rounded up the actual answer is less than the estimate. If the numbers were rounded down the product is greater than the estimate.   **35.** about 15,000,000   **37.** 5 mph
39. $\dfrac{17}{14}$ or $1\dfrac{3}{14}$ **41.** 27   **43.** 2

Pages 286–288   Lesson 6-3

5. d   **7.** $\dfrac{1}{2}$ **9.** $\dfrac{4}{9}$ **11.** -24 **13.** $\dfrac{3}{8}$ **15.** $\dfrac{9}{25}$

17a. 3.8¢   **17b.** 28 years, 1.5 months   **19.** $\dfrac{1}{7}$

21. $-1\dfrac{1}{2}$ **23.** $-4\dfrac{1}{2}$ **25.** -3 **27.** -24 **29.** $\dfrac{9}{25}$

31. $6\dfrac{12}{25}$ **33.** $\dfrac{4}{15}$ **35.** $1\dfrac{1}{18}$ **37.** $\dfrac{4}{5}$ **39.** $\dfrac{11}{36}$

41. $\dfrac{7}{2} \times \dfrac{6}{3}$ or $\dfrac{6}{2} \times \dfrac{7}{3}$; use the greatest factors as numerators and the least as denominators.

43. $3\dfrac{3}{4}$ hours   **45a.** 6 picas   **45b.** $4\dfrac{1}{2}$ picas
45c. about 91 characters   **47.** 15   **49.** $2^3 \cdot 3^2$
51. $-20xy$ **53.** 15

Pages 291–293   Lesson 6-4

7. no   **9.** yes   **11.** -1 **13.** $-3\dfrac{3}{4}$ **15.** $\dfrac{20}{57}$ **17.** $\dfrac{11}{18}$

19. $-\dfrac{3}{7}$ **21.** $\dfrac{5}{8}$ **23.** $\dfrac{2}{3}$ **25.** $\dfrac{y}{x}$ **27.** $-1\dfrac{2}{25}$ **29.** -10

31. $-1\dfrac{1}{2}$ **33.** 27   **35.** $34\dfrac{2}{7}$ **37.** $-1\dfrac{1}{2}$ **39.** $6\dfrac{2}{3}$

---

41. $-2\dfrac{5}{8}$ **43.** $-2\dfrac{2}{7}$ **45.** $12\dfrac{2}{3}$ **47.** 21 uniforms plus extra fabric remaining   **49.** 8 days   **51.** $\dfrac{2}{5}$ **53.** 6
55. II   **57.** $13h + 23$

Page 294   Lesson 6-5A

1. 24 squares; 24 hundredths or 0.24
3. $0.4 \times 0.2 = 0.08$, each square represents one one-hundredth   **5.** 5

Pages 297–299   Lesson 6-5

5. 1.44   **7.** 0.7   **9.** 0.62   **11.** -90 **13.** 25.12
15. Inverse, multiplication   **17.** 70 pounds
19. 1.4147   **21.** -5.5875 **23.** -6.592 **25.** 0.32
27. -17 **29.** 1460   **31.** 5.6   **33.** -15 **35.** 1.21
37. -1.16 **39.** Identity, multiplication
41. Associative, multiplication   **43.** Commutative, multiplication   **45.** -2.5; -2.6 **47.** -10; -10.2
49. 0.1; 0.2   **51.** Muesli, muesli 18.1¢ cents per oz; frosted corn flakes 18.4¢ per oz.   **53.** \$3.2 billion
55. $2\dfrac{7}{12}$ **57.** integers, rationals   **59.** \$314
61. Commutative, addition

Page 299   Self Test

1. 0.875   **3.** $-0.\overline{72}$ **5.** 16   **7.** 3   **9.** -24

Pages 304–306   Lesson 6-6

5. 29.4, 33, 12   **7.** 100.4, 105, 78   **9.** 49.9, 45, 45
11. 6.0, 6.5, 3.6 and 7.2   **13.** 104.4, 105, 113
15. 0.92, 0.85, none   **17.** 85.4   **19.** 86   **21a.** 86
21b. no effect   **23a.** 90   **23b.** no effect   **25.** 91
27. Sample answer: 2, 4, 6, 8, 8, 10, 12, 14
31. -12.9824 **33.** $y \leq -24$ **35.** Multiplication property of zero

Page 307   Lesson 6-6B

1. 5.55, 5.75   **3.** -16.36, -15 **5.** 73.6, 49.7   **7.** The mean is affected more by very large or very small numbers. So the median is a better representation if the data has extreme values. If the data are more closely clustered, the mean is a good representation.

Pages 310–311   Lesson 6-7

5. -15

7. $-9 < f$

9. $\dfrac{18}{25}$

11. 6 cm   **13.** -0.5 **15.** $16\dfrac{1}{2}$ **17.** $x < -14.4$

19. $-6\dfrac{1}{4}$ **21.** -3.55 **23.** $z \leq -1.3$ **25.** $\dfrac{-7}{10}$

27. 19   **29.** $x < \dfrac{1}{5}$

31. $y < \dfrac{15}{32}$

35. \$16.66   **37.** 88.8   **39.** $1\dfrac{3}{14}$ **41.** -36

Pages 315–316   Lesson 6-8

7. no   **9.** yes, $\frac{1}{3}$; 1, $\frac{1}{3}$, $\frac{1}{9}$ **11.** yes, 2; 8, 16, 32
13. 2048   **15.** yes, $\frac{1}{2}$; $\frac{3}{2}$, $\frac{3}{4}$, $\frac{3}{8}$ **17.** yes, $\frac{1}{12}$; $\frac{1}{144}$, $\frac{1}{1728}$, $\frac{1}{20736}$ **19.** yes, $\frac{1}{2}$; $\frac{1}{32}$, $\frac{1}{64}$, $\frac{1}{128}$ **21.** yes, -2; 112, -224, 448   **23.** no   **25.** no   **27.** yes, $\frac{1}{4}$; $\frac{3}{32}$, $\frac{3}{128}$, $\frac{3}{512}$
29. 2, -4, 8, -16, 32   **31.** 80, 100, 125   **33.** 1
35a. 1.3   **35b.** 37,129,300   **37.** 40   **39.** 27
41. $\frac{1}{3^4}$ or $\frac{1}{81}$ **43.** B   **45.** 34

Pages 319–320   Lesson 6-9

5. yes   **7.** 8.490×10^3 **9.** 8.479×10^2 **11.** 61,000
13. 31,557,600 seconds in a year; (31,557,600)
(3.00×10^5) or about 9.47×10^{12} km   **15.** 5.9×10^{-3} **17.** 1.5×10^{-4} **19.** 4.98×10^{-8} **21.** 7.01×10^8 **23.** 0.0000056   **25.** 0.009001   **27.** 0.00005985
29. 1454   **31.** 0.00005   **33.** 700,080,003
35a. 2,000,000; 2×10^6 **35b.** 252,000; 2.52×10^5
35c. 0.000016575; 1.6575×10^{-5}
35d. 0.00000000192; 1.92×10^{-9} **37.** 0.000002 or
2.0×10^{-6} **39.** yes; $-\frac{1}{3}$, $\frac{1}{9}$, $-\frac{1}{27}$, $\frac{1}{81}$ **41.** $-2\frac{4}{5}$
43. $b + 3$

Page 321   Highlights

1. true   **3.** false; Each term in a geometric sequence increases or decreases by a common ratio.   **5.** false   **7.** true   **9.** false; It is possible to have a set of numerical data with no mode.

Pages 322–324   Study Guide and Assessment

11. 0.625   **13.** 0.425   **15.** $0.2\overline{6}$ **17.** 21   **19.** 9
21. $-\frac{8}{45}$ **23.** $-1\frac{1}{5}$ **25.** $\frac{64}{125}$ **27.** $\frac{3}{4}$ **29.** $-\frac{1}{9}$
31. 28.28   **33.** 0.0629   **35.** $0.291\overline{6}$ **37.** -47.7
39. 20.6; 20; 18, 21, and 25   **41.** 14.17; 14; 14   **43.** 9
45. -12 **47.** $-4 \geq n$ **49.** 5.3   **51.** no   **53.** no
55. no   **57.** 6.74×10^{-6} **59.** 5.81×10^3
61. 57,200   **63.** 202,000,000   **65.** 132¢ or \$1.32
67. 15.2 times

Chapter 7   Solving Equations and Inequalities

Pages 331–332   Lesson 7-1

5. 3 ounces   **7.** \$16,200   **9.** \$5   **11.** 729 people
13. 29.8.92 and 2.9.92   **15.** 0.0000006789
17. whole numbers, integers, rationals   **19.** D

Page 333   Lesson 7-2A

1. -3 **3.** 4   **5.** 2   **7.** Sample answer: You undo operations to find the value of x.

Pages 336–337   Lesson 7-2

5. 13   **7.** -4 **9.** -3 **11.** 6 ounces   **13.** 4   **15.** 17
17. -1.2 **19.** 84   **21.** 33   **23.** 51   **25.** -153
27. 22.5   **29.** -93 **31.** 16   **33.** Sample answer: $6x + 3 = 2$ **35.** 4, 5   **37.** 90   **39.** $>$ **41.** y^3
43. false

Pages 339–340   Lesson 7-3

3. $17 - 2x = 5$; 6   **5a.** $2682 = 4d - 354$
5b. 759 thousand   **7.** C   **9.** E   **11.** $\frac{n + 6}{7} = 5$; 29

13. $\frac{c}{-4} - 8 = -42$; 136   **15.** $5x - 10 = 145$; 31
17. 7 yd, 9 yd, 11 yd   **19a.** $2w + 50 = 130$
19b. 40 feet   **21.** 50   **23.** 72

Pages 343–344   Lesson 7-4

7. 11.31 cm   **9.** 10.99 m   **11.** 12.57 cm   **13.** 34.56 ft
15. 43.96 mm   **17.** 59.032 m   **19.** 8.164 yd
21. $7\frac{1}{3}$ ft   **23.** C   **25.** B   **27.** 0.65 m   **29.** 2 units
31a. 125°   **31b.** 55°   **31c.** 235°   **31d.** 125°
31e. 125°   **31f.** 55°   **31g.** 30°   **31h.** 95°
33a. The perimeter of the square is longer.
33b. Sample answer: The perimeter of a square is 4 times the length of one side. The circumference of a circle is approximately 3.14 times the diameter. $4 > 3.14$ **35.** 3.7 million   **37.** $2 \cdot 2 \cdot 5 \cdot 11 \cdot p \cdot q \cdot q$
39. $>$

Page 345   Lesson 7-5A

1. -8 **3.** 4   **5.** 1   **7.** Sample answer: No; the result is the same.

Pages 348–350   Lesson 7-5

5. Sample answer: Add 1 to each side. Subtract $3x$ from each side. The solution is 3.   **7.** Sample answer: Simplify the left side to $3a + 66$. Subtract $3a$ from each side. The solution is 4.   **9.** Sample answer: Add n to each side. Subtract 4 from each side.   The solution is 3.   **11.** 21   **13.** 2   **15.** -0.5
17. 5   **19.** all numbers   **21.** $\varnothing$ **23.** 2   **25.** all numbers   **27.** 2.48   **29.** 100 m by 130 m   **31.** 25 ft by 30 ft   **33.** $3y - 14 = y$; 7
35a. $199,000 - 700x$ **35b.** $165,000 + 3700x$
35c. $199,000 - 700x = 165,000 + 3700x$
35d. about 8 years   **37a.** \$20   **39.** 32   **41.** $\frac{2}{7}$
43. 52

Page 350   Self Test

1. \$569.19   **3.** -12 **5.** $4x - 12 = 18$; 7.5
7. 1.02 yd   **9.** 65 yards by 120 yards

Pages 353–354   Lesson 7-6

5. $x \leq 7$

7. $k > 2$

9. $m < -2$

11. Sample answer: If Heather can find CDs that cost no more than \$12.50 each, she can buy the shoes and two CDs.

13. $u \geq -4$

17. $j \geq -3$

21. $1.17 > t$

25. $x \le 4.5$

29. $c < -4$

31. $-5 < x < 1$ **33.** $s \ge 97$ **35.** 258   **37.** 8
39. $n + 3$

Pages 356–357   Lesson 7-7
5. $2n - 9 > 11; n > 10$ **7.** $55c + 35 \le 200; c \le 3$
9. $4x + 4 \ge 16; x \ge 3$ **11.** Sample answer: A stove
and a freezer weigh at least 260 kg. The stove
weighs 115 kg. What is the weight of the freezer?
13. $3(h - 1) \ge 18; h \ge 7$ **15.** 17, 18   **17.** $t \ge \frac{11}{3}$
19. 3, 4, or 5   **21.** $50 > c > 6$ **23.** $x > -5$
25a. $\frac{1.7}{10^{24}}$ **25b.** 0.0000000000000000000000017
27. $-126b$

Pages 360–361   Lesson 7-8
5. mL   **7.** kg   **9.** 0.040   **11.** 12.8 km   **13.** cm or m
15. L   **17.** 0.4   **19.** 9.4   **21.** 6   **23.** 8000   **25.** 1000
27. 316   **29.** 946 mL   **31.** 603,200 cm   **33.** 13 cm
35. 65 cm, 55 cm   **37.** 0.035 m   **41.** $b > 7$
43. 17.25   **45.** $-20\frac{4}{5}$

Page 363   Highlights
1. false   **3.** true   **5.** true   **7.** true

Pages 364–366   Study Guide and Assessment
9. 12   **11.** 60   **13.** -20 **15.** 0.2   **17.** 13   **19.** $250
21. 22.0 in.   **23.** 30.8 m   **25.** -2 **27.** -1 **29.** 14

31. $n > 6$

33. $r \le 98$

35. $c < 7.5$

37. $8n + 2 \ge 18; n \ge 2$ **39.** $x + (x + 1) > 47$;
$x > 23$ **41.** 0.006   **43.** 43,000   **45.** 300,000
47. 8800   **49.** 140.8 beats per minute   **51.** less
than 16.67 minutes

Chapter 8   Functions and Graphing

Pages 374–377   Lesson 8-1
5. $\{-1.3, 4, -2.4\}; \{1, -3.9, 3.6\}$ **7.** $\{(-1, 3), (0, 6), (4, -1), (7, 2)\}; D = \{-1, 0, 4, 7\}, R = \{3, 6, -1, 2\}$
9. $\{(-2, 1), (0, -1), (2, 2)\}; D = \{-2, 0, 2\}$,
$R = \{-1, 1, 2\}$ **11.** yes   **13.** no   **15.** $\{(8, 4), (9, 3), (10, 2), (11, 1), (12, 0)\}$ **17.** $\{-1, 4, 2, 1\}; \{6, 2, 36\}$
19. $\{1.4, -2, 4, 6\}; \{3, 9.6, 4, -2.7\}$
21. $\left\{-\frac{1}{2}, 4\frac{2}{3}, -12\frac{3}{8}\right\}; \left\{\frac{1}{3}, -17, 66\right\}$ **23.** $\{(-4, -2), (-2, 1), (0, 2), (1, -3), (3, 1)\}; D = \{-4, -2, 0, 1, 3\},$
$R = \{-2, 1, 2, -3\}$ **25.** $\{(5, 4), (2, 8), (-7, 9), (2, 12), (5, 14)\}; D = \{5, 2, -7\}, R = \{4, 8, 9, 12, 14\}$
27. $\{(-3, -2), (-2, -1), (0, 0), (1, 1)\};$

$D = \{-3, -2, 0, 1\}, R = \{-2, -1, 0, 1\}$ **29.** yes
31. no   **33.** yes   **35.** no   **37.** no   **39.** no   **41.** yes
43a. $6.95   **43b.** more than $30 and less than or
equal to $70   **43c.** Yes, each x-value (total price)
has exactly one y-value (shipping cost).

45. The six-pack has 124 mL more.   **47.** $14\frac{2}{9}$ ft
49. $-21y^5$ **51.** -4

Page 378   Lesson 8-2A
1. Sample answer: Generally, as shoe length
increases, armspan increases also.   **3.** Sample
answer: Because points appear to be scattered
around a general area.

Pages 381–384   Lesson 8-2
5. negative   **7.** positive   **9.** positive   **11.** none
13. negative   **15.** positive   **17.** no   **19.** negative
21. no   **23a.** The data show a positive
relationship.   **23b.** Sample answer: about 62 wpm
23c. Sample answer: between 3 and 4 weeks
23d. The more experience a student has, the more
words per minute he or she can key.

25a.

Relationship of Year and Cost per Year

25b. Yes; positive.
27a.

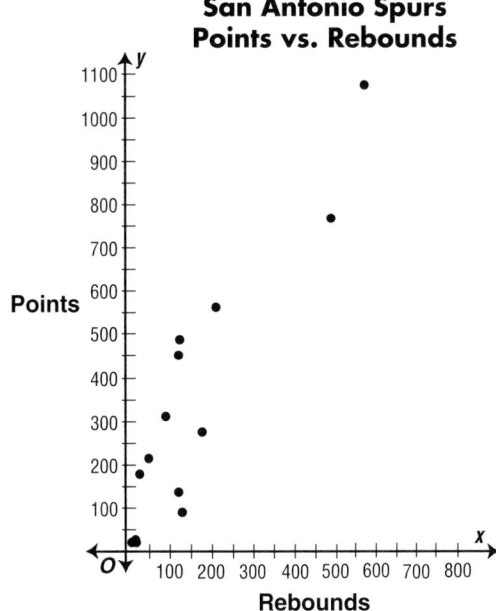

San Antonio Spurs Points vs. Rebounds

27b. yes; positive   **27c.** Sample answer: yes; you
could estimate by using the points in the scatter
plot.   **27d.** Sample answer: Yes; some players have

more opportunities to get rebounds because of the position they play for the team. **27e.** Sample answer: Yes, since more playing time provides a player with more opportunity to score points. **29.** {8, 4, 6, 5}; {1, 2, −4, −3, 0} **31.** 127.7; 129; 130 **33.** 2075

Pages 387–390 Lesson 8-3
7. c, d **9.** (−2, −5), (0, −2), (2, 1), (4, 4)
11. Sample answer: (0, 2.8), (−2, −7.2), (2, 12.8), (−1, −2.2). **13.** no

15.

17.

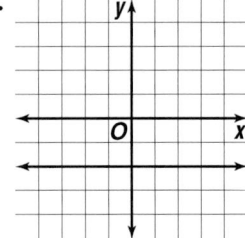

19a. $y = 0.5x$

19b.

x	y = 0.5x	y	(x, y)
0	y = 0.5(0)	0	(0, 0)
1	y = 0.5(1)	0.5	(1, 0.5)
2	y = 0.5(2)	1	(2, 1)
3	y = 0.5(3)	1.5	(3, 1.5)

19c. Sample answer: There is no such thing as a negative number of seedlings.

19d.

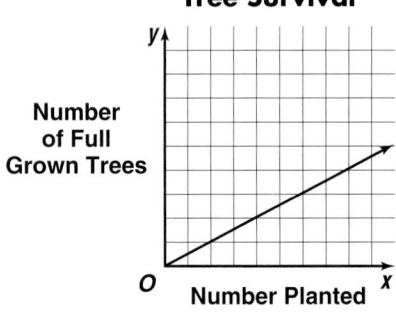

Tree Survival

Number of Full Grown Trees

Number Planted

21. a, d **23.–37.** Sample answers given.

23.{(0 ,0), (1, 2.5), (2, 5), (3, 7.5)}
25. {(0, 7), (1, 10), (2, 13), (3, 16)}
27. {(0, −4), (1, −5), (2, −6), (3, −7)}
29. {(0, 1), (1, 0), (2, −1), (3, −2)}
31. {(1, 2), (3, 4), (5, 6), (6, 7)}
33. {(0, 3), (3, 0.5), (4, 1), (6, 0)}
35. {(1, −1), (2, −2), (3, −3), (4, −4)}
37. {(−2, 0), (−2, 1), (−2, 2), (−2, 3)} **39.** no

41.

45.

49.

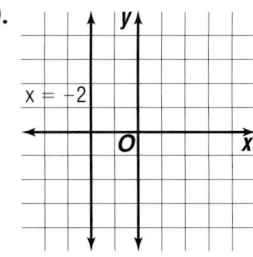

x = −2

51. $x = \frac{1}{4}y$; {(0, 0), (1, 4), (2, 8), (3, 12)};

53. 2 **55.** $y = −3x$ or $3x + y = 0$ **57a.** Sample answer: (1, 4), (2, 8), (4, 16), (6, 24), (8, 32)

57b.

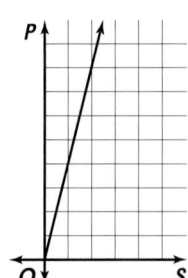

57c. Sample answer: Length cannot have a negative value. **59a.** (0, 32), (100, 212)

59b.

59c. Find the coordinates of other points on the graph.   **59d.** Sample answer: (20, 68)   **61.** $a > 11$ **63.** 4   **65.** 7^2

67.

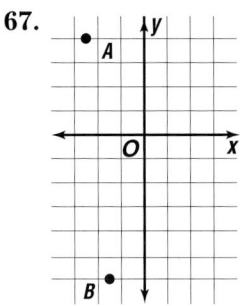

69. $14n$

Page 391   Lesson 8-3B

1. Each distance is the same.
3a. $\{7, 2, -1, -2, -1, 2, 7\}$
3b. $\{(-3, 7), (-2, 2), (-1, -1), (0, -2), (1, -1),$
$(2, 2), (3, 7)\}$;

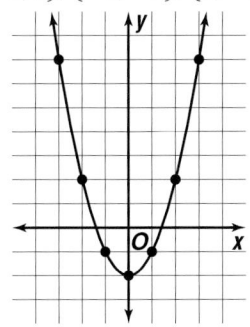

3c. Sample answer: They both open upward and have their vertex on the y-axis. The graphs pass through different points on the y-axis. The graph of $y = x^2 - 2$ is narrower.

Pages 394–395   Lesson 8-4

5. 10, 1, 2, 17; yes   **7.** 3   **9.** 6   **11.** 9 weeks   **13.** 8, 16, 28, 36; yes   **15.** 2, 2, 2, 2; yes   **17.** 5, 4, 2.5, 1.5; yes   **19.** 7, 11, 2, -14; yes   **21.** -47 **23.** $-1\frac{3}{4}$
25. 35   **27.** 4.25   **29.** $4b^2 - 2$ **31.** 118
33a. $\left\{-17\frac{7}{9}, 0, 22\frac{2}{9}, 37\right\}$

33b.

33c. Yes; No member of the domain is paired with more than one member of the range.
35a.   Sample answer:

x	y = 18x + 66.5	y	(x, y)
1	y = 18(1) + 66.5	84.5	(1, 84.5)
5	y = 18(5) + 66.5	156.5	(5, 156.5)
10	y = 18(10) + 66.5	246.5	(10, 246.5)
15	y = 18(15) + 66.5	336.5	(15, 336.5)
20	y = 18(20) + 66.5	426.5	(20, 426.5)

35b.

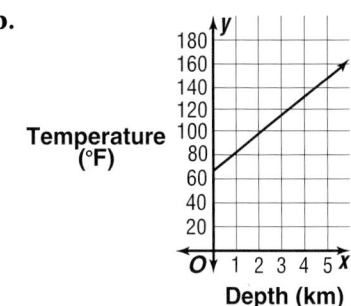

35c. 131.3°F   **37.** $\frac{9}{16}$

39. $a \le -6$;

Pages 398–399   Lesson 8-5
3a. about 78.2 m³   **3b.** about $12.76   **5.** about 4500 lb/in²   **7.** $1.11 a gallon   **9.** about 150 cubic meters   **11.** 156 pounds   **13.** 4   **15.** $-\frac{5}{12}$
17. negative

Page 399   Self Test
1. D = {4, 1, 3, 6}, R = {2, 3, 4}; yes   **3.** positive
5. Sample answer: (2, 6), (1, 2), (0, −2), (−1, −6)

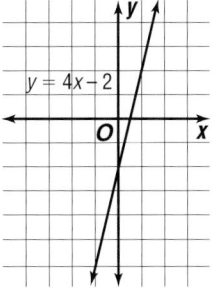

7. 50   **9.** −31

7. 1 **9.** $\frac{1}{2}$ **11.** -1 **13.** 4 **15.** -1 **17.** 0 **19.** $-\frac{1}{5}$
21. -2 **23.** Sample answer: Both graphs slant up
to the right. **25.** $-\frac{3}{2}$ **27.** $\frac{7}{4}$ **29.** 1 **31.** $\frac{3}{8}$
33. $-\frac{3}{4}$ **35.** $\frac{2}{3}$ **37.** 11,080 feet **39.** 89 minutes or
1 hour 29 minutes **41.** composite **43.** 12

3. The point should be on the line.

9. 1.5, 3 **11.** -6, 6

13. $\frac{2}{3}$, 6

15.
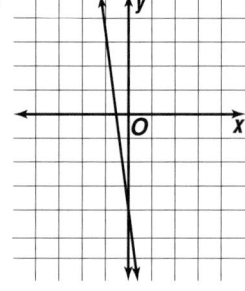

17. 5, -5 **19.** -2, -1 **21.** 2, -4
23.

27.

31.

35.

39.
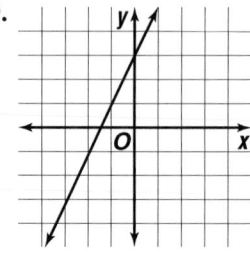

43. The x-intercept and y-intercept are both
zero. The line, therefore, passes through the
origin. Since two points are needed to graph a line,
$y = 2x$ cannot be graphed using only the
intercepts.

45a.

45b. 16 **45c.** The x-intercept is the time when the
plane lands. **47.** Gettysburg **49.** 13 **51.** $-5a$
53. associative, addition

1. The family of graphs is lines having a slope of
1.5. All are parallel and have different x- and
y-intercepts. **3.** Sample answer: $y = -5x$

7. (1, 1) **9.** (2, -6) **11.** (2, 1) **13a.** $2\ell + 2w = 12$;
$\ell = 2w$ **13b.** (4, 2) **13c.** The length of the frame
will be 4 feet and the width will be 2 feet.
15. (0, 0) **17.** (2, 1) **19.** no solution **21.** no
solution **23.** (5, 7) **25.** (-1, 3) **27.** (3, 9)
29. (-2, 8) **31.** (5, 7) **33.** infinitely many
35. $A = 3$, $B = 4$ **37a.** (300, 1500) **37b.** The
break-even point would be at point (300, 1500)
because that is the solution when the cost equals
the income. When 300 items are sold, the income is
$1500. **39a.** $y = 15x$, $y = 5x + 60$ **39b.** (6, 90)

39c. After 6 seconds, the dog will catch up to the prowler at 90 meters away from where the dog started.   **41.** 3, 4, or 5   **43.** A   **45.** 14

Page 417   Lesson 8-9A
1. Both graphs are shaded regions and the boundary is the same line.   **3.** The graph of $y = 2x + 5$ would not be shaded at all.

Pages 420–422   Lesson 8-9
7. yes   **9.** b, c   **11.** The region above the boundary.   **13.** The region to the right of the boundary.

15.   **17.**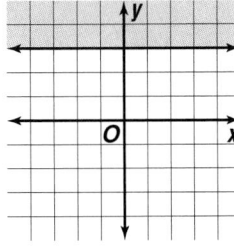

19. a   **21.** c   **23.** a, b, c   **25.** The region to the left of the boundary   **27.** The region to the left of the boundary.

29.   **33.**

37.   **41.**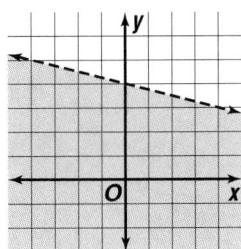

45a. $O < S - 3000$

45b.

45c. below the line   **45d.** No; this would mean that your cost was greater than your sales, which doesn't happen in a successful business.
45e. Sample answer: (6600, 3600), (7000, 4000)

47a.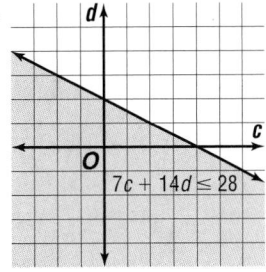

47b. Sample answer: (1, 1) means purchasing 1 cassette and 1 CD for a total cost of $21. (2, 1) means purchasing 2 cassettes and 1 CD for a total of $28.   **49.** $(-1, -1)$ **51.** $\frac{3}{5}, 6.02 \times 10^{-1}, 0.63$
53. 2   **55.** -12

Page 423   Highlights
1. function   **3.** *y*-intercept   **5.** relation
7. *x*-intercept

Pages 424–426   Study Guide and Assessment
9. {0, 6, 7, 9}; {17, 18, 19, 40}; yes   **11.** {6, 8, 15}; {2}; yes   **13.** $\left\{-2, -\frac{1}{2}, 4, 5\right\}$; {6, 3, 1}; yes
15. no   **17.** negative   **19.** no

21.

$(0, 2), (1, 5), (-1, -1), (2, 8)$

25.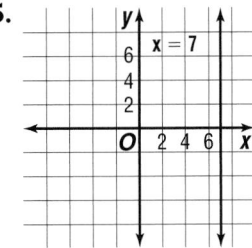

$(7, 0), (7, 2), (7, 4), (7, 6)$

29. 5   **31.** 4   **33.** $\frac{2}{5}$ **35.** -1 **37.** 0

39.   **43.**

45.

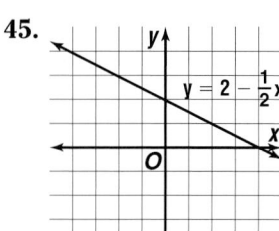

47. (0, 8) **49.** empty set **51.** empty set

53.

55.

57.

59a.

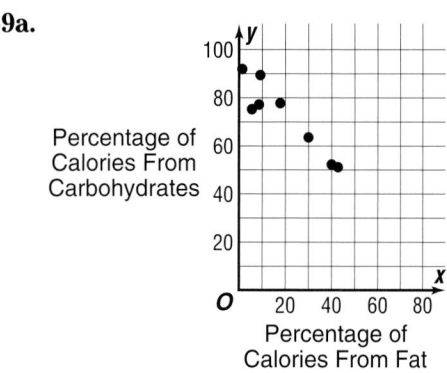

59b. negative **61.** approximately 36 goals scored

Chapter 9 Ratio, Proportion, and Percent

Pages 434–436   Lesson 9-1

5. $\frac{1}{5}$ **7.** $\frac{25}{1}$ **9.** $\frac{43}{46}$ **11.** \$0.38/can **13.** 0.25 in./hour
15. $\frac{1}{13}$ **17.** $\frac{4}{1}$ **19.** $\frac{13}{7}$ **21.** $\frac{2}{15}$ **23.** $\frac{5}{7}$ **25.** $\frac{3}{7}$
27. 24.2 miles/gallon **29.** 9.52 m/s **31.** \$7/ ticket
33a. 50 **33b.** 200 **33c.** 225 **33d.** 88 **33e.** 12
33f. 20 **35a.** the 32 oz bag at \$3.69 **35b.** Sample
answer: When the quality of the less expensive
product is not as good as that of the more
expensive product, it may be better to buy the
more expensive product. **37a.** All ratios are 2.
37b. The ratios approach the golden ratio.

39a. 34.8 mph **39b.** 5 ft/s

41.

43. 6,582,761; 391,000; no mode

45. $x < 28$;

Page 437–438   Lesson 9-2

3. 10 **5.** 21 **7.** 6 **9.** She gained \$800. **11.** −21
13. 5 **15.** 249.5 mph **17.** Sample answer: 180
19. 24 > 7

Page 439   Lesson 9-3A

1. Game A is fair. It is fair because there are the
same amount of odd and even sums. Game B is
unfair because there are many more even products
than odd products. **3.** Yes; there are the same
number of blue and red marbles.

Pages 442–443   Lesson 9-3

7. $\frac{1}{12}$ **9.** $\frac{1}{8}$ **11.** $\frac{1}{2}$ **13.** $\frac{7}{8}$ **15.** $\frac{5895}{7424}$ **17.** 0 **19.** $\frac{17}{24}$
21. 1 **23.** $\frac{1}{2}$ **25.** 0 **27.** $\frac{1}{4}$ **29.** $\frac{1}{36}$ **31.** 0 **33.** $\frac{7}{10}$
35. 1–humans have been to the Moon; 0–the Sun is
too hot. **37.** 4 **39.** −2 **41.** 15.4 m;
13.26 sq meters

Pages 446–447   Lesson 9-4

7. = **9.** 7.5 **11.** 2.1 **13.** $\frac{2.7}{m} = \frac{3}{7}$; 6.3 **15.** ≠
17. ≠ **19.** = **21.** 7 **23.** 42 **25.** 2.5 **27.** 16
29. 2.1 **31.** 10.2 **33.** $\frac{5}{6.15} = \frac{x}{8.00}$; 6.5 **35.** $\frac{8}{2} = \frac{z}{5}$;
20 **37.** $\frac{25}{5} = \frac{m}{25}$; 125 **39.** 1 **41.** 15 **43.** 6 and 8
45. 4 quarts **47.** $\frac{8}{3}$ **49.** 9.412 × 10⁶ **51.** $\frac{1}{4^3}$

Page 448   Lesson 9-4B

1. Sample answer: A larger sample will lead to a
more reliable prediction and result.

Pages 451–453   Lesson 9-5

7. 44% **9.** 3.2% **11.** 9 **13.** 12.5% **15.** 25%
17. 60% **19.** 225% **21.** 75.5% **23.** 106.72
25. 45% **27.** 35 **29.** 16 **31.** 80 **33.** 1200
35. 100 **37a.** 50 g **37b.** about 2375 mg or 2.375 g
39. 31,813; 8238; 18,779; 7279; 35,071 **41.** $\frac{3}{4}$
43. \$4.22 **45.** 5

Pages 455–457   Lesson 9-6

5. 2% **7.** Sample answer: High school cafeteria;
because teen's tastes are not representative of all
ages. **9.** 35% **11.** 125 **13.** No; because the
sample is too small and it is not random.
15. Sample answer: No; many of the common
two-letter words like *at*, *in*, *of*, and *on* do not
contain the letter E. **17a.** 52% **17b.** 60%

19. 58% **21.** D **23.** true

Page 457 Self Test
1. 60 ft/min **3.** 12 **5.** $\frac{9}{16}$ **7.** 1.8 **9.** 50 **11.** 105

Pages 460–461 Lesson 9-7
5. 37% **7.** 103% **9.** 7.2% **11.** $\frac{1}{4}$ **13.** $\frac{3}{1000}$

15. 0.245 **17a.** $\frac{2}{5}$ **17b.** 40% **19.** 3% **21.** 237%

23. 132% **25.** 0.04% **27.** $58\frac{1}{3}$% **29.** $233\frac{1}{3}$%

31. $55\frac{5}{9}$% **33.** 7.5% **35.** $\frac{5}{4}$ **37.** $\frac{69}{200}$ **39.** $\frac{3}{8}$

41. $\frac{49}{300}$ **43.** 0.75 **45.** 0.398 **47.** 0.004 **49.** 2.354

51. 22% **53.** 0.067, 16%, $\frac{1}{4}$ **55.** $\frac{1}{3}$ **57.** $\frac{4}{5}$ **59.** 1.5, -12 **61.** $\frac{5}{27}$ **63.** n^{14} **65.** C **67.** 8

Pages 464–466 Lesson 9-8
7. c **9.** $\frac{4}{5}$ **11.** 215 **13.** 1.5 **15.** 50% **17a.** 60%

17b. 40% **19.** c **21.** b **23.** 1 **25.** $\frac{3}{100}$ **27.** $\frac{1}{10}$

29. $1\frac{1}{5}$ **31.** 120 **33.** 98 **35.** 5.15 **37.** 20
39. 80% **41.** 50% **43.** 10% **45.** 10% **47.** 0.3%
49. 10%, 20%, 60%, 45%, 40% **51.** Sample answers: Connecticut 10%; Delaware 1%; Georgia 2%; Kentucky 1%; Maine 2%; Maryland 10%; New Hampshire 3%; New Jersey 5%; New York 10%; North Carolina 10%; Pennsylvania 15%; Plymouth 10%; Rhode Island 2%; South Carolina 10%; Tennessee 1%; Vermont 2%; Virginia 20% **53.** $\frac{16}{25}$

55. geometric; 1.5; 60.75, 91.125, 136.6875

57. 21;

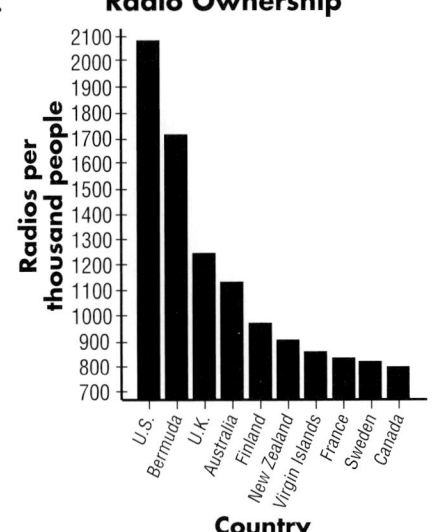

59.

Radio Ownership

Pages 469–471 Lesson 9-9
7. 200 **9.** 17.28 **11.** $47.25 **13.** 3.36 million
15. 27 **17.** 70 **19.** 31.5 **21.** 300% **23.** $1.\overline{6}$%
25. 120 **27.** 20 **29.** $70 **31.** $187.50 **33.** $42.50
35. $41.34 **37.** about 93 million **39.** They would

be paying the customers. **41a.** $264.72 **41b.** No; the price is the same either way. **41c.** You could not add the percents first. For example, if the price was $100, 10% + 33% = 43% off. Using this method the price would be 100 − 43 or $57. Taking 33% off and then 10% would have the price as $60.30.
43. The statements are not equivalent. For example, suppose an item costs $1.00 on day 1. According to the first statement, it would cost $2.08, not $1.00 × 2 or $2.00, on the fifth day.

Day 2   1.2 × 1.0 = $1.20
Day 3   1.2 × 1.20 = $1.44
Day 4   1.2 × 1.44 = 1.728 or $1.73
Day 5   1.2 × 1.73 = 2.076 or $2.08
45. positive **47a.** $3v - 298 = 1916$ **47b.** 738
49. 15 **51.** C

Pages 474–475 Lesson 9-10
7. decrease; 12% **9.** decrease; 7% **11.** decrease; 20% **13.** increase; 13% **15.** increase; 13%
17. increase; 10% **19.** increase; 7% **21.** increase; 136% **23a.** no; Suppose the price of the outfit is $100. The price would be $50 after the first reduction. After the second reduction, the price would be 50 − 0.3(50) = 50 − 15 or $35. A final price of $35 is $\frac{100 - 35}{100}$ or 65% off, not 80% off.
23b. 65% **25.** 20% **27.** $3.30 **29.** 125 **31.** Yes; it passes the vertical line test. **33.** A

Page 477 Highlights
1. true **3.** false **5.** true **7.** false

Pages 478–480 Study Guide and Assessment
9. $\frac{41}{45}$ **11.** $\frac{3}{5}$ **13.** $0.89 per pound

15. 57 miles/hour **17.** $\frac{13}{21}$ **19.** $\frac{8}{21}$ **21.** $\frac{11}{21}$ **23.** 2.5

25. $\frac{88.4}{3.4} = \frac{161.2}{g}$; 6.2 gal. **27.** 76% **29.** 4 **31.** 32

33. 0.56% **35.** 56.5% **37.** $\frac{13}{20}$ **39.** 2.35 **41.** a
43. a **45.** b **47.** 20 **49.** 4000 **51.** 31.25%
53. decrease; 1% **55.** decrease; 22%
57. decrease; 12% **59.** 28

Chapter 10   More Statistics and Probability

Pages 488–489 Lesson 10-1
7. 5, 6, 7, 8

5	4 7
6	3 7 8 9
7	1 5 7
8	5

$5|4 = 5.4$

9.

6	0 1 4 5 7 8
7	0 0 2 3
8	0 0

$6|0 = 60$

11.

0	5 6 7 9 9
1	1 2 4
2	1 3
3	
4	0 2 5

$4|0 = 40$

13.

9	2
10	4 8
11	1 2 2 7
12	3 6 8 9
13	3 8
14	7

$9|2 = 9.2$

15.

Women's		Men's
5	4	5
5 5	5
7 5 0	6
5 2 0 0 0	7	0 0 0 0 0 5
5 0	8	2 2 3
0	9
	10	0
	11	0
4	12	0 5
	13	3

$5|4 = 45$ $4|5 = 45$

15b. Answers will vary. Sample answer: Men's running shoes are more expensive than women's running shoes. The typical running shoe costs $70. **17.** Sample answer: Yes; Since the data can be read from a stem-and-leaf plot, you can estimate their sum and then divide by the number of leaves to estimate the mean.

19a.

Distress		No Distress
8 7 3	5
3	6	6 7 7 7 8 9
5 0 0	7	0 0 2 3 5 6 6 8 9
	8	1

$3|5 = 53$ $7|0 = 70$

21. $\frac{9}{20}$ **23.** D

Pages 492–494   Lesson 10-2
5. 13   **7.** 52

9a.

3	8 9
4	0 2 6
5	8
6	1 2 3 3
7	0
8	0 4
9	4

$3|8 = 38$

9b. 61.5   **9c.** 70, 42   **9d.** 28   **11.** 70; 65; 85, 45; 40   **13.** 21; 7; 13.5, 3; 10.5   **15.** 49; 208; 213, 200; 13   **17.** 4.1; 9.2; 9.75, 8.85; 0.9   **19.** 37,600,000; 43,100,000; 52,600,000; 33,100,000; 19,500,000   **21.** American League: 39; 42.5; 46, 38; 8; National League: 21; 40.5; 46.5, 38; 8.5   **23c.** The first set of data has a smaller interquartile range, thus the data in the first set are more tightly clustered around the median and the data in the second set are spread out over the range.   **25a.** participants 60; observers 71   **25b.** 51; 35   **25c.** Sample answer: The participants' ages are between 12 and 72. Their ages are spread out over the range of the data. The median age of the participants is 25 years.   **25d.** Sample answer: The observers are both younger and older than the participants, but the range of ages is very close to that of the participants. The median age of the observers is 46.5 while the median ages of the participants is 25.   **27.** >   **29.** C

Pages 499–501   Lesson 10-3
7a. 40   **7b.** 50   **7c.** 30   **7d.** 20   **7e.** 25, 55   **7f.** none   **9a.** 15   **9b.** 68   **9c.** 71   **9d.** 65   **9e.** 6   **9f.** no   **11a.** 50%   **11b.** 50%   **11c.** 25%   **11d.** 25%   **11e.** Scores were closer together between 80–83.   **11f.** The scores were spread further apart.   **13.** box-and-whisker plot   **15.** box-and-whisker plot   **17.** line graph or comparative graph   **19.** The box did not change. The weight 200 is an outlier.

21a.

21b. The medians are the same.   **21c.** The range and interquartile range are greater for the first period class.

23a.

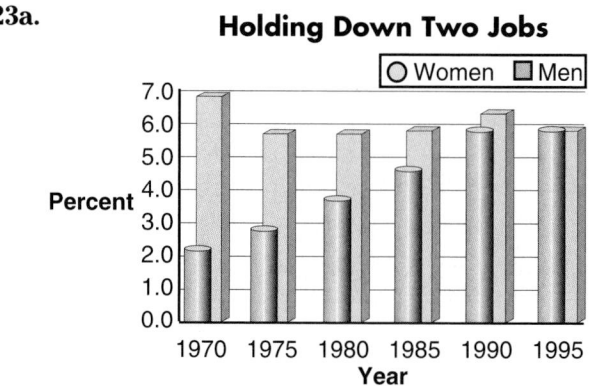

23b. Men's employment decreased from 1970 to 1975, but since 1975 men's employment has been relatively consistent.   **23c.** Women's employment increased steadily from 1970 to 1990 and has remained steady from 1990–95.   **23d.** Women are as likely as men to have two jobs.   **25.** {5, −1, 4.5}; {−3, 4, 0}   **27.** Sample answer: $7

1.

3.

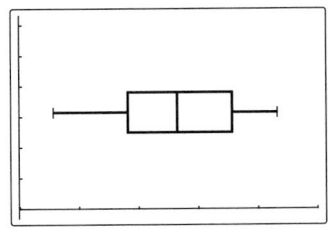

5a. about 1.1   **5b.** The second car is shorter in length and width.   **5c.** Yes; the actual number of cars sold in 1992 is 1.1 times greater than the number sold in 1991. However, the graph appears to be longer and wider, but both dimensions of the car should not have changed.   **7a.** Median; there are large and small values.   **7b.** Mean; the high values will raise the center of the data.   **9.** 172.6; mean   **11.** sports; mode
13.

	Company A	Company B
mean	$22,413.79	$21,935.48
median	$15,000.00	$20,000.00
mode	$15,000.00	$20,000.00
range	$85,000.00	$30,000.00

You would probably prefer to work for Company B. Both the median and mode salaries are greater, so the salary is greater for more employees.
15a. Sample answer: No; it only shows the population increases for 6 years.

15b.

17a. $58   **17b.** $25   **17c.** $35   **17d.** $15   **17e.** $20
17f. $12, $70   **17g.** $0 and $65   **17h.** $70   **19a.** $\frac{8}{11}$
19b. 1,965,955   **21.** $-6, 6$

5a. 12 outcomes   **5b.** $\frac{1}{12}$ **7.** 18 outcomes
9. 24 outcomes   **11.** 25 ways   **13.** 36 outcomes
15a. 8 outcomes   **15b.** $\frac{1}{8}$ **17.** 32 outcomes
19a. 2^{30} **19b.** $\frac{1}{2^{30}}$ **21a.** 10,000 extensions
21b. 4,860,000   **21c.** 2,880,000 new numbers
23. The median; it is close to three of the four actual prices.   **25.** -6 **27.** $h =$ hours, $3.50h < 20$, $h < 5.7$

1.

```
2 | 8
3 | 6 7 9
4 | 3 7 9
5 | 0 1 1 3
6 | 2        6|2 = 6.2
```

3. 2.7, 4.7   **5.** A comparative graph is probably the best way.   **7.** Brian would use the bottom graph because it makes the increases he has received appear smaller.   **9.** 24 outcomes

1. 120 words

5. combination   **7.** 2   **9.** 24   **11.** 120   **13.** 252
15. 56 programs   **17.** combination
19. permutation   **21.** combination   **23.** 720
25. 720   **27.** 362,880   **29.** 39,916,800   **31.** 35
33. 1   **35.** 45   **37.** 252   **39.** $1\frac{2}{3}$ **41.** 5005 teams
43. 6 ways   **45.** 10 students   **47.** 5005 combinations   **49.** 140,400,000 license plates
51. about $-15°$ C   **53.** C   **55.** $\frac{5}{4}$ **57.** $-9x$

5. 1:1   **7.** 5:1   **9.** 1:4   **11.** 4:1   **13.** 1:1   **15.** 1:2
17. 7:5   **19.** 5:1   **21.** 1:17   **23.** 7:11   **25.** 9:4
27. 1:9999   **29a.** less than 1 or none   **29b.** about 89 people   **31.** 40,320   **33.** $0.16, $0.31, $0.36, $0.40, $0.51, $0.55, $0.60   **35.** >

Pages 526–528   Lesson 10-8

5a. Sample answer: Select a marble out of a bag with 4 marbles the same color and one marble a different color. Then repeat three more times.
7a. Roll a die.   **9.** $25   **11a.** 15 games   **11b.** 240
13b. Change the second line to "For (N, 1, 50)".
15. 1:5   **17a.** $110 = \frac{1}{2}x + 20$ **17b.** $180   **19.** D

Page 529   Lesson 10-8B

1. The most frequently drawn color is probably the one with the greatest number of marbles in the bag.   **3.** yes; It is more likely to happen with fewer draws.

Pages 533–534   Lesson 10-9

5. dependent   **7.** $\frac{21}{380}$ **9.** $\frac{9}{190}$ **11.** $\frac{11}{625}$ or 1.76%

13. independent   **15.** independent   **17.** $\frac{1}{30}$

19. $\frac{2}{15}$ **21.** 0   **23.** $\frac{10}{91}$ **25.** $\frac{3}{91}$ **27.** $\frac{1}{91}$ **29a.** $\frac{3}{8}$

29b. $\frac{3}{10}$ **31.** $\frac{4993}{5000}$ or 99.86%   **33.** $10; about $40

35. 2;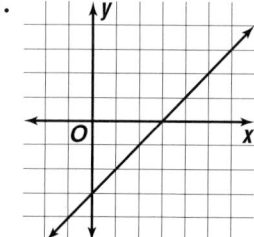

37. $2(n + 3)$

Pages 537–538   Lesson 10-10

5a. $\frac{1}{3}$ **5b.** $\frac{1}{3}$ **5c.** $\frac{2}{3}$ **7a.** inclusive; $\frac{3}{4}$

7b. exclusive; $\frac{5}{8}$ **7c.** exclusive; $\frac{5}{8}$ **7d.** inclusive; $\frac{1}{2}$

9. 0   **11.** $\frac{2}{15}$ **13.** $\frac{2}{221}$ **15.** $\frac{4}{9}$ **17.** about 10 times
19. about 35 times   **21.** about 20 times   **23.** 53%
25. independent

27.

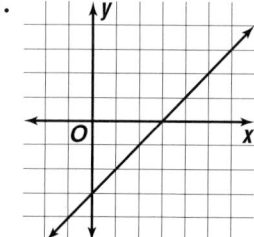

29. 300

Page 539   Highlights
1. C   **3.** I   **5.** A   **7.** G

Pages 540–542   Study Guide and Assessment
9.
9 | 5 8
10 | 7 7 8
11 | 0 5 9   *9|5 = 9.5*

11.
11 | 1 2 7
12 | 0 1
13 | 3
14 | 6 8   *11|1 = 111*

13. 30; 18; 10, 28; 18   **15.** 98; 217; 205, 240; 35

17a.

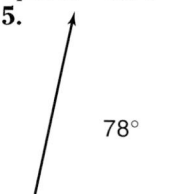

17b. 35   **17c.** none   **19.** 6.75; mean   **21.** favorite sport: soccer; 6.75 classes; 6.5 family members
23. 1024   **25.** 104   **27.** C   **29.** 120   **31.** 90   **33.** 10:3
35. 1:1   **37.** dependent   **39.** $\frac{3}{22}$ **41.** $\frac{3}{22}$
43. inclusive; $\frac{27}{36}$ **45.** $\frac{1}{2}$ **47.** 64

Chapter 11   Applying Algebra to Geometry

Pages 552–553   Lesson 11-1
7. plane   **9.** A   **11.** $\overrightarrow{CD}$ **13.** 63°, acute
15.

17. The three "pods" determine a plane.   **19.** plane
21. ray

23.   **25.**

27.   **29.**

31.   **33.**

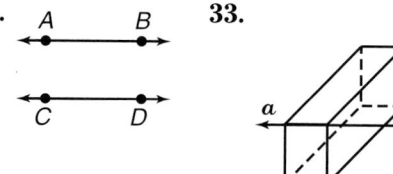

35. 155°; obtuse   **37.** 65°; acute   **39.** 15°; acute

41. acute;

43. acute;

45. right;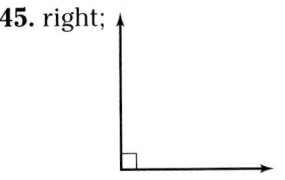

47. 3° **49.** 22 times **51.** = **53.** 50, 52 **55.** 15
57. A

Page 555   Lesson 11-1B
3. Cut in two equal pieces. **5.** They are the same.

Pages 558–560   Lesson 11-2
5. silicon **7a.** 100% **7b.** 360° **7c.** Housing: 31%,
111.6°; Food:19%, 68.4°; Transportation: 22%, 79.2°;
Insurance: 7%, 25.2°; Savings: 4%, 14.4°; Misc.: 17%,
61.2°; Totals, $2080, 100%, 360.0°

7d.   Monthly Budget

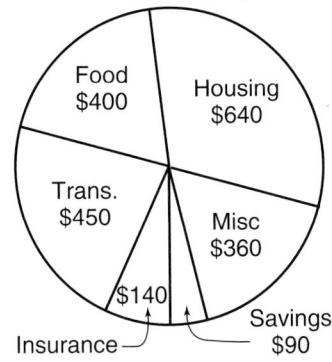

9.   Days in the Month

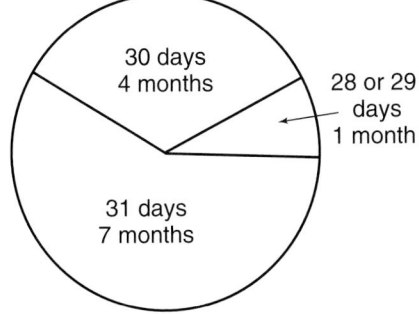

11a. $\frac{3}{8}$

**11b.   Probability
of
Winning or Losing**

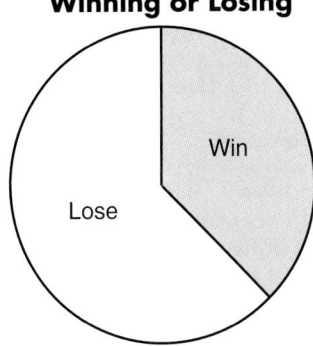

**13.   Areas of Oceans
of the World**

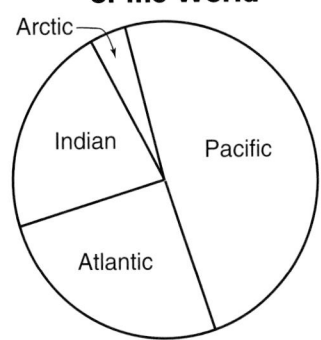

**15a.   Black Hills'
Reclamation Mix**

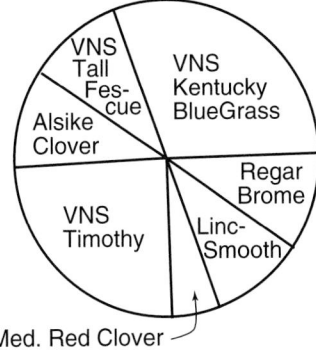

15b. 75 lb. VNS Kentucky Bluegrass;   25 lb each of
Regar Bromegrass, Lincoln Smooth Brome, Alsike
Clover, and VNS Tall Fescue; 12.5 lb Medium Red
Clover; 62.5 lb VNS Timothy

17a. $1000 × 0.05 × 1.5; $75; compounding
semiannually pays $1.90 more than simple
interest

17b. $10, 613.63

17c. $1212.04; The simple interest for 3 months is
$12, about 4 cents less.

17d. As the frequency of compounding increases,
the amount of interest increases.

17e. $340.10; The simple interest is $300, which is
$40.10 less.

19.

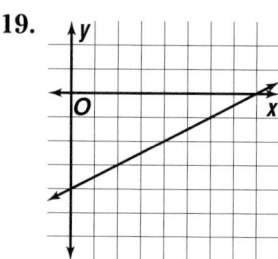

21. Sample answer: Joe earned 3 times as much as Sue. If Joe earned $25, how much did Sue earn?

Pages 564–566   Lesson 11-3
7. 40   **9.** 15°   **11.** $m\angle X = 60°, m\angle Y = 120°$
13. $\angle AFG, \angle GFB, \angle CGF, \angle FGD$
15. $m\angle AFE = 130°; m\angle EFB = 50°; m\angle BFG = 130°;$
$m\angle AFG = 50°; m\angle CGF = 130°; m\angle FGD = 50°;$
$m\angle DGH = 130°; m\angle CGH = 50°$ **17.** 105   **19.** 66
21. 55   **23.** 120°, 60°   **25.** 18°, 72°   **27.** 47°
29. 133°   **31.** 133°   **33.** $m\angle R = 32°; m\angle S = 58°$
35. $m\angle F = 118°; m\angle G = 62°$ **37.** 30   **39.** 55°
41. Yes; all the angles formed are right angles.

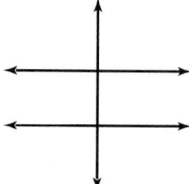

43. 135°   **45.** $\frac{5}{18}$ **47.** 5.678×10^4 **49.** $-7b$ **51.** 5

Page 567   Lesson 11-3B
1. It divides the difference of the y-coordinates by the difference of the x-coordinates.   **3.** yes
5. Draw two lines with the same slope.

Pages 570–572   Lesson 11-4
7. 117; obtuse   **9.** equilateral   **11.** 119°   **13.** 50; right   **15.** 36°; 54°; 90°   **17.** isosceles
19. equilateral   **21.** obtuse   **23.** 85°   **25.** 25°
27. 85°   **29.** 45°, 80°, 55°   **33.** 60°   **35.** 25°   **37.** $\frac{5}{13}$
39. $59\frac{1}{4}$ inches   **41.** D

Page 572   Self Test
1. acute   **3.** acute   **5.** 34°   **7.** 34°   **9.** $m\angle F = 100°;$ $m\angle G = 80°$ **11.** 100; obtuse

Pages 575–577   Lesson 11-5
5a. $\triangle ZYX$ **5b.** $\overline{ZY}$ **5c.** $\angle B$ **5d.** $\angle X$ **5e.** $\overline{YX}$
5f. $\overline{CA}$ **5g.** $\angle A$ **7.** c   **9.** $\triangle ABC \cong \triangle FED;$
$\angle A \cong \angle F, \angle B \cong \angle E, \angle C \cong \angle D, \overline{AB} \cong \overline{FE}, \overline{BC} \cong \overline{ED},$
$\overline{AC} \cong \overline{FD}$ **11.** $\triangle BAD \cong \triangle ABC; \angle BAD \cong \angle ABC,$
$\angle D \cong \angle C, \angle ABD \cong \angle BAC, \overline{BA} \cong \overline{AB}, \overline{AD} \cong \overline{BC},$
$\overline{BD} \cong \overline{AC}$ **13.** $\triangle DEF \cong \triangle HGF$ **15.** $\overline{MA}$ **17.** $\overline{RT}$
19. $\angle M$ **21.** 3   **23.** 8   **25.** $\triangle ABH \cong \triangle IJG;$
$\triangle ACG \cong \triangle IEF$ **27.** 5 ft   **29.** 95°; obtuse   **31.** 125
33. Liz–pack; Renee–snack bar; Pablo–plate lunch

Pages 581–583   Lesson 11-6
7a. $\dfrac{m\overline{AB}}{m\overline{AC}} = \dfrac{m\overline{BE}}{m\overline{CD}} = \dfrac{m\overline{AE}}{m\overline{AD}}$ **7b.** $\angle CAD$ **7c.** $\overline{CD}$
7d. 10   **9.** $\dfrac{1}{x} = \dfrac{2}{6}$; 3   **11.** 121 m   **13.** $\dfrac{3}{2} = \dfrac{x}{3}$; 4.5
15. $\dfrac{4}{5} = \dfrac{8}{x}$; 10   **17.** $\dfrac{3}{5} = \dfrac{4.5}{x}$; 7.5   **19a.** Yes; the ratio of the corresponding sides is 1:1.   **19b.** No; only those where the ratio of the corresponding sides is 1:1.   **21.** 2.7 ft   **23.** 32 m   **25.** 2^5 **27.** 47.9; 42.5; 54

Pages 586–588   Lesson 11-7
5. 115   **7.** quadrilateral, trapezoid; trapezoid
9. 6 m   **11.** 30°   **13.** $x = 60$; 60°, 120°   **15.** $x = 60$; 60°, 120°   **17.** square   **19.** rectangle
21. quadrilateral   **23.** sometimes   **25.** never
27. always   **29.** Yes; a rhombus is equilateral but may not be equiangular.   **29b.** Yes; a rectangle is equiangular but may not be equilateral.   **35.** 3 cm
37. 22.5 miles per gallon   **39.** inductive

Pages 592–593   Lesson 11-8
7. hexagon, not regular   **9.** 40°; 140°   **11.** 900°
13. 1800°   **15.** 2340°   **17.** 60°, 120°   **19.** 30°, 150°
21. 60°   **23.** 72 cm   **25.** 27 in.

27.

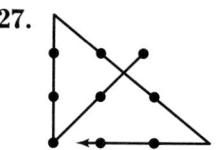

29a. trapezoid   **29b.** 45°, 45°, 135°, 135°
29c. square   **31.** $\dfrac{1}{5}$
33. $(1 \times 10^3) + (6 \times 10^2) + (4 \times 10^0)$

Page 594   Lesson 11-8B
1. It will tessellate if 360 is a multiple of the angle's measure.

Pages 597–599   Lesson 11-9
5. translation

7.

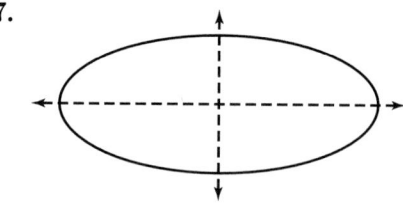

9. reflection   **11.** rotation   **13.** b

15.   **17.**

19.

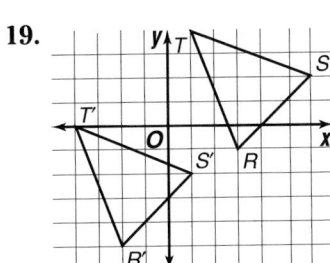

21. top to bottom: *all except* E L N O U Y Z
side to side: C D E F I J M P Q R S V W X
diagonal: F I L M N P Q R S U V W X Y
25. A

27. $t < -21$

1. b   **3.** g   **5.** a   **7.** i

9.

11.

13.

15.

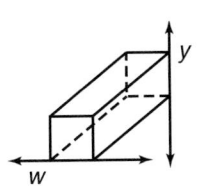

19. 48°   **21.** 48°   **23.** 90°   **25.** 92°; obtuse
27. $\angle Y$ **29.** $\overline{XZ}$ **31.** 2 in.   **33.** 70°; 110°   **35.** 144°

37.

39. reflection

41.

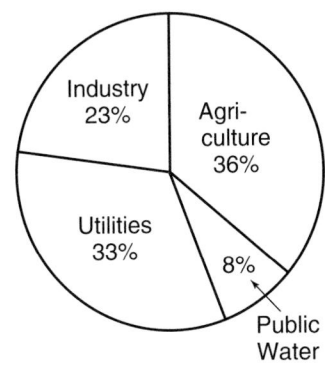

Water Use in U.S.

Industry 23%
Agriculture 36%
Utilities 33%
8%
Public Water

---

1. Sample answer: Count squares and half-squares.
3. Sample answer: Find the area of the rectangle and divide by 2.

5. 3; 4; 6 sq cm   **7.** 15 m and 5 m; 6 m; 60 m²
9. 216 cm²   **11a.** about 33,600 mi²   **11b.** The actual area is 35,936 mi².   **13.** 6 cm²   **15.** 90 cm²
17. 21.6 in²   **19.** 57 ft²   **21.** 9 in²   **23.** $6\frac{1}{2}$ ft²
25. 36.75 m²   **27.** $A_1 = \frac{1}{2} h_1 d$; $A_2 = \frac{1}{2} h_2 d$; $A_1 + A_2 = \frac{1}{2}$ $d(h_1 + h_2) = \frac{1}{2} d(d) = \frac{1}{2} d^2$ **29a.** 308 ft²   **29b.** 15-foot width   **31.** A shape as close as possible to a square gives the greatest area.   **33.** 5   **35.** 2.1
37. $x + 7$

1. Each section is a replica of the whole curve.
3. infinite

5. 452.4 in²   **7.** 49,087.4 ft²   **9.** 50.3 cm²
11. 120.8 m²   **13.** 475.3 cm²   **15.** 167.4 in²
17. 39.3 ft²   **19.** 7854.0 ft²   **21.** about 707 m²
23. about 26 times   **25.** about 13 mi²   **27.** C
29. $\frac{4}{9}$

5. $\frac{9}{50}$ **7.** about 9   **9.** $\frac{3}{8}$ **11.** $\frac{1}{6}$ **13.** $\frac{3}{4}$
15. about 38   **17.** 0.16   **19.** $\frac{1}{35}$ **21.** D   **23.** 0.1
25. $32 < w < 212$

1. The ratios should be similar.   **3.** More data makes the ratios closer.

5. 875 cm²
7b. 10 units; 10 units; 8 units; 10 units; 10 units
7c. Shapes where two squares share one side with another square and the rest of the squares share two sides with other squares are the greatest in perimeter.   **7d.** Square shapes have the least perimeter.   **7e.** 12; yes   **9.** 4 m by 12 m
11. $3 \times 3 \times 4$ **13.** $700 gain   **15.** A   **17.** $\frac{2}{5}$
19. 11 hours   **21.** B

7. triangular prism; 184 ft²   **9.** cylinder; 3166.7 m²
11. 282 in²   **13.** 132 cm²   **15.** 211.2 cm²
17. 636.2 in²   **19.** 264 in²   **21.** d   **23a.** 2250 in²
23b. No, as long as the entire surface of the face of the smaller cube is touching a face of the larger cube, the surface area of the tower will be the same.   **25.** about 6189 cm²   **27.** 512 in², 1200 in², 1800 in²; no

31.

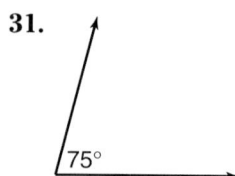

75°

33. 3,605

5. pyramid; 134.4 m² **7.** triangular pyramid; 57 in²
9. 66.4 ft² **11.** 404.6 m² **13.** 282.7 in²
15. 471.2 cm² **17.** 275 in² **19.** ≈ 25,780.8 ft²
21. 3160.1 sq ft **23.** 900 cm² **25.** $4x + 8 = 20$; 3
27. $-15xy$

Page 642   Self Test
1. 15m² **3.** 113.1 in² **5.** cylinder, 402.1 in²
7. ≈ 0.028 or about $\frac{3}{100}$ **9.** 47.1 in²

Page 643   Lesson 12-7A
1. triangular prism; cylinder **3.** same; 8 in.
5. cylinder

Pages 646–648   Lesson 12-7
7. 600 in³ **9.** 225 in² **11.** 28.3 ft³ **13.** 125 ft³
15. 512 cm³ **17.** 1608.5 ft³ **19.** 565.5 cm³
21. 24.5 ft³ **23.** 83.3 cm³ **25.** 4712.4 yd³
27. 18.6 m³ **29.** The volume is greater with $8\frac{1}{2}$ as
the height. **31.** 3.29 cm **33.** a **35a.** 128 ft³
35b. $\frac{5}{8}$ **37.** 4 mi **39.** 7 **41.** D

Pages 651–653   Lesson 12-8
5. 32 ft³ **7.** 160 ft³ **9.** 270.8 in³ **11.** 70 ft³
13. 160 m³ **15.** 150 in³ **17.** 9189.2 m³
19. 733 cm³ **21.** 50 cm³ **23.** 1885.0 m³ **25.** 44 m³
27. No; the customer gets only $\frac{1}{3}$ as much popcorn
for $\frac{1}{2}$ the price. **29.** 91,636,272 ft³ **31.** 15 ft³
33. 20% **35.** 55 cm; 184 cm²

Page 654   Lesson 12-8B
1. 1 unit² **3.** 1 unit³ **5.** 9 unit²
7.

Scale Factor	Length of a Side	Area of a Face	Volume
1	1	1	1
2	2	4	8
3	3	9	27

9. x^2 **11.** x^3

Page 655   Highlights
1. G **3.** H **5.** A **7.** C

Pages 656–658   Study Guide and Assessment
9. $3\frac{15}{16}$ in² **11.** 476 mm² **13.** 13.9 m² **15.** 8.0 m²
17. $3\frac{3}{8}$ ft² **19.** 7665 ft² **21.** 9.4 m² **23.** 3325 in³
25. 2770.9 ft³ **27.** 504 m³ **29.** 864 in² **31.** 198
33. about 1275 gallons

Pages 667–668   Lesson 13-1
7. 4 **9.** 9 **11.** -7 **13.** 11 **15.** 2 **17.** 3 **19.** -4
21. 10 **23.** -12 **25.** 20 **27.** 9 **29.** 14 **31.** -10
33. 51 **35.** 11 **37.** -5 **39.** a **41.** 3 **43.** Yes;
because when a negative number is squared the
result is positive. **45.** 20 feet **47.** 192 in³ **49.** $\frac{7}{25}$
51. 36 **53.** -20

Pages 670–671   Lesson 13-2
3. 18 people **5.** 100 cm **7.** 77 pieces **9.** 10 girls
11. 7 students
13.

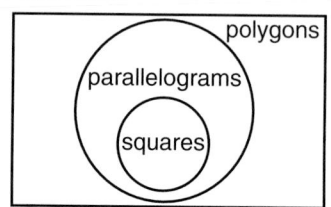

15. A rhombus is a quadrilateral with four
congruent sides. **17.** 26.5

Pages 674–675   Lesson 13-3
5. whole, integer, rational, real **7.** rational, real
9. 19, -19 **11.** 4.1, -4.1 **13.** whole number,
integer, rational, real **15.** rational, real
17. rational, real **19.** irrational, real
21. irrational, real **23.** rational, real **25.** 6, -6
27. 3.5, -3.5 **29.** 11.0, -11.0 **31.** 13.4, -13.4
33. 0.02, -0.02 **35.** 20, -20 **37.** 9, -9 **39.** 3, -3
41. Yes; sample answer: $\sqrt{2} \times \sqrt{2} = 2$ **43.** 2400 ft²
47. 4908.7 cm³ **49.** $-\frac{2}{5}$ **51.** 72

Pages 679–681   Lesson 13-4
5. $x^2 = 12^2 + 16^2$; 20 in. **7.** $30^2 = 21^2 + b^2$; 21.4 km
9. 5 **11.** 32.9 **13.** yes **15.** about 127.3 ft
17. $x^2 = 8^2 + 15^2$; 17 cm **19.** 26.5 m **21.** 19.6 ft
23. 20 **25.** 5.4 **27.** 5.7 **29.** 9.9 **31.** 20.5 **33.** 5
35. no **37.** yes **39.** no **41.** 12 m **43.** 5
45. 25 km **47.** 1.6 **49.** 5 **51.** $\frac{4}{9}$ **53.** 20

Page 681   Self Test
1. 6 **3.** -10 **5.** 12, -12 **7.** 6.3, -6.3 **9.** 16

Page 682   Lesson 13-4B
1. Use the procedure in the activity, drawing a
right triangle with legs 1 and 3 units long. Then use
a compass to find the distance $\sqrt{10}$ on the x-axis.
3. Use the procedure in the activity, drawing a
right triangle with legs 1 unit and 2 units long; or
use the procedure in the activity to find the
lengths $\sqrt{2}$ and $\sqrt{3}$ and use those lengths as the
legs of a right triangle. **5.** Draw a triangle with
one leg 2 units long and hypotenuse 4 units long.
Transfer $\sqrt{12}$ to the number line.

Pages 685–686   Lesson 13-5
3. 1.4 yd **5.** 9 m **7.** $a = 3$ m, $c \approx 4.2$ m
9. $a = 6$ m, $b = 10.4$ m **11.** 15 cm **13.** 21.2 cm

15. 7.4 m   **17.** 9.8 in.   **19.** 24 in.   **21.** $\frac{1}{8}$ mi
23. 2.315 cm   **25.** $a \approx 8.1$ cm, $c \approx 16.2$ cm
27. $a = 6$ in., $b \approx 10.4$ in.   **29.** $b = 5$ yd, $c \approx 7.1$ yd
31. $a \approx 4.0$ m, $b \approx 4.0$ m   **33.** $a = 5.5$ km, $b \approx 9.5$
km   **35.** about 5.8 yards   **37.** about 7.07 feet
39. 23.3 feet   **41.** 108°   **43.** C   **45.** $s \geq 50$

Page 687   Lesson 13-6A
1. The ratios are the same.   **3.** The ratios of the given angles are the same no matter how long the sides are.

Pages 690–692   Lesson 13-6
7. $\frac{24}{25}$ **9.** $\frac{7}{25}$ **11.** $\frac{24}{7}$ **13.** 0.3256   **15.** 0.4226
17. 80°   **19.** about 19 feet   **21.** 0.724, 0.690, 1.05
23. 0.6, 0.8, 0.75   **25.** 0.246, 0.969, 0.254
27. 0.5000   **29.** 0.9063   **31.** 0.9455   **33.** 53°
35. 44°   **37.** 61°   **39.** 10°   **41.** 42°   **43.** 19°
45. about 84.8°   **47.** 10 yd   **49.** 9 ways
51. deductive
53. -242;

Page 693   Lesson 13-6B
1. $\frac{1}{2}$ **3.** They are the same.   **5.** 3

Pages 696–697   Lesson 13-7
5. $\tan 39° = \frac{10}{x}$; 12.3   **7.** about 18.1 feet
9. $\sin 60° = \frac{x}{16}$; 13.9   **11.** $\cos 30° = \frac{9}{x}$; 10.4
13. $\sin x° = \frac{8}{10}$; 53.1   **15.** $\sin x° = \frac{15}{21.2}$; 45.0
17. 38.3 yards   **19.** about 118 feet
21. about 44 m   **23.** $a \approx 12.4$ yd, $b \approx 15.9$ yd
25. 37 feet   **27.** about 62°   **29.** 30; 62; 70.5; 48.5; 22
31. yes

Page 699   Highlights
1. square   **3.** rational   **5.** 30°–60°

Pages 700–702   Study Guide and Assessment
7. 3   **9.** -5 **11.** 15   **13.** 9   **15.** 15   **17.** -6
19. whole, integer, rational, real   **21.** rational, real
23. 14, -14 **25.** 3.9, -3.9 **27.** 20   **29.** 34   **31.** 16
33. 6 m   **35.** 3 cm   **37.** $\sin X = 0.882$, $\cos X =$
0.471, $\tan X = 1.875$ **39.** 84°   **41.** $\tan 35° = \frac{x}{15}$;
10.5 ft   **43.** $\sin 25° = \frac{15}{x}$; 35.5 in.   **45.** 28.4 ft
47. about 321.68 ft

Chapter 14   Polynomials

Pages 708–709   Lesson 14-1
5. yes; monomial   **7.** no   **9.** 3   **11.** 3   **13.** 35
15. 57   **17.** $38.82   **19.** yes, binomial   **21.** yes,
trinomial   **23.** yes, monomial   **25.** yes, trinomial
27. yes, monomial   **29.** yes, trinomial   **31.** 1
33. 3   **35.** 6   **37.** 5   **39.** 7   **41.** 5   **43.** -26
45. -18 **47.** -43 **49.** 97   **51.** -44 **53.** 82
55a. $2x + 2y + z + xy$ **55b.** 2   **59.** 156.8 m²
61. $104.37   **63.** yes; 3; 81, 243, 729   **65.** $x > 8$
67. 13

Page 710   Lesson 14-1B
1. A monomial uses just one size of tiles, a binomial uses two different sizes, and a trinomial uses three different sizes.   **3.** The degree of the polynomial is determined by the size of the largest algebra tile.

Pages 713–714   Lesson 14-2
7. $3x + 8$ **9.** $4x + 4$ **11.** $3x^2 + 3x + 2$
13. $6x + 24$ units   **15.** $4x + 14$ **17.** $11m - n$
19. $2x^2 - 2x + 9$ **21.** $17m + 3n$ **23.** $8x^2 - 3x + 3$
25. $4a^3 + 2a^2b - b^2 + b^3$ **27.** $4a + 6b + c$
29. $-2a - 3ab + 4b^2 + 2$ **31.** $4x^2 + x - 1$
33. $4a^2 - 3ab - 2b - 4b^2$; 0   **35.** $(4r + 3s)$
37. $4x + 2$ united inches   **39.** 7.5 cm²
41a. $t + h > 130$

41b. the area above the boundary line
41c. No points with $h > 100$ are included in the graph.
43. 15 ft

Pages 717–718   Lesson 14-3
9. $-2abc$ **11.** $-x^2 - 5x - 1$ **13.** $8m - 7n$
15. $2x + 2$ **17.** $2x - 3$ **19.** $x^2 + 3x - 3$
21. $4x - 6$ **23.** $x - 3$ **25.** $2x - 8$ **27.** $-3x + 5y$
29. $-2x^2 + 6x - 7$ **31.** $m^2 - 7m + 10$
33. $15m^2 - 2m + 10$ **35.** $9a + 3b - 7c$
37. $8y^2 + 4y - 5$ **39.** $-6x^3 - 4x^2y + 13xy^2$
41a. 184   **41b.** 370   **43.** $8a^2 + 2a + 1$ **45.** 90°;
right   **47.** every 12 days   **49.** II

Page 718   Self Test
1. yes, binomial; 3   **3.** yes, monomial; 12
5. $-x^2 + 12$ **7.** $2x + 6y$ units   **9.** $a + 13d$

Pages 722–723   Lesson 14-4
5. 2^{12} **7.** $3^3x^3y^3$ or $27x^3y^3$ **9.** x^2y^2 **11.** 9
13. $48x^6y^2$ **15.** y^4z^4 **17.** $16r^4s^4$ **19.** m^{10}
21. $-32x^{10}$ **23.** x^4y^{12} **25.** $12x^3$ **27.** $16x^4y^6$
29. t^{-12} **31.** $-2x^{-6}y^{-2}$ **33.** -36 **35.** 108
37. -1296 **39.** $\frac{1}{324}$ **41a.** The numbers in the
values columns are the same.   **41b.** The exponents are distributed across the factors.
41c. When you raise a product to a power, distribute the exponent over each factor.
43a. $1000(1.01^4)^6$ **43b.** $1269.73   **45.** 5280^3
47. 13 in.   **49.** A spinner divided into 10 sections, six of them being successes.

51.

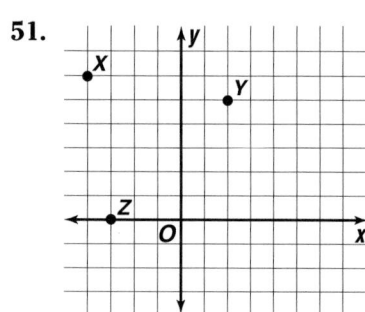

51a. II   **51b.** I   **51c.** none   **53.** $xy \div 5$

1. true   **3.** $x^2 + 5x$ **5.** $6x^2 + 3x$

7. $3x + 12$ **9.** $3x - 6$ **11.** $12x^2 + 28x$
13. 160 ft by 360 ft   **15.** $-2x - 16$ **17.** $6x^2 - 3x$
19. $a^2c + bc$ **21.** $p^2q^2 + 8pq$ **23.** $-18a + 2a^3$
25. $-15x^2 + 35x - 45$ **27.** $4c^4 + 28c^3 - 40c$
29. $-12x^5 + 48x^4 - 42x^2$ **31.** $-x^5 + x^4 - 3x^3 + 5x^2$
33. 2   **35a.** $12x^2 + 8x + 1$ **35b.** 96 in^2 **37.** $-32b^4$
39. Sample answer: Choose a scale that makes one product look much better than the other; omit some information or labels.   **41.** 13 times
43. 192.5 miles

7. $x + 3, 2x + 1; 2x^2 + 7x + 3$ **9.** $2x^2 + 5x + 3$
11. $10x^2 - x - 3$ **13.** $2x + 1, x + 2; 2x^2 + 5x + 2$
15. $2x + 3, 2x + 2; 4x^2 + 10x + 6$ **17.** $x^2 + 7x + 12$
19. $x^2 - 4x - 12$ **21.** $x^2 - 5x - 36$
23. $2x^2 - 5x - 12$ **25.** $x^2 + 6x + 9$
27. $10x^2 - 11x - 6$ **29.** $18x^2 - 3x + \frac{1}{8}$
31. $(x + 2)(x - 1); x^2 + x - 2$ **33a.** 14 in. by 27 in.   **33b.** 378 in^2 **33c.** 464 in^2 **35.** 90 ft
37. 3.14 m   **39.** $x < -14$

1. yes   **3.** no   **5.** no

1. sum   **3.** greatest   **5.** additive inverse

11. yes; binomial   **13.** yes; monomial   **15.** yes; binomial   **17.** 1   **19.** 2   **21.** 3   **23.** $5x^2 - 4x$
25. $x^2 - x$ **27.** $5x^2 + 3x + 1$ **29.** $4a - 15b$
31. $2a^2 - 3b^2 + c^2$ **33.** $16x^2 - x - 7$ **35.** a^6
37. p^6q^3 **39.** $4y^7z^3$ **41.** $-6a^8b^7$
43. $8d^2 - 20d$ **45.** $2a^5 + a^3 - 5a^2$
47. $-2g^4 - 12g^2 - 6g$
49. $2x + 1, x + 3; 2x^2 + 7x + 3$ **51.** $2x^2 + 3x + 1$
53. $2xy + 2y^2 + 2yz$ **55.** $3x^3$

2, 10, 35, 56 (t) Vince Streano/Tony Stone Images; **3** (t) Charles O'Rear/Westlight, (b) Chris Sorensen; **4** (bl) Lois Gervais/Westlight, (br) Bettmann; **5** (t) Bob Daemmrich/Stock Boston, (bl) Beverly Johnson by Francesco Scavullo August 1974 Vogue. Courtesy Vogue. ©1974 by the Conde Nast Publications, Inc.; (br) Carl Corey/Westlight; **6** (t) Duomo, (b) Mark Burnett; **7** Stacy Pick/Stock Boston; **8** Custom Medical Stock Photo; **9** Mark Steinmetz; **13** Joe Rimkus Jr./The Miami Herald; **14** Lester Lefkowitz/Tony Stone Images; **17** Don Smetzer/Tony Stone Images; **19** Jack Zehrt/FPG International; **20** (t) Wide World Photos, (b) FOXTROT ©Bill Amend. Reprinted with permission of UNIVERSAL PRESS SYNDICATE. All rights reserved.; **23** (l) David Allen/LGI Photo Agency, (r) David Allen/LGI Photo Agency; **28** Katrina De Leon/Still Life Stock; **30** E. Sampers/Gamma Liaison; **32** (t) David Young Wolff/Tony Stone Images, (b) Mark Burnett; **33** Doug Martin; **36** Matt Meadows; **38** Mark Burnett; **41** Bettmann; **43** Aaron Haupt; **44** Ben Van Hook/Duomo; **45** Photofest; **48** Owen Franken/Stock Boston; **50** Matt Meadows; **51** Murray & Associates/Tony Stone Images; **52** Mark Reinstein/FPG International; **55** Alexander Gardner, 1863/Bettmann Archive; **56** Chris Sorensen; **61-63** Mark Burnett; **64** (t) Mark Burnett, (bl) Eberhard E. Otto/FPG International, (br) Chromosohm/Joe Sohm/Photo Researchers; **65** FRANK & ERNEST, reprinted by permission of Newspaper Enterprise Association, Inc.; **65** (tl) Steve Gottlieb/FPG International, (b) Bettmann; **66** Buddy Mays/FPG International; **69** Mark Conlin; **73** Michael Keller/FPG International; **76** Robin Smith/Tony Stone Images; **79** Mark Steinmetz; **80** Jose Carrillo; **81** Steve Seiller/Ken Graham Agency; **84** AP/World Wide Photos; **86** Stephen Rose/Gamma International; **87** Stan Osolinski/FPG International; **90** Stan Osolinski/FPG International; **94** Biblioteca Universitaria, Bologna, Italy/Art Resource; **96** Comstock Inc./Gary Benson; **99** (t) Mark Conlin, (b) Stephen Rose/The Gamma Liaison Network; **102** Ken Graham/Ken Graham Agency; **106** Remi Benali/Gamma International; **107** John Terence Turner/FPG International; **108** (t) courtesy NASA, (b) Comstock Inc./Jack Clark; **112** Carl R. Sams II/Peter Arnold, Inc.; **116** (t) Hoffines/Young, (br) courtesy Perkins School for the Blind, (bl) Bettmann; **117** (t) Ron Rovtar/FPG International, (b) Bettmann; **118** Matt Meadows; **120** Nawrocki Stock Photo; **121** J. McGuire/Washington Stock Photo; **122** Henry Horenstein/Viesti; **127** Bonnie Sue/Photo Researchers; **129, 131** Mark Burnett; **132** Nawrocki Stock Photo; **133** (t) Sharon M. Kurgis, (b) Deborah Denker/Liaison International; **134** Frank Whitney/The Image Bank; **135** Fred Bavendam/Peter Arnold, Inc.; **136** Nawrocki Stock Photo; **138** Mark Burnett; **141** Michael Newman/PhotoEdit; **142** Marcel Ehrhard/Liaison International; **144** (t) Gerard Lacz/Peter Arnold, Inc., (b) ZIGGY ©1978 ZIGGY & FRIENDS, INC. Dist. by UNIVERSAL PRESS SYNDICATE. Reprinted with permission. All rights reserved.; **145** Nawrocki Stock Photo; **148** Mark Burnett; **149** Uniphoto; **153** Frank Spinelli/Liaison International; **154** Charlie Trainor/The Miami Herald; **155** Michael Simpson/FPG International; **160** Mark Burnett; **164** Bettmann; **167** Mark Steinmetz; **168** (br) Frank Driggs Collection, (bl) Painting by Abraham Anderson, c. 1889, National Portrait Gallery,

Smithsonian/Bettmann Archives; **169** (t) Comstock Inc./Comstock Inc., (c) Ken Frick, (br) Photofest, (bl) Charles O'Rear/Westlight; **172** Bridgeman/Art Resource, NY; **175** Aaron Haupt; **178** Comstock Inc./Comstock Inc.; **179** (t) Alinari/SEAT/Art Resource, NY, (b) THE FAR SIDE ©1985 FARWORKS, INC. Dist. by UNIVERSAL PRESS SYNDICATE. Reprinted with permission. All rights reserved.; **181** (t) Pam Francis/Liaison International, (b) Lawrence Migdale; **183** Lynn Stone; **185** Rosemary Weller/Tony Stone Images; **187** Photofest; **188** Dick Luria/FPG International; **190** S. Feld/H. Armstrong Roberts, Inc.; **191** Elizabeth Simpson/FPG International; **192** Aaron Haupt; **196** courtesy GCI/Atlanta; **199** James Blank/FPG International; **201** NASA/Peter Arnold, Inc.; **204** Mike Theiler/Reuters/Bettmann Archive; **207** Paul Richards/UPI/Bettmann Newsphotos; **208, 211** Manfred Kage/Peter Arnold, Inc.; **212** Howard Hughes Medical Institute/Peter Arnold, Inc.; **214** Arthur Tilley/Tony Stone Images; **219** Comstock Inc./Robert Houser; **222** *American Heritage* magazine; **223** (c) Ken Huang/The Image Bank, (br) Mark Steinmetz, (bl) Studiohio; **224** Everett Collection; **227** Steven Alexander/Uniphoto Picture Agency; **228** courtesy Ingrid Proctor-Fridia; **230** Mark Burnett; **233** Robert Landau/Westlight; **236** Eugene Gebhardt/FPG International; **237** Bettmann; **241** G. Schwartz/FPG International; **242** David W. Hamilton/The Image Bank; **245** Frank Saragnese/FPG International; **249, 250** Bettmann; **252** Mark Burnett; **255** PEANUTS reprinted by permission of United Feature Syndicate, Inc.; **256** Mark Steinmetz; **258** A. T. Willett/The Image Bank; **261** Studiohio; **262** Shinichi Kanno/FPG International; **264** Mark Steinmetz; **268** First Image; **271** Mark Burnett; **272** (t) Uniphoto, (bl, bc) Bettmann, (br) courtesy International Business Machines Corporation; **273** (c) Arthur Tilley/FPG International, (bl) Photofest, (br) Kazu Studio Ltd./FPG International; **274** Mark Burnett; **278** (l) David Zosel, (r) NASA; **281** Michal Heron/The Stock Market; **283** Mark Sennet/Onyx, **285** Richard Bowditch; **287** Mark Steinmetz; **290** Bob Daemmrich; **292** Aaron Haupt; **293** Dave Bartruff; **295** Aaron Haupt; **298** Elaine Shay; **300** Will Vinton/Photofest; **301** U.S. Representative/Westlight; **303** James Schnepf/The Gamma Liaison Network; **305** Chromosohm/Sohm/The Stock Market; **306** Jim Pickerell/Westlight; **308** (t) Terence Gili/FPG International, (b) Mark Steinmetz; **310** Photofest; **312** Matt Meadows; **314** Long Photography; **315** Mark Burnett; **317** Manfred Gottschalk/Westlight; **319** Paul W. Nesbit; **325** Richard Pasley/Viesti Associates; **328** (t, br) FPG International, (bl) Science & Society Picture Library; **329** Photofest; **331** Matthew Klein/Photo Researchers; **334** E. Nagele/FPG International; **336** Mark Steinmetz; **338** Focus on Sports; **340** from *Not Strictly By the Numbers* ©Carolina Biological Supply Company); **342-3** Jack Zehrt/FPG International; **349** Alan Pitcairn from Grant Heilman; **352** Larry Lefever from Grant Heilman; **354** Warren Morgan/Westlight; **355** Mark Steinmetz; **357** courtesy John Deere Precision Farming; **358** Joachim Messerschmidt/FPG International; **360** Dr.E.R. Degginger/Color-Pic; **362** James Marshall/The Stock Market; **366** (l) The Smithsonian Institution, (r) From the Al Rose Collection, Hogan Jazz

Archive, Tulane University; **369** Flip Chalfant/The Image Bank; **370** (t) Ben Van Hook/ Black Star Publishing, (b) Archive Photos; **371** (c) Cindy Lewis, (bl) Bettmann, (br) Michael R. Brown/Gamma-Liaison; **372** Comstock Inc./Comstock Inc.; **374** MAK-I; **375** Fred Hirschmann/ Ken Graham Agency; **377** Henry Groskinsky/Peter Arnold, Inc.; **378** Aaron Haupt; **381** Long Photography; **382** Uniphoto; **384** J.R. Schnelzer; **388** Comstock Inc./ M & C Werner; **394** Jose Carrillo; **395** Mark Stephenson/ Westlight; **397** Ken Skalski/Liaison International; **400** B. Busco/The Image Bank; **405** Mark Steinmetz; **406** Aaron Haupt; **408** Penny Gentieu/Tony Stone Images; **416** James Shaffer/ PhotoEdit; **419** Photofest; **422** Mark Conlin; **430** (l) Francois Gohier/Photo Researchers, (r) Larry Hamill; **431** (t) Anthro-Photo, (l) John Reader/Science Photo Library/Photo Researchers, (r) Francois Gohier/Photo Researchers; **432** Mark Burnett; **433** Aaron Haupt; **435** Eberhard Otto/FPG International; **436** John Mitchell/Photo Researchers; **437** Bettmann; **438** (t) Mark Steinmetz, (b) *People Weekly* 1995 Steve Liss; **439** Fred Lyon/Photo Researchers; **441** Matt Meadows; **444** De Keerle/Sygma; **445** Mark Steinmetz; **446** Doug Martin; **448** D. Robert Franz/Masterfile; **449** Mark Burnett; **452** (t) Ralph Clevenger/Westlight, (b) Mark Steinmetz; **454** Mark Burnett; **455** W. Hodges/Westlight; **459** Rosemary Weller/Tony Stone Images; **461** Uniphoto; **462** Aaron Haupt; **466** Cindy Lewis Photography; **469** Lisa Quinones/Black Star; **470** (t) Larry Mulvehill/Photo Researchers, (b) Aldo Pavan/Gamma Liaison Network; **471** Auguste Renoir. *The Reader,* 1874-1876. Musee d'Orsay, Paris, France. Erich Lessing/Art Resource, NY; **475** Geof du Feu/Masterfile; **476** Brian Bahr/Allsport; **480** courtesy Vehicle Research & Testing Center; **481** Mark Steinmetz; **482, 528** Mark Burnett; **483** KS Studio; **483** Rod Joslin; **483** (t) Mark Steinmetz, (b) Mark Burnett; **484** Bettmann; **485** (l) Glencoe files, (c) Bettmann, (r) Mark Steinmetz; **486** Glencoe files; **488** (l) Mark Burnett, (r) Ed Bock/The Stock Market; **493** David R. Frazier Photolibrary; **495** Trent Steffler/ David R. Frazier Photolibrary; **500** Mark Steinmetz; **502** Matt Meadows; **503** Nawrocki Stock Photo; **504** Barbara Campbell/ Gamma Liaison International; **509** Mark Burnett; **510** David M. Grossman; **514** Conway Trefethen/FPG International; **515** Tim Davis/Photo Researchers; **516** Larry Kunkel/FPG International; **517** Nawrocki Stock Photo; **518** Mark Burnett; **519** courtesy Dr. Carol Malloy; **520, 521** Mark Steinmetz; **521** (t) Mark Burnett from the Paul Baumann collection; **523** (t) Cecil Fox/Science Source/Photo Researchers, (b) PEANUTS reprinted by permission of United Feature Syndicate, Inc.; **524** Corbis Media; **526** James Levin/FPG International; **527** Ken Biggs/The Stock Market; **529** Mark Burnett from the Paul Baumann collection; **530** Thayer Syme/FPG International; **532** Mark Steinmetz; **533** Index Stock Photography; **538** Mark Steinmetz; **546** (l) Bettmann, (c) Stock Newport/Daniel Forster, (r) Thomas Zimmermann/FPG International; **547** (t) G. Whiteley/ Photo Researchers, (bl) Stock Newport/Bob Grieser, (br) Stock Newport/Daniel Forster; **548** David Zalubowski/AP Colorado;

550 Bettmann; **551** Mathias Oppersdorff/Photo Researchers; **552** Mark Burnett; **556** Tracy Borland; **557** Bettmann; **564** Mark Steinmetz; **567, 568** David R. Frazier Photolibrary; **570** David Gorman; **576** Carl Purcell/Photo Researchers; **578** Mark Steinmetz; **580** V.A.P. Neal/Photo Researchers; **583** Mark C. Burnett/Photo Researchers; **584** Janet Adams; **591** Defense Dept.; **594** Pictures Unlimited; **595** David R. Frazier Photolibrary; **596** (l) PhotoDisc, (lc) Matt Meadows, (c) Aaron Haupt, (rc) Mark Steinmetz, (r) Mark Burnett; **597** (l) Dr.E.R. Degginger/Color-Pic, (r) Harald Sund/Image Bank; **599** FOXTROT ©Bill Amend. Reprinted with permission of UNIVERSAL PRESS SYNDICATE. All rights reserved.; **600** Mark Burnett; **605** Scott Camazine/Photo Researchers; **606** Jose Luis Banus-March/FPG International; **607** David A. Barnes/FPG International; **608** (l) Bridgeman/Art Resource, NY, (r) Scala/Art Resource, NY; **609** (t) Uniphoto, (l) Bettmann, (c) Superstock, (r) Photofest; **612** Mark Burnett; **615** Mark Burnett; **618, 619** Mark Steinmetz; **621** Mark Segal/Tony Stone Images; **622** FRANK & ERNEST, reprinted by permission of Newspaper Enterprise Association, Inc.; **623** Mark Steinmetz; **624** Mark Burnett; **629** Werner Muller/Peter Arnold, Inc.; **630** Ben Mitchell/The Image Bank; **634** Mark Burnett; **638** Marvin E. Newman/The Image Bank; **641** courtesy Dr. Michael R. Doss; **645** Gerard Photography; **647** E.R.Degginger/ Photo Researchers; **650** Bernard Roussel/The Image Bank; **651** Mark Burnett; **653** William E. Ferguson; **658** David M. Grossman; **659** Ancient Art & Architecture Collection; **662** (t) George Obremski/The Image Bank, (l) Glencoe files, (r) Scala/Art Resource, NY; **663** (bl) Peter Beney/The Image Bank, (br) Bettmann; **664** Superstock; **665** Mark Burnett; **667** Cliff Feulner/The Image Bank; **669** Telegraph Colour Library/FPG International; **670, 672** David R. Frazier Photolibrary; **674** Index; **675** Doug Martin; **678** Leverett Bradley/FPG International; **682** Mark Burnett; **683** Stock Montage; **684** Pictures Unlimited; **685** Mark E. Gibson/Mark E. Gibson Photography; **687** Morton & White Photographic; **690** James Blank/FPG International; **692** John Elk III/Stock Boston; **695** Bill Lea; **697** Aaron Haupt/David R. Frazier Photolibrary; **698** (t) Courtesy Sullivan Gray Bruck Architects, Inc., (b) courtesy Darlene Leshnock; **702** Ron Thomas/FPG International; **704** (l) Mark Burnett, (c) Bettmann, (r) Superstock; **705** (l) Mark Burnett, (r) courtesy Bryan Berg; **706** Tom Carroll/FPG International; **707** Photo Researchers; **708** John Kaprielian/Photo Researchers; **709** Audrey Gibson/Mark E. Gibson Photography; **709** Craig K. Lorenz/Photo Researchers; **712** Ronald Sheridan/ Ancient Art & Architecture Collection; **715** Steve Allen/Gamma Liaison; **716** Beth Ava/FPG International; **720** Superstock; **723** Porterfield-Chickering/Photo Researchers; **724** Index; **725** Glencoe files; **726** Franc Shor/©National Geographic Society; **727** Mark Burnett; **728** Index; **729** Ed Braverman/FPG International; **731** Superstock; **736** Tony Quinn Photography; **Cover** Gus Chan; **Cover** Tony Stone Images; **Cover** Tony Stone Images

APPLICATIONS INDEX

INDEX

Tangent, 688–691, 695–697, 701
Technology, *see* Graphing calculator; Modeling; Spreadsheet; Technology Tips
Technology Tips, 12, 177, 235, 276, 318, 450, 491, 666, 689, 719
Temperature, *see* Celsius and Fahrenheit
Terminating decimal, 225, 275
Terms, 27, 258
Tessellation, 594
Tests, *see* Assessment
Test-taking tips, 115, 327, 429, 545, 661
Theoretical probability, 511, 529
Think About It, 7, 12, 17, 18, 23, 24, 27, 33, 37, 42, 43, 67, 104, 171, 172, 197, 201, 202, 226, 252, 281, 341, 352, 372, 386, 401, 406, 458, 536, 548, 556, 562, 563, 574, 580, 589, 596, 613, 614, 629, 632, 634, 638, 645
30°–60° right triangle, 683–686, 701
Timelines, *see* Putting It into Perspective
Transformation, 595–598
Translation, 595, 604
Transversal, 563–566, 602
Trapezoid, 585, 587
  area of, 335, 614–616

Tree diagram, 509, 511, 541
Triangle, 568–572
  acute, 569–571, 603
  angles, 569–577
  area of, 311, 610–611, 613–614, 616
  congruent, 573–577, 603
  equilateral, 569–571, 596
  isosceles, 569–571
  obtuse, 569–571, 603
  right, 569–571, 603
  scalene, 569–571
  similar, 578–583, 603
  special, 683–686
Triangular prism, 632–633
Triangular pyramid, 638
Trigonometric ratios, 688–691, 694–697, 701
Trigonometry, 688
Trinomials, 706–708, 734
Truncate, 275
Twin primes, 188

Unit rate, 433–435, 478
Upper quartile, 491–494, 540

Variable, 16–18, 32–35
  defining, 42
Venn diagram, 669, 671, 673
Vertex, 75, 549, 589, 638

Vertical angles, 561–563, 602
Vertical line test, 373–374
Volume, 643–648
  of a circular cylinder, 645–648, 657
  of a cone, 650–652, 658
  of a cube, 177–178, 391
  of a prism, 645–648, 657
  of a pyramid, 649–652, 658, 686

Whole numbers, 224
Windchill factor, 97
Working drawings, 653

x-**axis,** 36–40
x-**coordinate,** 37–40
x-**intercept,** 406–409

y-**axis,** 36–40
y-**coordinate,** 37–40
y-**intercept,** 406–409

Zero pair, 82

Number and Operations

$+$	plus or positive
$-$	minus or negative

$\left.\begin{array}{l} a \cdot b \\ a \times b \\ ab \text{ or } a(b) \end{array}\right\}$ a times b

$\div$	divided by
$\pm$	positive or negative
$=$	is equal to
$\neq$	is not equal to
$<$	is less than
$>$	is greater than
$\leq$	is less than or equal to
$\geq$	is greater than or equal to
$\approx$	is approximately equal to
$\%$	percent
$a{:}b$	the ratio of a to b, or $\frac{a}{b}$

Geometry and Measurement

$\cong$	is congruent to
$\sim$	is similar to
$^\circ$	degree(s)
$\overleftrightarrow{AB}$	line AB
$\overline{AB}$	segment AB
$\overrightarrow{AB}$	ray AB
$\llcorner$	right angle
$\perp$	is perpendicular to
$\parallel$	is parallel to
AB	length of $\overline{AB}$, distance between A and B
$\triangle ABC$	triangle ABC
$\angle ABC$	angle ABC
$\angle B$	angle B
$m\angle ABC$	measure of angle ABC
$\odot C$	circle C
$\overset{\frown}{AB}$	arc AB
π	pi $\left(\text{approximately } 3.14159 \text{ or } \frac{22}{7}\right)$
(a, b)	ordered pair with x-coordinate a and y-coordinate b
$\sin A$	sine of angle A
$\cos A$	cosine of angle A
$\tan A$	tangent of angle A

Algebra and Functions

a'	a prime
a^n	a to the nth power
a^{-n}	$\frac{1}{a^n}$ (one over a to the nth power)
$\lvert x \rvert$	absolute value of x
$\sqrt{x}$	principal (positive) square root of x
$f(n)$	function, f of n

Probability and Statistics

$P(A)$	the probability of event A
$n!$	n factorial
$P(n, r)$	permutation of n things taken r at a time
$C(n, r)$	combination of n things taken r at a time